Applied Mathematical Sciences
Volume 114

Editors
J.E. Marsden L. Sirovich F. John (deceased)

Advisors
M. Ghil J.K. Hale T. Kambe
J. Keller K. Kirchgässner
B.J. Matkowsky C.S. Peskin
J.T. Stuart

Springer
New York
Berlin
Heidelberg
Barcelona
Budapest
Hong Kong
London
Milan
Paris
Santa Clara
Singapore
Tokyo

Applied Mathematical Sciences

1. *John:* Partial Differential Equations, 4th ed.
2. *Sirovich:* Techniques of Asymptotic Analysis.
3. *Hale:* Theory of Functional Differential Equations, 2nd ed.
4. *Percus:* Combinatorial Methods.
5. *von Mises/Friedrichs:* Fluid Dynamics.
6. *Freiberger/Grenander:* A Short Course in Computational Probability and Statistics.
7. *Pipkin:* Lectures on Viscoelasticity Theory.
8. *Giacoglia:* Perturbation Methods in Non-linear Systems.
9. *Friedrichs:* Spectral Theory of Operators in Hilbert Space.
10. *Stroud:* Numerical Quadrature and Solution of Ordinary Differential Equations.
11. *Wolovich:* Linear Multivariable Systems.
12. *Berkovitz:* Optimal Control Theory.
13. *Bluman/Cole:* Similarity Methods for Differential Equations.
14. *Yoshizawa:* Stability Theory and the Existence of Periodic Solution and Almost Periodic Solutions.
15. *Braun:* Differential Equations and Their Applications, 3rd ed.
16. *Lefschetz:* Applications of Algebraic Topology.
17. *Collatz/Wetterling:* Optimization Problems.
18. *Grenander:* Pattern Synthesis: Lectures in Pattern Theory, Vol. I.
19. *Marsden/McCracken:* Hopf Bifurcation and Its Applications.
20. *Driver:* Ordinary and Delay Differential Equations.
21. *Courant/Friedrichs:* Supersonic Flow and Shock Waves.
22. *Rouche/Habets/Laloy:* Stability Theory by Liapunov's Direct Method.
23. *Lamperti:* Stochastic Processes: A Survey of the Mathematical Theory.
24. *Grenander:* Pattern Analysis: Lectures in Pattern Theory, Vol. II.
25. *Davies:* Integral Transforms and Their Applications, 2nd ed.
26. *Kushner/Clark:* Stochastic Approximation Methods for Constrained and Unconstrained Systems.
27. *de Boor:* A Practical Guide to Splines.
28. *Keilson:* Markov Chain Models—Rarity and Exponentiality.
29. *de Veubeke:* A Course in Elasticity.
30. *Shiatycki:* Geometric Quantization and Quantum Mechanics.
31. *Reid:* Sturmian Theory for Ordinary Differential Equations.
32. *Meis/Markowitz:* Numerical Solution of Partial Differential Equations.
33. *Grenander:* Regular Structures: Lectures in Pattern Theory, Vol. III.
34. *Kevorkian/Cole:* Perturbation Methods in Applied Mathematics.
35. *Carr:* Applications of Centre Manifold Theory.
36. *Bengtsson/Ghil/Källén:* Dynamic Meteorology: Data Assimilation Methods.
37. *Saperstone:* Semidynamical Systems in Infinite Dimensional Spaces.
38. *Lichtenberg/Lieberman:* Regular and Chaotic Dynamics, 2nd ed.
39. *Piccini/Stampacchia/Vidossich:* Ordinary Differential Equations in $\mathbf{R}^n$.
40. *Naylor/Sell:* Linear Operator Theory in Engineering and Science.
41. *Sparrow:* The Lorenz Equations: Bifurcations, Chaos, and Strange Attractors.
42. *Guckenheimer/Holmes:* Nonlinear Oscillations, Dynamical Systems and Bifurcations of Vector Fields.
43. *Ockendon/Taylor:* Inviscid Fluid Flows.
44. *Pazy:* Semigroups of Linear Operators and Applications to Partial Differential Equations.
45. *Glashoff/Gustafson:* Linear Operations and Approximation: An Introduction to the Theoretical Analysis and Numerical Treatment of Semi-Infinite Programs.
46. *Wilcox:* Scattering Theory for Diffraction Gratings.
47. *Hale et al:* An Introduction to Infinite Dimensional Dynamical Systems—Geometric Theory.
48. *Murray:* Asymptotic Analysis.
49. *Ladyzhenskaya:* The Boundary-Value Problems of Mathematical Physics.
50. *Wilcox:* Sound Propagation in Stratified Fluids.
51. *Golubitsky/Schaeffer:* Bifurcation and Groups in Bifurcation Theory, Vol. I.
52. *Chipot:* Variational Inequalities and Flow in Porous Media.
53. *Majda:* Compressible Fluid Flow and System of Conservation Laws in Several Space Variables.
54. *Wasow:* Linear Turning Point Theory.
55. *Yosida:* Operational Calculus: A Theory of Hyperfunctions.
56. *Chang/Howes:* Nonlinear Singular Perturbation Phenomena: Theory and Applications.
57. *Reinhardt:* Analysis of Approximation Methods for Differential and Integral Equations.
58. *Dwoyer/Hussaini/Voigt (eds):* Theoretical Approaches to Turbulence.
59. *Sanders/Verhulst:* Averaging Methods in Nonlinear Dynamical Systems.
60. *Ghil/Childress:* Topics in Geophysical Dynamics: Atmospheric Dynamics, Dynamo Theory and Climate Dynamics.

(continued following index)

J. Kevorkian J.D. Cole

Multiple Scale and Singular Perturbation Methods

With 83 Illustrations

Springer

J. Kevorkian
Department of Applied Mathematics
University of Washington
Seattle, WA 98195
USA

J.D. Cole
Department of Mathematical Sciences
Rensselaer Polytechnic Institute
Troy, NY 12181
USA

Editors

J.E. Marsden
Control and Dynamical Systems, 104-44
California Institute of Technology
Pasadena, CA 91125
USA

L. Sirovich
Division of Applied Mathematics
Brown University
Providence, RI 02912
USA

Mathematics Subject Classification (1991): 34E10, 35B20, 76Bxx

Library of Congress Cataloging-in-Publication Data
Kevorkian, J.
Multiple scale and singular perturbation methods/J. Kevorkian, J.D. Cole.
p. cm. – (Applied mathematical sciences; v. 114)
Includes bibliographical references and index.
ISBN 0-387-94202-5 (hardcover:alk. paper)
1. Differential equations—Numerical solutions. 2. Differential equations—Asymptotic theory. 3. Perturbation (Mathematics) I. Cole, Julian D. II. Title. III. Series: Applied mathematical sciences (Springer-Verlag New York Inc.); v. 114.
QA1.A647 vol. 114
[QA371]
510 s—dc20
[515'.35] 95-49951

Printed on acid-free paper.

Production managed by Hal Henglein; manufacturing supervised by Jeffrey Taub.
Camera-ready copy prepared from the authors' TeX file.
Printed and bound by R.R. Donnelley & Sons, Harrisonburg, VA.
Printed in the United States of America.

9 8 7 6 5 4 3 2 1

ISBN 0-387-94202-5 Springer-Verlag New York Berlin Heidelberg SPIN 10424264

Preface

This book is a revised and updated version, including a substantial portion of new material, of our text *Perturbation Methods in Applied Mathematics* (Springer-Verlag, 1981). We present the material at a level that assumes some familiarity with the basics of ordinary and partial differential equations. Some of the more advanced ideas are reviewed as needed; therefore this book can serve as a text in either an advanced undergraduate course or a graduate-level course on the subject.

Perturbation methods, first used by astronomers to predict the effects of small disturbances on the nominal motions of celestial bodies, have now become widely used analytical tools in virtually all branches of science. A problem lends itself to perturbation analysis if it is "close" to a simpler problem that can be solved exactly. Typically, this closeness is measured by the occurrence of a small dimensionless parameter, ϵ, in the governing system (consisting of differential equations and boundary conditions) so that for $\epsilon = 0$ the resulting system is exactly solvable. The main mathematical tool used is asymptotic expansion with respect to a suitable asymptotic sequence of functions of ϵ.

In a regular perturbation problem, a straightforward procedure leads to a system of differential equations and boundary conditions for each term in the asymptotic expansion. This system can be solved recursively, and the accuracy of the result improves as ϵ gets smaller, for all values of the independent variables throughout the domain of interest. We discuss regular perturbation problems in the first chapter.

In a singular perturbation problem, also called a layer-type problem, there are one or more thin layers at the boundary or in the interior of the domain where the above procedure fails. Often, this failure is due to the fact that ϵ multiplies the highest derivative in the differential equation; therefore the leading approximation obeys a lower-order equation that cannot satisfy all the prescribed boundary conditions. Layer-type problems for ordinary differential equations are discussed in Chapter 2 and for partial differential equations in Chapter 3.

Regular perturbations also fail if the govening system is to be solved over an infinite domain and contains small terms with a cumulative effect. The two principal techniques for deriving asymptotic solutions that remain valid in the far field are multiple scale expansions and the method of averaging. These techniques are discussed in Chapters 4 and 5 for systems of ordinary differential equations. Applications of multiple scale methods to problems in partial differential equations appear in Chapter 6.

The aim of this book is to survey perturbation methods as currently used in various application areas. We introduce a particular topic by means of a simple illustrative example and then build up to more challenging problems. Whenever possible (and practical), we give the general theory for a procedure that applies to a broad class of problems. However, we do not consider rigorous proofs for the validity of our results; to do so would take us far afield from our stated aim. Also, in spite of the progress in this regard in recent years, rigorous justification of asymptotic validity remains generally out of reach except for simple, well-understood problems.

The basic ideas discussed in this book are, as is usual in scientific work, the contributions of many people. We have made some attempt to cite original sources, but we do not claim perfect historical accuracy, nor do we give a complete list of references. Rather, we have tried to present the state of the art in a systematic and unified manner. There are several excellent references that cover various aspects of layer-type problems in some detail; fewer are available on multiple scale methods. We present a comprehensive treatment of both types of problems, including recent developments in multiple scale and averaging methods not available in other reference work.

Summer 1995

J. Kevorkian
J.D. Cole

Contents

1

Introduction

1.1 Order Symbols, Uniformity

We will use the conventional order symbols O and o as a mathematical measure of the relative order of magnitude of various expressions. Although generalizations are straightforward, we will restrict attention to scalar functions $u(\mathbf{x}; \epsilon)$ of the n-vector independent variable $\mathbf{x} = (x_1, x_2, \ldots, x_n)$ and the scalar parameter ϵ. The x_i will vary over some specified domain D and ϵ will lie on the interval $I : 0 < \epsilon \leq \epsilon_1$.

1.1.1 Large O

For a given domain D and ϵ-interval I, the statement

$$u(\mathbf{x}; \epsilon) = O(v(\mathbf{x}; \epsilon)) \text{ in } I \tag{1.1.1}$$

means that for each $\mathbf{x}$ in D there exists a positive number $k(\mathbf{x})$ such that

$$|u(\mathbf{x}; \epsilon)| \leq k(\mathbf{x})|v(\mathbf{x}; \epsilon)| \tag{1.1.2}$$

for all ϵ in I. Similarly, we say that

$$u(\mathbf{x}; \epsilon) = O(v(\mathbf{x}; \epsilon)) \text{ as } \epsilon \to 0, \tag{1.1.3}$$

if for each $\mathbf{x}$ in D, there exists a positive number $k(\mathbf{x})$ and a neighborhood N of $\epsilon = 0$ such that (1.1.2) holds for all ϵ in the intersection of N with I. Notice that if u/v is defined, (1.1.2) implies that $|u/v|$ is bounded above by $k(\mathbf{x})$.

The statement (1.1.1) is said to be *uniformly valid* in D if k does not depend on the value of $\mathbf{x}$, i.e., one can find a finite constant k for which (1.1.2) holds for any $\mathbf{x}$ in D. Similarly, (1.1.3) is said to be uniformly valid in D if, in addition to k being constant, the neighborhood N is also independent of $\mathbf{x}$.

Consider the following one-dimensional ($n = 1$) examples defined on $D : 0 < x < 1$ and $I : 0 < \epsilon \leq \epsilon_1 < 1$.

(i) $$x + \epsilon = O(1) \text{ in } I, \text{ uniformly in } D. \tag{1.1.4}$$

Clearly, (1.1.4) is true since (1.1.2) holds for any x in D and ϵ in I with the choice $k = 2$.

(ii) $$e^{-\sin \epsilon x} = O(e^{-2\epsilon x/\pi}) \text{ in } I, \text{ uniformly in } D. \tag{1.1.5}$$

This follows from the fact that $0 < 2z/\pi < \sin z < 1$ for all z in $0 < z < \pi/2$. Since $0 < \epsilon x < 1$ always, the inequality in (1.1.2) holds for all x in D and all ϵ in I with the choice $k = 1$.

(iii) $$\frac{1}{x+\epsilon} = O(1) \text{ in } I. \tag{1.1.6}$$

This is true because $(x+\epsilon)^{-1} < 1/x$ for all x in D and all ϵ in I; thus, $k(x) = 1/x$. But it is clear that the statement (1.1.6) is *not* uniformly valid in D because there is no finite constant for which the required inequality holds for all x in D so long as x is allowed to approach the origin. If, however, we restrict x to lie in the interval $\tilde{D} : 0 < \delta \le x < 1$, we can choose $k = 1/\delta$ and we see that (1.1.6) is uniformly valid in $\tilde{D}$. For similar reasons, the statement

(iv) $$\frac{\epsilon}{x(1-x)} = O(\epsilon) \text{ in } I \tag{1.1.7}$$

is true but not uniformly in D; it is uniformly valid in $D^* : 0 < \delta_1 \le x \le \delta_2 < 1$ with

$$k = \max\left(\frac{1}{\delta_1(1-\delta_1)}, \frac{1}{\delta_2(1-\delta_2)}\right).$$

(v) $$\sin\frac{x}{\epsilon} = O(x) \text{ as } \epsilon \to 0. \tag{1.1.8}$$

Here, even though the limit as $\epsilon \to 0$ of $\sin(x/\epsilon)$ does not exist for any $x \ne 0$, it is still true that $|\sin(x/\epsilon)| \le 1$ for any x in D. Therefore, (1.1.8) holds with $k(x) = 1/x$ and the statement is not uniformly valid. However, the statement $\sin(x/\epsilon) = O(1)$ as $\epsilon \to 0$ is uniformly valid in D.

Now let $n = 2$, $D_2 : 0 < x_1 < \infty; 0 < x_2 < \infty$, and I as above, i.e., $0 < \epsilon \le \epsilon_1 < 1$. Consider the two functions $u = x_1\epsilon^\alpha$ and $v = x_2\epsilon^\beta$, where α and β are constants with $\alpha \ge \beta$. We see that

(vi) $$x_1\epsilon^\alpha = O(x_2\epsilon^\beta) \text{ in } I \tag{1.1.9}$$

since the choice $k(x_1, x_2) = x_1/x_2$, which is finite in D_2, satisfies (1.1.2) because $\epsilon^\alpha \le \epsilon^\beta$. As k becomes infinite for $x_1 \to \infty$ or $x_2 \to 0$, the ordering (1.1.9) is not uniformly valid in D_2. In order to have uniform validity, we need to restrict x_1 and x_2 to lie in $\overline{D}_2 : 0 < x_1 \le X_1 < \infty; 0 < X_2 \le x_2 < \infty$, in which case (1.1.2) is satisfied with $k = X_1/X_2 = \text{const}$.

This example also points out the fact that the statement $u = O(v)$ does not necessarily imply that u and v are "of the same order of magnitude." In fact, for $\alpha > \beta$ we see that $x_1\epsilon^\alpha < x_2\epsilon^\beta$ for any pair (x_1, x_2) in D_2 if ϵ is sufficiently small. Moreover, $(x_1\epsilon^\alpha/x_2\epsilon^\beta) \to 0$ as $\epsilon \to 0$ in this case. Thus, the O symbol only provides a *one-sided* bound. One way to characterize two functions u and v that are of the same order of magnitude is to have $u = O(v)$ and $v = O(u)$. In this case, if $\lim_{\epsilon\to 0}(u/v)$ exists it is neither zero nor infinity. We therefore introduce the notation

$$u = O_s(v) \tag{1.1.10}$$

to indicate $u = O(v)$ and $v = O(u)$. Thus, the u and v functions in (1.1.9) do not satisfy (1.1.10), whereas those used in (1.1.4) do.

1.1.2 Small o

For a given domain D, the statement

$$u(\mathbf{x};\epsilon) = o(v(\mathbf{x};\epsilon)) \text{ as } \epsilon \to 0 \tag{1.1.11}$$

means that for each point $\mathbf{x}$ in D and any given $\delta > 0$, there exists an ϵ-interval $I(\mathbf{x},\delta) : 0 < \epsilon \leq \epsilon_1(\mathbf{x},\delta)$ such that

$$|u(\mathbf{x};\epsilon)| \leq \delta|v(\mathbf{x};\epsilon)| \tag{1.1.12}$$

for all ϵ in I. The inequality (1.1.12) indicates that $|u|$ becomes arbitrarily small compared to $|v|$ as $\epsilon \to 0$. Note also that $u = o(v)$ always implies $u = O(v)$ (the converse is not true). Often, the notation

$$u \ll v \tag{1.1.13}$$

is used to indicate (1.1.11).

We say that $u = o(v)$ as $\epsilon \to 0$ uniformly in D if ϵ_1 depends only on δ but not on $\mathbf{x}$.

The following examples illustrate ideas. Consider first the same functions and domain used in (vi), i.e., $u = x_1\epsilon^\alpha$; $v = x_2\epsilon^\beta$ with $\alpha > \beta$ in the domain $D_2 : 0 < x_1 < \infty, 0 < x_2 < \infty$. We have

(vii) $$x_1\epsilon^\alpha = o(x_2\epsilon^\beta) \text{ as } \epsilon \to 0, \tag{1.1.14}$$

because for any given $\delta > 0$, (1.1.12) holds as long as ϵ is in the neighborhood $0 < \epsilon \leq (\delta x_2/x_1)^{1/(\alpha-\beta)}$. This neighborhood depends on x_1 and x_2 and shrinks to zero if either $x_2 \to 0$ or $x_1 \to \infty$. Therefore, (1.1.14) is not uniformly valid in D_2. However, if we restrict x_1 and x_2 to $\overline{D}_2 : 0 < x_1 \leq X_1 < \infty; 0 < X_2 \leq x_2 < \infty$ for constants X_1 and X_2, then (1.1.14) is uniformly valid. The neighborhood of $\epsilon = 0$ that we need is $0 < \epsilon \leq (\delta X_2/X_1)^{1/(\alpha-\beta)}$, and it does not depend on x_1 and x_2.

Let D_Δ be the triangular domain $0 < x_1 < \infty$; $0 < x_2 < x_1$. For any *arbitrarily large* positive constant β, we have

$$e^{(x_2-x_1)/\epsilon} = o(\epsilon^\beta) \text{ as } \epsilon \to 0. \tag{1.1.15}$$

Note first that $(x_2 - x_1)/\epsilon < 0$. To verify (1.1.15), it suffices to show that

$$\lim_{\epsilon\to 0} \frac{e^{-\alpha/\epsilon}}{\epsilon^\beta} = 0 \tag{1.1.16}$$

for any $\alpha > 0$ and any β. (The result is trivially true for $\beta < 0$.) We have

$$\frac{e^{-\alpha/\epsilon}}{\epsilon^{\beta}} = \frac{e^{-\alpha/\epsilon}}{e^{\beta \log \epsilon}} = e^{-(\alpha+\beta\epsilon \log \epsilon)/\epsilon}.$$

Now, $\lim_{\epsilon\to 0} \epsilon \log \epsilon = \lim_{\epsilon\to 0} \frac{\log \epsilon}{1/\epsilon} = \lim_{\epsilon\to 0} \frac{1/\epsilon}{-1/\epsilon^2} = 0$ using L'Hospital's rule. Therefore,

$$\lim_{\epsilon\to 0} \frac{\alpha + \beta\epsilon \log \epsilon}{\epsilon} = \infty,$$

and (1.1.16) follows. The statement (1.1.15) is not uniformly valid in D_Δ, but it is uniformly valid in the subdomain $\overline{D}_\Delta : 0 < X_1 \leq x_1 < \infty; 0 < x_2 \leq x_1 - X_1$.

A function such as $e^{-\alpha/\epsilon}$ that tends to zero faster than any algebraic power of ϵ as $\epsilon \to 0$ is said to be *transcendentally small*. Henceforth, we shall use the abbreviated notation TST to refer to a transcendentally small term.

Various operations such as addition, multiplication, and integration may be performed with the O and o relations (see Problem 1). In general, differentiation of order relations with respect to ϵ or $\mathbf{x}$ is not permissible. For these and further results, the reader may consult [1.2].

Problems

1. Show the following:
   a. $O(O(u)) = O(u)$.
   b. $O(o(u)) = o(O(u)) = o(u)$.
   c. $O(u)O(v) = O(uv)$.
   d. $O(u)o(v) = o(u)o(v) = o(uv)$.
   e. $O(u) + O(u) = O(u) + o(u) = O(u)$.
   f. $o(u) + o(u) = o(u)$.
2. Consider $u(\mathbf{x}; \epsilon) = (1 - x_1^2 + x_2^2 + \epsilon)^{-1}$. Let $D : x_1^2 + x_2^2 = 1 - \Delta$, where $0 < \Delta = \text{const}$. Show that $u(\mathbf{x}; \epsilon) = O_s(1)$ as $\epsilon \to 0$, uniformly in D.
3. Let $u(\mathbf{x}; \epsilon) = O(v(\mathbf{x}; \epsilon))$ as $\epsilon \to 0$. Show that

$$\int_0^\epsilon u(\mathbf{x}; t)dt = O(\int_0^\epsilon |v(\mathbf{x}; t)|dt),$$

   and give an example.
4. Show that
   a. $\epsilon^\alpha \log \epsilon = o(1)$ as $\epsilon \to 0$ for any $\alpha > 0$.
   b. $e^{\alpha/\epsilon} = o(e^{\beta/\epsilon})$ as $\epsilon \to 0^+$ for $0 < \alpha < \beta$.
   c. $u = o(v)$ implies $|u|^\alpha = o(|v|^\alpha)$ for $\alpha > 0$.

1.2 Asymptotic Expansion of a Given Function

1.2.1 Asymptotic Sequence and Asymptotic Expansion

Consider a sequence of functions $\{\phi_n(\epsilon)\}, n = 1, 2, \ldots$. Such a sequence is called an asymptotic sequence if

$$\phi_{n+1}(\epsilon) = o(\phi_n(\epsilon)) \text{ as } \epsilon \to 0 \tag{1.2.1}$$

for each $n = 1, 2, \ldots$.

The following are some examples of asymptotic sequences (as $\epsilon \to 0$):

$$\phi_n(\epsilon) = \epsilon^{n-1} \quad n = 1, 2, \ldots \tag{1.2.2}$$

$$\phi_1 = \log\epsilon;\ \phi_2 = 1;\ \phi_3 = \epsilon\log\epsilon;\ \phi_4 = \epsilon$$
$$\phi_5 = \epsilon^2\log^2\epsilon;\ \phi_6 = \epsilon^2\log\epsilon;\ \phi_7 = \epsilon^2 \ldots. \tag{1.2.3}$$

Notice that the definition (1.2.1) does not preclude having one or more of the starting terms in an asymptotic sequence being infinite as the $\log\epsilon$ term in (1.2.3).

Here again, various operations, such as multiplication of two sequences or integration, can be used to generate a new sequence. Differentiation with respect to ϵ may not lead to a new asymptotic sequence. For more details, see [1.2].

Let $u(\mathbf{x}; \epsilon)$ be defined in some domain D of $\mathbf{x}$ and some neighborhood of $\epsilon = 0$. Let $\{\phi_n(\epsilon)\}$ be a given asymptotic sequence. The series $\sum_{n=1}^{N}\phi_n(\epsilon)u_n(\mathbf{x})$ is called the *asymptotic expansion* of $u(\mathbf{x}; \epsilon)$ to N terms (N may be a finite integer or infinity) as $\epsilon \to 0$ with respect to the sequence $\{\phi_n(\epsilon)\}$ if

$$u(\mathbf{x}; \epsilon) - \sum_{n=1}^{M}\phi_n(\epsilon)u_n(\mathbf{x}) = o(\phi_M) \text{ as } \epsilon \to 0 \tag{1.2.4}$$

for each $M = 1, 2, \ldots, N$.

If $N = \infty$, the following notation is generally used to indicate an asymptotic expansion

$$u(\mathbf{x}; \epsilon) \sim \sum_{n=1}^{\infty}\phi_n(\epsilon)u_n(\mathbf{x}) \text{ as } \epsilon \to 0. \tag{1.2.5}$$

A definition of an asymptotic expansion equivalent to (1.2.4) is

$$u(\mathbf{x}; \epsilon) - \sum_{n=1}^{M}\phi_n(\epsilon)u_n(\mathbf{x}) = O(\phi_{M+1}) \text{ as } \epsilon \to 0 \tag{1.2.6}$$

for each $M = 1, 2, \ldots, N - 1$. An asymptotic expansion is said to be *uniformly valid* in D if the order relations in (1.2.4) or (1.2.6) hold uniformly in D.

1.2.2 Asymptotic Expansion of an Explicitly Defined Function

Once the function $u(\mathbf{x}; \epsilon)$ is given and the asymptotic sequence $\{\phi_n(\epsilon)\}$ is specified, we can define each of the $u_n(\mathbf{x})$ uniquely by repeated application of definition

(1.2.4). Thus,

$$u_1(\mathbf{x}) = \lim_{\epsilon \to 0} \frac{u(\mathbf{x}; \epsilon)}{\phi_1(\epsilon)}, \tag{1.2.7a}$$

$$u_2(\mathbf{x}) = \lim_{\epsilon \to 0} \frac{u(\mathbf{x}; \epsilon) - \phi_1(\epsilon)u_1(\mathbf{x})}{\phi_2(\epsilon)}, \tag{1.2.7b}$$

$$u_k(\mathbf{x}) = \lim_{\epsilon \to 0} \frac{u(\mathbf{x}; \epsilon) - \sum_{n=1}^{k-1} \phi_n(\epsilon)u_n(\mathbf{x})}{\phi_k(\epsilon)}. \tag{1.2.7c}$$

For example, $u(x; \epsilon) = (x + \epsilon)^{-1/2}$ has the asymptotic expansion

$$(x + \epsilon)^{-1/2} \sim \sum_{n=1}^{\infty} \frac{(-1)^{n-1}}{2^{n-1}(n-1)!} \Pi_{k=1}^{n} |2k - 3| \frac{\epsilon^{n-1}}{x^{(2n-1)/2}} \tag{1.2.8}$$

as $\epsilon \to 0$, with respect to the sequence $1, \epsilon, \epsilon^2, \ldots$. Equation (1.2.8) is also the Taylor series expansion of $(x + \epsilon)^{-1/2}$ around $\epsilon = 0$, and it is a convergent series for $\epsilon < |x|$. Note that the expansion (1.2.8) is not uniformly valid in any x-interval that has $x = 0$ as a limit point.

Consider next the function

$$u = e^{-x/\epsilon} - \frac{\epsilon e^{-x}}{x + \epsilon} \equiv f(x; \epsilon) \tag{1.2.9}$$

on $0 \leq x \leq 1$, $0 < \epsilon \ll 1$. If we fix x and apply the limit process defined by (1.2.7) with $\phi_n = \epsilon^{n-1}$, we find the following expansion for u:

$$\begin{aligned} f &= -\epsilon \frac{e^{-x}}{x} + \epsilon^2 \frac{e^{-x}}{x^2} - \epsilon^3 \frac{e^{-x}}{x^3} + O(\epsilon^4) \\ &= \sum_{n=1}^{N} \epsilon^n h_n(x) + O(\epsilon^{N+1}). \end{aligned} \tag{1.2.10}$$

The term $e^{-x/\epsilon}$ in (1.2.9) is transcendentally small and makes no contribution to any $O(\epsilon^N)$ term in (1.2.10).

Clearly, the expansion (1.2.10) is not uniformly valid in any subinterval with $x = 0$ as the left limit point. In fact, this expansion is singular at $x = 0$, and it does not provide a good approximation of (1.2.9) no matter how small ϵ is chosen if x is allowed to become arbitrarily small. This is seen in Fig. 1.2.1, where we compare (1.2.9) with (1.2.10) for the choice $\epsilon = 0.1$.
However, if we restrict x to lie in the subinterval $0 < x_0 \leq x \leq 1$, then (1.2.10) is uniformly valid there. The source of the nonuniformity near $x = 0$ is easily traced to the expansion of the denominator $x + \epsilon$ in the second term. This expansion is based on the limit $\epsilon \to 0$ with x fixed and is thus incorrect if $x = O_s(\epsilon)$ or smaller.

It is natural to seek another expansion that adequately approximates (1.2.9), for x small. Since the nonuniformity occurs for $x = O_s(\epsilon)$, and since the combination x/ϵ appears in the first term of (1.2.9), one is led to the change of variable $x^* = x/\epsilon$

$$u = e^{-x^*} - \frac{e^{-\epsilon x^*}}{x^* + 1} \equiv g(x^*; \epsilon). \tag{1.2.11}$$

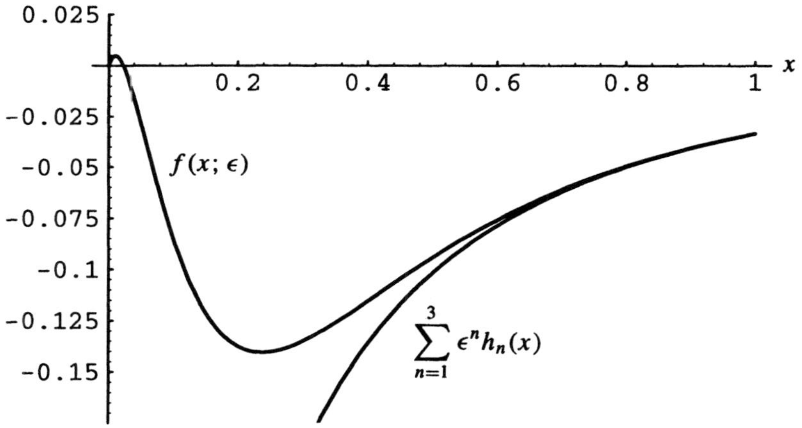

FIGURE 1.2.1. Exact Solution and Outer Expansion

With $x^* = x/\epsilon$, f and g define the *same* function u, i.e., $f(x;\epsilon) = g(x/\epsilon;\epsilon)$ or $f(\epsilon x^*;\epsilon) = g(x^*;\epsilon)$. However, the asymptotic expansion of g with x^* fixed and $\epsilon \to 0$ is quite different from (1.2.10). For this limit process, it is easy to see that we obtain the following expansion with respect to the sequence ϵ^{n-1}:

$$g = e^{-x^*} - \frac{1}{1+x^*} + \frac{\epsilon x^*}{1+x^*} - \frac{\epsilon^2 x^{*2}}{2(1+x^*)} + \frac{\epsilon^3 x^{*3}}{6(1+x^*)} + O(\epsilon^4)$$

$$\equiv \sum_{n=1}^{N} \epsilon^{n-1} g_n(x^*) + O(\epsilon^{N+1}). \tag{1.2.12}$$

We note that this expansion gives a good approximation for u for small x. In particular, the right-hand side of (1.2.12) vanishes at $x^* = x = 0$ *to all orders in* ϵ, and this conforms with the exact value of u at $x = 0$. However, (1.2.12) fails to be uniformly valid for $x^* \to \infty$, as seen in Fig. 1.2.2. Therefore, the two expansions (1.2.10) and (1.2.12) have mutually exclusive domains of validity. Depending on the magnitude of x compared with ϵ, one expansion or the other should be used.

We refer to (1.2.10) as the *outer expansion* of u (because it is valid away from the boundary point $x = 0$), and the expansion (1.2.12) is called the *inner expansion* of u (because it is valid near $x = 0$). In Sec. 1.4, we discuss in detail the sense in which outer and inner expansions (such as (1.2.10), (1.2.12)) are approximations of the exact expression such as (1.2.9) from which they arise. For the moment, we note only the following curious property of these two expansions. If we express the outer expansion in terms of the x^* variable and re-expand the result, the series we obtain agrees with the series that results from evaluating the inner expansion

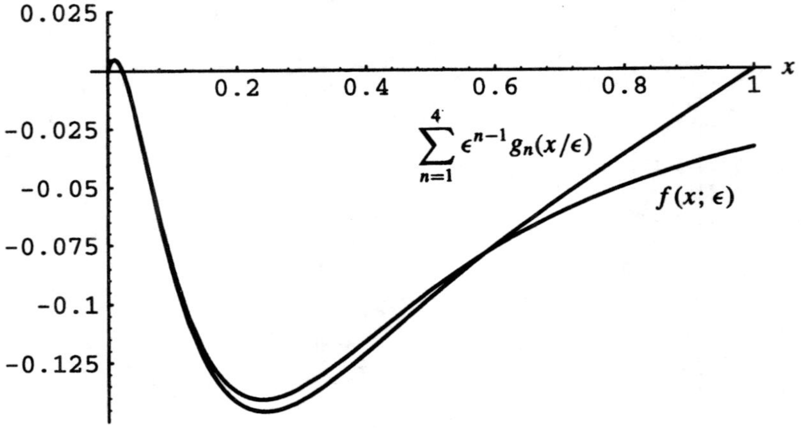

FIGURE 1.2.2. Exact Solution and Inner Expansion

for x^* large. To show this, first replace x by ϵx^* in (1.2.10)

$$f = -\frac{\epsilon e^{-\epsilon x^*}}{\epsilon x^*} + \frac{\epsilon^2 e^{-\epsilon x^*}}{\epsilon^2 x^{*2}} - \frac{\epsilon^3 e^{-\epsilon x^*}}{\epsilon^3 x^{*3}} + \ldots$$
$$= e^{-\epsilon x^*}\left(-\frac{1}{x^*} + \frac{1}{x^{*2}} - \frac{1}{x^{*3}} + \ldots\right),$$

then expand $e^{-\epsilon x^*}$ for ϵ small, and denote the resulting series $\widetilde{f}$

$$\widetilde{f} = \left(1 - \epsilon x^* + \frac{\epsilon^2 x^{*2}}{2} - \frac{\epsilon^3 x^{*3}}{6} + \ldots\right)\left(-\frac{1}{x^*} + \frac{1}{x^{*2}} - \frac{1}{x^{*3}} + \ldots\right). \tag{1.2.13}$$

On the other hand, if x^* is large, the e^{-x^*} term in (1.2.12) will be transcendentally small compared to all of the others, and we can write this expression as

$$\widetilde{g} = \frac{1}{x^*(1+\frac{1}{x^*})}[-1 + \epsilon x^* - \frac{\epsilon^2 x^{*2}}{2} + \frac{\epsilon^2 x^{*3}}{6} + \ldots] + \text{T.S.T.},$$

where T.S.T. denotes transcendentally small terms (see (1.1.15)). Expanding $\frac{1}{1+1/x^*}$ for large x^* gives

$$\widetilde{g} = \left(\frac{1}{x^*} - \frac{1}{x^{*2}} + \frac{1}{x^{*3}} + \ldots\right)\left(-1 + \epsilon x^* - \epsilon^2 \frac{x^{*2}}{2}\right.$$
$$\left. -\epsilon \frac{x^{*3}}{6} + \ldots\right) + \text{T.S.T.}, \tag{1.2.14}$$

and this product for $\widetilde{g}$ is identical with the product for $\widetilde{f}$.

The expansion (1.2.10) is not valid near $x = 0$, and $\tilde{f}$ approximates f near $x = 0$. Similarly, the expansion (1.2.12) is not valid for x^* near infinity, and $\tilde{g}$ approximates g near $x^* = \infty$. Therefore, it is indeed curious that $\tilde{f}$ and $\tilde{g}$ agree! The reason for this agreement will be discussed in Sec. 2.1.

1.2.3 Asymptotic Expansion for the Root of an Algebraic Equation

Sometimes a function $u = f(\mathbf{x}; \epsilon)$ is defined implicitly as the root of a certain algebraic equation $R(\mathbf{x}, u; \epsilon) = 0$ that cannot be solved explicitly for $f(\mathbf{x}; \epsilon)$ for arbitrary $\epsilon \neq 0$. If one is interested only in the solution of $R = 0$ for ϵ small, and if $R(\mathbf{x}, u; 0) = 0$ is solvable, it is useful to construct the asymptotic expansion of $f(\mathbf{x}; \epsilon)$ as $\epsilon \to 0$.

Consider the following example:

$$R(x, u; \epsilon) \equiv x - u + \epsilon \sin u = 0. \tag{1.2.15}$$

We see that for $\epsilon = 0$, we have $u = x$, and we wish to derive the next term in the asymptotic expansion of f. We set

$$f(x; \epsilon) \equiv x + \phi_2(\epsilon) f_2(x) + o(\phi_2(\epsilon)) \text{ as } \epsilon \to 0. \tag{1.2.16}$$

We are not given ϕ_2 and so we need to make a reasonable choice for this function as part of the calculation. Substituting (1.2.16) into (1.2.15) gives

$$x - (x + \phi_2(\epsilon) f_2(x) + o(\phi_2)) + \epsilon \sin(x + \phi_2(\epsilon) f_2(x) + o(\phi_2)) = 0 \text{ as } \epsilon \to 0. \tag{1.2.17}$$

Using the Taylor series for the sine function, we have $\sin u = \sin(x + \phi_2(\epsilon) f_2(x) + o(\phi_2)) = \sin x + O_s(\phi_2)$. Therefore, (1.2.17) simplifies to

$$-\phi_2(\epsilon) f_2(x) + \epsilon \sin x = o(\phi_2) \text{ as } \epsilon \to 0, \tag{1.2.18}$$

because $\epsilon O_s(\phi_2) = O_s(\epsilon\phi_2) = o(\phi_2)$ as $\epsilon \to 0$.

The choice of ϕ_2 affects the relative importance of the two terms on the left-hand side of (1.2.18). First, we note that the choice $\phi_2 \ll \epsilon$ gives a contradiction because (1.2.18) implies $\epsilon \ll \phi_2$. If, on the other hand, we choose a ϕ_2 such that $\epsilon \ll \phi_2$, (1.2.18) implies that $f_2 = 0$. In fact, we find a nontrivial $f_2(x)$ only if $\phi_2 = O_s(\epsilon)$. This does not define ϕ_2 uniquely but merely specifies its strict order of magnitude. For example, we may choose $\phi_2 = \epsilon, 5\epsilon, 100\epsilon, \epsilon + 2\epsilon^2, \sin\epsilon, \log(1 + \epsilon)$, etc. Clearly, the *simplest* choice is $\phi_2 = \epsilon$ for which $f_2 = \sin x$.

We compute the next term as follows. Let

$$f(x; \epsilon) = x + \epsilon \sin x + \phi_3(\epsilon) f_3(x) + o(\phi_3) \text{ as } \epsilon \to 0. \tag{1.2.19}$$

Now,

$$\sin u = \sin x + \epsilon \sin x \cos x + O_s(\phi_3) + O_s(\epsilon^2),$$

and substitution of (1.2.19) into (1.2.15) and simplification gives

$$-\phi_3(\epsilon) f_3(x) + \epsilon^2 \sin x \cos x = o(\phi_3) + O_s(\epsilon^3).$$

Using the same reasoning as before, we conclude that $\phi_3 = O_s(\epsilon^2)$. We choose $\phi_3 = \epsilon^2$ for simplicity and find $f_3 = \sin x \cos x$. This process can be continued indefinitely. Thus, we have found the three-term asymptotic expansion

$$f(x;\epsilon) = x + \epsilon \sin x + \epsilon^2 \sin x \cos x + o(\epsilon^2) \tag{1.2.20}$$

for the function defined by (1.2.15) with respect to the sequence $1, \epsilon, \epsilon^2, \ldots$ as $\epsilon \to 0$, and this expansion is uniformly valid for all x.

Had we chosen the sequence $c_1, c_2\epsilon, c_3\epsilon^2, \ldots$ for nonzero constants $c_1, c_2, c_3, \ldots$ the final result (12.20) would have been the same (because we would have computed $f_1 = (x/c_1)$, $f_2 = (\sin x/c_2)$, $f_3 = (\sin x \cos x/c_3)$, etc.). Suppose instead that we choose[1] the sequence $1, (\epsilon/(1+\epsilon)), (\epsilon^2/(1+\epsilon)^2), \ldots$. It is easily seen that the three-term expansion for $f(x;\epsilon)$ is now

$$f(x;\epsilon) = x + \frac{\epsilon}{1+\epsilon}\sin x + \frac{\epsilon^2}{(1+\epsilon)^2}(\sin x \cos x + \sin x) + o(\epsilon^2/(1+\epsilon)^2). \tag{1.2.21}$$

In fact, re-expanding (1.2.21) with respect to the sequence $1, \epsilon, \epsilon^2, \ldots$ gives (1.2.20). We say that two expansions are *asymptotically equivalent* with respect to a given sequence $\{\phi_n\}$ to a given $O(\phi_N(\epsilon))$ if their difference is $o(\phi_N(\epsilon))$. Thus, the two expansions (1.2.20) and (1.2.21) are asymptotically equivalent to $O(\epsilon^2)$.

Once we recognize that the appropriate asymptotic sequence $\{\phi_n(\epsilon)\}$ is $\{\epsilon^{n-1}\}$, $n = 1, 2, \ldots$, we may postulate the following form for the asymptotic expansion for u:

$$f(x;\epsilon) \sim \sum_{n=1}^{\infty} \epsilon^{n-1} f_n(x). \tag{1.2.22}$$

Substituting this into the defining relation (1.2.15) and expanding $\sin(\sum_{n=1}^{\infty} \epsilon^{n-1} f_n(x))$ to as many terms as desired gives a series in powers of ϵ that must vanish identically. Therefore, each coefficient of ϵ^n must vanish, and this sequentially gives f_1, f_2, etc.

To illustrate how nonuniformities may arise in the above process and how to deal with these, consider the following modified form of example (1.2.15):

$$R(x, u; \epsilon) \equiv x - u + \epsilon \sin u - \frac{\epsilon c}{u-1}, \tag{1.2.23}$$

where c is a constant.

Assuming an expansion of the form (1.2.22), we find the following result using the same procedure as before:

$$f_1(x) = x, \tag{1.2.24a}$$

$$f_2(x) = \sin x - \frac{c}{x-1}, \tag{1.2.24b}$$

[1] In some problems, a particular choice of asymptotic sequence, e.g., $\left\{\left(\epsilon/(1+\epsilon)\right)^n\right\}$ instead of $\{\epsilon^n\}$, may result in better numerical accuracy for certain values of ϵ and a given number of terms.

$$f_3(x) = \sin x \cos x - c\frac{\cos x}{x-1} + c\frac{\sin x}{(x-1)^2} - \frac{c^2}{(x-1)^3}. \quad (1.2.24c)$$

We note the singularity in this result at $x = 1$ if $c \neq 0$. A consequence of this singularity is that the asymptotic expansion (1.2.24) is not uniformly valid in any domain for which the point $x = 1$ is a limit point. Like (1.2.10), the series defined by (1.2.24) is the outer expansion of f as $\epsilon \to 0$, in this case for values of x fixed away from $x = 1$.

It is easy to pinpoint the source of the nonuniformity in this case. The expansion (1.2.24) is based on the premise that

$$\epsilon\frac{c}{u-1} = O(\epsilon) \text{ as } \epsilon \to 0, \quad (1.2.25)$$

and this, in particular, implies that $f = x + O(\epsilon)$ as $\epsilon \to 0$. But suppose we let $x \to 1$ at some rate that depends on ϵ, for example, by setting

$$x = 1 + x^*\phi(\epsilon), \quad (1.2.26)$$

where x^* is fixed and $\phi(\epsilon) = o(1)$ as $\epsilon \to 0$. It then follows from (1.2.24)–(1.2.25) that

$$f = 1 + O(\phi) + O\left(\frac{\epsilon}{\phi}\right) \text{ as } \epsilon \to 0,$$

which violates the premise (1.2.25) on which our calculation is based! In particular, for $x \approx 1$, it is incorrect to assume that $f = x + O(\epsilon)$ as we have done; we must determine the appropriate rate at which $f \to 1$ as $x \to 1$. To accomplish this, we set

$$u = 1 + u^*\psi(\epsilon), \quad (1.2.27)$$

where u^* is fixed and $\psi(\epsilon) = o(1)$ as $\epsilon \to 0$, and proceed to determine the order of the unknown functions $\phi(\epsilon)$ and $\psi(\epsilon)$ by re-examining (1.2.23) in this limit. Substitution of (1.2.26)–(1.2.27) into (1.2.23) gives

$$x^*\phi(\epsilon) - u^*\psi(\epsilon) + \epsilon \sin[1 + u^*\psi(\epsilon)] - \frac{\epsilon}{\psi(\epsilon)}\frac{c}{u^*} = 0 \quad (1.2.28)$$

and since $\epsilon \ll \epsilon/\psi(\epsilon)$ for any $\psi(\epsilon) \ll 1$, we may ignore the third term in (1.2.28) to leading order. The *most terms* that could remain in (1.2.28) to leading order are the first, second, and fourth terms. In order for this to occur, we must have $\phi = O_s(\psi) = O_s(\epsilon/\psi)$. For simplicity, we choose $\phi = \psi = \epsilon/\psi$ and conclude that $\phi = \psi = \sqrt{\epsilon}$. With this choice, we substitute $x = 1 + \epsilon^{1/2}x^*$, $u = 1 + \epsilon^{1/2}u^*$ into the defining relation (1.2.23) and divide by $\epsilon^{1/2}$ to obtain

$$R^*(x^*, u^*; \epsilon) \equiv x^* - u^* + \epsilon^{1/2}\sin(1 + \epsilon^{1/2}u^*) - \frac{c}{u^*} = 0. \quad (1.2.29)$$

This criterion, that the limiting expression upon rescaling has the maximal number of terms to leading order, is prevalent in various contexts of perturbation analysis. It is sometimes called the *principle of least degeneracy* or, more simply, rescaling for the *richest* limiting equation.

We may regard $R^* = 0$ as a new relation that defines the function $u^* = g(x^*; \epsilon)$ implicitly, and we proceed as before to compute its asymptotic expansion. To leading order, i.e., with $g = g_1 + o(1)$, we have the quadratic equation

$$g_1^2 - x^* g_1 + c = 0 \tag{1.2.30}$$

that has the two solutions

$$g_1^+ = \frac{1}{2}(x^* + \sqrt{x^{*^2} - 4c}); \quad g_1^- = \frac{1}{2}(x^* - \sqrt{x^{*^2} - 4c}). \tag{1.2.31}$$

These are real for all x^* if $c \leq 0$, and for $x^* \geq 2\sqrt{c}$ if $c > 0$.

The next term in the asymptotic expansion of g is $O_s(\epsilon^{1/2})$. Substituting

$$g(x^*; \epsilon) = g_1(x^*) + \epsilon^{1/2} g_2(x^*) + o(\epsilon^{1/2}) \tag{1.2.32}$$

into (1.2.29) gives

$$x^* - g_1 - \frac{c}{g_1} + \epsilon^{1/2}\left(s + c\frac{g_2}{g_1^2} - g_2\right) = o(\epsilon^{1/2}),$$

where $s = \sin 1 = 0.841471$. Setting the coefficient of the $\epsilon^{1/2}$ term equal to zero gives

$$g_2^+ = \frac{s}{1 - c/g_1^{+^2}}; \quad g_2^- = \frac{s}{1 - c/g_1^{-^2}}. \tag{1.2.33}$$

Therefore, the two inner expansions for u are given by

$$u = 1 + \epsilon^{1/2} u^* = \begin{cases} 1 + \epsilon^{1/2} g_1^+ + \epsilon g_2^+ + o(\epsilon) & (1.2.34\text{a}) \\ 1 + \epsilon^{1/2} g_1^- + \epsilon g_2^- + o(\epsilon). & (1.2.34\text{b}) \end{cases}$$

The expansion (1.2.32) to $O(\epsilon^{1/2})$ is invariant under the transformation $x^* \to -x^*$ if we replace $g_1^+ \to -g_1^-$. Therefore, it suffices to restrict attention to $x^* > 0$.

If $c > 0$, g_2^+ and g_2^- become singular when the two roots g_1^+ and g_1^- coalesce to $\sqrt{c}$, i.e., at $x^* = 2\sqrt{c}$. Therefore, if $c > 0$ the expansions (1.2.34) are not uniformly valid in any domain that contains the point $x^* = 2\sqrt{c}$ or $x = \frac{2\sqrt{c}-1}{\epsilon^{1/2}}$ as a limit point. The appropriate expansion near this point is left as an exercise (see Problem 3).

Notice that as $x^* \to \infty$, $g_1^+ = O(x^*)$ (and as $x^* \to -\infty$, $g_1^- = O(x^*)$). Therefore, the expansion (1.2.34a) is also not uniformly valid for large x^*. But, in this case, the expansion (1.2.24) is valid and can be used to approximate u.

To explore further the connection between (1.2.24) and (1.2.34a), let us focus first on (1.2.34a) and choose $x^* > 0$ (i.e., $x > 1$). We re-expand (1.2.24) and (1.2.34a) for values of x for which each becomes nonuniform as we did in (1.2.13) and (1.2.14). We set $x = 1 + \epsilon^{1/2} x^*$ in (1.2.24), expand the result holding x^* fixed, and collect terms according to their powers of ϵ to find

$$u = (1 + \epsilon^{1/2} x^*) + \epsilon\left[\sin(1 + \epsilon^{1/2} x^*) - \frac{c}{\epsilon^{1/2} x^*}\right]$$

$$+\epsilon^2 \Bigg[\sin(1+\epsilon^{1/2}x^*)\cos(1+\epsilon^{1/2}x^*)$$

$$-\frac{c\cos(1+\epsilon^{1/2}x^*)}{\epsilon^{1/2}x^*} + \frac{c\sin(1+\epsilon^{1/2}x^*)}{\epsilon x^{*2}} - \frac{c^2}{\epsilon^{3/2}x^{*3}}\Bigg] + \dots$$

$$= 1 + \epsilon^{1/2}\left(x^* - \frac{c}{x^*} - \frac{c^2}{x^{*3}}\right) + \epsilon\left(s + \frac{sc}{x^{*2}}\right) + \dots, \tag{1.2.35}$$

where ... denotes $o(\epsilon)$ as $\epsilon \to 0$ with x^* fixed. On the other hand, if we evaluate (1.2.34a) for $x^* \to \infty$, where it fails to be uniformly valid, we first calculate

$$g_1^+ = x^* - \frac{c}{x^*} - \frac{c^2}{x^{*2}} + O(x^{*-3}) \text{ as } x^* \to \infty,$$

$$g_2^+ = s + \frac{sc}{x^{*2}} + O(x^{*-4}) \text{ as } x^* \to \infty.$$

Therefore, as $x^* \to \infty$ the expansion (1.2.34) for u becomes

$$u = 1 + \epsilon^{1/2}\left(x^* - \frac{c}{x^*} - \frac{c^2}{x^{*2}} + O(x^{*-3})\right) + \epsilon\left(s + \frac{sc}{x^{*3}} + O(x^{*-4})\right)$$

$$+ o(\epsilon) \text{ as } \epsilon \to 0,\ x^* \to \infty. \tag{1.2.36}$$

We see that (1.2.35) and (1.2.36) *agree up to all the terms we have retained.* Again, the basis for this correspondence will become evident when we discuss matching of asymptotic expansions in Sec. 2.1.

At this point, it is reasonable to regard (1.2.24) and (1.2.34a) (or (1.2.24) and (1.2.34b) if $x^* < 0$) as asymptotic expansions of the *same root* of (1.2.23) valid in different x-domains. In addition to this root, we have the second root having the expansion (1.2.34b) or (1.2.34a) if $x^* < 0$. This expansion is uniformly valid as $x^* \to \infty$ and predicts that $u \downarrow 1$ in this limit.

1.2.4 Asymptotic Expansion for a Definite Integral

A less trivial definition of a function involves an integral representation. Consider, for example, the error function

$$\operatorname{erf}\lambda = \frac{2}{\sqrt{\pi}}\int_0^\lambda e^{-t^2}dt. \tag{1.2.37a}$$

Since $\int_0^\infty e^{-t^2}dt = \sqrt{\pi}/2$, we may write

$$\operatorname{erf}\lambda = 1 - \frac{2}{\sqrt{\pi}}\int_\lambda^\infty e^{-t^2}dt, \tag{1.2.37b}$$

and by setting $t^2 = \tau$ this becomes

$$\operatorname{erf}\lambda = 1 - \frac{1}{\sqrt{\pi}}\int_{\lambda^2}^\infty e^{-\tau}\tau^{-1/2}d\tau. \tag{1.2.37c}$$

The form (1.2.37c) is chosen because upon integration by parts we find

$$\operatorname{erf}\lambda = 1 - \frac{1}{\sqrt{\pi}}\left[\frac{e^{-\lambda^2}}{\lambda} - \frac{1}{2}\int_{\lambda^2}^{\infty} e^{-\tau}\tau^{-3/2}d\tau\right],$$

and this suggests repeating the process in order to generate an expansion in increasing powers of $\epsilon \equiv \lambda^{-1}$. If such an expansion were asymptotic in the limit $\lambda \to \infty$ ($\epsilon \to 0$), it would be useful for numerical evaluation of erf λ for λ large.

Defining

$$F_n(\lambda) = \int_{\lambda^2}^{\infty} e^{-\tau}\tau^{-(2n+1)/2}d\tau;\ n = 0, 1, 2, \ldots \tag{1.2.38}$$

and integrating $F_n(\lambda)$ by parts results in the recursion relation

$$F_n(\lambda) = \frac{e^{-\lambda^2}}{\lambda^{2n+1}} - \frac{(2n+1)}{2}F_{n+1}(\lambda);\ n = 0, 1, 2, \ldots. \tag{1.2.39}$$

This can be used to calculate the following *exact* result for F_0

$$F_0(\lambda) = e^{-\lambda^2}\left[\frac{1}{\lambda} - \frac{1}{2\lambda^3} + \frac{1\cdot 3}{2^2\lambda^5} + \ldots + \frac{(-1)^{n-1}1\cdot 3\cdot 5\ldots(2n-3)}{2^{n-1}\lambda^{2n-1}}\right]$$

$$+(-1)^n\frac{1\cdot 3\cdot 5\ldots(2n-1)}{2^n}F_n(\lambda);\ n = 1, 2, \ldots. \tag{1.2.40}$$

Equation (1.2.40) provides a formal series in ascending powers of λ^{-1} together with an exact expansion for the remainder if the series is truncated after n terms. To show that the bracketed expansion in (1.2.40) is the asymptotic expansion of F_0, we must verify that (1.2.6) is satisfied, i.e., that

$$F_0(\lambda) - e^{-\lambda^2}\sum_{n=1}^{M}(-1)^{n-1}\frac{1\cdot 3\cdot 5\ldots(2n-3)}{2^{n-1}\lambda^{2n-1}} = O(\lambda^{-(2M-1)}e^{-\lambda^2}) \tag{1.2.41}$$

as $\lambda \to \infty$.

According to (1.2.40), the above reduces to showing that $S_M(\lambda)$ defined by

$$S_M(\lambda) = \lambda^{2M-1}e^{\lambda^2}\frac{(-1)^M 1\cdot 3\cdot 5\ldots(2M-1)}{2^M}F_M(\lambda) \tag{1.2.42}$$

tends to zero as $\lambda \to \infty$. This is easily accomplished once we note that

$$F_M(\lambda) \le \frac{1}{\lambda^{2M+1}}\int_{\lambda^2}^{\infty} e^{-\tau}d\tau = \frac{e^{-\lambda^2}}{\lambda^{2M+1}}. \tag{1.2.43}$$

Therefore,

$$|S_M(\lambda)| \le \frac{1\cdot 3\cdot 5\ldots(2M-1)}{2^M\lambda^2},$$

and hence $S_M = o(1)$ as $\lambda \to \infty$.

We note that the asymptotic expansion

$$\operatorname{erf} \lambda \sim 1 - \frac{e^{-\lambda^2}}{\sqrt{\pi}} \sum_{n=1}^{\infty} \frac{(-1)^{n-1} 1 \cdot 3 \cdot 5 \ldots (2n-3)}{2^{n-1} \lambda^{2n-1}} \tag{1.2.44}$$

is *divergent* because the absolute value of the coefficients of λ^{-2n+1} in the series in (1.2.44) becomes large as n increases. Actually, (1.2.40) provides an exact expression for the error resulting from using M terms of the expansion (1.2.44) to approximate erf λ . It is easily verified that for any fixed λ there is an optimal value M_0 of M in the sense that the absolute value of the error decreases as the number M of terms retained increases, as long as $M < M_0$. But, if we insist on retaining $M > M_0$ terms, the absolute value of the error will increase with M. Moreover, M_0 depends on λ and M_0 increases as λ increases, whereas the absolute value of the error for the series with M_0 terms decreases as λ increases. The above features are typical of divergent asymptotic expansions.

The reader may verify that for $\lambda = 2$ the series in (1.2.44) gives the best accuracy if five terms are used and that the absolute value of the error in this case is only 6.43×10^{-5}, which is remarkable since $\lambda = 2$ is not a large number.

In general, we do not need to have an exact result such as (1.2.40) in order to determine when an asymptotic expansion begins to diverge. We need only monitor the absolute value of each successive term in the expansion; the optimal cutoff value M_0 occurs when we calculate the smallest absolute value for the added term. This argument presumes that the absolute value of each added term decreases monotonically up to $M = M_0$ and increases monotonically thereafter. It is possible, but rather unlikely, to encounter an example where the absolute value of each added term oscillates about some mean value.

Functions defined in integral form also occur naturally in the solution of linear partial differential equations by transform techniques. Various methods, such as stationary phase and steepest descents, have been developed for calculating the asymptotic behavior of the integral representation of the solution. A discussion of this topic is beyond the scope of this book and the reader is referred to standard texts (for example, [1.1], [1.2] and [1.5]).

Problems

1. Calculate the asymptotic expansion to $O(\epsilon^2)$ of the function

$$u(x;\epsilon) = \sin\sqrt{1-\epsilon x} \tag{1.2.45}$$

for a fixed $x < \infty$ with respect to the sequence $1, \epsilon, \epsilon^2, \ldots$. Show that the result is not uniformly valid on $0 \le x < \infty$. Show, however, that the approximation $\sin\left(1 - \frac{\epsilon}{2} - \frac{\epsilon^2}{8}\right) x$ is uniformly valid to $O(\epsilon)$ for all x in $0 \le x \le X(\epsilon)$ where $X(\epsilon) = O(\epsilon^{-1})$ as $\epsilon \to 0$.

2. Consider the cubic

$$R(x, y, u; \epsilon) \equiv \epsilon u^3 + u^2 - f^2(x, y) = 0, \tag{1.2.46}$$

where $f(x, y)$ is prescribed for all x and y.

a. Show that if $f^2 < 4/27\epsilon^2$, $R = 0$ has three real roots: $u^+(x, y; \epsilon)$, $u^-(x, y; \epsilon)$, and $\widetilde{u}(x, y; \epsilon)$ where $u^+ > 0$, $u^- < 0$, and $\widetilde{u} < u^-$.

b. Calculate the expansions of these roots for $\epsilon \to 0$ in the form

$$u^{\pm}(x, y; \epsilon) = \pm f + \epsilon(-\frac{1}{2} f^2) + \epsilon^2(\pm\frac{5}{8} f^3) + O(\epsilon^3), \quad (1.2.47)$$

$$\widetilde{u}(x, y; \epsilon) = -\frac{1}{\epsilon} + \epsilon f^2 + \epsilon^3 2 f^4 + O(\epsilon^5). \quad (1.2.48)$$

c. Show that for the choice $f = x \tanh y$, the above expansions are not uniformly valid as $|x| \to \infty$. What is the correct scaling for x, y, and u in $R = 0$ to calculate uniformly valid results as $|x| \to \infty$?

3. Calculate the asymptotic expansion for the roots of (1.2.29) with $c > 0$, when $x^* \approx 2\sqrt{c}$.

4. One often encounters an integral representation for a function $u(t)$ that becomes singular in the limit as $t \to t_0$ (see Problem 2 of Sec. 2.4). The asymptotic expansion of such a function in the limit $\epsilon \equiv (t_0 - t) \to 0$ can still be derived. As an example, consider the integral

$$u(t) = \int_0^t \frac{\sin(t-\tau)d\tau}{[1 - \sin\tau + \tau\cos\tau]^2}, \quad (1.2.49)$$

which is singular as $t \to \pi/2$. To compute the asymptotic expansion of u as $\epsilon = \pi/2 - t \to 0$, we change the variable of integration from τ to $\sigma = \pi/2 - \tau$.

5. Show that u may be written in the form

$$\widetilde{u}(\epsilon) \equiv u\left(\frac{\pi}{2} - \epsilon\right) = \cos\epsilon \int_\epsilon^{\pi/2} \frac{\sin\sigma}{D(\sigma)} d\sigma - \sin\epsilon \int_\epsilon^{\pi} \frac{\cos\sigma}{D(\sigma)} d\sigma, \quad (1.2.50)$$

where

$$D(\sigma) = [1 - \cos\sigma + \left(\frac{\pi}{2} - \sigma\right)\sin\sigma]^2.$$

6. Show that the integrands in (1.2.50) have the following expansions:

$$\frac{\sin\sigma}{D(\sigma)} = \frac{4}{\pi^2\sigma} + O(1) \text{ as } \sigma \to 0, \quad (1.2.51)$$

$$\frac{\cos\sigma}{D(\sigma)} = \frac{4}{\pi^2\sigma^2} + \frac{8}{\pi^3\sigma} + O(1) \text{ as } \sigma \to 0, \quad (1.2.52)$$

where we have exhibited the singular terms only. Subtracting out the singular parts of the integrands in (1.2.36) and then adding these gives the following *exact* alternate form for $\widetilde{u}(\epsilon)$:

$$\widetilde{u}(\epsilon) = \cos\epsilon \int_\epsilon^{\pi/2} \left[\frac{\sin\sigma}{D(\sigma)} - \frac{4}{\pi^2\sigma}\right] d\sigma + \frac{4}{\pi^2}\cos\epsilon \int_\epsilon^{\pi/2} \frac{d\sigma}{\sigma}$$
$$- \sin\epsilon \int_\epsilon^{\pi/2} \left[\frac{\cos\sigma}{D(\sigma)} - \frac{4}{\pi^2\sigma^2} - \frac{8}{\pi^3\sigma}\right] d\sigma$$

$$- \frac{4}{\pi^2} \sin \epsilon \int_{\epsilon}^{\pi/2} \frac{d\sigma}{\sigma^2} - \frac{8}{\pi^3} \sin \epsilon \int_{\epsilon}^{\pi/2} \frac{d\sigma}{\sigma}$$

$$= \cos \epsilon \int_{\epsilon}^{\pi/2} F(\sigma) d\sigma - \sin \epsilon \int_{\epsilon}^{\pi/2} G(\sigma) d\sigma$$

$$+ \frac{4}{\pi^2} \left[\log \frac{\pi}{2} - \log \epsilon \right] \cos \epsilon + \frac{4}{\pi^2} \left(\frac{1}{\pi} - \frac{1}{\epsilon} \right) \sin \epsilon$$

$$+ \frac{8}{\pi^3} \left(\log \epsilon - \log \frac{\pi}{2} \right) \sin \epsilon, \tag{1.2.53}$$

where

$$F(\sigma) = \frac{\sin \sigma}{D(\sigma)} - \frac{4}{\pi^2 \sigma} = O(1) \text{ as } \sigma \to 0$$

$$G(\sigma) = \frac{\cos \sigma}{D(\sigma)} - \frac{4}{\pi^2 \sigma^2} - \frac{8}{\pi^3 \sigma} = O(1) \text{ as } \sigma \to 0.$$

Now, the two integrals in (1.2.53) exist as $\epsilon \to 0$ since $F(\sigma)$ and $G(\sigma)$ are regular at $\sigma = 0$.

7. Show that the asymptotic expansion for $\tilde{u}(\epsilon)$ to three terms is given by

$$\tilde{u}(\epsilon) = -\frac{4}{\pi^2} \log \epsilon + \left[\int_0^{\pi/2} F(\sigma) d\sigma + \frac{4}{\pi^2} \left(\log \frac{\pi}{2} - 1 \right) \right]$$

$$+ \frac{8}{\pi^3} \epsilon \log \epsilon + O(\epsilon) \text{ as } \epsilon \to 0. \tag{1.2.54}$$

8. Consider the initial value problem

$$\frac{d^2 u}{dt^2} + u = \frac{1}{t}; \quad \pi \le t < \infty \tag{1.2.55}$$

with zero initial conditions: $u(\pi) = \frac{du}{dt}(\pi) = 0$.

a. Write the exact solution

$$u(t) = \int_{\pi}^{t} \frac{\sin(t - \tau)}{\tau} d\tau \tag{1.2.56}$$

in the form

$$u(t) = a \sin t + b \cos t - \int_{t}^{\infty} \frac{\sin(t - \tau)}{\tau} d\tau, \tag{1.2.57}$$

where

$$a = \int_{\pi}^{\infty} \frac{\cos \tau}{\tau} d\tau = -0.073668$$

$$b = -\int_{\pi}^{\infty} \frac{\sin \tau}{\tau} d\tau = -1.815937.$$

b. Consider the function

$$F_n(t) = -\int_{t}^{\infty} \frac{\sin(t - \tau)}{\tau^n} d\tau; \quad n = 1, 2, \ldots. \tag{1.2.58}$$

Thus, we are interested in $F_1(t)$. Integrate this by parts twice to derive the recursion relation

$$F_n(t) = \frac{1}{t^n} - n(n+1)F_{n+2}(t), \quad n = 1, 2, \ldots, \tag{1.2.59}$$

and use this to derive the exact result

$$F_1(t) = \sum_{n=1}^{N} \frac{(2n-2)!(-1)^{n-1}}{t^{2n-1}} + (2N)!(-1)^N F_{2N+1}(t); \quad N = 1, \ldots. \tag{1.2.60}$$

c. Show that the series in (1.2.60) gives the asymptotic expansion of F_1 as $t \to \infty$ and that the series diverges.

d. Observe that

$$u = a \sin t + b \cos t + \sum_{n=1}^{\infty} \frac{c_n}{t^n} \tag{1.2.61}$$

is formally a general solution of (1.2.55). Substitute into (1.2.55) and show that $c_{2n} = 0$, $c_{2n-1} = (2n-2)!(-1)^{n-1}$ as in (1.2.60).

9. Consider the nonlinear equation

$$\frac{d^2u}{dt^2} - \sin u = -\frac{1}{2}, \quad 0 \le t < \infty \tag{1.2.62}$$

with zero initial conditions $u(0) = \frac{du}{dt}(0) = 0$.

a. Multiply (1.2.62) by du/dt to derive the energy integral

$$\frac{1}{2}\left(\frac{du}{dt}\right)^2 + \cos u + \frac{1}{2}u = 1. \tag{1.2.63}$$

b. Use (1.2.63) to show that $u(t)$ is the inverse of the expression

$$t = -\int_0^u \frac{d\eta}{\sqrt{2 - \eta - 2\cos\eta}}. \tag{1.2.64}$$

c. Argue that $u \to -t^2/4$ as $t \to \infty$ and write (1.2.63) for $u < 0$ in the form

$$t = 2z^{1/2} + \kappa - \int_z^\infty \left[\frac{1}{\sqrt{2 + \eta - 2\cos\eta}} - \frac{1}{\sqrt{\eta}}\right] d\eta, \tag{1.2.65}$$

where $z = -u$ and κ is the constant

$$\kappa = \int_0^\infty \left[\frac{1}{\sqrt{2 + \eta - 2\cos\eta}} - \frac{1}{\sqrt{\eta}}\right] d\eta \approx -1.72.$$

Show that (1.2.65) has the expansion

$$t = 2z^{1/2} + \kappa + 2z^{-1/2} - \frac{3}{2}z^{-3/2} + z^{-3/2}\sin z + O(z^{-5/2}) \quad \text{as } z \to \infty. \tag{1.2.66}$$

d. Invert (1.2.66) to compute the following asymptotic expansion for u

$$u = -\frac{t^2}{4} + \frac{\kappa t}{2} + \left(2 - \frac{\kappa^2}{4}\right) + \frac{1}{t^2}\left[4\sin\left(\frac{t}{2} - \frac{\kappa}{2}\right)^2 - 2\right] + O(t^{-3}) \text{ as } t \to \infty. \tag{1.2.67}$$

10. Show that

$$e^{-x^2}\int_0^x e^{s^2}ds = \frac{1}{2x} + o(x^{-1}) \text{ as } x \to \infty. \tag{1.2.68}$$

Hint: Split the integral into one over $(0, a)$ plus one over (a, x), where a is a constant. In the integral over (a, x), change the variable integration for s to $t = \sqrt{s}$ and then integrate by parts.

1.3 Regular Expansions for Ordinary and Partial Differential Equations

In this section, we study functions $u(\mathbf{x}; \epsilon)$ that are defined as the solutions of ordinary or partial differential equations that involve the parameter ϵ. We restrict attention to the class of problems, often called *regular perturbation* problems, where the asymptotic expansion of the solution can be directly derived in a form that remains uniformly valid throughout the domain of interest. We point out via examples that many cases of interest do not fall in this category and that an asymptotic expansion derived by a given limit process often fails somewhere in the domain of interest. Such singular perturbation problems will be our focus for the major part of this book.

To fix ideas, let L and M be given differential operators. For simplicity, we assume L to be linear and consider the differential equation

$$L(u) + \epsilon M(u) = 0 \tag{1.3.1}$$

in some domain D with initial and/or boundary conditions that do not involve ϵ. We will later consider examples where L is nonlinear and where the initial or boundary data depend on ϵ. Suppose that $u_0(\mathbf{x})$, the solution of

$$L(u_0) = 0, \tag{1.3.2}$$

satisfying the given initial and/or boundary data, is known. We assume a solution of the perturbed problem (1.3.1) in the form

$$u(\mathbf{x}; \epsilon) = u_0(\mathbf{x}) + \phi_1(\epsilon)u_1(\mathbf{x}) + o(\phi_1), \tag{1.3.3}$$

where $\phi_1(\epsilon) = o(1)$ as $\epsilon \to 0$ but is otherwise unknown. Substituting the perturbation expansion (1.3.3) in (1.3.1) gives

$$\phi_1(\epsilon)L(u_1) + o(\phi_1) + \epsilon M(u_0) = 0, \tag{1.3.4}$$

because $L(u_0 + \phi_1 u_1) = L(u_0) + \phi_1 L(u_1)$ for a linear operator, and $L(u_0) = 0$. If $\phi_1 = O_s(\epsilon)$, say $\phi_1 = \epsilon$, u_1 obeys the linear inhomogeneous equation

$$L(u_1) = -M(u_0) \tag{1.3.5}$$

subject to zero initial and/or boundary data. Because (1.3.5) is inhomogeneous, one finds a nontrivial solution $u_1(\mathbf{x})$.

The other choice, $\epsilon \ll \phi_1$, is not of interest because it gives $L(u_1) = 0$, and this, along with the vanishing of the initial and/or boundary values, usually implies $u_1 = 0$. The third alternative, $\phi_1 \ll \epsilon$, leads to an inconsistent condition; it requires that we set $M(u_0) = 0$, an algebraic relation that is not true in general.

Once u_1 is calculated using (1.3.5), we modify (1.3.3) to include the next higher-order term and proceed to derive the equation it obeys. Or we anticipate the structure of the expansion, say, $\phi_1 = \epsilon$, $\phi_2 = \epsilon^2, \ldots$ and solve the sequence of inhomogeneous equations that result from (1.3.1)

$$L(u_i) = f_i(\mathbf{x});\ i = 0, 1, 2, \ldots, \tag{1.3.6}$$

where $f_0 = 0$, and each f_i for $i > 0$ is a function of the previously calculated solutions. The examples presented in this section illustrate these ideas.

1.3.1 Ordinary Differential Equation Examples

A first-order equation

Consider the first-order ordinary differential equation

$$u' + 2xu - \epsilon u^2 = 0 \tag{1.3.7}$$

for $\epsilon \to 0$ on $0 \le x < \infty$, where $' \equiv d/dx$, and we impose the initial condition

$$u(0) = 1. \tag{1.3.8}$$

The unperturbed problem satisfies $u_0' + 2xu_0 = 0$ and $u_0(0) = 1$. The solution is easily found:

$$u_0(x) = e^{-x^2}. \tag{1.3.9}$$

Substituting the expansion

$$u(x;\epsilon) = u_0(x) + \phi_1(\epsilon)u_1(x) + o(\phi_1) \tag{1.3.10}$$

into (1.3.7) gives

$$\phi_1(\epsilon)[u_1' + 2xu_1] - \epsilon e^{-2x^2} = o(\phi_1). \tag{1.3.11}$$

The initial condition (1.3.8) implies that $u_1(0) = 0$. If $\epsilon \ll \phi_1$, u_1 obeys the homogeneous equation $u_1' + 2xu_1 = 0$ with zero boundary condition, and we find $u_1(x) = 0$. The choice $\phi_1 \ll \epsilon$ in (1.3.11) gives the inconsistent requirement $e^{-2x^2} = 0$. A nontrivial $u_1(x)$ results only if $\phi_1 = O_s(\epsilon)$, and we choose $\phi_1 = \epsilon$

for simplicity. We then have $u_1' + 2xu_1 = e^{-2x^2}$ with $u_1(0) = 0$. The solution is

$$u_1(x) = e^{-x^2} \int_0^x e^{-s^2} ds. \tag{1.3.12}$$

The expansion

$$u(x; \epsilon) = e^{-x^2} + \epsilon e^{-x^2} \int_0^x e^{-s^2} ds + o(\epsilon)$$

is uniformly valid in $D : 0 \le x < \infty$. This follows from the fact that $e^{-x^2} < 1$ and $\int_0^x e^{-s^2} ds < \int_0^\infty e^{-s^2} ds = \sqrt{\pi}/2$ on D. Hence $0 \le u_1(x) < \sqrt{\pi}/2$ on D.

Perturbed oscillator

Consider next an oscillator with a nearly constant frequency modeled by

$$u'' + (1 - \epsilon e^{-ax})u = 0, \tag{1.3.13}$$

where a is a positive constant. We take the initial conditions to be: $u(0) = 0$, $u'(0) = 1$. Proceeding as in the previous example, we look for an expansion in the form

$$u(x; \epsilon) = u_0(x) + \phi_1(\epsilon)u_1(x) + o(\phi_1), \tag{1.3.14}$$

and we find that u_0 satisfies $u_0'' + u_0 = 0$ with $u_0(0) = 0$ and $u_0'(0) = 1$. Thus,

$$u_0(x) = \sin x. \tag{1.3.15}$$

Unless we choose $\phi_1(\epsilon) = O_s(\epsilon)$, we find no consistent result for u_1; we pick $\phi_1 = \epsilon$ for simplicity. We then find the equation

$$u_1'' + u_1 = e^{-ax} \sin x$$

with initial conditions $u_1(0) = u_1'(0) = 0$. The solution is easily found:

$$u_1(x) = \frac{e^{-ax} + 1}{4 + a^2} \sin x + \frac{2(e^{-ax} - 1)}{a(4 + a^2)} \cos x. \tag{1.3.16}$$

The expansion to $O(\epsilon)$ we have calculated for u is uniformly valid for all x in $0 \le x < \infty$ since, for all x in this domain, we have

$$u_1 \le \frac{1 + e^{-ax}}{4 + a^2} + \frac{2(1 - e^{-ax})}{a(4 + a^2)} \le \frac{1}{4 + a^2} + \frac{2}{a(4 + a^2)} = \frac{a + 2}{a(4 + a^2)}.$$

The reader should not be lulled into a false sense of confidence in the efficacy of this approach. In general, a small perturbation to the harmonic oscillator equation will introduce a cumulative perturbation that *cannot* be uniformly described by a regular expansion such as (1.3.14) if x is allowed to become large.

To show this, we replace the exponentially decaying term ϵe^{-ax} in (1.3.13) by a term that does not decay as $x \to \infty$. For example, consider

$$u'' + (1 + \epsilon u^2)u = 0 \tag{1.3.17}$$

with the same initial conditions: $u(0) = 0$, $u'(0) = 1$.

We find $u_0 = \sin x$ as before, but u_1 now obeys

$$u_1'' + u_1 = -u_0^2 = -\sin^3 x = -\frac{3}{4}\sin x + \frac{1}{4}\sin 3x. \tag{1.3.18}$$

The forcing term, $\frac{3}{4}\sin x$, has the same frequency as the homogeneous solution and gives rise to a resonant response. In fact, the solution for u_1 satisfying $u_1(0) = u_1'(0) = 0$ is found to be

$$u_1(x) = -\frac{9}{32}\sin x - \frac{1}{32}\sin 3x + \frac{3}{8}x\cos x, \tag{1.3.19}$$

where the last term is due to the resonant forcing. Clearly, the expansion $u_0 + \epsilon u_1$ is uniformly valid in any interval $0 \le x \le X(\epsilon)$ so long as $X(\epsilon) = O(1)$ as $\epsilon \to 0$. The term $(3/8)x\cos x$ oscillates with an amplitude that increases linearly with x. Consequently, its contribution to ϵu_1 will become $O(1)$ and make the expansion $u_0 + \epsilon u_1$ nonuniform if $X(\epsilon) = O(\epsilon^{-1})$. Thus, the cumulative effect of the small term ϵu^3 in (1.3.17) is not correctly taken into account in a regular expansion. We will discuss the appropriate expansion for this type of nonuniformity in Chapter 4.

Perturbed two-point boundary-value problems

A two-point boundary-value problem on a finite domain involving a small parameter ϵ usually has a regular expansion if ϵ does not multiply a term that becomes large along the solution.

Consider the following example:

$$(xu')' - \epsilon u = 0 \tag{1.3.20}$$

on $1 \le x \le 2$ with $u(1) = 1$ and $u(2) = 2$. The unperturbed problem has a singularity at $x = 0$, but this is outside our domain, and we find

$$u_0(x) = \frac{\log x}{\log 2} + 1. \tag{1.3.21}$$

The next term in the expansion $u = u_0 + \epsilon u_1$ obeys $(xu_1)' = u_0$ with zero boundary conditions: $u_1(1) = u_1(2) = 0$. We find

$$u_1(x) = \left[\frac{2}{(\log 2)^2} - \frac{3}{\log 2}\right]\log x + \frac{2}{\log 2} - 1 + \frac{x\log x}{\log 2} + \left(1 - \frac{2}{\log 2}\right)x. \tag{1.3.22}$$

Again, since $u_1(x)$ is bounded in $1 \le x \le 2$, the expansion $u_0 + \epsilon u_1$ is uniformly valid to $O(\epsilon)$ in this interval.

A regular expansion procedure may fail for one of several possible reasons. We saw in the previous example that for a problem over an unbounded domain, the cumulative effect of a small term may not be relegated to higher order. For problems over a bounded domain, one of the reasons a regular expansion may fail is if ϵ multiplies a term in the differential equation that becomes large somewhere in the domain or its boundary.

For example, consider

$$\epsilon u'' + u' = 2 \tag{1.3.23}$$

with $u(0) = 0$ and $u(1) = 1$. The exact solution is easily found:

$$u = 2x + \frac{1 - e^{-x/\epsilon}}{e^{-1/\epsilon} - 1} \equiv f(x;\epsilon). \tag{1.3.24}$$

In the limit $\epsilon \to 0$ with x fixed not equal to zero, we have $f(x;0) = f_0(x) = 2x - 1$. It is easily seen that $f - 2x + 1$ is transcendentally small; hence the outer expansion in this limit consists of the one term $2x - 1$. We note immediately that f_0 does not satisfy the boundary condition at the origin, but the right boundary condition is satisfied since $f_0(1) = 1$.

If we attempt to derive this result from the governing differential equation (1.3.23) without recourse to the exact solution, we are faced with a dilemma. Since ϵ multiplies the second derivative term in (1.3.23), the leading approximation $f_0(x) \equiv f(x;0)$ satisfies the first-order equation $f_0' = 2$; its solution $f_0(x) = 2x + c$ involves only one integration constant c and cannot satisfy both boundary conditions. Evidently, the correct choice is to have f_0 satisfy the right boundary condition $f_0(1) = 1$, but this choice is not directly obvious without knowledge of the exact solution.

One might argue that a transformation of independent variable $x \to x^* = x/\epsilon$ would circumvent the difficulty of having ϵ multiply the highest derivative. In fact, the exact expression becomes

$$u = 2\epsilon x^* + \frac{1 - e^{-x^*}}{e^{-1/\epsilon} - 1} \equiv g(x^*;\epsilon), \tag{1.3.25}$$

and in the limit $\epsilon \to 0$, x^* fixed, we find $g_0(x^*) \equiv g(x^*, 0) = e^{-x^*} - 1$. This expression satisfies the boundary condition at the origin but not the one at $x = 1$.

As pointed out in connection with example (1.2.9), f_0 is valid for $x \neq 0$ and g_0 is valid for $x^* \neq \infty$. Therefore, we should not expect $f_0(0) = 0$ or $g_0(\infty) = 1$. However, we again note the curious identity $f_0(0) = g_0(\infty)$ that will be explained when we discuss matching in Sec. 2.1.

We can also calculate g_0 by solving an appropriate limiting differential equation in terms of the x^* variable. Upon substitution of $u = g(x^*;\epsilon)$ into (1.3.23) and the boundary conditions, we find

$$\frac{d^2g}{dx^{*^2}} + \frac{dg}{dx^*} = 2\epsilon \tag{1.3.26}$$

with $g(0;\epsilon) = 0$ and $g(\frac{1}{\epsilon};\epsilon) = 1$. Thus, g_0 satisfies $d^2g_0/dx^{*^2} + dg_0/dx^* = 0$ with $g_0(0) = 0$.

If we also require g_0 to satisfy the right boundary condition, i.e., $g_0(\infty) = 1$, we obtain the *incorrect* result $g_0 = 1 - e^{-x^*}$. As we shall see later on, we can only require $g_0(0) = 0$ to find $g_0 = k(1 - e^{-x^*})$. The *matching condition* $f_0(0) = g_0(\infty)$ then gives the correct result $k = -1$.

In summary, we see that for this example neither the outer limit ($\epsilon \to 0$, x fixed $\neq 0$) nor the inner limit ($\epsilon \to 0$, x^* fixed $\neq \infty$) individually defines an approximation that is valid throughout $0 \leq x \leq 1$. This feature characterizes a *singular perturbation* problem. In contrast, the asymptotic expansion for a *regular* *perturbation* problem valid throughout the domain can be obtained by a single limit process.

The outer limit fails near $x = 0$ because it presupposes that $\epsilon f'' = O(\epsilon)$. However, as is easily seen using the exact solution, we have $\epsilon f''(x; 0) = O(\epsilon^{-1})$. On the other hand, the inner limit fails near $x^* = \infty$ because it presupposes that d^2g/dx^* and dg/dx^* are both $O(1)$ and dominate over the 2ϵ term on the right-hand side of (1.3.26). However, both of these terms are transcendentally small and negligible compared to 2ϵ if $x^* \to \infty$.

A singular perturbation problem need not be characterized by ϵ multiplying the highest derivative term in the governing equation. Examples are given in Secs. 2.4 and 3.3.

1.3.2 A Perturbed Eigenvalue Problem

A variety of physically interesting problems involve calculating the eigenvalues and eigenfunctions of a perturbed linear self-adjoint operator. A particularly simple example that illustrates ideas is that of the small amplitude transverse vibrations of a string over an elastic support with a small linear but spatially dependent restoring force. In appropriate dimensionless variables (see section 3.1 of [1.4]), we study the wave equation

$$u_{tt} - u_{xx} + \epsilon u \sin x = 0; \quad 0 \le x \le \pi. \tag{1.3.27}$$

Thus, the elastic support exerts a restoring force proportional to $\sin x$. We assume the string is fixed at the endpoints $x = 0, x = \pi$

$$u(0, t; \epsilon) = u(\pi, t; \epsilon) = 0, \tag{1.3.28}$$

and take arbitrary initial conditions for the displacement and velocity

$$u(x, 0; \epsilon) = f(x), \tag{1.3.29a}$$

$$u_t(x, 0; \epsilon) = g(x). \tag{1.3.29b}$$

Separation of variables leads to an eigenvalue problem just as for the unperturbed case. We assume a solution of the form

$$u(x, t; \epsilon) = X(x; \epsilon)T(t; \epsilon) \tag{1.3.30}$$

and find, upon substitution into (1.3.27), that

$$-\frac{1}{T}\frac{d^2T}{dt^2} = -\frac{1}{X}\frac{d^2X}{dx^2} + \epsilon \sin x = \lambda = \text{constant}. \tag{1.3.31}$$

Solutions bounded in t require that $\lambda > 0$, so $X(x; \epsilon)$ obeys the perturbed eigenvalue problem

$$L(X) \equiv -\frac{d^2X}{dx^2} + \epsilon(\sin x)X = \lambda X, \tag{1.3.32a}$$

$$X(0; \epsilon) = 0; \tag{1.3.32b}$$

$$X(\pi; \epsilon) = 0. \tag{1.3.32c}$$

Let us first review the theory of self-adjoint linear operators. (For example, see section 5.5 of [1.3]).

The linear operator L in (1.3.32a) is self-adjoint in the following sense. Let $u(x)$ and $v(x)$ be any two solutions of the eigenvalue problem, i.e., $L(u) = \lambda u$ or $L(v) = \lambda v$. Then

$$(u, L(v)) = (L(u), v), \tag{1.3.33}$$

where the inner product of two functions $\alpha(x)$ and $\beta(x)$ defined on $(0, \pi)$ is given by

$$(\alpha, \beta) \equiv \int_0^\pi \alpha(x)\beta(x)dx. \tag{1.3.34}$$

To prove (1.3.33), consider $(u, L(v))$ for our example (1.3.32a). We have

$$(u, L(v)) = -\int_0^\pi u(x)v''(x)dx + \epsilon \int_0^\pi u(x)v(x) \sin x dx.$$

Integrating the first term on the right-hand side by parts gives

$$(u, L(v)) = \int_0^\pi u'(x)v'(x)dx + \epsilon \int_0^\pi u(x)v(x) \sin x dx.$$

The boundary contributions vanish because u satisfies the homogeneous boundary conditions $u(0) = u(\pi) = 0$. Integrating by parts again and using the fact that $v(0) = v(\pi) = 0$ gives

$$(u, L(v)) = -\int_0^\pi u''(x)v(x)dx + \epsilon \int_0^\pi u(x)v(x) \sin x dx = (L(u), v).$$

An immediate consequence of (1.3.33) is that two eigenfunctions u_m and u_n associated respectively with the *distinct* eigenvalues λ_m and λ_n are orthogonal. More precisely, let u_m and u_n satisfy $L(u_m) = \lambda_m u_m$, $L(u_n) = \lambda_n u_n$, and the homogeneous boundary conditions $u_m(0) = u_m(\pi) = u_n(0) = u_n(\pi) = 0$. If λ_m and λ_n are distinct, then $(u_m, u_n) = 0$.

To prove this orthogonality condition, we note

$$(u_m, L(u_n)) = (u_m, \lambda_n u_n) = \lambda_n (u_m, u_n).$$

But since L is self-adjoint, we also have

$$(u_m, L(u_n)) = (L(u_m), u_n) = \lambda_m (u_m, u_n).$$

We have shown that $\lambda_n(u_m, u_n) = \lambda_m(u_m, u_n)$. Therefore, if $\lambda_m \neq \lambda_n$, we must have $(u_m, u_n) = 0$.

The above ideas apply to operators more general than the one in (1.3.32a) and to more independent variables. For the case at hand, the eigenvalues of the unperturbed problem are $\lambda_n^{(0)} = n^2$, $n = 1, 2, \ldots$. The associated orthogonal eigenfunctions are $c_n \sin nx$, where the c_n are arbitrary constants. It is convenient to set each $c_n = \left(\frac{2}{\pi}\right)^{1/2}$ and to work with the normalized set of eigenfunctions

$$\xi_n^{(0)}(x) = \left(\frac{2}{\pi}\right)^{1/2} \sin nx$$

for which $(\xi_n^{(0)}, \xi_n^{(0)}) = 1$.

For the perturbed problem (1.3.32), we assume that the eigenvalues $\lambda_n(\epsilon)$ and eigenfunctions $\xi_n(x;\epsilon)$ have the regular expansions

$$\lambda_n(\epsilon) = n^2 + \epsilon\lambda_n^{(1)} + O(\epsilon^2), \tag{1.3.35a}$$

$$\xi_n(x;\epsilon) = \left(\frac{2}{\pi}\right)^{1/2} \sin nx + \epsilon\xi_n^{(1)}(x) + O(\epsilon^2). \tag{1.3.35b}$$

Substituting these expansions into (1.3.32) gives

$$\frac{d^2\xi_n^{(1)}}{dx^2} + n^2\xi_n^{(1)} = -\left(\frac{2}{\pi}\right)^{1/2} \lambda_n^{(1)} \sin nx + \left(\frac{2}{\pi}\right)^{1/2} \sin x \sin nx \tag{1.3.36}$$

with $\xi_n^{(1)}(0) = \xi_n^{(1)}(\pi) = 0$, for each $n = 1, 2, \ldots$.

The general solution of (1.3.36) is easily found in this case:

$$\xi_n^{(1)} = A_n \sin nx + B_n \cos nx + \frac{\lambda_n^{(1)}}{n(2\pi)^{1/2}} x \cos nx$$

$$+ \frac{1}{(2n-1)(2\pi)^{1/2}} \cos(n-1)x + \frac{1}{(2n+1)(2\pi)^{1/2}} \cos(n+1)x. \tag{1.3.37}$$

The homogeneous boundary conditions $\xi_n^{(1)}(0) = \xi_n^{(1)}(\pi) = 0$ determine B_n and $\lambda_n^{(1)}$, but A_n is arbitrary

$$B_n = -\frac{4n}{(2\pi)^{1/2}(4n^2-1)}, \tag{1.3.38a}$$

$$\lambda_n^{(1)} = \frac{8n^2}{\pi(4n^2-1)}. \tag{1.3.38b}$$

The indeterminacy of the A_n is a direct consequence of the fact that an eigenfunction has an arbitrary constant multiplier. We may fix this constant by normalizing the perturbed eigenfunctions as we did the $\xi_n^{(0)}$. Thus, if we require $(\xi_n, \xi_n) = 1$, we have the expansion

$$(\xi_n, \xi_n) = (\xi_n^{(0)}, \xi_n^{(0)}) + 2\epsilon(\xi_n^{(0)}, \xi_n^{(1)}) + O(\epsilon^2) = 1.$$

We therefore set

$$(\xi_n^{(0)}, \xi_n^{(1)}) = 0.$$

Multiplying the expression in (1.3.37) for $\xi_n^{(1)}$ by $\xi_n^{(0)} = \left(\frac{2}{\pi}\right)^{1/2} \sin nx$ and integrating the result from 0 to π gives

$$(\xi_n^{(1)}, \xi_n^{(0)}) = \left(\frac{\pi}{2}\right)^{1/2} A_n + \frac{\lambda_n^{(1)}}{n\pi} \int_0^\pi x \cos nx \sin nx dx = 0.$$

Evaluating the definite integral and using (1.3.38b) for $\lambda_n^{(1)}$ gives

$$A_n = \left(\frac{2}{\pi}\right)^{3/2} \frac{1}{4n^2 - 1}. \tag{1.3.39}$$

In many applications, the complexity of the operator L puts an explicit result such as (1.3.37) out of reach. A less direct solution of $\xi_n^{(1)}$ is still possible if we express this function in a series of the unperturbed eigenfunctions $\xi_n^{(0)}$, a Fourier sine series in this case. We assume

$$\xi_n^{(1)}(x) = \left(\frac{2}{\pi}\right)^{1/2} \sum_{j=1}^{\infty} a_{nj} \sin jx \tag{1.3.40}$$

and substitute this into (1.3.36) to find

$$\left(\frac{2}{\pi}\right)^{1/2} \sum_{j=1}^{\infty} (-j^2 + n^2) a_{nj} \sin jx = -\left(\frac{2}{\pi}\right)^{1/2} \sin nx + \left(\frac{2}{\pi}\right)^{1/2} \sin x \sin nx.$$

We now multiply this expression by $\left(\frac{2}{\pi}\right)^{1/2} \sin kx$ and integrate the result from 0 to π to obtain

$$a_{nk}(-k^2 + n^2) = -\frac{2}{\pi}\lambda_n^{(1)} \int_0^\pi \sin nx \sin kx dx + \frac{2}{\pi}\int_0^\pi \sin x \sin nx \sin kx dx, \tag{1.3.41}$$

where we have used orthogonality to simplify the left-hand side.

If $k \neq n$, the integral multiplying $\lambda_n^{(1)}$ in (1.3.41) vanishes and we find

$$a_{nk} = \frac{2}{\pi(n^2 - k^2)} \int_0^\pi \sin x \sin nx \sin kx dx.$$

Evaluating this integral gives

$$a_{nk} = \begin{cases} \frac{1}{\pi(n^2-k^2)}\left[\frac{(-1)^{k+n}+1}{(k+n)^2-1} - \frac{(-1)^{k-n}+1}{(k-n)^2-1}\right] & \text{if } k \neq n-1, k \neq n+1 \\ 0 & \text{if } k = n-1 \text{ or } k = n+1. \end{cases} \tag{1.3.42}$$

If $k = n$, the left-hand side of (1.3.41) vanishes and this equation reduces to

$$0 = -\lambda_n^{(1)} + \frac{2}{\pi}\int_0^\pi \sin x \sin^2 nx dx.$$

Evaluating the integral gives the previously derived expression, (1.3.38b), for $\lambda_n^{(1)}$. At this point, the a_{nn} are arbitrary. We fix these coefficients by imposing the normalization condition $(\xi_n^{(0)}, \xi_n^{(1)}) = 0$, which now takes the form

$$\int_0^\pi \left[\left(\frac{2}{\pi}\right)^{1/2} \sin nx \cdot \left(\frac{2}{\pi}\right)^{1/2} \sum_{j=1}^{\infty} a_{nj} \sin jx \right] dx = 0.$$

Therefore, we must set $a_{nn} = 0$. The reader can verify that the expression (1.3.40) for $\xi_n^{(1)}$ in terms of the coefficients a_{nj} that we have just derived is the Fourier sine series of the expression in (1.3.37). A discussion of the eigenvalue problem for a general self-adjoint operator is given in section 8.2.2 of [1.4].

To complete the solution of the original initial value problem, we express $u(x, t; \epsilon)$ in the series of eigenfunctions ξ_n

$$u(x, t; \epsilon) = \sum_{n=1}^{\infty} \rho_n(t; \epsilon)\xi_n(x; \epsilon). \tag{1.3.43}$$

Substituting this into (1.3.27) and noting that ξ_n satisfies (1.3.32a) gives

$$\frac{d^2\rho_n}{dt^2} + \lambda_n \rho_n = 0.$$

Therefore,

$$\rho_n(t, \epsilon) = \alpha_n(\epsilon) \sin \lambda_n^{1/2} t + \beta_n(\epsilon) \cos \lambda_n^{1/2} t. \tag{1.3.44}$$

Since we have determined the λ_n and ξ_n to $O(\epsilon)$, it is appropriate to expand the α_n and β_n in terms of ϵ and to retain only terms up to $O(\epsilon)$. Thus, we set

$$\alpha_n(\epsilon) = \alpha_n^{(0)} + \epsilon\alpha_n^{(1)} + O(\epsilon^2), \tag{1.3.45a}$$

$$\beta_n(\epsilon) = \beta_n^{(0)} + \epsilon\beta_n^{(1)} + O(\epsilon^2), \tag{1.3.45b}$$

and we obtain the following expansion for u correct to $O(\epsilon)$

$$\begin{aligned} u(x, t; \epsilon) &= \sum_{n=1}^{\infty} [\alpha_n^{(0)} \sin \omega_n^{(1)}(\epsilon)t + \beta_n^{(0)} \cos \omega_n^{(1)}(\epsilon)t]\xi_n^{(0)}(x) \\ &+ \epsilon \sum_{n=1}^{\infty} \left\{ \left[\alpha_n^{(1)} \sin \omega_n^{(1)}(\epsilon)t + \beta_n^{(1)} \cos \omega_n^{(1)}(\epsilon)t\right] \xi_n^{(0)}(x) \right. \\ &+ \left. \left[\alpha_n^{(0)} \sin \omega_n^{(1)}(\epsilon)t + \beta_n^{(0)} \cos \omega_n^{(1)}(\epsilon)t\right] \xi_n^{(1)}(x) \right\} + O(\epsilon^2), \end{aligned} \tag{1.3.46}$$

where

$$\omega_n^{(1)}(\epsilon) = [n^2 + \epsilon\lambda_n^{(1)} + O(\epsilon^2)]^{1/2} = n + \epsilon \frac{4n}{\pi(4n^2 - 1)} + O(\epsilon^2).$$

It is important to note that in approximating $\sin \lambda_n^{1/2} t$ and $\cos \lambda_n^{1/2} t$ by $\sin \omega_n^{(1)} t$ and $\cos \omega_n^{(1)} t$, respectively, we do not expand these further to avoid nonuniformities for t large. For example, the expansion

$$\sin \lambda_n^{1/2}(\epsilon)t = \sin nt + \epsilon \frac{4n}{\pi(4n^2 - 1)} t \cos nt + O(\epsilon^2)$$

is uniformly valid to $O(\epsilon)$ in any interval $I(\epsilon)$: $0 \leq t \leq T(\epsilon)$ so long as $T(\epsilon) = O(1)$ as $\epsilon \to 0$. It fails to be uniform if $T(\epsilon) = O(\epsilon^{-1})$. On the other hand, the approximation $\sin \lambda_n^{1/2}(\epsilon)t = \sin \omega_n^{(1)}(\epsilon)t + O(\epsilon^2)$ is uniformly valid to $O(\epsilon)$ in $I(\epsilon)$ with $T(\epsilon) = O(\epsilon^{-1})$. (See Problem 1 of Sec. 1.2, where a similar example is discussed.)

Applying the initial conditions (1.3.29) to the expansion (1.3.46) gives

$$\sum_{n=1}^{\infty} \beta_n^{(0)}\xi_n^{(0)}(x) + \epsilon \sum_{n=1}^{\infty}(\beta_n^{(1)}\xi_n^{(0)}(x) + \beta_n^{(0)}\xi_n^{(1)}(x)) + O(\epsilon^2) = f(x), \quad (1.3.47a)$$

$$\sum_{n=1}^{\infty} \omega_n^{(1)}\alpha_n^{(0)}\xi_n^{(0)}(x) + \epsilon \sum_{n=1}^{\infty} \omega_n^{(1)}(\alpha_n^{(1)}\xi_n^{(0)}(x) + \alpha_n^{(0)}\xi_n^{(1)}(x)) + O(\epsilon^2) = g(x). \quad (1.3.47b)$$

Expanding $\omega_n^{(1)}(\epsilon)$ in (1.3.47b) gives

$$\sum_{n=1}^{\infty} n\alpha_n^{(0)}\xi_n^{(0)} + \epsilon \sum_{n=1}^{\infty} \left[n\alpha_n^{(1)}\xi_n^{(0)}(x) + \left(n + \frac{4n}{\pi(4n^2-1)}\right)\alpha_n^{(0)}\xi_n^{(1)}(x)\right]$$
$$+O(\epsilon^2) = g(x). \quad (1.3.48)$$

The condition (1.3.47a) to $O(1)$ shows that $\sum_{n=1}^{\infty} \beta_n^{(0)}\xi_n^{(0)}(x)$ is just the Fourier sine series of $f(x)$. Thus,

$$\beta_n^{(0)} = \left(\frac{2}{\pi}\right)^{1/2} \int_0^{\pi} f(x) \sin nx dx. \quad (1.3.49)$$

Since there are no $O(\epsilon)$ terms on the right-hand side of (1.3.47a), the series multiplied by ϵ on the left-hand side must vanish. When we use (1.3.40) for $\xi_n^{(1)}$ in this series, we find

$$\left(\frac{2}{\pi}\right)^{1/2} \sum_{n=1}^{\infty}(\beta_n^{(1)} \sin nx + \beta_n^{(0)} \sum_{j=1}^{\infty} a_{nj} \sin jx) = 0.$$

Multiplying this expression by $\left(\frac{2}{\pi}\right)^{1/2}$ sin kx and integrating the result from 0 to π gives

$$\beta_k^{(1)} = -\sum_{n=1}^{\infty} \beta_n^{(0)} a_{nk}, \quad (1.3.50)$$

which defines $\beta_k^{(1)}$ since $\beta_n^{(0)}$ and a_{nk} are known.

Repeating these steps for (1.3.48) defines $\alpha_n^{(0)}$ and $\alpha_n^{(1)}$:

$$n\alpha_n^{(0)} = \left(\frac{2}{\pi}\right)^{1/2} \int_0^{\pi} g(x) \sin nx dx, \quad (1.3.51a)$$

$$k\alpha_k^{(1)} = -\sum_{n=1}^{\infty} \left(n + \frac{4n}{\pi(4n^2-1)}\right) \alpha_n^{(0)} a_{nk}, \quad (1.3.51b)$$

and this completes the solution to $O(\epsilon)$.

A characteristic feature of the linear problem (1.3.27) is that the modal amplitudes $\rho_n(t)$ obey *decoupled* oscillator equations if we express the solution in terms of the *perturbed* eigenfunctions $\xi_n(x;\epsilon)$. Moreover, the frequency $\omega_n(\epsilon)$ of oscillation for each mode is known once the eigenvalue problem (1.3.32) has been calculated. This allows us to express the solution in the form (1.3.46) that remains uniformly valid for t in the interval $0 \le t \le T(\epsilon) = O(\epsilon^{-1})$.

It is also possible to express the solution for u in terms of the *unperturbed* eigenfunctions $\sin nx$ at the price of not having the modes decouple and not knowing the frequency a priori. To illustrate this, let us assume a solution of (1.3.27) for $u(x,t;\epsilon)$ in the form

$$u(x,t;\epsilon) = \left(\frac{2}{\pi}\right)^{1/2} \sum_{n=1}^{\infty} \gamma_n(t;\epsilon) \sin nx, \tag{1.3.52}$$

which automatically satisfies the two boundary conditions (1.3.28). Substituting this series into (1.3.27) and using orthogonality immediately gives the *coupled* linear system

$$\ddot{\gamma}_k + k^2\gamma_k + \epsilon \sum_{n=1}^{\infty} b_{kn}\gamma_n = 0; \quad k = 1, 2, \ldots, \tag{1.3.53}$$

where

$$b_{kn} = \begin{cases} \frac{1}{\pi}\left[\frac{(-1)^{k+n}+1}{(k+n)^2-1} - \frac{(-1)^{k-n}+1}{(k-n)^2-1}\right] & \text{if } k \ne n-1 \text{ and } k \ne n+1 \\ 0, & \text{if } k = n-1 \text{ or } k = n+1. \end{cases} \tag{1.3.54}$$

A regular perturbation expansion of (1.3.53) leads to terms in γ_k proportional to $t \sin nt$ and $t \cos nt$ to $O(\epsilon)$ and is therefore not uniformly valid to $O(\epsilon)$ if $T(\epsilon) = O(\epsilon^{-1})$. In fact, the solution of (1.3.52) with appropriate initial conditions reproduces the result we would get by expanding $\sin \omega_n^{(1)}(\epsilon)t$ and $\cos \omega_n(\epsilon)t$ in (1.3.46).

Although it is possible to derive a uniformly valid perturbation expansion of the solution of (1.3.53) for $T = O(\epsilon^{-1})$ using a multiple-scale or averaging procedure, as we shall see in Chaps. 4 and 5, this approach is not efficient for a linear problem such as (1.3.27). The expansion (1.3.43) based on perturbed eigenfunctions is significantly more elegant and direct.

If the perturbation term is nonlinear, one can no longer derive a perturbed eigenvalue problem such as (1.3.32). For example, if instead of $\epsilon u \sin x$ in (1.3.27) we have the term ϵu^2 we cannot separate variables. We can, however, look for a solution of the form (1.3.52) and derive a coupled weakly nonlinear system of oscillator equations for the γ_n. (See Problem 5.) Again, this coupled system of equations must be solved using multiple scales or averaging in order to ensure uniformity for t large.

1.3.3 A Boundary-Perturbation Problem

One often needs to solve a partial differential equation on a domain that is uniformly close to a simple domain over which the exact solution is known. For example, consider Laplace's equation

$$u_{rr} + \frac{1}{r}u_r + \frac{1}{r^2}u_{\theta\theta} = 0 \tag{1.3.55}$$

over the planar, nearly circular domain D: $r \leq 1 + \epsilon f(\theta)$. Here ϵ is a small parameter and f is smooth on $0 \leq \theta \leq 2\pi$, i.e., f and $df/d\theta$ are continuous on $0 \leq \theta \leq 2\pi$ with $f(0) = f(2\pi)$. We consider the general linear boundary condition for $u(r, \theta; \epsilon)$

$$\alpha u(1 + \epsilon f(\theta), \theta; \epsilon) + \beta u_r(1 + \epsilon f(\theta), \theta; \epsilon) = g(\theta), \tag{1.3.56}$$

where α and β are arbitrary constants and g is a prescribed smooth function on $0 \leq \theta \leq 2\pi$. Thus, for the special case $\beta = 0$ we have the Dirichlet problem for D (see Problem 7), whereas $\alpha = 0$ gives Neumann's problem. Strictly speaking, the second term on the left-hand side of (1.3.56) should be β times the derivative of u normal to the actual boundary, $r = 1 + \epsilon f(\theta)$, instead of the unperturbed boundary, $r = 1$. The two choices differ to $O(\epsilon)$ only, and there is no essential difference in the analysis for the simpler choice in (1.3.56).

If we assume the regular expansion

$$u(r, \theta; \epsilon) = u^{(0)}(r, \theta) + \epsilon u^{(1)}(r, \theta) + O(\epsilon^2), \tag{1.3.57}$$

(1.3.56) gives

$$\alpha u^{(0)}(1 + \epsilon f(\theta), \theta) + \alpha\epsilon u^{(1)}(1 + \epsilon f(\theta), \theta)$$

$$+\beta u_r^{(0)}(1 + \epsilon f(\theta), \theta) + \beta\epsilon u_r^{(1)}(1 + \epsilon f(\theta), \theta) + O(\epsilon^2) = g(\theta).$$

Expanding the arguments gives

$$\alpha u^{(0)}(1, \theta) + \alpha u_r^{(0)}(1, \theta)\epsilon f(\theta) + \alpha\epsilon u^{(1)}(1, \theta)$$

$$+\beta u_r^{(0)}(1, \theta) + \beta u_{rr}^{(0)}(1, \theta)\epsilon f(\theta) + \beta\epsilon u_r^{(1)}(1, \theta) + O(\epsilon^2) = g(\theta).$$

Therefore, $u^{(0)}(r, \theta)$ and $u^{(1)}(r, \theta)$ satisfy the following boundary conditions:

$$\alpha u^{(0)}(1, \theta) + \beta u_r^{(0)}(1, \theta) = g(\theta) \equiv g^{(0)}(\theta) \tag{1.3.58a}$$

$$\alpha u^{(1)}(1, \theta) + \beta u_r^{(1)}(1, \theta) = -[\alpha u_r^{(0)}(1, \theta) + \beta u_{rr}^{(0)}(1, \theta)]f(\theta) \equiv g^{(1)}(\theta). \tag{1.3.58b}$$

Because of linearity, $u^{(0)}$ and $u^{(1)}$ each satisfy Laplace's equation (1.3.55). Once the unperturbed problem for $u^{(0)}$ is known, the right-hand side of (1.3.58b) is available and $u^{(1)}$ is formally governed by the same problem as $u^{(0)}$.

One approach for solving (1.3.55) is to separate variables. The assumption that u consists of terms of the form $R(r)\Theta(\theta)$ leads to the following condition:

$$r^2\frac{d^2R}{dr^2} + r\frac{dR}{dr} = -\frac{d^2\Theta}{d\theta^2} = \text{constant}. \tag{1.3.59}$$

The solution must be 2π-periodic in θ; this requires that the separation constant in (1.3.59) be n^2, where $n = 0, 1, 2, \ldots$. With this choice, R obeys a Cauchy equation with solutions proportional to r^n and r^{-n}. We discard solutions proportional to r^{-n} as they become singular at the origin, and we obtain the following Fourier series in θ for $u^{(i)}$:

$$u^{(i)}(r, \theta) = \frac{a_0^{(i)}}{2} + \sum_{n=1}^{\infty} r^n (a_n^{(i)} \cos n\theta + b_n^{(i)} \sin n\theta); \; i = 0, 1, \ldots, \quad (1.3.60)$$

where the unknown $a^{(i)}$ are to be determined from the boundary conditions (1.3.58).

Substituting (1.3.60) into (1.3.58) gives

$$\alpha \frac{a_0^{(i)}}{2} + \sum_{n=1}^{\infty} [(\alpha + n\beta) a_n^{(i)} \cos n\theta + (\alpha + n\beta) b_n^{(i)} \sin n\theta] = g^{(i)}(\theta); \; i = 0, 1, \ldots.$$

Thus, $(\alpha + n\beta)a_n^{(i)}$ and $(\alpha + n\beta)b_n^{(i)}$ are the Fourier coefficients of $g^{(i)}$. Orthogonality defines

$$(\alpha + n\beta)a_n^{(i)} = \frac{1}{\pi} \int_0^{2\pi} g^{(i)}(\theta) \cos n\theta d\theta; \quad i = 0, 1, \ldots \quad (1.3.61a)$$

$$(\alpha + n\beta)b_n^{(i)} = \frac{1}{\pi} \int_0^{2\pi} g^{(i)}(\theta) \sin n\theta d\theta; \quad i = 0, 1, \ldots. \quad (1.3.61b)$$

Here $g^{(0)}$ is given, and we compute the right-hand side of (1.3.58b) to obtain $g^{(1)}$ in the form

$$g^{(1)}(\theta) = -f(\theta)\{\sum_{n=1}^{\infty} [n\alpha + \beta n(n-1)][a_n^{(0)} \cos n\theta + b_n^{(0)} \sin \theta]\}. \quad (1.3.62)$$

The solution for $u^{(1)}$ is given by (1.3.60) and (1.3.61) with $i = 1$, etc.

The original problem, (1.3.55)–(1.3.56), thus reduces to a sequence of formally identical problems for each of the $u^{(i)}$; each of these satisfies Laplace's equation in the interior of the unit disc with the general boundary condition (1.3.58), where $g^{(i)}(\theta)$ is known in terms of the previously computed solutions for $u^{(0)}$, $u^{(1)}, \ldots u^{(i-1)}$.

It is important to review the question of the existence and uniqueness for each $u^{(i)}$. As is well known, e.g. see section 2.5.3 of [1.4], the Dirichlet problem ($\beta = 0$) is unique, and the Neumann problem ($\alpha = 0$) is unique to within an arbitrary constant. Our series solution confirms the results for these two special cases. In fact, with $\beta = 0$, (1.3.61) defines each of the $a_n^{(i)}$, $n = 0, 1, \ldots$ and $b_n^{(i)}$, $n = 1, 2, \ldots$ uniquely. However, if $\alpha = 0$, we see from (1.3.61a) for $n = 0$ that a_0 is arbitrary. Thus, for any solution $u^{(i)}$ of the Neumann problem $u^{(i)} - a_0^{(i)}/2$ is also a solution for any constant $a_0^{(i)}$.

For the general boundary-value problem ($\alpha \neq 0$, $\beta \neq 0$), we see that solutions are unique if α and β have the same sign. This is a special case of the more general uniqueness result for Laplace's equation in an arbitrary domain with the boundary condition $\alpha u + \beta \frac{\partial u}{\partial n} = g$, where α, β, and g may vary along the boundary and $\partial u/\partial n$ denotes the outward normal derivative to the boundary. One can show that solutions are unique if α and β have the same sign for all points on the boundary. However, if α and β have different signs, solutions may not exist, and even if they exist they may not be unique. To show this, it suffices to restrict attention to the case where α and β are nonzero constants with opposite signs, choosing $|\frac{\alpha}{\beta}| = m =$ integer.

Consider first the case where $g^{(i)}$ is orthogonal to both $\cos m\theta$ and $\sin m\theta$, i.e., the Fourier series of $g^{(i)}(\theta)$ does not contain $\cos m\theta$ and $\sin m\theta$. Then the right-hand sides of (1.3.61a) and (1.3.61b) both vanish for $n = m$. The left-hand sides also both vanish because the factor $(\alpha + m\beta) = 0$. In this case, a solution exists but is not unique since a_m and b_m are arbitrary. If $g^{(i)}$ is not orthogonal to either $\cos m\theta$ or $\sin m\theta$ (or both), the right-hand side of either (1.3.61a) or (1.3.61b) (or both) will be nonzero, whereas both left-hand sides are zero. Therefore, either a_m or b_m (or both) will be undefined, and a solution will not exist.

The reader can verify that our solution procedure easily generalizes to the case where f, g, α, and β also depend on ϵ. The series solution (1.3.60) is also appropriate for the more general linear boundary condition where α and β in (1.3.56) are given 2π-periodic functions of θ. In this case, the boundary condition leads to a linear system of algebraic equations for the $a_n^{(i)}$ and $b_n^{(i)}$. This system can be solved in principle if α and β have the same sign for all θ if one approximates (1.3.60) by truncating the series after a finite number of terms N. (See Problem 8.)

Problems

1. Show that the first-order equation

$$\epsilon \frac{du}{dx} + u = \frac{\epsilon[x(\epsilon - 1) + \epsilon^2]e^{-x}}{(x + \epsilon)^2} \tag{1.3.63}$$

with boundary condition $u(0; \epsilon) = 0$ has (1.2.9) as its exact solution.

a. Derive the outer expansion (1.2.10) by solving the appropriate algebraic equations that result from (1.3.63) for each of the $h_n(x)$, $n = 1, 2, 3$.

b. Now express (1.3.63) in terms of $x^* = x/\epsilon$ to find

$$\frac{du^*}{dx^*} + u = \frac{[x^*(\epsilon - 1) + \epsilon]e^{-\epsilon x^*}}{(x^* + 1)^2}, \tag{1.3.64}$$

with boundary condition $u^*(0; \epsilon) = 0$, where $u^*(x^*; \epsilon) \equiv u(\epsilon x^*; \epsilon)$.

c. Derive the inner expansion (1.2.12) by solving the appropriate differential equations that result from (1.3.64) for each of the $g_n(x^*)$, $n = 1, 2, 3$.

2. Generalize (1.3.13) to the case

$$u'' + (1 - \epsilon f(x))u = 0 \tag{1.3.65}$$

for a given $f(x)$ on $0 \le x < \infty$, and initial conditions $u(0) = 0$, $u'(0) = 1$. Show that a necessary condition that the regular expansion of the solution be uniformly valid to $O(\epsilon)$ on $0 \le x < \infty$ is to have $\int_0^x f(\xi)d\xi$ bounded on $0 \le x < \infty$. Show that this condition is not sufficient by giving an example of a function f where $\int_0^x f(\xi)d\xi$ is bounded but where the regular expansion of the solution is not uniformly valid to $O(\epsilon)$ on $0 \le x < \infty$.

3. Calculate the regular expansion to $O(\epsilon)$ for

$$x^2u'' - (2x + \epsilon x^2)u' + 2u = 0, \tag{1.3.66}$$

$$u(-1; \epsilon) = 1; \tag{1.3.67a}$$

$$u(1; \epsilon) = 0. \tag{1.3.67b}$$

4. Does (1.3.20) with the boundary conditions $u(\epsilon) = 1, u(1) = 2$ have a regular expansion?
5. Consider the weakly nonlinear eigenvalue problem

$$-X'' + \epsilon x^2 X^3 = \lambda X, \tag{1.3.68}$$

$$X(0; \epsilon) = 0, \tag{1.3.69a}$$

$$X(\pi; \epsilon) = 0. \tag{1.3.69b}$$

Assume that λ_n and X_n have expansions as in (1.3.35) to calculate $\lambda_n^{(1)}$ and the Fourier coefficients a_{nj} of $\xi_n^{(1)}$.
6. Generalize the vibrating string problem in (1.3.27) to have a weak *nonlinear* restoring force term

$$u_{tt} - u_{xx} + \epsilon u^3 = 0 \tag{1.3.70}$$

and keep the same boundary (1.3.28) and initial (1.3.29) conditions.

Show that a separable solution $u = X(x; \epsilon)T(t; \epsilon)$ does not exist. Assume instead that

$$u(x, t; \epsilon) = \sum_{n=1}^{\infty} b_n(t; \epsilon) \sin nx. \tag{1.3.71}$$

Show that this implies that

$$u^3(x, t; \epsilon) = \sum_{n=1}^{\infty} f_n(a_1, a_2, \ldots) \sin nx, \tag{1.3.72}$$

where

$$f_n = \frac{3}{4} \sum_{j=1}^{\infty} \sum_{k=1}^{j} b_{n+j+1} b_{j-k+1} b_k - \frac{1}{4} \sum_{j=1}^{n-2} \sum_{k=1}^{n-j-1} b_{n-j-k} b_j b_k. \tag{1.3.73}$$

Thus, the b_n obey the coupled weakly nonlinear system

$$\frac{d^2 b_n}{dt^2} + n^2 b_n + \epsilon f_n(a_1, a_2, \ldots) = 0. \tag{1.3.74}$$

7. For the special case $\alpha = 1, \beta = 0$ in (1.3.56), use Poisson's formula to express $u^{(0)}$ and $u^{(1)}$ in integral form

$$u^{(0)}(r, \theta) = \frac{1 - r^2}{2\pi} \int_0^{2\pi} \frac{g(\xi)}{1 + r^2 - 2r\cos(\theta - \xi)} d\xi, \tag{1.3.75}$$

$$u^{(1)}(r, \theta) = \frac{1 - r^2}{2\pi} \int_0^{2\pi} \frac{h(\xi)}{1 + r^2 - 2r\cos(\theta - \xi)} d\xi, \qquad (1.3.76)$$

where

$$h(\xi) = -\frac{\partial u^{(0)}}{\partial r}(1^-, \xi) f(\xi). \qquad (1.3.77)$$

Verify that expanding the integrand for $u^{(0)}$ in a power series in r and then integrating term by term gives the series solution (1.3.60), where $a_n^{(0)}$ and $b_n^{(0)}$ are defined by (1.3.61) with $\alpha = 1$ and $\beta = 0$.

Calculate $u^{(1)}(r, \theta)$ explicitly for the case where f and g have finite Fourier series

$$f(\theta) = f_1 \sin\theta + f_2 \sin 2\theta, \qquad (1.3.78a)$$

$$g(\theta) = \frac{\lambda_0}{2} + \lambda_1 \cos\theta + \lambda_2 \cos 2\theta + \nu_1 \sin\theta + \nu_2 \sin 2\theta, \qquad (1.3.78b)$$

for constants $f_1, f_2, \lambda_0, \lambda_1, \lambda_2, \nu_1$, and ν_2.

8. Consider the generalization of the problem discussed in Sec. 1.3.3. for which α and β are smooth functions of θ and have the same sign for all θ on $0 \leq \theta \leq 2\pi$. Assume that α and β have finite Fourier series

$$\alpha(\theta) = \frac{\gamma_0}{2} + \gamma_1 \cos\theta + \gamma_2 \cos 2\theta + \delta_1 \cos\theta + \delta_2 \cos 2\theta \qquad (1.3.79a)$$

$$\beta(\theta) = \frac{\rho_0}{2} + \rho_1 \cos\theta + \rho_2 \cos 2\theta + \mu_1 \sin\theta + \mu_2 \sin 2\theta. \qquad (1.3.79b)$$

Assume also that f and g have the finite Fourier series given by (1.3.78).

Ignore third and higher harmonics in the solutions and calculate $a_0^{(i)}$, $a_1^{(i)}$, $a_2^{(i)}$, $b_1^{(i)}$, and $b_2^{(i)}$ for $i = 0, 1$.

References

1.1. G.F. Carrier, M. Krook, and C.E. Pearson, *Functions of a Complex Variable, Theory and Technique*, McGraw-Hill Book Company, New York, 1966.

1.2. A. Erdelyi, *Asymptotic Expansions*, Dover Publications, New York, 1956.

1.3. R. Haberman, *Elementary Applied Partial Differential Equations*, Second Edition, Prentice-Hall, Englewood Cliffs, NJ, 1987.

1.4. J. Kevorkian, *Partial Differential Equations: Analytical Solution Techniques*, Chapman & Hall, New York, London, 1990, 1993.

1.5. J.D. Murray, *Asymptotic Analysis*, Springer-Verlag, New York, 1984.

2

Limit Process Expansions for Ordinary Differential Equations

In this chapter, a series of simple examples are considered, some model and some physical, in order to demonstrate the application of various techniques concerning limit process expansions. In general, we expect analytic dependence of the exact solution on the small parameter ϵ, but one of the main tasks in the various problems is to discover the nature of this dependence by working with suitable approximate differential equations. Another problem is to systematize as much as possible the procedures for discovering these expansions.

The main unifying features of problems having two or more limit process expansions is that certain terms in the governing differential equation will change their orders of magnitude depending on the domain in x. Often (but not in all cases), the highest derivative in the differential equation will be multiplied by the small parameter ϵ, and this term will be small everywhere except near special points, e.g., boundary points.

In physical problems, ϵ is considered dimensionless and is found by expressing the entire problem in suitable dimensionless coordinates. Physical problems have an advantage from the point of view of perturbation procedures: very often the general nature of the solution is known, and this simplifies the task of finding the appropriate limit process expansions.

2.1 The Linear Oscillator

As a first example that illustrates ideas, we consider a case for which the exact solution is easily found: the response of a linear spring-mass-damping system, initially at rest, to an impulse I_0 (see Fig. 2.1.1).

The equation and initial conditions are

$$M\frac{d^2Y}{dT^2} + B\frac{dY}{dT} + KY = I_0\delta(T), \tag{2.1.1a}$$

$$Y(0^-) = \frac{dY(0^-)}{dT} = 0, \tag{2.1.1b}$$

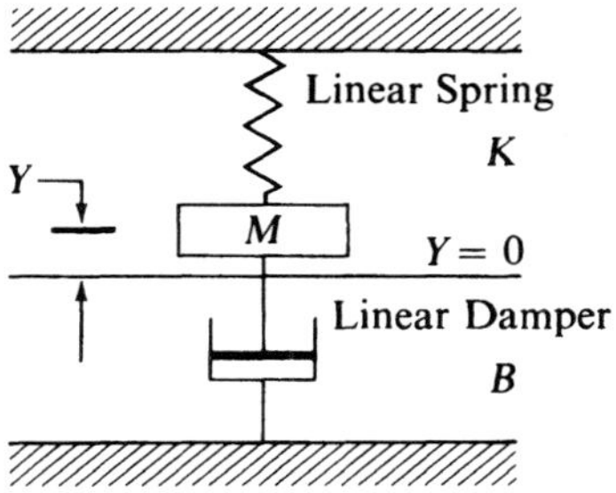

FIGURE 2.1.1. Spring-Mass-Damping System

where δ is the Dirac delta function.

Problem (2.1.1) can be replaced by an equivalent one, (2.1.2), by considering an impulse-momentum balance across $T = 0$ or by integrating Equation (2.1.1) from $T = 0^-$ to $T = 0^+$:

$$M\frac{d^2Y}{dT^2} + B\frac{dY}{dT} + KY = 0, \quad T > 0, \tag{2.1.2a}$$

$$Y(0^+) = 0, \tag{2.1.2b}$$

$$\frac{dY(0^+)}{dT} = \frac{I_0}{M}. \tag{2.1.2c}$$

The solution defined by this problem is the fundamental solution of this linear equation.

2.1.1 Dimensionless Variables

Before proceeding with the perturbation analysis, it is crucial to choose dimensionless variables that are appropriate for the limiting case to be studied. Two such limiting cases are of interest for the linear oscillator.

Small damping (cumulative perturbation)

If B is small, we expect the motion to be a weakly damped oscillation close to the free simple harmonic oscillation of the system—the solution of (2.1.2) with $B = 0$. For the introduction of dimensionless coordinates, a suitable time scale is $\sqrt{M/K}$, the reciprocal of the natural frequency of free undamped motion, since this scale remains in the limit $B \to 0$. The length scale A, a measure of the amplitude, can be chosen arbitrarily, and this choice will not affect the resulting dimensionless differential equation since it is linear. Actually, we will choose A in a form convenient for normalizing the initial velocity.

Setting

$$t^* = \frac{T}{(M/K)^{1/2}}, \quad y = \frac{Y}{A}, \tag{2.1.3}$$

we find

$$\frac{d^2y}{dt^{*2}} + 2\epsilon^* \frac{dy}{dt^*} + y = 0, \tag{2.1.4}$$

where

$$\epsilon^* = \frac{B}{2(MK)^{1/2}}.$$

In these variables $y(0^+) = 0, dy(0^+)/dt^* = 1$ if we set $A = I_0/(MK)^{1/2}$.

We see that the solution involves the one parameter ϵ^*, and small damping corresponds to ϵ^* small. The exact solution is easily found:

$$y(t^*; \epsilon) = \frac{e^{-\epsilon^* t^*}}{\sqrt{1-\epsilon^{*2}}} \sin(\sqrt{1-\epsilon^{*2}}t^*). \tag{2.1.5}$$

A regular perturbation expansion of (2.1.5), i.e., $\epsilon^* \to 0$ with t^* fixed and finite, is

$$y = \sin t^* - \epsilon^* t^* \sin t^* + O(\epsilon^{*^2}) + O(\epsilon^{*^2} t^*) + O(\epsilon^{*^2} t^{*^2}). \tag{2.1.6}$$

This result also follows if we assume the expansion

$$y = g_1(t^*) + \epsilon^* g_2(t^*) + \ldots \tag{2.1.7}$$

and solve the equations that result for g_1 and g_2 (see Problem 1). As discussed in connection with the example (1.3.17), the expansion (2.1.6) is uniformly valid to $O(\epsilon^*)$ only if t^* is in the interval $0 \leq t^* \leq T_0 = O(1)$.

Small mass (singular perturbation)

A singular problem is associated with approximations of (2.1.1) for small values of the mass M. The difficulty near $T = 0$ arises from the fact that the limit equation with $M = 0$ is first order, so that the initial conditions, (2.1.2b) and (2.1.2c), cannot both be satisfied. The loss of an initial or boundary condition in a problem leads, in general, to the occurrence of a boundary layer.

We discuss this problem first using physical reasoning. The general nature of the solution for small values of M is sketched in Figure 2.1.2 with each solid curve corresponding to a fixed value of M. After a short time interval, it can be expected that the motion of the system is described by the limit form of (2.1.1a) with $M = 0$:

$$B\frac{dY}{dT} + KY = I_0\delta(T). \tag{2.1.8}$$

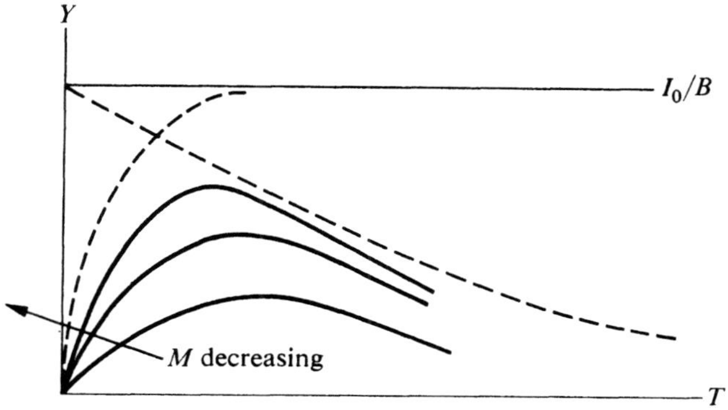

FIGURE 2.1.2. Solution Curves, Varying M

The initial condition in velocity is lost, and the effect of the impulse is to introduce a jump in the initial displacement from $Y(0^-) = 0$ to

$$Y(0^+) = \frac{I_0}{B}. \tag{2.1.9}$$

The solution is

$$Y = \frac{I_0}{B} e^{-KT/B}. \tag{2.1.10}$$

We see that the solution decays exponentially after the short initial interval in which the displacement increases infinitely rapidly from 0 to I_0/B.

In order to describe the motion during the initial instants, we remark that inertia is certainly dominant at $T = 0$ (impulse-momentum balance). Due to the large initial velocity, damping is immediately important, whereas the restoring force of the spring is not; the spring must be deflected before its influence is felt. Thus, in the initial instants, (2.1.1a) can be approximated by

$$M\frac{d^2Y}{dT^2} + B\frac{dY}{dT} = I_0\delta(t), \quad Y(0^-) = 0, \quad \frac{dY}{dT}(0^-) = 0 \tag{2.1.11}$$

with the solution

$$Y(t) = \frac{I_0}{B}\{1 - e^{-BT/M}\}. \tag{2.1.12}$$

This solution shows the approach of the deflection in a very short time ($M \to 0$) to the starting value for the decay solution (2.1.10). The curves are shown dashed in Figure 2.1.2 and give an overall picture of the motion.

Following our physical considerations, we aim to construct suitable asymptotic expansions for expressing these physical ideas and to show how to join these expansions. The method uses expansions valid after a short time (away from the initial point) and expansions valid near the initial point.

For the expansion valid away from the initial point, we find that natural variables are those based on a time scale for decay (B/K) and on an amplitude linear in I_0. Let

$$t = \frac{K}{B}T, \quad y = B\frac{Y}{I_0}$$

so that (2.1.2) for $y(t;\epsilon)$ reads

$$\epsilon\frac{d^2y}{dt^2} + \frac{dy}{dt} + y = 0, \tag{2.1.13}$$

where $\epsilon = MK/B^2$, with initial conditions

$$y(0;\epsilon) = 0, \quad \frac{dy}{dt}(0;\epsilon) = \frac{1}{\epsilon}. \tag{2.1.14}$$

The exact solution is

$$y(t;\epsilon) = \frac{1}{\sqrt{1-4\epsilon}}\left\{\exp[-(1-\sqrt{1-4\epsilon})\frac{t}{2\epsilon}] - \exp[-(1+\sqrt{1-4\epsilon})\frac{t}{2\epsilon}]\right\}. \tag{2.1.15}$$

2.1.2 *Singular Perturbation Problem*

In this section, we use the simple model described by (2.1.13)–(2.1.14) to study the outer expansion valid for $t > 0$, the inner expansion valid for $t \approx 0$, and the connection between these two expansions.

Outer and inner expansions of exact solution

First, let us use the exact solution (2.1.15) to calculate the outer expansion defined by the limit $\epsilon \to 0$ with t fixed $\neq 0$. For ϵ small, the ϵ-dependent factors in (2.1.15) have the expansions

$$-(1-\sqrt{1-4\epsilon})/2\epsilon = -1 - \epsilon + O(\epsilon^2),$$
$$-(1+\sqrt{1-4\epsilon})/2\epsilon = -\frac{1}{\epsilon} + 1 + O(\epsilon),$$
$$\frac{1}{\sqrt{1-4\epsilon}} = 1 + 2\epsilon + O(\epsilon^2).$$

Therefore,

$$y = (1 + 2\epsilon + \ldots)[e^{-t-\epsilon t+\ldots} - e^{-t/\epsilon+t+\ldots}]. \tag{2.1.16}$$

For t fixed $\neq 0$ and $\epsilon \to 0$, the second term in the bracketed expansion in (2.1.16) is transcendentally small, whereas the first term has the expansion $e^{-t}(1-\epsilon t+\ldots)$. Collecting terms of $O(1)$ and $O(\epsilon)$ then gives the outer expansion to $O(\epsilon)$ in the form

$$y = e^{-t} + \epsilon(2 - t)e^{-t} + O(\epsilon^2). \tag{2.1.17}$$

This result is not uniformly valid for $t \to 0$; in fact, it violates both initial conditions. Also, (2.1.17) is not uniformly valid for $t \to \infty$ because of the presence of the $-\epsilon t e^{-t}$ term.

To compute an approximation valid for small t, we must not ignore $e^{-t/\epsilon}$. Accordingly, we introduce the rescaled inner time $t^* = t/\epsilon$ and express (2.1.16) as follows

$$y = (1 + 2\epsilon + \ldots)[e^{-\epsilon t^* - \epsilon^2 t^* + \ldots} - e^{-t^* + \epsilon t^* + \ldots}].$$

For fixed $t^* \neq \infty$ and $\epsilon \to 0$, we compute

$$y = (1 - e^{-t^*}) + \epsilon[(2 - t^*) - (2 + t^*)e^{-t^*}] + \ldots. \tag{2.1.18}$$

This approximation predicts $y \to 1 + \epsilon(2 - t^*) + \ldots$ as $t^* \to \infty$ and is not uniformly valid as $t^* \to \infty$.

Extended domains of validity of outer and inner expansions

We now show that the outer expansion (2.1.17) and the inner expansion (2.1.18) are each valid in a wider domain of the $t\epsilon$-plane than the nominal domains inherent in the defining limit processes.

Consider the outer limit $h_1(t) = e^{-t}$. It is calculated via the limit process

$$\lim_{\substack{\epsilon \to 0 \\ t \text{ fixed} \neq 0}} y(t; \epsilon) = h_1(t). \tag{2.1.19a}$$

The second term, $h_2(t) = (2 - t)e^{-t}$, is calculated using

$$\lim_{\substack{\epsilon \to 0 \\ t \text{ fixed} \neq 0}} \frac{y(t; \epsilon) - h_1(t)}{\epsilon} = h_2(t), \tag{2.1.19b}$$

and similar limits define the higher-order terms (see (1.2.7)).

Actually, we can show that (2.1.17) is valid in a more general sense by allowing t to either remain fixed or tend to zero at some maximal rate as $\epsilon \to 0$.

To establish more precisely the domain of validity of (2.1.17) in the $t\epsilon$-plane, we set $t = \eta(\epsilon)t_\eta$ for some *fixed* $t_\eta > 0$ and some function $\eta(\epsilon)$ that remains bounded as $\epsilon \to 0$. Thus, if $\eta(\epsilon) \ll 1$, t tends to zero "at the rate" $\eta(\epsilon)$ as $\epsilon \to 0$. If $\eta = O_s(1)$, then the limiting value of t is fixed as in the outer limit. In order for the outer limit $h_1(t)$ to be valid in this extended sense for a given $\eta(\epsilon)$, we must have

$$\lim_{\substack{\epsilon \to 0 \\ t_\eta \text{ fixed} \neq 0}} y(\eta t_\eta; \epsilon) - h_1(\eta t_\eta) = 0. \tag{2.1.20}$$

It follows from (2.1.16) that (2.1.20) holds as long as $e^{-\eta t_\eta/\epsilon}$ is transcendentally small, as postulated earlier. This implies that any $\eta(\epsilon)$ such that $\epsilon|\log\epsilon| \ll \eta(\epsilon)$ is allowable. However, if η equals the critical value $\eta_1 \equiv \epsilon|\log\epsilon|$, then $e^{-\eta_1 t_\eta/\epsilon} =$

$O_s(\epsilon^{t_\eta})$; this does vanish as $\epsilon \to 0$, but it is not transcendentally small. The class of all possible functions $\eta(\epsilon)$ that are bounded as $\epsilon \to 0$ and satisfy the requirement that $e^{-\eta t_\eta/\epsilon}$ be transcendentally small is called the *extended domain of validity* of h_1; this class can be expressed in the form

$$\epsilon|\log\epsilon| \ll \eta(\epsilon) \underline{\ll} 1, \tag{2.1.21}$$

evocative of a half-open interval. Here, we have introduced the notation

$$\eta(\epsilon) \underline{\ll} \lambda(\epsilon) \text{ if } \eta \ll \lambda \text{ or } \eta = O_s(\lambda). \tag{2.1.22}$$

We can also give a pictorial representation of this extended domain as in Figure 2.1.3. The left boundary of the shaded region corresponds to the critical curve $t = \epsilon|\log\epsilon|t_\eta$ and the right boundary to the curve $t = \eta_0(\epsilon)t_\eta$ for some $\eta_0 = O_s(1)$. The actual shaded region is not significant per se; in particular, the specific choices of the η_1 and η_0 functions are not relevant, as we are interested only in the limiting behaviors of η_1 and η_0 as $\epsilon \to 0$.

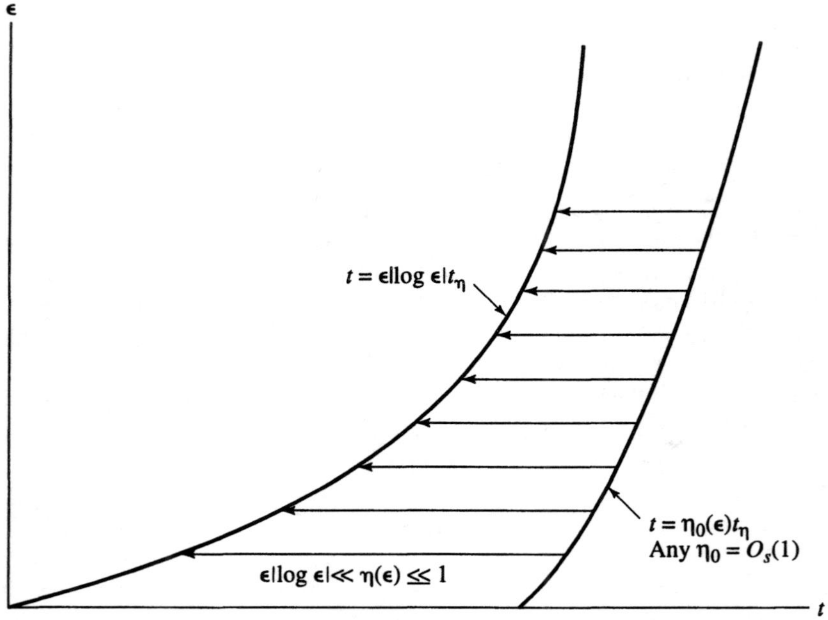

FIGURE 2.1.3. Extended Domain of Validity for h_1

The extended domain of validity of the two-term outer expansion $h_1(t) + \epsilon h_2(t)$ is still given by (2.1.21) because

$$\lim_{\substack{\epsilon \to 0 \\ t_\eta \text{ fixed} \neq 0}} \frac{y(\eta t_\eta; \epsilon) - h_1(\eta t_\eta) - \epsilon h_2(\eta t_\eta)}{\epsilon} = 0 \tag{2.1.23}$$

for any η in the class defined by (2.1.21).

Now consider the inner limit $g_1(t^*) = 1 - e^{-t^*}$. It follows from the limit process

$$\lim_{\substack{\epsilon \to 0 \\ t^* \text{ fixed} \neq \infty}} y(\epsilon t^*; \epsilon) = g_1(t^*). \tag{2.1.24a}$$

Similarly, the second term $g_2(t^*) = (2 - t^*) - (2 + t^*)e^{-t^*}$ in the inner expansion (2.1.18) obeys

$$\lim_{\substack{\epsilon \to 0 \\ t^* \text{ fixed} \neq \infty}} \frac{y(\epsilon t^*; \epsilon) - g_1(t^*)}{\epsilon} = g_2(t^*). \tag{2.1.24b}$$

We can show that the inner limit is also valid in an extended domain of validity. To do so, we again set $t = \eta t_\eta$, i.e. $t^* = \eta t_\eta / \epsilon$ for some fixed, finite, non-negative t_η, and look for the class of functions $\eta(\epsilon)$ for which (2.1.24a) remains valid. In other words, for what functions $\eta(\epsilon)$ is the following true?

$$\lim_{\substack{\epsilon \to 0 \\ t_\eta \text{ fixed} \neq \infty}} y(\eta t_\eta; \epsilon) - g_1(\eta t_\eta/\epsilon) = 0. \tag{2.1.25}$$

It follows from (2.1.16) that the above holds for any η in the following extended domain of validity for g_1

$$\epsilon \underline{\ll} \eta \ll 1, \tag{2.1.26}$$

and this domain is sketched in Figure 2.1.4.

The left boundary $\eta = O_s(\epsilon)$ corresponds to the inner limit, whereas the requirement $\eta \ll 1$ ensures that $e^{-t} \to 1$ in (2.1.16). Actually, it is possible to extend (2.1.26) to the "left," i.e., to have $\eta \ll \epsilon$. In this case, each term in (2.1.25) vanishes individually. But such an extension is of no interest as we shall be concerned only with the intersection of the two sets (2.1.21) and (2.1.26) that lie to the right of $\eta = \epsilon$.

The domain of validity of the two-term inner expansion (2.1.18) is the set of $\eta(\epsilon)$ for which the limit

$$\lim_{\substack{\epsilon \to 0 \\ t_\eta \text{ fixed} \neq \infty}} \frac{y(\eta t_\eta; \epsilon) - g_1(\eta t_\eta/\epsilon) - \epsilon g_2(\eta t_\eta/\epsilon)}{\epsilon} = 0 \tag{2.1.27}$$

holds.

Expanding $y(\eta t_\eta; \epsilon)$ for η and ϵ small and dividing by ϵ gives

$$\frac{y(\eta t_\eta; \epsilon)}{\epsilon} = \frac{1}{\epsilon} - \frac{e^{-\eta t_\eta/\epsilon}}{\epsilon} - \frac{\eta}{\epsilon} t_\eta - \frac{\eta}{\epsilon} t_\eta e^{-\eta t_\eta/\epsilon} + 2 - 2e^{-\eta t_\eta/\epsilon}$$

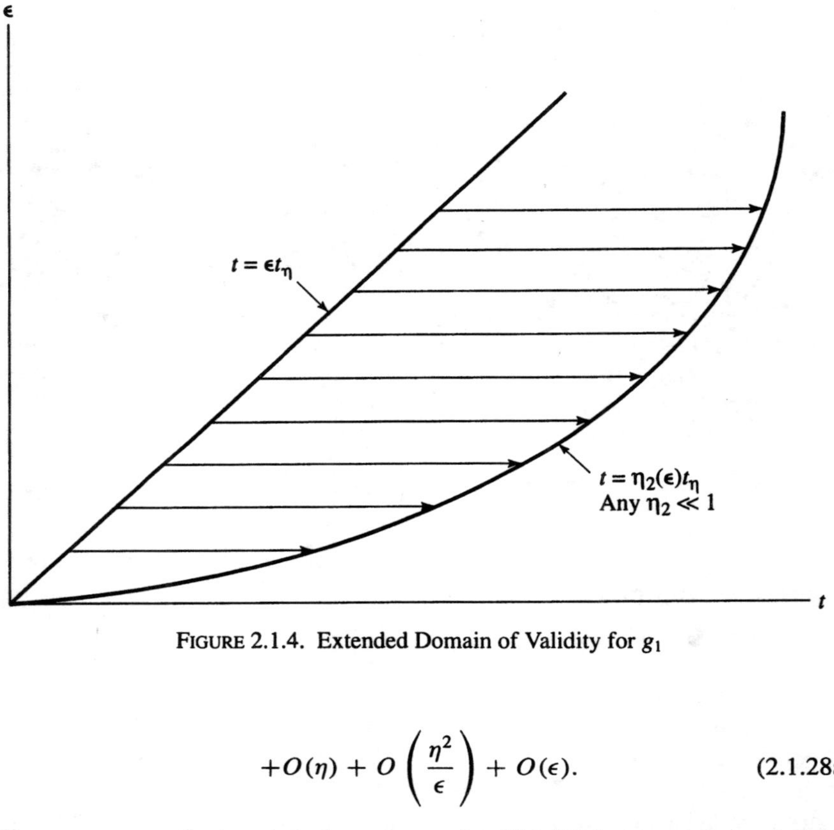

FIGURE 2.1.4. Extended Domain of Validity for g_1

$$+O(\eta) + O\left(\frac{\eta^2}{\epsilon}\right) + O(\epsilon). \tag{2.1.28a}$$

When we express the two-term inner expansion (2.1.18) in terms of t_η and divide by ϵ, we find

$$\frac{1}{\epsilon} g_1(\eta t_\eta/\epsilon) + g_2(\eta t_\eta/\epsilon) = \frac{1}{\epsilon} - \frac{e^{-\eta t_\eta/\epsilon}}{\epsilon} + 2 - \frac{\eta t_\eta}{\epsilon}$$

$$-2e^{-\eta t_\eta/\epsilon} - \frac{\eta}{\epsilon} t_\eta e^{-\eta t_\eta/\epsilon}. \tag{2.1.28b}$$

Subtracting (2.1.28b) from (2.1.28a) shows that (2.1.27) is satisfied if $\eta \ll 1$, $\eta^2/\epsilon \ll 1$, and $\epsilon \ll 1$. The crucial requirement is $\eta^2/\epsilon \ll 1$ as $\epsilon \ll 1$ automatically, and $\eta \ll 1$ is already a requirement for the validity of the inner limit. Therefore, we must restrict η further by requiring $\eta \ll \epsilon^{1/2}$, and we find that the extended domain of validity of the two-term inner expansion is

$$\epsilon \lll \eta \ll \epsilon^{1/2}. \tag{2.1.29}$$

Matching of outer and inner expansions in the overlap domain

We are now in a position to explain the curious and seemingly paradoxical result we first observed in Chapter 1 (see (1.2.13)–(1.2.14) and (1.2.35)–(1.2.36)). We had found that when the outer expansion was expressed in terms of the inner variable $x^* = x/\epsilon$ and re-expanded, it agreed with the inner expansion evaluated for large x^*.

We first note that, in the previous example, the extended domains of validity of the outer and inner expansions to $O(1)$ and $O(\epsilon)$ *overlap* in the sense that for each order in ϵ there exists a set of η that belongs to *both* extended domains (see Figure 2.1.5).

For the outer and inner limits of this example, the overlap domain is

$$\epsilon|\log\epsilon| \ll \eta \ll 1,$$

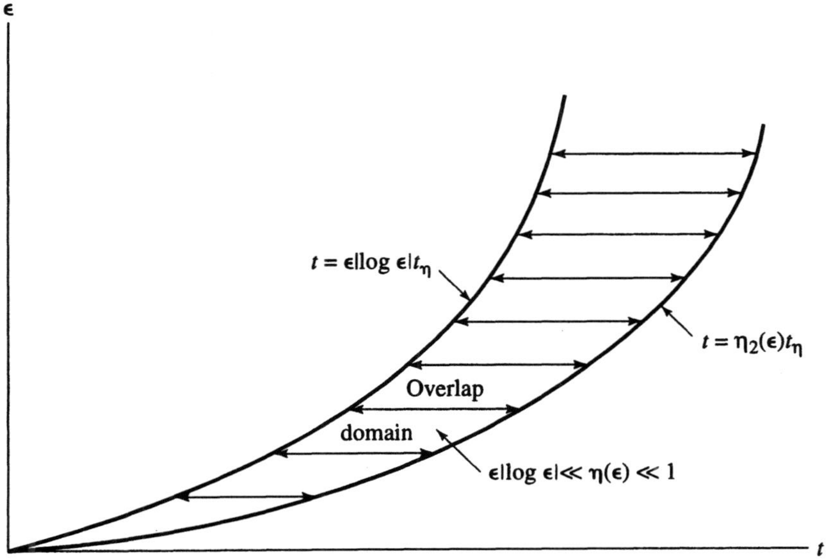

FIGURE 2.1.5. Overlap Domain for h_1 and g_1

and for the two-term outer and inner expansions, it is

$$\epsilon|\log\epsilon| \ll \eta \ll \epsilon^{1/2}.$$

Subtracting (2.1.20) from (2.1.25) gives the *direct matching condition* to $O(1)$

$$\lim_{\substack{\epsilon\to 0\\ t_\eta \text{ fixed } \neq 0, \neq\infty}} (h_1(\eta t_\eta) - g_1(\eta t_\eta/\epsilon)) = 0 \tag{2.1.30a}$$

for all $\eta(\epsilon)$ in the overlap domain $\epsilon|\log\epsilon| \ll \eta \ll 1$. Similarly, the matching condition to $O(\epsilon)$ follows from (2.1.23) and (2.1.27)

$$\lim_{\substack{\epsilon\to 0\\ t_\eta \text{ fixed } \neq 0, \neq\infty}} \frac{h_1(\eta t_\eta) + \epsilon h_2(\eta t_\eta) - g_1(\eta t_\eta/\epsilon) - \epsilon g_2(\eta t_\eta/\epsilon)}{\epsilon} = 0 \tag{2.1.30b}$$

for all η in the overlap domain $\epsilon|\log\epsilon| \ll \eta \ll \epsilon^{1/2}$.

Note that in the overlap domain the outer expansion is being evaluated for *small* t since t is replaced by ηt_η with $\eta \ll 1$. This yields the same result as that we would obtain by setting $t = \epsilon t^*$ and re-expanding. Similarly, in the overlap domain, the inner expansion is being evaluated for large t^* because we replace t^* by $\eta t_\eta/\epsilon$ and $\eta/\epsilon \to \infty$ as $\epsilon \to 0$ in the overlap domain. In effect, expressing the outer and inner expansions in terms of t_η in the overlap domain simply corresponds to evaluating the outer expansion for t small and the inner expansion for t^* large. Although it is often possible to carry out the matching by this simple scheme, in this book we will always take the more systematic approach of expressing both expansions in terms of the matching variable t_η to make sure that terms ignored are indeed negligible in the overlap domain and to exhibit this domain explicitly.

When the exact solution is not available but one is able to construct the outer and inner expansions by solving the differential equations associated with the respective limit processes, the direct matching condition will be used to verify overlap and to determine unknown constants. We will discuss numerous examples of matching in the absence of an exact solution for singular perturbation problems. Next, we illustrate ideas for the example at hand.

Limiting equations, distinguished limits

The equation governing our example is (2.1.13), with initial conditions given by (2.1.14). We may consider various limits in which t approaches the origin in the $t\epsilon$-plane at varying rates. As before, we study all such limits by introducing $t_\eta = t/\eta(\epsilon)$, where $\eta(\epsilon) \leqq 1$ and t_η is fixed and positive. In terms of t_η, (2.1.13) becomes

$$\frac{\epsilon}{\eta^2}\frac{d^2\overline{y}}{dt_\eta^2} + \frac{1}{\eta}\frac{d\overline{y}}{dt_\eta} + \overline{y} = 0, \tag{2.1.31}$$

where $\overline{y}(t_\eta;\epsilon) \equiv y(\eta t_\eta;\epsilon)$.

Four cases evidently arise, each yielding a different limiting equation that would govern the dominant term of the corresponding asymptotic expansion.

(i) Outer limit. If $\eta = O_s(1)$, the limiting equation represents a balance between the damping and spring forces (the second and third terms in (2.1.31)). For simplicity, we choose $\eta = 1$ and obtain the outer limiting equation

$$\frac{d\overline{y}}{dt_\eta} + \overline{y} = 0. \tag{2.1.32}$$

We construct the outer expansion from (2.1.13) using the limit process $\epsilon \to 0$, $t =$ fixed $\neq 0$. Because (2.1.32) is derived for a choice of η having a *definite* order as $\epsilon \to 0$, we refer to its solution as a *distinguished* limit.The corresponding expansion will be a limit process expansion in the sense that each term in the expansion is defined by a limit process where $\epsilon \to 0$ with a fixed independent variable, which is t in this case (see (2.1.19)).

(ii) Inner limit (initial layer limit). With $\eta = O_s(\epsilon)$ the first two terms in (2.1.31) are of the same order and dominate in comparison with the third term. Again, this is a distinguished limit because η has a definite order as $\epsilon \to 0$, and we adopt the simple choice $\eta = \epsilon$ for which the inner limiting equation is

$$\frac{d^2\overline{y}}{dt_\eta^2} + \frac{d\overline{y}}{dt_\eta} = 0. \tag{2.1.33}$$

The associated limit process expansion has t^* fixed as $\epsilon \to 0$; see (2.1.24). Equation (2.1.33) corresponds to (2.1.11), derived earlier using physical reasoning, and the initial conditions (2.1.14) can be satisfied.

(iii) Intermediate limit. If $\epsilon \ll \eta \ll 1$, i.e., $\epsilon/\eta \to 0$, the damping term alone dominates

$$\frac{d\overline{y}}{dt_\eta} = 0. \tag{2.1.34}$$

This limit is *not* distinguished because it consists of an "open interval" of classes of η intermediate to the outer and inner limits. Equation (2.1.34), derived from this limit, can satisfy neither the initial conditions nor the expected behavior of the solution for t large. We see that this limit is in fact superfluous because it is contained in both the outer and inner limits in the sense that the solution $y =$ const. of (2.1.34) is a special case of the solution of (2.1.32) with $t_\eta \to 0$ and of (2.1.33) with $t_\eta \to \infty$.

(iv) Inner-inner limit. This final case corresponds to $\eta \ll \epsilon$ or $\epsilon/\eta \to \infty$ and yields the inertia-dominated limiting equation

$$\frac{d^2\overline{y}}{dt_\eta^2} = 0. \tag{2.1.35}$$

Again, this is not a distinguished limit and is superfluous, as it is contained in (2.1.33) ($t_\eta \to 0$). The expansion associated with this limit does satisfy both initial conditions but is valid only in a very small time interval $t \leq k\eta(\epsilon)$, $k =$ const., around $t = 0$.

In conclusion, we remark that distinguished limits are important because they correspond to a specific ordering of the magnitudes of terms appearing in the governing equation. We shall show in examples throughout this chapter that distinguished limits have common overlap domains of validity.

Limit process expansions

We assume an outer expansion in the form

$$y(t;\epsilon) = \gamma_1(\epsilon)h_1(t) + \gamma_2(\epsilon)h_2(t) + o(\gamma_2). \tag{2.1.36}$$

The equations derived by repeated application of the outer limit to (2.1.13), or by equating terms of the same order when (2.1.36) is substituted into (2.1.13), are

$$\frac{dh_1}{dt} + h_1 = 0, \tag{2.1.37a}$$

$$\frac{dh_2}{dt} + h_2 = \begin{cases} -\frac{d^2h_1}{dt^2} & \text{if } (\epsilon\gamma_1/\gamma_2) = O_s(1) \\ 0 & \text{if } (\epsilon\gamma_1/\gamma_2) \ll 1. \end{cases} \tag{2.1.37b}$$

The initial conditions for this set of equations, as well as the orders of the various $\gamma_i(\epsilon)$, are unknown and have to be found by matching with the inner expansion.

The solutions of (2.1.37) are

$$h_1(t) = A_1 e^{-t}, \tag{2.1.38a}$$

$$h_2(t) = A_2 e^{-t} - A_1 t e^{-t}, \tag{2.1.38b}$$

where A_1 and A_2 are constants. The term $-A_1 te^{-t}$ would be missing if it turned out that $(\gamma_1\epsilon/\gamma_2) \ll 1$.

We next express (2.1.13) and the initial condition (2.1.14) in terms of the inner variable $t^* = t/\epsilon$ that we found to be significant because $\eta = \epsilon$ gives a distinguished iimit. Although we already know that t/ϵ is significant from the exact solution, we proceed as though this is not available, as is the case for most problems of interest. We find

$$\frac{d^2g}{dt^{*2}} + \frac{dg}{dt^*} + \epsilon g = 0 \tag{2.1.39}$$

with initial conditions

$$g(0;\epsilon) = 0, \tag{2.1.40a}$$

$$\frac{dg}{dt^*}(0;\epsilon) = 1, \tag{2.1.40b}$$

where $g(t^*;\epsilon) \equiv y(\epsilon t^*;\epsilon)$.

Consider the following asymptotic expansion associated with the inner limit $\epsilon \to 0$, t^* fixed $\neq \infty$:

$$g(t^*;\epsilon) = \mu_1(\epsilon)g_1(t^*) + \mu_2(\epsilon)g_2(t^*) + o(\mu_2), \tag{2.1.41}$$

and the associated sequence of approximate equations that result from (2.1.39):

$$\frac{d^2 g_1}{dt^{*2}} + \frac{dg_1}{dt^*} = 0, \tag{2.1.42a}$$

$$\frac{d^2 g_2}{dt^{*2}} + \frac{dg_2}{dt^*} = \begin{cases} -g_1 & \text{if } (\epsilon\mu_1/\mu_2) = O_s(1) \\ 0 & \text{if } (\epsilon\mu_1/\mu_2 \ll 1. \end{cases} \tag{2.1.42b}$$

The initial condition (2.1.40b) fixes $\mu_1(\epsilon) = 1$ since

$$\frac{dg(0;\epsilon)}{dt^*} = \mu_1(\epsilon)\frac{dg_1(0)}{dt^*} + \mu_2(\epsilon)\frac{dg_2(0)}{dt^*} + \ldots = 1.$$

Thus, the initial conditions associated with (2.1.42) are

$$g_1(0) = 0, \tag{2.1.43a}$$

$$\frac{dg_1(0)}{dt^*} = 1. \tag{2.1.43b}$$

$$g_2(0) = 0, \tag{2.1.44a}$$

$$\frac{dg_2(0)}{dt^*} = 0. \tag{2.1.44b}$$

The solution of the inner limit equation (2.1.42a) is thus

$$g_1(t^*) = 1 - e^{-t^*}, \tag{2.1.45}$$

as found from the exact solution earlier (see (2.1.18)).

In view of the fact that both g_2 and dg_2/dt^* vanish initially, we need to have a nonzero right-hand side for (2.1.42b) to find a nonzero g_2. Accordingly, we must have $(\epsilon\mu_1/\mu_2) = O_s(1)$ and, since $\mu_1 = 1$, we choose $\mu_2 = \epsilon$, and (2.1.42b) reads

$$\frac{d^2 g_2}{dt^{*2}} + \frac{dg_2}{dt^*} = -(1 - e^{-t^*}).$$

The solution satisfying (2.1.42b) is

$$g_2(t^*) = (2 - t^*) - (2 + t^*)e^{-t^*}, \tag{2.1.46}$$

in agreement with (2.1.18). The inner expansion can be computed in this way to any order.

At this point, it is worthwhile to give a formal definition of a *limit process expansion* (see (1.2.4)). For a given function $f(x;\epsilon)$, we say that $\sum_{n=1}^{N} f_n(x^*)\mu_n(\epsilon)$ is a limit process expansion for fixed $x^* = x/s(\epsilon)$ as $\epsilon \to 0$ of $f(x;\epsilon)$ with respect to the asymptotic sequence $\{\mu_n(\epsilon)\}$ if for a given $s(\epsilon)$ and each $M = 1, 2, \ldots, N$

$$\lim_{\epsilon\to 0} \frac{f(s(\epsilon)x^*;\epsilon) - \sum_{n=1}^{M} f_n(x^*)\mu_n(\epsilon)}{\mu_M(\epsilon)} = 0. \tag{2.1.47}$$

Thus, the outer and inner expansions

$$h_1 + \epsilon h_2 = e^{-t} + \epsilon(2-t)e^{-t}, \tag{2.1.48a}$$

$$g_1 + \epsilon g_2 = (1 - e^{-t^*}) + \epsilon[(2-t^*) - (2+t^*)e^{-t^*}] \tag{2.1.48b}$$

are each limit process expansions of $y(t; \epsilon)$ for $N = 2$ and $s = 1$ and $s = \epsilon$ respectively.

Matching

The unknown constants A_1, A_2 ... as well as the asymptotic sequence γ_1, γ_2, ... are to be determined by matching the outer and inner expansions in their common domain of validity. The matching conditions to $O(1)$ and $O(\epsilon)$ are given by (2.1.30a) and (2.1.30b), respectively. To apply (2.1.30a), we express h_1 and g_1 in terms of t_η and expand for η small

$$\gamma_1(\epsilon)h_1(\eta t_\eta) = \gamma_1(\epsilon)A_1(1 - \eta t_\eta + O(\eta^2)), \tag{2.1.49a}$$

$$g_1(\eta t_\eta)/\epsilon) = 1 - e^{-\eta t_\eta/\epsilon}. \tag{2.1.49b}$$

The dominant term in (2.1.49a) is $\gamma_1(\epsilon)A_1$; it must match with the dominant term in (2.1.49b). Therefore, γ_1 cannot depend on ϵ since A_1 is also independent of ϵ. Thus we pick $\gamma_1 = 1$ and $A_1 = 1$ (or $\gamma_1 = 1/A_1$, as the end result is the same). The next term in (2.1.49a), $-\eta t_\eta$, must be made to vanish as nothing will match it to this order. So we must have $\eta \ll 1$. The exponential term in (2.1.49b) must be made transcendentally small because no other term in either expansion to any $O(\epsilon^M)$ will match with it. So we must have $\epsilon|\log\epsilon| \ll \eta$.

In summary, the outer and inner expansions match to $O(1)$ with $\gamma_1 = 1$, $A_1 = 1$ for η in the overlap domain $\epsilon|\log\epsilon| \ll \eta(\epsilon) \ll 1$.

For the matching to $O(\epsilon)$, we use (2.1.30b) and write

$$\frac{1}{\epsilon}h_1(\eta t_\eta) + \frac{\gamma_2}{\epsilon}h_2(\eta t_\eta) = \frac{1}{\epsilon} - \frac{\eta}{\epsilon}t_\eta + O(\frac{\eta^2}{\epsilon}) + \frac{\gamma_2}{\epsilon}A_2 + O(\eta), \tag{2.1.50a}$$

$$\frac{1}{\epsilon}g_1(\eta t_\eta/\epsilon) + g_2(\eta t_\eta/\epsilon) = \frac{1}{\epsilon} + 2 - \frac{\eta}{\epsilon}t_\eta + \text{T.S.T.} \tag{2.1.50b}$$

The singular terms $(1/\epsilon)$ and $(-\eta t_\eta/\epsilon)$ in each expansion match, so they cancel out identically in (2.1.30b). The only term that will match the 2 in (2.1.50b) is $(\gamma_2/\epsilon)A_2$ in (2.1.50a). To do so, we must basically have $\gamma_2 = \epsilon$ and $A_2 = 2$. Finally, the term of order (η^2/ϵ) in (2.1.50a) vanishes if $\eta \ll \epsilon^{1/2}$. This, combined with the condition $\epsilon|\log\epsilon| \ll \eta$, which ensures that e^{-t^*} is transcendentally small, gives the overlap domain $\epsilon|\log\epsilon| \ll \eta(\epsilon) \ll \epsilon^{1/2}$ for the matching to $O(\epsilon)$.

We note that all the assumptions that were made prior to the matching are justified a posteriori: it was necessary to set $\gamma_2 = \epsilon$ in order to obtain $A_2 = 2$. The choice $\epsilon \ll \gamma_2$ would have required that we set $A_2 = 0$, leaving no term to match the constant contribution 2 from g_2. In this example, we were able to match the leading $O(1)$ terms in each expansion to $O(1)$ and the two-term expansions to $O(\epsilon)$. Failure of the matching would have required that these assumptions be abandoned or modified. For example, in some problems a homogeneous solution (corresponding here to the choice $\gamma_1\epsilon \ll \gamma_2$) might be needed to carry out the matching. Also, to match to a given order, the number of terms that are required in one of the expansions might exceed those in the other. Examples of these and other possibilities will be discussed as we study progressively more complicated problems.

Uniformly valid composite expansion, general asymptotic expansion

At this point, it is natural to look for an approximation of the solution that remains uniformly valid on $0 \leq t \leq T$, where T is finite. We shall consider the question of allowing $T \to \infty$ later on.

Our study of the respective domains of validity of the outer and inner expansions indicates that a function $u(t; \epsilon)$ is a uniformly valid approximation to the exact solution $y(t; \epsilon)$ to some order in ϵ if it yields the outer and inner expansion to the same order under the corresponding limit process. Thus, $u(t; \epsilon)$ must satisfy *both* (2.1.19) and (2.1.24) in order to be a uniformly valid approximation to $O(\epsilon)$.

A useful approach for constructing $u(t; \epsilon)$, once the outer and inner expansions are known, is to add these two expansions and subtract those terms that are common (cancel out in the matching). For our example, the outer and inner expansions to order ϵ are

$$h_1 + \epsilon h_2 = e^{-t} + \epsilon(2 - t)e^{-t} + O(\epsilon^2),$$
$$g_1 + \epsilon g_2 = (1 - e^{-t^*}) + \epsilon[(2 - t^*) - (2 + t^*)e^{-t^*}] + O(\epsilon^2),$$

and the common part is $1 + \epsilon(2 - t^*)$. Therefore, we propose

$$\begin{aligned} u(t; \epsilon) &= (h_1 + g_1 - 1) + \epsilon[h_2 + g_2 - (2 - t^*)] \\ &= (e^{-t} - e^{-t^*}) + \epsilon[(2 - t)e^{-t} - (2 + t^*)e^{-t^*}] \end{aligned} \tag{2.1.51}$$

as a uniformly valid approximation to $O(\epsilon)$. It is easily verified that u contains both the outer and inner expansions to $O(\epsilon)$ in the sense that when u is used instead of y, (2.1.19) and (2.1.24) both hold.

The expansion (2.1.51) has the form

$$u(t; \epsilon) = F_1(t; \epsilon) + \epsilon F_2(t; \epsilon) + \ldots, \tag{2.1.52}$$

where

$$F_1(t; \epsilon) = h_1(t) + g_1^*(t/\epsilon),$$
$$F_2(t; \epsilon) = h_2(t) + g_2^*(t/\epsilon).$$

Here g_1^* and g_2^* consist of the inner minus common terms to each order. They are referred to as the *inner layer* (or *boundary-layer*) *correction terms*, and they decay exponentially in the outer domain. The approximation (2.1.52) is an asymptotic expansion of $y(t; \epsilon)$ in the following general sense (see (2.1.4)).

For a given function of $f(t; \epsilon)$, the expansion $\sum_{n=1}^{N} F_n(t; \epsilon)$ is a *general asymptotic expansion* as $\epsilon \to 0$ of $f(t; \epsilon)$ with respect to the asymptotic sequence

$\{\mu_n(\epsilon)\}$ if for each $M = 1, 2, \ldots N$

$$\lim_{\epsilon \to 0} \frac{f(t;\epsilon) - \sum_{n=1}^{M} F_n(t;\epsilon)}{\mu_M(\epsilon)} = 0. \tag{2.1.53}$$

Thus, (2.1.51) is a uniformly valid (on $0 \le t \le T < \infty$) generalized asymptotic expansion of $y(t;\epsilon)$, but it is not a limit process expansion, as each F_n involves both t and t^* (or t and ϵ) simultaneously. A form such as (2.1.51) may be assumed *a priori* in many singular perturbation problems. The $g_i^*(t/\epsilon)$ should have the property of correcting the generally incorrect boundary value given by the h_i and should decay exponentially away from the boundary ($t = 0$ in this case).

Finally, note that in this example the first term of the uniformly valid approximation gives a good description of the physical phenomenon for small M. In physical variables, we have

$$y \cong \frac{I_0}{B}\{e^{-KT/B} - e^{-BT/M}\}.$$

The motion shows a rapid rise to a peak at $T \cong (M/B)\log(B^2/KM)$ and an eventual decay.

Modified outer expansion: uniformity at $t = \infty$

If we are interested in extending the domain of uniform validity of the outer expansion to $t = \infty$, we must avoid expanding the first exponential in (2.1.16). Because there is a cumulative effect of inertia that weakly shifts the time scale of decay, we can choose a new outer time scale

$$t^+ = t(1 + \epsilon a_1 + \epsilon^2 a_2 + \ldots) \tag{2.1.54}$$

instead of t, and a limit process where $\epsilon \to 0$ with t^+ fixed. It is evident from (2.1.16) that $1 + \epsilon a_1 + \epsilon^2 a_2 + \ldots$ is just the expansion of $(1 - \sqrt{1-4\epsilon})/2\epsilon$ in powers of ϵ.

Next we will show that the constants a_k can be found, without knowledge of the exact solution, by enforcing uniform validity at $t = \infty$ on the modified outer expansion. This now has the form

$$y(t;\epsilon) = h_1^+(t^+) + \epsilon h_2^+(t^+) + \ldots. \tag{2.1.55}$$

Note that

$$\frac{dy}{dt} = \frac{dh_1^+}{dt^+} + \epsilon\left\{\frac{dh_2^+}{dt^+} + a_1\frac{dh_1^+}{dt^+}\right\} + \ldots$$

$$\frac{d^2y}{dt^2} = \frac{d^2h_1^+}{dt^{+2}} + \epsilon\left\{\frac{d^2h_2^+}{dt^{+2}} + 2a_1\frac{d^2h_1^+}{dt^{+2}}\right\} + \ldots.$$

Thus, replacing (2.1.37), we have

$$\frac{dh_1^+}{dt^+} + h_1^+ = 0, \tag{2.1.56}$$

$$\frac{dh_2^+}{dt^+} + h_2^+ = -\frac{d^2h_1^+}{dt^{+2}} - a_1 \frac{dh_1^+}{dt^+}. \tag{2.1.57}$$

With $h_1^+ = A_1^+ e^{-t^+}$, the right-hand side of (2.1.57) becomes $-A_1^+(1 - a_1)e^{-t^+}$. Previously, the term $\epsilon t e^{-t}$, which caused trouble as $T \to \infty$, arose from the right-hand side of (2.1.37b), in particular, from the forcing term proportional to e^{-t}, which is a solution of the homogeneous equation. Now, such a homogeneous solution can be eliminated by the choice $a_1 = 1$ that makes the right-hand side of (2.1.57) equal to zero.

The outer expansion that results,

$$h_1^+(t^+) + \epsilon h_2^+(t^+) = A_1^+ e^{-t^+} + \epsilon A_2^+ t^+ e^{-t^+},$$

is now uniform near infinity. A similar idea of "strained coordinates" will be discussed in Sec. 4.1 for periodic solutions.

Problems

1. Calculate the regular expansion of (2.1.5) to $O(\epsilon^{*^2})$ and identify the nonuniform contributions for t large due to the expansions of $e^{-\epsilon^* t^*}$ as well as $\sin\sqrt{1-\epsilon^2}t^*$. Show that this result also follows from the solution of the initial value problems that one finds for g_1, g_2, and g_3 when (2.1.7) is substituted into (2.1.4).
2. What is the overlap domain for the matching to $O(\epsilon^3)$ of (1.2.10) and (1.2.12)?
3. What is the overlap domain for the matching to $O(\epsilon)$ of (1.2.24) and (1.2.34)?
4. Compute $a_2 = 2$ by carrying out the modified outer expansion (2.1.55) to $O(\epsilon^2)$ and verify that your result is the term of order ϵ^2 in the expansion for $(1 - \sqrt{1-4\epsilon})/2\epsilon$.

2.2 Linear Singular Perturbation Problems with Variable Coefficients

The ideas of the previous section are now applied to some further linear examples with the aim of showing how variable coefficients in the equation can affect the nature of the expansion.

2.2.1 General Problem

Consider first a general boundary-value problem for a finite interval, $0 \le x \le 1$, say, for

$$\epsilon \frac{d^2y}{dx^2} + a(x)\frac{dy}{dx} + b(x)y = 0, \tag{2.2.1}$$

with the boundary conditions

$$y(0) = A, \quad y(1) = B. \tag{2.2.2}$$

In general, a, b, A, and B can depend in a regular way on ϵ, but here it is assumed that they are independent of ϵ.

The first question to be discussed is the existence of a solution to the boundary-value problem for some range of $\epsilon > 0$ in the neighborhood of $\epsilon = 0$. A general result is that the solution to the boundary-value problem specified by (2.2.1) and (2.2.2) exists and is unique if the corresponding problem with zero boundary conditions ($y(0) = y(1) = 0$) has only the trivial solution $y = 0$. That is, in order to apply the ideas of the previous section here, we have to be sure that no eigenvalues exist for sufficiently small ϵ. If in fact eigenvalues exist as $\epsilon \to 0$, the structure of the solution is much more complicated, although it is still possible to find asymptotic results.

The conditions under which boundary-layer theory can be applied are easily deduced from a canonical form of (2.2.1). Under the transformation

$$y(x) = \exp\left[-\frac{1}{2\epsilon}\int_9^x a(\xi)d\xi\right] w(x), \tag{2.2.3}$$

(2.2.1) takes the form ($' = d/dx$)

$$\epsilon\frac{d^2w}{dx^2} - \left[\frac{a^2(x)}{4\epsilon} + \frac{a'(x)}{2} - b(x)\right] w(x) = 0. \tag{2.2.4}$$

It is only necessary to discuss the lowest eigenvalue of (2.2.4) and the corresponding eigenfunction since its existence (or nonexistence) implies the existence (or nonexistence) of the infinite discrete spectrum. The qualitative shape of the eigenfunctions for $y(x)$ will be the same as that for $w(x)$ if we assume that

$$\int_0^x a(\xi)d\xi < \infty.$$

Then, the fundamental eigenfunctions will necessarily have the qualitative shape shown in Figure 2.2.1. It can be seen from (2.2.4) that $d^2w/dx^2 > 0$ when $w > 0$ if

$$\frac{a^2(x)}{4\epsilon} + \frac{a'(x)}{2} - b(x) > 0, \quad 0 \le x \le 1, \tag{2.2.5}$$

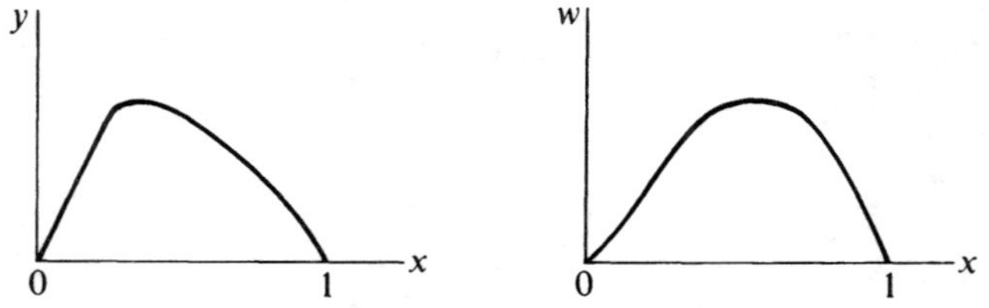

FIGURE 2.2.1. Shape of Eigenfunctions

and in this case it is not possible to have an eigenfunction of the shape in Figure 2.2.1. Thus, sufficient conditions for the existence of simple boundary layers as $\epsilon \to 0$ are merely that on $0 \leq x \leq 1$

$$a(x) \neq 0, \quad |a'(x)| < \infty, \quad b(x) < \infty. \tag{2.2.6}$$

Under these conditions, it is always possible to find sufficiently small ϵ so that the inequality in (2.2.5) is satisfied. In all the examples that we consider in this section, the inequality in (2.2.5) will hold.

Assume now that $a(x)$, $b(x)$ are such that the solution to the boundary-value problem exists for ϵ sufficiently small. In the limit problem ($\epsilon = 0$), the equation reduces to first order and, in general, both boundary conditions cannot be satisfied by the reduced equation. We therefore expect a boundary layer to exist at either end of the domain or possibly in the interior. As we saw previously, the idea of the boundary layer is that the higher-order terms of (2.2.1) dominate the behavior of the solution in the boundary layer. Thus, we have

$$\epsilon \frac{d^2 y_{BL}}{dx^2} + a_{BL} \frac{dy_{BL}}{dx} = 0 \tag{2.2.7}$$

where a_{BL} is the value of a at the boundary-layer location.

Exponential decay (rather than growth) is essential for boundary-layer behavior. Thus, if $a(x) > 0$ in $0 \leq x \leq 1$, the solutions of (2.2.7) can be expected to decay exponentially near $x = 0$ ($\epsilon > 0$), and the boundary layer occurs there; if $a(x) < 0$, the boundary layer occurs near $x = 1$. The case where $a(x)$ changes sign in the interval $0 \leq x \leq 1$ is evidently more complicated; examples of this situation are discussed in Sections 2.2.4 and 2.2.5.

We can make a few remarks about the general form of the outer expansion. Asssuming that $a(x) > 0$, $0 \leq x \leq 1$, we see that the outer expansion, valid away from $x = 0$, must proceed in powers of ϵ, that is,

$$y(x; \epsilon) = h_0(x) + \epsilon h_1(x) + \epsilon^2 h_2(x) + \dots. \tag{2.2.8}$$

The various h_i all satisfy first-order differential equations and the boundary conditions

$$h_0(1) = B, \quad h_i(1) = 0, \quad i = 1, 2, 3, \dots. \tag{2.2.9}$$

The sequence in powers of ϵ is necessary to ensure that the various $h_i, i = 1, 2, \dots,$ satisfy nonhomogeneous differential equations. If other orders of ϵ were used, the corresponding h_j would be identically zero. Thus, a sequence of equations for the outer expansion is obtained:

$$a(x) \frac{dh_0}{dx} + b(x) h_0 = 0 \tag{2.2.10a}$$

$$a(x) \frac{dh_1}{dx} + b(x) h_1 = -\frac{d^2 h_0}{dx^2} \tag{2.2.10b}$$

$$\vdots$$

$$a(x)\frac{dh_i}{dx} + b(x)h_i = -\frac{d^2h_{i-1}}{dx^2}, \qquad i = 2, 3, \ldots. \tag{2.2.10c}$$

The solution of (2.2.10a), if we take account of the boundary condition, is

$$h_0(x) = B \exp\left[\int_x^1 \frac{b(\xi)}{a(\xi)}\, d\xi\right]. \tag{2.2.11}$$

In order for a simple boundary layer to exist at $x = 0$, the first term $h_0(x)$ should be defined on $0 \le x \le 1$; thus, it is assumed that the integral in (2.2.11) exists and that $h_0(x)$ takes a value as $x \to 0$.

$$h_0(0) = B \exp\left[\int_0^1 \frac{b(\xi)}{a(\xi)}\, d\xi\right] = C \quad \text{(say).} \tag{2.2.12}$$

In general, $h_0(0) \ne A$, so that a boundary layer exists at $x = 0$ (see Figure 2.2.2).

Assuming now that the variable coefficients have regular expansions near $x = 0$,

$$\begin{aligned} a(x) &= a^{(0)} + a^{(1)}x + \ldots, \qquad a^{(0)} > 0 \\ b(x) &= b^{(0)} + b^{(1)}x + \ldots, \end{aligned} \tag{2.2.13}$$

we can construct a boundary-layer expansion. The suitable inner variable is

$$x^* = \frac{x}{\epsilon} \tag{2.2.14}$$

since the basic equation behaves near the boundary $x = 0$ in a way that is essentially the same as the constant-coefficient equation of the previous section. The orders in ϵ of the terms in the asymptotic sequence for the inner expansion are found, strictly, from the condition of matching with the outer expansion. For the functions considered here, a power series in ϵ is adequate:

$$y(x; \epsilon) = g_0(x^*) + \epsilon g_1(x^*) + \ldots. \tag{2.2.15}$$

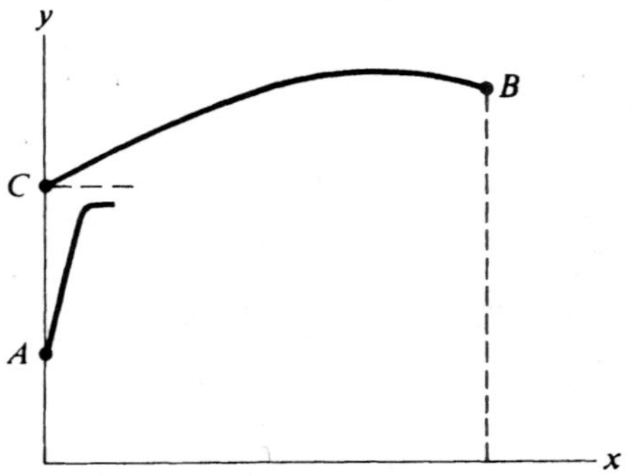

FIGURE 2.2.2. Boundary Layer at $x = 0$

The coefficients in (2.2.1) are also expressed in terms of the inner coordinate x^* and thus have the expansion

$$\begin{aligned} a(x) &= a(\epsilon x^*) = a^{(0)} + \epsilon a^{(1)} x^* + \dots \\ b(x) &= b(\epsilon x^*) = b^{(0)} + \epsilon b^{(1)} x^* + \dots . \end{aligned} \tag{2.2.16}$$

These expansions are useful for the inner limit process $\epsilon \to 0$, x^* fixed. Thus, the equations satisfied by the g_i are

$$\frac{d^2 g_0}{dx^{*2}} + a^{(0)} \frac{dg_0}{dx^*} = 0, \tag{2.2.17}$$

$$\frac{d^2 g_1}{dx^{*2}} + a^{(0)} \frac{dg_1}{dx^*} = -b^{(0)} g_0(x^*) - a^{(1)} x^* \frac{dg_0}{dx^*}. \tag{2.2.18}$$

The boundary conditions at $x^* = 0$ are

$$g_0(0) = A, \tag{2.2.19}$$

$$g_1(0) = g_2(0) = \dots = 0. \tag{2.2.20}$$

The solution of (2.2.17) satisfying the boundary condition (2.2.19) is

$$g_0(x^*) = Ae^{-a^{(0)}x^*} + B_0(1 - e^{-a^{(0)}x^*}). \tag{2.2.21}$$

The constant B_0 is found from matching with the first term of the outer expansion.

To carry out this matching to $O(1)$, we introduce the intermediate variable $x_\eta = x/\eta(\epsilon)$, where $\eta(\epsilon)$ belongs to the class that specifies the overlap domain, and we express h_0 and g_0 in terms of x_η. We find

$$h_0(\eta x_\eta) = B \exp\left[\int_0^1 \frac{b(\xi)}{a(\xi)} d\xi\right] + O(\eta), \tag{2.2.22a}$$

$$g_0\left(\frac{\eta x_\eta}{\epsilon}\right) = B_0 + \text{T.S.T.}, \tag{2.2.22b}$$

where the exponential terms in (2.2.22b) are transcendentally small as long as $\epsilon|\log\epsilon| \ll \eta$, exactly as in the previous example.

The matching condition to $O(1)$ (see (2.1.30a))

$$\lim_{\substack{\epsilon\to 0\\ x_\eta \text{ fixed}}} (h_0(\eta x_\eta) - g_0(\eta x_\eta/\epsilon)) = 0 \tag{2.2.23}$$

is therefore satisfied if we choose

$$B_0 = B \exp\left[\int_0^1 \frac{b(\xi)}{a(\xi)} d\xi\right] \tag{2.2.24}$$

for all η in the overlap domain $\epsilon|\log\epsilon| \ll \eta \ll 1$.

A uniformly valid approximation to order unity is obtained, as before, by adding h_0 and g_0 and subtracting the common part (2.2.24). Thus, the uniformly valid $(0 \leq x \leq 1)$ approximation to $O(1)$ is

$$u(x; \epsilon) = B \exp\left[\int_0^1 \frac{b(\xi)}{a(\xi)} d\xi\right] + \left\{A - B \exp\left[\int_0^1 \frac{b(\xi)}{a(\xi)} d\xi\right]\right\} e^{-a^{(0)}x^*}. \tag{2.2.25}$$

Higher approximations can be computed. We will carry out such a computation in the next section.

2.2.2 *Analytic Coefficients*

In this subsection, we consider the following special case of (2.2.1):

$$\epsilon \frac{d^2y}{dx^2} + (1 + \alpha x)\frac{dy}{dx} + \alpha y = 0, \qquad 0 \leq x \leq 1,\ \alpha = \text{const.} > -1 \tag{2.2.26}$$

with boundary conditions

$$y(0) = 0, \qquad y(1) = 1. \tag{2.2.27}$$

The outer expansion is

$$y(x; \epsilon) = h_0(x) + \epsilon h_1(x) + \ldots, \tag{2.2.28}$$

so that

$$(1 + \alpha x)\frac{dh_0}{dx} + \alpha h_0 = 0, \qquad h_0(1) = 1; \tag{2.2.29}$$

$$(1 + \alpha x)\frac{dh_1}{dx} + \alpha h_1 = -\frac{d^2h_0}{dx^2}, \qquad h_1(1) = 0. \tag{2.2.30}$$

The boundary conditions for the outer expansion are taken at the right-hand end of the interval, since the boundary layer occurs at $x = 0$. The solutions of (2.2.29) and (2.2.30) are easily obtained as

$$h_0(x) = \frac{1 + \alpha}{1 + \alpha x}, \tag{2.2.31}$$

$$h_1(x) = -\alpha\left\{\frac{1}{(1 + \alpha)(1 + \alpha x)} - \frac{1 + \alpha}{(1 + \alpha x)^3}\right\}. \tag{2.2.32}$$

Note that $h_0(0) = 1 + \alpha > 0$. The inner expansion, valid near the boundary, is

$$y(x; \epsilon) = g_0(x^*) + \epsilon g_1(x^*) + \ldots, \qquad x^* = \frac{x}{\epsilon}. \tag{2.2.33}$$

The basic equation (2.2.26) becomes

$$\frac{1}{\epsilon}\left\{\frac{d^2g_0}{dx^{*2}} + \epsilon\frac{d^2g_1}{dx^{*2}} + \ldots\right\} + \frac{1}{\epsilon}\{1 + \alpha\epsilon x^*\}\left\{\frac{dg_0}{dx^*} + \epsilon\frac{dg_1}{dx^*} + \ldots\right\} + \alpha g_0$$

$$+\epsilon\alpha g_1 + \ldots = 0 \tag{2.2.34}$$

so that we have

$$\frac{d^2 g_0}{dx^{*2}} + \frac{dg_0}{dx^*} = 0, \qquad g_0(0) = 0; \tag{2.2.35}$$

$$\frac{d^2 g_1}{dx^{*2}} + \frac{dg_1}{dx^*} = -\alpha x^* \frac{dg_0}{dx^*} - \alpha g_0, \qquad g_1(0) = 0. \tag{2.2.36}$$

The constants of integration for the boundary-layer solutions are found from the boundary condition at $x^* = 0$ and by matching with the outer expansion. The solution of (2.2.35) is

$$g_0(x^*) = B_0(1 - e^{-x^*}). \tag{2.2.37}$$

The matching condition to $O(1)$ is (2.2.23), and we have

$$h_0(\eta x_\eta) = 1 + \alpha + o(\eta) \tag{2.2.38a}$$

$$g_0(\eta x_\eta/\epsilon) = B_0 + \text{T.S.T.} \tag{2.2.38b}$$

Thus, we must set $B_0 = 1 + \alpha$. The boundary layer rises from $y = 0$ at $x = 0$ asymptotically to the value $1 + \alpha > 0$. Using this value for B_0, we find the following solution for g_1:

$$g_1(x^*) = C_1 + B_1 e^{-x^*} - \alpha(\alpha + 1)\left\{(x^* - 1) - \frac{x^{*2}}{2} e^{-x^*}\right\}, \tag{2.2.39}$$

and the boundary condition at $x^* = 0$ requires

$$B_1 = -C_1 - \alpha(\alpha + 1).$$

The matching condition to $O(\epsilon)$ is (see (2.1.30b))

$$\lim_{\substack{\epsilon \to 0 \\ x_\eta \text{ fixed}}} \frac{1}{\epsilon}\{h_0(\eta x_\eta) + \epsilon h_1(\eta x_\eta) - g_0(\eta x_\eta/\epsilon) - \epsilon g_1(\eta x_\eta/\epsilon)\} = 0 \tag{2.2.40}$$

for an overlap domain in η to be determined. We now expand h_0 to higher order and ignore transcendentally small terms in g_0 and g_1 to calculate

$$\frac{1}{\epsilon} h_0(\eta x_\eta) = \frac{(1 + \alpha)}{\epsilon} - \frac{\eta}{\epsilon}\alpha(1 + \alpha)x_\eta + O(\eta^2/\epsilon),$$

$$h_1(\eta x_\eta) = -\frac{\alpha}{1 + \alpha} + \alpha(1 + \alpha) + O(\eta),$$

$$\frac{1}{\epsilon} g_0(\eta x_\eta/\epsilon) = \frac{1 + \alpha}{\epsilon} + \text{T.S.T.},$$

$$g_1(\eta x_\eta/\epsilon) = C_1 - \alpha(\alpha + 1)\left(\frac{\eta x_\eta}{\epsilon} - 1\right) + \text{T.S.T.}$$

The singular terms $(1 + \alpha)/\epsilon$ (in h_0/ϵ and g_0/ϵ) and $-\alpha(1 + \alpha)\eta x_\eta/\epsilon$ (in h_0/ϵ and g_1) are matched. In order that the constant terms match, we must set

$$C_1 + \alpha(\alpha + 1) = -\frac{\alpha}{1 + \alpha} + \alpha(1 + \alpha)$$

or

$$C_1 = -\frac{\alpha}{1+\alpha}.$$

The requirement that $\eta^2/\epsilon \to 0$ implies $\eta \ll \sqrt{\epsilon}$. Thus, as in the example of Sec. 2.1, the overlap domain for the matching to $O(\epsilon)$ is

$$\epsilon|\log\epsilon| \ll \eta(\epsilon) \ll \sqrt{\epsilon}.$$

We refer to the terms in the outer and inner expansions that match in the overlap region as the common part (cp). Thus,

$$\text{cp} = (1+\alpha) - \epsilon\left(\frac{\alpha}{1+\alpha} + \alpha(1+\alpha)[x^* - 1]\right) \tag{2.2.41}$$

written in terms of the inner variable x^*. Adding the first two terms of the inner and outer expansions and subtracting the common part yields the uniformly valid expansion to $O(\epsilon)$:

$$u(x;\epsilon) = (1+\alpha)\left\{\frac{1}{1+\alpha x} - e^{-x^*}\right\} - \epsilon\left\{\frac{\alpha}{1+\alpha}\left[\frac{1}{1+\alpha x} - e^{-x^*}\right]\right.$$

$$\left. -\alpha(1+\alpha)\left[\frac{1}{(1+\alpha x)^3} - e^{-x^*}\right] - \alpha(\alpha+1)\frac{x^{*2}}{2}e^{-x^*}\right\}. \tag{2.2.42}$$

Note that the inner expansion contributes only transcendentally small terms away from the boundary. The uniformly valid expansion again has the form of a composite expression

$$y(x;\epsilon) = \sum_{k=0} \epsilon^k \left\{h_k(x) + f_k\left(\frac{x}{\epsilon}\right)\right\}. \tag{2.2.43}$$

This form could have been taken as the starting point for the expansion of (2.2.26) with the requirements that $h_0 + f_0$ satisfy both boundary conditions, ($h_i + f_i$, $i \neq 0$) satisfy zero boundary conditions, and all f_i are transcendentally small away from the boundary.

2.2.3 Nonanalytic Coefficients

This example is chosen to illustrate the effect of coefficients that are not analytic at the boundary point in modifying the form of the expansion. The basic assumption about the coefficients is that the first term of the outer expansion exists so that $h_0(x)$ is defined on $0 \le x \le 1$:

$$\epsilon\frac{d^2y}{dx^2} + \sqrt{x}\frac{dy}{dx} - y = 0, \qquad 0 \le x \le 1, \tag{2.2.44}$$

$$y(0) = 0, \qquad y(1) = e^2. \tag{2.2.45}$$

In order to see if the solution to the boundary-value problem exists, we check the coefficient of $w(x)$ in (2.2.4). With

$$a(x) = \sqrt{x}, \qquad b(x) = -1$$

we have

$$\frac{a^2(x)}{4\epsilon} + \frac{a'(x)}{2} - b(x) = \frac{x}{4\epsilon} + \frac{1}{4\sqrt{x}} + 1 > 0.$$

Thus, the solution to this particular problem exists for all $\epsilon > 0$.

The outer expansion ($\epsilon \to 0$, x fixed $\neq 0$) is

$$y(x, \epsilon) = h_0(x) + \epsilon h_1(x) + O(\epsilon^2)$$

with the following equations and boundary conditions at $x = 1$:

$$\sqrt{x}\,\frac{dh_0}{dx} - h_0 = 0, \qquad h_0(1) = e^2, \tag{2.2.46}$$

$$\sqrt{x}\,\frac{dh_1}{dx} - h_1 = -\frac{d^2h_0}{dx^2}, \qquad h_1(1) = 0. \tag{2.2.47}$$

The solutions are

$$h_0(x) = e^{2\sqrt{x}}, \tag{2.2.48}$$

$$h_1(x) = e^{2\sqrt{x}}\left[-\frac{1}{2x} + \frac{2}{\sqrt{x}} - \frac{3}{2}\right], \tag{2.2.49}$$

$$\epsilon^2 h_2 = O\left(\frac{\epsilon^2}{x^{5/2}}\right) \quad \text{as } x \to 0. \tag{2.2.50}$$

We see that h_0 does not satisfy the boundary condition at $x = 0$ and that h_1, $h_2 \ldots$ have singularities at $x = 0$. Clearly, an inner expansion near $x = 0$ is needed to complete the description of the solution.

To investigate the nature of the distinguished limit corresponding to a boundary layer, we set $x = \delta(\epsilon)x^*$ for some as yet unspecified $\delta(\epsilon)$. The original equation (2.2.44) for $y^*(x^*; \epsilon) = y(\delta(\epsilon)x^*; \epsilon)$ becomes (see (2.1.31))

$$\frac{\epsilon}{\delta^2}\frac{d^2y^*}{dx^{*2}} + \frac{1}{\delta^{1/2}}\sqrt{x^*}\,\frac{dy^*}{dx^*} - y^* = 0. \tag{2.2.51}$$

Because the second derivative term must survive in the inner limit, and because the second term always dominates over the third for $\delta \ll 1$, we must set $\epsilon/\delta^2 = 1/\delta^{1/2}$ to obtain a distinguished limit. This gives $\delta(\epsilon) = \epsilon^{2/3}$ and

$$x^* = \frac{x}{\epsilon^{2/3}}. \tag{2.2.52}$$

The dominant boundary-layer equation is

$$\frac{d^2g_0}{dx^{*2}} + \sqrt{x^*}\,\frac{dg_0}{dx^*} = 0, \qquad g_0(0) = 0. \tag{2.2.53}$$

The solution, if we take account of the boundary condition at $x^* = 0$, is

$$g_0(x^*) = C_0 \int_0^{x^*} \exp\left(-\frac{2}{3}\zeta^{3/2}\right) d\zeta. \tag{2.2.54}$$

As $x^* \to \infty$, the integral defining g_0 approaches a constant

$$k = \int_0^{\infty} \exp\left(-\frac{2}{3}\zeta^{3/2}\right) d\zeta \tag{2.2.55}$$

that can be expressed in terms of the Gamma function. The integral in (2.2.54) can be transformed to an incomplete Gamma function. The approach of g_0 to its asymptotic value is exponential. In fact, if we write

$$\int_0^{x^*} \exp\left(-\frac{2}{3}\zeta^{3/2}\right) d\zeta = \int_0^{\infty} \exp\left(-\frac{2}{3}\zeta^{3/2}\right) d\zeta - \int_{x^*}^{\infty} \exp\left(-\frac{2}{3}\zeta^{3/2}\right) d\zeta \tag{2.2.56}$$

and use repeated integrations by parts of the second term on the right-hand side, we find

$$g_0 = C_0 \left\{k + \left[-\frac{1}{x^{*1/2}} + \frac{1}{2x^{*2}} - \frac{1}{x^{*7/2}} + O(x^{*-5})\right] \exp\left(-\frac{2}{3}x^{*3/2}\right)\right\}. \tag{2.2.57}$$

The matching to $O(1)$ can now be carried out. Introducing the matching variable

$$x_\eta = \frac{x}{\eta(\epsilon)}$$

for some class of functions $\eta(\epsilon)$ that define the overlap domain that is to be determined, we see that

$$g_0\left(\frac{\eta x_\eta}{\epsilon^{2/3}}\right) = kC_0 + \text{T.S.T.}$$

$$h_0(\eta x_\eta) = 1 + O(\sqrt{\eta}).$$

In order that $g_0 - kC_0$ be transcendentally small in the overlap domain, it is necessary that $\epsilon^{-M} e^{-\eta^{3/2}/\epsilon} \to 0$ as $\epsilon \to 0$ for any finite positive M, i.e., $\epsilon^{2/3}|\log\epsilon|^{2/3} \ll \eta$. Thus, we have matching to $O(1)$ in the overlap domain

$$\epsilon^{2/3}|\log\epsilon|^{2/3} \ll \eta(\epsilon) \ll 1$$

by setting

$$C_0 = \frac{1}{k}, \qquad k = \int_0^{\infty} e^{-(2/3)\zeta^{3/2}} d\zeta = \left(\frac{2}{3}\right)^{1/3} \Gamma\left(\frac{2}{3}\right) = 1.17\ldots.$$

For matching to higher orders, it is necessary to calculate further terms in the inner expansion, which is evidently in the form

$$y(x, \epsilon) = \sum_{i=0} \epsilon^{i/3} g_i(x^*). \tag{2.2.58}$$

Each g_i satisfies the boundary condition, $g_i(0) = 0$, and obeys

$$\frac{d^2 g_i}{dx^{*2}} + \sqrt{x^*}\frac{dg_i}{dx^*} = g_{i-1}, \qquad i = 1, 2, \ldots$$

or

$$\frac{d}{dx^*}\left[e^{(2/3)x^{*3/2}}\frac{dg_i}{dx^*}\right] = e^{(2/3)x^{*3/2}} g_{i-1}, \qquad i = 1, 2, \ldots . \tag{2.2.59}$$

Using (2.2.59) one can calculate each g_i by quadrature. Only the asymptotic behavior of the g_i as $x^* \to \infty$ is needed for matching. This is easily found using (2.2.59) directly rather than attempting to expand the exact solution by repeated integration by parts. For example, we have

$$\frac{d^2 g_1}{dx^{*2}} + \sqrt{x^*}\frac{dg_1}{dx^*} = 1 + \text{T.S.T.} \tag{2.2.60}$$

Clearly, as $x^* \to \infty$, dg_1/dx^* dominates over d^2g_1/dx^{*2}. Therefore, to leading order, the behavior of (2.2.60) must be $\sqrt{x^*}dg_1/dx^* \approx 1$, which implies that $g_1 \sim 2x^{*1/2}$ as $x^* \to \infty$. We confirm that $d^2g_1/dx^{*2} = O(x^{*-3/2})$, and therefore $d^2g_1/dx^{*2} = o(\sqrt{x^*}dg_1/dx^*)$ as $x^* \to \infty$.

We are thus led to seek an expansion for g_1 in the form

$$g_1 = 2x^{*1/2} + K_1 + \sum_{n=1}^{\infty}\frac{C_{n/2}}{x^{*n/2}} \quad \text{as } x^* \to \infty. \tag{2.2.61}$$

Substituting the series (2.2.61) into (2.2.60) then defines all the $C_{n/2}$. The constant K_1, which is obviously the limiting value of $g_1 - 2x^{*1/2}$ as $x^* \to \infty$, will be determined by the matching. Since we have imposed only one boundary condition on (2.2.60), g_1 must involve an arbitrary constant. Thus, K_1 is a function of this constant.

Using this procedure, we calculate the following expansions for the g_i as $x^* \to \infty$:

$$g_0 = 1 + \text{T.S.T.}, \tag{2.2.62}$$

$$g_1 = 2x^{*1/2} + K_1 - \frac{1}{2x^*} + O(x^{*-5/2}), \tag{2.2.63}$$

$$g_2 = 2x^* + 2K_1x^{*1/2} + K_2 + \frac{1}{x^{*1/2}} - \frac{K_1}{2x^*} + O(x^{*-3/2}), \tag{2.2.64}$$

$$g_3 = \frac{4}{3}x^{*3/2} + 2K_1x^* + 2K_2x^{*1/2} + K_3 + \frac{K_1}{x^{*1/2}} - \frac{K_2}{2x^*} + O(x^{*-3/2}), \tag{2.2.65}$$

where K_2, K_3, etc. are arbitrary constants to be determined by matching.

In order that the inner and outer expansions match to some specified order $\gamma(\epsilon) \ll 1$, we must have

$$\lim_{\substack{\epsilon\to 0\\ x_\eta \text{ fixed}}} \frac{h_0(\eta x_\eta) + \epsilon h_1(\eta x_\eta) - \sum_{j=0}^{3}\epsilon^{j/3}g_j(\eta x_\eta/\epsilon^{2/3})}{\gamma(\epsilon)} = 0 \tag{2.2.66}$$

for all $\eta(\epsilon)$ in some overlap domain. In preparation for this matching, we expand h_0, h_1, and $g_0, \ldots, g_3$ in terms of x_η for η small to find

$$h_0 = 1 + 2\eta^{1/2}x_\eta^{1/2} + 2\eta x_\eta + \frac{4}{3}\eta^{3/2}x_\eta^{3/2} + \frac{2}{3}\eta^2 x_\eta^2 + \frac{4}{15}\eta^{5/2}x_\eta^{5/2} + O(\eta^3), \tag{2.2.67}$$

$$h_1 = -\frac{1}{2\eta x_\eta} + \frac{1}{\eta^{1/2}x_\eta^{1/2}} + \frac{3}{2} + O(\eta^{1/2}), \tag{2.2.68}$$

$$g_0 = 1 + \text{T.S.T.}, \tag{2.2.69}$$

$$g_1 = \frac{2\eta^{1/2}x_\eta^{1/2}}{\epsilon^{1/3}} + K_1 - \frac{\epsilon^{2/3}}{2\eta x_\eta} + O(\epsilon^{5/3}\eta^{-5/2}), \tag{2.2.70}$$

$$g_2 = \frac{2\eta x_\eta}{\epsilon^{2/3}} + 2K_1\frac{\eta^{1/2}x_\eta^{1/2}}{\epsilon^{1/3}} + K_2 + \frac{\epsilon^{1/3}}{\eta^{1/2}x_\eta^{1/2}} - \frac{K_1\epsilon^{2/3}}{2\eta x_\eta} + O(\epsilon\eta^{-3/2}), \tag{2.2.71}$$

$$g_3 = \frac{4}{3}\frac{\eta^{3/2}x_\eta^{3/2}}{\epsilon} + 2K_1\frac{\eta x_\eta}{\epsilon^{2/3}} + 2K_2\frac{\eta^{1/2}x_\eta^{1/2}}{\epsilon^{1/3}} + K_3 + \frac{K_1\epsilon^{1/3}}{\eta^{1/2}x_\eta^{1/2}} - \frac{K_2}{2}\frac{\epsilon^{2/3}}{\eta x_\eta} + O(\epsilon\eta^{-3/2}). \tag{2.2.72}$$

We see that the first four terms in the expansion of h_0 match with the leading terms of g_0, g_1, g_2, and g_3, respectively. Moreover, the first three terms in h_1 match with corresponding terms in g_1, g_2, and g_3 if we set $K_3 = 3/2$.

Let $\gamma(\epsilon)$ in (2.2.66) belong to an asymptotic sequence of functions $\{\gamma_n(\epsilon)\}$. Eventually, we must proceed in the matching to functions $\gamma_m(\epsilon) \ll \epsilon^{1/3}$. At this point, the term $\epsilon^{1/3}K_1/\gamma_m(\epsilon)$ arising from g_1 will become singular. Evidently, this term has no counterpart in the outer expansion. Therefore, we must set $K_1 = 0$. A similar argument requires that $K_2 = 0$ because with $\gamma_k \ll \epsilon^{2/3}$ the term $\epsilon^{2/3}K_2/\gamma_k$ arising in g_2 becomes singular. This completes the determination of K_1, K_2, and K_3. However, as shown next, the matching can only be carried out with $\gamma = \epsilon^{2/3}$ but not with $\gamma = \epsilon$, without further terms.

With $\gamma = \epsilon^{2/3}$, the leading unmatched term in h_0 is $(2/3)\eta^2 x_\eta^2$. Thus, in the matching condition (2.2.66) this term becomes $O(\eta^2\epsilon^{-2/3})$. It will vanish, if we choose $\eta \ll \epsilon^{1/3}$. The leading unmatched term in the inner expansion arises from g_1 and is $O(\epsilon^{5/3}\eta^{-5/2})$; its contribution in (2.2.66) is $O(\epsilon^{4/3}\eta^{-5/2})$. Therefore, it will vanish if $\epsilon^{8/15} \ll \eta$, and this establishes the nonempty overlap domain

$$\epsilon^{8/15} \ll \eta \ll \epsilon^{1/3} \tag{2.2.73}$$

for the matching to $O(\epsilon^{2/3})$.

It is natural to attempt the matching to $O(\epsilon)$ since both inner and outer expansions have been carried to $O(\epsilon)$. We will show that this fails unless we include g_4 and g_5 in (2.2.66).

Note first that with $\gamma = \epsilon$ the most critical unmatched term in the outer expansion is still $(2/3)\eta^2 x_\eta^2$ because its contribution is $O(\eta^2/\epsilon)$ in the matching condition (2.2.66). In order for this contribution to vanish, we must choose $\eta \ll \epsilon^{1/2}$. Next, consider the most critical unmatched term in the inner expansion, the $O(\epsilon^{5/3}\eta^{-5/2})$ term in g_1. In order for this contribution to vanish in (2.2.66), we must have $\epsilon/\eta^{5/2} \ll 1$, i.e., $\epsilon^{2/5} \ll \eta$, which contradicts the prior requirement that $\eta \ll \epsilon^{1/2}$. Thus, the outer and inner expansions to $O(\epsilon)$ *do not overlap*; we need to relax one of the constraints, $\eta \ll \epsilon^{1/2}$ or $\epsilon^{2/5} \ll \eta$. To relax the requirement $\epsilon^{2/5} \ll \eta$, we would have to include $\epsilon^2 h_2$ in (2.2.66) because $\epsilon^2 h_2 = O(\epsilon^2 x^{-5/2}) = O(\epsilon^2\eta^{-5/2})$ and this is precisely the critical term in $\epsilon^{1/3}g_1$. But, now the most critical term in the inner expansion becomes the $O(\epsilon^{5/3}\eta^{-3/2})$ in $\epsilon^{2/3}g_2$. In order for this to vanish, we must have $\epsilon^{4/9} \ll \eta$, which still contradicts $\eta \ll \epsilon^{1/2}$! We therefore conclude that we cannot relax the requirement $\epsilon^{2/5} \ll \eta$, but we must instead avoid having $\eta \ll \epsilon^{1/2}$.

We recall that each successive term in the expansion (2.2.67) for h_0 matches with the *leading* term in the successive g_i's. Thus, we expect the two unmatched terms $\frac{2}{3}\eta^2 x_\eta^2$ and $\frac{4}{15}\eta^{5/2}x_\eta^{5/2}$ in h_0 to match with the leading term in g_4 and g_5, respectively. That this is indeed the case can be verified easily once we calculate the following expansions of g_4 and g_5 for x^* large:

$$\begin{aligned} g_4 &= \frac{2}{3}x^{*2} + \frac{4}{3}K_1 x^{*3/2} + 2K_2 x^* + O(x^{*1/2}), \\ g_5 &= \frac{4}{15}x^{*5/2} + \frac{2}{3}K_1 x^{*2} + O(x^{*3/2}). \end{aligned} \tag{2.2.74}$$

Now that the first six terms in h_0 have been matched, the remainder in (2.2.66) arising from h_0 is $O(\eta^3\epsilon^{-1})$ and will vanish if $\eta \ll \epsilon^{1/3}$ as in (2.2.73). We have already shown that the requirement $\epsilon^{2/5} \ll \eta$ ensures that no contributions from the inner expansion remain unmatched. Our overlap domain for the matching to $O(\epsilon)$ is

$$\epsilon^{2/5} \ll \eta \ll \epsilon^{1/3}.$$

The uniformly valid composite expansion to $O(\epsilon)$ is

$$\begin{aligned} y = h_0 + \epsilon h_1 + \sum_{j=0}^{5}\epsilon^{j/3}g_j(x^*) - \Bigg[1 + \epsilon^{1/3}\left(2x^{*1/2} - \frac{1}{2x^*}\right) \\ + \epsilon^{2/3}\left(2x^* + \frac{1}{x^{*1/2}}\right) + \epsilon\left(\frac{4}{3}x^{*3/2} + \frac{3}{2}\right) + \epsilon^{4/3}\left(\frac{2}{3}x^{*2}\right) \\ + \epsilon^{5/3}\left(\frac{4}{15}x^{*5/2}\right)\Bigg] + O(\epsilon^2). \end{aligned} \tag{2.2.75}$$

The need to include g_4 and g_5 in this example became apparent in the course of determining the overlap region. Had we simply ignored the higher-order terms in h_0, we would have incorrectly concluded that outer and inner expansions to $O(\epsilon)$ match to $O(\epsilon)$ also.

2.2.4 Interior Layer

This example and the following one are designed to illustrate some of the complications that may arise if $a(x)$ in (2.2.1) changes sign in the domain of interest. We first consider the case where $a'(x)$ is positive throughout the interval with $a(0) = 0$, and we choose an equation for which the exact solution can be explicitly calculated. We consider

$$\epsilon \frac{d^2y}{dx^2} + x\frac{dy}{dx} - y = 0, \qquad -1 \le x \le 1,\ 0 < \epsilon \ll 1, \tag{2.2.76}$$

$$y(-1) = 1; \qquad y(1) = 2. \tag{2.2.77}$$

Noting that $y = Cx$, where C is a constant, is one solution [1] of (2.2.76), we can calculate the other linearly independent solution in the form $y = x\int^x (e^{-s^2/2\epsilon}/s^2)ds$. Integrating the second solution by parts gives a more convenient representation, and we have the general solution in the form

$$y = C_1x + C_2\left(e^{-x^2/2\epsilon} + \frac{x}{\epsilon}\int_{-1}^{x} e^{-s^2/2\epsilon}ds\right). \tag{2.2.78}$$

Imposing the boundary conditions (2.2.77) gives

$$C_1 = -1 + \frac{3e^{-1/2\epsilon}}{2e^{-1/2\epsilon} + I(\epsilon)/\epsilon} \tag{2.2.79}$$

$$C_2 = \frac{3}{2e^{-1/2\epsilon} + I(\epsilon)/\epsilon}, \tag{2.2.80}$$

where

$$I(\epsilon) = \int_{-1}^{1} e^{-s^2/2\epsilon}ds = 2\int_0^{\infty} e^{-s^2/2\epsilon}ds - 2\int_1^{\infty} e^{-s^2/2\epsilon}ds. \tag{2.2.81}$$

The integral over $(1, \infty)$ is transcendentally small, as can be seen by changing the integration variable from s to $u = s^2/2\epsilon$. Thus,

$$I(\epsilon) = \sqrt{2\pi\epsilon} + \text{T.S.T.} \tag{2.2.82}$$

Equations (2.2.79)–(2.2.80) then give

$$C_1 = -1 + \text{T.S.T.} \tag{2.2.83}$$

$$C_2 = 3\sqrt{\frac{\epsilon}{2\pi}} + \text{T.S.T.} \tag{2.2.84}$$

We now examine the solution (2.2.78) in various regions of the interval $-1 \le x \le 1$. With the change of variable $s = \sqrt{\epsilon}\sigma$, we can write the solution

$$y = -x + \frac{3x}{\sqrt{2\pi}}\int_{-1/\sqrt{\epsilon}}^{x/\sqrt{\epsilon}} E^{-\sigma^2/2}d\sigma + \text{T.S.T.} \quad \text{for } x \ne 0 \tag{2.2.85a}$$

---

[1] For this example $y = Cx$ is also the outer solution *to all orders* ϵ^M. The second solution is found by assuming the form $y = xw(x;\epsilon)$, a standard method.

$$y = 3\sqrt{\frac{\epsilon}{2\pi}} + \text{T.S.T.} \quad \text{for } x = 0. \tag{2.2.85b}$$

Therefore, if $x < 0$, the result in (2.2.85a) reduces as $\epsilon \to 0$ to

$$y \sim -x + \text{T.S.T.} \tag{2.2.86}$$

and if $x > 0$, we have

$$y \sim 2x + \text{T.S.T.} \tag{2.2.87}$$

The exact solution thus shows that there are no boundary layers at either endpoint, and two branches of the outer solution, each satisfying the appropriate boundary condition, hold on each half of the interval $-1 \leq x \leq 1$. Clearly, if $x = O(\sqrt{\epsilon})$, the above approximations are not valid and we have to examine (2.2.78) more carefully. We note that (2.2.76) has a distinguished limit with $\epsilon \to 0, x^* = x/\sqrt{\epsilon}$ fixed, and that in this limit the exact differential equation

$$\frac{d^2y}{dx^{*2}} + x^* \frac{dy}{dx^*} - y = 0 \tag{2.2.88}$$

must be solved. This distinguished limit follows directly from the solution (2.2.78) written in terms of x^* as

$$y = -\epsilon^{1/2}x^* + \frac{3\epsilon^{1/2}}{\sqrt{2\pi}}\left[e^{-x^{*2}/2} + x^* \int_{-1/\sqrt{\epsilon}}^{x^*} e^{-\sigma^2/2}d\sigma\right] + \text{T.S.T.}$$

Now, letting $\epsilon \to 0$ with x^* fixed, we obtain

$$y \sim \epsilon^{1/2}\left[-x^* + \frac{3}{\sqrt{2\pi}}\left(e^{-x^{*2}/2} + x^* \int_{-\infty}^{x^*} e^{-\sigma^2/2}d\sigma\right)\right]. \tag{2.2.89}$$

Equation (2.2.89) describes a "corner-layer" solution valid near $x = 0$. We verify that this corner layer matches to the left with $y = -x$ and to the right with $y = 2x$. Thus, the uniformly valid solution to $O(1)$ is, in fact, the corner-layer solution given by (2.2.89), and the result is sketched in Figure 2.2.3.

The reader can easily verify that the above results also follow systematically by seeking boundary layer, interior, and outer solutions to (2.2.76). In particular, the fact that no boundary layers are possible at the endpoints $x = \pm 1$ follows from the observation that the corresponding boundary-layer equations give exponential *growth* in the interior of the $-1 < x < 1$ interval. Thus, one must use the two outer expansions (2.2.86) and (2.2.87). Then, it remains to smooth the resulting corner at $x = 0$, and a solution of the form (2.2.89) with two arbitrary constants can be calculated for (2.2.88). Matching to the left and right then determines these two constants.

2.2.5 Two Boundary Layers

This example illustrates an interesting collection of difficulties and is one where techniques we have used so far *fail* to yield a unique result. We simply change

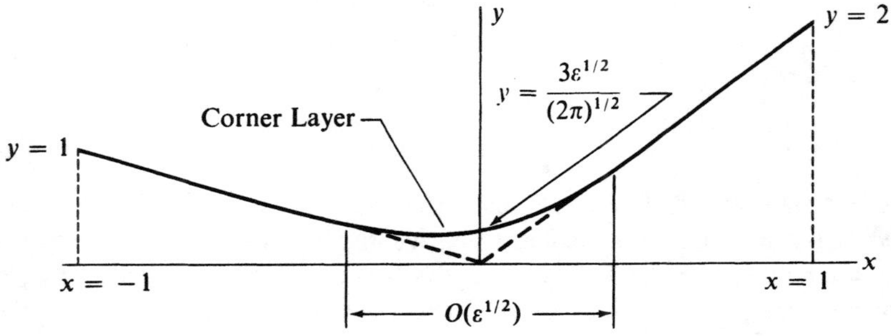

FIGURE 2.2.3. Corner Layer at $x = 0$

the signs of the second and third terms in Equation (2.2.76) and consider the boundary-value problem

$$\epsilon y'' - xy' + y = 0, \qquad -1 \le x \le 1; \; 0 < \epsilon \ll 1, \tag{2.2.90}$$

$$y(-1) = 1; \qquad y(1) = 2. \tag{2.2.91}$$

Again, an exact solution can be found explicitly since $y = Cx$ solves Equation (2.2.90). We write the exact solution in the form

$$y = C_1 x + C_2 \left[e^{x^2/2\epsilon} - \frac{x}{\epsilon} \int_{-1}^{x} e^{s^2/2\epsilon} ds \right]. \tag{2.2.92}$$

Imposing the boundary conditions gives

$$C_1 = -1 + \frac{3e^{1/2\epsilon}}{2e^{1/2\epsilon} - J(\epsilon)/\epsilon} \tag{2.2.93}$$

and

$$C_2 = \frac{3}{2e^{1/2\epsilon} - J(\epsilon)/\epsilon}, \tag{2.2.94}$$

where

$$J(\epsilon) = \int_{-1}^{1} e^{s^2/2\epsilon} ds. \tag{2.2.95}$$

To approximate $J(\epsilon)$, we write (2.2.95) in the form

$$J(\epsilon) = 2e^{1/2\epsilon} \int_{0}^{1} e^{-(1-s^2)/2\epsilon} ds \tag{2.2.96}$$

and then change the variable of integration, setting $(1 - s^2)/2 = u$ to obtain

$$J(\epsilon) = 2e^{1/2\epsilon} \int_0^{1/2} \frac{e^{-u/\epsilon}}{\sqrt{1-2u}} du. \quad (2.2.97)$$

The asymptotic expansion for J can now be directly calculated by Laplace's method[1] as

$$J(\epsilon) = 2\epsilon e^{1/2\epsilon}[1 + \epsilon + 3\epsilon^2 + 15\epsilon^3 + 105\epsilon^4 + O(\epsilon^5)]. \quad (2.2.98)$$

The asymptotic expansions for C_1 and C_2 thus become

$$C_1 = -\frac{3}{2\epsilon} + \frac{7}{2} + 9\epsilon + 63\epsilon^2 + O(\epsilon^3) \quad (2.2.99)$$

$$C_2 = e^{-1/2\epsilon}\left[-\frac{3}{2\epsilon} + \frac{9}{2} + 9\epsilon + 63\epsilon^2 + O(\epsilon^3)\right]. \quad (2.2.100)$$

We will now show that the solution consists of a boundary layer at $x = -1$

$$y = -\frac{1}{2} + \frac{3}{2}e^{-(x+1)/\epsilon} + \ldots, \quad (2.2.101)$$

a boundary layer at $x = 1$

$$y = \frac{1}{2} + \frac{3}{2}e^{(x-1)/\epsilon} + \ldots, \quad (2.2.102)$$

and an outer solution

$$y = \frac{x}{2} + \ldots. \quad (2.2.103)$$

The interesting feature of the outer solution, valid at all interior points $|x| \neq 1$, is the fact that it does not satisfy either boundary condition.

Consider first the behavior of the exact solution near $x = -1$. We introduce the boundary layer variable $x^* = (x+1)/\epsilon$ and write the solution (2.2.92) in the form

$$y = C_1(-1+\epsilon x^*) + C_2 e^{1/2\epsilon}\left[e^{-x^*+\epsilon x^{*2}/2} - \left(x^* - \frac{1}{\epsilon}\right)\int_{-1}^{-1+\epsilon x^*} e^{-(1-s^2)/2\epsilon} ds\right]. \quad (2.2.104)$$

The integral appearing in (2.2.104) can be decomposed as

$$\int_{-1}^{-1+\epsilon x^*} e^{-(1-s^2)/2\epsilon} ds = \int_{-1}^{0} e^{-(1-s^2)/2\epsilon} ds + \int_0^{-1+\epsilon x^*} e^{-(1-s^2)/2\epsilon} ds$$
$$= \frac{J(\epsilon)}{2} e^{-1/2\epsilon} + \int_0^{-1+\epsilon x^*} e^{-(1-s^2)/2\epsilon} ds. \quad (2.2.105)$$

[1]This method is essentially based on the observation that for $\epsilon \to 0$ the integrand is negligible for all u except in some neighborhood of $u = 0$. Therefore, we expand $1/\sqrt{1-2u}$ for u small and integrate the result term by term. For more details and a proof of asymptotic validity, see section 6.2 of [2.2].

To calculate the asymptotic behavior of the integral appearing on the right-hand side of (2.2.105), we first make the change of variable $s + 1 - \epsilon x^* = u$ to obtain

$$K(x^*;\epsilon) = -\int_0^{-1+\epsilon x^*} e^{-(1-s^2)/2\epsilon} ds = e^{-x^*+\epsilon x^{*2}/2}\int_0^{1-\epsilon x^*} e^{-(2u-u^2)/2\epsilon + ux^*} du.$$

Then we set $(2u - u^2)/2 = \sigma$ and develop the integrand near $\sigma = 0$, which is the point giving the dominant contribution. We find

$$K(x^*;\epsilon) = e^{-x^*+\epsilon x^{*2}/2}\int_0^{1/2+O(\epsilon^2)} e^{-\sigma/\epsilon}[1+(x^*+1)\sigma+O(\sigma^2)]d\sigma. \quad (2.2.106)$$

This result can now be directly evaluated by Laplace's method:

$$K(x^*;\epsilon) = \epsilon e^{-x^*+\epsilon x^{*2}/2}[1 + (x^* + 1)\epsilon + O(\epsilon^2)].$$

Substituting for K, C_1, and C_2 into (2.2.104) then shows that in the limit $\epsilon \to 0$, x^* fixed, we have

$$y = -\frac{1}{2} + \frac{3}{2}e^{-x^*} + O(\epsilon). \quad (2.2.107a)$$

A similar calculation near $x = 1$ with $x^{**} = (x - 1)/\epsilon$ leads to

$$y = \frac{1}{2} + \frac{3}{2}e^{x^{**}} + O(\epsilon). \quad (2.2.107b)$$

Thus, the boundary-layer solution at the left end tends to $y = -\frac{1}{2}$ as $x^* \to \infty$ and the boundary layer at the right end tends to $y = +\frac{1}{2}$ as $x^{**} \to -\infty$.

To determine the asymptotic limit of the exact solution away from the ends, we use the expression for y given by (2.2.92) and write the integral appearing there as

$$\int_{-1}^{x} e^{s^2/2\epsilon} ds = \frac{J(\epsilon)}{2} + e^{x^2/2\epsilon}\int_0^x e^{-(x^2-s^2)/2\epsilon} ds.$$

Using the same procedure as in (2.2.98) to evaluate the asymptotic expansion for the integral on the right-hand side, and some algebra, we find

$$y = C_1 x + C_2\left\{-xe^{1/2\epsilon}\left[1 + \epsilon + 3\epsilon^2 + 15\epsilon^3 + 105\epsilon^4 + O(\epsilon^5)\right]\right.$$
$$\left. - \frac{\epsilon}{x^2}e^{x^2/2\epsilon}\left(1 + \frac{3\epsilon}{x^2} + \frac{15\epsilon^2}{x^4} + \frac{105\epsilon^3}{x^6} + O(\epsilon^4)\right)\right\}.$$

Since $C_2 = O(e^{-1/2\epsilon}/\epsilon)$, the second series multiplying C_2 is transcendentally small for $\epsilon \to 0$, $|x|$ fixed $\neq 1$, and we find

$$y = x\{C_1 - C_2 e^{1/2\epsilon}[1 + \epsilon + 3\epsilon^2 + 15\epsilon^3 + 105\epsilon^4 + O(\epsilon^5)]. \quad (2.2.108)$$

Now, using the expansions we have calculated for C_1 and C_2 shows

$$y = \frac{x}{2} + \text{T.S.T.} \quad (2.2.109)$$

Thus, the uniformly valid solution to $O(1)$ consists of the outer limit plus left and right boundary corrections. Using (2.2.109), (2.2.107a), and (2.2.107b) and taking into account the common parts, we find

$$y = \frac{x}{2} + \frac{3}{2}e^{-(x+1)/\epsilon} + \frac{3}{2}e^{(x-1)/\epsilon} + O(\epsilon), \tag{2.2.110}$$

and this is sketched in Figure 2.2.4.

Let us now pretend that we do not have an exact solution and proceed to calculate and match outer and boundary-layer expansions.

For an outer expansion in the form

$$y = \sum_{i=0}^{\infty} \epsilon^i h_i(x), \tag{2.2.111}$$

we see that the h_i obey $-xh_i' + h_i = h_{i-1}''$. But since $h_0(x) = c_0 x$, we have $h_0'' = 0$ and hence $h_1 = c_1 x$, etc. Thus, the outer expansion to $O(\epsilon^N)$, for any integer N, has the degenerate form

$$y = C(\epsilon)x = (c_0 + \epsilon c_1 + \epsilon^2 c_2 + \ldots)x, \tag{2.2.112}$$

where the c_i are constants independent of ϵ.

If we assume a boundary layer at $x = -1$, we need an expansion

$$y = g_0^{(L)}(x^*) + \epsilon g_1^{(L)}(x^*) + \ldots, \tag{2.2.113}$$

where $x^* = (x+1)/\epsilon$.

The functions $g_0^{(L)}$ and $g_1^{(L)}$ obey

$$\frac{d^2 g_0^{(L)}}{dx^{*2}} + \frac{dg_0^{(L)}}{dx^*} = 0 \tag{2.2.114a}$$

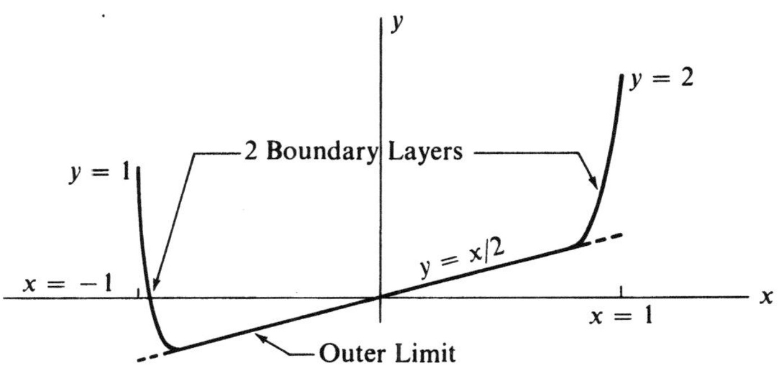

FIGURE 2.2.4. Two Boundary Layers

$$\frac{d^2 g_1^{(L)}}{dx^{*2}} + \frac{dg_1^{(L)}}{dx^*} = -g_0^{(L)} + x^* \frac{dg_0^{(L)}}{dx^*} \tag{2.2.114b}$$

with boundary conditions $g_0^{(L)}(0) = 1$, $g_1^{(L)}(0) = 0$, etc.

The solution of (2.2.114a) subject to $g_0^{(L)}(0) = 1$ is

$$g_0^{(L)}(x^*) = A_0 + (1 - A_0)e^{-x^*},$$

where A_0 is to be determined by matching.

Similarly, with $x^{**} = (x - 1)/\epsilon$, we consider a boundary layer near $x = 1$ in the form

$$y = g_0^{(R)}(x^{**}) + \epsilon g_1^{(R)}(x^{**}) + \dots. \tag{2.2.115}$$

Now, $g_0^{(R)}$ and $g_1^{(R)}$ obey

$$\frac{d^2 g_0^{(R)}}{dx^{**2}} - \frac{dg_0^{(R)}}{dx^{**}} = 0 \tag{2.2.116a}$$

$$\frac{d^2 g_1^{(R)}}{dx^{**2}} - \frac{dg_1^{(R)}}{dx^{**}} = -g_0^{(R)} + x^{**} \frac{dg_0^{(R)}}{dx^{**}} \tag{2.2.116b}$$

with boundary conditions $g_0^{(R)}(0) = 2$, $g_1^{(R)}(0) = 0$, etc.

Solving (2.2.116a) with $g_0^{(R)}(0) = 2$ gives

$$g_0^{(R)} = B_0 + (2 - B_0)e^{x^{**}}.$$

The matching of the outer limit $c_0 x$ with $g_0^{(L)}$ gives

$$-c_0 = A_0,$$

and matching $c_0 x$ with $g_0^{(R)}$ gives

$$c_0 = B_0.$$

Thus, to $O(1)$ the two boundary layer limits may be expressed in terms of c_0, but there is no criterion left to determine c_0 itself! If $c_0 = 1$ there is no left boundary layer, and if $c_0 = 2$ there is no right boundary layer; all other values of c_0 predict two boundary layers. Let us examine whether this indeterminacy can be resolved by including the next higher-order terms.

It is easily found that (2.2.114b), subject to $g_1^{(L)}(0) = 0$, gives

$$g_1^{(L)} = c_0(x^* - 1) + 2(1 + c_0)x^* e^{-x^*} + (1 + c_0)\frac{x^{*2}}{2} e^{-x^*} + A_1 + (c_0 - A_1)e^{-x^*} \tag{2.2.117a}$$

and (2.2.116b), subject to $g_1^{(R)}(0) = 0$, gives

$$g_1^{(R)} = c_0(x^{**} + 1) - 2(2 - c_0)x^{**}e^{x^{**}} + (2 - c_0)\frac{x^{**2}}{2} e^{x^{**}} + B_1 - (B_1 + c_0)e^{x^{**}}. \tag{2.2.117b}$$

Matching the left boundary layer and outer expansions to $O(\epsilon)$ gives

$$A_1 = c_0 - c_1.$$

Similarly, matching the right boundary layer and outer expansions to $O(\epsilon)$ gives

$$B_1 = (c_1 - c_0).$$

This procedure may be repeated to any order ϵ^N; each of the unknown boundary layer constants A_N, B_N is found in terms of the unknown outer constants $c_0, \ldots c_N$. In particular, to $O(1)$ we have the one-parameter family of uniformly valid "solutions" (see (2.2.110))

$$y = c_0 x + (1 + c_0)e^{-(x+1)/\epsilon} + (2 - c_0)e^{(x-1)/\epsilon} + O(\epsilon). \tag{2.2.118}$$

Our procedure of matching outer and boundary-layer expansions does not determine the correct value $c_0 = 1/2$ for the unknown parameter.

This behavior was first pointed out in [2.1] in the more general setting (2.2.1). It has subsequently received much attention in the literature, where it is often referred to as "boundary layer resonance." In this context, *resonance* is associated with the behavior of eigenvalues of (2.2.1) and is a somewhat misleading characterization. Since the unknown constant c_0 specifies the outer limit, we shall refer to "outer limit indeterminacy" instead. In [2.6] Grasman and Matkowsky give conditions under which the general problem (2.2.1), with arbitrary boundary conditions (2.2.2), will exhibit an indeterminate outer expansion.[1] Their proposal for resolving the indeterminacy will be discussed after we take note of a simple symmetry argument that may be invoked for the special example (2.2.90).

Symmetry

Since (2.2.90) is invariant under the transformation $x \to -x$, solutions would be even functions of x if the boundary values $y(-1)$ and $y(1)$ were equal, in which case we would have $c_0 = 0$. If the boundary values (2.2.91) are arbitrary, it is still possible to use a symmetry argument to determine the c_i for the special example (2.2.90). For example, with $y(-1) = Y^-$ and $y(1) = Y^+$, where Y^- and Y^+ are arbitrary constants, we would have computed

$$y = c_0 x + (Y^- + c_0)e^{-(x+1)/\epsilon} + (Y^+ - c_0)e^{(x-1)/\epsilon} + O(\epsilon) \tag{2.2.119}$$

instead of (2.1.118). Now, the exact solution gives $c_0 = (Y^+ - Y^-)/2$.

Lagerstrom pointed out (see exercise 2.3 in [2.8]) that the change of dependent variable $w = y - Kx$ for an appropriate constant K will transform (2.2.90), and arbitrary boundary values Y^- and Y^+, to a problem whose solution is *even* in x. We see that if y satisfies (2.2.90) then w satisfies the *same* equation, and the boundary conditions become $w(-1) = Y^- + K$ and $w(1) = Y^+ - K$. Therefore, solutions for w are even in x if $Y^- + K = Y^+ - K$, i.e., $K \equiv (Y^+ - Y^-)/2$. Since $w = y - (Y^+ - Y^-)x/2$ is even, we must have

$$[c_0 - (Y^+ - Y^-)/2]x + (Y^- + c_0)e^{-(x+1)/\epsilon} + (Y^+ - c_0)e^{(x-1)/\epsilon}$$

[1]Note that in the general case the outer expansion need not have the degenerate form $y = (c_0 + \epsilon c_1 + \epsilon^2 c_2 + \ldots)x$; each term $h_i(x)$ in (2.2.111) will have the form $h_i = c_i k_i(x)$ with a different $k_i(x)$ for each i.

$$= [c_0 - (Y^+ - Y^-)/2](-x) + (Y^- + c_0)e^{(x-1)/\epsilon} + (Y^+ - c_0)e^{-(x+1)/\epsilon},$$

and this is true only if $c_0 = (Y^+ - Y^-)/2$. Similarly, when the terms of order ϵ are retained in the uniform result, we find $c_1 = 0$, etc.

The above symmetry argument only applies to problems where the governing equation may be transformed to one that is even in x. Such a transformation does not exist for the general problem (2.2.1) for which it is possible to have an indeterminate outer expansion.

A Variational Principle

Consider the general problem (see 2.2.90)

$$\epsilon y'' + a(x;\epsilon)y' + b(x;\epsilon)y = 0, \tag{2.2.120}$$

$$y(-1;\epsilon) = A(\epsilon), \tag{2.2.121a}$$

$$y(1;\epsilon) = B(\epsilon). \tag{2.2.121b}$$

We have normalized x so that the solution domain is $-1 \le x \le 1$, with no loss of generality, and we are interested in the case where $a(x;\epsilon)$ changes sign in this interval.

Let us construct a Lagrangian $L(x, y, y'; \epsilon)$ for which our governing equation (2.2.120) is the Euler-Lagrange equation, i.e., (2.2.120) is obtained from

$$\frac{d}{dx}\left(\frac{\partial L}{\partial y'}\right) - \frac{\partial L}{\partial y} = 0. \tag{2.2.122}$$

For a given differential equation (2.2.120), the associated Lagrangian is *not unique* (see Problem 3). One Lagrangian may be obtained by first transforming (2.2.120) to the form

$$(p(x;\epsilon)y')' + q(x;\epsilon)y = 0.$$

This is accomplished by multiplying (2.2.120) by the integrating factor

$$\alpha(x;\epsilon) = \exp\left[\frac{1}{\epsilon}\int_0^x a(t;\epsilon)dt\right]$$

to find $p = \epsilon\alpha$ and $q = \alpha b$. Now, identifying $\frac{\partial L}{\partial y'} = py'$ and $\frac{\partial L}{\partial y} = -qy$ gives the Lagrangian

$$L = \frac{1}{2}(\epsilon y'^2 - b(x;\epsilon)y^2)\exp\left[\frac{1}{\epsilon}\int_0^x a(t;\epsilon)dt\right]. \tag{2.2.123}$$

An alternate Lagrangian is derived in Problem 3.

As is well known, an *exact* solution of (2.2.122) passing through two fixed points $y(-1;\epsilon) = A(\epsilon)$ and $y(1;\epsilon) = B(\epsilon)$ is an *extremal* for the variational principle

$$\delta I = 0 \tag{2.2.124a}$$

with fixed endpoints

$$\delta y(-1;\epsilon) = \delta y(1;\epsilon) = 0, \tag{2.2.124b}$$

where

$$I = \int_{-1}^{1} L(x, y, y'; \epsilon)dx. \tag{2.2.124c}$$

This variational principle is called *Hamilton's principle* in dynamics and *Fermat's principle* in optics (see, for example, section 2.1 of [2.5]).

Stated otherwise, for *arbitrary* small variations of y, subject to the boundary condition (2.2.121), I is stationary along a solution of (2.2.122). In particular, let $y = f(x; c_0, \epsilon)$ be a one-parameter family of functions for varying c_0 satisfying $f(-1; c_0, \epsilon) = A(\epsilon)$, $f(1; c_0, \epsilon) = B(\epsilon)$ for *all* c_0, and satisfying (2.2.122) for one value of $c_0 = \bar{c}_0$. It then follows that $\bar{c}_0$ is defined by the solution of the algebraic condition: $\frac{\partial I}{\partial c_0}(\bar{c}_0; \epsilon) = 0$, where

$$I(c_0, \epsilon) = \int_{-1}^{1} L(x, f(x; c_0, \epsilon), f'(x; c_0, \epsilon); \epsilon)dx. \tag{2.2.125}$$

In [2.6] it is proposed to apply this property of *exact* solutions of (2.2.120) to *asymptotic* solutions involving a parameter, as in (2.2.118), in order to isolate the correct value $\bar{c}_0$. Before proceeding with the calculations for the specific example of (2.2.90)–(2.2.91), it is important to keep in mind that the expression (2.2.118) is not the exact solution for any c_0, nor does it satisfy the boundary conditions (2.2.91); transcendentally small errors are present in each of these two categories, and their effects may be important.

For the example problem (2.2.90), the Lagrangian (2.2.123) becomes

$$L = \frac{1}{2}(\epsilon y'^2 - y^2)\exp(-x^2/2\epsilon),$$

and when we use (2.2.118) for y, we find

$$\begin{aligned}
2L =& c_0^2(\epsilon - x^2)\exp(-x^2/2\epsilon) - 2c_0(1+c_0)(1+x)\exp\left[-\frac{(x+1)^2+1}{2\epsilon}\right] \\
&+ 2c_0(2-c_0)(1-x)\exp\left[-\frac{(x-1)^2+1}{2\epsilon}\right] \\
&+ (1+c_0)^2\left(\frac{1}{\epsilon} - 1\right)\exp\left[-\frac{(x+2)^2}{2\epsilon}\right] \\
&- 2(1+c_0)(2-c_0)\left(\frac{1}{\epsilon} + 1\right)\exp\left[-\frac{x^2+4}{2\epsilon}\right] \\
&+ (2-c_0)^2\left(\frac{1}{\epsilon} - 1\right)\exp\left[-\frac{(x-2)^2}{2\epsilon}\right].
\end{aligned} \tag{2.2.126}$$

The various integrals that arise in (2.2.125) for I can be evaluated explicitly. We have

$$I_1 \equiv \int_{-1}^{1} (\epsilon - x^2) \exp\left(-\frac{x^2}{2\epsilon}\right) dx = 2\epsilon \exp\left(-\frac{1}{2\epsilon}\right) \tag{2.2.127a}$$

$$\begin{aligned} I_2 &\equiv \int_{-1}^{1} (1+x) \exp\left[-\frac{(x+1)^2+1}{2\epsilon}\right] dx \\ &= \int_{-1}^{1} (1-x) \exp\left[-\frac{(x-1)^2+1}{2\epsilon}\right] dx \\ &= \epsilon \left[\exp\left(-\frac{1}{2\epsilon}\right) - \exp\left(-\frac{3}{2\epsilon}\right)\right] \end{aligned} \tag{2.2.127b}$$

$$\begin{aligned} I_3 &\equiv \int_{-1}^{1} \exp\left[-\frac{(x+2)^2}{2\epsilon}\right] dx = \int_{-1}^{1} \exp\left[-\frac{(x-2)^2}{2\epsilon}\right] dx \\ &= \sqrt{\frac{\pi\epsilon}{2}} \left[\operatorname{erf}\left(\frac{3}{\sqrt{2\epsilon}}\right) - \operatorname{erf}\left(\frac{1}{\sqrt{2\epsilon}}\right)\right] \\ &= \exp(-\frac{1}{2\epsilon})[\epsilon + O(\epsilon^2)] - \exp(-\frac{3}{2\epsilon})\left[\frac{\epsilon}{3} + O(\epsilon^2)\right] \end{aligned} \tag{2.2.127c}$$

$$\begin{aligned} I_4 &\equiv \int_{-1}^{1} \exp\left(-\frac{x^2+4}{2\epsilon}\right) dx = \sqrt{2\pi\epsilon} \exp\left(-\frac{2}{\epsilon}\right) \operatorname{erf}\left(\frac{1}{\sqrt{2\epsilon}}\right) \\ &= \sqrt{2\pi\epsilon} \exp\left(-\frac{2}{\epsilon}\right) + O\left(\epsilon \exp\left(-\frac{5}{2\epsilon}\right)\right), \end{aligned} \tag{2.2.127d}$$

where erf( ) denotes the error function defined by (see (1.2.37))

$$\operatorname{erf} z = \frac{2}{\sqrt{\pi}} \int_0^z e^{-s^2} ds. \tag{2.2.127e}$$

In (2.2.127c) and (2.2.127d), the asymptotic behavior for I_3 and I_4 follows from the general result (1.2.44). Therefore, the dominant contribution to I is due to I_3, and we have

$$\begin{aligned} I = \int_{-1}^{1} L dx &= \frac{1}{2}\left[(1+c_0)^2 + (2-c_0)^2\right] \exp\left(-\frac{1}{2\epsilon}\right) \\ &\quad + O\left(\epsilon \exp\left(-\frac{1}{2\epsilon}\right)\right). \end{aligned} \tag{2.2.128}$$

The fact that I itself is transcendentally small is not relevant, as we could have included the factor $\exp\left(\frac{1}{2\epsilon}\right)$ in L (and hence in I) without affecting the equivalence of (2.2.122) and (2.2.120); the crucial issue is to identify the *leading* term in I. Using this leading term, we see that $\frac{\partial I}{\partial c_0} = 0$ gives

$$(1 + c_0) - (2 - c_0) = 0, \quad \text{or} \quad c_0 = 1/2.$$

The reader can find the corresponding results for the general case of (2.2.120) in [2.6]. It has been pointed out that for this general case, where the outer expansion is nondegenerate, the variational principle (2.2.124) *fails to higher order*. The details, and references to earlier works, are given in [2.15], where a modified variational principle that is valid to any order in ϵ is proposed. However, the calculations of higher-order terms are very tedious.

The two approaches we have used so far—symmetry and a variational principle—make use of *global* properties of the solution to fix the undetermined constant c_0. The idea discussed next is more appealing and applies to other classes of problems where matching outer and inner expansion fails to define the asymptotic solution.

Inclusion of transcendentally small terms[1]

Let us augment the conventional outer expansion with transcendentally small correction terms to remove the indeterminacy of the outer expansion $C(\epsilon)$. For our example, let us construct the general asymptotic expansion

$$y = C(\epsilon)x + \delta_1(\epsilon)f_1(x;\epsilon) + \delta_2(\epsilon)f_2(x;\epsilon) + \ldots \tag{2.2.129}$$

to approximate the solution of (2.2.90) away from the boundaries. We expect to determine $C(\epsilon)$ as well as the transcendentally small functions $\delta_1(\epsilon), \delta_2(\epsilon), \ldots$ from the matching.

Since δ_1 is transcendentally small, the correction term $\delta_1 f_1$ comes into play only after we have taken account of the usual outer expansion to *all orders*. Also, for reasons that will become clear, we need to allow f_1 to depend on ϵ.

Substituting (2.2.129) into (2.2.90) gives

$$\epsilon\delta_1 f_1'' - \delta_1 x f_1' + \delta_1 f_1 = O(\delta_2). \tag{2.2.130}$$

If we were to ignore the term of order $\epsilon\delta_1$ relative to the two $O(\delta_1)$ terms, we would find f_1 linear in x and make no progress. It is crucial to include this term and thus regard the augmented part of the outer expansion, $\delta_1 f_1 + \delta_2 f_2 + \ldots$, as a general asymptotic expansion (see (2.1.52)). In the present case, since (2.2.90) is linear, the governing equation for f_1 is just (2.2.90)

$$\epsilon f_1'' - x f_1' + f_1 = 0. \tag{2.2.131}$$

In general, for a nonlinear governing equation, $f_1, f_2, \ldots$ would *still obey linear equations*. More importantly, we do not need the exact solution of f_1 in general; we only need its asymptotic behavior near the boundaries in order to determine c_0 and δ_1.

We now write the two linearly independent solutions of f_1 as $f_{11}(x) = x$ and $f_{12}(x;\epsilon) = e^{-x^2/2\epsilon} + \frac{x}{\epsilon}\int_0^x \exp(s^2/2\epsilon)ds$, (see (2.2.92)). Calculations analogous to those leading to (2.2.107a) show that f_{12} has the following behavior near $x =$

[1] This discussion is based on some lecture notes and ongoing research by Professor A.D. MacGillivray.

-1:

$$f_{12} = e^{1/2\epsilon}[-\epsilon e^{-x^*} + O(\epsilon^2) + O(e^{-1/\epsilon})]. \tag{2.2.132a}$$

Similarly, near $x = 1$, we have

$$f_{12} = e^{1/2\epsilon}[-\epsilon e^{x^{**}} + O(\epsilon^2) + O(e^{-1/\epsilon})]. \tag{2.2.132b}$$

Let us now match the augmented outer expansion to $O(\delta_1)$ with the $O(1)$ boundary-layer limits. The augmented outer expansion has the form

$$y = (C(\epsilon) - D_1\delta_1(\epsilon))x - D_1\delta_1(\epsilon) f_{12}(x; \epsilon), \tag{2.2.133}$$

where D_1 is an arbitrary constant. For the matching with the left boundary-layer limit, we introduce, as usual, the matching variable $x_{\eta_L} = (x+1)/\eta_L(\epsilon)$ for a class of $\eta_L(\epsilon) \ll 1$. Thus, $x^* = \eta_L x_{\eta_L}/\epsilon$ and $x = -1 + \eta_L x_{\eta_L}$. The augmented outer expansion expressed in terms of x_{η_L} and expanded for $\epsilon \to 0$, with x_{η_L} fixed, becomes

$$y = -c_0 + D_1\delta_1(\epsilon)\epsilon e^{1/2\epsilon} e^{-\eta_L x_{\eta_L}/\epsilon} + \ldots \tag{2.2.134}$$

and the left boundary-layer limit gives

$$y = A_0 + (1 - A_0)e^{-\eta_L x_{\eta_L}/\epsilon}.$$

Therefore, we must have $A_0 = -c_0$ as before, and

$$D_1\delta_1(\epsilon)\epsilon e^{1/2\epsilon} = 1 - A_0 = 1 + c_0. \tag{2.2.135}$$

For matching with the right boundary-layer limit, we introduce $x_{\eta_R} = (x - 1)/\eta_R(\epsilon)$ for a class of $\eta_R(\epsilon) \ll 1$. Now, $x^{**} = \eta_R x_{\eta_R}/\epsilon$, $x = 1 + \eta_R x_{\eta_R}$, and the augmented outer expansion gives

$$y = c_0 + D_1\delta_1(\epsilon)e^{1/2\epsilon}\epsilon e^{\eta_R x_{\eta_R}/\epsilon} + \ldots. \tag{2.2.136}$$

The right boundary layer limit gives

$$y = B_0 + (2 - B_0)e^{\eta_R x_{\eta_R}/\epsilon}.$$

We conclude that $B_0 = c_0$ as before, and

$$D_1\delta(\epsilon)\epsilon e^{1/2\epsilon} = (2 - B_0) = 2 - c_0. \tag{2.2.137}$$

The two conditions (2.2.135) and (2.2.137) define c_0 and $D_1\ \delta_1$ as follows:

$$c_0 = 1/2; \qquad D_1\delta_1(\epsilon) = \frac{3}{2}\epsilon^{-1}e^{-1/2\epsilon}.$$

Thus, δ_1 is indeed transcendentally small, and we see that augmenting the outer expansion by such a term does provide the condition that was missing in the usual matching.

Use of transcendentally small (or large) terms in asymptotic expansions is of much current interest in various contexts. The earliest application of this idea is given by Lange in [2.14]. He shows that augmenting conventional asymptotic expansions to include transcendentally small terms resolves the difficulty of spurious solutions pointed out in [2.3] for a class of nonlinear singular perturbation problems.

Problems

1. Consider the following boundary-value problem on $0 \leq x \leq 1$ that can be solved exactly

$$\epsilon y'' + y' - xy = 0, \tag{2.2.138}$$

$$y(0) = 0, \qquad y(1) = e^{1/2}. \tag{2.2.139}$$

a. Change dependent variable $y \rightarrow z$ according to $y = ze^{-x/2\epsilon}$ and independent variable $x \rightarrow \xi$

$$\xi = \frac{4\epsilon x + 1}{4\epsilon^{4/3}} \tag{2.2.140}$$

to transform the governing equation to Airy's equation

$$\frac{d^2 z}{d\xi^2} - \xi z = 0 \tag{2.2.141}$$

with solution

$$z = \xi^{1/2}\left[AI_{1/3}\left(\frac{2}{3}\xi^{3/2}\right) + BK_{1/3}\left(\frac{2}{3}\xi^{3/2}\right)\right], \tag{2.2.142}$$

where $I_{1/3}$ and $K_{1/3}$ are modified Bessel functions of the first and second kind of order $1/3$. Thus,

$$y = e^{-x/2\epsilon}\sqrt{1 + 4\epsilon x}\left[MI_{1/3}\left(\frac{(4\epsilon x + 1)^{3/2}}{12\epsilon^2}\right)\right.$$

$$\left. + NK_{1/3}\left(\frac{(4\epsilon x + 1)^{3/2}}{12\epsilon^2}\right)\right], \tag{2.2.143}$$

where A, B and M, N are arbitrary constants.

b. Evaluate M, N using the given boundary conditions, then use asymptotic properties of the Bessel functions to show that the uniformly valid solution to $O(1)$ is

$$y = e^{x^2/2}e^{-x/\epsilon}. \tag{2.2.144}$$

c. Match outer and inner limits to confirm this result.

2. Calculate the uniformly valid solution to $O(\epsilon)$ on $0 \leq x \leq 1$ for

$$\epsilon y'' + x^{-1/2}y' - y = 0 \tag{2.2.145}$$

with

$$y(0) = 0, \quad y(1) = e^{2/3}. \tag{2.2.146}$$

3. As an alternate derivation of a Lagrangian for (2.2.120), introduce the transformation

$$w = y \exp\left[\frac{1}{2\epsilon}\int_0^x a(t;\epsilon)dt\right], \tag{2.2.147}$$

which removes the first derivative term [see Equations (2.2.3) and (2.2.4)] and gives the following equation for w

$$\epsilon w'' + \left(b - \frac{a^2}{4\epsilon} - \frac{a'}{2}\right) w = 0. \tag{2.2.148}$$

Now interpret the above as the equation for an oscillator where ϵ is the mass, w is the displacement, x is the time, and the oscillator is acted on by a linear time-dependent spring with spring constant k

$$k = b - \frac{a^2}{4\epsilon} - \frac{a'}{2}. \tag{2.2.149}$$

Clearly, the kinetic energy is

$$T = \frac{\epsilon w'^2}{2} \tag{2.2.150}$$

and the potential energy V, defined by

$$\frac{\partial V}{\partial w} = w\left(b - \frac{a^2}{4\epsilon} - \frac{a'}{2}\right), \tag{2.2.151}$$

is

$$V = \frac{w^2}{2}\left(a - \frac{a^2}{4\epsilon} - \frac{a'}{2}\right). \tag{2.2.152}$$

Thus, a Lagrangian for the oscillator equation is

$$L_1 = T - V = \frac{1}{2}\left[\epsilon w'^2 - w^2\left(b - \frac{a^2}{4\epsilon} - \frac{a'}{2}\right)\right]. \tag{2.2.153}$$

Show that the Euler-Lagrange equation resulting from L_1 gives the oscillator equation.

Now transform w to y in L_1 and show that the Lagrangian

$$\widetilde{L}(x, y, y'; \epsilon) = \frac{1}{2}\left[\epsilon y'^2 + ayy' + \left(\frac{a^2}{2\epsilon} + \frac{a'}{2} - b\right) y'\right]$$

$$\cdot \exp\left[\frac{1}{\epsilon}\int_0^x a(t;\epsilon)dt\right] \tag{2.2.154}$$

that results will also give Equation (2.2.120) as its Euler-Lagrange equation. Thus, the Lagrangian corresponding to a given differential equation is not unique.

Note that $\widetilde{L}$ is more complicated than the L used in Equation (2.2.123). It must follow that the Euler-Lagrange equation corresponding to $L^* = \widetilde{L} - L$ is identically satisfied for any y.

In our case

$$L^* = \left[\frac{ayy'}{2} + \left(\frac{a^2}{4\epsilon} + \frac{a'}{4}\right)y^2\right]\exp\left[\frac{1}{\epsilon}\int_0^x a(t;\epsilon)dt\right]. \qquad (2.2.155)$$

Verify that

$$\frac{d}{dx}\left[\frac{\partial L^*}{\partial y'}\right] - \frac{\partial L^*}{\partial y} = 0$$

for any y.

Show that in general, for a Lagrangian that is linear in y' of the form

$$L^*(x, y, y'; \epsilon) = A(x, y; \epsilon)y' + B(x, y; \epsilon), \qquad (2.2.156)$$

the Euler-Lagrange equation is identically satisfied for any y as long as

$$\frac{\partial A}{\partial x} = \frac{\partial B}{\partial y}, \qquad (2.2.157)$$

and this is the case for L^* in our problem.

4. Calculate the uniformly valid solution to $O(1)$ for

$$\epsilon y'' + (x - \frac{1}{2})y' + \epsilon y = 0, \qquad (2.2.158)$$

$$y(0) = 1, \qquad y(1) = -1, \qquad (2.2.159)$$

and

$$\epsilon y'' + (\frac{1}{4} - x^2)y' + 2xy = 0, \qquad (2.2.160)$$

$$y(-1) = 1, \qquad y(1) = 2. \qquad (2.2.161)$$

5. If all terms in (2.2.76) or (2.2.90) have the same sign, we have

$$\epsilon y'' + xy' + y = 0, \quad -1 \le x \le 1, \quad 0 < \epsilon \ll 1, \qquad (2.2.162)$$

$$y(-1) = 1, \quad y(1) = 2. \qquad (2.2.163)$$

Derive the exact solution in the form

$$y(x;\epsilon) = e^{(1-x^2)/2\epsilon}R(x;\epsilon), \qquad (2.2.164)$$

where

$$R(x;\epsilon) = \frac{3}{2} + \frac{1}{2}\frac{\int_0^x \exp(s^2/2\epsilon)ds}{\int_0^1 \exp(s^2/2\epsilon)ds}. \qquad (2.2.165)$$

Show that $1 \le R \le 2$ for all x in $-1 \le x \le 1$. Therefore, the solution is exponentially large in the interior, reaching a maximum value $y(0;\epsilon) = \frac{3}{2}e^{1/2\epsilon}$.

2.3 Model Nonlinear Example for Singular Perturbations

In this section, a model nonlinear example is studied that illustrates the following points: (1) A consistent study of boundary layers in the general sense enables the correct limit ($\epsilon = 0$) solutions to be isolated, and (2) a wide variety of phenomena can occur even in a simple-looking nonlinear problem. The example is

$$\epsilon \frac{d^2y}{dx^2} + y\frac{dy}{dx} - y = 0, \qquad 0 \le x \le 1, \tag{2.3.1}$$

$$y(0) = A, \qquad y(1) = B.$$

Here A and B are not considered dependent on ϵ. The main problem of interest is the study of the dependence of the solutions on the boundary values A and B. Since the problem is nonlinear, the dependence on boundary conditions is nontrivial and can change the qualitative nature of the solution.

Actually, we can use a symmetry argument to cut in half the range of values of A and B that one need consider. We note that (2.3.1) is invariant under the transformation

$$y \leftrightarrow -y, \quad x \leftrightarrow (1-x), \quad A \leftrightarrow -B, \quad B \leftrightarrow -A.$$

Therefore, if $y = f(x, A, B)$ is a solution of (2.3.1) subject to the boundary conditions $y(0) = A$, $y(1) = B$, then $y = -f(1-x, -B, -A)$ is also a solution, and this solution satisfies the boundary conditions $y(0) = -B$, $y(1) = -A$.

Thus, the solution corresponding to a given point A, B generates a solution for the "reflected point" $-B, -A$. This reflected solution is obtained by the transformation $y \to -y$, $x \to 1-x$, i.e., by a reflection about the x axis followed by a reflection about the line $x = \frac{1}{2}$ in the xy-plane. We need therefore only consider values of A and B on one side of the line $B = -A$, and we will take $B \ge -A$. Regions of the A, B plane with reflected solutions ($B < -A$) will be labeled with the subscript R.

The outer limit ($\epsilon \to 0$, x fixed) satisfies the equation

$$h\frac{dh}{dx} - h = 0. \tag{2.3.2}$$

Two branches appear in the limit solution:

$$h = 0 \tag{2.3.3a}$$

$$h = x + c, \qquad c = \text{const.} \tag{2.3.3b}$$

Only the branch $h = x + c$ has a chance of satisfying an arbitrary end condition. Note that since the outer limit (2.3.3) is linear in x it is an exact solution of (2.3.1).

Therefore, as in the examples of Secs. 2.2.4–2.2.5, the outer expansion in powers of ϵ has the degenerate form

$$y(x;\epsilon) = (c_0 + \epsilon c_1 + \epsilon^2 c_2 + \dots) + x. \tag{2.3.4}$$

In this problem, it is not clear a priori where boundary layers will occur, so various possibilities must be examined. Two outer solutions h_R and h_L are possible, depending on whether the boundary condition on the right or left is satisfied, respectively,

$$h_R(x) = x + B - 1, \tag{2.3.5}$$

$$h_L(x) = x + A. \tag{2.3.6}$$

These solutions take the values

$$\begin{aligned} h_R(0) &= B - 1, \\ h_L(1) &= A + 1 \end{aligned} \tag{2.3.7}$$

at the other end of the interval. We see that if $B - 1 = A$ (i.e., $h_R(0) = A$ or $h_L(1) = B$) the outer limit $h_R(x) = h_L(x)$ is an *exact solution* to the boundary-value problem (2.3.1).

If $A \neq B - 1$, we expect a boundary or interior layer to occur somewhere on the interval $(0, 1)$. A study of the possible boundary layers is now made with the aim of determining where these may occur and of what types they may be.

If y is not small, the simplest type of boundary layer can occur over a scale of order ϵ in x. In such a boundary layer, the derivative terms in (2.3.1) are dominant. The corresponding asymptotic expansion is of the form

$$y(x;\epsilon) = g(x^*) + \epsilon g_1(x^*) + \dots, \quad x^* = \frac{x - x_d}{\epsilon}. \tag{2.3.8}$$

Here x_d gives the location of the layer. Since (2.3.1) is autonomous, the choice of x_d does not alter the equations governing $g, g_1, \dots$. We have

$$\frac{d^2 g}{dx^{*2}} + g\frac{dg}{dx^*} = 0, \tag{2.3.9}$$

which has a first integral

$$\frac{dg}{dx^*} + \frac{g^2}{2} = C = \text{const.} \tag{2.3.10}$$

Choosing $C < 0$ in (2.3.10) leads to a result that cannot match with the outer solution. Thus, we set $C = \beta^2/2 > 0$ and write the solution of (2.3.10) as

$$g(x^*) = \beta \tanh\frac{\beta}{2}(x^* + k) \tag{2.3.11}$$

or

$$g(x^*) = \beta \coth\frac{\beta}{2}(x^* + k), \tag{2.3.12}$$

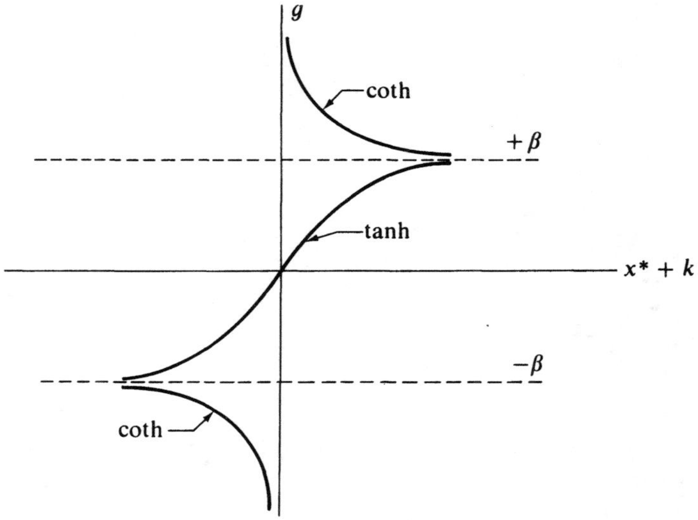

FIGURE 2.3.1. Solution Curves for Boundary Layers

where k is a second constant of integration. A sketch of these solutions is given in Figure 2.3.1, where the abscissa is $x^* + k$. Thus, varying k results in a translation of the curves along the x^* axis.

The tanh solution, (2.3.11), increases to its asymptotic value β as $x^* \to \infty$ and decreases to $-\beta$ as $x^* \to -\infty$. The approach is exponential:

$$g(x^*) \sim \beta\{1 - 2e^{-\beta(x^*+k)} + \ldots\} \quad \text{as } x^* \to \infty. \tag{2.3.13}$$

The coth solution (2.3.12) decreases from infinity at $x^* = -k$ to its asymptotic value β as $x^* \to \infty$ and increases from $-\infty$ at $x^* = -k$ to its asymptotic value $-\beta$ as $x^* \to -\infty$.

Segments of these solutions can be used as boundary layers that match in a simple way with the outer solutions (2.3.5) or (2.3.6). Evidently, the matching conditions are

$$h_{\mathrm{R}}(x_d) = g(\infty) \quad \text{or} \quad h_{\mathrm{L}}(x_d) = g(-\infty). \tag{2.3.14}$$

This type of simplified matching to order unity follows directly from the consideration of suitable limits in terms of a matching variable $x_\eta = \frac{x - x_d}{\eta(\epsilon)}$ as discussed earlier.

Before carrying out the discussion of the possibilities for various boundary conditions, notice that another distinguished limit exists for (2.3.1) if y is allowed to be small. The fact that scaling y may give a different distinguished limit is a consequence of the nonlinearity. In fact, if $\overline{y} = y/\sqrt{\epsilon}$ and $\overline{x} = (x - x_0)/\sqrt{\epsilon}$, the equation for $\overline{y}$ is free of ϵ. Considered as a local solution derived from the exact equation by means of the asymptotic expansion

$$y(x;\epsilon) = \sqrt{\epsilon} f(\overline{x}) + \epsilon f_1(\overline{x}) + \ldots, \quad \overline{x} = \frac{x - x_0}{\sqrt{\epsilon}}, \tag{2.3.15}$$

$f(\overline{x})$ should be an important element in some approximations. This statement is based on the idea that distinguished limits are always significant. The equation for f is the exact equation (2.3.1) with $\epsilon = 1$

$$\frac{d^2 f}{d\overline{x}^2} + f\frac{df}{d\overline{x}} - f = 0. \tag{2.3.16}$$

Thus, the local solution (2.3.15) can be calculated only if the boundary conditions appropriate for f simplify the solution of (2.3.16).

Next, consider the range of values of A and B for which solutions can be composed of h and g functions, that is, of outer solutions and boundary layers of order ϵ in thickness. The situation is represented on the (A, B) diagram of Figure 2.3.2. The line $B = A + 1$ represents solutions with no boundary layer where the exact solution is the outer limit $h_R = h_L = x + A = x + B - 1$. If $A > B - 1$, the outer solution $h_R = x + B - 1$ satisfies the right-hand boundary condition and takes a positive value if $B - 1 > 0$. A boundary layer at $x = 0$ descending to $B - 1$ can then be used to complete the solution as shown in Figure 2.3.3a. Thus, the triangular domain $A > B - 1 > 0$ consists of a left boundary layer descending by a coth solution to the h_R outer limit. This is abbreviated by (LBL $\downarrow$ coth) in the diagram. Such a boundary layer [see Figure 2.3.1] matched to h_R is

$$g_L(x^*) = (B - 1)\coth\left(\frac{B-1}{2}\right)(x^* + k), \quad x^* = \frac{x}{\epsilon}. \tag{2.3.17}$$

The value of k is chosen to satisfy the boundary condition at $x^* = 0$.

$$A = (B - 1)\coth\left(\frac{B-1}{2}k\right), \tag{2.3.18}$$

i.e., $k = [2/(B - 1)]\coth^{-1}[A/(B - 1)]$. Note that the lower boundary of this domain brings us to the limiting case $B = 1$, where the boundary-layer solution decays algebraically as $x^* \to \infty$, and we put this case aside temporarily. Corresponding to region I, we have region I_R consisting of a right boundary layer ascending by a coth solution for values of A and B such that $B < A + 1 < 0$. It is easy to verify that the choice of h and g functions is unique here. For example, for region I, if we had chosen h_L we would have needed a right boundary layer rising to a positive value, but no such boundary-layer solution is available.

Next consider the region to the left of the line $B = A + 1$, i.e., $A < B - 1$ but with B still greater than unity. It is still possible to fit in a tanh-type boundary layer at the left end to match h_R, provided that $|A| < B - 1$. The restriction $|A| < B - 1$ ensures that the asymptotic value of the tanh solution as $x^* \to \infty$ can be equal to A. Thus, region II is defined by $0 \leq |A| < B - 1$ with a left tanh boundary layer that rises from A to $B - 1$ as shown in Figure 2.3.3b for $A > 0$ and Figure 2.3.3c

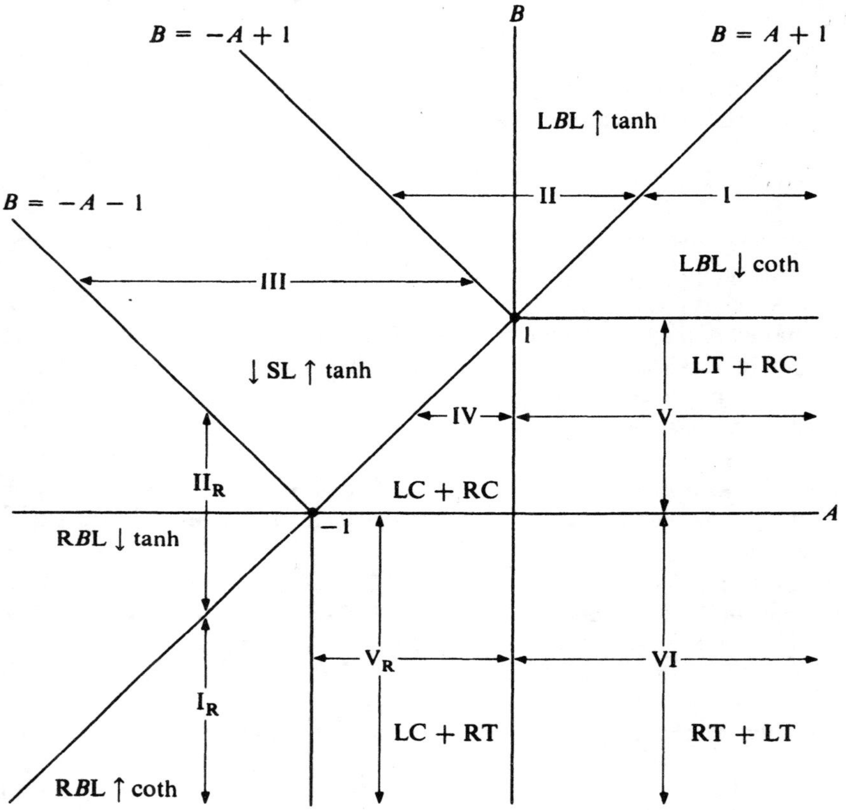

FIGURE 2.3.2. Possible Solutions in the AB-Plane

for $A < 0$. In region II_R with $|B| < |A+1|$, $A + 1 < 0$ we have a right tanh boundary layer that descends from B to $A + 1$.

A case between II and II_R has $B > A + 1$, but a tanh boundary layer at the end cannot provide a sufficient rise (or descent) to match the end condition. There is, however, the possibility of using the tanh solution at an interior point x_d and matching both as $x^* \to \infty$ and $x^* \to -\infty$. The boundary layer is, so to speak, pushed off the ends and appears in the interior as a shock layer. This is the case III in Figure 2.3.2 and has $B > A + 1$, $-(B + 1) < A < 1 - B$. The left and right boundary conditions are satisfied by outer solutions $h_{\text{L}} = A + x$, $h_{\text{R}} = x - 1 + B$. As seen in Figure 2.3.3d, the tanh solution (2.3.11) matches to values $\pm B$ symmetric about $y = 0$ as $x^* \to \pm\infty$. Thus, this solution can serve as

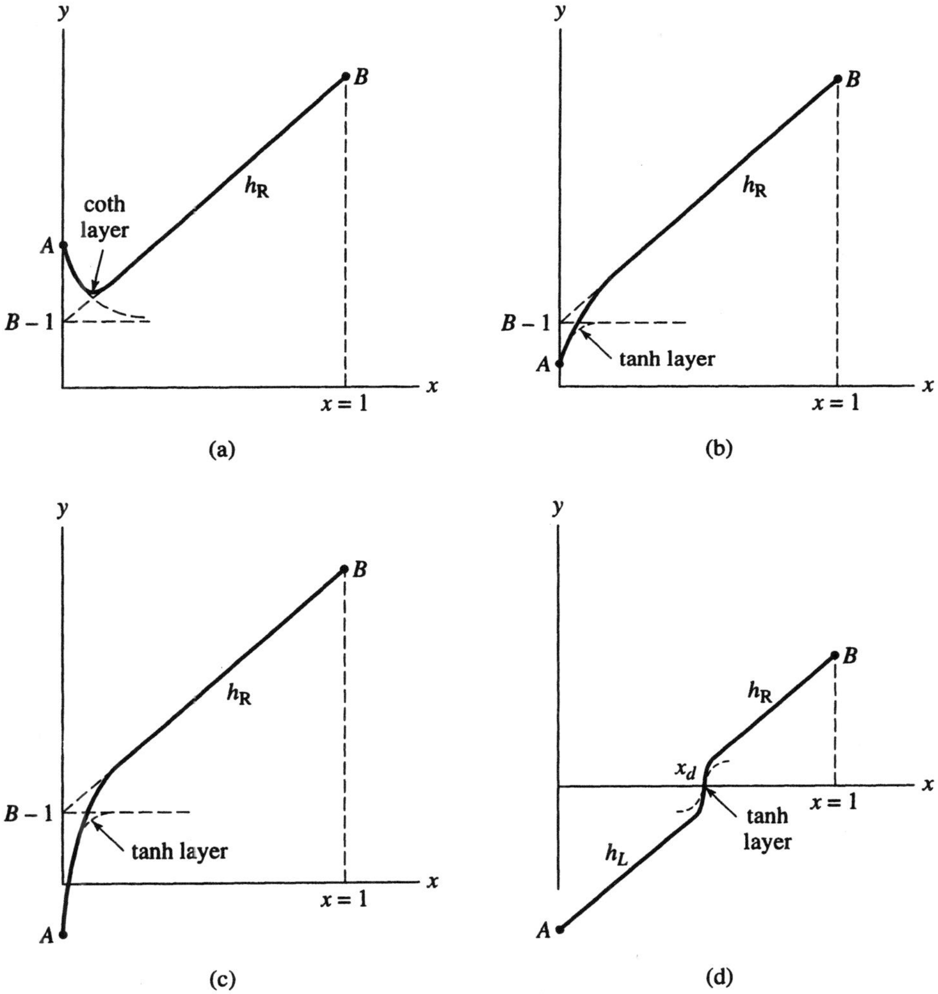

FIGURE 2.3.3. Solutions with $O(\epsilon)$ Layers. (a) Region I, (b) Region II, $A > 0$, (c) Region II, $A < 0$, (d) Region III

a shock layer centered at $x = x_d$, where x_d is defined by the symmetry condition $h_L(x_d) = -h_R(x_d)$, i.e.,

$$A + x_d = -B - x_d + 1, \tag{2.3.19}$$

which gives

$$x_d = \frac{1 - A - B}{2}. \tag{2.3.20}$$

Matching to $O(1)$ requires $\beta = -x_d$ in (2.3.11) but does not determine k, which represents an $O(\epsilon)$ shift in the shock location in this case. The inner solution is

$$g(x^*) = \frac{B - A - 1}{2} \tanh \frac{B - A - 1}{4} (x^* + k), \tag{2.3.21}$$

where

$$x^* = \frac{x - (1 - A - B)/2}{\epsilon}.$$

The possibilities for boundary layers of order ϵ in thickness are now exhausted, but large parts of the AB- plane are still inaccessible; for example, $A > 0, B < 0$. A hint of the kind of solutions needed is obtained by considering the special case $A = 0, 0 < B < 1$. In this example, outer solutions of different branches can be used to satisfy the end conditions

$$h_L(x) = 0, \quad h_R = x + B - 1. \tag{2.3.22}$$

These solutions intersect in a corner at $x = x_c = 1 - B$. A smooth solution over the full interval can be found if we can exhibit a corner-layer solution centered about $x = x_c$ and matching with h_L and h_R of (2.3.22). Such a corner-layer solution, if it exists, must be contained in the solutions of (2.3.16). The matching conditions for h_L and h_R are such that

$$f(\overline{x}) \to 0 \quad \text{as } \overline{x} \to -\infty; \quad f(\overline{x}) \to \overline{x} \quad \text{as } \overline{x} \to \infty. \tag{2.3.23}$$

To determine whether such solutions exist, we study the phase plane of (2.3.16). Setting

$$v = \frac{df}{d\overline{x}}, \tag{2.3.24}$$

(2.3.16) becomes

$$\frac{dv}{df} = -\frac{f(v - 1)}{v}. \tag{2.3.25}$$

The diagram of the integral curves of (2.3.25) is Figure 2.3.4. Along any path, the direction of increasing $\overline{x}$ is indicated by an arrow, as found from (2.3.24). It is clear that the paths that approach $v = 1 = df/d\overline{x}$ are capable of matching of the type $f(\overline{x}) \to \overline{x}$ as $\overline{x} \to +\infty$.

The exceptional path labeled f_{RC}, which starts from the origin, has a chance also to satisfy $f(\overline{x}) \to 0$ as $\overline{x} \to -\infty$, since the origin is a singular point. The nature

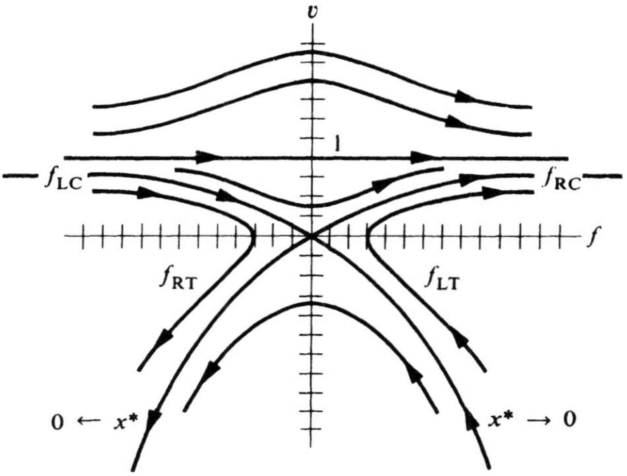

FIGURE 2.3.4. Phase Plane of Various Solutions

of the singularity is found from linearization of (2.3.25), which gives

$$v^2 - f^2 = \text{const.}$$

Since f_{RC} passes through the origin, the constant is equal to zero, so that along f_{RC}

$$v = f + \ldots \text{ as } f \to 0.$$

The integration of (2.3.24) with $v = f$ shows that

$$f_{RC} = k_0 e^{\bar{x}} + \ldots \text{ as } \bar{x} \to -\infty. \tag{2.3.26}$$

That is, the matching condition as $\bar{x} \to -\infty$ is satisfied with an exponential approach. Here f_{RC} is called a right-corner solution and can be used together with (2.3.22) to complete the solution for $A = 0$, $0 < B < 1$. The reflected solution for $B = 0$, $-1 < A < 0$ will involve the left-corner solution, labeled f_{LC} in Figure 2.3.4.

The combination of these cases has solutions with both left and right corners and occurs in the triangular region IV of Figure 2.3.2. In this region $B - 1 < A < 0$, $0 < B < A + 1$. The outer solution has three pieces:

$$h_L(x) = x + A, \qquad 0 \le x \le -A;$$
$$h_m(x) = 0, \qquad -A < x \le 1 - B;$$
$$h_R(x) = x - 1 + B, \qquad 1 - B \le x \le 1.$$

Here f_{LC} provides the match between h_{L} and h_m, and f_{RC} the match between h_m and h_{R} as seen in Figure 2.3.5a.

The other two exceptional paths in the phase plane of Figure 2.3.4 are also necessary to complete the coverage of the AB-plane. Consider, for example, that $B = 0$, $A > 0$. The outer solution satisfying the right-boundary condition is $h_R = 0$. The special solution $f_{\text{LT}}(\overline{x})$ with $\overline{x} = x/\sqrt{\epsilon}$ can match to this as $\overline{x} \to \infty$ with exponential approach. This can be seen from a discussion of the behavior near $(v = f = 0)$ analogous to that for f_{RC} [see (2.3.25)]. However, it is not reasonable to expect to satisfy a boundary condition where y is $O(1)$ with a transition layer where y is $O(\sqrt{\epsilon})$. Therefore, we must try to match this transition layer to a thinner boundary layer around $x = 0$. In order to study this matching, we need to know the behavior of f_{LT} as $\overline{x} \to 0$. In this case, the behavior of f_{LT} as $\overline{x} \to 0$ can be obtained from the complete integral of (2.3.25) taken along the exceptional path. Equation (2.3.25) can be written

$$\left(1 + \frac{1}{v-1}\right) dv + f\,df = 0$$

so that the first integral representing paths through the origin is

$$v + \log|1 - v| + \frac{f^2}{2} = 0. \tag{2.3.27}$$

We see that as $-v$, f approach infinity, the dominant terms in (2.3.27) balance with

$$v \to -\frac{f^2}{2},$$

and in this limit (2.3.24) gives

$$d\overline{x} \to -\frac{2\,df}{f^2}.$$

Thus, $f(\overline{x})$ has the algebraic decay

$$f(\overline{x}) = \frac{2}{\overline{x} + k_0} + \cdots. \tag{2.3.28}$$

It is clear that for matching we must have $f(\overline{x})$ becoming large as $\overline{x} \to 0$ so that y can become $O(1)$. Thus the constant of integration $k_0 = 0$ and

$$f(\overline{x}) = \frac{2}{\overline{x}} + \cdots, \qquad \overline{x} \to 0. \tag{2.3.29}$$

Next we consider an $O(\epsilon)$ boundary layer at $x = 0$, an inner layer, as in (2.3.9) but in order to have algebraic decay we choose the constant of integration in (2.3.10) to be zero. Thus

$$\frac{dg}{dx^*} + \frac{g^2}{2} = 0. \tag{2.3.30}$$

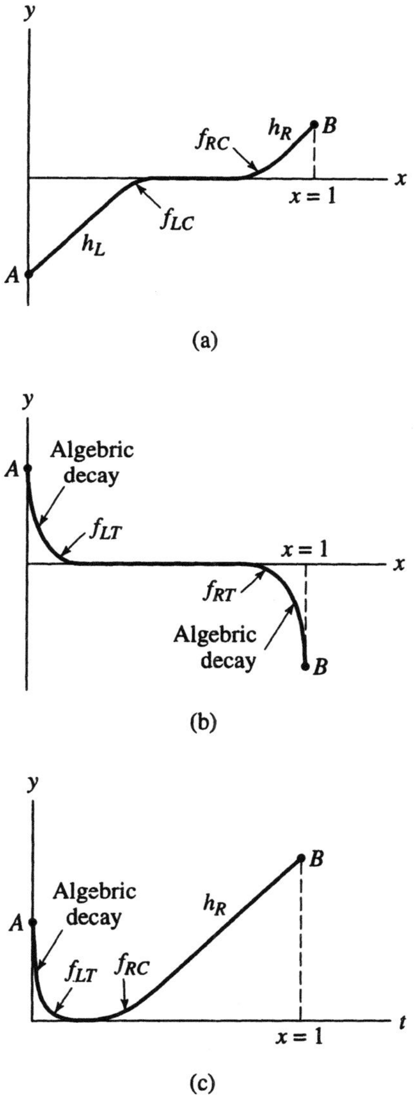

FIGURE 2.3.5. Solutions with Corner and Transition Layers. (a) Region IV, (b) Region VI, (c) Region V

The appropriate solution satisfying the boundary condition $g = A$ at $x^* = 0$ is

$$g(x^*) = \frac{2}{x^* + 2/A}. \tag{2.3.31}$$

Now the $O(1)$ matching can be discussed using the limit $\epsilon \to 0$, x_η fixed where $x_\eta = x/\eta(\epsilon)$, and η belongs to an overlap domain that is contained in $\epsilon \ll \eta \ll \sqrt{\epsilon}$.

In order to demonstrate that this matching is really valid and that the singularity of $f(\overline{x})$ does not produce too high a singularity in $f_1(\overline{x})$, it is useful to work out both the next term in the expansion of f and the first term of f_1 (see (2.3.15)). The occurrence of a log term in the integral (2.3.27) indicates that a log term appears in the expansion of $f(\overline{x})$ as $\overline{x} \to 0$. By substitution in the basic equation (2.3.27), the following expansion can be verified:

$$f(\overline{x}) = \frac{2}{x} + \frac{2}{3}\overline{x}\log\overline{x} + k\overline{x} + O(\overline{x}^2\log\overline{x}). \tag{2.3.32}$$

Here k is to be regarded as a known constant that could be found, for example, from the complete numerical integration of (2.3.27) for $f(\overline{x})$. The boundary conditions used to specify $f(\overline{x})$ are $f(\overline{x}) \to 0$ as $\overline{x} \to \infty$, $f(\overline{x}) \to 2/\overline{x}$ as $\overline{x} \to 0$ [see (2.3.29)].

Next, by substituting the expansion (2.3.15) in the original equation (2.3.1), we find that f_1 satisfies

$$\frac{d^2 f_1}{d\overline{x}^2} + f\frac{df_1}{d\overline{x}} + \left(\frac{df}{d\overline{x}} - 1\right) f_1 = 0. \tag{2.3.33}$$

This linear equation is the variational equation of the original equation (2.3.1). We can remark that whenever the basic equation is nonlinear, the equations for the higher approximations are linear. In general, these linear equations have variable coefficients that depend on the earlier terms in the expansion. In this particular case, it is easy to verify that there is a solution of (2.3.33) such that $f_1(\overline{x}) \to$ (const.)$e^{-\overline{x}}$ as $\overline{x} \to \infty$. Now as $\overline{x} \to 0$ this solution can be expressed as some linear combinations of the two independent solutions. As $\overline{x} \to 0$, (2.3.33) can be approximated by using the asymptotic form of f:

$$\frac{d^2 f_1}{d\overline{x}^2} + \left(\frac{2}{\overline{x}} + \dots\right)\frac{df_1}{d\overline{x}} - \left(\frac{2}{\overline{x}^2} + \dots\right) f_1 = 0. \tag{2.3.34}$$

By seeking solutions of the form $f_1 \sim \overline{x}^\alpha$, we find the indicial equation

$$\alpha(\alpha - 1) + 2\alpha - 2 = (\alpha + 2)(\alpha - 1) = 0.$$

The appropriate root is $\alpha = -2$. Thus, as $\overline{x} \to 0$, f_1 has the expansion

$$f_1 = \frac{k_1}{\overline{x}^2} + \dots. \tag{2.3.35}$$

The constant k_1 is to be found when matching is carried out to a sufficiently high order.

Next, we need to consider a corresponding expansion with leading term $g(x^*)$ as $x^* \to \infty$. Because of the occurrence of the log term in the expansion (2.3.32) of $f(\overline{x})$ it turns out that a log term has to be included in the inner expansion. The occurrence of this term would really not be discovered until matching to higher order was attempted. The expansion is

$$y(x;\epsilon) = g(x^*) + \epsilon \log \epsilon g_{11}(x^*) + \epsilon g_1(x^*) + \dots \tag{2.3.36}$$

and on substitution in the basic equation we find the following sequence of equations and boundary conditions:

$$\frac{d^2 g}{dx^{*2}} + g\frac{dg}{dx^*} = 0; \qquad g(0) = A, \tag{2.3.37}$$

$$\frac{d^2 g_{11}}{dx^{*2}} + g\frac{dg_{11}}{dx^*} + \frac{dg}{dx^*}g_{11} = 0; \qquad g_{11}(0) = 0, \tag{2.3.38}$$

$$\frac{d^2 g_1}{dx^{*2}} + g\frac{dg_1}{dx^*} + \frac{dg}{dx^*}g_1 = g; \qquad g_1(0) = 0. \tag{2.3.39}$$

We have already found $g(x^*)$ so that straightforward integration yields the following solutions:

$$g(x^*) = \frac{2}{x^* + (2/A)}, \tag{2.3.40}$$

$$g_{11}(x^*) = \frac{C_{11}}{3}\left(x^* + \frac{2}{A}\right) - \frac{8}{3A^3}\frac{C_{11}}{(x^* + (2/A))^2}, \tag{2.3.41}$$

$$g_1(x^*) = \frac{2}{3}\left(x^* + \frac{2}{A}\right)\log\left(x^* + \frac{2}{A}\right) + C_1\left(x^* + \frac{2}{A}\right)$$

$$- \frac{(16C_1/A^4) + (32/3A^4)\log(2/A)}{(x^* + (2/A))^3}. \tag{2.3.42}$$

The constants C_{11} and C_1 would be found in higher-order matching.

However, there are no arbitrary constants to be found in matching to order unity because both $f(\overline{x})$ and $g(x^*)$ are completely defined; f was defined by conditions of exponential decay at infinity and singular behavior at the origin, g by algebraic decay at infinity and $g(0) = A$. Thus, matching to order unity here is a verification that, in a certain sense, g is contained in f. In order to match, we introduce $x_\eta = x/\eta(\epsilon)$ so that

$$\overline{x} = \frac{x}{\sqrt{\epsilon}} = \frac{\eta}{\sqrt{\epsilon}}x_\eta \to 0, \qquad x^* = \frac{x}{\epsilon} = \frac{\eta}{\epsilon}x_\eta \to \infty. \tag{2.3.43}$$

For the transition-layer expansion and the inner expansion to match to $O(1)$, we want

$$\lim_{\substack{\epsilon\to 0\\ x_\eta \text{ fixed}}} \left\{\sqrt{\epsilon} f\left(\frac{\eta x_\eta}{\sqrt{\epsilon}}\right) + \epsilon f_1\left(\frac{\eta x_\eta}{\sqrt{\epsilon}}\right) + \dots - g\left(\frac{\eta x_\eta}{\sqrt{\epsilon}}\right)\right.$$

$$-\epsilon\log\epsilon g_{11}\left(\frac{\eta x_\eta}{\epsilon}\right)-g_1\left(\frac{\eta x_\eta}{\epsilon}\right)-\ldots\Bigg\}=0. \tag{2.3.44}$$

Substituting from the various expansions just constructed, we find

$$\lim_{\substack{\epsilon\to 0\\ x_\eta\text{ fixed}}}\left\{\sqrt{\epsilon}\left[\frac{2}{(\eta/\sqrt{\epsilon})x_\eta}+O\left(\frac{\eta}{\sqrt{\epsilon}}\log\eta\right)+O\left(\frac{\eta}{\sqrt{\epsilon}}\log\epsilon\right)+O\left(\frac{\epsilon^2}{\eta^2}\right)\right]\right.$$

$$\left.-\left[\frac{2}{(\eta x_\eta/\epsilon)+A}+O(\eta\log\epsilon)+O(\eta\log\eta)\right]\right\}=0. \tag{2.3.45}$$

The dominant terms cancel and the other terms all vanish if $\eta(\epsilon)$ is in the class of $\epsilon \ll \eta \ll \sqrt{\epsilon}/\log\epsilon$. Thus, the $O(\sqrt{\epsilon})$ transition layer can be continued with this inner layer to satisfy an $O(1)$ boundary condition.

In essence, our discussion of all the possible solutions is now complete. All solutions with corner layers or transition layers match to an outer solution $h = 0$. The case $A > 0$, $B < 0$ demands f_{LT}, f_{RT} as shown in Figure 2.3.5b, whereas the case $A > 0$, $0 < B < 1$ demands f_{LT}, f_{RC} as shown in Figure 2.3.5c.

Thus, the systematic use of boundary-layer theory and matching can successfully cope with the wide variety of problems that can arise in a nonlinear case. While it is true that the full equation must be integrated to find f_{LT}, f_{RT}, f_{LC}, f_{RC}, the boundary conditions are canonical so that this integration can be done once for all problems.

Problems

1. Study the solutions to order unity (as $\epsilon \to 0$) for all A and B independent of ϵ, of the boundary-value problem with $y(0) = A$, $y(1) = B$ governed by the following equations:
   a. $\epsilon yy'' + yy' = -x.$ (2.3.46)
   b. $\epsilon y'' + \frac{1}{2}y^2y' - y = 0.$ (2.3.47)
2. Solve the following boundary-value problem exactly by differentiating the equation. Are the nonunique solutions that result derivable by perturbations for $0 < \epsilon \ll 1$?

$$\epsilon(y'')^2 + (x - \frac{1}{2})y' - y = 0, \tag{2.3.48}$$

$$y(0) = y(1) = 1. \tag{2.3.49}$$

3. Solve the following nonlinear problem (which is a model of the shock structure for an isothermal shock for large Mach number) exactly on $-\infty < x < \infty$:

$$y'' + \frac{\epsilon y'}{y^2} - y' = 0, \tag{2.3.50}$$

$$y(-\infty) = 1, \qquad y(\infty) = \epsilon = \text{const.} > 0. \tag{2.3.51}$$

Next, study the asymptotic behavior of the solution as $\epsilon \to 0$ for $x < 0$, and $x \approx 0$ using the differential equation. Verify your results by comparison with the exact solution.

2.4 Singular Boundary Problems

In this section, two problems are discussed in which the expansions in terms of a small parameter are singular, not because of a lowering of the order of the equation in the limit but rather because of a difficulty associated with the behavior near a boundary point. Nevertheless, the same method as used in previous sections enables the expansions of the solutions to be found. Different asymptotic expansions valid in different regions are constructed, and the matching of these expansions in an overlap domain enables all unknown constants to be found. Thus, a uniformly valid approximation can be constructed.

2.4.1 Periodic Collision Orbits in the Problem of Two Fixed Force-Centers

Consider the planar motion of a point mass in the gravitational field of two *fixed* centers of attraction. This is a classical example of an *integrable* dynamical system and is discussed in various texts in dynamics (e.g., see section 64 of [2.4]).

Using suitable dimensionless variables, the two fixed centers have masses $1-\mu$ and μ and can be located at $x = 0$ and $x = 1$, respectively. The equations of motion will then become

$$\frac{d^2x}{dt^2} = -\frac{(1-\mu)x}{r^3} - \frac{\mu(x-1)}{r_1^3} \tag{2.4.1a}$$

$$\frac{d^2y}{dt^2} = -\frac{(1-\mu)y}{r^3} - \frac{\mu y}{r_1^3}, \tag{2.4.1b}$$

where

$$r = \sqrt{(x^2+y^2)}, \qquad r_1 = \sqrt{(x-1)^2+y^2}$$

are the distances from the particle to these two centers as shown in Figure 2.4.1.

Actually, the system (2.4.1) is dynamically inconsistent if the points μ and $1-\mu$ are to be regarded as masses that move under their mutual gravitational influence. In fact, if two point masses are initially at rest some distance apart, they will be attracted along the line joining them to eventually collide. More generally, for arbitrary initial positions and velocities, the particles would describe a Keplerian orbit about the center of mass (e.g., see sections 3.7–3.8 of [2.5] or sections 36–37 of [2.4]). The particle P, with a negligible mass, would then move in the given gravitational field of the orbiting pair μ and $1-\mu$ of gravitational centers. This is the restricted three-body problem, also well known in dynamics (e.g., see

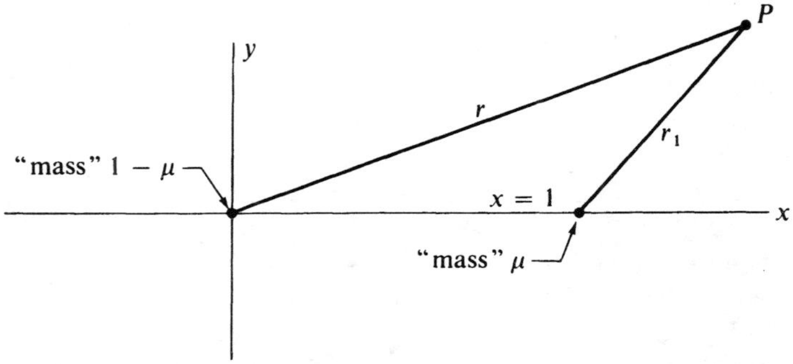

FIGURE 2.4.1. Planar Motion in the Field of Two Fixed Centers

[2.16] and Section 4.3). In spite of the physical inconsistency of (2.4.1), it is a useful mathematical model of the restricted three-body problem, particularly for studying perturbation solutions for $\mu \ll 1$.

In this section, we focus on the very special case of solutions of (2.4.1) that have $y = 0$ and $0 \leq x \leq 1$. This is a singular perturbation problem for $\mu \to 0$ even though μ does not multiply the highest derivative term. Instead, the term multiplied by μ becomes singular at $x = 1$, $y = 0$. A slightly more general version (where solutions have $y = O(\mu)$) was discussed in [2.13] in preparation for the more realistic analysis given in [2.12] for motion in the restricted three-body problem. This work has been extended to trajectories with minimal energy [2.11], to three dimensions [2.10], and to periodic orbits in a rotating frame [2.7]. In addition, there are numerous other references that deal with dynamical systems where a particle approaches two or more gravitational centers, as in cometary orbits and interplanetary spacecraft trajectories.

All such problems are characterized by the following features:

1. For motion away from the small mass μ one has a perturbed Keplerian ellipse relative to the dominant gravitational center. This outer expansion fails when the solution approaches the small mass μ.
2. Motion close to the small mass μ is a perturbed Keplerian hyperbola relative to this mass. This inner expansion fails when the solution is at a large distance from the small mass μ.
3. In order to determine the constants that define the *leading* term in the inner expansion, one needs to match with the *two-term* outer expansion defining the perturbed Keplerian ellipse.
4. Once the inner limit is known, one can use matching again to compute the leading Keplerian orbit that describes the solution after close passage to the small mass.

5. Using the above results, a composite expansion can be constructed. This expansion describes the motion from the initial time through close passage and beyond so long as the particle does not again pass close to the small mass.

Detailed results can be found in [2.7], [2.10]–[2.13], and the references cited therein.

For the model problem (2.4.1) it is well known that there exists a family of periodic solutions that are confocal ellipses with foci at $x = 0$ and $x = 1$. If the speed at any point on this elliptic orbit is denoted by v, one can show that $v = \sqrt{v_{1-\mu}^2 + v_\mu^2}$, where $(v_{1-\mu})$, (v_μ) is the speed corresponding to the same elliptic orbit but with the point mass $(1-\mu)$, (μ) as the only center of attraction. In the limit as the eccentricity of this family of orbits tends to unity, we have the periodic double collision orbit in the unit interval.

For this limiting case, (2.4.1) reduces to

$$\frac{d^2x}{dt^2} = -\frac{(1-\mu)}{x^2} + \frac{\mu}{(x-1)^2}, \qquad 0 \le x \le 1, \tag{2.4.2}$$

and we wish to study the solution for the case $\mu \ll 1$.

Equation (2.4.2) has the energy integral

$$\frac{1}{2}\left(\frac{dx}{dt}\right)^2 - \frac{(1-\mu)}{x} - \frac{\mu}{1-x} = h = \text{const.}, \tag{2.4.3}$$

which can be used to express t as a function of x by quadrature. The result involves elliptic integrals and is not very instructive. The qualitative nature of the collision orbits can be determined from the phase-plane path of solutions for various values of h given in Figure 2.4.2. Setting the right-hand side of (2.4.2) equal to zero gives the equilibrium point

$$x_e = \frac{(1-\mu) - \sqrt{\mu - \mu^2}}{1 - 2\mu} = 1 - \mu^{1/2} + O(\mu), \tag{2.4.4}$$

which is a saddle point.

Substituting $x = x_e$ and $dx/dt = 0$ into (2.4.3) gives the value of h for the trajectories passing through the saddle point

$$h_e = -1 - 2\sqrt{\mu - \mu^2} = -1 + O(\mu^{1/2}). \tag{2.4.5}$$

For $h < h_e$, the motion consists of a single collision periodic orbit relative to the mass point $1 - \mu$ or μ. In either case, the trajectory does not go beyond a certain maximum distance from the center of attraction. For $h > h_e$, the motion spans the entire interval with alternate collisions at $x = 0$ and $x = 1$, and in our calculations later, we take $h = 0 > h_e$.

Using (2.4.3), we can calculate t as a function of x by quadrature, and this result is given in [2.13]. Here we will derive the asymptotic representation of this solution from (2.4.2), ignoring the existence of the exact integral (2.4.3).

For $\mu \to 0$, (2.4.2) defines a singular perturbation problem because the outer expansion, which neglects the term $\mu/(x-1)^2$ to first order, is in error in a

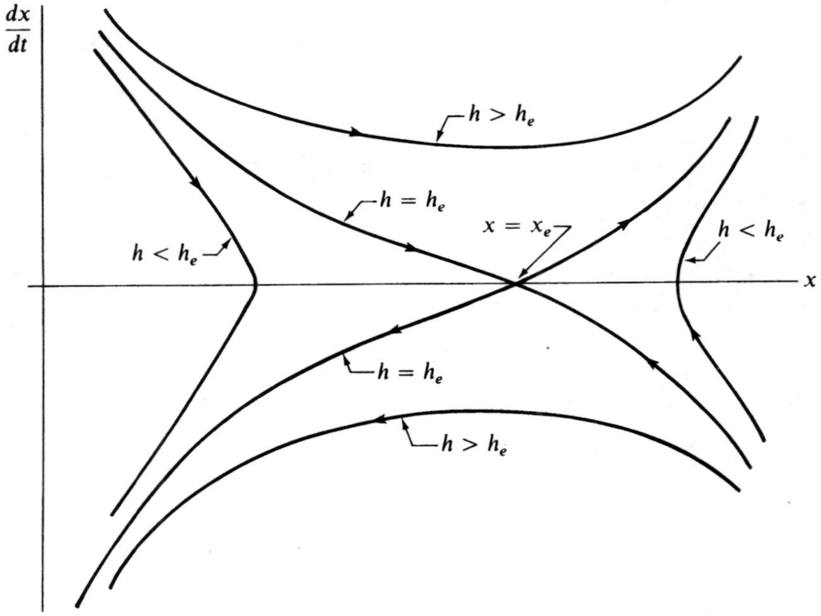

FIGURE 2.4.2. Phase-Plane Trajectories

neighborhood of $x = 1$. In fact, it is clear that regardless of the size of μ, there is a neighborhood of $x = 1$ in which this term is dominant. Since the nonuniformity location is a priori known, it is convenient to choose x as the independent variable and regard t as the dependent variable. Denoting derivatives with respect to x by primes, (2.4.2) transforms to

$$-\frac{t''}{t'^3} = -\frac{(1-\mu)}{x^2} + \frac{\mu}{(x-1)^2}. \tag{2.4.6}$$

We now assume an outer expansion

$$t(x;\mu) = t_0(x) + \mu t_1(x) + O(\mu^2) \tag{2.4.7}$$

and derive the following equations for t_0 and t_1:

$$-\frac{t_0''}{t_0'^3} = -\frac{1}{x^2}, \tag{2.4.8}$$

$$-\frac{t_1''}{t_0'^3} + \frac{3t_1't_0''}{t_0'^4} = \frac{1}{x^2} + \frac{1}{(x-1)^2}. \tag{2.4.9}$$

The foregoing equations are integrable and simplify to

$$\left(\frac{1}{2t_0'^2}\right)' = -\frac{1}{x^2} = \left(\frac{1}{x}\right)', \tag{2.4.10}$$

$$-\left(\frac{t_1'}{t_0'^3}\right)' = \frac{1}{x^2} + \frac{1}{(x-1)^2} = \left(-\frac{1}{x} - \frac{1}{(x-1)}\right)', \tag{2.4.11}$$

which, of course, is a consequence of the existence of the exact integral (2.4.3)

$$\frac{1}{2t'^2} - \frac{(1-\mu)}{x} - \frac{\mu}{1-x} = h.$$

Integrating (2.4.10) and (2.4.11) with $h = 0$ and $t(0) = 0$ gives

$$\sqrt{2}t(x;\mu) = \frac{2}{3}x^{3/2} + \mu\left[\frac{2}{3}x^{3/2} + x^{1/2} - \frac{1}{2}\log\frac{1+x^{1/2}}{1-x^{1/2}}\right] + O(\mu^2). \tag{2.4.12}$$

Because of the logarithmic singularity at $x = 1$, the expansion (2.4.12) is not uniformly valid there, and we must consider an inner expansion centered at $x = 1$. We introduce inner variables

$$x_\alpha = \frac{1-x}{\mu^\alpha}; \qquad t_\beta = \frac{t-\tau(\mu)}{\mu^\beta}, \tag{2.4.13}$$

where the constants α and β are to be determined by an analysis of the order of magnitude of the various terms in the differential equation written in inner variables. Here $\tau(\mu)$ is the half-period and will be determined from the matching.

Equation (2.4.2) transforms to

$$-\frac{d^2x_\alpha}{dt_\beta^2} = -\mu^{2\beta-\alpha}\frac{(1-\mu)}{(1-\mu^\alpha x_\alpha)^2} + \frac{\mu^{2\beta-3\alpha+1}}{x_\alpha^2} = \frac{\mu^{2\beta-3\alpha+1}}{x_\alpha^2} + O(\mu^{2\beta-\alpha}). \tag{2.4.14}$$

Since the attraction of the mass at $x = 1$ must be taken into account to first order, we must set $2\beta - 3\alpha + 1 = 0$, and this gives a relationship between β and α. For initial conditions that correspond to trajectories spanning the unit interval, the velocities of the inner and outer expansions must also match. The velocity dx/dt, calculated from the outer expansion, is $O(1)$ in the matching region. According to the inner expansion, the velocity $dx/dt = O(\mu^{\alpha-\beta})$. Therefore, we must set $\alpha = \beta$ to conclude that $\alpha = \beta = 1$, and we have the inner variables

$$x^* = \frac{1-x}{\mu}, \qquad t^* = \frac{t-\tau(\mu)}{\mu}.$$

Since $t = \tau + \mu t^*$, we need only calculate t^* as a function of x^* to $O(1)$ in order to obtain $t(x, \mu)$ to $O(\mu)$. Thus, it suffices to study the solution of the following limiting differential equation (x^*, t^* fixed, $\mu \to 0$) that ensues from (2.4.2):

$$-\frac{d^2x^*}{dt^{*2}} = \frac{1}{x^{*2}}. \tag{2.4.15a}$$

This has the integral

$$\frac{1}{2}\left(\frac{dx^*}{dt^*}\right)^2 - \frac{1}{x^*} = h^*. \tag{2.4.15b}$$

We calculate h^* by matching the velocities given by the inner and outer limits. According to (2.4.12) (or the energy integral with $h = 0$), the outer expansion for the kinetic energy $(dx/dt)^2/2$ is $(dx/dt)^2/2 = 1/x + O(\mu)$. Equation (2.4.15b) gives $(dx/dt)^2/2 = (dx^*/dt^*)^2/2 = h^* + 1/x^* + \dots$. To match, we require the outer limit at $x \to 1$ to agree with the inner limit as $x^* \to \infty$. The above simplified matching, which can be justified using a matching variable, gives $h^* = 1$. We can now solve (2.4.15).

Restricting attention to the half-period with positive velocity, we have

$$-\frac{\sqrt{2}}{\mu}\int_t^\tau ds = \int_{x^*}^0 \sqrt{\frac{\xi}{1+\xi}}\, d\xi. \tag{2.4.16}$$

Setting $\tau = \tau_0 + \mu\tau_1 + \dots$, (2.4.16) integrates to

$$\frac{\sqrt{2}}{\mu}[t - (\tau_0 + \mu t_1 + \dots)] = -\sqrt{x^*(1+x^*)} + \log(\sqrt{x^*} + \sqrt{1+x^*}). \tag{2.4.17}$$

If we indicate the outer and inner expansions for $\sqrt{2}t(x, \mu)$ by

$$\sqrt{2}t(x, \mu) = h_0(x) + \mu h_1(x) + O(\mu^2), \tag{2.4.18}$$

$$\sqrt{2}t(x, \mu) = g_0(x^*) + \mu g_1(x^*) + O(\mu^2), \tag{2.4.19}$$

h_0 and h_1 are defined by (2.4.12), and (2.4.17) gives

$$g_0(x^*) = \sqrt{2}\tau_0, \tag{2.4.20}$$

$$g_1(x^*) = \sqrt{2}\tau_1 - \sqrt{x^*(1+x^*)} + \log(\sqrt{x^*} + \sqrt{1+x^*}). \tag{2.4.21}$$

The matching condition to $O(\mu)$ [with $x_\eta = (1-x)/\eta(\mu)$] is

$$\lim_{\substack{\mu\to 0\\ x_\eta \text{ fixed}}} \frac{1}{\mu}\left[h_0(1-\eta x_\eta) + \mu h_1(1-\eta x_\eta) - g_0\left(\frac{\eta x_\eta}{\mu}\right) - \mu g_1\left(\frac{\eta x_\eta}{\mu}\right)\right] = 0. \tag{2.4.22}$$

We calculate

$$h_0 + \mu h_1 = \frac{2}{3} - \eta x_\eta + \mu\left[\frac{5}{3} - \frac{1}{2}\log 2 + \frac{1}{2}\log\frac{\eta x_\eta}{2}\right] + O(\eta^2), \tag{2.4.23}$$

$$g_0 + \mu g_1 = \sqrt{2}\tau_0 + \mu\left[\sqrt{2}\tau_1 - \frac{\eta x_\eta}{\mu} - \frac{1}{2} + \log 2 + \frac{1}{2}\log\frac{\eta x_\eta}{\mu}\right] + O\left(\frac{\mu^2}{\eta}\right). \tag{2.4.24}$$

Thus, we must set

$$\sqrt{2}\tau_0 = \frac{2}{3}, \tag{2.4.25}$$

$$\sqrt{2}\tau_1 = \frac{13}{6} - \log 4 + \log \mu^{1/2}, \tag{2.4.26}$$

and all singular terms match. The neglected terms will vanish in (2.4.22) as long as $\eta^2/\mu \to 0$ and $\mu/\eta \to 0$, and this determines the overlap domain

$$\mu \ll \eta \ll \mu^{1/2}. \tag{2.4.27}$$

The composite expansion, uniformly valid to order μ on $0 \le x \le 1$ with $dx/dt > 0$, is

$$\begin{aligned}\sqrt{2}t =& \frac{2}{3}x^{3/2} + \mu\left[\frac{2}{3}x^{3/2} + x^{1/2} - \frac{1}{2}\log\frac{1+x^{1/2}}{1-x^{1/2}}\right] \\ &+ \mu[x^* - \sqrt{x^*(1+x^*)} + \log(\sqrt{x^*} + \sqrt{1+x^*}) \\ &- \log 2 + \frac{1}{2} - \frac{1}{2}\log x^*] + O(\mu^2).\end{aligned} \tag{2.4.28}$$

If we denote the solution (2.4.28) for $0 \le t \le \tau$ by

$$t = f(x, \mu), \tag{2.4.29a}$$

then the solution for $\tau \le t \le 2\tau$ is

$$t = 2\tau - f(x, \mu) \tag{2.4.29b}$$

by symmetry; periodicity then defines the solution for all times.

2.4.2 *A Model Example for the Stokes-Oseen Problem*

The mathematical problem of low Reynolds number flows past an object is outlined in Section 3.2.2. A model example for this flow was proposed and discussed in a seminar given by P.A. Lagerstrom in 1960. An early version of this study is included in the lecture notes [2.9]. A detailed analysis and a list of more recent references can be found in [2.8].

The model is a singular boundary-value problem in the sense that the form of the expansion comes not from a distinguished limit but from the behavior of the solution near the boundary points.

Consider the equation

$$\frac{d^2u}{dr^2} + \frac{1}{r}\frac{du}{dr} + u\frac{du}{dr} = 0 \tag{2.4.30}$$

with boundary conditions

$$u(\epsilon) = 0, \tag{2.4.31}$$

$$u(\infty) = 1. \tag{2.4.32}$$

Actually, a slightly more general version is studied in [2.8] where (2.4.30) reads

$$\frac{d^2u}{dr^2} + \frac{k}{r}\frac{du}{dr} + \alpha u\frac{du}{dr} + \beta\left(\frac{du}{dr}\right)^2 = 0, \tag{2.4.33}$$

where α and β are arbitrary non-negative constants and k is any real number.

Although the above vaguely resembles the radial momentum equation for a viscous incompressible flow, no physical correspondence is intended, as the relation of the model to the Navier-Stokes equations is strictly qualitative.

In this section, we will only study the special case (2.4.30) for which it is easy to give a rigorous demonstration of the validity of the asymptotic expansions and matching.

We want the behavior of the solution $u(r; \epsilon)$ as $\epsilon \to 0$. The problem also has an analogy with a steady cylindrically symmetric heat-flow problem outside a cylinder of radius ϵ. The temperature u at infinity equals 1 and the cylinder surface temperature is maintained at $u = 0$. The nonlinear term represents a heat source with strength/area proportional to $u(du/dr)$. In these coordinates, as $\epsilon \to 0$ the size of the cold cylinder shrinks to zero. The general shape of the expected solution is shown in Figure 2.4.3. From this, the first term of the limiting solution connected with the outer limit ($\epsilon \to 0$, r fixed) can be intuitively guessed as

$$u \to 1. \tag{2.4.34}$$

That is, the zero-size cold cylinder does not disturb the temperature field at all. Away from $r = \epsilon$, one might expect only small perturbations to this solution. Thus, an outer expansion of the form

$$u(r; \epsilon) = 1 + \mu_1(\epsilon)h_1(r) + \mu_2(\epsilon)h_2(r) + \dots \tag{2.4.35}$$

is assumed with the idea of satisfying the boundary conditions at infinity and matching to an inner expansion near $r = \epsilon$. The first term of (2.4.35) is not a good approximation in some neighborhood of $r = \epsilon$, and the orders in the asymptotic sequence $\mu_i(\epsilon)$ are not known a priori.

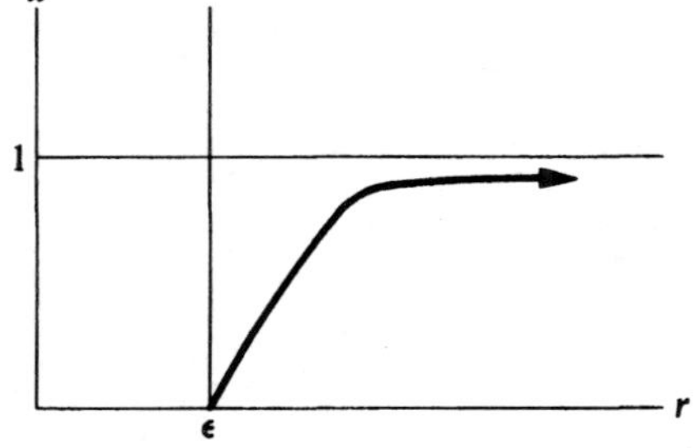

FIGURE 2.4.3. Stokes-Oseen Model Solution

The equations satisfied by h_1 and h_2 are

$$\frac{d^2h_1}{dr^2} + \left(\frac{1}{r} + 1\right)\frac{dh_1}{dr} = 0, \tag{2.4.36}$$

$$\frac{d^2h_2}{dr^2} + \left(\frac{1}{r} + 1\right)\frac{dh_2}{dr} = \begin{cases} 0 & \text{if } (\mu_1^2/\mu_2) \ll 1 \\ -h_1\frac{dh_1}{dr} & \text{if } (\mu_1^2/\mu_2) = O_s(1). \end{cases} \tag{2.4.37}$$

The boundary condition at infinity becomes

$$h_1(\infty) = 0, \qquad h_2(\infty) = 0. \tag{2.4.38}$$

An h_2 that is significantly different from h_1 appears only if $\mu_2 = O_s(\mu_1^2)$, and we can assume that $\mu_2 = \mu_1^2$ with the option of inserting h_1 terms of various orders larger than μ_1^2.

The solutions for h_1 and h_2 satisfying the condition at infinity are easily found. Equation (2.4.36) can be written

$$\frac{d}{dr}\left(re^r\frac{dh_1}{dr}\right) = 0,$$

so that

$$h_1(r) = A_1E_1(r), \tag{2.4.39}$$

where

$$E_1(r) = \int_r^\infty \frac{e^{-\rho}}{\rho}\,d\rho. \tag{2.4.40}$$

Here E_1 is the well-known exponential integral (sometimes denoted by $-E_i(-r)$) and has the following expansion (useful for matching) as $r \to 0$.

$$E_1(r) = -\log r - \gamma + r + O(r^2), \qquad \gamma = \text{Euler's const.} = 0.577215\ldots. \tag{2.4.41}$$

Similarly, we have

$$\frac{d}{dr}\left(re^r\frac{dh_2}{dr}\right) = A_1^2E_1(r). \tag{2.4.42}$$

Defining

$$E_n(r) = \int_r^\infty \frac{e^{-\rho}}{\rho^n}\,d\rho, \tag{2.4.43}$$

we can easily show that

$$\int_r^\infty E_n(\rho)d\rho = -rE_n(r) + E_{n-1}(r). \tag{2.4.44}$$

Hence, (2.4.42) becomes

$$\frac{dh_2}{dr} = -A_1^2\frac{e^{-2r}}{r} + A_1^2e^{-r}E_1(r) - A_2\frac{e^{-r}}{r}$$

and

$$h_2(r) = A_2E_1(r) + A_1^2\{2E_1(2r) - e^{-r}E_1(r)\}. \tag{2.4.45}$$

Use has been made of the result

$$\int_r^\infty e^{-\rho}E_1(\rho)d\rho = e^{-r}E_1(r) - E_1(2r). \tag{2.4.46}$$

The expansion of $h_2(r)$ as $r \to 0$ is, thus,

$$\begin{aligned} h_2(r) = &- (A_2 + A_1^2)\log r - (A_2 + A_1^2)\gamma - A_1^2 2\log 2 - A_1^2 r\log r \\ &+ [A_2 + (3-\gamma)A_1^2]r + O(r^2\log r). \end{aligned} \tag{2.4.47}$$

Now an inner expansion has to be constructed that can take care of the boundary condition $u = 0$ on $r = \epsilon$. A suitable inner coordinate is

$$r^* = \frac{r}{\epsilon},$$

and the limit process has r^* fixed as $\epsilon \to 0$. The form of this expansion is

$$u(r;\epsilon) = \nu_0(\epsilon)g_0(r^*) + \nu_1(\epsilon)g_1(r^*) + \nu_2(\epsilon)g_2 + \dots. \tag{2.4.48}$$

Again, choose $\nu_1 = \epsilon\nu_0^2$ so that the equation for g_1 has a forcing term; other terms similar to g_0 but of order intermediate to ν_0, ν_1 can be inserted in the expansion if necessary. For g_0 we have

$$\frac{d^2g_0}{dr^{*2}} + \frac{1}{r^*}\frac{dg_0}{dr^*} = 0, \qquad g_0(1) = 0, \tag{2.4.49}$$

so that

$$g_0 = B_0\log r^*.$$

Then, we have

$$\frac{d^2g_1}{dr^{*2}} + \frac{1}{r^*}\frac{dg_1}{dr^*} = -g_0\frac{dg_0}{dr^*} = -B_0^2\frac{\log r^*}{r^*}, \qquad g_1(1) = 0. \tag{2.4.50}$$

Integration of (2.4.50) yields

$$g_1(r^*) = B_1\log r^* - B_0^2(r^*\log r^* - 2r^* + 2). \tag{2.4.51}$$

For matching the inner and outer expansions, we introduce the matching variable $r_\eta = r/\eta(\epsilon)$ for a class of functions $\eta(\epsilon)$ contained in $\epsilon \ll \eta \ll 1$, and we consider the limit as $\epsilon \to 0$ with r_η fixed. In this limit $r = \eta r_\eta \to 0$, $r^* = (\eta/\epsilon)r_\eta \to \infty$. The first-order matching condition is

$$\lim_{\substack{\epsilon\to 0\\ r_\eta \text{ fixed}}}\left[1 + \mu_1(\epsilon)h_1(\eta r_\eta) + \dots - \nu_0(\epsilon)g_0\left(\frac{\eta}{\epsilon}r_\eta\right) + \dots\right] = 0$$

or

$$\lim_{\substack{\epsilon\to 0\\ r_\eta \text{ fixed}}}\left[1 + O(\mu_1\log\eta) - \nu_0(\epsilon)B_0\log\left(\frac{1}{\epsilon}\right) + O(\nu_0\log\eta)\right] = 0. \tag{2.4.52}$$

If we choose

$$\nu_0(\epsilon) = \frac{1}{\log(1/\epsilon)}, \qquad B_0 = 1, \tag{2.4.53}$$

the first terms are matched since $\mu_1 \to 0$ and $\nu_0 \to 0$.

Matching to the next order demands that

$$\lim_{\substack{\epsilon\to 0\\ r_\eta \text{ fixed}}} \frac{1}{\delta(\epsilon)}\left[1 + \mu_1(\epsilon)h_1(\eta r_\eta) + \mu_1^2(\epsilon)h_2(\eta r_\eta) + \ldots - \frac{1}{\log(1/\epsilon)} g_0\left(\frac{\eta}{\epsilon} r_\eta\right)\right.$$

$$\left. - \frac{\epsilon}{\log^2(1/\epsilon)} g_1\left(\frac{\eta}{\epsilon} r_\eta\right) + \ldots\right] = 0 \tag{2.4.54}$$

for some suitable $\delta(\epsilon) \to 0$. Writing out (2.4.54) and omitting the terms that have already been matched, we have

$$\frac{1}{\delta}\left\{ \mu_1(\epsilon)[A_1(-\log \eta r_\eta) - A_1\gamma + A_1\eta r_\eta + O(\eta^2)]\right.$$

$$+ \mu_1^2(\epsilon)[-(A_1^2 + A_2)\log \eta r_\eta - (A_1^2 + A_2)\gamma - A_1^2 2\log 2$$

$$- A_1^2 \eta r_\eta \log \eta r_\eta + O(\eta)] - \frac{1}{\log(1/\epsilon)}\left[\log \eta r_\eta\right]$$

$$- \frac{\epsilon}{\log^2(1/\epsilon)}\left[B_1 \log\left(\frac{\eta r_\eta}{\epsilon}\right) - \frac{\eta r_\eta}{\epsilon}\log\frac{\eta r_\eta}{\epsilon} + \frac{2\eta r_\eta}{\epsilon} - 2\right]$$

$$\left. + \ldots\right\} \to 0. \tag{2.4.55}$$

If we choose a sequence of functions $\delta(\epsilon)$ that vanish successively faster as $\epsilon \to 0$, we will eventually have $\delta(\epsilon) = O_s(\mu_1(\epsilon))$. At this point, the term $-(\mu_1/\delta)\log \eta r_\eta$, contributed by h_1 to (2.4.55), becomes singular unless we choose

$$\mu_1(\epsilon) = \frac{1}{\log(1/\epsilon)}, \qquad A_1 = -1 \tag{2.4.56}$$

so that this term matches with what remains from $\nu_0 g_0$. Next, we see that the term $-A_1\gamma$, now equal to γ, has no corresponding term in the inner expansion. This difficulty is easily resolved by inserting a term

$$\nu^+(\epsilon)g^+(r^*) = \nu^+(\epsilon)B^+ \log r^*$$

between g_0 and g_1 in the inner expansion, i.e., $\nu_1 \ll \nu^+ \ll \nu_0$. In fact, in order that $\nu^+ g^+$ generates a constant that matches with the γ term in h_1, we must choose

$$\nu^+ = \frac{1}{\log^2(1/\epsilon)}, \qquad B^+ = \gamma. \tag{2.4.57}$$

Rewriting (2.4.55) with the $\gamma^+ g^+$ term added and omitting the $\log \eta r_\eta$ terms already matched, we have

$$\frac{1}{\delta}\left\{\frac{1}{\log(1/\epsilon)}\left[\gamma - \eta r_\eta + O(\eta^2)\right] + \frac{1}{\log^2(1/\epsilon)}\left[(-1 + A_2)\log \eta r_\eta\right.\right.$$
$$\left. - (1 + A_2)\gamma - 2\log 2 - \eta r_\eta \log \eta r_\eta + O(\eta)\right] - \frac{\gamma}{\log^2(1/\epsilon)}\log\frac{\eta r_\eta}{\epsilon}$$
$$- \frac{\epsilon}{\log^2(1/\epsilon)}\left[B_1 \log\frac{\eta r_\eta}{\epsilon} - \left(\frac{\eta r_\eta}{\epsilon}\log\frac{\eta r_\eta}{\epsilon}\right) + \frac{2\eta r_\eta}{\epsilon} - 2\right]$$
$$\left. + \ldots\right\} \to 0. \tag{2.4.58}$$

The next term to match is the $\log \eta r_\eta$ term in h_2. We see that this term matches the $\log \eta r_\eta$ contribution of $\gamma^+ g^+$ if we set

$$A_2 = -\gamma - 1. \tag{2.4.59}$$

This procedure can evidently be continued, with the appropriate insertion of terms of intermediate order in the inner expansion. Summarizing the results, we see that the terms and orders of inner and outer expansions are as shown in the following table.

Order	Term
Outer	
1	1
$\mu_1 = \frac{1}{\log(1/\epsilon)}$	$h_1 = -E_1(r)$
$\mu_2 = \frac{1}{\log^2(1/\epsilon)}$	$h_2 = -(1 + \gamma)E_1(r) + E_1(2r) - e^{-r}E_1(r)$
Inner	
$\nu_0 = \frac{1}{\log(1/\epsilon)}$	$g_0 = \log r^*$
$\nu^+ = \frac{1}{\log^2(1/\epsilon)}$	$g^+ = \gamma \log r^*$
$\nu_1 = \frac{\epsilon}{\log^2(1/\epsilon)}$	$g_1 = B_1 \log r^* - (r^* \log r^* - 2r^* + 2)$

The term with g_1 is transcendentally small compared with the g_0 term; it does not enter the matching until further terms such as

$$\left[\frac{(\gamma^2 - 2\log 2)}{\log^2(1/\epsilon)}\right]$$

are matched by the introduction of an intermediate order $\upsilon^{++} = 1/[\log^3(1/\epsilon)]$. This serves to determine A_3 associated with the log term of h_3, etc. The effect of the nonlinearity never appears in the inner expansion, but this effect is in the far field of the outer expansion. The outer expansion contains the inner expansion. In the analogy with viscous flow, the inner or Stokes flow is not adequate for finding the solution and evaluating the skin friction $(du/dr)_0$, but the outer or Oseen flow is.

In this example, we can give a proof that our guess of the first term in the outer expansion is really correct. We can write (2.4.30) as

$$\frac{d}{dr}\left(r\frac{du}{dr}\right)+u\left(r\frac{du}{dr}\right)=0, \qquad u(\epsilon)=0,\ u(\infty)=1. \tag{2.4.60}$$

If we regard (2.4.60) as a linear problem for $r(du/dr)$ and integrate, the problem can be formulated as an integral equation:

$$u(r;\epsilon)=1-\frac{G(r;\epsilon)}{G(\epsilon;\epsilon)}, \tag{2.4.61}$$

where

$$G(r;\epsilon)=\int_r^\infty \frac{[\exp - \int_\epsilon^\rho u(\sigma;\epsilon)d\sigma]}{\rho}\,d\rho.$$

Equation (2.4.60) is invariant under a group of transformations if $ru \sim$ const. and can be reduced to the first-order system

$$\frac{dt}{ds}=\frac{t(1-s)}{t+s}, \qquad \frac{dr}{r}=\frac{dt}{t(1-s)}, \tag{2.4.62}$$

where

$$s=ru, \qquad t=r^2\frac{du}{dr}.$$

From a study of the integral curves of (2.4.62), we can conclude that the only possible solution of (2.4.60) satisfying the boundary condition has $s \geq 0$, and hence $u \geq 0$. These phase-plane considerations also can be used to prove the existence of a unique solution. It follows from (2.4.61) that

$$0 \leq u \leq 1, \qquad \epsilon \leq r \leq \infty. \tag{2.4.63}$$

Thus, we have

$$G(\epsilon;\epsilon)\geq \int_\epsilon^\infty \frac{e^{-(\rho-\epsilon)}}{\rho}\,d\rho \geq \int_\epsilon^\infty \frac{e^{-\rho}}{\rho}\,d\rho \quad \text{or} \quad G(\epsilon;\epsilon)\geq E_1(\epsilon). \tag{2.4.64}$$

Now we can write

$$G(r;\epsilon)=\int_r^{r_0} \frac{[\exp - \int_\epsilon^\rho u(\sigma;\epsilon)d\sigma]}{\rho}\,d\rho+\int_{r_0}^\infty \frac{[\exp - \int_\epsilon^\rho u(\sigma;\epsilon)d\sigma]}{\rho}\,d\rho.$$

Thus, we have

$$G(r;\epsilon)\leq \log\frac{r_0}{r}+\frac{1}{r_0u(r_0;\epsilon)}\int_{r_0}^\infty u(r;\epsilon)\exp\left[-\int_\epsilon^\rho u(\sigma;\epsilon)d\sigma\right]d\rho \tag{2.4.65}$$

if we use the fact that $u(r)$ is monotonic $(r > 0)$.

Integrating gives

$$G(r;\epsilon) \leq \log\frac{r_0}{r} + \frac{1}{r_0 u(r_0;\epsilon)}\left[\exp - \int_\epsilon^{r_0} u(\rho;\epsilon)d\rho\right],$$

$$G(r;\epsilon) \leq \log\frac{r_0}{r} + \frac{1}{r_0 u(r_0;\epsilon)}.$$

Now, for any given δ and all $\epsilon < \delta$, it follows that $u(r_0;\epsilon) > u(r_0,\delta)$, so that

$$G(r;\epsilon) \leq \log\frac{r_0}{r} + \frac{1}{r_0 u(r_0;\epsilon)}, \qquad \epsilon < \delta. \tag{2.4.66}$$

Thus, in (2.4.61) the outer limit $\epsilon \to 0$, r fixed shows that

$$\frac{G(r;\epsilon)}{G(\epsilon;\epsilon)} \leq \frac{1}{E_1(\epsilon)}\left\{\log\frac{r_0}{r} + \frac{1}{r_0 u(r_0;\delta)}\right\} \to 0, \qquad \epsilon \to 0. \tag{2.4.67}$$

The problem of finding the viscous, incompressible flow past a circular cylinder relies on considerations such as these, although a rigorous proof has not been provided for that case. The considerations given earlier can actually be extended to demonstrate the overlapping of the two expansions used.

Problems

1. Consider the initial value problem for (2.4.2):

$$x(0;\mu) = 1 - \mu^{1/2}, \tag{2.4.68}$$

$$\frac{dx}{dt}(0;\mu) = \sqrt{2(2-c^2)}\mu^{1/4}, \tag{2.4.69}$$

where c is a constant with $0 \leq c \leq \sqrt{2}$. These initial conditions correspond to starting near the equilibrium point with a small positive velocity. What are the appropriate inner variables? Calculate the inner and outer expansions to order μ and match to find the half-period for this case. Compare your result with the exact expression for the half-period.

2. Consider the following "collision" problem:

$$\frac{d^2y}{dt^2} + y - \frac{\epsilon}{(y-1)^2} = 2\sin t \tag{2.4.70}$$

with zero initial conditions $y(0;\epsilon) = \frac{dy}{dt}(0;\epsilon) = 0$. Note that for $\epsilon = 0$, y describes the linear resonant behavior

$$y(t;0) \equiv h_0(t) = \sin t - t\cos t. \tag{2.4.71}$$

But, as $h_0 \to 1$ (collision), the term ignored in (2.4.70) becomes singular. Thus, the outer expansion

$$y(t;\epsilon) = h_0(t) + \epsilon h_1(t) + O(\epsilon^2) \tag{2.4.72}$$

obtained in the limit $\epsilon \to 0$, t fixed cannot be valid when $h_0 \to 1$, i.e., $t \to \pi/2$. We wish to calculate the uniformly valid asymptotic expansion of the solution $y(t;\epsilon)$ over $0 \le t \le T(\epsilon)$, where $T(\epsilon)$ is the collision time, i.e., $y(T(\epsilon);\epsilon) = 1$.

a. Show that the outer expansion is given by

$$y(t;\epsilon) = \sin t - t\cos t + \epsilon \int_0^t \frac{\sin(t-\tau)}{[1-\sin\tau + \tau\cos\tau]^2}\,d\tau + O(\epsilon^2) \tag{2.4.73}$$

and refer to Problem 4 of Sec. 1.2 for the singular behavior of $h_1(t)$ as $t \to \pi/2$.

b. For t close to $T(\epsilon)$, define the new rescaled variables $y^*(t^*;\epsilon) = (1-y)/\epsilon$ where $t^* = [t - T(\epsilon)]/\epsilon$. Now consider the inner expansion

$$y^*(t^*;\epsilon) = g_0(t^*) + O(\epsilon) \tag{2.4.74}$$

and show that g_0 has the first integral

$$\frac{1}{2}\left(\frac{dg_0}{dt^*}\right)^2 - \frac{1}{g_0} = c_0 = \text{const.} \tag{2.4.75}$$

c. Match the outer and inner expansions for (dy/dt) to $O(1)$ to show that $c_0 = \pi^2/8$.

d. Integrate the result in (2.4.75) with $c_0 = \pi^2/8$ to calculate

$$t^* = -\frac{2}{\pi} g_0\left(1 + \frac{8}{\pi^2 g_0}\right)^{1/2} + \frac{16}{\pi^3}\log\left\{\left[g_0\left(1+\frac{8}{\pi^2 g_0}\right)\right]^{1/2} + g_0^{1/2}\right\} + \frac{8}{\pi^3}\log\frac{\pi^2}{8}. \tag{2.4.76}$$

e. Use (2.4.76) to compute the following asymptotic behavior for g_0 as $t^* \to -\infty$:

$$g_0 = -\frac{\pi}{2}t^* + \frac{4}{\pi^2}\log(-t^*) + \frac{4}{\pi^2}\left(\log\frac{\pi^3}{4} - 1\right) + O\left[\frac{1}{t^*}\log(-t^*)\right]. \tag{2.4.77}$$

f. Match the outer and inner expansions of y to $O(\epsilon)$ to obtain

$$T(\epsilon) = \frac{\pi}{2} + \frac{8}{\pi^3}\epsilon\log\epsilon + \epsilon C + o(\epsilon), \tag{2.4.78}$$

where the constant C is

$$C = \frac{8}{\pi^3}\left(2 - \log\frac{\pi^4}{8}\right) - \frac{2}{\pi}\int_0^{\pi/2}\left\{\frac{\sin\sigma}{[1-\cos\sigma + (\frac{\pi}{2}-\sigma)\sin\sigma]^2} - \frac{4}{\pi^2\sigma}\right\}d\sigma. \tag{2.4.79}$$

3. Consider (2.4.33) for the special case $k = 0$, $\alpha = 1$, $\beta = 0$ and calculate the exact solution. Show that the outer limit, $\epsilon \to 0$, r fixed, is not $u = 1$. Show

also that the incorrect assumption of an outer expansion of the form of (2.4.35) cannot be made to match with the inner expansion.

4. Parallel the discussion of Section 2.4.2 for the case $k = 2, \alpha = 1, \beta = 1$ in (2.4.33).

2.5 Higher-Order Example: Beam String

In this section, an elementary example of a higher-order equation is constructed in order to show that the ideas of the previous sections have a natural and general validity.

The engineering theory of an elastic beam with tension that supports a given load distribution leads to the following differential equation, when it is assumed that the deflection W is small (linearized theory):

$$EI\frac{d^4W}{dX^4} - T\frac{d^2W}{dX^2} = P(X), \qquad 0 \le X \le L. \tag{2.5.1}$$

Here

E is the constant modulus of elasticity,
I is the constant moment of inertia of cross section about the neutral axis, and
T is the constant external tension,
X is the coordinate along the beam,
W is the deflection of the neutral axis, and
$P(X)$ is the external load per length on the beam.

In the engineering theory of bending, see, for example, [2.17], a model is made for the deformation of the beam under load in which plane cross sections of the beam remain plane under load. The tension and compression forces due to bending that act along the beam are computed by Hooke's law from the stretching of the fibers; the neutral axis is unstressed. Adjacent sections exert a bending moment M on each other proportional to the beam curvature. For small deflections, we have

$$M(X) = -EI\frac{d^2W}{dX^2}. \tag{2.5.2}$$

A vertical shear V at these sections produces a couple to balance the bending moment:

$$V = \frac{dM}{dX}. \tag{2.5.3}$$

The effect of this shear in supporting the external load is expressed in (2.5.1) by the fourth derivative term. The load is carried by the tension in the structure in the usual way, in which a string or cable supports an external load, $[T(d^2W)/(dX^2)]$. When the deflection is known, the stresses of interest can be calculated.

Singular perturbation problems arise when the effect of bending rigidity is relatively small in comparison to the tension. In general, when a more complicated model of a physical phenomenon (for example, beam vs. string) is constructed, the order of the differential equations is raised. Correspondingly, the nature of the boundary conditions at the ends is more complicated; due to the higher order of the equations, more conditions are needed. For the string problem, for example, it is sufficient to prescribe the deflection. In a beam problem, the mode of support must also be given. The loss of a boundary condition in passage from the beam-string to the string implies the existence of a boundary layer near the support, a local region in which bending rigidity is important. A similar phenomenon can occur under a region of rapid change of the load or near a concentrated load. Various types of boundary conditions can be used to represent the end of a beam of which the following are the most common and important:

1. Pin-end: no restoring moment M applied at the end, $d^2W/dX^2 = 0$; deflection prescribed, for example, $W = 0$.
2. Built-in end: slope at end prescribed, for example, $dW/dX = 0$; deflection prescribed, for example, $W = 0$;
3. Free end: no bending moment exerted on end, $d^2W/dX^2 = 0$; no shear exerted on end, $dW/dX = 0$.

Consider now the typical problem for a beam with built-in ends under the distributed load $P(X)$. The problem can be expressed in suitable dimensionless coordinates by measuring lengths in terms of L and using a characteristic load density $\mathcal{P}$ so that

$$P(X) = \mathcal{P}p(x), \qquad 0 \le x \le 1, \tag{2.5.4}$$

where $x = X/L$. The deflection is conveniently measured in terms of that characteristic of the string alone,

$$w(x) = \left(\frac{T}{\mathcal{P}L^2}\right) W(X), \tag{2.5.5}$$

and the resulting dimensionless equation and boundary conditions are

$$\epsilon \frac{d^4w}{dx^4} - \frac{d^2w}{dx^2} = p(x), \qquad 0 \le x \le 1; \tag{2.5.6a}$$

$$w(0) = \frac{dw}{dx}(0) = w(1) = \frac{dw}{dx}(1) = 0. \tag{2.5.6b}$$

The small parameter ϵ of the problem is

$$\epsilon = \frac{EI}{TL^2} \tag{2.5.7}$$

and measures the relative importance of the bending rigidity in comparison to the tension.

Next, we construct the inner and outer expansions. For the outer expansion ($\epsilon \to 0$, x fixed), we expect the first term to be independent of ϵ, and we write

$$w(x;\epsilon) = h_0(x) + \nu_1(\epsilon)h_1(x) + \nu_2(\epsilon)h_2(x) + \dots. \tag{2.5.8}$$

The corresponding differential equations are

$$-\frac{d^2h_0}{dx^2} = p(x),$$

$$-\frac{d^2h_1}{dx^2} = \begin{cases} -\dfrac{d^4h_0}{dx^4} = p''(x) & \text{if } \epsilon/\nu_1 = O_s(1) \\ 0 & \text{if } \epsilon/\nu_1 \ll 1. \end{cases} \tag{2.5.9}$$

There are no boundary conditions for the h_i, but the constants of integration in the solutions must be obtained by matching with the boundary layers at each end. The solution for $h_0(x)$ is, thus,

$$h_0(x) = B_0 + A_0 x - \int_0^x (x-\lambda)p(\lambda)d\lambda. \tag{2.5.10}$$

For purposes of matching later, it is useful to have the series expansions of $h_0(x)$ near $x = 0$ and $x = 1$.

Near $x = 0$,

$$\begin{aligned} h_0(x) &= h_0(0) + xh_0'(0) + \frac{x^2}{2}h_0''(0) + \dots \\ &= B_0 + A_0 x - p(0)\frac{x^2}{2!} - p'(0)\frac{x^3}{3!} + O(x^4). \end{aligned} \tag{2.5.11}$$

Near $x = 1$,

$$\begin{aligned} h_0(x) = B_0 + A_0 - \int_0^1 (1-\lambda)p(\lambda)d\lambda + \left\{A_0 - \int_0^1 p(\lambda)d\lambda\right\}(x-1) \\ - p(1)\frac{(x-1)^2}{2!} - p'(1)\frac{(x-1)^3}{3!} + O((x-1)^4). \end{aligned} \tag{2.5.12}$$

A suitable boundary-layer coordinate x^* is chosen by the requirement that the bending and tension terms are of the same order of magnitude near $x = 0$. This gives

$$x^* = \frac{x}{\sqrt{\epsilon}}. \tag{2.5.13}$$

The corresponding asymptotic expansion near $x = 0$ is

$$w(x;\epsilon) = \mu_0(\epsilon)g_0(x^*) + \mu_1(\epsilon)g_1(x^*) + \dots. \tag{2.5.14}$$

Equation (2.5.6) thus becomes

$$\frac{1}{\epsilon}\left\{\mu_0\frac{d^4g_0}{dx^{*4}} + \mu_1\frac{d^4g_1}{dx^{*4}} + \dots\right\} - \frac{1}{\epsilon}\left\{\mu_0\frac{d^2g_1}{dx^{*2}} + \mu_1\frac{d^2g_1}{dx^{*2}} + \dots\right\}$$
$$= p(0) + \sqrt{\epsilon}x^*p'(0) + \dots. \tag{2.5.15}$$

Two possibilities arise: either $\mu_0/\epsilon \to \infty$ or $\mu_0/\epsilon \to 1$. It can be shown that, in general, the second possibility does not allow matching, and thus only the first is considered here. The effect of the external load, then, does not appear in the first boundary-layer equation but enters at first only through the matching. Of course, it has to be verified that the assumption $\mu_0/\epsilon \to \infty$ is correct after μ_0 is found from the matching. Thus, we have

$$\frac{d^4 g_0}{dx^{*2}} - \frac{d^2 g_0}{dx^{*2}} = 0. \tag{2.5.16}$$

Both boundary conditions at $x^* = 0$ are to be satisfied by g_0:

$$\frac{dg_0}{dx^*} = g_0 = 0, \qquad \text{at } x^* = 0. \tag{2.5.17}$$

Using the fact that exponential growth (e^{x^*}) cannot match as $x^* \to \infty$, and taking into account the boundary conditions, we obtain a solution with one arbitrary constant,

$$g_0(x^*) = C_0\{x^* - 1 + e^{-x^*}\}. \tag{2.5.18}$$

For matching near $x^* = 0$, we introduce

$$x_\eta = \frac{x}{\eta(\epsilon)} \tag{2.5.19}$$

for a class of functions $\eta(\epsilon)$ contained in $\sqrt{\epsilon} \ll \eta(\epsilon) \ll 1$ so that

$$x^* = \frac{\eta}{\sqrt{\epsilon}} x_\eta \to \infty, \qquad x = \eta x_\eta \to 0. \tag{2.5.20}$$

Matching to $O(\sqrt{\epsilon})$ near $x = 0$ takes the form

$$\lim_{\substack{\epsilon\to 0\\ x_\eta \text{ fixed}}} \frac{1}{\sqrt{\epsilon}} \left\{ h_0(\eta x_\eta) + \nu_1(\epsilon) h_1(\eta x_\eta) + \ldots - \mu_0(\epsilon) g_0\left(\frac{\eta}{\sqrt{\epsilon}} x_\eta\right) \right.$$
$$\left. - \mu_1(\epsilon) g_1\left(\frac{\eta x_\eta}{\sqrt{\epsilon}}\right) + \ldots \right\} = 0. \tag{2.5.21}$$

Using the expansion (2.5.11) and (2.5.18), we find that the matching condition to $O(\sqrt{\epsilon})$ is

$$\lim_{\substack{\epsilon\to 0\\ x_\eta \text{ fixed}}} \left\{ \frac{B_0}{\sqrt{\epsilon}} + A_0 \frac{\eta x_\eta}{\sqrt{\epsilon}} + \ldots - \frac{\mu_0(\epsilon)}{\sqrt{\epsilon}} C_0 \left(\frac{\eta}{\sqrt{\epsilon}} x_\eta - 1 + \exp \frac{-\eta x_\eta}{\sqrt{\epsilon}} \right) \right.$$
$$\left. + \ldots + \frac{\nu_1(\epsilon)}{\sqrt{\epsilon}} h_1(\eta x_\eta) \right\} = 0. \tag{2.5.22}$$

The term linear in x_η dominates g_0, so that matching is only possible if

$$B_0 = 0 \tag{2.5.23}$$

and

$$\mu_0 = \sqrt{\epsilon}, \qquad A_0 = C_0. \tag{2.5.24}$$

This verifies the fact that $\mu_0/\epsilon \to \infty$.

Another point can be noticed from (2.5.22). The term $O(1)$ in g_0 cannot be matched except by a suitable h_1. That is, we must have $\nu_1(\epsilon) = \sqrt{\epsilon}$, and h_1 satisfies the equation of an unloaded string (see (2.5.9)):

$$\frac{d^2 h_1}{dx^2} = 0. \tag{2.5.25}$$

The solution is

$$h_1 = B_1 + A_1 x. \tag{2.5.26}$$

Completing the matching to order $\sqrt{\epsilon}$ in (2.5.22), we have

$$B_1 = -C_0, \tag{2.5.27}$$

and the overlap domain is $\sqrt{\epsilon}|\log \epsilon| \ll \eta(\epsilon) \ll 1$.

The final determination of the unknown constants depends on the application of similar considerations at the other end of the beam. Summarizing, for the outer expansion we have thus far

$$w(x;\epsilon) = h_0(x) + \sqrt{\epsilon} h_1(x) + \ldots,$$

where

$$h_0(x) = C_0 x - \int_0^x (x-\lambda)p(\lambda)d\lambda, \qquad h_1(x) = -C_0 + A_1 x. \tag{2.5.28}$$

The first approximation h_0 satisfies a zero-deflection boundary condition at $x = 0$ as might have been expected from physical consideration. Applying the same reasoning at $x = 1$, we find, for example, that

$$C_0 = \int_0^1 (1-\lambda)p(\lambda d\lambda \equiv -M^{(1)}. \tag{2.5.29}$$

Here $M^{(1)}$ represents the total moment of the applied load about $x = 1$. The result of (2.5.29) is now verified by detailed matching, and the unknown constant A_1 is also found.

For the boundary layer near $x = 1$, the coordinate

$$x^+ = \frac{(x-1)}{\sqrt{\epsilon}} \tag{2.5.30}$$

is used, and the boundary-layer expression is

$$w(x;\epsilon) = \sqrt{\epsilon} f_0(x^+) + \ldots. \tag{2.5.31}$$

The equation for f_0 is the same as that for g_0, so that the solution satisfying the boundary condition at $x = 1$, $x^+ = 0$ is

$$f_0(x^+) = D_0\{x^+ + 1 - e^{x^+}\}. \tag{2.5.32}$$

Exponential growth as $x^+ \to -\infty$ is ruled out near $x = 1$.

The matching variable x_ξ is defined by

$$x_\xi = \frac{x-1}{\xi(\epsilon)} < 0, \tag{2.5.33}$$

where

$$x = 1 + \xi x_\xi \to 1, \qquad x^+ = \left(\frac{\xi}{\sqrt{\epsilon}}\right) x_\xi \to -\infty.$$

The expansion (2.5.12) of h_0 near $x = 1$ is now

$$h_0(x) = C_0 + M^{(1)} + (C_0 - k)(x - 1) + \dots, \tag{2.5.34}$$

where

$$k = \int_0^1 p(\lambda)d\lambda = \text{ total load on beam.}$$

The matching condition to $O(\sqrt{\epsilon})$ near $x = 1$ is, thus,

$$\lim_{\substack{\xi \to 0 \\ x_\xi \text{ fixed}}} \frac{1}{\sqrt{\epsilon}} \Big\{ C_0 + M^{(1)} + (C_0 - k)\xi x_\xi + \dots + \sqrt{\epsilon}(-C_0 + A_1 + A_1 \xi x_\xi) + \dots$$

$$-\sqrt{\epsilon} D_0 \left(\frac{\xi}{\sqrt{\epsilon}} x_\xi + 1 - \exp\left(\frac{\xi x_\xi}{\sqrt{\epsilon}}\right)\right) + \dots \Big\} = 0. \tag{2.5.35}$$

The terms of $O(\epsilon^{-1/2})$ must vanish, so we have

$$C_0 = -M^{(1)},$$

and the vanishing of the terms linear in x_ξ gives

$$C_0 - k = D_0, \tag{2.5.36}$$

that is,

$$D_0 = \int_0^1 (1 - \lambda) p d\lambda - \int_0^1 p\, d\lambda = -\int_0^1 \lambda p(\lambda) d\lambda \equiv -M^{(0)}. \tag{2.5.37}$$

Further, the matching of the constant term in f_0 yields

$$A_1 = C_0 + D_0 = -[M^{(1)} + M^{(0)}]. \tag{2.5.38}$$

Thus, finally, the three expansions are fully determined to the orders considered.

$$\begin{aligned} w(x; \epsilon) &= -\sqrt{\epsilon} M^{(1)} \{x^* - 1 + e^{-x^*}\} + \dots \text{ near } x = 0, \\ w(x; \epsilon) &= -M^{(1)} x - \int_0^x (x - \lambda) p(\lambda) d\lambda \\ &\quad + \sqrt{\epsilon}\{M^{(1)} - (M^{(0)} + M^{(1)})x\} + \dots \text{ away from the ends,} \\ w(x; \epsilon) &= -\sqrt{\epsilon} M^{(0)} \{x^+ + 1 - e^{x^+}\} + \dots \text{ near } x = 1. \end{aligned} \tag{2.5.39}$$

The uniformly valid approximation to $O(\sqrt{\epsilon})$ is constructed, as before, by adding all three expansions and subtracting the common part, which has canceled out identically in the matching. Thus, we have

$$\begin{aligned} w_{\text{uv}} = -M^{(1)} x - \int_0^x (x - \lambda) p(\lambda) d\lambda + \sqrt{\epsilon}\{M^{(0)} e^{(x-1)/\sqrt{\epsilon}} - M^{(1)} e^{-x/\sqrt{\epsilon}} \\ + M^{(1)} - (M^{(0)} + M^{(1)})x\} + o(\sqrt{\epsilon}). \end{aligned} \tag{2.5.40}$$

The uniformly valid expansion is again recognized as having the form of a composite expansion,

$$\sum_{n=0} \epsilon^{n/2}\{h_n(x) + G_n(x^*) + F_n(x^+)\},$$

where the F_n and G_n decay exponentially.

From this expansion, the deflection curve, bending moment, and stresses are easily calculated. For example, the bending moment distribution near $x = 0$, proportional to d^2w/dx^2, comes only from the boundary-layer term. Near $x = 0$, the moment (and stress) decay exponentially:

$$M = -\frac{EI\mathcal{P}}{T}\frac{d^2w}{dx^2} = \frac{M^{(1)}}{\sqrt{\epsilon}}\frac{EI\mathcal{P}}{T}e^{-x/\sqrt{\epsilon}}$$

$$= M^{(1)}\sqrt{\frac{EI}{T}}L\mathcal{P}\exp\left(-X\sqrt{\frac{T}{EI}}\right). \tag{2.5.41}$$

Problem

1. In [2.17] (pp. 277ff), the deflection theory of suspension bridges is discussed, and the differential equation for the additional cable deflection $w(X)$ (or beam deflection) over that due to dead load is obtained. This equation is of the form studied in this section,

$$EI\frac{d^4W}{dX^4} - (T+\tau)\frac{d^2W}{dX^2} = P - Q\frac{\tau}{T}, \qquad 0 \le x \le L, \tag{2.5.42}$$

where

$$\begin{aligned} T &= \text{dead load cable tension,} \\ \tau &= \text{increase in cable tension due to live load,} \\ Q(X) &= \text{dead load per length, and} \\ P(X) &= \text{live load per length.} \end{aligned}$$

The increase of the main-cable tension depends on the stretching of the cable and is thus related to W. Assuming a linear elasticity for the cable (and small-cable slopes), we find that

$$\frac{TL_c}{E_cA_c}\tau = \int_0^L W(X)Q(X)dX, \tag{2.5.43}$$

where L_c, E_c, and A_c are the original cable length, modulus of elasticity, and cross-sectional area, respectively.

According to the usual boundary conditions, the truss (or beam) is considered pin-ended.

Using boundary-layer theory, calculate the deflection $W(X)$ due to a uniform dead load, $Q = \text{const.}$, and a concentrated live load, $P = P_0S(X - L/2)$, at the center. Use either matched or composite expansions. Indicate what kind of problem must be solved to find the additional tension.

References

2.1. R.C. Ackerberg and R.E. O'Malley Jr., "Boundary layer problems exhibiting resonance," *Stud. Appl. Math.*, **49**, 1970, pp. 277–295.

2.2. G.F. Carrier, M. Krook, and C.E. Pearson, *Functions of a Complex Variable, Theory and Technique*, McGraw-Hill Book Company, New York, 1966.

2.3. G.F. Carrier and C.E. Pearson, *Ordinary Differential Equations*, Blaisdell, Waltham, MA, 1968.

2.4. H.C. Corben and P. Stehle, *Classical Mechanics*, 2nd ed. Wiley, New York, 1960.

2.5. H. Goldstein, *Classical Mechanics*, 2nd ed., Addison-Wesley, Reading, MA, 1980.

2.6. J. Grasman and B.J. Matkowsky, "A variational approach to singularly perturbed boundary value problems for ordinary and partial differential equations with turning points," *SIAM J. Appl. Math.*, **32**, 1977, pp. 588–597.

2.7. J. Kevorkian and J.E. Lancaster, "An asymptotic solution for a class of periodic orbits of the restricted three-body problem," *Astron. J.*, **73**, 1968, pp. 791–806.

2.8. P.A. Lagerstrom, *Matched Asymptotic Expansions, Ideas and Techniques*, Springer-Verlag, 1988.

2.9. P.A. Lagerstrom, "Méthodes asymptotiques pour l'étude des équations de Navier-Stokes," *Lecture Notes*, Institut Henri Poincaré, Paris, 1961. Translated by T.J. Tyson, California Institute of Technology, Pasadena, California, 1965.

2.10. P.A. Lagerstrom and J. Kevorkian, "Nonplanar earth-to-moon trajectories in the restricted three-body problem," *AIAA J.*, **4**, 1966, pp. 149–152.

2.11. P.A. Lagerstrom and J. Kevorkian, "Earth-to-moon trajectories with minimal energy," *J. Mécanique*, **2**, 1963, pp. 493–504.

2.12. P.A. Lagerstrom and J. Kevorkian, "Earth-to-moon trajectories in the restricted three-body problem," *J. Mécanique*, **2**, 1963, pp. 189–218.

2.13. P.A. Lagerstrom and J. Kevorkian, "Matched conic approximation to the two fixed force-center problem," *Astron. J.*, **68**, 1963, pp. 84–92.

2.14. C.G. Lange, "On spurious solutions of singular perturbation problems," *Stud. Appl. Math.*, **68**, 1983, pp. 227–257.

2.15. R. Srinivasan, "A variational principle for the Ackerberg-O'Malley resonance problem," *Stud. Appl. Math.*, **79**, 1988, pp. 271–289.

2.16. V. Szebehely, *Theory of Orbits, the Restricted Problem of Three Bodies*, Academic Press, New York, 1967.

2.17. T. von Karman, and M. Biot, *Mathematical Methods in Engineering*, McGraw-Hill Book Co., New York, 1940.

3

Limit Process Expansions for Partial Differential Equations

In this chapter, the methods developed in Chapter 2 are applied to partial differential equations. The plan is the same as for the cases of ordinary differential equations discussed earlier. First, we discuss the very simplest case in which a singular perturbation problem arises; that of a second-order equation that becomes a first-order one in the limit $\epsilon \to 0$. Following this, various more complicated physical examples of boundary-layer theory in fluid mechanics are discussed. The final section deals with a variety of physical examples for singular boundary-value problems.

3.1 Limit Process Expansions for Second-Order Partial Differential Equations

In this section, a study is made of the simplest problems for partial differential equations that lead to boundary layers; that is, problems that are singular in the sense of Section 2.2. We base the discussion as much as possible on the mathematical situation. The simplest nontrivial case is that of a second-order equation that drops to a first-order one as the small parameter $\epsilon \to 0$. It is clear that some of the boundary data cannot be satisfied by the limit equation, so that boundary layers (in general) occur. The analogous case of a first-order partial differential equation reducing to a zero-order equation as $\epsilon \to 0$, is, by the theory of characteristics for first-order equations, equivalent to a problem in ordinary differential equations and is not discussed here.

Consider

$$\epsilon \left\{ \alpha_{11} \frac{\partial^2 u}{\partial x^2} + 2\alpha_{12} \frac{\partial^2 u}{\partial x \partial y} + \alpha_{22} \frac{\partial^2 u}{\partial y^2} \right\} = a \frac{\partial u}{\partial x} + b \frac{\partial u}{\partial y} \tag{3.1.1}$$

with constant coefficients. In the case where the coefficients are functions of (x, y), the solutions can be expected to behave locally in the same way as the constant coefficient equation. However, nonlinearities, especially in the lower-order operator, usually introduce new effects.

A boundary- or initial-value problem for (3.1.1), which leads to a unique solution $u(x, y; \epsilon)$, is considered. The kind of boundary-value problem that makes sense for (3.1.1) depends on the type of the equation that is a property only of the highest-order differential operator appearing in that equation:

$$L_2 u \equiv \alpha_{11} \frac{\partial^2 u}{\partial x^2} + 2\alpha_{12} \frac{\partial^2 u}{\partial x \partial y} + \alpha_{22} \frac{\partial^2 u}{\partial y^2}. \tag{3.1.2}$$

The classification of the operator L_2 and some significant properties are summarized next. For a detailed discussion, see chapter 4 of [3.15].

I *Elliptic Type*

No real characteristics exist. A point-disturbance solution of $L_2 u = 0$ influences the entire space. One boundary condition for u, its normal derivative, or a linear combination may be prescribed on a closed boundary. The simplest case and canonical form is the Laplace equation $u_{xx} + u_{yy} = 0$.

II *Hyperbolic Type*

Real characteristic curves exist in the xy-plane and form a coordinate system. A point-disturbance solution of $L_2 u = 0$ influences a restricted part of the domain; that bounded by the characteristic curves of the (x, y) space. The direction of propagation (future) is not necessarily implicit in the equation and must be assigned. Two initial conditions on a spacelike arc define a solution in a domain bounded by characteristics. One boundary condition is assigned on a timelike arc. A characteristic boundary-value problem assigns a compatible boundary condition on a characteristic curve. The simplest case is the wave equation $u_{xx} - u_{yy} = 0$.

III *Parabolic Type*

This is a type intermediate to I and II for which the real characteristics coalesce. Half the space is influenced by a point disturbance. One initial condition on a characteristic arc or one boundary condition is prescribed. The simplest case is $u_{xx} = u_y$.

If $\phi(x, y) =$ const. is the equation of a family of characteristic curves for L_2, the equation of these curves is

$$\alpha_{11} \left(\frac{\partial \phi}{\partial x} \right)^2 + 2\alpha_{12} \left(\frac{\partial \phi}{\partial x} \right) \left(\frac{\partial \phi}{\partial y} \right) + \alpha_{22} \left(\frac{\partial \phi}{\partial y} \right)^2 = 0 \tag{3.1.3}$$

or, if the slope of a characteristic $\zeta = dy/dx = -\phi_x/\phi_y$ (on $\phi =$ const.) is introduced, we have

$$\alpha_{11}\zeta^2 - 2\alpha_{12}\zeta + \alpha_{22} = 0. \tag{3.1.4}$$

The classification of (3.1.1) thus depends only on the discriminant, and we have three possibilities:

I. Elliptic: $\alpha_{12}^2 - \alpha_{11}\alpha_{22} < 0$, no real roots,

II. Hyperbolic: $\alpha_{12}^2 - \alpha_{11}\alpha_{22} > 0$, two real directions $\zeta_\pm$,
III. Parabolic: $\alpha_{12}^2 - \alpha_{11}\alpha_{22} = 0$, double root $\zeta = \alpha_{12}/\alpha_{11}$.
In the next three subsections, a separate discussion is given for each type of equation.

3.1.1 Linear Elliptic Equations, $\alpha_{12}^2 - \alpha_{11}\alpha_{22} < 0$

Since α_{11}, α_{22} must be of the same algebraic sign, let $\alpha_{11} > 0, \alpha_{22} > 0$, and note that

$$|\alpha_{12}| < \sqrt{\alpha_{11}\alpha_{22}}; \qquad \alpha_{11}\alpha_{22} > 0. \tag{3.1.5}$$

Consider first an interior boundary-value problem (Dirichlet problem) with $u = u_B(P_B)$, a prescribed continuous function of position on a closed smooth boundary curve (Fig. 3.1.1). Here u_B is independent of ϵ.

This set of boundary conditions defines a unique regular solution $u(x, y; \epsilon)$ at all points interior to and on the boundary. As $\epsilon \to 0$ with (x, y) fixed in the interior of the domain, the exact solution of (3.1.1) approaches the outer limit, which can be thought of as the first term of an outer expansion (valid off the boundary):

$$\lim_{\substack{\epsilon\to 0\\ x,y \text{ fixed}}} u(x, y; \epsilon) = u_0(x, y). \tag{3.1.6}$$

Here $u_0(x, y)$ satisfies the limit equation of first order,

$$a\frac{\partial u_0}{\partial x} + b\frac{\partial u_0}{\partial y} = 0. \tag{3.1.7}$$

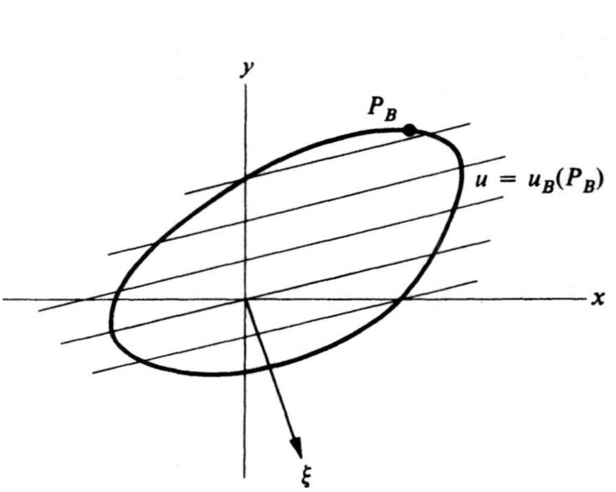

FIGURE 3.1.1. Boundary-Value Problem

Solutions of (3.1.7) are calculated along the characteristic curves:

$$\xi = bx - ay = \text{const.}$$

These are defined parametrically by $dx/ds = a$, $dy/ds = b$, $du/ds = 0$ in accordance with the familiar interpretation of (3.1.7) as a derivative in the direction with slope b/a.

We call the lines $\xi = \text{const.}$ *subcharacteristics* of the original equation (3.1.1). The main underlying structure of the solution is given by these subcharacteristics since $u_0 = u_0(\xi)$. The subcharacteristic curves are sketched in Figure 3.1.1. It is clear that the boundary condition on one side of the domain is sufficient to define $u_0(x, y) = u_0(\xi)$ in the whole domain. Also, $u_0(\xi)$, in general, does not satisfy the boundary condition on the other side of the domain, so that a boundary layer is needed.

In order to study this boundary layer and its matching to $u_0(\xi)$, it is convenient to introduce an orthogonal coordinate system (ξ, η) based on the subcharacteristic ξ. The coordinate η is chosen here arbitrarily for convenience. Different choices of η do not affect the essential boundary-layer character but may influence the form of higher-order corrections:

$$\xi = bx - ay, \qquad \eta = ax + by. \tag{3.1.8}$$

The transformation formulas are

$$\begin{aligned}
\frac{\partial u}{\partial x} &= b\frac{\partial u}{\partial \xi} + a\frac{\partial u}{\partial \eta}, \qquad \frac{\partial u}{\partial y} = -a\frac{\partial u}{\partial \xi} + b\frac{\partial u}{\partial \eta},\\
\frac{\partial^2 u}{\partial x^2} &= b^2\frac{\partial^2 u}{\partial \xi^2} + 2ab\frac{\partial^2 u}{\partial \xi\partial \eta} + a^2\frac{\partial^2 u}{\partial \eta^2},\\
\frac{\partial^2 u}{\partial x\partial y} &= -ab\frac{\partial^2 u}{\partial \xi^2} + (b^2 - a^2)\frac{\partial^2 u}{\partial \xi\partial \eta} + ab\frac{\partial^2 u}{\partial \eta^2},\\
\frac{\partial^2 u}{\partial y^2} &= a^2\frac{\partial^2 u}{\partial \xi^2} - 2ab\frac{\partial^2 u}{\partial \xi\partial \eta} + b^2\frac{\partial^2 u}{\partial \eta^2}.
\end{aligned} \tag{3.1.9}$$

The original equation (3.1.1) now reads

$$\epsilon\left[A_{11}\frac{\partial^2 u}{\partial \xi^2} + 2A_{12}\frac{\partial^2 u}{\partial \xi\partial \eta} + A_{22}\frac{\partial^2 u}{\partial \eta^2}\right] = \frac{\partial u}{\partial \eta}, \tag{3.1.10}$$

where the coefficients A_{ij} are given by

$$\begin{aligned}
(a^2 + b^2)A_{11} &= \alpha_{11}b^2 - 2\alpha_{12}ab + \alpha_{22}a^2,\\
(a^2 + b^2)A_{12} &= \alpha_{11}ab + \alpha_{12}(b^2 - a^2) - \alpha_{22}ab,\\
(a^2 + b^2)A_{22} &= \alpha_{11}a^2 + 2\alpha_{12}ab + \alpha_{22}b^2.
\end{aligned} \tag{3.1.11}$$

Since the equation is still elliptic, we know that the discriminant

$$A_{12}^2 - A_{11}A_{22} < 0. \tag{3.1.12}$$

We also note the identity (to be used later on)

$$(a^2 + b^2)A_{22} = (\sqrt{\alpha_{11}}\, a \pm \sqrt{\alpha_{22}}\, b)^2 \mp 2ab(\sqrt{\alpha_{11}\alpha_{22}} \mp \alpha_{12}) > 0 \quad (3.1.13)$$

if we choose the upper sign for $ab < 0$ and the lower for $ab > 0$.

$O(\epsilon)$ Boundary layer

The original domain has an image in the $\xi\eta$-plane as shown in Figure 3.1.2. We denote the upper and lower boundaries of the image domain by $\eta = \eta_B^+(\xi)$ and $\eta = \eta_B^-(\xi)$, respectively. Let the assigned boundary values be denoted $u = u_U(\xi)$ on the upper part of the domain between A and B and by $u = u_L(\xi)$ on the lower part.

In the region of the boundary layer, we expect $\partial/\partial\eta$ to be large, so that an appropriate boundary-layer coordinate has the form

$$\eta^* = \frac{\eta - \eta_B(\xi)}{\delta(\epsilon)} \quad (3.1.14)$$

for fixed ξ and $\delta \ll 1$. The associated limit process is $\epsilon \to 0$, η^*, ξ fixed. The largest order terms on the left-hand side of (3.1.10) are then $O(\epsilon/\delta^2)$, and the term on the right is $O(1/\delta)$. Thus, $\epsilon/\delta^2 = 1/\delta$ for dominant balance, i.e., the boundary layer has thickness $\delta = \epsilon$, and (3.1.14) reads

$$\eta^* = \frac{\eta - \eta_B(\xi)}{\epsilon}. \quad (3.1.15)$$

The boundary-layer solution is represented as the first term of an asymptotic expansion:

$$u(x, y; \epsilon) = u_{BL}(\xi, \eta^*) + \ldots . \quad (3.1.16)$$

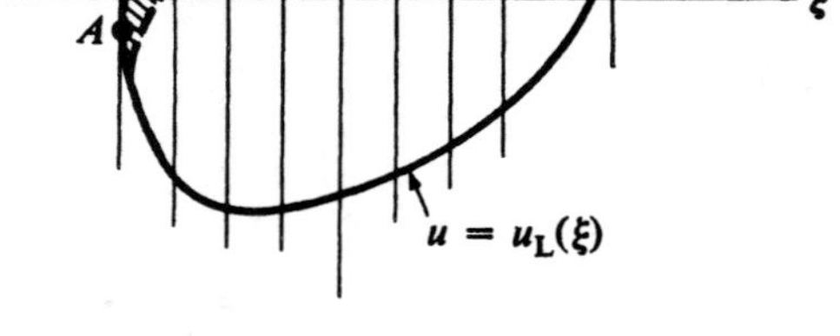

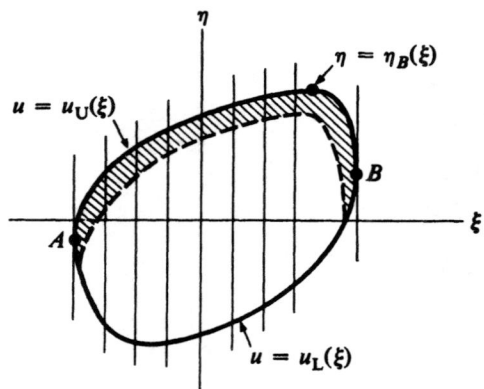

FIGURE 3.1.2. Subcharacteristic Coordinates

Note that, under the transformation from (ξ, η) to (ξ, η^*), we have

$$\frac{\partial u}{\partial \eta} \to \frac{1}{\epsilon}\frac{\partial u}{\partial \eta^*}, \qquad \frac{\partial u}{\partial \xi} \to -\frac{1}{\epsilon}\left(\frac{d\eta_B}{d\xi}\right)\frac{\partial u}{\partial \eta^*} + \dots,$$

so that the boundary-layer equation derived from (3.1.10) is

$$K(\xi)\frac{\partial^2 u_{BL}}{\partial \eta^{*2}} = \frac{\partial u_{BL}}{\partial \eta^*}, \tag{3.1.17}$$

where

$$K(\xi) = A_{11}\left(\frac{d\eta_B}{d\xi}\right)^2 - 2A_{12}\left(\frac{d\eta_B}{d\xi}\right) + A_{22}. \tag{3.1.18}$$

Thus, the boundary-layer equation is an ordinary differential equation in this case. The assumed orders of magnitude are correct provided $(d\eta_B/d\xi)$ is finite. The important case where $d\eta_B/d\xi = \infty$ along an arc, that is, the case of subcharacteristic boundary, will be discussed later. The location of the boundary layer is now decided by the criterion that it must match with the outer limit, $u_0(\xi)$.

The solution of (3.1.17) is

$$u_{BL}(\xi, \eta^*) = A(\xi) + B(\xi)e^{\eta^*/K(\xi)}. \tag{3.1.19}$$

In order that (3.1.19) match with the outer limit $u_0(\xi)$, the exponent η^*/K must be *negative* in the boundary layer. Thus, the location of the boundary layer depends on the sign of $K(\xi)$. If $K > 0$, we must have the boundary layer along $\eta = \eta_B^+(\xi)$ because η_-^* is negative in this case. Conversely, if $K < 0$ the boundary layer is located along $\eta = \eta_B^-(\xi)$. The quadratic in $(d\eta_B/d\xi)$ that results from setting the right-hand side of (3.1.18) equal to zero has no real roots because this expression is just the characteristic condition (3.1.4) expressed in terms of the A_{ij}. Therefore, $K(\xi)$ does not vanish, and sign $K =$ sign $A_{22} > 0$ according to (3.1.13). Exponential decay occurs as $\eta^* \to -\infty$; the boundary layer appears on the upper boundary in Figure 3.1.2.

With matching, the solution is thus

$$u_{BL}(\xi, \eta^*) = u_L(\xi) + \{u_U(\xi) - u_L(\xi)\}e^{\eta^*/K(\xi)}, \quad u_0(\xi) = u_L(\xi). \tag{3.1.20}$$

The boundary-layer solution in this case is also the first term of a uniformly valid composite expansion.

The location of the boundary layer in the original figure (Figure 3.1.1) depends on the orientation of the coordinates (ξ, η), that is, on the signs of (a, b). A qualitative sketch of the possibilities is given in Figure 3.1.3.

Subcharacteristic boundary, $O(\sqrt{\epsilon})$ boundary layer

It remains to discuss the case where a segment of the boundary is subcharacteristic, say $\xi = \xi_s =$ const., as shown in Figure 3.1.4.

The outer solution carries the constant $u_L(\xi_s)$ as the right boundary is approached, so that the boundary condition $u = u_s(\eta)$ is violated there except for the very special case $u_s(\eta) =$ const. $= u_L(\xi_s)$. In general, it can be expected that

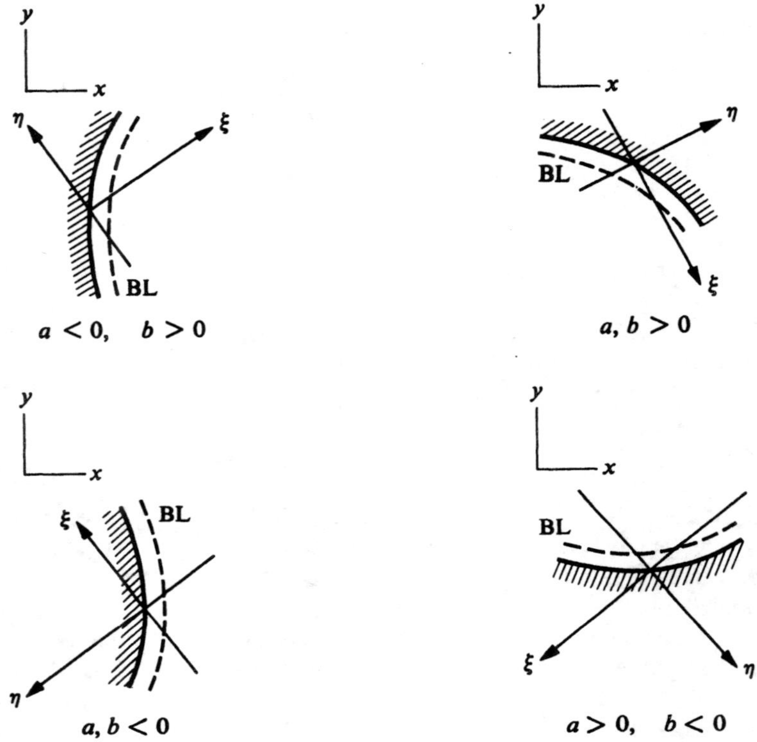

FIGURE 3.1.3. Various Locations of Boundary Layers

$\partial/\partial\xi$ is large near $\xi = \xi_s$, so that a suitable boundary-layer coordinate (balancing $\partial^2/\partial\xi^2$ and $\partial/\partial\eta$) is

$$\xi^* = \frac{\xi - \xi_s}{\sqrt{\epsilon}}, \tag{3.1.21}$$

and the boundary-layer thickness is $O(\sqrt{\epsilon})$. The subcharacteristic boundary-layer solution is the first term of an asymptotic expansion

$$u(\xi, \eta; \epsilon) = u^*_{BL}(\xi^*, \eta) + \ldots \tag{3.1.22}$$

associated with the limit ($\epsilon \to 0, \xi^*, \eta$ fixed). Thus, (3.1.10) yields

$$A_{11}\frac{\partial^2 u^*_{BL}}{\partial \xi^{*2}} = \frac{\partial u^*_{BL}}{\partial \eta}. \tag{3.1.23}$$

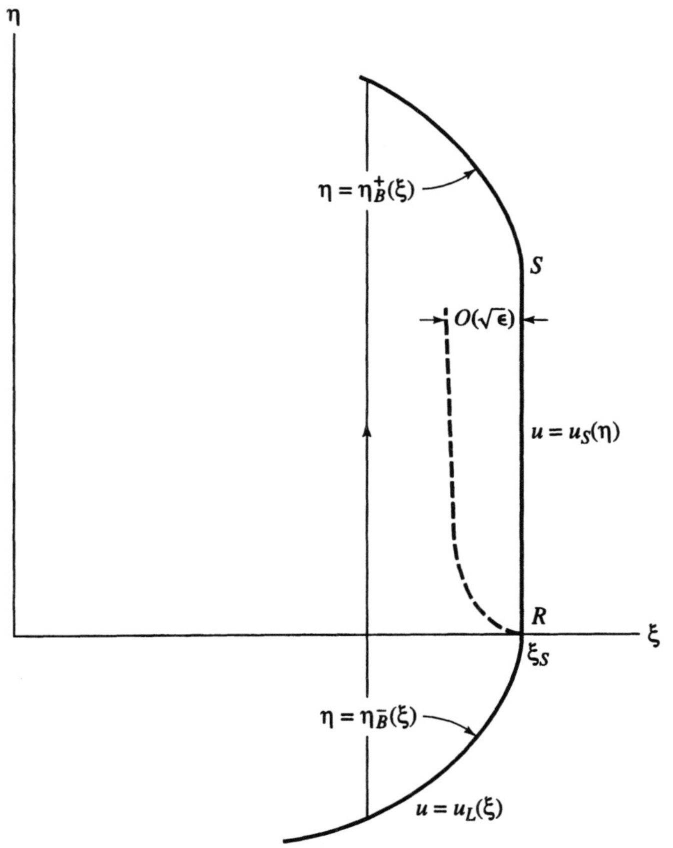

FIGURE 3.1.4. Boundary Layer on a Subcharacteristic

Here A_{11} = const. and the subcharacteristic boundary-layer equation is the diffusion equation, a partial differential equation in this case. Also, $A_{11} > 0$ just as $A_{22} > 0$ [see (3.1.13)] so that η is the timelike coordinate.

The solution of (3.1.23) is to be found satisfying the boundary condition

$$u^*_{BL}(0, \eta) = u_s(\eta). \tag{3.1.24}$$

Also, the solution must match with the outer limit, $u_0(\xi)$. This matching can be carried out in terms of the variable $\xi_\delta = (\xi - \xi_s)/\delta(\epsilon)$ for a class of functions $\delta(\epsilon)$

in $\sqrt{\epsilon} \ll \delta(\epsilon) \ll 1$. As usual, the matching gives the result that the boundary-layer limit as $\xi^* \to -\infty$ must equal the outer limit as $\xi \to \xi_s$

$$u^*_{BL}(-\infty, \eta) = u_L(\xi_s) = \text{const.} \tag{3.1.25}$$

and this gives the second boundary condition for the diffusion equation (3.1.23) on $-\infty < \xi < 0$. To completely define the solution for u^*_{BL}, we also need to specify an "initial condition." This follows from the requirement that along a fixed subcharacteristic $\xi^* = \text{const.}$ in the subcharacteristic boundary layer $u^*_{BL} \to u_L$ as η approaches the lower boundary.

To fix ideas, suppose the lower boundary $\eta = \eta_B^-(\xi)$ has the following behavior as $\xi \to \xi_s$ (near the point R in Figure 3.1.4):

$$\eta_B^-(\xi) = c(\xi_s - \xi)^\gamma + o((\xi_s - \xi)^\gamma) \text{ as } \xi \to \xi_s, \tag{3.1.26}$$

where for the situation shown in Figure 3.1.4, c is a negative constant and γ is a positive constant. If the arc $\eta_B^-(\xi)$ joins smoothly with the subcharacteristic arc $\xi = \xi_s$, $(d\eta_B^-/d\xi) \to \infty$ as $\xi \to \xi_s$, and we must have $0 < \gamma < 1$. If, however, there is a discontinuity in slope at the point $\xi = \xi_s$, $\eta = 0$ we have $\gamma \geq 1$. Expressing (3.1.26) in terms of ξ^* gives

$$\eta_B^- = c(-\xi^*)^\gamma \epsilon^{\gamma/2} + \ldots \tag{3.1.27}$$

Therefore, to leading order

$$u^*_{BL}(\xi^*, \eta_B^-) = u^*_{BL}(\xi^*, 0) + \ldots \tag{3.1.28}$$

On the other hand, for a fixed ξ^*, we have

$$u_L(\xi) = u_L(\xi_s + \sqrt{\epsilon}\xi^*) = u_L(\xi_s) + \ldots \tag{3.1.29}$$

Identifying u^*_{BL} in (3.1.28) with u_L in (3.1.29) gives the initial condition

$$u^*_{BL}(\xi^*, 0) = u_L(\xi_s). \tag{3.1.30}$$

Various representations can be used to solve (3.1.23) subject to (3.1.24), (3.1.25), and (3.1.30). For example, see section 1.4 of [3.15]. A convenient form that follows from use of Green's function is

$$u^*_{BL}(\xi^*, \eta) = u_s(0^+) + [u_s(0^+) - u_L(\xi_s)] \, \text{erf}\left(\frac{\xi^*}{2\sqrt{A_{11}\eta}}\right)$$

$$+ \int_0^\eta \dot{u}_s(\bar\eta)\text{erfc}\left(\frac{-\xi^*}{2\sqrt{A_{11}(\eta - \bar\eta)}}\right) d\bar\eta, \tag{3.1.31}$$

where $\text{erfc}(z) = 1 - \text{erf}(z)$. Thus, for $u_s(\eta) = \text{const.} = u_s(0^+) \neq u_L(\xi_s)$ the integral in (3.1.31) vanishes and the subcharacteristic boundary-layer solution is given simply by

$$u^*_{BL}(\xi^*, \eta) = u_s(0^+) + [u_s(0^+) - u_L(\xi_s)] \, \text{erf}\left(\frac{\xi - \xi_s}{2\sqrt{\epsilon A_{11}\eta}}\right). \tag{3.1.32}$$

Of course, if the boundary data at $\xi - \xi_s, \eta = 0$ is continuous ($u_s(0^+) = u_L(\xi_s)$) this result reduces to $u^*_{BL} = u_s(0^+)$.

Discontinuous boundary data

An $O(\sqrt{\epsilon})$ layer centered along an *interior* subcharacteristic is also appropriate if the boundary data on η_B^- is discontinuous. For example, assume $u_L(\xi_0^-) \neq u_L(\xi_0^+)$ at some point ξ_0 along the lower arc AB of Figure 3.1.2. Since the solution of an elliptic equation is continuous in the interior, we must introduce an $O(\sqrt{\epsilon})$ layer centered at the subcharacteristic $\xi = \xi_0$. With $\xi^* = (\xi - \xi_0)/\sqrt{\epsilon}$, and $u = u^*(\xi^*, \eta) + \ldots$, we see that u^* obeys (3.1.23). Now, the left and right matching conditions are $u^*(-\infty, \eta) = u_L(\xi_0^-)$ and $u^*(\infty, \eta) = u_L(\xi_0^+)$, respectively. The initial condition ($\eta = 0$) is $u^*(\xi^*, 0) = u_L(\xi_0^-)$ if $\xi^* < 0$ and $u^*(\xi^*, 0) = u_L(\xi_0^+)$ if $\xi^* > 0$. The solution is in the form (3.1.32) with $u_L \to u_L(\xi_0^-)$ and $u_s \to \frac{1}{2}[u_L(\xi_0^+) + u_L(\xi_0^-)]$.

$O(\epsilon)$ local layer

We note that our result (3.1.31) or (3.1.32) is not defined for $\eta < 0$. Now, as long as the boundary data are continuous at R (i.e., $u_L(\xi_s) = u_s(0^+)$) everything is fine and the outer limit joins continuously with the side boundary layer along $\eta = 0$. But, suppose the boundary data are discontinuous, $u_L(\xi_s) \neq u_s(0^+)$. We must then introduce a local layer centered at $\xi = \xi_s, \eta = 0$. Clearly, variations with respect to ξ and η are equally important and we must rescale as

$$\overline{\xi} = \frac{\xi - \xi_s}{\epsilon}; \quad \overline{\eta} = \frac{\eta}{\epsilon}$$

to study (3.1.10) in a domain of order ϵ centered at R.

It is easily seen that the exact equation (3.1.10) results for $u = \overline{u}(\overline{\xi}, \overline{\eta})$ in this small domain:

$$A_{11}\frac{\partial^2 \overline{u}}{\partial \overline{\xi}^2} + 2A_{12}\frac{\partial^2 \overline{u}}{\partial \overline{\xi}\partial \overline{\eta}} + A_{22}\frac{\partial^2 \overline{u}}{\partial \overline{\eta}^2} = \frac{\partial \overline{u}}{\partial \overline{\eta}}. \tag{3.1.33}$$

In terms of the $\overline{\xi}$, $\overline{\eta}$ variables, the domain in which (3.1.33) is to be solved is bounded on the right by two *straight* lines that follow upon rescaling from the actual boundaries: $\xi = \xi_s, \eta > 0$ and $\eta = \eta_B^-(\xi), \xi < \xi_s$. The boundary condition is that $\overline{u}$ = constant on each of these straight lines. We find the side boundary condition:

$$\overline{u}(0, \overline{\eta}) = u_s(0^+) \quad \text{on } \overline{\eta} > 0. \tag{3.1.34}$$

The boundary condition on the lower arc depends on γ, (see (3.1.26)). For $0 < \gamma < 1$, the limiting form of the lower boundary is a vertical line, and we have

$$\overline{u}(0, \overline{\eta}) = u_L(\xi_s) \quad \text{on } \overline{\eta} < 0. \tag{3.1.35a}$$

If $\gamma = 1$, the lower boundary is the straight line $\overline{\eta} = -c\overline{\xi}$, and we find

$$\overline{u}(\overline{\xi}, -c\overline{\xi}) = u_L(\xi_s) \quad \text{on } \overline{\xi} < 0. \tag{3.1.35b}$$

Finally, for $\gamma > 1$ the lower boundary is the horizontal line $\overline{\eta} = 0$, and we have

$$\overline{u}(\overline{\xi}, 0) = u_L(\xi_s) \quad \text{on } \overline{\xi} < 0. \tag{3.1.35c}$$

We confirm our earlier statement that for continuous boundary data, $u_L(\xi_S) = u_s(0^+)$, the outer limit is valid for $\overline{\eta} \leq 0, \overline{\xi} \leq 0$. We note that with $u_L(\xi_S) = u_s(0^+)$ the two boundary values (3.1.34) and (3.1.35) are identical constants and the solution of (3.1.33) in this case is just $\overline{u}(\overline{\xi}, \overline{\eta}) = \text{const.} = u_L(\xi_S) = u_s(0^+)$. This is just the outer limit at R and the side boundary layer evaluated at $\eta = 0^+$. Thus, a local layer with $O(\epsilon)$ radius centered at R is needed only if $u_L(\xi_S) \neq u_s(O^+)$. In this case, one needs to solve the exact problem (3.1.33) subject to the constant boundary data (3.1.34), (3.1.35). The boundary condition at infinity follows from matching with the subcharacteristic boundary-layer solution and the outer limit.

An $O(\epsilon)$ local layer is also needed near S (see Figure 3.1.4) if the boundary data are discontinuous there. Arguments analogous to those used for R show that (3.1.33) with appropriate rescaled local variables holds near S, and constant boundary values must be imposed along straight lines representing the rescaled boundaries near S.

In all our discussion so far, we have confined attention to the solution to leading order. Local layers of $O(\epsilon)$ radius may have to be introduced (where none are needed for the $O(1)$ solution) if one considers higher-order terms.

Mixed boundary conditions

The approach we have used generalizes to problems where u is prescribed on part of the boundary and an outward derivative is prescribed on the remainder of the boundary. However, some difficulties, not encountered so far, may occur. To illustrate these, consider the following special case of (3.1.1):

$$\epsilon(u_{xx} + u_{yy}) = u_y \tag{3.1.36}$$

to be solved in the interior of the unit circle, $x^2 + y^2 < 1$, subject to the mixed boundary condition

$$u(x, \sqrt{1 - x^2}; \epsilon) = 1, \tag{3.1.37a}$$

$$u_y(x, -\sqrt{1 - x^2}; \epsilon) = 0. \tag{3.1.37b}$$

The unique[1] solution is $u = 1$. But, let us see what follows by matching outer and inner limits.

Consider the outer expansion

$$u(x, y; \epsilon) = u_0(x, y) + \epsilon u_1(x, y) + \dots. \tag{3.1.38}$$

[1] We first transform (3.1.36) to $w_{xx} + w_{yy} = \frac{1}{4\epsilon^2} w$ by setting $u = we^{y/2\epsilon}$. Because the coefficient $1/4\epsilon^2$ is non-negative, it follows, as for Laplace's equation, that $w \equiv 0$ inside the domain if it vanishes on part of the boundary and the normal derivative vanishes on the remainder. Uniqueness is a consequence of the fact that for two solutions w_1, w_2 the difference $w = w_1 - w_2$ satisfies a zero boundary condition.

The outer limit $u_0(x, y)$ satisfies $\partial u_0/\partial y = 0$. Therefore,

$$u_0 = f(x).$$

To specify $f(x)$ further, we note that u_1 satisfies

$$\frac{\partial u_1}{\partial y} = \frac{d^2 f}{dx^2}. \tag{3.1.39}$$

Since $\partial u_1/\partial y$ must vanish on the lower half-circumference, we must have $f''(x) = 0$, i.e.,

$$u_0 = A_0 x + B_0 \tag{3.1.40}$$

for constant A_0 and B_0.

A boundary layer along the upper half-circumference can now be constructed satisfying the given boundary condition $u = 1$ and matching with (3.1.40) for *any* A_0 and B_0. To see this, we introduce the boundary-layer coordinate

$$y^* = \frac{y - \sqrt{1 - x^2}}{\epsilon},$$

where $y^* < 0$ in the interior of the unit circle. For the boundary-layer expansion

$$u = u_{BL}(x, y^*) + \dots,$$

we find that u_{BL} obeys

$$\frac{\partial^2 u_{BL}}{\partial y^{*2}} - (1 - x^2)\frac{\partial u_{BL}}{\partial y^*} = 0. \tag{3.1.41}$$

Therefore, the solution that satisfies the boundary condition $u_{BL}(x, 0) = 1$ and matches with (3.1.40) is

$$u_{BL}(x, y^*) = (A_0 x + B_0) + (1 - A_0 x - B_0)e^{(1-x^2)y^*}. \tag{3.1.42}$$

As (3.1.42) contains the outer limit, it is the uniformly valid solution to $O(1)$ in the interior of the unit circle. But, we are unable to specify $A_0 = 0$ and $B_0 = 1$ by any matching considerations.

For this example, the evenness of solutions with respect to x requires $A_0 = 0$, leaving B_0 undefined. To determine B_0, one could use a variational principle or an augmented outer expansion including exponentially small terms as discussed in Sec. 2.2.5. A less trivial example for a mixed boundary-value problem that also results in an unspecified outer limit is outlined in Problem 2.

Nonparallel subcharacteristics

We expect more general elliptic problems to have a structure similar to that encountered so far. For example, for the strictly linear equations (3.1.1) for which $a = a(x, y)$ and $b = b(x, y)$, the subcharacteristics are no longer parallel straight lines but form a one-parameter family of curves defined by the first-order differential equation $dy/dx = b(x, y)/a(x, y)$. If this equation has no singular points in the domain of interest, the behavior of solutions is the same as for the case of constant a and b. If singular points do occur, then certain complications may arise. An example is discussed next; a second example is outlined in Problem 4.

We want to calculate the solution to $O(1)$ for the interior Dirichlet problem

$$\epsilon(u_{xx} + u_{yy}) = xu_x + yu_y \tag{3.1.43a}$$

in the unit circle $x^2 + y^2 \leq 1$ with prescribed boundary values for u.

It is more convenient to transform the problem to polar coordinates (r, θ), and we have to satisfy

$$\epsilon\left(u_{rr} + \frac{u_r}{r} + \frac{u_{\theta\theta}}{r^2}\right) = ru_r \tag{3.1.43b}$$

inside $0 \leq r \leq 1$, with boundary condition

$$u(1, \theta) = f(\theta) \tag{3.1.44}$$

prescribed.

The outer limit is $u_r = 0$, i.e., $u_0 = A(\theta)$. This means that u is constant along rays $y/x =$ constant. It follows by consideration of the solution near $r = 0$, where there is no singularity for $\epsilon > 0$, that A must be a constant, say $A = \alpha$.

To satisfy the boundary condition at $r = 1$, we introduce a boundary layer there. It is easily seen that

$$r^* = \frac{r-1}{\epsilon} \tag{3.1.45}$$

is the appropriate boundary-layer variable. Hence, (3.1.43b) in the limit $\epsilon \to 0$, r^* fixed, reduces to

$$\frac{\partial^2 u}{\partial r^{*2}} - \frac{\partial u}{\partial r^*} = O(\epsilon), \tag{3.1.46}$$

and we have the boundary-layer limit

$$u_{BL} = a(\theta) + [f(\theta) - a(\theta)]e^{r^*}. \tag{3.1.47}$$

Matching with the outer limit u_0 shows that

$$a(\theta) = \alpha = \text{const.}. \tag{3.1.48}$$

but there is no information on the value of this constant.

This is the analog, for partial differential equations, of the problem we encountered in Sec. 2.2.5 and is also discussed in [3.8]. A boundary layer is possible everywhere along the circle $r = 1$, and matching does not determine the unknown constant α.

Motivated by the resolution of the difficulty for the ordinary differential equation case, we seek a variational principle for which (3.1.43) is the associated Euler equation. We will then use the variational principle to calculate the unknown constant α.

Thus, we seek a Lagrangian $L(x, y, u, u_x, u_y)$ such that (3.1.43a) corresponds to Euler's equation (see sec. IV.10 of [3.6])

$$\frac{\partial}{\partial x}\left(\frac{\partial L}{\partial u_x}\right)+\frac{\partial}{\partial y}\left(\frac{\partial L}{\partial u_y}\right)-\frac{\partial L}{\partial u}=0 \tag{3.1.49}$$

associated with the variational principle

$$\delta J \equiv \delta\left[\iint\limits_{\sqrt{x^2+y^2}\leq 1} L(x, y, u, u_x, u_y)dxdy\right]=0 \tag{3.1.50}$$

with u prescribed on the boundary.

For the Laplacian operator, interpreted as defining the equilibrium deflection u of a membrane, it is well known that $L = (u_x^2 + u_y^2)/2$, and the variational principle (3.1.50) is the principle of least potential energy.

Guided by this result, we attempt to calculate a Lagrangian associated with (3.1.43b). As in Sec. 2.2.5, we eliminate the first derivative term on the right-hand side by introducing the transformation

$$v(r, \theta) = u(r, \theta)e^{-r^2/4\epsilon}.$$

Equation (3.1.43b) becomes

$$\epsilon\left(v_{rr}+\frac{v_r}{r}+\frac{v_{\theta\theta}}{r^2}\right)+\left(1-\frac{r^2}{4\epsilon}\right)v=0. \tag{3.1.51}$$

Now, since the Lagrangian associated with the Laplacian operator is known, we can easily extend the definition to include the added linear term in v in (3.1.51). We find

$$L(r, \theta, v, v_r, v_\theta)=\frac{\epsilon}{2}\left(v_r^2+\frac{v_\theta^2}{r^2}\right)-\frac{1}{2}\left(1-\frac{r^2}{4\epsilon}\right)v^2. \tag{3.1.52a}$$

Thus, (3.1.51) is the Euler equation associated with

$$\delta\left\{\frac{1}{2}\int_0^1\int_0^{2\pi}\left[\epsilon\left(v_r^2+\frac{v_\theta^2}{r^2}\right)-\left(1-\frac{r^2}{4\epsilon}\right)v^2\right]rdr\,d\theta\right\}=0. \tag{3.1.52b}$$

Hence, (3.1.43b) is the Euler equation for the variational principle

$$\delta J = 0, \quad \delta u = 0 \quad \text{on } r = 1, \tag{3.1.53a}$$

where

$$J=\frac{1}{2}\int_0^1\int_0^{2\pi}\left[\epsilon\left(u_r^2+\frac{u_\theta^2}{r^2}\right)+\left(\frac{r^2u^2}{2\epsilon}-u^2-ruu_r\right)\right]e^{-r^2/2\epsilon}r\,dr\,d\theta. \tag{3.1.53b}$$

Again, we point out that the form of the Lagrangian is not unique [see Problem 2.2.3], and (3.1.53b) is somewhat simpler than the expression given in [3.8].

It is easily verified that the Euler equation

$$\frac{\partial}{\partial r}\left(r\frac{\partial L_1}{\partial u_r}\right)+\frac{\partial}{\partial \theta}\left(r\frac{\partial L_1}{\partial u_\theta}\right)-\frac{\partial}{\partial u}(rL_1)=0 \tag{3.1.54a}$$

with

$$L_1(r,\theta,u,u_r,u_\theta)=\left[\frac{\epsilon}{2}\left(u_r^2+\frac{u_\theta^2}{r^2}\right)+\frac{1}{2}\left(\frac{r^2u^2}{2\epsilon}-u^2-ruu_r\right)\right]e^{-r^2/2\epsilon} \tag{3.1.54b}$$

gives (3.1.43b).

What we need to do now is evaluate L_1 along the candidate solution

$$u=\alpha+[f(\theta)-\alpha]e^{-(1-r)/\epsilon}, \tag{3.1.55}$$

which, for an appropriate α, is uniformly valid to $O(1)$ in the interior of the unit circle. Knowing L_1 to leading order then defines J to leading order and α is the solution of

$$\frac{\partial J}{\partial \alpha}=0. \tag{3.1.56}$$

The result is

$$\alpha=\frac{1}{2\pi}\int_0^{2\pi}f(\theta)d\theta, \tag{3.1.57}$$

i.e., α is the average value of the boundary data. The calculations are left as an exercise [Problem 5].

The situation when the singular point is not at the center and for a more general boundary is explored in Problem 6.

3.1.2 Linear Hyperbolic Equations, $\alpha_{12}^2-\alpha_{11}\alpha_{22}>0$

Typically, linear hyperbolic equations describe perturbations about a constant solution for a quasilinear hyperbolic system (see Sec. 6.2.1). The essential features are illustrated by considering L_2 to be the simple wave operator and t a coordinate corresponding to time. Assuming a suitable length scale and normalizing the time by (length/signal speed), a dimensionless equation can be written in the form

$$\epsilon\left\{\frac{\partial^2 u}{\partial x^2}-\frac{\partial^2 u}{\partial t^2}\right\}=a\frac{\partial u}{\partial x}+b\frac{\partial u}{\partial t}. \tag{3.1.58}$$

Here, the small parameter ϵ is somewhat artificial. In the context of the leading perturbation to a constant solution for a hyperbolic system, one would derive (3.1.58) (with $\epsilon=1$), and ϵ would then measure the nonlinear higher-order terms that have been ignored. One could, of course, introduce far field variables $\tilde{x}=\epsilon x$ and $\tilde{t}=\epsilon t$ to obtain (3.1.58) (after dropping the tildes). However, as we shall see in Chapter 6, it is crucial to account for the small nonlinear perturbation terms in order to correctly describe the solution in the far field. A second possible rescaling that would introduce the ϵ in (3.1.58) corresponds to numerically large coefficients multiplying u_x and u_t. Rescaling these as (a/ϵ) and (b/ϵ) gives (3.1.58). At any rate, it is mathematically instructive to study the rescaled equations in the form (3.1.58) to illustrate various features of the solution.

Equation (3.1.58) has real characteristics:

$$r = t - x; \qquad s = t + x. \tag{3.1.59}$$

The characteristics serve to define the region of influence, propagating into the future, of a disturbance at a point Q (Figure 3.1.5). For the specification of a boundary-value problem for (3.1.58), the number of boundary conditions to be imposed on an arc depends on the nature of the arc with respect to the characteristic directions of propagation. For example, on $t = 0$, typical initial conditions (u, u_t) must be given in order to find the solution for $t > 0$. For $x = 0$, one boundary condition (for example, u) must be given to define the solution (signal propagating from the boundary) for $x > 0$. Along $t = 0$, two characteristics lead from the boundary into the region of interest, but for $x = 0$, only one does. Generalizing this idea, the directions of an arc, with respect to the characteristic directions and the future directions, can be classified as timelike, spacelike, or characteristic. (See Figure 3.1.6, where the characteristics have arrows pointing to the future.)

One boundary condition is specified on the timelike arc corresponding to one characteristic leading into the adjacent region in which the solution is defined. Two initial conditions are given on the spacelike arc corresponding to the two characteristics leading into the adjacent domain. When the boundary curves are characteristic, only one condition can be prescribed, and the characteristic relations must hold. The characteristic initial-value problem prescribes one condition on AB and on AC to define the solution in $ABCD$.

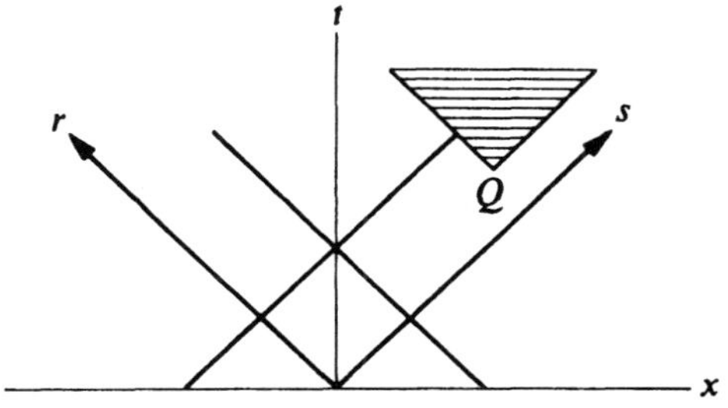

FIGURE 3.1.5. Characteristic Coordinate System

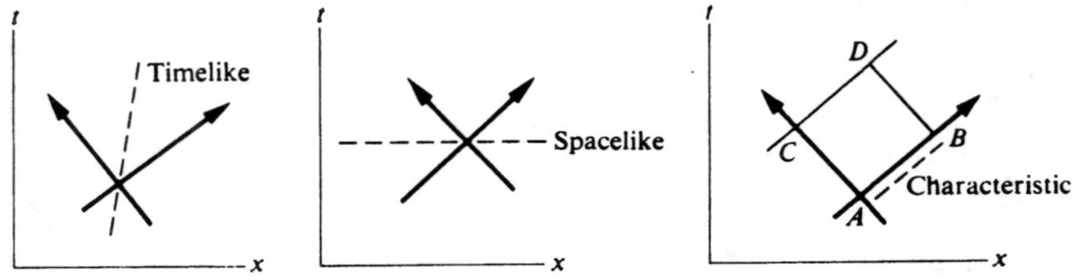

FIGURE 3.1.6. Directions of Arcs

Now consider the initial-value problem in $-\infty < x < \infty$ for (3.1.58):

$$u(x, 0) = F(x), \tag{3.1.60}$$

$$u_t(x, 0) = G(x). \tag{3.1.61}$$

(See Figure 3.1.7.)

According to the general theory, the solutions at a point $F(x, t)$ can depend only on that part of the initial data that can send a signal to P, the part cut out of the initial line by the backward running characteristics through P, $(x_1 < x < x_2)$. Now, consider what happens as $\epsilon \to 0$. In particular, consider the behavior of the

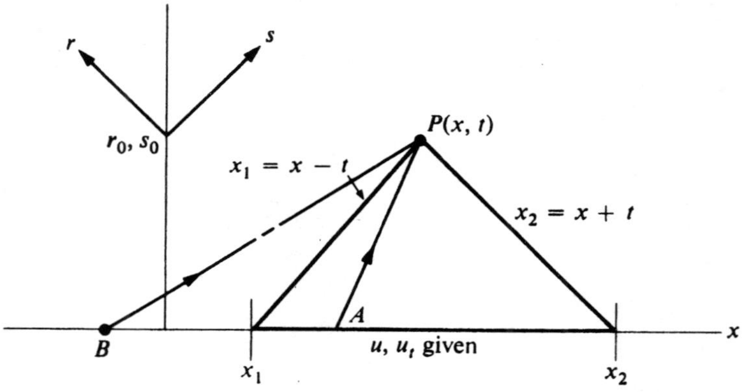

FIGURE 3.1.7. Initial-Value Problem

limit equation

$$a\frac{\partial u}{\partial x}+b\frac{\partial u}{\partial t}=0. \tag{3.1.62}$$

The solution in this case depends only on the data connected to P along the subcharacteristic

$$bx-at=\text{const.} \tag{3.1.63}$$

The general solution of (3.1.62) has the form

$$u(x,t)=f\left(x-\frac{a}{b}t\right). \tag{3.1.64}$$

Now, if the subcharacteristic originates at point A between x_1 and x_2, that is, if

$$\left|\frac{b}{a}\right|>1, \tag{3.1.65}$$

it is reasonable to conceive of (3.1.64) as a limiting form of the exact solution. In this case, as $\epsilon\to 0$, it is only the data at point A that affects the solution in the limit. However, if

$$\left|\frac{a}{b}\right|>1, \tag{3.1.66}$$

the subcharacteristic to P lies outside the usual domain of influence and originates at B. The speed of the disturbances associated with the subcharacteristics is greater than that of the characteristics. In this case, one cannot expect the solution (3.1.64) to the limit problem to be a limit of the exact solution of (3.1.58). This strange behavior is connected with the fact that condition (3.1.66) makes the solutions $u(x,t;\epsilon)$ unstable.[1] For stability, it is necessary that the subcharacteristics be timelike. These points about stability are easily demonstrated from the characteristic form of (3.1.58) and the rules for the propagation of jumps. If the characteristic coordinates (r,s) given by (3.1.59) are introduced, then (3.1.58) becomes (see Fig. 3.1.7)

$$-4\epsilon\frac{\partial^2 u}{\partial r\,\partial s}=(b-a)\frac{\partial u}{\partial r}+(b+a)\frac{\partial u}{\partial s}. \tag{3.1.67}$$

Consider now the propogation along ($r=r_0=$ const.) of a jump in the derivative $(\partial u/\partial r)$. Let

$$\kappa=\left[\frac{\partial u}{\partial r}\right]_{r=r_0}\equiv\frac{\partial u}{\partial r}(r_0^+,s)-\frac{\partial u}{\partial r}(r_0^-,s). \tag{3.1.68}$$

[1] In particular, the small perturbation assumption, which is implicit in the derivation of the linear problem (3.1.58), is violated if u is unstable.

Assuming that u and $\partial u/\partial s$ are continuous across $r = r_0$, we can evaluate (3.1.67) at r_0^+ and r_0^- and form the difference to obtain

$$-4\epsilon \frac{\partial \kappa}{\partial s} = (b - a)\kappa. \tag{3.1.69}$$

The solution has the form

$$\kappa = \kappa_0 \exp\left[-\frac{b-a}{4\epsilon}(s - s_0)\right]. \tag{3.1.70}$$

A jump across a characteristic $r = r_0$ propagates to infinity along that characteristic. If

$$b - a > 0, \qquad \text{we have exponential decay and stability;}$$

if

$$b - a < 0, \qquad \text{we have exponential growth and instability.}$$

A parallel discussion for jumps in $\partial u/\partial s$ across a characteristic $s = s_0$ gives the following results:

$$b + a > 0 \quad \text{implies exponential decay and stability;}$$

$$b + a < 0 \quad \text{implies exponential growth and instability.}$$

Combining these relations, we see that for stability we must have

$$\frac{b}{|a|} > 1. \tag{3.1.71}$$

Thus, we restrict further discussion to the stable case for which (3.1.71) is satisfied and study first the initial-value problem specified by (3.1.60)–(3.1.61) as $\epsilon \to 0$.

Initial-value problem

Since the limit solution (3.1.64) can only satisfy one initial condition, the existence of an initial boundary layer, analogous to that discussed in Sec. 2.1.2, can be expected. An initially valid expansion can be expressed in the coordinates (t^*, x), where

$$t^* = \frac{t}{\delta(\epsilon)} \tag{3.1.72}$$

for an appropriate $\delta(\epsilon) \ll 1$. The associated limit process has $(x, t^*$ fixed) and consists of a "vertical" approach to the initial line. The expansion has the form

$$u(x, t; \epsilon) = U_0(x, t^*) + \beta_1(\epsilon)U_1(x, t^*) + \dots. \tag{3.1.73}$$

In order to satisfy the initial conditions independent of ϵ, we need

$$\beta_1(\epsilon) = \delta,$$

and then we obtain

$$\frac{\partial u}{\partial t}(x, t; \epsilon) = \frac{1}{\delta}\frac{\partial U_0}{\partial t^*} + \frac{\partial U_1}{\partial t^*} + \dots. \tag{3.1.74}$$

Thus, to take care of the initial conditions (3.1.60)–(3.1.61), we need

$$U_0(x, 0) = F(x), \quad U_1(x, 0) = U_2(x, 0) = \dots = 0, \tag{3.1.75}$$

$$\frac{\partial U_0}{\partial t^*}(x, 0) = 0, \quad \frac{\partial U_1}{\partial t^*}(x, 0) = G(x), \quad \frac{\partial U_2}{\partial t^*}(x, 0) = \dots = 0. \tag{3.1.76}$$

Using the inner expansion in (3.1.58), we find

$$\epsilon\left\{\frac{\partial^2 U_0}{\partial x^2} + \delta\frac{\partial^2 U_1}{\partial x^2} + \dots - \frac{1}{\delta^2}\frac{\partial^2 U_0}{\partial t^{*2}} - \frac{1}{\delta}\frac{\partial^2 U_1}{\partial t^{*2}} + \dots\right\}$$

$$= a\frac{\partial U_0}{\partial x} + \dots + \frac{b}{\delta}\frac{\partial U_0}{\partial t^*} + b\frac{\partial U_1}{\partial t^*} + \dots.$$

A second-order equation results for U_0 only if $\delta = O_s(\epsilon)$, and we choose $\delta = \epsilon$ for simplicity. With this choice, the following sequence of approximate equations is obtained:

$$\frac{\partial^2 U_0}{\partial t^{*2}} + b\frac{\partial U_0}{\partial t^*} = 0, \tag{3.1.77}$$

$$\frac{\partial^2 U_1}{\partial t^{*2}} + b\frac{\partial U_1}{\partial t^*} = -a\frac{\partial U_0}{\partial x}. \tag{3.1.78}$$

In accordance with the general ideas of the first part of this section, the boundary-layer equations are ordinary differential equations, since the boundary layer does not occur on a subcharacteristic. This must be true for any hyperbolic initial value problem, since a spacelike arc can never be subcharacteristic. Equations (3.1.75) and (3.1.76) provide the initial conditions for the initial-layer equations, and the solutions are easily found:

$$U_0(x, t^*) = F(x), \tag{3.1.79}$$

$$U_1(x, t^*) = \left[G(x) + \frac{a}{b}F'(x)\right]\left[\frac{1 - e^{-bt^*}}{b}\right] - \frac{a}{b}t^* F'(x). \tag{3.1.80}$$

Thus, finally, we have the initially valid expansion (3.1.73)

$$u(x, t; \epsilon) = F(x) + \epsilon\left\{\left[G(x) + \frac{a}{b}F'(x)\right]\left[\frac{1 - e^{-bt^*}}{b}\right] - \frac{a}{b}t^* F'(x)\right\} + \dots. \tag{3.1.81}$$

These solutions contain persistent terms as well as typical decaying terms of boundary layer type with a time scale $t = O(\epsilon)$. The behavior of (3.1.81) provides initial conditions for an outer expansion.

Next, we construct the outer expansion, based on the limit process ($\epsilon \to 0$, x, t fixed), the first term of which is the limit solution (3.1.64). The orders of the various terms are evident from the orders in (3.1.81). Thus, we have

$$u(x, t; \epsilon) = u_0(x, t) + \epsilon u_1(x, t) + \epsilon^2 u_2(x, t) + \dots. \tag{3.1.82}$$

In the outer expansion, the higher-order derivatives are small and the lower-order operator dominates. The sequence of equations that approximates (3.1.58) is

$$a\frac{\partial u_0}{\partial x} + b\frac{\partial u_0}{\partial t} = 0, \tag{3.1.83}$$

$$a\frac{\partial u_1}{\partial x} + b\frac{\partial u_1}{\partial t} = \left(\frac{\partial^2 u_0}{\partial x^2} - \frac{\partial^2 u_0}{\partial t^2}\right). \tag{3.1.84}$$

The general solutions can all be expressed in terms of arbitrary functions:

$$u_0 = f(\xi), \quad \xi = x - \frac{a}{b}t. \tag{3.1.85}$$

The equation for u_1 becomes

$$a\frac{\partial u_1}{\partial x} + b\frac{\partial u_1}{\partial t} = \left(1 - \frac{a^2}{b^2}\right) f''(\xi). \tag{3.1.86}$$

This is easily solved by introducing $\tau = t$ and ξ as coordinates, and the result is

$$u_1(x, t) = \left(1 - \frac{a^2}{b^2}\right) \frac{t}{b} f''(\xi) + f_1(\xi). \tag{3.1.87}$$

Thus, the outer expansion is

$$u(x, t; \epsilon) = f(\xi) + \epsilon \left[\left(1 - \frac{a^2}{b^2}\right) \frac{t}{b} f''(\xi) + f_1(\xi)\right] + \dots. \tag{3.1.88}$$

The arbitrary functions, f, f_1, etc., in the outer expansion (3.1.88) must be determined by matching with (3.1.81). We introduce the matching variable

$$t_\eta = \frac{t}{\eta(\epsilon)} \tag{3.1.89}$$

for a class of functions $\eta(\epsilon)$ contained in $\epsilon \ll \eta \ll 1$, and we consider the limit $\epsilon \to 0$ with x and t_η fixed. Thus, we have $t = \eta t_\eta \to 0$, $t^* = (\eta/\epsilon)t_\eta \to \infty$ in the matching domain. The initially valid expansion (3.1.81) becomes

$$u(x, t; \epsilon) = F(x) + \epsilon \left\{\frac{G(x)}{b} + \frac{a}{b^2}F'(x) - \frac{a}{b}\frac{\eta}{\epsilon}t_\eta F'(x)\right\} + O(\epsilon^2) + \text{T.S.T.} \tag{3.1.90}$$

In the outer expansion, we have

$$f(\xi) = f\left(x - \frac{a}{b}t\right) = f\left(x - \frac{a}{b}\eta t_\eta\right) = f(x) - \frac{a}{b}\eta t_\eta f'(x) + O(\eta^2).$$

Therefore, the outer expansion has the following form:

$$u(x,t;\epsilon) = f(x) - \frac{a}{b}\eta t_\eta f'(x) + \epsilon f_1(x) + O(\epsilon\eta) + O(\epsilon^2). \tag{3.1.91}$$

The terms of order one in (3.1.90) and (3.1.91) can be matched, and then the terms of order η are matched identically; terms of order ϵ can also be matched. Thus, we have

$$f(x) = F(x), \tag{3.1.92}$$

$$f_1(x) = \frac{G(x)}{b} + \frac{a}{b^2}F'(x), \tag{3.1.93}$$

as long as $\epsilon|\log\epsilon| \ll \eta \ll 1$.

The final result expresses the outer expansion in terms of the given initial values:

$$u(x,t;\epsilon) = F\left(x - \frac{a}{b}t\right) + \epsilon\left\{\frac{1}{b}\left(1 - \frac{a^2}{b^2}\right)tF''\left(x - \frac{a}{b}t\right)\right.$$
$$\left. + \frac{a}{b^2}F'\left(x - \frac{a}{b}t\right) + \frac{1}{b}G\left(x - \frac{a}{b}t\right)\right\} + O(\epsilon^2). \tag{3.1.94}$$

We see that, after a little while, the solution is dominated by the given initial value of u, which propagates along the subcharacteristic. We also note that the outer expansion is not uniformly valid in the far field ($t = O(\epsilon^{-1})$). As mentioned earlier, the governing equation is itself not generally valid in the far field, as one must account for the small nonlinearities that have been ignored in deriving (3.1.88). A detailed discussion of solutions uniformly valid for large x and t is given in Chapter 6.

Signaling problems

We next consider a signaling or radiation problem in which boundary conditions are prescribed on a timelike arc and propagate into the quiescent medium in $x > 0$. For the first problem, the boundary condition is prescribed at $x = 0$, and we have to distinguish two cases according to the slope of the subcharacteristics, that is, whether they run into or out of the boundary $x = 0$ (see Figure 3.1.8). The subcharacteristics are given by

$$\xi = x - \frac{a}{b}t = \text{const.}, \tag{3.1.95}$$

and for $a > 0$ (outgoing) and $a < 0$ (incoming) the boundary condition is

$$u(0,t) = F(t), \quad t > 0. \tag{3.1.96}$$

There is a real discontinuity in the function and in its derivative along the characteristic curve $x = t$, but the intensity of the jump decays exponentially ($b > |a|$) according to the considerations of (3.1.70). The solution is identically zero for $x > t$.

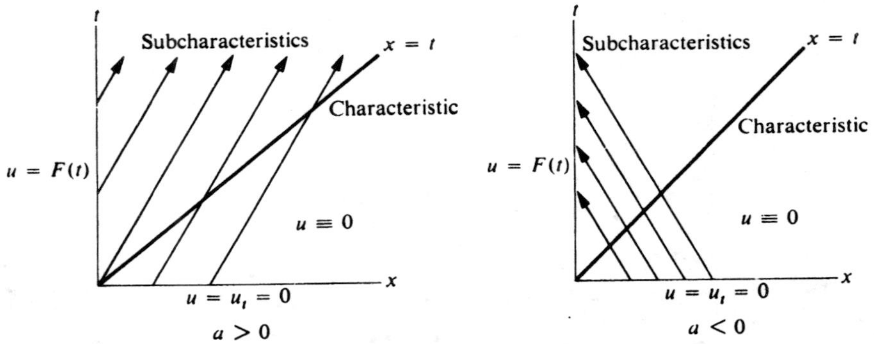

FIGURE 3.1.8. Signaling Problems

$a > 0$: *Outgoing subcharacteristics*

The outer solution is an asymptotic expansion of the same form as before:

$$u(x, t; \epsilon) = u_0(x, t) + \epsilon u_1(x, t) + \dots . \tag{3.1.97}$$

The sequence of equations satisfied by u_i is the same as before. The solution of (3.1.83) can be written in the form

$$u_0 = f(\zeta), \quad \zeta = t - \frac{b}{a}x. \tag{3.1.98}$$

The boundary condition (3.1.96) can be satisfied by identifying $f = F$, so that the first term in the outer expansion is

$$u_0 = \begin{cases} 0, & t < \frac{b}{a}x, \\ F\left(t - \frac{b}{a}x\right), & t > \frac{b}{a}x. \end{cases} \tag{3.1.99}$$

This solution, however, has a discontinuity on the particular subcharacteristic through the origin. Such a discontinuity is not permitted in the solution to the exact problem (3.1.58) with $\epsilon > 0$, since any discontinuities can appear only on characteristics. Thus, to obtain a uniformly valid solution, a suitable interior layer must be introduced on the particular subcharacteristic $\zeta = 0$, which supports the discontinuity in the outer solution. In order to derive the interior-layer equations, we consider a limit process in which (x^*, t^*) are fixed, where

$$x^* = \frac{x - (a/b)t}{\delta(\epsilon)}, \tag{3.1.100}$$

$$t^* = t, \tag{3.1.101}$$

and try to choose $\delta(\epsilon) \ll 1$ so that a meaningful problem results. Here, $\delta(\epsilon)$ is the measure of thickness of the interior layer. The expansion has the form

$$u(x, t; \epsilon) = U_0(x^*, t^*) + \mu(\epsilon)U_1(x^*, t^*) + \dots. \tag{3.1.102}$$

The first term is of order one so that it can match to (3.1.99). The derivatives of u have the form

$$\frac{\partial u}{\partial x}(x, t; \epsilon) = \frac{1}{\delta}\frac{\partial U_0}{\partial x^*} + \frac{\mu}{\delta}\frac{\partial U_1}{\partial x^*} + \dots$$

$$\frac{\partial u}{\partial t}(x, t; \epsilon) = \frac{-(a/b)}{\delta}\frac{\partial U_0}{\partial x^*} + \frac{\partial U_0}{\partial t^*} - \frac{(a/b)}{\delta}\mu\frac{\partial U_1}{\partial x^*} + \dots.$$

The operator on the right-hand side of (3.1.58) now has the form

$$a\frac{\partial u}{\partial x} + b\frac{\partial u}{\partial t} = b\left[\frac{\partial U_0}{\partial t^*} + \mu(\epsilon)\frac{\partial U_1}{\partial t^*} + \dots\right]. \tag{3.1.103}$$

The dominant terms of the wave operator are proportional to $\partial^2 U_0/\partial x^{*2}$ so that (3.1.58) becomes

$$\frac{\epsilon}{\delta^2}\left\{\frac{\partial^2 U_0}{\partial x^{*2}} + \dots - \frac{a^2}{b^2}\frac{\partial^2 U_0}{\partial x^{*2}} + \dots\right\} = b\frac{\partial U_0}{\partial t^*} + \dots. \tag{3.1.104}$$

The distinguished limiting case, which results in a nontrivial equation, has

$$\delta = \sqrt{\epsilon}. \tag{3.1.105}$$

The interior-layer thickness here is an order of magnitude larger than in the initial boundary layer. The interior-layer equation is then a partial differential equation with one coordinate along a subcharacteristic. From (3.1.104), we obtain

$$\kappa\frac{\partial^2 U_0}{\partial x^{*2}} = \frac{\partial U_0}{\partial t^*}, \tag{3.1.106}$$

where $\kappa = (1 - a^2/b^2)/b > 0$. Here $\kappa > 0$ assures that $t^* = t$ is a positive timelike direction and that (3.1.106) is an ordinary diffusion equation that smooths the discontinuity of the outer expansion on $\zeta = 0$. The boundary conditions for (3.1.106) have to come from matching with the outer expansion.

Matching for this problem is carried out in terms of x_η with x_η, t fixed as $\epsilon \to 0$, where

$$x_\eta = \frac{x - (a/b)t}{\eta(\epsilon)}, \tag{3.1.107}$$

and η is contained in the class $\sqrt{\epsilon} \ll \eta \ll 1$. Thus, in the matching domain

$$x - \frac{a}{b}t = \eta x_\eta \to 0, \quad x^* = \frac{x - (a/b)t}{\sqrt{\epsilon}} = \frac{\eta}{\sqrt{\epsilon}}x_\eta \to \pm\infty.$$

Under this limit, the leading terms in the outer and interior-layer expansions behave as follows:

$$\text{Outer limit:} \qquad u(x, t; \epsilon) = u_0 + \ldots \to \begin{cases} 0, & t < \frac{b}{a}x \\ F(0^+), & t > \frac{b}{a}x \end{cases} \tag{3.1.108}$$

$$\text{Interior layer:} \qquad u(x, t; \epsilon) = U_0(x^*, t) + \ldots \to U_0(\pm\infty, t) + \ldots . \tag{3.1.109}$$

For matching, the terms in (3.1.108) and (3.1.109) must be the same, and this provides the boundary conditions illustrated in Figure 3.1.9 for the interior-layer equation (3.1.106). Initial conditions are chosen here consistent with the boundary conditions and in such a way that the physical process represents the resolution of the discontinuity at $\zeta = 0$. A rigorous treatment of the initial conditions demands a discussion of initially valid expansions, which is not given here. The main point here is the nature of the interior-layer equation. The solution corresponding to the conditions stated is

$$U_0(x^*, t) = \frac{F(0^+)}{2} \operatorname{erfc}\left(\frac{x^*}{2\sqrt{\kappa t}}\right). \tag{3.1.110}$$

The discontinuity of the outer solution is replaced here by the diffusive solution of the heat equation.

$a < 0$: Incoming subcharacteristics

The qualitative difference between this case and the case just discussed is striking. If an outer expansion of the form of (3.1.97) is contemplated, the only reasonable solution of the form $f(t - (b/a)x)$ is zero, since disturbances now propagate along the subcharacteristics from the quiescent region to the boundary.

Thus, we have

$$u_0 = 0. \tag{3.1.111}$$

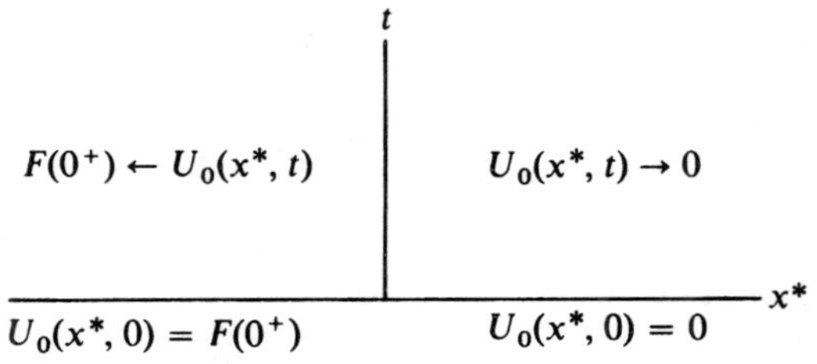

FIGURE 3.1.9. Initial Condition and Limiting Values of U_0

The discontinuity occurs at the boundary $x = 0$, and the boundary layer occurs at $x = 0$. The boundary-layer equations should be ordinary differential equations, since again in this case the boundary layer is not on a subcharacteristic. To derive these equations, consider (x^*, t^*) fixed, where

$$x^* = \frac{x}{\delta(\epsilon)}, \quad t^* = t, \quad \text{as } \epsilon \to 0. \tag{3.1.112}$$

The expansion is of the usual form,

$$u(x, t; \epsilon) = U_0(x^*, t^*) + \nu_1(\epsilon)U_1(x^*, t^*) + \dots. \tag{3.1.113}$$

The basic equation (3.1.58) takes the form

$$\frac{\epsilon}{\delta^2}\frac{\partial^2 U_0}{\partial x^{*2}} + \dots = \frac{a}{\delta}\frac{\partial U_0}{\partial x^*} + b\frac{\partial U_0}{\partial t^*} + \dots. \tag{3.1.114}$$

Again, $\delta = \epsilon$ is a distinguished case for the boundary layer not on a subcharacteristic, and U_0 obeys the ordinary differential equation

$$\frac{\partial^2 U_0}{\partial x^{*2}} = a\frac{\partial U_0}{\partial x^*}. \tag{3.1.115}$$

The solution satisfying the boundary condition (3.1.96) is

$$U_0(x^*, t^*) = F(t^*)e^{ax^*}, \quad a < 0. \tag{3.1.116}$$

The boundary layer here is just a region of exponential decay adjacent to the boundary.

Many of the results we have derived in this section can be obtained after laborious calculations from the asymptotic behavior of exact solutions. For a discussion, see Sec. 10.1 of [3.44].

3.1.3 Burgers' Equation, a Quasilinear Parabolic Problem

In this section, we illustrate some of the consequences of having a quasilinear lower-order operator, and we choose an equation that can be solved exactly as our model.

Cole-Hopf transformation

The equation

$$u_t + uu_x - \epsilon u_{xx} = 0 \tag{3.1.117}$$

was originally proposed by Burgers to model turbulence. It was later shown in [3.5] that (3.1.117) is derivable from the Navier-Stokes equations in the limit of a weak shock layer. It was also observed in [3.5], and independently in [3.11], that (3.1.117) can be transformed to the diffusion equation, and we discuss this next.

Consider the transformation of dependent variable $u \to v$ defined by

$$u(x,t;\epsilon) = -2\epsilon \frac{v_x(x,t;\epsilon)}{v(x,t;\epsilon)}. \tag{3.1.118}$$

We then find

$$\begin{aligned} u_t &= -2\epsilon \frac{v_{xt}}{v} + 2\epsilon \frac{v_x v_t}{v^2}, \\ u_x &= -2\epsilon \frac{v_{xx}}{v} + 2\epsilon \frac{v_x^2}{v^2}, \\ u_{xx} &= -2\epsilon \frac{v_{xxx}}{v} + 6\epsilon \frac{v_x v_{xx}}{v^2} - 4\epsilon \frac{v_x^3}{v^3}, \end{aligned}$$

and using these expressions to compute the left-hand side of (3.1.117) gives

$$u_t + uu_x - \epsilon u_{xx} = \frac{2\epsilon}{v^2}(v_t - \epsilon v_{xx}) - \frac{2\epsilon}{v}(v_t - \epsilon v_{xx})_x.$$

Therefore, *any* solution $v(x,t;\epsilon)$ of

$$\frac{2\epsilon}{v^2}(v_t - \epsilon v_{xx}) - \frac{2\epsilon}{v}(v_t - \epsilon v_{xx})_x = 0 \tag{3.1.119}$$

is also a solution of Burgers' equation when (3.1.118) is used to compute $u(x,t;\epsilon)$. In particular, if $v(x,t)$ satisfies the *linear* diffusion equation,

$$v_t - \epsilon v_{xx} = 0, \tag{3.1.120}$$

it also solves (3.1.119) trivially and can be used to find a solution of Burgers' equation (3.1.117).

Initial-value problem on $-\infty < x < \infty$, exact solution

To solve the initial-value problem

$$u(x,0;\epsilon) = F(x)$$

for (3.1.117), we first need to compute the corresponding initial condition $v(x,0;\epsilon)$ for (3.1.120). It follows from (3.1.118) that $v(x,0;\epsilon)$ obeys the *linear* first-order ordinary differential equation

$$v_x(x,0;\epsilon) + \frac{F(x)}{2\epsilon} v(x,0;\epsilon) = 0.$$

Integrating this gives

$$v(x,0;\epsilon) = \exp\left(-\frac{1}{2\epsilon}\int_0^x F(s)ds\right) \equiv G(x;\epsilon), \tag{3.1.121}$$

where, with no loss of generality, we have set $v(0,0;\epsilon) = 1$. This arbitrary constant will cancel out in the final result for u.

Now, using the well-known result (e.g., see Sec. 1.3 of [3.15]) for the solution of the initial-value problem for v:

$$v(x, t; \epsilon) = \frac{1}{\sqrt{4\pi\epsilon t}} \int_{-\infty}^{\infty} G(s; \epsilon) e^{-(x-s)^2/4\epsilon t} ds, \tag{3.1.122}$$

we calculate u from Equation (3.1.118) in the form

$$u(x, t; \epsilon) = \frac{\int_{-\infty}^{\infty} G(s; \epsilon) \frac{(x-s)}{t} e^{-(x-s)^2/4\epsilon t} ds}{\int_{-\infty}^{\infty} G(s; \epsilon) e^{-(x-s)^2/4\epsilon t} ds}. \tag{3.1.123}$$

This gives u explicitly for a given $F(x; \epsilon)$. If $F(x; \epsilon)$ is piecewise constant the integrals in (3.1.123) can be evaluated explicitly. The details can be found in Sec. 5.3.6 of [3.15], and we summarize the results next.

Consider first the initial-value problem where

$$F(x; \epsilon) = \begin{cases} u_1 = \text{const.} & \text{if } x < x_0 \\ u_2 = \text{const.} & \text{if } x > x_0. \end{cases} \tag{3.1.124}$$

It is easily seen that the transformation

$$\overline{x} = \frac{x - x_0 - (u_1 + u_2)t/2}{2/(u_1 - u_2)}, \tag{3.1.125a}$$

$$\overline{t} = \frac{t}{4/(u_1 - u_2)^2}, \tag{3.1.125b}$$

$$\overline{u} = \frac{2u - (u_1 + u_2)}{u_1 - u_2} \tag{3.1.125c}$$

leaves Burgers' equation invariant and transforms (3.1.124) to

$$\overline{u}(\overline{x}, 0; \epsilon) = \begin{cases} 1, & \text{if } \overline{x} < 0 \\ -1 & \text{if } \overline{x} > 0. \end{cases} \tag{3.1.126}$$

Thus, if $u_1 > u_2$, the larger initial value is mapped to 1 and the smaller initial value is mapped to -1, with the positive and negative x axes mapping, respectively, to the positive and negative $\overline{x}$ axes.

Similarly, the transformation

$$\overline{x} = \frac{x - x_0 - (u_1 + u_2)t/2}{2/(u_2 - u_1)}, \tag{3.1.127a}$$

$$\overline{t} = \frac{t}{4(u_2 - u_1)^2}, \tag{3.1.127b}$$

$$\overline{u} = \frac{2u - (u_1 + u_2)}{u_2 - u_1} \tag{3.1.127c}$$

also leaves Burgers' equation invariant and transforms (3.1.124) to

$$\overline{u}(\overline{x}, 0; \epsilon) = \begin{cases} -1 & \text{if } \overline{x} < 0 \\ 1 & \text{if } \overline{x} > 0. \end{cases} \tag{3.1.128}$$

This transformation is appropriate if $u_2 > u_1$ since it maps the larger initial value to 1 and the smaller to -1, while mapping the positive and negative x axes, respectively, to the positive and negative $\overline{x}$ axes.

Thus, with no loss of generality, we need only study the canonical initial conditions (3.1.126) and (3.1.128).

The exact solution for (3.1.126) is found in the form (dropping overbars for simplicity)

$$u(x,t;\epsilon) = \frac{e^{-x/\epsilon}\operatorname{erfc}\left(\frac{x-t}{2\sqrt{\epsilon t}}\right) - \operatorname{erfc}\left(-\frac{x+t}{2\sqrt{\epsilon t}}\right)}{e^{-x/\epsilon}\operatorname{erfc}\left(\frac{x-t}{2\sqrt{\epsilon t}}\right) + \operatorname{erfc}\left(-\frac{x+t}{2\sqrt{\epsilon t}}\right)} \tag{3.1.129a}$$

when the integrals occurring in (3.1.123) are evaluated. Similarly, the solution corresponding to (3.1.128) is

$$u(x,t;\epsilon) = \frac{e^{-x/\epsilon}\operatorname{erfc}\left(\frac{t-x}{2\sqrt{\epsilon t}}\right) - \operatorname{erfc}\left(\frac{x+t}{2\sqrt{\epsilon t}}\right)}{e^{-x/\epsilon}\operatorname{erfc}\left(\frac{t-x}{2\sqrt{\epsilon t}}\right) + \operatorname{erfc}\left(\frac{x+t}{2\sqrt{\epsilon t}}\right)}. \tag{3.1.129b}$$

It is easily seen using the asymptotic behavior of error functions that the outer limit of (3.1.129a) ($\epsilon \to 0$ $x \neq 0$, t fixed) is

$$u_0 = \lim_{\substack{\epsilon\to 0\\ x\neq 0,\ t \text{ fixed}}} u(x,t;\epsilon) = \begin{cases} 1 & \text{if } x < 0 \\ -1 & \text{if } x > 0. \end{cases} \tag{3.1.130}$$

The discontinuity at $x = 0$ is smoothed by the shock layer shock-layer obtained in the limit $\epsilon \to 0$, $x^* = x/\epsilon$ and t fixed

$$u_0^* = \lim_{\substack{\epsilon\to 0\\ x^*, t \text{ fixed}}} u(\epsilon x^*, t; \epsilon) = -\tanh\frac{x^*}{2}. \tag{3.1.131}$$

Thus, (3.1.131) gives the uniformly valid solution to $O(1)$ and is shown in Figure 3.1.10.

The outer limit that results from (3.1.129b) is

$$u_0(x,t) = \begin{cases} -1 & \text{if } x < -t \\ x/t & \text{if } -t < x < t \\ 1 & \text{if } x > t. \end{cases}$$

The discontinuities in u_x and u_t encountered at $x = t$ are smoothed by the corner-layer expansion

$$u = 1 - 2\left(\frac{\epsilon}{\pi t}\right)^{1/2} \frac{e^{-x_c^2/4t}}{\operatorname{erfc}\left(-\frac{x_c}{2\sqrt{t}}\right)} + O(\epsilon), \tag{3.1.132}$$

where $x_c = (x - t)/\sqrt{\epsilon}$. Again, using the asymptotic behavior of erfc(z), we see that $u_c \to 1$ as $x_c \to \infty$ and $u_c \to x/t$ as $x_c \to -\infty$. A similar expression is found near the corner $x = -t$. The resulting solution is shown in Figure 3.1.11.

The integrals in (3.1.123) can also be explicitly evaluated for certain other more complicated forms of F. Also, one can choose a particular solution of the diffusion

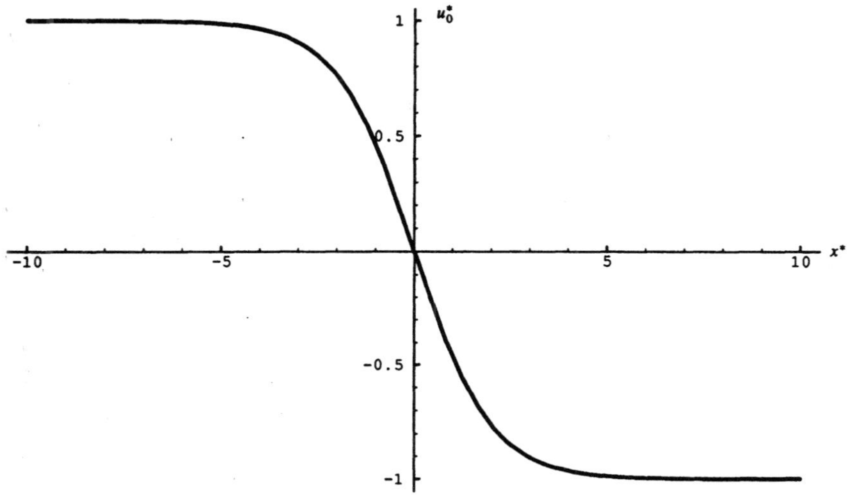

FIGURE 3.1.10. Shock Layer for Burgers' Equation

equation (3.1.120) and then derive the corresponding solution for u directly from (3.1.118). Details can be found in chapter 4 of [3.44]. We will show later that perturbation solutions of (3.1.117) give the correct limiting expressions that we have obtained from the exact solution.

Boundary-value problems, exact solution

Suppose now that we wish to solve Burgers' equation on the semi-infinite interval $0 \leq x < \infty$ with a prescribed value $u(0, t; \epsilon) = g(t)$ if $t > 0$ at the left end. Let us take the initial condition $u(x, 0) = 0$ for simplicity.

The Cole-Hopf transformation (3.1.118) implies that v still satisfies the linear diffusion equation (3.1.120) with initial condition $v(x, 0) = c =$ constant, and the *linear homogeneous* boundary condition

$$2\epsilon v_x(0, t; \epsilon) + g(t)v(0, t; \epsilon) = 0. \tag{3.1.133}$$

If $g(t) =$ constant $= a$, the explicit solution for v is easily found (e.g., see problem 10b of [3.15]). In fact, it is shown in [3.5] that the solution for Burgers' equation for $a = 1$ is

$$u(x, t; \epsilon) = \frac{\operatorname{erfc}\left(\frac{x-t}{2\sqrt{\epsilon t}}\right)}{e^{1/2\epsilon(x-t/2)}\operatorname{erf}\left(\frac{x}{2\sqrt{\epsilon t}}\right) + \operatorname{erfc}\left(\frac{x-t}{2\sqrt{\epsilon t}}\right)}. \tag{3.1.134}$$

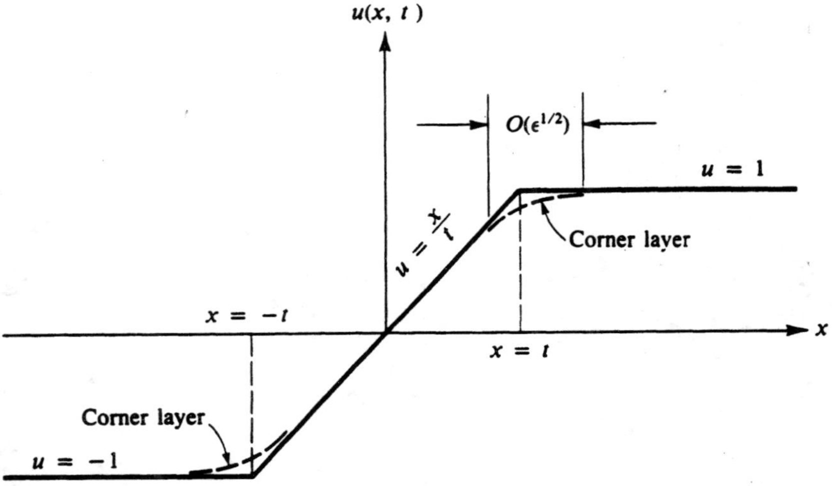

FIGURE 3.1.11. Corner Layers for Burgers' Equation

To extend this result to arbitrary a, we note that the transformation $u \to (u/a)$, $x \to ax$, $t \to a^2 t$ leaves Burgers' equation invariant and simplifies the left boundary condition to equal 1. Therefore, the solution for $a \neq 1$ follows from (3.1.134).

It is also possible to derive an explicit result for an arbitrary initial condition as long as $g(t) = \text{constant}$. However, for variable g an explicit result appears out of reach. One approach (see Sec. 1.6.3 of [3.15]) is to solve the problem for v for an unknown boundary condition $v(0, t; \epsilon) = k(t; \epsilon)$ and then use (3.1.118) to derive the following integral equation for k (for the choice $v(x, 0; \epsilon) = k(0^+; \epsilon)$):

$$g(t)k(t; \epsilon) = 2\sqrt{\frac{\epsilon}{\pi}} \int_0^t \frac{\dot{k}(\tau; \epsilon)d\tau}{\sqrt{t - \tau}}. \tag{3.1.135}$$

Once k is determined, e.g., by solving (3.1.135) numerically, the solution for $v(x, t; \epsilon)$ is given by (see (3.1.131))

$$v(x, t; \epsilon) = k(0^+; \epsilon) + \int_0^t \dot{k}(\tau; \epsilon) \operatorname{erfc}\left(\frac{x}{2\sqrt{\epsilon(t - \tau)}}\right) d\tau, \tag{3.1.136}$$

and u follows from (3.1.118).

Explicit exact solutions for Burgers' equation on a finite domain are also possible as long as the boundary conditions are constant (see Problem 13). We will not study this class of solutions here.

Weak solutions for $\epsilon = 0$

Since the outer limit ($\epsilon \to 0$, x, t fixed) of Burgers' equation is quasilinear, subcharacteristics may cross and solutions may no longer exist in the strict sense. We therefore briefly review the idea of a *weak* solution of

$$u_t + uu_x = 0. \tag{3.1.137}$$

This is a solution of (3.1.137) everywhere in some domain of the xt plane except along a specific trajectory where u or its first partial derivatives may be discontinuous. The reader will find a detailed discussion for general systems in Sec. 5.3 of [3.15].

In order to handle possible discontinuities, we need to broaden our definition of a solution and consider an *integral conservation law* as the governing equation instead of (3.1.137). Consider the conservation law

$$\frac{d}{dt}\int_{x_1}^{x_2} u(x,t)dx = \frac{1}{2}[u^2(x_1,t) - u^2(x_2,t)] \tag{3.1.138}$$

for two fixed points x_1 and x_2. It is easily seen that if u, u_x, and u_t are continuous, (3.1.138) implies (3.1.137). However, (3.1.138) *still makes sense* if u, u_x, or u_t are discontinuous in the interval (x_1, x_2). In this case, (3.1.138) implies that the discontinuity path $x_s(t)$ is governed by

$$\frac{dx_s}{dt} = \frac{u(x_s^+, t) + u(x_s^-, t)}{2}. \tag{3.1.139}$$

Note that (3.1.139) implies that an infinitesimal discontinuity, $u(x_s^+, t) \approx u(x_s^-, t)$, travels at the subcharacteristic speed u of (3.1.137). In addition to (3.1.139), an allowable discontinuity must satisfy the *entropy condition*

$$u(x_s^+, t) \le \frac{dx_s}{dt} \le u(x_s^-, t), \tag{3.1.140}$$

so called because its generalization to the case of compressible flow ensures that the entropy cannot decrease downstream of a physically consistent shock.

It is also important to note that the conservation law (3.1.138) is the *primitive* governing equation from which (3.1.137) follows for smooth solutions and for which (3.1.139) governs discontinuities. Equation (3.1.138) specifies a definite conservation law in the sense that the time rate of change of the quantity $\int_{x_1}^{x_2} u dx$ is balanced by a certain *net flux* through the boundaries; here, the flux is given by $\frac{u^2}{2}$.

It is possible to construct any number of different conservation laws, all of which imply (3.1.137) for smooth solutions. In fact, given any two functions $p(u)$ and $q(u)$, subject to the constraint $p'(u) - uq'(u) = 0$, the integral conservation law

$$\frac{d}{dt}\int_{x_1}^{x_2} p(u(x,t))dx = q(u(x_1,t)) - q(u(x_2,t)) \tag{3.1.141}$$

reduces, for smooth solutions, to (3.1.137). The shock condition for (3.1.141) is

$$\frac{dx_s}{dt} = \frac{[p]}{[q]}, \tag{3.1.142}$$

where $[p]$ denotes $p(u(x_s^+, t)) - p(u(x_s^-, t))$. For each choice of p and q satisfying $p' - uq' = 0$, (3.1.142) gives a different shock condition.

In general, the correct conservation law follows either from physical considerations (mass conservation, momentum conservation, etc.) or knowledge of the limiting form, as $\epsilon \to 0$, of the exact solution of a higher-order governing equation. We will show next that for Burgers' equation we must use (3.1.138) as it corresponds to the limiting form as $\epsilon \to 0$ of exact solutions.

Initial-value problem, straight shock layer, asymptotic solution

One way to generate a discontinuity of the solution of the lower-order problem (3.1.137) is by considering discontinuous initial data. Let us study the case where $u(x, 0; \epsilon)$ is constant on either side of a point of discontinuity. As discussed earlier, we may take the case (3.1.126) with no loss of generality. Later, we will study the case (3.1.128).

The leading term u_0 of the outer expansion obeys (3.1.137), and the subcharacteristics of this equation are solutions of

$$\frac{dt}{ds} = 1, \tag{3.1.143a}$$

$$\frac{dx}{ds} = u, \tag{3.1.143b}$$

$$\frac{du_0}{ds} = 0. \tag{3.1.143c}$$

That is, u_0 is a constant equal to its initial value along the straight subcharacteristics with speed $\frac{dx}{dt} = u_0$. Since u_0 is initially piecewise constant, we find the two families (parameterized in terms of ξ) of intersecting subcharacteristics

$$u_0 = -1 \quad \text{on} \quad x = \xi - t; \ \ \xi > 0$$

$$u_0 = 1 \quad \text{on} \quad x = \xi + t; \ \ \xi < 0.$$

To avoid a two-valued solution in the triangular region $-t < x < t$ of the xt plane, we insert a shock that starts at $x = 0$, $t = 0$ and obeys (see (3.1.139)) $(dx_s/dt) = \frac{1}{2}(-1 + 1) = 0$. Thus, we need a stationary shock $x_s = 0$, as is obvious from symmetry. The shock prevents crossing of subcharacteristics from either side, and we find the outer limit (3.1.130) that we derived earlier from the exact solution.

We now introduce a shock layer around $x = 0$, and it is easily seen that this layer must have $O(\epsilon)$ thickness for a distinguished limit. Thus, (3.1.117) becomes

$$\epsilon \frac{\partial u^*}{\partial t} + u^* \frac{\partial u^*}{\partial x^*} = \frac{\partial^2 u^*}{\partial x^{*2}}, \tag{3.1.144}$$

where $x^* = x/\epsilon$ and $u^*(x^*, t; \epsilon) = u(\epsilon x^*, t; \epsilon)$. Expanding $u^* = u_0^* + \epsilon u_1^* + \dots$, we see that the inner limit u_0^* obeys

$$\frac{\partial^2 u_0^*}{\partial x^{*2}} - u_0^* \frac{\partial u_0^*}{\partial x^*} = 0. \tag{3.1.145}$$

The general solution of (3.1.145) is (see (2.3.11)–(2.3.12))

$$u_0^*(x^*, t) = -c_0^* \tanh \frac{c_0^*}{2}(x^* + k) \tag{3.1.146a}$$

or

$$u_0^*(x^*, t) = -c_0^* \coth \frac{c_0^*}{2}(x^* + k), \tag{3.1.146b}$$

where c_0^* and k are functions of t. Matching with the outer limit $u_0 = \pm 1$ as $x^* \to \mp\infty$ shows that we must use the tanh solution and that $c_0^* = 1$. *The matching to $O(1)$ does not determine k* (see (2.3.21)). Consideration of u_1^*, the next term in the inner expansion, does not help. In fact, with $u_0^* = -\tanh(x^* + k)/2$, we find

$$u_1^* = c_1^* \operatorname{sech}^2 \frac{1}{2}(x^* + k) + c_2^*[x^* \operatorname{sech}^2 \frac{1}{2}(x^* + k) + \frac{1}{2} \sinh(x^* + k) \operatorname{sech}^2 \frac{1}{2}(x^* + k)] \tag{3.1.147}$$

where c_1^* and c_2^* are unknown functions of t^*. Matching requires $u_1^* \to 0$ as $|x^*| \to \infty$. Since $\frac{1}{2} \sinh(x^* + k) \operatorname{sech}^2 \frac{1}{2}(x^* + k) \to \pm 1$ as $x^* \to \pm\infty$, we must set $c_2^* = 0$. Thus, $u_1^* = c_1^* \operatorname{sech}^2(x^* + k)/2$, but because $\operatorname{sech}^2(x^* + k)/2$ decays exponentially as $|x^*| \to \infty$, c_1^* is a second arbitrary function!

Of course, for this example, symmetry considerations directly demand $k = 0$. In fact, we know that (3.1.117), together with the initial data (3.1.126), is invariant under the transformation $x \to -x, t \to t, u \to -u$. Therefore, the exact solution must be an odd function of x, $u(x, t; \epsilon) = -u(-x, t; \epsilon)$, and this immediately implies $k = 0$, and we recover the limiting result (3.1.131) obtained from the exact solution.

Initial-value problem, corner layer, asymptotic solution

The solution of the subcharacteristic equations (3.1.143) for the initial condition (3.1.130) is found in the form (3.1.131). The constant solutions $u = -1$ if $x < -t$ and $u = 1$ if $x > t$ follow directly. The solution $u = x/t$ (centered fan) in the triangular region $-t < x < t$ can be calculated in various ways. One approach is to artificially smooth the initial data over a small interval $-\delta < x < \delta$ centered at $x = 0$, calculate the solution for the smooth data, then take the limit $\delta \to 0$. A second approach is to look for a similarity solution. One could also insert N straight shocks with finite jumps centered at the origin to discover that the entropy condition (3.1.140) is violated unless $N \to \infty$ and the jump across each shock is infinitesimal, i.e., we end up with the family of characteristics $x/t =$ constant.

The discontinuity in u_x and u_t along $x = \pm t$ is smoothed by a corner-layer solution. Consider the ray $x = t$. We are looking for a solution that has $u \approx 1$ in some neighborhood of $x = t$. Therefore, we look for a distinguished limit for

Burgers' equation for the scaling

$$u_c = \frac{u-1}{\alpha(\epsilon)}; \qquad x_c = \frac{x-t}{\beta(\epsilon)},$$

where $\alpha \ll 1$ and $\beta \ll 1$. Equation (3.1.117) transforms to

$$\alpha \frac{\partial u_c}{\partial t} + \frac{\alpha^2}{\beta} u_c \frac{\partial u_c}{\partial x_c} = \frac{\epsilon \alpha}{\beta^2} \frac{\partial^2 u_c}{\partial x_c^2}.$$

For a distinguished limit, we must have $\alpha = \beta = \sqrt{\epsilon}$ to find the full equation. As noted in earlier examples for ordinary differential equations, a corner layer requires an exact solution of the full governing equation (see (2.3.16)). The exact solution we need here is just (3.1.132) expressed in terms of u_c:

$$u_c = -\frac{2}{\sqrt{\pi t}} \frac{e^{-x_c^2/4t}}{\operatorname{erfc}\left(-\frac{x_c}{2\sqrt{t}}\right)}. \tag{3.1.148}$$

Curved shock layer, asymptotic solution

Consider the piecewise constant initial data

$$u(x, 0; \epsilon) = \begin{cases} 1 & \text{if } x < 0 \\ -1 & \text{if } 0 < x < 1 \\ 0 & \text{if } 1 < x < \infty. \end{cases} \tag{3.1.149}$$

The subcharacteristics are the family of straight lines shown in Figure 3.1.12.

Clearly, we need to introduce a shock at the origin, and as the values of u_0 on either side of this shock are ± 1 we have a stationary shock $x_s = 0$ for $0 \le t \le 1$.

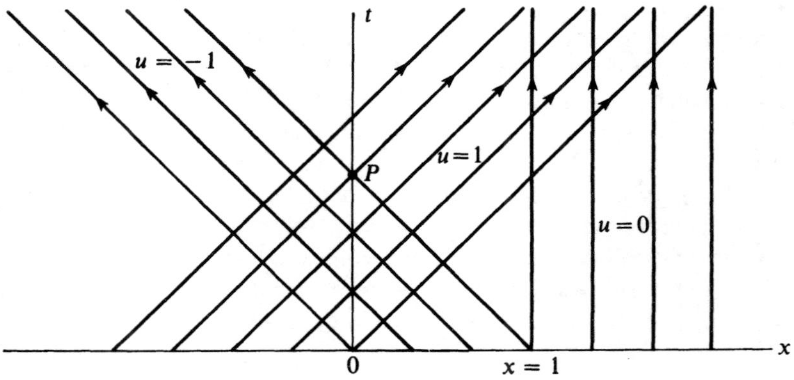

FIGURE 3.1.12. Subcharacteristics for (3.1.149)

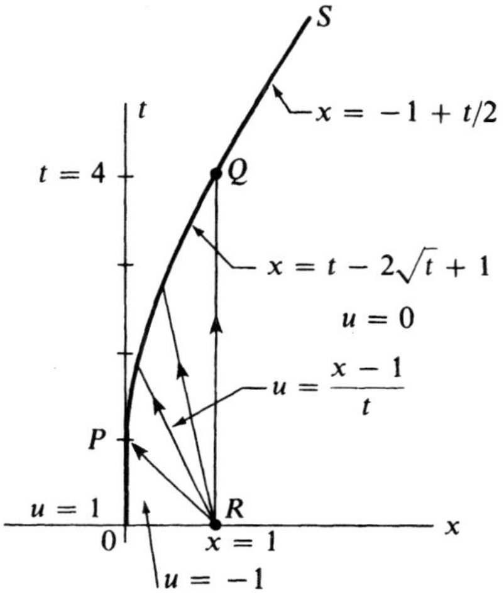

FIGURE 3.1.13. Outer Limit for (3.1.149)

Next, we insert the centered fan $u_0 = (x - 1)/t$ for $1 - t \leq x \leq 1$, as shown in Figure 3.1.13.

Above the point P, $(t > 1)$, the shock must curve because the values of u to the right are no longer constant. Using the shock condition (3.1.139), we find that

$$\frac{dx_s}{dt} = \frac{1}{2}\left[\frac{x_s - 1}{t} + 1\right], \tag{3.1.150}$$

and we need to solve this for the initial condition $x_s = 0$ at $t = 1$. The result is easily found:

$$x_s = t - 2\sqrt{t} + 1 \qquad 1 \leq t \leq 4. \tag{3.1.151}$$

The reason that (3.1.151) is valid only up to $t = 4$ is because above $Q : x = 1$, $t = 4$, we have $u(x_s^+, t) = 0$. Therefore, the shock remains straight from then on and is governed by $dx_s/dt = 1/2$, i.e., $x_s = -1 + t/2$.

The shock layers appropriate along the straight segments OP and QS have been discussed, and we concentrate on the shock layer along the curved arc PQ. Introduce the inner variable $x^* = (x - x_s(t))/\epsilon$ and expand $u(x, t; \epsilon) = u_0^*(x^*, t) + \epsilon u_1^*(x^*, t) + \ldots$ to calculate the following governing equations for u_0^* and u_1^*:

$$\frac{\partial^2 u_0^*}{\partial x^{*2}} + (\dot{x}_s(t) - u_0^*)\frac{\partial u_0^*}{\partial x^*} = 0, \tag{3.1.152a}$$

$$\frac{\partial^2 u_1^*}{\partial x^{*2}} + \dot{x}_s(t)\frac{\partial u_1^*}{\partial x^*} - \frac{\partial}{\partial x^*}(u_0^* u_1^*) = \frac{\partial u_0^*}{\partial t}. \tag{3.1.152b}$$

The solution of (3.1.152a) for u_0^* is (see (3.1.146a))

$$u_0^* = \dot{x}_s(t) - c_0^* \tanh \frac{c_0^*}{2}(x^* + k), \tag{3.1.153}$$

where c_0^* and k are functions of t. Again, we discard the coth solution as it is not appropriate for an interior layer.

To carry out the matching, we introduce the variable $x_\eta = (x - x_s(t))/\eta(\epsilon)$ for a class of $\eta(\epsilon)$ contained in $\epsilon \ll \eta \ll 1$. Thus, $x \to x_0$ and $x^* \to \pm\infty$ in the matching. The leading term of the outer expansion resulting from u_0 becomes

$$u = \frac{x_s(t) - 1}{t} + O(\eta), \tag{3.1.154a}$$

whereas the leading term of the inner expansion gives

$$u = \dot{x}_s(t) - c_0^*(t) + O(\epsilon) + \text{ T.S.T.} \tag{3.1.154b}$$

The contribution of (3.1.153) to (3.1.154b) is transcendentally small as long as $\epsilon|\log\epsilon| \ll \eta$. The $O(\epsilon)$ term we have ignored in (3.1.154b) comes from ϵu_1^*.

Matching to $O(1)$ demands that (3.1.154a) and (3.1.154b) agree and this determines c_0^*:

$$c_0^* = \dot{x}_s(t) - \frac{x_s - 1}{t} = \frac{1}{\sqrt{t}}. \tag{3.1.155}$$

Again, the matching to $O(1)$ does not determine k, and now we do not have a symmetry condition to calculate k. In this simple case, it is possible to derive an explicit exact solution and determine k from this result (Problem 12).

Boundary-value problem, asymptotic solution

The structure of asymptotic solutions for initial and boundary-value problems for Burgers' equation can be illustrated with the following simple special case on $0 \le x < \infty, 0 \le t < \infty$, with

$$u(x, 0; \epsilon) = A = \text{ constant} \tag{3.1.156a}$$

$$u(0, t; \epsilon) = B = \text{ constant} \tag{3.1.156b}$$

for varying values of the constants A and B taken to be independent of ϵ.

The simplest choice is $A = B$, in which case the *exact* solution is $u = A = B$. This gives us a starting point for studying various possible combinations of A and B as we did in the example of Section 2.3. We consider first the case (see Figure 3.1.14):

(I) $0 < B < A$. In this case, the outer limit consists of a centered fan $u = x/t$ between the two uniform states $u = A$, $u = B$, as shown in Figure 3.1.15.

To smooth out the discontinuities in u_x along the rays $x = Bt$, $x = At$, we introduce corner layers similar to the ones given by (3.1.148).

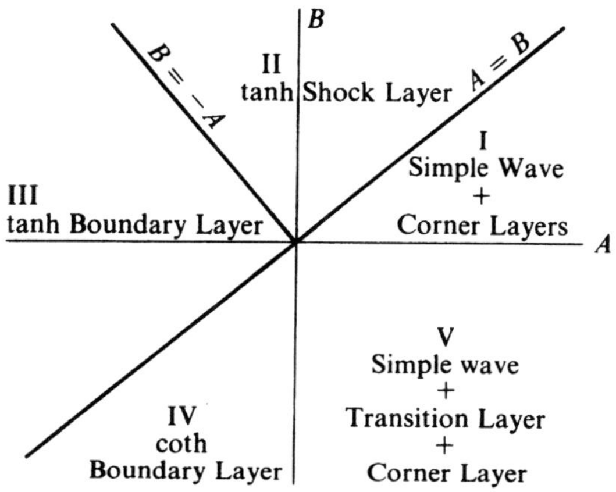

FIGURE 3.1.14. Possible Solutions in the AB Plane

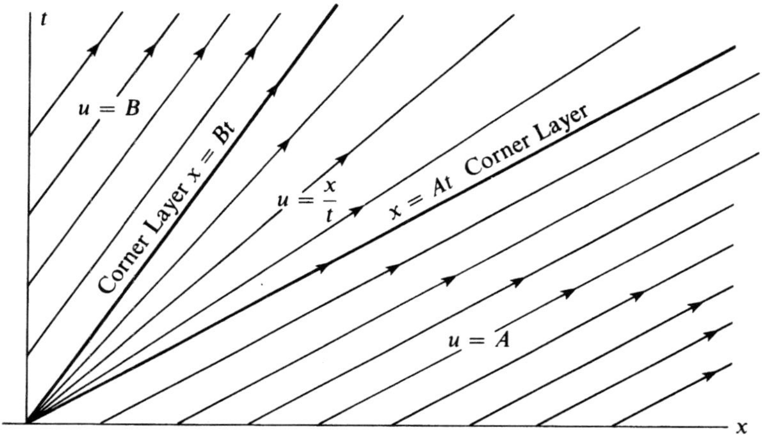

FIGURE 3.1.15. Solutions for $0 < B < A$

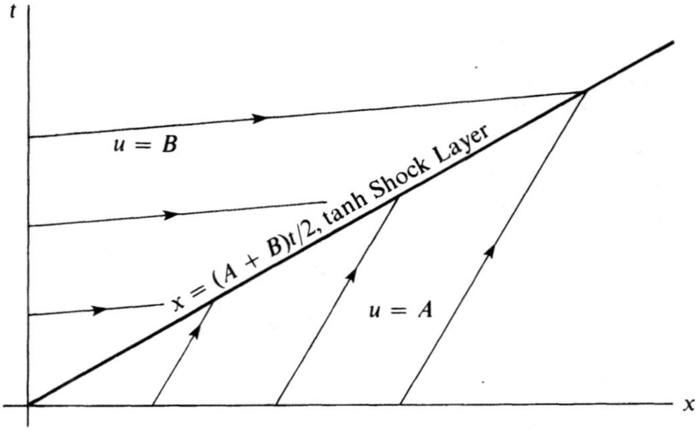

FIGURE 3.1.16. Solution for $0 < |A| < B$

(II) $0 < |A| < B$. In this case, the subcharacteristics flowing out of the x and t axes intersect and we must introduce a shock with slope $dx/dt = (A+B)/2 > 0$. The shock divides the first quadrant of the xt plane into two regions of uniform u, as shown in Figure 3.1.16.

To smooth out the discontinuity in u along the shock, we introduce a shock layer of order ϵ thickness along $x = (A + B)t/2$.

As discussed earlier, this is a solution of Equation (3.1.152a) with $\dot{x}_s = \text{const.} = b$, i.e.,

$$u_0^* = b - c_0^* \tanh c_0^*(x^* + k)/2 \tag{3.1.157a}$$

or

$$u_0^* = b - c_0^* \coth c_0^*(x^* + k)/2. \tag{3.1.157b}$$

As will become evident, the coth branch of the solution will also be needed in certain cases.

In the present case, we use (3.1.157a) with $b = (A + B)/2$; then matching to $O(1)$ requires that

$$u_0^*(\infty, t^*) = A; \quad u_0^*(-\infty, t^*) = B. \tag{3.1.158}$$

These two conditions are satisfied by setting $c_0^* = B - (A + B)/2 = (B - A)/2$. Again, the value of k is not determined by the matching. As pointed out in [2.8] (see page 216), the appropriate phase shift that follows from the exact solution (3.1.134) for the case $A = 0$ is $k = -\log\sqrt{2}$.

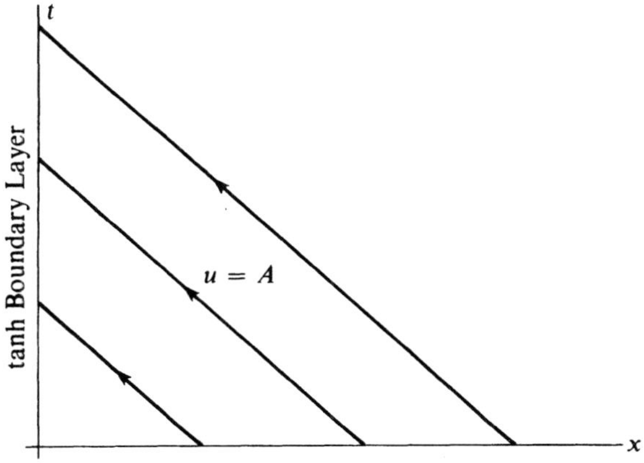

FIGURE 3.1.17. Solution for $A < 0, |B| < -A$

(III) $A < 0$, $|B| < -A$. In this case, the characteristics emerging from the x axis intersect the t axis, and no shock can exist in the first quadrant ($t > 0$, $x > 0$), as shown in Figure 3.1.17.

Thus, we need a boundary layer along $x = 0$ to satisfy the condition $u(0, t) = B$. Such a boundary layer can be constructed using the tanh solution, (3.1.157a), with $b = 0$.

The boundary condition $u(0, t) = B$ requires that

$$B = -c_0^* \tanh c_0^* \frac{k}{2}, \tag{3.1.159a}$$

and the matching condition gives

$$A = -c_0^*. \tag{3.1.159b}$$

Solving this for c_0^* and k, we obtain

$$c_0^* = -A; \qquad k = -\frac{2}{A} \tanh^{-1} \frac{B}{A}. \tag{3.1.160}$$

Note that $\tanh^{-1}(B/A)$ is real only if $|B/A| \leq 1$, so that $B = A < 0$ is the limit beyond which we cannot use a tanh solution. However, we do have the coth solution, (3.1.157b), and this is needed in case IV.

(IV) $A < 0$, $B < 0$, $|B/A| > 1$. Now, the outer solution is as in case III. Using a coth boundary layer, we calculate

$$c_0^* = -A, \tag{3.1.161a}$$

$$k = -\frac{2}{A} \coth^{-1}(B/A). \tag{3.1.161b}$$

These cases exhaust the possibilities of using layers of order ϵ thickness, but we have not been able to handle the case $A > 0$, $B < 0$.

(V) $A > 0$, $B < 0$. Now, the outer solution consists of a uniform region and a centered fan, as sketched in Figure 3.1.18.

Thus, on the boundary $x = 0$, the outer limit, $u = 0$, disagrees with the boundary condition $u = B < 0$. It is clear that neither the tanh nor the coth solutions can be used, as they do not increase from a negative value to zero as $x^* \to \infty$.

The situation here is analogous to case (V) of Section 2.3. We need to introduce a transition layer of thickness $\sqrt{\epsilon}$ at the origin

$$x_T = \frac{x}{\sqrt{\epsilon}}, \tag{3.1.162a}$$

$$u_T = u/\sqrt{\epsilon}. \tag{3.1.162b}$$

Again, we find that u_T satisfies the full equation, and we match this solution with a thinner ($O(\epsilon)$) boundary layer at the origin (see the discussion at the end of Section 2.3). Of course, in addition to the above, we need a corner-layer solution along $x = At$.

Finally, we note that since Burgers' equation is invariant under the transformation $u \to -u$, $x \to \overline{X} - x$ for any $\overline{X}$, the solution corresponding to a right

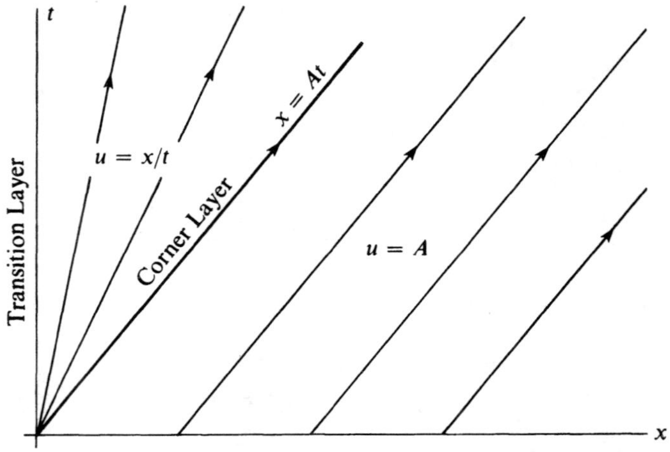

FIGURE 3.1.18. Solution for $A > 0$, $B < 0$

boundary can be derived by symmetry from the above discussion of the case of a left boundary.

Problems

1. Consider steady-state heat conduction in the rectangular region $0 \leq x \leq L_1$, $-L_2 \leq Y \leq L_2$. Assume that the temperature is prescribed along the edges $X = 0, L_1$ and that the edges $Y = \pm L_2$ are insulated. We are interested in the limiting case $(L_2/L_1) \ll 1$. If we normalize X with respect to L_1 and Y with respect to L_2, we need to solve ($\epsilon = L_2/L_1$)

$$\epsilon^2 u_{xx} + u_{yy} = 0 \tag{3.1.163}$$

in $0 \leq x \leq 1$; $-1 \leq y \leq 1$, with boundary conditions

$$u(0, y; \epsilon) = f(y), \tag{3.1.164a}$$

$$u(1, y; \epsilon) = g(y), \tag{3.1.164b}$$

$$u_y(x, 1; \epsilon) = u_y(x, -1; \epsilon) = 0. \tag{3.1.165}$$

a. Construct an outer expansion in the form

$$u(x, y; \epsilon) = u_0(x, y) + \epsilon^2 u_1(x, y) + \ldots \tag{3.1.166}$$

and by requiring u_0 and u_1 to satisfy (3.1.165) show that

$$u_0 = a_0 x + b_0(1 - x), \tag{3.1.167}$$

where a_0 and b_0 are undetermined constants.

b. Introduce appropriate boundary layers along the edges $x = 0$ and $x = 1$, and show that matching with the outer limit (3.1.167) gives

$$a_0 = \frac{1}{2}\int_{-1}^{1} g(y)dy, \tag{3.1.168a}$$

$$b_0 = \frac{1}{2}\int_{-1}^{1} f(y)dy. \tag{3.1.168b}$$

c. Solve the problem exactly and verify your results in parts (a) and (b).

2. Consider the simple elliptic equation

$$\epsilon(u_{xx} + u_{yy}) = u_y \tag{3.1.169}$$

for $0 < \epsilon \ll 1$ in the interior of the unit square $0 < x < 1, 0 < y < 1$, with mixed boundary conditions

$$u_y(x, 0) = 0, \tag{3.1.170a}$$

$$u(1, y) = -1, \tag{3.1.170b}$$

$$u(0, y) = 1, \tag{3.1.170c}$$

$$u(x, 1) = 0. \tag{3.1.170d}$$

a. Show that the outer limit is

$$u_0 = A_0 x + B_0 \tag{3.1.171}$$

and the boundary-layer limit is

$$u_{BL} = (A_0 x + B_0)(1 - e^{y^*}), \tag{3.1.172}$$

where $y^* = (y-1)/\epsilon$.

b. Since the transformation $x \to 1-x, y \to y, u \to -u$ leaves the equation and boundary condition invariant, we must have

$$u(x, y; \epsilon) = -u(1-x, y; \epsilon). \tag{3.1.173}$$

Use this condition to conclude that $A_0 = -2B_0$. Thus,

$$u_0 = B_0(1-2x), \tag{3.1.174}$$

$$u_{BL} = B_0(1-2x)(1-e^{y^*}). \tag{3.1.175}$$

c. Show that side boundary layers of $O(\sqrt{\epsilon})$ thickness can also be introduced along $x = 0$ and $x = 1$, and derive these in the form

$$u^*_{BL} = -1 + (B_0 - 1)\,\text{erf}\left(\frac{x-1}{2\sqrt{\epsilon y}}\right), \tag{3.1.176}$$

$$u^{**}_{BL} = 1 + (B_0 - 1)\,\text{erf}\left(\frac{x}{2\sqrt{\epsilon y}}\right). \tag{3.1.177}$$

d. Show that a possible solution that is uniformly valid to $O(1)$ everywhere inside the unit square (except in $O(\epsilon)$ neighborhoods around $x = 1, y = 1$ and $x = 0$, $y = 1$) is given by

$$u_{uv} = B_0(1-2x)(1-e^{\frac{y-1}{\epsilon}})+(B_0-1)\left[\text{erf}\left(\frac{x-1}{2\sqrt{\epsilon y}}\right)+\text{erf}\left(\frac{x}{2\sqrt{\epsilon y}}\right)\right]. \tag{3.1.178}$$

e. Use (3.1.178) in conjunction with an appropriate variational principle to show that $B_0 = 1$, i.e., there are no side boundary layers.

3. Discuss in detail the solution of (3.1.169) to $O(1)$ inside the annular region $1/2 < r < 1, 0 < \theta \leq 2\pi$ with boundary condition

$$u(1/2, \theta) = 1 \tag{3.1.179a}$$

$$u(1, \theta) = \sin\theta. \tag{3.1.179b}$$

Here $x = r\cos\theta$ and $y = r\sin\theta$.

4. Consider the problem

$$\epsilon(u_{xx} + u_{yy}) = yu_x + xu_y \tag{3.1.180}$$

in the interior of the unit circle $x^2 + y^2 < 1$, with boundary condition

$$u(1, \theta) = f(\theta). \tag{3.1.181}$$

Now the subcharacteristics have a saddle point singularity at the origin. Determine the locations of boundary and interior layers and derive the solution in each of these regions.

5. Carry out the details of the calculations leading to (3.1.57).
6. Study (3.1.43) inside a simply connected bounded domain containing the origin. Let the boundary be given by

$$r = r_B(\theta) \tag{3.1.182}$$

and assume that the boundary condition is specified in the form

$$u(r_B(\theta), \theta) = f(\theta). \tag{3.1.183}$$

a. Assume that the point $P : r = 1, \theta = 0$, is a unique closest point to the origin and that near this point

$$r_B(\theta) = 1 + a\theta^2 + \dots, \quad \text{as } \theta \to 0; \ a > 0. \tag{3.1.184}$$

Thus, at P, r_B has a first-order contact with the unit circle. Show that in this case

$$\alpha = f(0). \tag{3.1.185}$$

b. Let the boundary have *two* distinct points ($P_1 : r = 1, \theta = 0$ and $P_2 : r = 1, \theta = \pi$) nearest the origin. Also assume that r_B has the following behavior near these points:

$$r_B(\theta) = 1 + a\theta^{2p} + \dots \text{ as } \theta \to 0; \ a > 0, \tag{3.1.186a}$$

$$r_B(\theta) = 1 + b(\theta - \pi)^{2q} + \dots \text{ as } \theta \to \pi; \ b > 0. \tag{3.1.186b}$$

Show that if $p \neq q$

$$\alpha = \begin{cases} f(0); & \text{if } p > q \\ f(\pi); & \text{if } p < q \end{cases} \tag{3.1.187a}$$

and that if $p = q$

$$\alpha = \frac{\sqrt{b} f(0) + \sqrt{a} f(\pi)}{\sqrt{a} + \sqrt{b}}. \tag{3.1.187b}$$

7. Consider the wave equation

$$\epsilon(u_{xx} - u_{tt}) = au_x + bu_t, \tag{3.1.188}$$

$$u(x, 0; \epsilon) = 0, \tag{3.1.189a}$$

$$u_t(x, 0; \epsilon) = V(x). \tag{3.1.189b}$$

a. Calculate the exact solution in the form

$$u(x, t; \epsilon) = \frac{1}{2} e^{-bt/2\epsilon} \int_{-t}^{t} V(x + \zeta) e^{-a\zeta/2\epsilon} I_0(g(\zeta)) d\zeta, \tag{3.1.190}$$

where I_0 is the modified Bessel function of the first kind of order zero, and

$$\zeta = \xi - x, \ g(\zeta) = \sqrt{(b^2 - a^2)(t^2 - \zeta^2)}/2\epsilon.$$

b. Use the asymptotic behavior

$$I_0(z) \sim (2\pi z)^{-1/2} e^z \text{ as } z \to \infty \tag{3.1.191}$$

to show that the outer limit ($\epsilon \to 0$, x, t fixed) is

$$u_0 = \frac{\epsilon}{b} V\left(x - \frac{a}{b}t\right). \tag{3.1.192}$$

c. Show that the inner expansion ($\epsilon \to 0$, x, $t^* = \frac{t}{\epsilon}$ fixed) is

$$u^*(x, t^*; \epsilon) = \sum_{n=0}^{\infty} \epsilon^{n+1} \frac{V^{(n)}(x)}{n!} \int_{-t^*}^{t^*} \tau^n g(t^*, \tau) d\tau, \tag{3.1.193}$$

where

$$g(t^*, \tau) = \frac{1}{2} e^{-\frac{bt^* + a\tau}{2}} I_0\left(\frac{1}{2}\sqrt{(b^2 - a^2)(t^{*^2} - \tau^2)}\right). \tag{3.1.194}$$

8. Consider the wave equation

$$\epsilon\left(\frac{\partial}{\partial t} + c_1 \frac{\partial}{\partial x}\right)\left(\frac{\partial}{\partial t} + c_2 \frac{\partial}{\partial x}\right) u + \left(\frac{\partial}{\partial t} + b\frac{\partial}{\partial x}\right) u = 0, \tag{3.1.195}$$

where $0 < \epsilon \ll 1$, c_1, c_2, and b are constants with $c_1 > 0$ and $c_2 < 0$.
a. Show that the stability condition is

$$c_2 < b < c_1. \tag{3.1.196}$$

b. For the stable case, study the signaling problem

$$u(x, 0; \epsilon) = 0, \tag{3.1.197a}$$

$$u_t(x, 0; \epsilon) = 0, \tag{3.1.197b}$$

$$u(0, t; \epsilon) = F(t) \text{ for } t > 0. \tag{3.1.197c}$$

In particular, for $b > 0$ show that the outer limit is

$$u_0 = \begin{cases} 0 & \text{if } t < x/b \\ F(t - \frac{x}{b}) & \text{if } t > x/b \end{cases} \tag{3.1.198}$$

and that there is an interior layer along the $x = bt$ subcharacteristic with leading term

$$U_0^*(x^*, t) = \frac{F(0^+)}{2} \operatorname{erfc}\left(\frac{x^*}{2\sqrt{\kappa t}}\right), \tag{3.1.199}$$

where $\kappa = (c_1 - b)(b - c_2)$ and $x^* = (x - bt)/\sqrt{\epsilon}$.

Show that for $b < 0$, the outer limit is $u_0 = 0$ and the boundary-layer limit is given by

$$U_0(x^*, t) = F(t) \exp(-bx^*/c_1 c_2), \tag{3.1.200}$$

where $x^* = x/\epsilon$.

c. For the special case

$$F = \begin{cases} 1 & \text{if } 0 < t \le 1 \\ 0 & \text{for } t > 1, \end{cases} \tag{3.1.201}$$

study the behavior of the signal for t large.

9. An incompressible fluid (density ρ, specific heat c, thermal conductivity k) flows through a circular grid of radius L located at $x = 0$. The velocity of the fluid is assumed always to be U in the $+x$ direction. The temperature of the grid is maintained at $T = T_B = \text{const.}$, and the temperature of the fluid at upstream infinity is T_∞. Thus, the differential equation for the temperature field is

$$\kappa \left\{ \frac{\partial^2 T}{\partial x^2} + \frac{\partial^2 T}{\partial r^2} + \frac{1}{r}\frac{\partial T}{\partial r} \right\} = U \frac{\partial T}{\partial x}, \tag{3.1.202}$$

where $\kappa = k/\rho c$. Write the problem in suitable dimensionless coordinates. [Use $x^* = x/L$, $r^* = r/L$, $T(x, r) = T_\infty + (T_B - T_\infty)\theta(x^*, r^*)$.] Study the behavior of the solution for ϵ small, where $\epsilon = \kappa/UL$. Find the outer solution and the necessary boundary layers. In particular, show that the rate of heat transfer to the fluid is independent of k as $\epsilon \to 0$.

Discuss the validity of these solutions for the regions where $x^* \to \infty$.

10. Consider heat transfer to a viscous incompressible fluid flowing steadily in a circular pipe of radius R. The equation for the temperature distribution is

$$\rho c u(r) \frac{\partial T}{\partial x} = k \left\{ \frac{\partial^2 T}{\partial r^2} + \frac{1}{r}\frac{\partial T}{\partial r} + \frac{\partial^2 T}{\partial x^2} \right\}, \tag{3.1.203}$$

where ρ = fluid density, c = fluid specific heat = const., k = thermal conductivity = const. For laminar flow, the velocity distribution in the pipe is parabolic:

$$\frac{u(r)}{U} = 1 - \left(\frac{r}{R}\right)^2. \tag{3.1.204}$$

Let the temperature be raised at the wall from the constant value T_0 for $x < 0$ to T_1 for $x > 0$ (see Fig. 3.1.19).

For $\epsilon = (k/\rho c U R) \ll 1$, construct a suitable boundary-layer theory. Note that $\epsilon = 1/(\text{Re Pr})$, where Re is the Reynolds number and Pr is the Prandtl number. Thus, show that $Q(\ell)$, the heat transferred to the fluid in the length ℓ, is approximately

$$\frac{Q}{\{[k(T_i - T_0)]/R\}(2\pi R\ell)} = -\frac{3}{\Gamma(1/3)}\left(\frac{3}{4\epsilon}\right)^{1/3}\left(\frac{R}{\ell}\right)^{1/3}, \tag{3.1.205}$$

where the minus sign indicates that heat is flowing from the pipe into the flow. *Hint*: Use similarity methods or Laplace transforms to solve the boundary-layer equation. Indicate the mathematical problem to be solved for the next

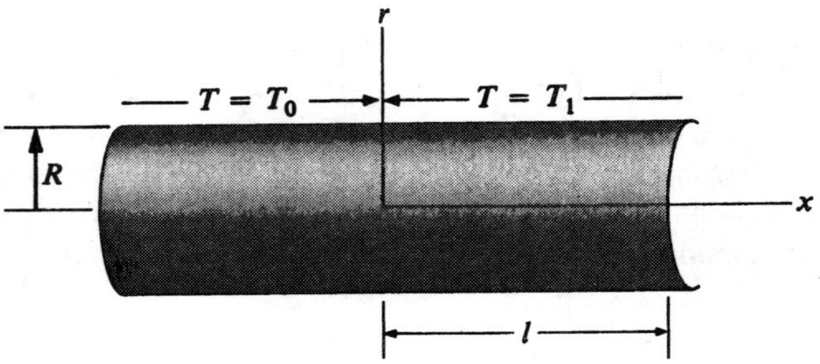

FIGURE 3.1.19. Heat Transfer in Pipe Flow

higher-order approximation. Is there a singularity of heat transfer at $x = 0$ in the higher approximation?

11. Calculate the solution to $O(1)$ of

$$\epsilon u_{xx} = u_t + \cos t u_x, \quad 0 \le x,\ 0 \le t \tag{3.1.206}$$

with initial condition $u(x, 0) = 0$ and boundary condition $u(0, t) = 1$. Note the locations of shock and boundary layers, and calculate the inner limit in these layers. Also, note the points near which exact local solutions are needed.

12. Calculate the exact solution of Burgers' equation for the initial data (3.1.149). Use this result to derive the interior-layer limit that is appropriate near the shock and obtain the expression for the shift $k(t)$.

13. Consider the following boundary-value problem for Burgers' equation on $0 \le x \le 1$:

$$u_t + uu_x = \epsilon u_{xx}, \tag{3.1.207}$$

$$u(0, t) = 1 \quad \text{if } t > 0, \tag{3.1.208a}$$

$$u(1, t) = 0 \quad \text{if } t > 0, \tag{3.1.208b}$$

$$u(x, 0) = 0. \tag{3.1.209}$$

a. Introduce appropriate shock and boundary layers and calculate the uniformly valid solution to $O(1)$.
b. Derive the exact solution and verify your results in part (a).

3.2 Boundary-Layer Theory in Viscous, Incompressible Flow

The original physical problem from which the ideas of mathematical boundary layer theory originated was the problem of viscous, incompressible flow past an object. The aim was to explain the origin of the resistance in a slightly viscous fluid. By the use of physical arguments, Prandtl, in [3.34], deduced that for small values of the viscosity a thin region near the solid boundary (where the fluid is brought to rest) is described by approximate boundary-layer equations, and the flow outside this region is essentially inviscid. These ideas find a natural mathematical expression in terms of the ideas of singular perturbation problems discussed in Chapter 2. In this section, we show how the external inviscid flow is associated with an outer limit process and the boundary layer with an inner limit process. The boundary condition of no slip is lost, and the order of the equations is lowered in the outer limit, so that the problem is indeed singular in the terminology used previously.

In order to illustrate these ideas explicitly, the entire discussion should be carried out in dimensionless variables. Consider uniform flow with velocity U past an object with characteristic length L (Figure 3.2.1). Given the fluid density, ρ, and viscosity coefficient, μ, there is one overall dimensionless number, the Reynolds number, Re:

$$\mathrm{Re} = UL/\nu, \quad \nu = \mu/\rho. \tag{3.2.1}$$

Pressure does not enter a dimensionless parameter since the level of pressure has no effect on the flow. That is, the Mach number is always zero. The limit processes are all concerned with Re $\to \infty$ or

$$\epsilon = 1/\mathrm{Re} \to 0. \tag{3.2.2}$$

Now, make all velocities dimensionless with U, all lengths with L, and the pressure with ρU^2. The full problem is expressed by the continuity and momentum equations, written in an invariant vector form as follows.

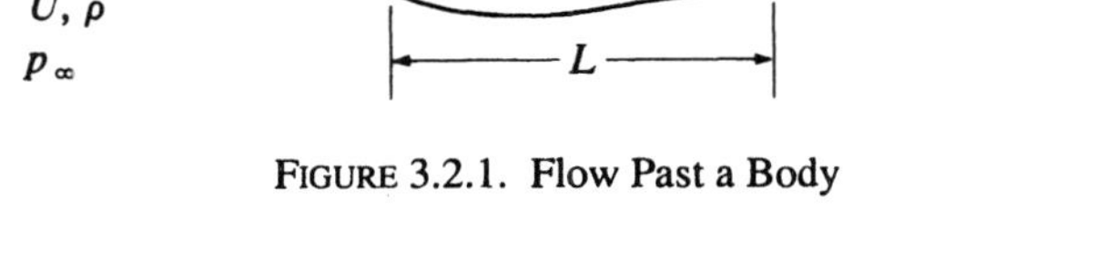

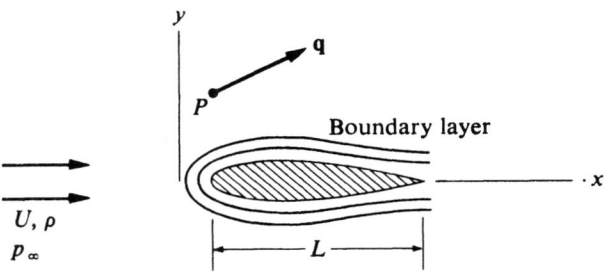

FIGURE 3.2.1. Flow Past a Body

Navier-Stokes equations

$$\text{div}\,\mathbf{q} = 0, \tag{3.2.3a}$$

$$\underbrace{\mathbf{q}\cdot\text{grad}\,\mathbf{q} \equiv \text{grad}\left(\frac{q^2}{2}\right) - (\mathbf{q}\times\boldsymbol{\omega})}_{\text{transport or inertia}} = \underset{\text{pressure}}{-\text{grad}\,p} \underset{\text{viscous body force}}{- \epsilon\,\text{curl}\,\boldsymbol{\omega}}, \tag{3.2.3b}$$

where

$$\boldsymbol{\omega} = \text{vorticity} = \text{curl}\,\mathbf{q}. \tag{3.2.3c}$$

Vorticity represents the angular velocity of a fluid element. Also, it can be shown that the viscous force per area on a surface is

$$\tau_v = -\epsilon(\boldsymbol{\omega}\times\mathbf{n}), \tag{3.2.4}$$

where $\mathbf{n}$ is the unit outward normal (dimensionless). The uniform flow boundary condition is

$$\mathbf{q}(\infty) = \mathbf{i}, \tag{3.2.5}$$

and the no-slip condition on the body is

$$\mathbf{q} = \mathbf{q}_b = 0. \tag{3.2.6}$$

The outer expansion is carried out, keeping the representative point P fixed and letting $\epsilon \to 0$. The expansion has the form of an asymptotic expansion in terms of a suitable sequence $\alpha_i(\epsilon)$:

$$\mathbf{q}(P;\epsilon) = \mathbf{q}_0(P) + \alpha_1(\epsilon)\mathbf{q}_1(P) + \dots, \tag{3.2.7a}$$

$$p(P;\epsilon) = p_0(P) + \alpha_1(\epsilon)p_1(P) + \dots. \tag{3.2.7b}$$

Here $(\mathbf{q}_0, p_0)$ represents an inviscid flow. In many cases of interest, this flow is irrotational. This fact can be demonstrated by considering the equation for vorticity propagation obtained by taking the curl of (3.2.3b),

$$\text{curl}\,(\mathbf{q}\times\boldsymbol{\omega}) = \epsilon\,\text{curl}\,\text{curl}\,\boldsymbol{\omega}. \tag{3.2.8}$$

For plane flow, the vorticity vector is normal to the xy plane and can be written

$$\boldsymbol{\omega} = \omega(P)\mathbf{k}. \tag{3.2.9}$$

Thus, (3.2.8) can be written

$$\mathbf{q}\cdot\text{grad}\,\omega = \epsilon\Delta\omega, \tag{3.2.10}$$

where Δ denotes the Laplacian ($\Delta \equiv \text{div grad}$).

The physical interpretation of (3.2.10) is that the vorticity is transported along the streamlines but diffuses (like heat) due to the action of viscosity. The solid boundary is the only source of vorticity; as $\epsilon \to 0$, the diffusion is small and is confined to a narrow boundary layer close to the body (except if the flow separates). Under the outer limit, (3.2.10) becomes

$$\mathbf{q} \cdot \operatorname{grad} \omega = 0, \tag{3.2.11}$$

so that vorticity is constant along a streamline. Here, $\mathbf{q} \cdot \operatorname{grad}$ is the operator of differentiation along a streamline. For uniform flow, $\omega = 0$ at ∞, and hence $\omega = 0$ throughout. We obtain, in terms of the outer expansion,

$$\omega_0 = \operatorname{curl} \mathbf{q}_0 = 0. \tag{3.2.12}$$

Thus, the basic outer flow $(\mathbf{q}_0, p_0)$ is a potential flow. For example, in Cartesian components, $\mathbf{q}_0 = u_0\mathbf{i} + v_0\mathbf{j}$, and the components can be expressed by an analytic function:

$$u_0 - i v_0 = F'(z), \quad z = x + iy. \tag{3.2.13}$$

The problem is thus purely kinematic. Integration of the limit form of (3.2.3) along a streamline yields Bernoulli's law:

$$p_0 = \frac{1}{2}(1 - q_0^2), \quad p_0(\infty) = 0 \tag{3.2.14}$$

and accounts for all the dynamics of potential flow. In addition, stream (ψ) and potential (ϕ) functions exist:

$$\phi + i\psi = F(z), \quad u_0 = \phi_x = \psi_y, \quad v_0 = \phi_y = -\psi_x. \tag{3.2.15}$$

An important question that cannot be answered by the current approach is what potential flow $F(z)$ to choose for the problem of high Re flow past a given body. Real flows tend to separate toward the rear of a closed body and to generate a viscous wake. This implies that the correct limiting potential flow separates from the body. Furthermore, if Re is sufficiently high, turbulence sets in, so that a description under the steady Navier-Stokes equations is not valid. Thus, for our purposes we consider the simplest potential flow, for example, that which closes around the body. The approximation is understood to be valid only in a region of limited extent near the nose of the body.

Now, with respect to the higher-order terms of the outer expansion, the following observation can be made: every $(\mathbf{q}_i, p_i)$ is a potential flow. This follows by induction from the fact that the viscous body force is zero in a potential flow:

$$\operatorname{curl} \omega_0 = 0 \quad \text{or} \quad \Delta_{(x,y)}\mathbf{q}_0 = 0. \tag{3.2.16}$$

The inner expansion is derived from a limit process in which a representative point P^* approaches the boundary as $\epsilon \to 0$. The boundary layer in this problem is along a streamline of the inviscid flow, a subcharacteristic of the full problem. Characteristic surfaces in general are the loci of possible discontinuities, and streamlines of an inviscid flow can support a discontinuity in vorticity. In the inviscid limit in which the external flow is potential flow, this discontinuity is only at the solid surface where the tangential velocity jumps (if the boundary condition of zero velocity on the surface is enforced). Now the vorticity equation, (3.2.10), has the same structure as the general partial differential equation discussed in Section 3.1. The boundary layer resolving the vorticity occurs on a subcharacteristic and, hence, should be of $O(\sqrt{\epsilon})$ in thickness. Thus, symbolically,

$$P^* = \frac{P - P_b}{\sqrt{\epsilon}}, \quad P_b = \text{point of the boundary}, \tag{3.2.17}$$

is held fixed as $\epsilon \to 0$. This order of magnitude can also be checked explicitly.

Before considering flow past a body, however, it is worthwhile to outline the simpler problem of purely radial flow in a wedge-shaped sector (Figure 3.2.2). For this simple geometry, the Navier-Stokes equations simplify sufficiently to allow an exact solution to be constructed. For inflow, there is a sink at the origin, and the solutions are well behaved as Re $\to \infty$ in the sense that boundary layers form near the walls. For outflow, however, the solutions can exhibit a much more complicated structure, including regions of backflow. The limit solutions for this case as Re $\to \infty$ may have vortex sheets in the interior of the channel. Only the case of inflow is considered here.

3.2.1 Radial Viscous Inflow

The mass flux per unit width of channel, Q = mass/sec-length, is prescribed. The overall Reynolds number is, thus,

$$\text{Re} = \frac{Q}{\mu} = \frac{1}{\epsilon}. \tag{3.2.18}$$

By dimensional reasoning, the radial velocity and pressure must be of the form

$$\text{outward radial velocity} = \frac{Q}{\rho r} f(\theta), \tag{3.2.19}$$

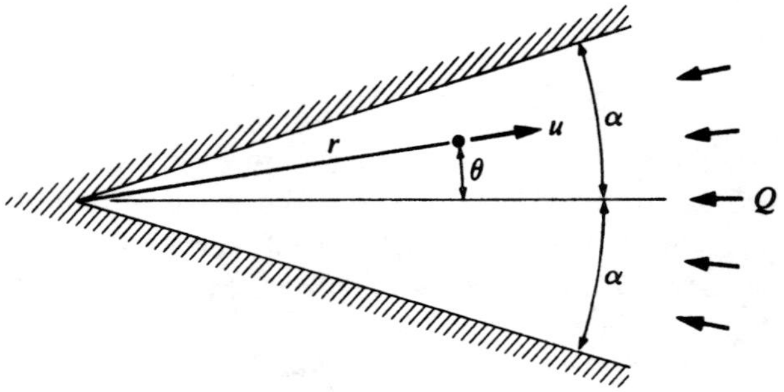

FIGURE 3.2.2. Viscous Sink Flow in a Sector

$$\text{pressure} = \frac{Q^2}{\rho r^2} g(\theta). \tag{3.2.20}$$

Thus, the full Navier-Stokes equations become ordinary differential equations:

$$\text{radial momentum, } -f^2(\theta) = 2g(\theta) + \epsilon \frac{d^2 f}{d\theta^2}, \tag{3.2.21}$$

$$\text{tangential momentum, } 0 = -\frac{dg}{d\theta} + 2\epsilon \frac{df}{d\theta}. \tag{3.2.22}$$

Mass conservation leads to the normalization

$$-\int_{-\alpha}^{\alpha} f(\theta) d\theta = 1, \tag{3.2.23}$$

while the condition of no slip at the walls is

$$f(\pm\alpha) = 0. \tag{3.2.24}$$

The exact solution may be derived in terms of elliptic functions but is not very instructive.

In order to study the solution as $\epsilon \to 0$, outer and inner expansions, as indicated earlier, are constructed. The outer expansion, associated with the limit $\epsilon \to 0, \theta$ fixed, represents a sequence of potential flows.

Outer expansion

$$\begin{aligned} f(\theta;\epsilon) &= F_0(\theta) + \gamma_1(\epsilon)F_1(\theta) + \gamma_2(\epsilon)F_2(\theta) + \dots, \\ g(\theta;\epsilon) &= G_0(\theta) + \gamma_1(\epsilon)G_1(\theta) + \gamma_2(\epsilon)G_2(\theta) + \dots. \end{aligned} \tag{3.2.25}$$

The limit ($\epsilon = 0$) form of (3.2.21) and (3.2.22) yields

$$-F_0^2 = 2G_0, \tag{3.2.26}$$

$$0 = -\frac{dG_0}{d\theta}. \tag{3.2.27}$$

The no-slip condition is given up so that $F_0 = G_0 = \text{const.}$, and the normalization (3.2.23) yields

$$F_0 = -\frac{1}{2\alpha}, \quad G_0 = -\frac{1}{8\alpha^2}. \tag{3.2.28}$$

Inner expansion

Now, in order to have a balance of viscous forces and inertia near the walls ($\theta = \pm\alpha$), it is necessary that the viscous layer have a thickness $O(\sqrt{\epsilon})$. An inner limit $\epsilon \to 0, \theta^* = (\theta \pm \alpha)/\sqrt{\epsilon}$ fixed, is considered. It follows that the inner expansion is of the form (valid near each wall)

$$f(\theta;\epsilon) = f_0(\theta^*) + \beta(\epsilon) f_1(\theta^*) + \dots, \quad g(\theta;\epsilon) = g_0(\theta^*) + \beta(\epsilon) g_1(\theta^*) + \dots$$

$$\theta^* = (\theta \pm \alpha)/\sqrt{\epsilon}. \tag{3.2.29}$$

The equations of motion reduce to

$$O(1): \quad -f_0^2 = 2g_0 + \frac{d^2 f_0}{d\theta^{*2}}, \tag{3.2.30}$$

$$O\left(\frac{1}{\sqrt{\epsilon}}\right): \quad 0 = -\frac{dg_0}{d\theta^*}. \tag{3.2.31}$$

The solutions to these boundary-layer equations should satisfy the no-slip condition, so that

$$f_0(0) = 0. \tag{3.2.32}$$

The other boundary conditions for (3.2.30) and (3.2.31) are found by matching with the outer solution. Equation (3.2.31) states that, to this order, there is no pressure gradient across the thin viscous layer adjacent to the wall. The matching thus fixes the level of pressure in the boundary layer.

We introduce the matching variable

$$\theta_\eta = \frac{(\theta \pm \alpha)}{\eta(\epsilon)} \tag{3.2.33}$$

for a class of $\eta(\epsilon)$ contained in the interval $\sqrt{\epsilon} \ll \eta \ll 1$ as $\epsilon \to 0$. In this limit, we have

$$\theta = \mp\alpha + \eta\theta_\eta \to \mp\alpha, \tag{3.2.34}$$

$$\theta^* = \frac{\eta}{\sqrt{\epsilon}}\theta_\eta \to \pm\infty. \tag{3.2.35}$$

It is assumed that the inner and outer expansions match directly in an overlap domain.

Consider now only the lower wall $\theta = -\alpha$; the solution at the upper wall is found by symmetry. The matching of pressures to first order is

$$\lim_{\substack{\epsilon\to 0\\ \theta_\eta \text{ fixed}}} \left[G_0(-\alpha + \eta\theta_\eta) + \ldots - g_0\left(\frac{\eta}{\sqrt{\epsilon}}\theta_\eta\right) + \ldots\right] = 0. \tag{3.2.36}$$

In this case, we know that

$$G_0 = g_0 = \text{const.} = -\frac{1}{8\alpha^2}. \tag{3.2.37}$$

Thus, (3.2.30) becomes

$$\frac{d^2 f_0}{d\theta^{*2}} + f_0^2(\theta^*) = \frac{1}{4\alpha^2}. \tag{3.2.38}$$

Matching of the velocities now gives

$$\lim_{\substack{\epsilon\to 0\\ \theta_\eta \text{ fixed}}} \left\{F_0(-\alpha + \eta\theta_\eta) - f_0\left(\frac{\eta}{\sqrt{\epsilon}}\theta_\eta\right)\right\} = 0. \tag{3.2.39}$$

The velocity of the potential flow at the wall is matched to the velocity of the boundary-layer flow at infinity in this case:

$$f_0(\infty) \to F_0(-\alpha) = -1/2\alpha. \tag{3.2.40}$$

Thus, the solution of (3.2.38) for $0 \le \theta^* < \infty$ must be found satisfying the conditions (3.2.32) and (3.2.40). It is clear that the boundary condition (3.2.40) is consistent with (3.2.38).

The existence of the solution to (3.2.38) and the form near infinity are easily seen from the phase plane of (3.2.38). Let

$$w_0 = df_0/d\theta^*. \tag{3.2.41}$$

Equation (3.2.38) then becomes

$$\frac{dw_0}{df_0} = \frac{(1/4\alpha^2) - f_0^2}{w_0}. \tag{3.2.42}$$

The paths of the integral curves are indicated in Figure 3.2.3. The arrows indicate the direction of increasing θ^* according to (3.2.41). The singularity at $w_0 = 0$, $f_0 = -1/2\alpha$ is a saddle point whose paths are

$$w_0^2 - \frac{1}{\alpha}\left(f_0 + \frac{1}{2\alpha}\right)^2 = \text{const.} \tag{3.2.43}$$

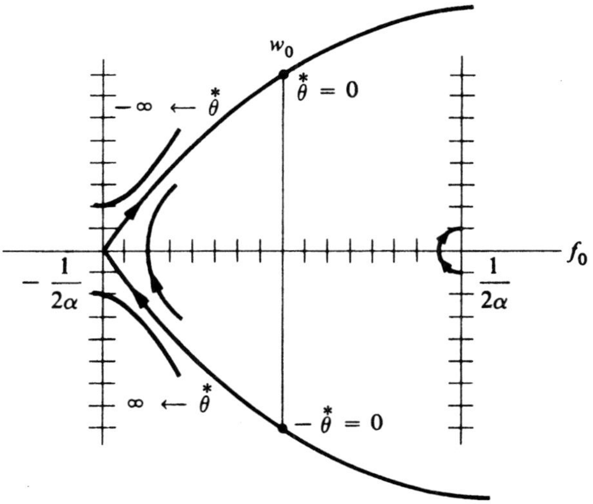

FIGURE 3.2.3. Phase Plane for (3.2.42)

Two exceptional paths,

$$w_0 = \pm\frac{1}{\sqrt{\alpha}}\left(f_0 + \frac{1}{2\alpha}\right), \tag{3.2.44}$$

enter the saddle points, the $(-)$ sign corresponding to the boundary layer at the lower wall. The value of θ^* along the path is found by integration of (3.2.42) along the path from $f_0 = 0$, $\theta^* = 0$. Near the singular point, integration of (3.2.41) shows that

$$f_0 = -\frac{1}{2\alpha} + k_0 e^{-\theta^*/\sqrt{\alpha}} + \dots, \tag{3.2.45}$$

where k_0 is known from integration along the path. Equation (3.2.45) shows that the boundary layer approaches its limiting value with an error that is transcendentally small as long as $\epsilon^{1/2}|\log\epsilon| \ll \eta$.

The need for higher-order terms arises because of the mass-flow defect in the boundary layer. The first term of a uniformly valid ($-\alpha \le \theta \le \alpha$) composite expansion of the form

$$f(\theta;\epsilon) = \mathcal{F}_0(\theta;\epsilon) + \gamma_1(\epsilon)\mathcal{F}_1(\theta;\epsilon) + \dots \tag{3.2.46}$$

can be found by adding f_0 for both walls to F_0 and subtracting the common part $-1/2\alpha$:

$$\mathcal{F}_0(\theta;\epsilon) = f_0\left(\frac{\theta+\alpha}{\sqrt{\epsilon}}\right) + f_0\left(\frac{\theta-\alpha}{\sqrt{\epsilon}}\right) + \frac{1}{2\alpha}. \tag{3.2.47}$$

The mass-flow integral (3.2.23) is

$$-\int_{-\alpha}^{\alpha}\left[f_0\left(\frac{\theta+\alpha}{\sqrt{\epsilon}}\right) + f_0\left(\frac{\theta-\alpha}{\sqrt{\epsilon}}\right) + \frac{1}{2\alpha}\right]d\theta$$
$$\to 1 - 2\sqrt{\epsilon}\int_0^{\infty}\left[\frac{1}{2\alpha} + f_0(\theta^*)\right]d\theta^* + \dots.$$

The error is thus $O(\sqrt{\epsilon})$, and this has to be made up by the next term in the outer expansion. Hence, we have $\gamma_1(\epsilon) = \sqrt{\epsilon}$ and

$$\int_{-\alpha}^{\alpha} F_1(\theta)d\theta = -2\int_0^{\infty}\left[\frac{1}{2\alpha} + f_0(\theta^*)\right]d\theta^*. \tag{3.2.48}$$

Once the (F_1, G_1) are found, the next boundary-layer terms (f_1, g_1) can be constructed and the procedure repeated.

We next apply these ideas to flow past a body.

3.2.2 Flow Past a Body

Consider steady plane flow past a body of the general form indicated in Figure 3.2.4. In order to carry out the boundary layer and outer expansions, it is convenient to choose a special coordinate system. In a very interesting paper [3.13], Kaplun discussed the choice of "optimal" coordinates. He shows that it is possible to

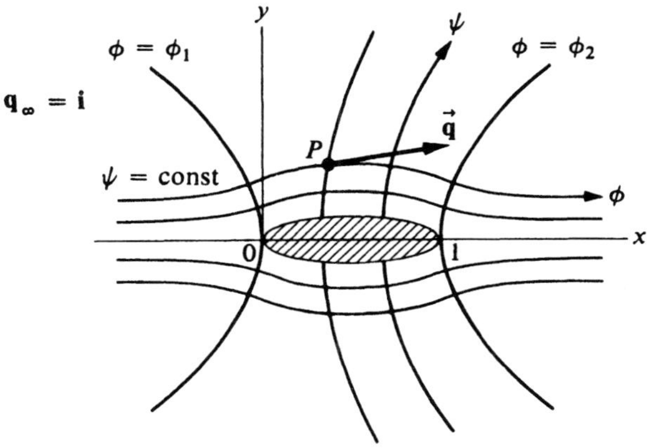

FIGURE 3.2.4. Streamline Coordinates

find certain coordinates in which the boundary-layer equations and solutions are uniformly valid first approximations in the *entire flow field*, including the so-called flow due to displacement thickness. However, the construction of such coordinates is, in general, just as difficult as it is to proceed directly in any convenient system. The first approximation to the skin friction is independent of the coordinates. Here we express the Navier-Stokes equations in terms of a network of potential lines (ϕ = const.) and streamlines (ψ = const.), which represent the idealized inviscid flow around the object.

This choice of coordinates at least has the advantage of allowing the ideas of boundary-layer theory to be expressed independently of the body shape and of having a simple representation of the first-order outer flow. Thus, $\psi = 0$ is always the bounding streamline along which the boundary layer appears. In addition to the basic definitions (3.2.13), (3.2.14), and (3.1.15), we note the following expressions for the velocity components, (q_ϕ, q_ψ), vorticity, etc., in the *viscous* flow. The results follow from general vector formulas in orthogonal curvilinear coordinates (see Figure 3.2.5):

$$dz = \frac{dF}{F'}, \quad (dx)^2 + (dy)^2 = \frac{(d\phi)^2 + (d\psi)^2}{|F'|^2} = \frac{(d\phi)^2 + (d\psi)^2}{q_0^2}, \tag{3.2.49}$$

$$\omega = q_0^2 \left\{ \frac{\partial}{\partial \phi}\left(\frac{q_\psi}{q_0}\right) - \frac{\partial}{\partial \psi}\left(\frac{q_\phi}{q_0}\right) \right\}, \quad \boldsymbol{\omega} = \omega \mathbf{k}, \tag{3.2.50}$$

$$\mathbf{q} \times \boldsymbol{\omega} = (q_\phi \mathbf{i}_\phi + q_\psi \mathbf{i}_\psi) \times (\omega \mathbf{k}) = \omega q_\psi \mathbf{i}_\phi - \omega q_\phi \mathbf{i}_\psi, \tag{3.2.51}$$

$$\nabla = \left(q_0 \frac{\partial}{\partial \phi}, \ q_0 \frac{\partial}{\partial \psi} \right), \tag{3.2.52}$$

$$\operatorname{curl} \omega = \mathbf{i}_\phi q_0 \frac{\partial \omega}{\partial \psi} - \mathbf{i}_\psi q_0 \frac{\partial \omega}{\partial \phi}. \tag{3.2.53}$$

Thus, the basic Navier-Stokes equations (3.2.3a,b) become

$$\text{continuity,} \quad \frac{\partial}{\partial \phi}\left(\frac{q_\phi}{q_0}\right) + \frac{\partial}{\partial \psi}\left(\frac{q_\psi}{q_0}\right) = 0; \tag{3.2.54}$$

$$\phi\text{-momentum,} \quad q_0 \frac{\partial}{\partial \phi}\left(\frac{q_\phi^2 + q_\psi^2}{2}\right) - q_\psi \omega = -q_0 \frac{\partial p}{\partial \phi} - \epsilon q_0 \frac{\partial \omega}{\partial \psi}; \tag{3.2.55}$$

$$\psi\text{-momentum,} \quad q_0 \frac{\partial}{\partial \psi}\left(\frac{q_\phi^2 + q_\psi^2}{2}\right) + q_\phi \omega = -q_0 \frac{\partial p}{\partial \psi} + \epsilon q_0 \frac{\partial \omega}{\partial \phi}. \tag{3.2.56}$$

The viscous stress is now $\tau_v = \epsilon q_0^2 (\partial/\partial\psi)(q_\phi/q_0)$. From the form of these equations, it seems clear that a small simplification can be achieved by measuring the velocities at a point P relative to the inviscid velocity q_0 at that point. Let

$$w_\phi = \frac{q_\phi}{q_0}, \qquad w_\psi = \frac{q_\psi}{q_0}. \tag{3.2.57}$$

It follows that, for the vorticity, we have

$$\omega = q_0^2 \left\{ \frac{\partial w_\psi}{\partial \phi} - \frac{\partial w_\phi}{\partial \psi} \right\}, \tag{3.2.58}$$

and (3.2.54), (3.2.55), and (3.2.56) become

$$\frac{\partial w_\phi}{\partial \phi} + \frac{\partial w_\psi}{\partial \psi} = 0, \tag{3.2.59}$$

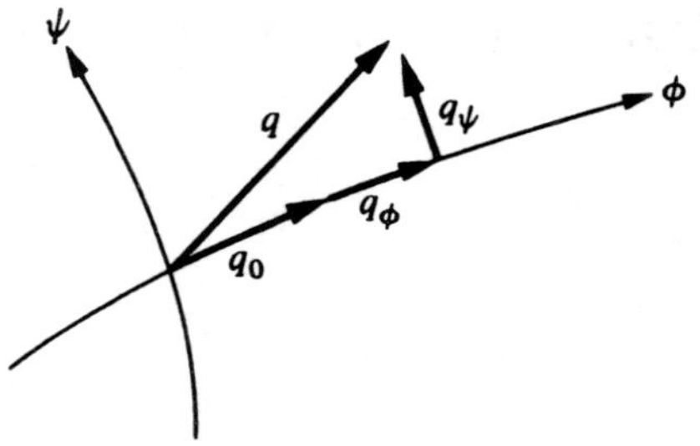

FIGURE 3.2.5. Detail of Velocity Components

$$w_\phi \frac{\partial w_\phi}{\partial \phi} + w_\psi \frac{\partial w_\phi}{\partial \psi} + (w_\phi^2 + w_\psi^2)\frac{\partial}{\partial \phi}(\log q_0) = -\frac{1}{q_0^2}\frac{\partial p}{\partial \phi}$$
$$+\epsilon\left\{\frac{\partial^2 w_\phi}{\partial \phi^2} + \frac{\partial^2 w_\phi}{\partial \psi^2} - 2\frac{\partial \log q_0}{\partial \psi}\left(\frac{\partial w_\psi}{\partial \phi} - \frac{\partial w_\phi}{\partial \psi}\right)\right\}, \quad (3.2.60a)$$

$$w_\phi \frac{\partial w_\psi}{\partial \phi} + w_\psi \frac{\partial w_\psi}{\partial \psi} + (w_\phi^2 + w_\psi^2)\frac{\partial}{\partial \psi}(\log q_0) = -\frac{1}{q_0^2}\frac{\partial p}{\partial \psi}$$
$$+\epsilon\left\{\frac{\partial^2 w_\psi}{\partial \phi^2} + \frac{\partial^2 w_\psi}{\partial \psi^2} + 2\frac{\partial \log q_0}{\partial \phi}\left(\frac{\partial w_\psi}{\partial \phi} - \frac{\partial w_\phi}{\partial \psi}\right)\right\}. \quad (3.2.60b)$$

The viscous surface stress, correspondingly, is $\tau_v = \epsilon q_0^2(\partial w_\phi/\partial \psi)$. The domain of the problem is sketched in Figure 3.2.6. The body occupies the slit $\psi = 0, \phi_1 < \phi < \phi_2$, for symmetric flow. For unsymmetric flow $\phi_2(\psi = 0^+) \neq \phi_2(\psi = 0^-)$. The boundary conditions are as follows:

uniform flow at infinity upstream: $w_\psi \to 0, \ w_\phi \to 1, \ \phi \to -\infty$; (3.2.61)

no slip at the body surface: $w_\phi = w_\psi = 0, \ \psi = 0, \ \phi_1 < \phi < \phi_2$. (3.2.62)

Now, in order to construct the expansions, we consider first the outer or Euler limit ($\epsilon \to 0, \phi, \psi$ fixed). This represents inviscid and, in this case, irrotational flow around the object. The limit flow is, thus,

$$w_\phi \to 1, \quad w_\psi \to 0, \quad p = \frac{1}{2}(1 - q_0^2) \ \text{(Bernoulli equation)}. \quad (3.2.63)$$

As an outer expansion, we have the limit flow as the first term and corrections due to the inner solution appearing as higher terms. The general form of the outer

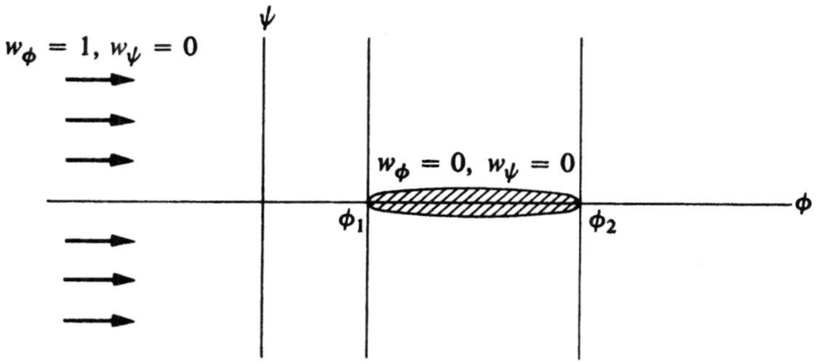

FIGURE 3.2.6. Boundary-Value Problem in the (ϕ, ψ) plane

expansion is, thus,

$$w_\phi(\phi, \psi; \epsilon) = 1 + \beta(\epsilon) w_\phi^{(1)}(\phi, \psi) + \dots, \tag{3.2.64}$$

$$w_\psi(\phi, \psi; \epsilon) = \beta(\epsilon) w_\psi^{(1)}(\phi, \psi) + \dots, \tag{3.2.65}$$

$$p(\phi, \psi; \epsilon) = \frac{1}{2}(1 - q_0^2) + \beta(\epsilon) p^{(1)}(\phi, \psi) + \dots. \tag{3.2.66}$$

All corrections with superscript (1) vanish at upstream infinity. However, other boundary conditions for these correction terms cannot be found without discussing the inner viscous boundary layer. To construct the boundary layer and correction equations, we consider an inner-limit process and associated expansion where

$$\psi^* = \frac{\psi}{\delta(\epsilon)}, \quad \phi \text{ fixed as } \epsilon \to 0. \tag{3.2.67}$$

The expansion has the following form:

$$w_\phi(\phi, \psi; \epsilon) = W_\phi(\phi, \psi^*) + \dots, \tag{3.2.68}$$

$$w_\psi(\phi, \psi; \epsilon) = \delta(\epsilon) W_\psi(\phi, \psi^*) + \dots, \tag{3.2.69}$$

$$p(\phi, \psi; \epsilon) = P(\phi, \psi^*) + \dots. \tag{3.2.70}$$

The form is deduced from the following considerations, in addition to those that indicated that the boundary layer occupies a thin region close to $\psi = 0$. The first term of the expression for the velocity component w_ϕ along the streamline is of order one, so that it can be matched to the outer expansion (3.2.64). The first term in the expansion for w_ψ must then be of the order $\delta(\epsilon)$ so that a nontrivial continuity equation results:

$$\frac{\partial W_\phi}{\partial \phi} + \frac{\partial W_\psi}{\partial \psi} = 0. \tag{3.2.71}$$

The first term in the pressure is also $O(1)$ in order to match to (3.2.66). Under the boundary-layer limit, the inviscid velocity field, which occurs in the coefficient, approaches the surface distribution of inviscid velocity according to

$$q_0(\phi, \psi) = q_0(\phi, \delta(\epsilon)\psi^*) = q_0(\phi, 0) + \delta(\epsilon)\psi^* \frac{\partial q_0}{\partial \psi}(\phi, 0) + \dots \tag{3.2.72}$$

or

$$q_0(\phi, \psi) = q_B(\phi) + O(\delta(\epsilon)),$$

where $q_B(\phi)$ is the inviscid surface velocity distribution. The inertia and pressure terms in (3.2.60a) are both of $O(1)$, while $\epsilon(\partial^2 W_\phi / \partial \psi^2)$ is of the order ϵ/δ^2. The distinguished limiting case has

$$\delta = \sqrt{\epsilon}. \tag{3.2.73}$$

It is only this case that allows a nontrivial system of boundary-layer equations capable of satisfying the boundary conditions and being matched to the outer flow.

With this assumption, the first approximation momentum equations are

$$W_\phi \frac{\partial W_\phi}{\partial \phi} + W_\psi \frac{\partial W_\phi}{\partial \psi^*} + \frac{W_\phi^2}{q_B(\phi)} \frac{dq_B}{d\phi} = -\frac{1}{q_B^2} \frac{\partial P}{\partial \phi} + \frac{\partial^2 W_\phi}{\partial \psi^{*2}}, \tag{3.2.74}$$

$$0 = -\frac{1}{q_B^2} \frac{\partial P}{\partial \psi^*}. \tag{3.2.75}$$

Equation (3.2.75) tells us that the layer is so thin that the pressure does not vary across the layer and, rather, that

$$P = P(\phi). \tag{3.2.76}$$

Hence, the pressure is easily matched to the pressure in the outer solution.

The matching is carried out in terms of the variable

$$\psi_\eta = \frac{\psi}{\eta(\epsilon)}, \tag{3.2.77}$$

with ϕ fixed, where η belongs to $\sqrt{\epsilon} \ll \eta \ll 1$. It then follows that

$$\psi = \eta\psi_\eta \to 0, \quad \psi^* = \frac{\eta}{\sqrt{\epsilon}} \psi_\eta \to \infty. \tag{3.2.78}$$

Thus, the representative point approaches the wall but not as fast as it does in the distinguished limit. It is sufficient to consider only positive ψ to illustrate the ideas.

Matching of pressure (see (3.2.66), (3.2.70)) takes the form

$$\lim_{\substack{\epsilon \to 0 \\ \psi_\eta \text{ fixed}}} \{\frac{1}{2}(1 - q_0^2(\phi, \eta\psi_\eta)) + \beta(\epsilon) p^{(1)}(\phi, \eta\psi_\eta) + \ldots - P(\phi) - \ldots\} = 0.$$

Hence, to first order, we have

$$P(\phi) = \frac{1}{2}(1 - q_B^2(\phi)) \equiv P_B(\phi). \tag{3.2.79}$$

The pressure distribution on the body is that of the inviscid flow, if we neglect the boundary layer. Thus, the system of boundary-layer equations (3.2.71), (3.2.74) is

$$\frac{\partial W_\phi}{\partial \phi} + \frac{\partial W_\psi}{\partial \psi^*} = 0, \tag{3.2.80a}$$

$$W_\phi \frac{\partial W_\phi}{\partial \phi} + W_\psi \frac{\partial W_\phi}{\partial \psi^*} = \frac{1 - W_\phi^2}{q_B} \frac{dq_B}{d\phi} + \frac{\partial^2 W_\phi}{\partial \psi^{*2}}, \tag{3.2.80b}$$

a system for (W_ϕ, W_ψ). The boundary conditions to be satisfied are no slip,

$$W_\phi(\phi, 0) = W_\psi(\phi, 0) = 0, \quad \phi_1 < \phi < \phi_2, \tag{3.2.81}$$

and matching.

The system (3.2.80) is parabolic, so that only the interval $\phi_1 < \phi < \phi_2$ need be considered at first. The next quantity to be matched is the velocity component along a streamline, which also contains an $O(1)$ term. Inner and outer expansions (Equations (3.2.68) and (3.2.64)) must match in terms of ψ_η, so that

$$\lim_{\substack{\epsilon\to 0\\ \psi_\eta \text{ fixed}}} \{1 + \beta(\epsilon) w_\phi^{(1)}(\phi, \eta\psi_\eta) + \ldots - W_\phi(\phi, (\eta/\sqrt{\epsilon})\psi_\eta) - \ldots\} = 0. \quad (3.2.82)$$

Thus, the following boundary condition is obtained:

$$\lim_{\psi^*\to\infty} W_\phi(\phi, \psi^*) = 1. \quad (3.2.83)$$

This is usually interpreted by saying that the velocity at the outer edge of the boundary layer is that of the inviscid flow adjacent to the body. Since the system (3.2.80) is parabolic, there is no upstream influence, so that the solution again must match the undisturbed flow:

$$W_\phi(\phi_1, \psi^*) = 1. \quad (3.2.84)$$

The conditions (3.2.84), (3.2.83), and (3.2.81) serve to define a unique solution in the strip $\phi_1 < \phi < \phi_2$. The solution downstream of the body, $\phi > \phi_2$, should really be discussed also. The boundary-layer equations and expansion are the same, but the boundary conditions corresponding to the wake are different. Now, the upstream boundary-layer solution just calculated provides initial conditions on $\phi = \phi_2$ for $-\infty < \psi^* < \infty$; and the initial-value problem can be solved to find the flow downstream.

Assume now that the solution of (3.2.80) has been found for all $\phi > \phi_1$, so that W_ϕ, W_ψ are known functions. The matching of the normal component of velocity W_ψ along the potential lines can be discussed next, and this provides a boundary condition that defines the correction in the outer flow due to the presence of the boundary layer. We require

$$\lim_{\substack{\epsilon\to 0\\ \psi_\eta \text{ fixed}}} \left\{\beta(\epsilon) w_\psi^{(1)}(\phi, \eta\psi_\eta) + \ldots - \sqrt{\epsilon} W_\psi \left(\phi, \frac{\eta\psi_\eta}{\sqrt{\epsilon}}\right) - \ldots\right\} = 0. \quad (3.2.85)$$

Matching is achieved to first order, provided the limits exist, if first of all

$$\beta(\epsilon) = \sqrt{\epsilon} \quad (3.2.86)$$

and

$$w_\psi^{(1)}(\phi, 0) = W_\psi(\phi, \infty), \quad \phi > \phi_1. \quad (3.2.87)$$

Equation (3.2.87) has the form of a boundary condition for $w_\psi^{(1)}$, which can be interpreted as an effective thin body added to the original body; it defines the flow due to displacement thickness. The limit in (3.2.87) exists since the solutions for $[1 - W_\phi]$ can be shown to decay exponentially as $\psi \to \infty$ and

$$W_\psi(\phi, \psi) = \int_0^\psi \left[-\frac{\partial W_\phi}{\partial \phi}(\phi, \lambda)\right] d\lambda. \quad (3.2.88)$$

Thus, the outer flow $w_\phi^{(1)}$, $w_\psi^{(1)}$, $p^{(1)}$ can, in principle, be computed, and further matching of p, w_ϕ can be used to define the second-order boundary layer. Various local nonuniformities can develop, such as near sharp, leading, or trailing edges, corners, etc., that make it unwise to attempt to carry the procedure very far. Analogous procedures can be carried out for compressible flow where the energy balance of the flow must also be considered. The subject is discussed with a point of view similar to that given here by P.A. Lagerstrom in [3.16]. Further complications occur when different types of interaction with shock waves in the outer flow have to be considered, but the ideas behind these methods seem capable of handling all cases that arise.

The quantity of most physical interest from the boundary-layer theory is the skin friction on the surface, which now is represented by

$$\tau_v = \sqrt{\epsilon} q_B^2(\phi)(\partial W_\phi/\partial\psi^*)(\phi, 0). \tag{3.2.89}$$

When the boundary-layer solution is found, $(\partial W_\phi/\partial\psi^*)$ $(\phi, 0)$ can be calculated, and the estimate of the skin friction is obtained. The classical result is given here—the skin friction coefficient is proportional to $1/\sqrt{\text{Re}}$.

Unfortunately, no elementary solutions of the boundary-layer system (3.2.80) under boundary conditions (3.2.81), (3.2.84) exist. The only cases in which substantial simplifications can be achieved are cases of similarity when the problem can be reduced to ordinary differential equations. Otherwise, numerical integration of the system must be relied on, although some rough approximate methods can also be derived.

The cases of similarity can either be interpreted as local approximations or as solutions that are really asymptotic to solutions of the Navier-Stokes equations in a sense different from having $\epsilon \to 0$. That is, the characteristic length L used to define ϵ really drops out of the problem, and the expansion is really in terms of the coordinates (x, y) or (ϕ, ψ). For example, consider the flow past a semi-infinite body generated by a source at $z = 1$ in a free stream (see Figure 3.2.7). The inviscid flow and coordinates are given by

$$\phi + i\psi = z + \log(z - 1) - i\pi = -\frac{z^2}{2} + \ldots, \quad \text{as } z \to 0. \tag{3.2.90}$$

Thus, near the origin we have the stagnation-point flow

$$\phi = \frac{y^2 - x^2}{2}, \quad \psi = -xy. \tag{3.2.91}$$

The body is, at $x = 0$,

$$q_B = \frac{\partial\phi}{\partial y} = y = \sqrt{2\phi}. \tag{3.2.92}$$

The boundary-layer equations (3.2.80) are, for this case,

$$\frac{\partial W_\phi}{\partial\phi} + \frac{\partial W_\psi}{\partial\psi^*} = 0, \tag{3.2.93a}$$

$$W_\phi \frac{\partial W_\phi}{\partial\phi} + W_\psi \frac{\partial W_\phi}{\partial\psi^*} = \frac{1 - W_\phi^2}{2\phi} + \frac{\partial^2 W_\phi}{\partial\psi^{*2}}, \tag{3.2.93b}$$

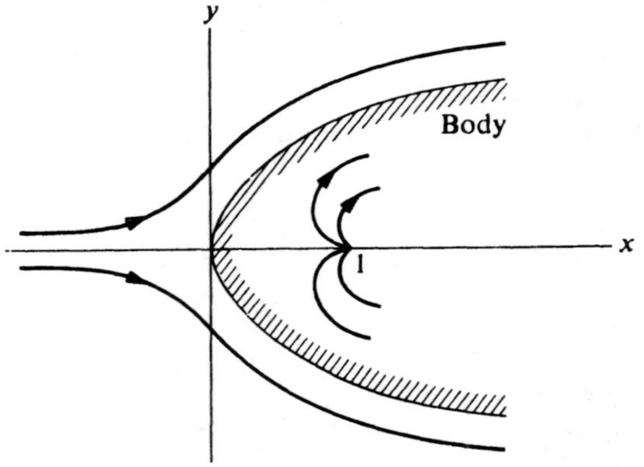

FIGURE 3.2.7. Source Half-Body

with boundary conditions

$$W_\phi(\phi, 0) = W_\psi(\phi, 0) = 0, \quad W_\phi(\phi, \infty) = 1. \tag{3.2.94}$$

The system (3.2.93)–(3.2.94) has similarity, which permits the problem to be reduced to ordinary differential equations. The form is

$$W_\phi = F(\eta), \tag{3.2.95}$$

where

$$\eta = \frac{\psi^*}{2\sqrt{\phi}}, \tag{3.2.96}$$

$$W_\psi = \frac{1}{\sqrt{\phi}} G(\eta), \tag{3.2.97}$$

$$\frac{dG}{d\eta} - \eta \frac{dF}{d\eta} = 0, \quad G(0) = F(0) = 0, \tag{3.2.98}$$

$$\frac{d^2 F}{d\eta^2} + 2(\eta F - G)\frac{dF}{d\eta} + 2(1 - F^2) = 0, \quad F(\infty) = 1. \tag{3.2.99}$$

According to (3.2.89), the skin friction is obtained once this system is solved from

$$\tau_v = \sqrt{\phi\epsilon}\, F'(0). \tag{3.2.100}$$

The existence of the solution to the problem posed in (3.2.98)–(3.2.99), as well as to the more general class of self-similar problems in which $q_B = c\phi^m$ (the form (3.2.95)–(3.2.97) is the same), is proved by H. Weyl in [3.43]. For a special case of the stagnation-point flow, the similar solution can be interpreted as the local solution near the origin. It turns out that in this case the solution to the boundary-layer equation (3.2.98), (3.2.99) is also a solution to the full Navier-Stokes equations. The parameter ϵ really drops out of the local solution, since the local solution cannot depend on the length L. The length L drops out of the similarity variable $\psi^*/\sqrt{\phi}$ when dimensional coordinates are reintroduced as follows:

$$\psi^* = UL\sqrt{\epsilon}\Psi, \quad \phi = UL\Phi, \tag{3.2.101}$$

where Φ, Ψ are dimensional:

$$\frac{\psi^*}{\sqrt{\phi}} = \frac{UL}{\sqrt{UL}}\sqrt{\frac{\mu}{\rho UL}}\frac{\Psi}{\sqrt{\Phi}} = \Psi\sqrt{\frac{\mu}{\rho\Phi}}. \tag{3.2.102}$$

Similar considerations apply to the velocity components to show that expansion is really in terms of Φ or (X) and is valid near the origin.

The same remarks apply to another classical case that is usually discussed, namely, the flow past a semi-infinite flat plate, in which case we have

$$\phi = x, \quad \psi = y, \tag{3.2.103}$$

$$q_B(\phi) = 1. \tag{3.2.104}$$

The similarity form (3.2.96)–(3.2.97) is the same, and the equations are a simplified version of (3.2.98) and (3.2.99) with the $(1 - F^2)$ term missing. There is no characteristic length L in the problem, so that the parameter ϵ is artificial. If an arbitrary length is used for L (and this can be done), it must drop out of the answer. When similarity is combined with the artificial expansion in terms of ϵ, the expansion corresponding to boundary-layer theory becomes an expansion in terms of the space coordinates. For example, in dimensional coordinates (X, Y), the boundary-layer expansion, (3.2.68)–(3.2.70), and outer expansions, (3.2.64)–(3.2.66), take the form

$$\begin{aligned}
\frac{q_x}{U} &= U_0(\zeta) + \sqrt{\frac{\nu}{UX}}U_1(\zeta) + \dots \\
\frac{q_y}{U} &= \sqrt{\frac{\nu}{UX}}V_0(\zeta) + \dots, \quad \text{where } \zeta = \frac{Y}{\sqrt{X}}\sqrt{\frac{U}{\nu}}, \\
\frac{p - p_\infty}{\rho U^2} &= \sqrt{\frac{\nu}{UX}}P_1(\zeta) + \dots, \quad \text{(boundary layer)},
\end{aligned} \tag{3.2.105}$$

$$\begin{aligned}
\frac{q_x}{U} &= 1 + \sqrt{\frac{\nu}{UX}}u_1(\zeta) + \dots \\
\frac{q_y}{U} &= \sqrt{\frac{\nu}{UX}}v_1(\zeta) + \dots \\
\frac{p - p_\infty}{\rho U^2} &= \sqrt{\frac{\nu}{UX}}p_1(\zeta) + \dots \quad \text{(outer expansion)}.
\end{aligned} \tag{3.2.106}$$

These expansions are seen to be valid for small ν/UX and are thus nonuniform near the nose, where a more complete treatment of the Navier-Stokes equations is needed. However, the skin friction has a singularity only like $1/\sqrt{x}$, which is integrable at the nose. This indicates that probably a first approximation to the total drag can be found as $\epsilon \to 0$.

A general result can be proved: if a problem with a parameter has similarity, then the approximate solution in terms of this parameter cannot be uniformly valid, unless the approximate solution turns out to be the exact solution (as in the stagnation-point case). By similarity, here we mean the fact, for example, that if a solution depends on coordinates and a parameter $(x, y; \epsilon)$, the solution must depend on two combinations of these due to invariance. In the case of the semi-infinite flat plate, the Navier-Stokes solution $u(x, y; \epsilon) = fn(x/\epsilon, y/\epsilon)$ and the boundary layer solution is not uniformly valid. The proof of this theorem, as well as much detailed discussion of expansions for both ϵ small and ϵ large in special problems for the Navier-Stokes equations, is given in [3.18].

3.3 Singular Boundary Problems

Just as we found for ordinary differential equations discussed in Section 2.5, there are problems for partial differential equations in which various asymptotic expansions are constructed in different regions, but where the order of the system does not change in the limit of vanishing of the small parameter. In these problems, the form of the expansions is dominated by the boundary conditions and usually, in one limit or other, a region degenerates to a line or a point, and may thus be singular. Narrow domains, slender bodies, and disturbances of small spatial extent are examples. In these cases, it is often useful, although not always necessary, to construct different expansions in different regions. Expansions valid near the singularity can be matched with expansions that are valid far away. Several such examples are now considered.

3.3.1 One-Dimensional Heat Conduction

In many examples, the geometrical shape of the domain of the problem introduces a small parameter. For such thin domains, it is often possible to introduce various asymptotic expansions based on the limit $\epsilon \to 0$. The terms in these asymptotic expansions can correspond to simplified models for the physical process. In this section, one-dimensional heat conduction is derived from a three-dimensional equation. In Section 3.3.2, elastic shell theory is derived from the three-dimensional linear elasticity equations, and there are many other examples. In general, the boundary conditions for the simplified equations have to be derived from matching with a more complicated boundary layer involving more independent variables.

Problem formulation

Consider steady heat conduction in a long rod of circular cross section whose shape is given by

$$S(X, R) = 0 = R - BF(X/L), \quad 0 \le X \le L \tag{3.3.1}$$

(see Figure 3.3.1). Assume that the side of the rod is insulated, so that $\partial T/\partial n = 0$, and assume that the temperature $T(X, R)$ is prescribed on the ends and is written in the form

$$T(0, R) = T^*\phi\left(\frac{R}{B}\right), \quad T(L, R) = T^*\psi\left(\frac{R}{B}\right), \tag{3.3.2}$$

so that heat flows down the rod. We are interested in the case where $B/L \ll 1$. Here T^* is a characteristic temperature, and the equation for steady heat flow with constant thermal properties is Laplace's equation with axial symmetry;

$$\frac{\partial^2 T}{\partial R^2} + \frac{1}{R}\frac{\partial T}{\partial R} + \frac{\partial^2 T}{\partial X^2} = 0. \tag{3.3.3}$$

The boundary condition on the insulated surface $S = 0$ can be expressed as

$$\text{grad}T \cdot \text{grad}S = 0 \quad \text{on } S = 0$$

or

$$\frac{\partial T}{\partial R} = \frac{B}{L}F'\left(\frac{X}{L}\right)\frac{\partial T}{\partial X} \text{ on } R = BF\left(\frac{X}{L}\right). \tag{3.3.4}$$

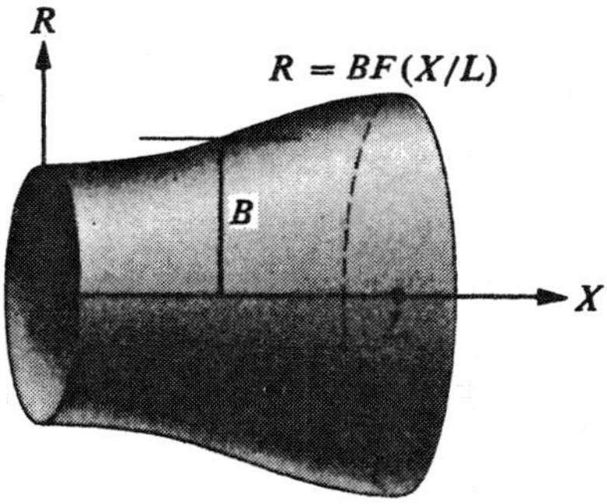

FIGURE 3.3.1. Quasi-One-Dimensional Heat Conduction

The entire problem can be expressed in the following suitable dimensionless coordinates:

$$r = \frac{R}{B}, \quad x = \frac{X}{L}, \quad \theta(x, r; \epsilon) = \frac{T(X, R)}{T*},$$

where $\epsilon = B/L$. In terms of these variables, (3.3.3) becomes

$$\frac{\partial^2\theta}{\partial r^2} + \frac{1}{r}\frac{\partial\theta}{\partial r} + \epsilon^2\frac{\partial^2\theta}{\partial x^2} = 0, \tag{3.3.5}$$

and the problem is specified by the following boundary conditions:

ends $$\theta(0, r; \epsilon) = \phi(r), \theta(1, r; \epsilon) = \psi(r), \tag{3.3.6}$$

side $$\frac{\partial\theta}{\partial r}(x, F(x); \epsilon) = \epsilon^2 F'(x)\frac{\partial\theta}{\partial x}(x, F(x); \epsilon). \tag{3.3.7}$$

Outer expansion

We now assume that a limiting solution, independent of ϵ, appears as $\epsilon \to 0$ and represents θ by the following asymptotic expansion, which we expect to be valid away from the ends of the rod:

$$\theta(x, r; \epsilon) = \theta_0(x, r) + \epsilon^2\theta_1(x, r) + \dots. \tag{3.3.8}$$

The corresponding limit process has $\epsilon \to 0$, (x, r) fixed.

In general, an arbitrary order could be chosen for the θ_0 term, but matching would show it to be of $O(1)$. The second term is of order ϵ^2, so that a nonhomogeneous equation for θ_1 results. Terms of intermediate order could be inserted if needed. The sequence of equations approximating (3.3.5) is

$$\frac{\partial^2\theta_0}{\partial r^2} + \frac{1}{r}\frac{\partial\theta_0}{\partial r} = 0, \tag{3.3.9}$$

$$\frac{\partial^2\theta_1}{\partial r^2} + \frac{1}{r}\frac{\partial\theta_1}{\partial r} = -\frac{\partial^2\theta_0}{\partial x^2}. \tag{3.3.10}$$

All subsequent equations are of the form of (3.3.10). The boundary condition (3.3.7) has the expansion

$$\frac{\partial\theta_0}{\partial r}(x, F(x)) + \epsilon^2\frac{\partial\theta_1}{\partial r}(x, F(x)) + \dots = \epsilon^2 F'(x)\left\{\frac{\partial\theta_0}{\partial x}(x, F(x)) + \dots\right\}$$

so that, on the insulated boundary, we have

$$\frac{\partial\theta_0}{\partial r}(x, F(x)) = 0, \tag{3.3.11}$$

$$\frac{\partial\theta_1}{\partial r}(x, F(x)) = F'(x)\frac{\partial\theta_0}{\partial x}(x, F(x)). \tag{3.3.12}$$

The solution of (3.3.9) for θ_0 is

$$\theta_0(x, r) = A_0(x) + B_0(x)\log r. \tag{3.3.13}$$

If we require finite temperature at the axis, then $B_0 = 0$, and the basic approximation is a one-dimensional temperature distribution:

$$\theta_0(x, r) = A_0(x). \tag{3.3.14}$$

This distribution automatically satisfies the boundary condition (3.3.11). In this case, further information about $A_0(x)$ cannot be found without considering the equation for θ_1 and its boundary condition. Equation (3.3.10) is now

$$\frac{\partial^2 \theta_1}{\partial r^2} + \frac{1}{r}\frac{\partial \theta_1}{\partial r} = -\frac{d^2 A_0}{\partial x^2}, \tag{3.3.15}$$

which, if we disregard the log r term, has the solution

$$\theta_1(x, r) = A_1(x) - \frac{r^2}{4}\frac{d^2 A_0}{dx^2}. \tag{3.3.16}$$

Now the boundary condition (3.3.12) on the insulated surface becomes

$$-\frac{F(x)}{2}\frac{d^2 A_0}{dx^2} = F'(x)\frac{dA_0}{dx} \tag{3.3.17}$$

or

$$\frac{d}{dx}\left(F^2(x)\frac{dA_0}{dx}\right) = 0. \tag{3.3.18}$$

Information about A_1 is found from the equation for θ_2, etc.

Remembering that $F(x)$ is proportional to the radius of a cross section, we see that (3.3.18) is the equation for one-dimensional heat conduction. It arises here as a formal consequence of the insulation boundary condition. For the uniform accuracy of this approximation over the center section of the rod, $F(x)$ has to be sufficiently smooth. To determine $A_0(x)$ from (3.3.18) uniquely, we need two conditions. We will see next that $A_0(0)$ and $A_0(1)$ are defined by matching with boundary layers at $x = 0$ and $x = 1$, respectively.

Boundary layer expansion, matching

Near $x = 0$, the only distinguished limit that preserves enough structure in (3.3.5) to allow for boundary conditions and matching is one in which $x^* = x/\epsilon$ is fixed. Thus, consider the following asymptotic expansion that is valid near $x = 0$:

$$\theta(x, r; \epsilon) = \vartheta(x^*, r) + \dots, \qquad x^* = x/\epsilon. \tag{3.3.19}$$

Then the full equation results for ϑ:

$$\frac{\partial^2 \vartheta}{\partial r^2} + \frac{1}{r}\frac{\partial \vartheta}{\partial r} + \frac{\partial^2 \vartheta}{\partial x^{*2}} = 0, \tag{3.3.20}$$

but the boundary condition on the insulated surface is somewhat simplified. Equation (3.3.7) becomes

$$\frac{\partial \vartheta}{\partial r}(x^*, F(\epsilon x^*)) + \dots = \epsilon^2 F'(\epsilon x^*)\frac{1}{\epsilon}\frac{\partial \vartheta}{\partial x^*}(x^*, F(\epsilon x^*)) + \dots. \tag{3.3.21}$$

Thus, as $\epsilon \to 0$, we have

$$\frac{\partial \vartheta}{\partial r}(x^*, r_0) = 0, \tag{3.3.22}$$

where $r_0 = F(0)$. Again the assumption that F is smooth has been used. It can be seen from (3.3.21) that the next term in the boundary layer expansion (3.3.19) is $O(\epsilon)$, but this is not considered here. At the end $x = 0$, we have

$$\vartheta(0, r) = \phi(r), \quad 0 \le r \le r_0. \tag{3.3.23}$$

Thus, the problem to be solved is that of heat flow in an insulated semi-infinite cylinder. The extent in the x^* direction is infinite, since $x^* \to \infty$, $x = 0$ in the matching. The matching condition here takes the simple form

$$\vartheta(\infty, r) = A_0(0). \tag{3.3.24}$$

It remains to be shown that $\vartheta(x^*, r) \to$ const. as $x^* \to \infty$ and to evaluate the constant. The solution to the problem for ϑ can be expressed, by separation of variables, in terms of functions like

$$e^{-\lambda x^*} J_0(\lambda r), \quad \lambda \ge 0,$$

where J_0 is the Bessel function of the first kind of order zero. The transcendental equation for the eigenvalues λ_n follows from (3.3.22):

$$\lambda_n J_0'(\lambda_n r_0) = 0. \tag{3.3.25}$$

There is an infinite set of roots starting with $\lambda_0 = 0, \lambda_1, \lambda_2, \lambda_3, \ldots$, and an infinite complete set of eigenfunctions. Thus, we may represent the solution for ϑ in the form

$$\vartheta(x^*, r) = a_0 + \sum_{n=1}^{\infty} a_n e^{-\lambda_n x^*} J_0(\lambda_n r). \tag{3.3.26}$$

From the equation for $J_0(\lambda r)$,

$$\frac{d}{dr}\left\{r \frac{dJ_0}{dr}\right\} + \lambda^2 r J_0(\lambda r) = 0, \tag{3.3.27}$$

it follows by integration from 0 to r_0 that

$$\int_0^{r_0} J_0(\lambda_n r) r \, dr = 0. \tag{3.3.28}$$

Thus, the constant a_0 is determined from

$$\int_0^{r_0} \vartheta(x^*, r) r \, dr = \frac{r_0^2}{2} a_0 = \int_0^{r_0} \vartheta(0, r) r \, dr$$

or

$$a_0 = \frac{2}{r_0^2} \int_0^{r_0} \phi(r) r \, dr. \tag{3.3.29}$$

Thus, the matching condition (3.3.24) states that the (weighted) average temperature at the end should be used as the boundary condition for the one-dimensional heat flow:

$$A_0(0) = (2/r_0^2) \int_0^{r_0} \phi(r) r \, dr, \quad r_0 = F(0). \tag{3.3.30}$$

Similar considerations apply near $x = 1$, so that we have

$$A_0(1) = (2/r_1^2) \int_0^{r_1} \psi(r) r \, dr, \quad r_1 = F(1), \tag{3.3.31}$$

and the net heat flow can then be calculated to $O(\epsilon)$.

The other coefficients in the series (3.3.26) can be calculated from the usual orthogonality properties of the eigenfunctions:

$$\int_0^{r_0} J_0(\lambda_n r) J_0(\lambda_m r) r \, dr = \begin{cases} 0, & n \neq m, \\ \gamma_m^2, & n = m. \end{cases} \tag{3.3.32}$$

The correctness of the one-dimensional approximation depends to a large extent on the type of boundary conditions. Problems 1 and 2 illustrate this point.

3.3.2 *Elastic-Shell Theory, Spherical Shell*

By an *elastic shell*, we mean a thin region of elastic material that responds to a load in a special way due to its geometrical properties. The theory of elastic shells can be derived in a systematic way, by the use of perturbation expansions, from the three-dimensional equations of elasticity. This is not the method usually followed in various books. Rather, shell equations are derived from overall assumptions about the total forces and moments acting on an infinitesimal element. These forces and moments are often thought of as averages across the cross section of a shell, or corresponding strain-energy methods are used (see [3.23], [3.25] and [3.41]).

Perturbation theory corroborates the approximate equations in certain cases and further provides a method for incorporating the boundary layers that inevitably arise. If some stage of the approximation corresponds to simplified shell theory, one cannot expect to satisfy full-elasticity boundary conditions.

Problem formulation

The basic small parameter ϵ of shell theory is the thickness over a characteristic length, say the sphere radius. The calculations here are based wholly on linear elasticity theory, that is, on small strains. Thus, the loads that are applied must be thought of as being sufficiently small so that the structure remains in the linear elastic range. Since the loads then occur linearly in the problem, they need not be considered in the perturbation scheme; all results are proportional to the loads. However, if large deformations or nonlinearities are to be considered, then the mutual dependence of load (made dimensionless with an elastic modulus) and ϵ is of vital importance.

In this section, we consider a simple special example of shell theory, namely, a segment of a spherical shell fastened rigidly around the edges and loaded by axisymmetric pressure forces on the inner surface (see Figure 3.3.2). The problem is sufficiently general to illustrate all the essential features of shell theory. First, the outer expansion valid away from the boundary is constructed and is shown to contain the membrane theory of thin shells. Then, the various boundary layers that must be added are discussed briefly.

The exact boundary-value problem demands a solution of the full equations for elasticity (written, for example, in terms of the displacements (q_r, q_θ, q_ϕ)) in polar coordinates (r, θ, ϕ) subject to the boundary conditions of prescribed stresses on the free surfaces and zero displacements on the fixed edges:

$$T_{rr}(a+t,\theta) = T_{r\theta}(a+t,\theta) = 0, \quad 0 \le \theta \le \nu;$$

$$T_{rr}(a-t,\theta) = -p(\theta), \quad T_{r\theta}(a-t,\theta) = 0, \quad 0 \le \theta \le \nu; \tag{3.3.33}$$

$$q_r(r,\nu) = q_\theta(r,\nu) = 0, \quad a-t \le r \le a+t.$$

Away from the edge ($\theta = \nu$), it can be expected that the solution to this problem should behave something like that of the full sphere under pressure. An exact solution of the latter problem is available for uniform pressure (e.g., see page 142 of [3.25]). A study of this exact solution shows that the deflections have an expansion starting with $O(1/\epsilon)$ terms and that the hoop stresses $T_{\theta\theta}, T_{\phi\phi}$ are also $O(1/\epsilon)$, as would be expected from an overall force balance. These facts can be used to start the expansion of our problem corresponding to membrane theory, which is valid away from the edge $\theta = \nu$.

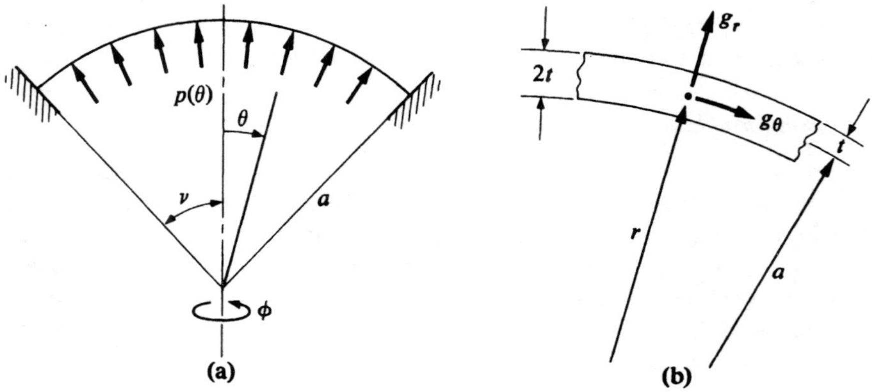

FIGURE 3.3.2. (a) Spherical Segment, (b) Detail of Thickness

Our procedure is to first construct the membrane-theory expansion assumed valid away from the boundary and then to construct the necessary boundary layer. A convenient starting point is the stress-equilibrium equations (e.g., see page 91 of [3.25]) (div $T = 0$):

$$\frac{\partial T_{rr}}{\partial r} + \frac{1}{r}\frac{\partial T_{r\theta}}{\partial \theta} + \frac{1}{r}(2T_{rr} - T_{\theta\theta} - T_{\phi\phi} + T_{r\theta}\cot\theta) = 0, \tag{3.3.34}$$

$$\frac{\partial T_{r\theta}}{\partial r} + \frac{1}{r}\frac{\partial T_{\theta\theta}}{\partial \theta} + \frac{1}{r}[(T_{\theta\theta} - T_{\phi\phi})\cot\theta + 3T_{r\theta}] = 0. \tag{3.3.35}$$

The stress components are related to the strain components with the help of the elastic constants (λ, μ):

$$T_{rr} = \lambda\Delta + 2\mu\frac{\partial q_r}{\partial r} \tag{3.3.36a}$$

$$T_{\theta\theta} = \lambda\Delta + 2\mu\left(\frac{1}{r}\frac{\partial q_\theta}{\partial \theta} + \frac{q_r}{r}\right), \tag{3.3.36b}$$

$$T_{\theta\theta} = \lambda\Delta + 2\mu\left(\frac{\cot\theta}{r}q_\theta + \frac{q_r}{r}\right), \tag{3.3.36c}$$

$$\frac{1}{\mu}T_{r\theta} = \frac{\partial q_\theta}{\partial r} - \frac{q_\theta}{r} + \frac{1}{r}\frac{\partial q_r}{\partial \theta}, \tag{3.3.36d}$$

$$T_{r\phi} = T_{\theta\phi} \equiv 0, \tag{3.3.36e}$$

where the dilatation Δ is

$$\Delta = \text{div}\,\mathbf{q} = \frac{\partial q_r}{\partial r} + \frac{2}{r}q_r + \frac{1}{r}\frac{\partial q_\theta}{\partial \theta} + \frac{\cot\theta}{r}q_\theta. \tag{3.3.37a}$$

Note also that

$$E = \text{modulus of elasticity} = \frac{\mu(3\lambda + 2\mu)}{\lambda + \mu}, \tag{3.3.37b}$$

$$\overline{\sigma} = \text{Poisson ratio} = \frac{\lambda}{2(\lambda + \mu)}. \tag{3.3.37c}$$

Inner expansions (membrane theory)

Now consider a limit process $\epsilon \to 0$ and coordinates fixed inside the shell (r^*, θ), where

$$r^* = \frac{r - a}{t} = \frac{r/a - 1}{\epsilon}, \quad -1 \le r^* \le 1, \quad \frac{\partial}{\partial r} = \frac{1}{\epsilon}\frac{\partial}{\partial r^*}. \tag{3.3.38}$$

The stresses in a thin shell thus have an asymptotic expansion of the form

$$\begin{aligned} T_{rr}(r, \theta; \epsilon) &= T(r^*, \theta) + \ldots, \quad T_{r\theta}(r, \theta; \epsilon) = S(r^*, \theta) + \ldots; \\ T_{\theta\theta}(r, \theta; \epsilon) &= \frac{H(\theta)}{\epsilon} + \ldots, \quad T_{\phi\phi}(r, \theta; \epsilon) = \frac{J(\theta)}{\epsilon} + \ldots. \end{aligned} \tag{3.3.39}$$

If, in fact, $O(1/\epsilon)$ stresses were allowed in T_{rr}, $T_{r\theta}$, the equilibrium equations would immediately show that these stresses are functions of θ only, and the boundary conditions would then show that these terms are zero.

The $O(1/\epsilon)$ hoop stress terms are here assumed to depend only on θ; this mirrors their behavior for the full sphere, where they are constant across the thickness. This can also be proved by a detailed consideration of the displacement equations and expansions. It is sufficient here to show consistency with the equations and boundary conditions. The equilibrium equations (3.3.34) and (3.3.35) have $O(1/\epsilon)$ terms, which are

$$\frac{\partial T}{\partial r^*} - (H + J) = 0, \tag{3.3.40}$$

$$\frac{\partial S}{\partial r^*} + \frac{dH}{d\theta} + (H - J)\cot\theta = 0. \tag{3.3.41}$$

Here we have used $r/a = 1 + \epsilon r^*$, $a/r = 1 - \epsilon r^* + \dots$. These approximate stress-balance equations can be integrated at once to yield

$$T(r^*, \theta) = \tau(\theta) + \{H(\theta) + J(\theta)\}r^*, \tag{3.3.42}$$

$$S(r^*, \theta) = \sigma(\theta) - r^*(dH/d\theta) - r^*(H - J)\cot\theta, \tag{3.3.43}$$

where $\tau(\theta)$, $\sigma(\theta)$ are functions of integration. Now, applying the boundary conditions (3.3.33), we have four relations:

$$0 = T(1, \theta) = \tau(\theta) + H(\theta) + J(\theta),$$

$$-p(\theta) = T(-1, \theta) = \tau - (H + J),$$

$$0 = S(1, \theta) = \sigma - \frac{dH}{d\theta} - (H - J)\cot\theta,$$

$$0 = S(-1, \theta) = \sigma + \frac{dH}{d\theta} + (H - J)\cot\theta.$$

Elimination from this system provides the basic equations for H, J:

$$H + J = \frac{p}{2} = -\tau, \tag{3.3.44}$$

$$\frac{dH}{d\theta} + (H - J)\cot\theta = 0 = \sigma \tag{3.3.45}$$

or, in terms of H itself,

$$\frac{d}{d\theta}(H\sin^2\theta) = \frac{p}{2}\sin\theta\cos\theta. \tag{3.3.46}$$

The integral of (3.3.46), which has H bounded as $\theta \to 0$, is

$$H(\theta) = \frac{1}{\sin^2\theta}\int_0^\theta \frac{p(\alpha)}{2}\sin\alpha\cos\alpha\, d\alpha. \tag{3.3.47}$$

Here, J follows from (3.3.44):

$$J(\theta) = \frac{p(\theta)}{2} - \frac{1}{\sin^2\theta}\int_0^\theta \frac{p(\alpha)}{2}\sin\alpha\cos\alpha\, d\alpha. \tag{3.3.48}$$

For the special case where $p = \text{const.}$, the classical result is obtained:

$$H = J = p/4 = \text{const.} \tag{3.3.49}$$

Since τ, σ are given by (3.3.44) and (3.3.45), we now have the $O(1)$ distribution of shear and normal stress across the section:

$$T(r^*, \theta) = -\frac{p(\theta)}{2}(1 - r^*), \tag{3.3.50}$$

$$S(r^*, \theta) = 0. \tag{3.3.51}$$

It is of interest now, and essential for matching later, to obtain the form of the deflection that corresponds to this distribution of stresses. As for the full sphere, the dominant terms of the deflection are $O(1/\epsilon)$ and are functions of θ alone. The justification is similar to that given before. Thus, tentatively assume that

$$\frac{q_r}{a} = \frac{u(\theta)}{\epsilon} + u_1(r^*, \theta) + \ldots, \tag{3.3.52}$$

$$\frac{q_\theta}{a} = \frac{v(\theta)}{\epsilon} + v_1(r^*, \theta) + \ldots. \tag{3.3.53}$$

The stresses produced by this set of displacements are now studied. From (3.3.37a), we have

$$\Delta = \frac{1}{\epsilon}\left\{\frac{\partial u_1}{\partial r^*} + 2u\frac{dv}{d\theta} + v\cot\theta\right\} + \ldots \tag{3.3.54}$$

so that the expansion for T_{rr} starts out as

$$T_{rr} = \frac{1}{\epsilon}\left\{\lambda\left(2u + \frac{dv}{d\theta} + v\cot\theta\right) + (\lambda + 2\mu)\frac{\partial u_1}{\partial r^*}\right\} + \ldots. \tag{3.3.55}$$

However, this term must be identically zero since no T_{rr} of $O(1/\epsilon)$ occurs in the problem. Thus, we have

$$\frac{\partial u_1}{\partial r^*} = -\frac{\lambda}{\lambda + 2\mu}\left(2u + \frac{dv}{d\theta} + v\cot\theta\right). \tag{3.3.56}$$

There is, in consequence, a dilatation of $O(1/\epsilon)$ corresponding to the general stretching of the shell:

$$\Delta = \frac{1}{\epsilon}\frac{2\mu}{\lambda + 2\mu}\left\{2u + \frac{dv}{d\theta} + v\cot\theta\right\} + \ldots. \tag{3.3.57}$$

Equations (3.3.44) and (3.3.45) can now be expressed as equations for the displacement of the shell:

$$T_{\theta\theta} + T_{\phi\phi} = 2(\lambda + \mu)\Delta - 2\mu\left(\frac{\partial q_r}{\partial r} + \frac{q_r}{r}\right)$$

$$= 2(\lambda + \mu)\frac{1}{\epsilon}\left\{\frac{2\mu}{\lambda + 2\mu}\right\}\left\{2u + \frac{dv}{d\theta} + v\cot\theta\right\} - \frac{2\mu}{\epsilon}\left(\frac{\partial u_1}{\partial r^*}\right)$$

or

$$H + J = \frac{2\mu(3\lambda + 2\mu)}{\lambda + 2\mu}\left\{2u + \frac{dv}{d\theta} + v\cot\theta\right\} + \dots. \tag{3.3.58}$$

Similarly, we obtain

$$T_{\phi\phi} - T_{\theta\theta} = \frac{2\mu}{\epsilon}\left(v\cot\theta - \frac{dv}{d\theta}\right) + \dots$$

or

$$J - H = 2\mu\left(v\cot\theta - \frac{dv}{d\theta}\right) + \dots. \tag{3.3.59}$$

Thus, (3.3.44) directly becomes

$$\frac{dv}{d\theta} + v\cot\theta + 2u = \frac{p(\theta)}{4\mu}\frac{\lambda + 2\mu}{3\lambda + 2\mu} \tag{3.3.60}$$

and, after a little elimination, (3.3.45) becomes

$$\frac{du}{d\theta} - v = \frac{1}{2\mu}\frac{\lambda + \mu}{3\lambda + 2\mu}\frac{dp(\theta)}{d\theta}. \tag{3.3.61}$$

Equations (3.3.60) and (3.3.61) form the basic system of equations for the shape of the shell and are identical to the membrane equations mentioned on page 584 of [3.25].

The equation for the tangential displacement alone, from (3.3.60) and (3.3.61), is

$$\frac{d^2v}{d\theta^2} + \cot\theta\frac{dv}{d\theta} + (2 - \csc^2\theta)v = -\frac{1}{4\mu}\frac{dp}{d\theta}. \tag{3.3.62}$$

A particular solution corresponding to a rigid displacement is a solution of the homogeneous equations

$$v_p = A\sin\theta, \quad u_p = -A\cos\theta. \tag{3.3.63}$$

The complete solution for the case $p =$ const. is, thus,

$$u = u_\infty - A\cos\theta, \quad v = A\sin\theta, \tag{3.3.64}$$

where the constant u_∞ is the radial displacement of the full sphere under uniform pressure:

$$u_\infty = \frac{1}{8\mu}\frac{\lambda + 2\mu}{3\lambda + 2\mu}p. \tag{3.3.65}$$

It is clear that the solution represented by (3.3.64) cannot satisfy the boundary condition of no displacement at $\theta = \nu$, even with a particular choice of the rigid displacement A. In fact, no such rigid displacement of $O(1/\epsilon)$ is to be expected in this problem. Some kind of a boundary layer is needed near $\theta = \nu$.

Boundary-layer expansion

If displacements q_r, q_θ of the same order occur in a thin layer of $O(\epsilon)$ in thickness near $\theta = \nu$, plane-strain elasticity equations result. These are expressed in terms of r and $\theta^* = [\theta - \nu]/\epsilon$. However, it is easy to show that no solution of these plane-strain equations in the "elasticity" boundary layer exists that matches to the membrane expansion. For matching, we would need

$$\frac{q_r}{a} = \frac{u^*(r, \theta^*)}{\epsilon} + \dots, \qquad \frac{q_\theta}{a} = \frac{v^*(r, \theta^*)}{\epsilon} + \dots, \tag{3.3.66}$$

and $u^* \to u_\infty$, $v^* \to 0$, $\theta^* \to \infty$ $(A = 0)$.

Thus, some intermediate boundary layer must be constructed, and its width must be greater than that of the elasticity layer. That is, a boundary-layer expansion is sought, in which

$$\overline{\theta} = \frac{\theta - \nu}{\delta(\epsilon)}, \qquad r^* = \frac{(r/a) - 1}{\epsilon}, \tag{3.3.67}$$

are fixed where $\epsilon \ll \delta(\epsilon)$. Note that

$$\cot\theta = \cot\nu - \delta(1 + \cot^2\nu)\overline{\theta}. \tag{3.3.68}$$

Returning to the stress equations, we assume an asymptotic expansion of the form

$$T_{rr}(r, \theta; \epsilon) = \tau(r^*, \overline{\theta}) + \dots, \tag{3.3.69}$$

$$T_{\theta\theta}(r, \theta; \epsilon) = (1/\epsilon)h(r^*, \overline{\theta}) + \dots, \tag{3.3.70}$$

$$T_{\phi\phi}(r, \theta; \epsilon) = (1/\epsilon)g(r^*, \overline{\theta}) + \dots, \tag{3.3.71}$$

$$T_{r\theta}(r, \theta; \epsilon) = \beta(\epsilon)\sigma(r^*, \overline{\theta}) + \dots. \tag{3.3.72}$$

The orders of the hoop stresses are in accord with overall equilibrium ideas, the order of the normal stress with the boundary conditions, and the order $\beta(\epsilon)$ of the shear is here undetermined. A large shear can be expected to be produced if substantial bending takes place near the boundary. Other possibilities should be investigated and ruled out. This assumption leads to an expansion capable of being matched. The dominant equations of stress equilibrium are, thus,

$$\frac{1}{\epsilon}\frac{\partial\tau}{\partial r^*} + \frac{\beta(\epsilon)}{\delta(\epsilon)}\frac{\partial\sigma}{\partial\overline{\theta}} - \frac{1}{\epsilon}(h + g) = 0, \tag{3.3.73}$$

$$\frac{\beta}{\epsilon}\frac{\partial\sigma}{\partial r^*} + \frac{1}{\epsilon\delta}\frac{\partial h}{\partial\overline{\theta}} = 0. \tag{3.3.74}$$

In order for (3.3.74) to yield a nontrivial result, we need

$$\beta(\epsilon) = \frac{1}{\delta(\epsilon)}. \tag{3.3.75}$$

Then, the distinguished limit of (3.3.73) occurs for $1/\epsilon = \beta/\delta$ or

$$\delta = \sqrt{\epsilon}, \quad \beta(\epsilon) = 1/\sqrt{\epsilon}, \tag{3.3.76}$$

fixing the order of the boundary-layer thickness and shear stress. The resulting equations include all the terms of the "elasticity" boundary layer and thus have at least the possibility of matching to an elasticity boundary layer.

Rewriting the basic equations (3.3.73) and (3.3.74), we have

$$\frac{\partial \tau}{\partial r^*} + \frac{\partial \sigma}{\partial \overline{\overline{\theta}}} - (h + g) = 0,$$

$$\frac{\partial \sigma}{\partial r^*} + \frac{\partial h}{\partial \overline{\overline{\theta}}} = 0.$$

Next, consider the displacement field corresponding to the assumed orders of stress in (3.3.70)–(3.3.72):

$$\frac{q_r}{a} = \frac{U(\overline{\theta})}{\epsilon} + U_1(r^*, \overline{\theta}) + \ldots, \tag{3.3.77}$$

$$\frac{q_\theta}{a} = \frac{V(\overline{\theta}, r^*)}{\sqrt{\epsilon}} + \sqrt{\epsilon} V_1(r^*, \overline{\theta}) + \ldots. \tag{3.3.78}$$

It is necessary that $U = U(\overline{\theta})$ only, so that dilatation of $O(1/\epsilon^2)$ does not occur. Then, for the dilatation Δ, we have

$$\Delta = \frac{1}{\epsilon}\left\{\frac{\partial U_1}{\partial r^*} + 2U + \frac{\partial V}{\partial \overline{\overline{\theta}}}\right\} + \ldots, \tag{3.3.79}$$

and the expressions for the $O(1/\epsilon)$ components of hoop stress are

$$h = \lambda \frac{\partial U_1}{\partial r^*} + (\lambda + 2\mu)\frac{\partial V}{\partial \overline{\overline{\theta}}} + 2(\lambda + \mu)U, \tag{3.3.80}$$

$$g = \lambda \frac{\partial U_1}{\partial r^*} + \lambda \frac{\partial V}{\partial \overline{\overline{\theta}}} + 2(\lambda + \mu)U, \tag{3.3.81}$$

$$h + g = 2\lambda \frac{\partial U_1}{\partial r^*} + 2(\lambda + \mu)\left\{\frac{\partial V}{\partial \overline{\overline{\theta}}} + 2U\right\}. \tag{3.3.82}$$

Considering next the normal stress T_{rr}, we have

$$T_{rr} = \frac{1}{\epsilon}\left\{\lambda\left(\frac{\partial U_1}{\partial r^*} + 2U + \frac{\partial V}{\partial \overline{\overline{\theta}}}\right) + 2\mu \frac{\partial U_1}{\partial r^*}\right\} + \tau(r^*, \overline{\theta}) + \ldots. \tag{3.3.83}$$

Again, the $O(1/\epsilon)$ term in T_{rr} must vanish. This provides an expression for $\partial U_1/\partial r^*$ in terms of U, V and allows h, g to be expressed completely in terms of these quantities:

$$\frac{\partial U_1}{\partial r^*} = -\frac{\lambda}{\lambda + 2\mu}\left(\frac{\partial V}{\partial \overline{\overline{\theta}}} + 2U\right). \tag{3.3.84}$$

Thus, we have

$$h = 4\mu \frac{\lambda+\mu}{\lambda+2\mu}\frac{\partial V}{\partial\overline{\theta}} + 2\mu\frac{3\lambda+2\mu}{\lambda+2\mu}U, \tag{3.3.85}$$

$$h+g = 2\mu\frac{3\lambda+2\mu}{\lambda+2\mu}\left(\frac{\partial V}{\partial\overline{\theta}} + 2U\right). \tag{3.3.86}$$

A similar argument can be applied to the shear stress:

$$\frac{1}{\mu}T_{r\theta} = \frac{1}{\epsilon^{3/2}}\frac{\partial V}{\partial r^*} + \frac{1}{\sqrt{\epsilon}}\frac{\partial V_1}{\partial r^*} - \frac{V}{\sqrt{\epsilon}}$$

$$+(1-\epsilon r^* + \ldots)\left\{\frac{1}{\epsilon^{3/2}}\frac{dU}{d\overline{\theta}} + \frac{1}{\sqrt{\epsilon}}\frac{\partial U_1}{\partial\overline{\theta}}\right\} + \ldots.$$

The $O(1/\epsilon^{3/2})$ term must vanish:

$$\frac{\partial V}{\partial r^*} + \frac{\partial U}{\partial\overline{\theta}} = 0, \tag{3.3.87}$$

which is one of the basic differential equations for the shell deflection. Also, the $O(1/\sqrt{\epsilon})$ term is

$$\frac{\sigma(r^*,\overline{\theta})}{\mu} = \frac{\partial V_1}{\partial r^*} - V + \frac{\partial U_1}{\partial\overline{\theta}} - r^*\frac{dU}{d\overline{\theta}}. \tag{3.3.88}$$

The consequence of (3.3.87) is a linear variation of tangential displacement across the cross section

$$V(r^*,\overline{\theta}) = A(\overline{\theta}) - r^*\frac{dU}{d\overline{\theta}}, \tag{3.3.89}$$

and a corresponding linear variation of the hoop stresses from (3.3.85) and (3.3.86). Introducing some special notation, we have

$$h = h^{(0)}(\overline{\theta}) + r^* h^{(1)}(\overline{\theta}), \tag{3.3.90}$$

$$g = g^{(0)}(\overline{\theta}) + r^* g^{(1)}(\overline{\theta}), \tag{3.3.91}$$

where

$$h^{(0)} = 4\mu\frac{\lambda+\mu}{\lambda+2\mu}\frac{dA}{d\overline{\theta}} + 2\mu\frac{3\lambda+2\mu}{\lambda+2\mu}U,$$

$$h^{(1)} = -4\mu\frac{\lambda+\mu}{\lambda+2\mu}\frac{d^2U}{d\overline{\theta}^2},$$

$$h^{(0)} + g^{(0)} = 2\mu\frac{3\lambda+2\mu}{\lambda+2\mu}\left(\frac{dA}{d\overline{\theta}} + 2U\right),$$

$$h^{(1)} + g^{(1)} = -2\mu\frac{3\lambda+2\mu}{\lambda+2\mu}\frac{d^2U}{d\overline{\theta}^2}.$$

The tangential equilibrium equation (3.3.74) can now be integrated in the form

$$\sigma(r^*, \bar{\theta}) = \sigma^{(0)}(\bar{\theta}) - r^* \frac{dh^{(0)}}{d\bar{\theta}} - \frac{r^{*2}}{2} \frac{dh^{(1)}}{d\bar{\theta}}. \tag{3.3.92}$$

The boundary condition states that $\sigma(\pm 1, \theta) = 0$, so that we have

$$\frac{dh^{(0)}}{d\bar{\theta}} = 0 \tag{3.3.93}$$

and

$$\sigma^{(0)} = \frac{1}{2} \frac{dh^{(1)}}{d\bar{\theta}}. \tag{3.3.94}$$

The shear stress has a parabolic distribution across the thickness:

$$\sigma(r^*, \bar{\theta}) = (1 - r^{*2}) \frac{1}{2} \frac{dh^{(1)}}{d\bar{\theta}}. \tag{3.3.95}$$

A similar study of the radial equilibrium equation (3.3.73), using (3.3.94), (3.3.90), and (3.3.91), allows the basic differential equation for the shell deflection to be found, and from its solution all the stresses can also be found. Integration of (3.3.73) shows that

$$\tau(r^*, \bar{\theta}) = \tau^{(0)}(\bar{\theta}) + r^* \tau^{(1)}(\bar{\theta}) + r^{*2} \tau^{(2)}(\bar{\theta}) + r^{*3} \tau^{(3)}(\bar{\theta}), \tag{3.3.96}$$

where

$$\begin{aligned} \tau^{(1)} &= -\frac{1}{2} \frac{d^2 h^{(1)}}{d\bar{\theta}^2} + h^{(0)} + g^{(0)}, \\ \tau^{(2)} &= \frac{1}{2} \{h^{(1)} + g^{(1)}\}, \\ \tau^{(3)} &= \frac{1}{6} \frac{d^2 h^{(1)}}{d\bar{\theta}^2}. \end{aligned}$$

The boundary conditions at $r^* = \pm 1$ are

$$r^* = +1, \quad 0 = \tau^{(0)} + \tau^{(1)} + \tau^{(2)} + \tau^{(3)}, \tag{3.3.97}$$

$$r^* = -1, \quad -p(\nu) = \tau^{(0)} - \tau^{(1)} + \tau^{(2)} - \tau^{(3)}, \tag{3.3.98}$$

or

$$\tau^{(1)} + \tau^{(3)} = \frac{p(\nu)}{2} = -\frac{1}{3} \frac{d^2 h^{(1)}}{d\bar{\theta}^2} + h^{(0)} + g^{(0)}. \tag{3.3.99}$$

Equations (3.3.93) and (3.3.99) provide the basic systems of equations. Equation (3.3.93) states that $h^{(0)} = \text{const.}$ or

$$2(\lambda + \mu)(dA/d\bar{\theta}) + (3\lambda + 2\mu)U = (3\lambda + 2\mu)U_\infty, \tag{3.3.100}$$

where $U \to U_\infty$ as $\bar{\theta} \to -\infty$, for matching the constant u of the membrane solution as $\theta \to \nu$; $dA/d\bar{\theta} \to 0$, $\bar{\theta} \to -\infty$. Equation (3.3.99) is, from the definitions of $h^{(0)}, h^{(1)}, q^{(0)}$,

$$\frac{4}{3}\mu\frac{\lambda+\mu}{\lambda+2\mu}\frac{d^4U}{d\overline{\theta}^4}+2\mu\frac{3\lambda+2\mu}{\lambda+2\mu}\left(\frac{dA}{d\overline{\theta}}+2U\right)=\frac{p(\nu)}{2}. \tag{3.3.101}$$

For the special case of $p(\theta) = \text{const.} = p$, which is all that will be considered further, the elimination of $dA/d\overline{\theta}$ from (3.3.100) and (3.3.101) results in

$$\frac{d^4U}{d\overline{\theta}^4}+4\kappa^4U=\text{const.}=4\kappa^4U_\infty, \tag{3.3.102}$$

where

$$4\kappa^4=\frac{3}{4}\frac{(3\lambda+2\mu)(\lambda+2\mu)}{(\lambda+\mu)^2},\qquad U_\infty=\frac{1}{8\mu}\frac{\lambda+2\mu}{3\lambda+2\mu}p.$$

Equation (3.3.102) looks exactly like the equation of a beam on an elastic foundation and has oscillatory decaying solutions as $\overline{\theta}\to-\infty$. Discarding the solutions that grow as $\overline{\theta}\to-\infty$ and making $U(\overline{\theta})=0$ as $\overline{\theta}=0$ to approach the fixed boundary condition at $\theta=\nu$, we have

$$U(\overline{\vartheta})=U_\infty\{1-e^{\kappa\overline{\theta}}\cos\kappa\overline{\theta}\}+be^{\kappa\overline{\theta}}\sin\kappa\overline{\theta}, \tag{3.3.103}$$

where the constant b is arbitrary. Next, from (3.3.100), we calculate $A(\theta)$:

$$A(\overline{\theta})=A_\infty+\frac{3\lambda+2\mu}{4(\lambda+\mu)\kappa}\{(U_\infty+b)\cos\kappa\overline{\theta}+(U_\infty-b)\sin\kappa\overline{\theta}\}e^{\kappa\overline{\theta}} \tag{3.3.104}$$

and

$$\frac{dU}{d\overline{\theta}}=\kappa\{(U_\infty+b)\sin\kappa\overline{\theta}-(U_\infty-b)\cos\kappa\overline{\theta}\}e^{\kappa\overline{\theta}}. \tag{3.3.105}$$

Matching

The solution for $U(\overline{\theta})$, $V(\overline{\theta})$ should match to the elasticity boundary layer as $\overline{\theta}\to 0$, $\theta^*\to-\infty$. The entire matching process can be expressed in terms of a suitable limit as usual, but here the details are omitted. The behavior of the "bending-layer" solution as $\overline{\theta}\to 0$ is

$$U(\overline{\theta})\to(b-U_\infty)\kappa\overline{\theta}+b\kappa^2\overline{\theta}^2+(U_\infty+b)\frac{\kappa^2\overline{\theta}^3}{3}+\dots,$$

$$\frac{dU}{d\overline{\theta}}\to\kappa(b-U_\infty)+2b\kappa^2\overline{\theta}+(U_\infty+b)\kappa^3\overline{\theta}^2+\dots,$$

$$A\to A_\infty+\frac{3\lambda+2\mu}{4\kappa(\lambda+\mu)}\{U_\infty+b\}+2U_\infty\kappa\overline{\theta}+\dots, \tag{3.3.106}$$

and, from (3.3.89),

$$V\to A_\infty+\frac{3\lambda+2\mu}{4\kappa(\lambda+\mu)}\{V_\infty+b\}-r^*\kappa(b-U_\infty)+O(\overline{\theta}).$$

All attempts at matching this behavior to that of an "elasticity" boundary layer as $\theta^* \to -\infty$ fail, except if (3.3.106) is made to satisfy the boundary condition at $\overline{\theta} = 0$ exactly; that is, the elasticity boundary layer is included in (3.3.106). Thus, the arbitrary constants (A_∞, b) must be chosen so that $V = 0$ at $\overline{\theta} = 0$:

$$b - U_\infty = 0, \quad A_\infty + \frac{3\lambda + 2\mu}{4\kappa(\lambda + \mu)}(U_\infty + b) = 0, \tag{3.3.107}$$

and the bending boundary layer satisfies the fixed-edge boundary condition exactly. As far as (3.3.102) is concerned, this means that a fixed edge forces the boundary conditions. Thus, we have

$$U(\overline{\theta}) = \frac{dU}{d\overline{\theta}} = 0 \quad \text{at } \overline{\theta} = 0. \tag{3.3.108}$$

A consequence of this solution is that

$$V(r^*, \overline{\theta}) \to A_\infty = -\frac{3\lambda + 2\mu}{2(\lambda + \mu)\kappa} U_\infty \tag{3.3.109}$$

as $\overline{\theta} \to -\infty$. A term must be added to the membrane solution to match this deflection. This term can be a rigid displacement of $O(1/\sqrt{\epsilon})$. In particular, we could have

$$\begin{aligned} \frac{q_r}{a} &= \frac{u(\theta)}{\epsilon} - \frac{A_\infty \cos\theta}{\sqrt{\epsilon}} + \cdots \\ \frac{q_\theta}{a} &= \frac{v(\theta)}{\epsilon} + \frac{A_\infty \sin\theta}{\sqrt{\epsilon}} + \cdots, \end{aligned} \tag{3.3.110}$$

and this is the ultimate effect of the rigid boundary on the main part of the shell. The presence of an $O(1/\sqrt{\epsilon})$ term in q_r implies a higher-order bending layer, etc.

It should be noted in conclusion that the preceding considerations are not valid for a shallow shell where ν is small. To study a shallow shell, $\nu(\epsilon)$ must be assigned an order by studying the various limits. The interested reader will find further recent results and references in the survey paper [3.42] and in [3.9] which deals with cylindrical shells.

3.3.3 Radially Deforming Slender Body in a Uniform Stream

Problem formulation

Consider the incompressible flow of a perfect fluid past a body of revolution capable of arbitrary radial deformations. This is a generalization of classical slender-body theory for a rigid body of revolution. A discussion of the perturbation solution for the special case of a rigid body is given in section 3.3.4 of [3.15]. The body is held fixed in a flow that has a velocity U_∞ along the body axis at infinity. We denote the body length by L and let the deformations be defined by

$$R = R_{\max} F\left(\frac{X}{L}, \frac{T}{\tau}\right), \tag{3.3.111}$$

where capital letters denote dimensional variables and τ is the characteristic time for the deformations (see Figure 3.3.3).

We also assume that the body does not shed any vortices. The flow outside the body is then everywhere irrotational and is governed by Laplace's equation which, in cylindrical polar coordinates, has the form

$$\frac{\partial^2 \Phi}{\partial X^2} + \frac{\partial^2 \Phi}{\partial R^2} + \frac{1}{R}\frac{\partial \Phi}{\partial R} = 0. \tag{3.3.112}$$

Here $\Phi(X, R, T)$ is the velocity potential that determines the axial and radial components U and V of the flow field according to

$$U = \frac{\partial \Phi}{\partial X}, \tag{3.3.113a}$$

$$V = \frac{\partial \Phi}{\partial R}. \tag{3.3.113b}$$

We note that the time occurs only as a parameter in (3.3.112). Once the solution for Φ is known, one can calculate the pressure field from Bernoulli's equation

$$P - P_\infty = -\rho \frac{\partial \Phi}{\partial T} + \frac{1}{2}\rho\left[U_\infty^2 - \left(\frac{\partial \Phi}{\partial X}\right)^2 - \left(\frac{\partial \Phi}{\partial R}\right)^2\right],$$

where P_∞ is the ambient pressure of the undisturbed flow at infinity, P is the pressure in the flow field, and ρ is the constant density of the fluid.

The boundary conditions to be imposed are that the potential tends to the undisturbed free-stream value at infinity and that the component of the flow velocity normal to the body at the body surface must be equal to the normal component of the body surface velocity.

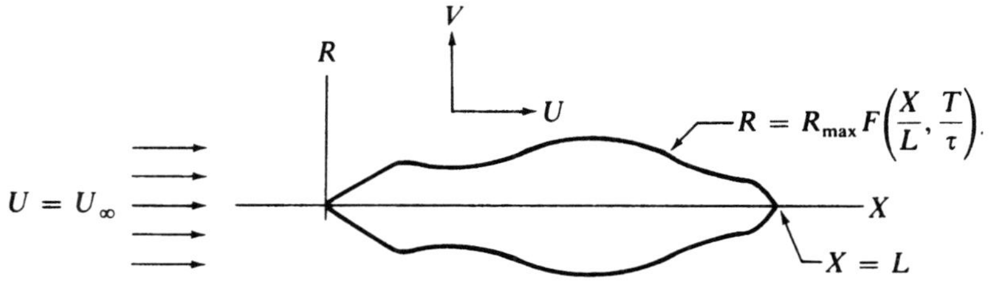

FIGURE 3.3.3. Flow Geometry

Now, the outward normal to the body $\mathbf{n}$ has components

$$\mathbf{n} = \left[-R_{\max}\frac{\partial F}{\partial X}, 1\right];$$

the velocity components of the body surface are defined by

$$\mathbf{V}_B = \left[0, R_{\max}\frac{\partial F}{\partial T}\right];$$

and the velocity components of the fluid are given by (3.3.113). Thus, the boundary condition

$$\mathbf{n}\cdot \operatorname{grad}\Phi = \mathbf{n}\cdot\mathbf{V}_B \quad \text{on } R = R_{\max}F \tag{3.3.114}$$

becomes

$$\frac{\partial \Phi}{\partial R}(X, R_{\max}F, T) = R_{\max}\left[\frac{\partial F}{\partial X}\frac{\partial \Phi}{\partial X} + \frac{\partial F}{\partial T}\right]. \tag{3.3.115}$$

To calculate the horizontal force on the body, we consider an annular element of thickness dX as shown in Figure 3.3.4.

The net pressure acting on the surface is $P - P_\infty$ and is normal to the surface. Thus, the horizontal component of the pressure is given by

$$P_H = (P - P_\infty)\sin\theta, \tag{3.3.116}$$

where θ is the angle between the tangent to the surface and the horizontal; i.e.,

$$\theta = \tan^{-1} R_{\max}\frac{\partial F}{\partial X}. \tag{3.3.117}$$

We note that P_H is negative, i.e., propulsive, if, say, $P > P_\infty$ and $(\partial F/\partial X) < 0$ as drawn in Figure 3.3.4.

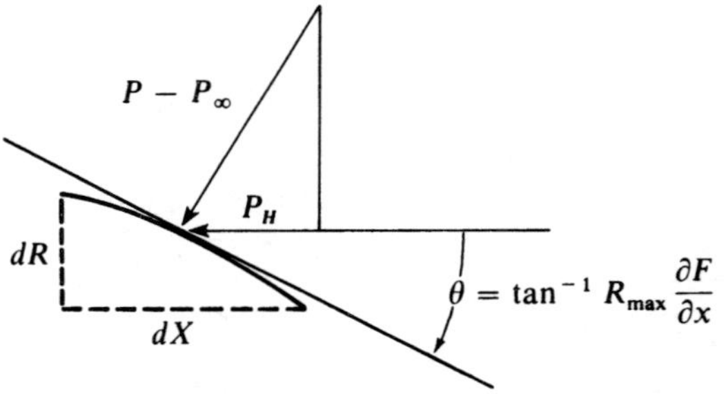

FIGURE 3.3.4. Pressure on the Surface

The horizontal force on the annulus is thus

$$dD = 2\pi(P - P_\infty)(\tan\theta)R_{\max}F\,dX \tag{3.3.118a}$$

and the total force D on the body is therefore

$$D(T) = 2\pi R_{\max}^2 \int_0^L (P - P_\infty)F\frac{\partial F}{\partial X}\,dX, \tag{3.3.118b}$$

where we have used the identity (3.3.117).

It is convenient to introduce dimensionless variables (denoted by lowercase letters) defined as follows

$$r = \frac{R}{L}, \quad x = \frac{X}{L}, \quad t = \frac{T}{\tau}, \quad \phi = \frac{\Phi}{LU_\infty}. \tag{3.3.119}$$

Then, the governing equation (3.3.112) becomes

$$\Delta\phi \equiv \phi_{xx} + \phi_{rr} + \frac{1}{r}\phi_r = 0. \tag{3.3.120}$$

The boundary condition on the surface reduces to

$$\phi_r[x, \epsilon F(x,t), t] = \epsilon\left\{\phi_x[x, \epsilon F(x,t), t]F_x(x,t) + \frac{1}{\delta}F_t(x,t)\right\}; \quad 0 \le x \le 1 \tag{3.3.121}$$

where

$$\epsilon = \frac{R_{\max}}{L} \ll 1$$

and

$$\delta = \frac{\tau U_\infty}{L} = O(1).$$

Thus, ϵ measures the slenderness of the body, while δ is the ratio of the two characteristic times in the problem, and we assume here that $\delta = O(1)$. The boundary condition at infinity is

$$\phi \to x \quad \text{as} \quad x^2 + r^2 \to \infty. \tag{3.3.122}$$

If we introduce the pressure coefficient, C_p, according to the usual convention

$$C_p = \frac{(P - P_\infty)}{\rho U_\infty^2},$$

Bernoulli's equation becomes

$$C_p = -\frac{1}{\delta}\phi_t + \frac{1}{2}[1 - \phi_x^2 - \phi_r^2]. \tag{3.3.123}$$

Finally, the force on the body can be expressed in terms of an axial force coefficient, C_D, nondimensionalizing (3.3.118b) as

$$C_D = 2\int_0^1 C_p(x,t)G_x(x,t)\,dx, \tag{3.3.124}$$

where

$$C_D = \frac{2D}{\rho U_\infty^2 \pi R_{\max}^2} \tag{3.3.125a}$$

and

$$G = F^2. \tag{3.3.125b}$$

Thus, the classical (rigid) slender-body theory corresponds to the special case $F_t = 0$.

For future reference, it is important to point out that if the potential ϕ is discontinuous with respect to the time (as would be the case if the velocity of surface deformations were discontinuous in time) (3.3.123) remains valid as long as we interpret ϕ_t in the sense of distributions. More precisely, if at some time $t = t_0$, ϕ has a finite jump discontinuity of the form

$$[\phi]_{t_0} \equiv \phi(x, r, t_0^+) - \phi(x, r, t_0^-), \tag{3.3.126a}$$

then (3.3.123) holds as long as we interpret

$$\phi_t = \{\phi_t\} + [\phi]_{t_0}\delta(t - t_0), \tag{3.3.126b}$$

where { } denotes the continuous part of ϕ_t and δ is the Dirac delta function.

This result follows directly by integration of the Euler equations

$$\text{grad } p = -\frac{\partial}{\partial t}\text{grad }\phi - \text{grad}\left[\frac{1}{2}(\text{grad }\phi)^2\right]$$

since it is easy to show that

$$\frac{\partial}{\partial t}\text{grad }\phi = \text{grad}\left[\left\{\frac{\partial \phi}{\partial t}\right\} + [\phi]_{t_0}\delta(t - t_0)\right].$$

Outer expansion, singularity at $r = 0$

At first glance, the problem defined by (3.3.112) subject to the boundary conditions (3.3.121), (3.3.122), appears to be a regular perturbation problem for $\epsilon \ll 1$. We show next that the outer expansion ($\epsilon \to 0$, x, r, t fixed) becomes singular because, for $\epsilon \to 0$, the boundary condition (3.3.121) is imposed at $r = 0$, where the solution of Laplace's equation has a singularity.

We look for an expansion for ϕ in the form

$$\phi(x, r, t; \epsilon) = x + \mu_1(\epsilon)\phi_1(x, r, t) + \dots, \tag{3.3.127}$$

where $\mu_1(\epsilon)$ is to be determined. Thus, ϕ_1 satisfies $\Delta\phi_1 = 0$ with zero boundary condition at infinity. If we now attempt to impose the surface boundary condition (3.3.121), we must have

$$\mu_1(\epsilon)\frac{\partial \phi_1}{\partial r}(x, \epsilon F(x, t), t) = \epsilon(F_x(x, t) + \frac{1}{\delta}F_t(x, t)) + \dots \tag{3.3.128}$$

on $0 \leq x \leq 1$. If $\epsilon \ll \mu_1$, we have $\frac{\partial \phi_1}{\partial r} = 0$ on the surface and conclude that $\phi_1(x, r, t) = 0$. The choice $\mu_1 \ll \epsilon$ is inconsistent, so that we must choose

$\mu_1 = O_s(\epsilon)$; say $\mu_1 = \epsilon$ for simplicity. Then, we have the surface boundary condition

$$\frac{\partial \phi_1}{\partial r}(x, 0, t) = F_x(x, t) + \frac{1}{\delta} F_t(x, t); \quad 0 \leq x \leq 1. \tag{3.3.129}$$

Now, the general solution for $\Delta\phi_1 = 0$, subject to (3.3.129) and $\phi_1 \to 0$ as $x^2 + r^2 \to \infty$, may be written in terms of a source distribution, of strength/unit distance $S_1(x, t)$, along the interval $0 \leq x \leq 1$ in the form (e.g., see Secs. 2.4.3 and 2.12.2 of [3.15]):

$$\phi_1(x, r, t) = -\frac{1}{4\pi} \int_0^1 \frac{S_1(\xi, t)}{\sqrt{(x-\xi)^2 + r^2}} d\xi. \tag{3.3.130}$$

Because the boundary condition is imposed at $r = 0$ instead of $r = \epsilon F$, $\frac{\partial \phi_1}{\partial r} = O(r^{-1})$ as $r \to 0$ (see (3.3.134)). Thus, the outer expansion (3.3.127) cannot hold near $r = 0$. Consequently, $\mu_1(\epsilon)$ need not equal ϵ and will have to be determined by matching with an appropriate inner expansion.

For any $\mu_1 \ll 1$, (3.3.130) results for ϕ_1 and we next investigate the behavior of this solution near $r = 0$ in preparation for matching.

To accomplish this, we note the identities

$$\frac{1}{\sqrt{(x-\xi)^2 + r^2}} = -\frac{\partial}{\partial \xi} \log[x - \xi + \sqrt{(x-\xi)^2 + r^2}], \quad \xi \leq x \tag{3.3.131a}$$

$$\frac{1}{\sqrt{(x-\xi)^2 + r^2}} = +\frac{\partial}{\partial \xi} \log[\xi - x + \sqrt{(x-\xi)^2 + r^2}], \quad \xi \geq x. \tag{3.3.131b}$$

Thus, (3.3.130) can be split into two parts as follows:

$$\begin{aligned}\phi_1(x, r, t) = {} & \frac{1}{4\pi} \int_0^x S_1(\xi, t) \frac{\partial}{\partial \xi} \log[x - \xi + \sqrt{(x-\xi)^2 + r^2}] d\xi \\ & - \frac{1}{4\pi} \int_x^1 S_1(\xi, t) \frac{\partial}{\partial \xi} \log[\xi - x + \sqrt{(x-\xi)^2 + r^2}] d\xi. \end{aligned} \tag{3.3.132}$$

Integrating by parts then gives the following *exact* result:

$$\begin{aligned}\phi_1(x, r, t) = {} & \frac{1}{2\pi} S_1(x, t) \log r - \frac{1}{4\pi} S_1(0, t) \log(x + \sqrt{x^2 + r^2}) \\ & - \frac{1}{4\pi} S_1(1, t) \log[1 - x + \sqrt{(1-x)^2 + r^2}] \\ & - \frac{1}{4\pi} \int_0^x \frac{\partial S_1}{\partial \xi} \log[x - \xi + \sqrt{(x-\xi)^2 + r^2}] d\xi \\ & + \frac{1}{4\pi} \int_x^1 \frac{\partial S_1}{\partial \xi} \log[\xi - x + \sqrt{(x-\xi)^2 + r^2}] d\xi. \end{aligned} \tag{3.3.133}$$

The limiting form of the above as $r \to 0$ is

$$\phi_1(x, r, t) = \frac{1}{2\pi} S_1(x, t) \log r - \frac{1}{4\pi} S_1(0, t) \log 2x$$

$$- \frac{1}{4\pi} S_1(1, t) \log 2(1 - x)$$
$$- \frac{1}{4\pi} \int_0^1 \frac{\partial S_1}{\partial \xi} \operatorname{sgn}(x - \xi) \log 2|x - \xi| d\xi$$
$$+ O(r^2). \tag{3.3.134}$$

Let us assume that near $x = 0$ and $x = 1$ the source strength tends to zero faster than $[\log x]^{-1}$ or $[\log(1 - x)]^{-1}$. This condition will, once S_1 is determined in terms of $F(x, t)$, impose a restriction on the allowable nose and tail shapes of the body.

Inner expansion, matching

Since the body radius is proportional to ϵ, we introduce the inner variable $r^* = r/\epsilon$ and inner limit $\epsilon \to 0$ with x, r^*, and t fixed. We expand $\phi^*(x, r^*, t; \epsilon) \equiv \phi(x, \epsilon r^*, t; \epsilon)$ as follows:

$$\phi^* = \phi_0^*(x, r^*, t) + \mu_1^*(\epsilon)\phi_1^*(x, r^*, t) + \dots . \tag{3.3.135}$$

The leading term ϕ_0^* satisfies

$$\frac{\partial^2 \phi_0^*}{\partial r^{*2}} + \frac{1}{r^*}\frac{\partial \phi_0^*}{\partial r^*} = 0 \tag{3.3.136}$$

subject to the surface boundary condition (see (3.3.121))

$$\frac{\partial \phi_0^*}{\partial r^*}(x, F(x, t), t) = \frac{\epsilon^2}{\delta}\frac{\partial F}{\partial t}(x, t) + \epsilon^2 \frac{\partial \phi_0^*}{\partial x}(x, F(x, t), t)\frac{\partial F}{\partial x}(x, t) + \dots .$$

Since $\delta = O_s(1)$, we find

$$\frac{\partial \phi_0^*}{\partial r^*}(x, F(x, t), t) = 0, \tag{3.3.137}$$

and the solution of (3.1.136) gives $\phi_0^* = B_0(x, t)$. Matching with the outer limit to $O(1)$ shows that $B_0 = x$.

To calculate ϕ_1^*, we substitute (3.3.135) into (3.3.120) and find that ϕ_1^* also obeys Equation (3.3.136).

Therefore,

$$\phi_1^*(x, r^*, t) = A_1(x, t) \log r^* + B_1(x, t). \tag{3.3.138}$$

The boundary condition on $r^* = F(x, t)$ to this order becomes

$$\mu_1^*(\epsilon)\frac{\partial \phi_1^*}{\partial r^*}(x, F(x, t), t) = \epsilon^2[F_x + F_t/\delta]. \tag{3.3.139}$$

Thus, we must choose $\mu_1^*(\epsilon) = \epsilon^2$, and using (3.3.138) in (3.3.139) gives (see (3.3.125b))

$$A_1(x, t) = \frac{1}{2}(G_x + G_t/\delta). \tag{3.3.140}$$

To determine $B_1(x, t)$, we must match with the outer expansion to $O(\epsilon^2)$. For this purpose, we introduce the matching variable

$$r_\eta = \frac{r}{\eta(\epsilon)} \tag{3.3.141}$$

for an appropriate class of functions $\eta(\epsilon)$ contained in $\epsilon \ll \eta \ll 1$. The outer expansion then becomes

$$\phi(x, r, t; \epsilon) = x + \mu_1(\epsilon)\left[\frac{S_1(x, t)}{2\pi}\log \eta r_\eta + T_1(x, t) + O(\eta^2)\right] + o(\mu_1), \tag{3.3.142}$$

where we have introduced the notation

$$T_1(x, t) = -\frac{1}{4\pi}\int_0^1 \frac{\partial S_1}{\partial \xi}\,\mathrm{sgn}\,(x - \xi)\log 2|x - \xi|d\xi. \tag{3.3.143}$$

The inner expansion, written in terms of r_η, gives

$$\phi(x, r, t; \epsilon) = x + \epsilon^2\left[A_1(x, t)\log\frac{\eta r_\eta}{\epsilon} + B_1(x, t)\right] + o(\epsilon^2). \tag{3.3.144}$$

We see that in order to match we must set

$$\mu_1(\epsilon) = \epsilon^2, \tag{3.3.145a}$$

$$A_1(x, t) = \frac{S_1(x, t)}{2\pi}, \tag{3.3.145b}$$

$$B_1(x, t) = T_1(x, t). \tag{3.3.145c}$$

However, the term $-A_1(x, t)\epsilon^2 \log \epsilon$ in the inner expansion cannot be matched.

As in the example of Section 2.5.2, matching this term requires introducing a homogeneous solution of order $\epsilon^2 \log \epsilon$ in the inner expansion. Thus, (3.3.135) must now be modified to read

$$\phi^* = x + A_1(x, t)\epsilon^2 \log \epsilon + \epsilon^2[A_1(x, t)\log r^* + B_1(x, t)] + o(\epsilon^2), \tag{3.3.146}$$

and the matching is demonstrated to $O(\epsilon^2)$ with the definitions in (3.3.145).

We note that the source strength S_1 is related to the body shape according to (3.3.145b) and (3.3.140) by

$$S_1(x, t) = \pi(G_x + G_t/\delta). \tag{3.3.147}$$

Since $G_t = 0$ at $x = 0$ and $x = 1$, we have the classical restrictions on the nose and tail shapes defined by

$$\lim_{x\to 0} G_x \log x = 0, \quad \lim_{x\to 1} G_x \log(1 - x) = 0. \tag{3.3.148}$$

For example, if the nose and tail are given by a power law $r \sim x^n$ or $(1 - x)^n$, we must restrict our attention to values of $n > \frac{1}{2}$.

We can now calculate the pressure coefficient C_p on the body by substituting (3.3.146), evaluated on $r^* = F(x, t)$, into (3.3.123). The result is easily derived in the form

$$C_p = -\frac{1}{2}(\epsilon^2 \log \epsilon) H^2(G) - \epsilon^2 \left\{ \frac{1}{4} H^2(G) \log G + H(T_1) + \frac{1}{8} \frac{[H(G)]^2}{G} \right\} + O(\epsilon^4 \log \epsilon), \tag{3.3.149a}$$

where H is the operator

$$H \equiv \frac{\partial}{\partial t} + \frac{\partial}{\partial x}. \tag{3.3.149b}$$

In the example of this section, the use of an inner expansion is not strictly necessary in the sense that the inner expansion is completely contained in the outer expansion [see (3.3.142) and (3.3.144)]. However, it is useful in making explicit the behavior near the boundary (particularly in the calculation of the pressure coefficient) and in emphasizing the nature of the different expansions as $\epsilon \to 0$ for a point fixed on the boundary as compared to a point fixed in space. For more complicated differential equations, the idea of local behavior near a singular line or point can be essential.

Force on the body

Consider now the axial force coefficient defined by (3.3.124). We encounter the following three integrals:

$$I_1(t) = \int_0^1 H^2(G) G_x dx, \tag{3.3.150a}$$

$$I_2(t) = \int_0^1 \left\{ H^2(G) \log G + \frac{1}{2G} [H(G)]^2 \right\} G_x dx, \tag{3.3.150b}$$

$$I_3(t) = \int_0^1 H(T_1) G_x dx, \tag{3.3.150c}$$

where T_1 is defined by (3.3.143), (3.3.145b), and (3.3.140) in the form

$$T_1(x, t) = -\frac{1}{4} \int_0^1 \frac{\partial}{\partial \xi} H(G) \operatorname{sgn}(x - \xi) \log 2|x - \xi| d\xi. \tag{3.3.151}$$

In evaluating these integrals, it is useful to note the identity for any two functions $A(x, t)$, $B(x, t)$:

$$H(AB) = AH(B) + BH(A). \tag{3.3.152}$$

Consider the following identity for I_1:

$$I_1 = \int_0^1 H^2(G) G_x dx = \int_0^1 H[H(G) G_x] dx - \frac{1}{2} \int_0^1 \frac{\partial}{\partial x} [H(G)]^2 dx. \tag{3.3.153a}$$

Since $H(G) = 2FH(F)$ and $F(0, t) = F(1, t) = 0$, the second integral on the right-hand side of (3.3.153a) vanishes, and the first integral reduces to

$$I_1 = \int_0^1 \frac{\partial}{\partial t}[H(G)G_x]dx \qquad (3.3.153b)$$

because $H(G)(\partial G/\partial x) = 0$ at $x = 0$ and $x = 1$. Similar calculations give the following results for I_2 and I_3:

$$I_2 = \int_0^1 \frac{\partial}{\partial t}[H(G)G_x \log G]dx, \qquad (3.3.154)$$

$$I_3 = \int_0^1 \frac{\partial}{\partial t}[T_1 G_x]dx. \qquad (3.3.155)$$

Thus, for the case where $G_t = 0$, i.e., a rigid body, we recover the classical result $C_D = 0$ (D'Alembert paradox).

Now consider a periodic deformation $G(x, t)$ with period λ, i.e.,

$$G(x, t + \lambda) = G(x, t),$$

and let G and its partial derivatives be continuous. It then follows that C_D is of the form

$$C_D = \frac{d}{dt}h(t, \epsilon), \qquad (3.3.156)$$

where $h(t, \epsilon)$ is the total contribution of all integrals in C_p.

Thus, C_D is also periodic with period λ and zero average, i.e., C_D can be represented in a Fourier series without a constant term in the form

$$C_D = \sum_{n=1}^{\infty} a_n(\epsilon) \sin\left[\frac{2n\pi t}{\lambda} + \beta_n(\epsilon)\right], \qquad (3.3.157)$$

where the a_n and β_n can be determined by quadrature once the deformation G is known. It is particularly important to note that the average value of C_D over one period is zero. In other words, *the net impulse over one period is zero.*

This situation is unaltered if we allow G_t to have any number of finite discontinuities over a period with G and G_x continuous. To prove this, we note that since the Bernoulli equation is still valid in this case (as long as time derivatives are interpreted in the sense of distributions), we have to consider in the calculation for C_D integrals of the form

$$I = \int_0^1 \frac{\partial}{\partial t}\left[K(x, t)\frac{\partial G}{\partial x}(x, t)\right]dx,$$

where both K and G_t have discontinuities. It is sufficient to consider the case of one discontinuity at $t = t_0$ in the interval $0 \le t \le \lambda$.

The impulse during one period is the integral of I, which has the form

$$J = \int_0^{\lambda}\int_0^1 \frac{\partial}{\partial t}\left[K(x, t)\frac{\partial G}{\partial x}(x, t)\right]dx\, dt. \qquad (3.3.158)$$

If we now split the integral as indicated in (3.3.126b), we have two contributions denoted by J_1 and J_2:

$$J = J_1 + J_2, \tag{3.3.159}$$

where

$$J_1 = \int_0^\lambda \int_0^1 \left\{ \frac{\partial}{\partial t} \left[K(x,t) \frac{\partial G}{\partial x} \right] \right\} dx\, dt, \tag{3.3.160a}$$

$$J_2 = \int_0^\lambda \int_0^1 [K(x,t)]_{t_0} \delta(t - t_0) \frac{\partial G}{\partial x} dx\, dt, \tag{3.3.160b}$$

and { } denotes the continuous part of the quantity.

We calculate J_1 in two parts as follows:

$$J_1 = \int_0^{t_0} \int_0^1 \left\{ \frac{\partial}{\partial t} \left[K(x,t) \frac{\partial G}{\partial x} \right] \right\} dx\, dt + \int_{t_0}^\lambda \int_0^1 \left\{ \frac{\partial}{\partial t} \left[K(x,t) \frac{\partial G}{\partial x} \right] \right\} dx\, dt. \tag{3.3.161}$$

Since the integrands are continuous in both open intervals, we can immediately integrate with respect to t and obtain

$$J_1 = \int_0^1 K(x, t_0^-) \frac{\partial G}{\partial x}(x, t_0) dx - \int_0^1 K(x, t_0^+) \frac{\partial G}{\partial x}(x, t_0) dx \tag{3.3.162a}$$

or

$$J_1 = -\int_0^1 [K(x,t)]_{t_0} \frac{\partial G}{\partial x}(x, t_0) dx. \tag{3.3.162b}$$

Integrating J_2 gives

$$J_2 = \int_0^1 [K(x,t)]_{t_0} \frac{\partial G}{\partial x}(x, t_0) dx. \tag{3.3.163}$$

Therefore, $J = 0$.

Actually, this conclusion regarding the impulse is valid even if the motion is nonperiodic. It is sufficient that the body return to some given configuration after an interval of time λ for the impulse during this interval to vanish.

To show an example of a deformation with nonvanishing impulse, let us consider only the leading term in the expansion for C_p. We then have

$$C_D = -\epsilon^2 \log \epsilon \frac{d}{dt} \int_0^1 H(G) G_x dx + O(\epsilon^2). \tag{3.3.164}$$

Assume now that G has the form

$$G(x,t) = f(x) g(t) \tag{3.3.165}$$

with $f(0) = f(1) = 0$. Thus, the body undergoes geometrically similar deformations. Equation (3.3.164) can then be easily integrated to give

$$C_D = -\epsilon^2 (\log \epsilon) \alpha^2 \frac{d(g^2)}{dt} + O(\epsilon^2), \tag{3.3.166a}$$

where α^2 is the positive constant

$$\alpha^2 = \int_0^1 \left[\frac{df}{dx}\right]^2 dx. \tag{3.3.166b}$$

We note that in order to have a propulsive force, ($C_D < 0$), we must let $dg^2/dt < 0$. Hence, the body must " collapse." For example, if we wish to maintain a constant propulsive force $C_D = C < 0$, we calculate (note: $C/\epsilon^2 \log\epsilon > 0$)

$$g(t) = \sqrt{1 - \frac{Ct}{\alpha^2\epsilon^2 \log\epsilon}}. \tag{3.3.167}$$

Thus, the body will have collapsed to a "needle" after an interval of time equal to $\alpha^2\epsilon^2 \log\epsilon/C$. It is also interesting to note (see Problem 3) that for a body starting from rest and moving under the influence of forces generated by periodic surface deformations, the best average velocity that can be achieved is a constant.

Nonuniformity near the nose

Let us now consider the local nonuniformity of the slender-body expansion near a blunt nose. We will consider the case of a rigid body for simplicity and outline a method based on a local solution for eliminating the difficulty. As we saw earlier [see the discussion following (3.3.148)], some difficulty occurs for a nose (or tail) that is so blunt that $F(x) \sim \sqrt{x}$, but if $F(x) \sim x^n$, $n < \frac{1}{2}$, the pressure force, at least in the first approximation, is integrable. Thus, we consider here a slender body whose shape function $F(x)$ has the following behavior near $x = 0$:

$$F(x) = (2ax)^{1/2}(1 + bx + \ldots), \tag{3.3.168a}$$

$$F'(x) = (a/2x)^{1/2} + O(x^{1/2}), \tag{3.3.168b}$$

$$F''(x) = -(2a)^{1/2}/4x^{3/2} + O(x^{-1/2}). \tag{3.3.168c}$$

The radius of curvature r_c at the nose is

$$r_c \to \frac{\epsilon^3(F'^3)}{\epsilon F''} \to \epsilon^2 a \quad \text{as } x \to 0. \tag{3.3.169}$$

The slender-body inner expansion for the potential is given by (3.3.146), but near the nose we find that the source strength is

$$A_1(x) = FF' = a + O(x). \tag{3.3.170}$$

Furthermore, in order to determine the function $B_1(x)$, we have to return to (3.3.134) and retain the boundary term $-(1/4\pi)S_1(0)\log 2x$, which was omitted for a sufficiently pointed nose. Now, the matching gives

$$B_1(x) = -\frac{1}{2}A_1(0)\log x + \ldots$$
$$-\frac{1}{2}\int_0^1 A_1'(\xi)\operatorname{sgn}(x-\xi)\log 2|x-\xi|\,d\xi \tag{3.3.171}$$

or, as $x \to 0$,

$$B_1(x) \to -(a/2)\log 2 - (a/2)\log x + \ldots. \tag{3.3.172}$$

Thus, the potential has the behavior

$$\phi = x + a\epsilon^2 \log\epsilon + \epsilon^2[a\log r^* - a\log 2 - (a/2)\log x + \ldots] \text{ as } x \to 0. \tag{3.3.173}$$

On the body surface, $r^* = (2ax)^{1/2}$ and ϕ is finite. However, the velocity is given by

$$u = \phi_x = 1 - a\epsilon^2/2x + \ldots \tag{3.3.174a}$$

$$v = \frac{1}{\epsilon}\phi_{r^*} = a\epsilon/r^* \tag{3.3.174b}$$

so that the local surface pressure coefficient, C_{p_b}, given by the Bernoulli equation (3.3.123) with $\phi_t = 0$, is

$$\begin{aligned} 2C_{p_b} &= 1 - (1 - a\epsilon^2/2x + \ldots)^2 - \left(\frac{a\epsilon}{\sqrt{2ax}} + \ldots\right)^2 \\ &= \epsilon^2 a/2x + O(\epsilon^4). \end{aligned} \tag{3.3.175}$$

The term of $O(\epsilon^2)$ shows large physically unrealistic compression and, in fact, the total force on the nose, which is proportional to $\int C_{p_b}FF'dx$, is infinite. In order to give a better representation of the flow near the nose, we can try to find a local expansion based on a limit process that preserves the structure of the flow near the nose. Since we are interested in the neighborhood of a point, both x and r must tend to zero in the limit, and it is clear that all terms in the basic equation (3.3.120) should be retained. Thus, the general form has

$$\overline{x} = \frac{x}{\alpha(\epsilon)}, \quad \overline{r} = \frac{r}{\alpha(\epsilon)}$$

in the limit. But considering that $r_b \sim \epsilon x^{1/2}$ as $x \to 0$, we see that $\alpha = \epsilon^2$ in order to keep the typical body structure near the surface. Thus, let

$$\overline{x} = \frac{x}{\epsilon^2}, \quad \overline{r} = \frac{r}{\epsilon^2}. \tag{3.3.176}$$

The asymptotic expansion of the potential near the nose is assumed in the form

$$\phi = x + \overline{\mu}_1(\epsilon)\overline{\phi}_1(\overline{x}, \overline{r}) + \ldots, \tag{3.3.177}$$

where $\overline{\phi}_1$ obeys (3.3.120).

As far as the first approximation goes, the body is represented by

$$\overline{r}_b = (2a\overline{x})^{1/2}. \tag{3.3.178}$$

Then the problem is one of flow past a paraboloid. The surface boundary condition (3.3.121) (with $\partial/\partial t = 0$) becomes

$$\frac{\overline{\mu}_1}{\epsilon^2}\frac{\partial \overline{\phi}_1}{\partial \overline{r}}(\overline{x}, \sqrt{2a\overline{x}}) + \ldots = \sqrt{\frac{a}{2\overline{x}}}\left[1 + \frac{\overline{\mu}_1}{\epsilon^2}\frac{\partial \overline{\phi}_1}{\partial \overline{x}}(\overline{x}, \sqrt{2a\overline{x}}) + \ldots\right] \tag{3.3.179}$$

from which we see that the proper choice for $\overline{\mu}_1$ is

$$\overline{\mu}_1 = \epsilon^2. \tag{3.3.180}$$

Note that the free-stream term x in (3.3.177) is just the same order ($x = \epsilon^2 \overline{x}$) as the $\overline{\mu}_1\overline{\phi}_1$ term. The potential for the paraboloid with the boundary condition (3.3.179) can be written

$$\overline{\phi}_1 = \frac{1}{2} a \log\left\{\left[\left(\overline{x} - \frac{a}{2}\right)^2 + \overline{r}^2\right]^{1/2} - \left(\overline{x} - \frac{a}{2}\right)\right\}. \tag{3.3.181}$$

There is no arbitrary constant here due to the form of (3.3.179). The x term in (3.3.177) is already matched, and the ϕ_1 term can be matched to the previously calculated inner expansion to remove the singularity at the nose and enable a uniformly valid approximation to be constructed. For the matching, we introduce

$$x_\eta = \frac{x}{\eta(\epsilon)}, \tag{3.3.182}$$

where $\eta(\epsilon)$ now belongs to a class contained in $\epsilon^2 \ll \eta \ll \epsilon$. In the limit $\epsilon \to 0$ with r^* and x_η fixed, we have

$$\overline{x} = \frac{\eta}{\epsilon^2} x_\eta \to \infty, \qquad x = \eta x_\eta \to 0, \qquad \overline{x} = \frac{r^*}{\epsilon} \to \infty. \tag{3.3.183}$$

Thus, we obtain

$$\left[\left(\overline{x} - \frac{a}{2}\right)^2 + \overline{r}^2\right]^{1/2} = \frac{\eta}{\epsilon^2} x_\eta \left(1 + \frac{1}{2}\frac{\epsilon^2}{\eta^2}\frac{r^{*2}}{x_\eta 2} + \ldots\right) \tag{3.3.184}$$

if $\epsilon^2 \ll \eta$, and therefore

$$\overline{\phi}_1 = \frac{a}{2}\log\frac{r^{*2}}{2\eta x_\eta} + \ldots. \tag{3.3.185}$$

By adding suitable constants that do not affect the velocity, it is seen that the potential in (3.3.181) matches with the $\log r^*$ and $\log x$ terms in (3.3.173). Thus, near the nose the pressure should be computed from the velocity components as found from (3.3.181):

$$\frac{\partial \phi}{\partial r} = \frac{a}{2}\frac{1}{\sqrt{(\overline{x} - a/2)^2 + \overline{r}^2} - (\overline{x} - a/2)} \cdot \frac{\overline{r}}{\sqrt{(\overline{x} - a/2)^2 + \overline{r}^2}}$$

$$\to \frac{(2a\overline{x})^{1/2}}{2\overline{x} + a} \text{ on the surface} \tag{3.3.186}$$

$$\frac{\partial\phi}{\partial x} = 1 + \frac{a}{2}\frac{1}{\sqrt{(\overline{x}-a/2)^2+\overline{r}^2}-(\overline{x}-a/2)}\left\{-1+\frac{\overline{x}-a/2}{\sqrt{(\overline{x}-a/2)^2+\overline{r}^2}}\right\}$$

$$\to \frac{\overline{x}}{\overline{x}+a/2} \text{ on the surface.} \tag{3.3.187}$$

This is a typical example of how a local solution, in this case flow past a paraboloid, can be used to improve the representation of the solution near a singularity. A composite expansion can be written by adding the local and outer expansions and subtracting the common part.

3.3.4 Low Reynolds Number Viscous Flow Past a Circular Cylinder

The basic references for this section are the papers [3.12], [3.14], and [3.17] by Kaplun and Lagerstrom. These (and [3.13]) are reprinted in [3.19], where the reader can find a comprehensive discussion by the editors.

Problem formulation

For this problem, the Navier-Stokes equations (3.2.3) are again considered to describe the flow. There is uniform flow at infinity, and the body is at the origin. Since the size of the body was used as the characteristic length in writing the system (3.2.3), the body diameter is one (see Figure 3.3.5). The boundary condition of no slip,

$$q = 0 \quad \text{on} \quad r = \sqrt{x^2+y^2} = \frac{1}{2}, \tag{3.3.188}$$

and conditions at infinity serve to define the problem. We are interested in a low Reynolds number, so that in Equation (3.2.3b) we have

$$\epsilon = \frac{1}{\text{Re}} = \frac{\nu}{UL} \to \infty. \tag{3.3.189}$$

The variables based on L are inner variables (Stokes variables in the notation of [3.12], [3.14], and [3.17]), since the boundary remains fixed in the limit. As it turns out, the inner problem, which is Stokes flow, cannot satisfy the complete boundary conditions at infinity, so that some suitable outer expansion, valid near infinity, must also be constructed. Both inner and outer expansions, which can be identified with the usual Stokes and Oseen flow approximations, respectively, are described here, and the matching is carried out. The model example corresponding to the kind of singular boundary-value problem that occurs here has already been discussed in Section 2.5.2.

The inner expansion is based on Re $\to$ 0, $\epsilon \to \infty$ in Equation (3.2.3), but if pressure is measured in units of ρU^2, both inertia and pressure terms drop out of the limiting momentum equations. There are not enough variables if continuity is to be considered, so that the physical pressure (difference from infinity) should be measured in terms of $U\mu/L$. Or let

$$p^*(x, y) = \frac{1}{\epsilon} p(x, y) = (\mathrm{Re}) p(x, y). \tag{3.3.190}$$

This is in accord with Stokes' idea of a balance between viscous stresses and pressure forces, at least near the body, for slow flow. Thus, in inner variables, the Navier-Stokes system can be written

$$\operatorname{div} \mathbf{q} = 0, \tag{3.3.191a}$$

$$\mathrm{Re}(\mathbf{q} \cdot \operatorname{grad}\mathbf{q}) = -\operatorname{grad} p^* - \operatorname{curl} \omega, \qquad \omega = \operatorname{curl} \mathbf{q}. \tag{3.3.191b}$$

Inner (Stokes) expansion

The inner expansion, associated with the limit process (Re $\to$ 0, x,y fixed), has the form

$$\mathbf{q}(x, y; \mathrm{Re}) = \alpha_0(\mathrm{Re})\mathbf{q}_0(x, y) + \alpha_1(\mathrm{Re})\mathbf{q}_1(x, y) + \dots, \tag{3.3.192a}$$
$$p^*(x, y; \mathrm{Re}) = \alpha_0(\mathrm{Re}) p_0^*(x, y) + \alpha_1(\mathrm{Re}) p_1^*(x, y) + \dots. \tag{3.3.192b}$$

Taking the limit of the Navier-Stokes equations (3.3.191) shows that the first term of the inner expansion satisfies the usual Stokes equations:

$$\operatorname{div} \mathbf{q}_0 = 0, \tag{3.3.193a}$$

$$0 = -\operatorname{grad} p_0^* - \operatorname{curl} \omega_0, \qquad \omega_0 = \operatorname{curl} \mathbf{q}_0. \tag{3.3.193b}$$

A fairly general discussion of the solutions to (3.3.193) can be given while taking account of the boundary conditions on the surface, so that the behavior at infinity can be ascertained. It is convenient to introduce the stream function $\psi(x, y)$, satisfying continuity identically, by

$$\mathbf{q} = \operatorname{curl} \boldsymbol{\psi}, \quad \boldsymbol{\psi} = \psi(x, y)\mathbf{k}, \quad q_x = \frac{\partial \psi}{\partial y}, \quad q_y = -\frac{\partial \psi}{\partial x}. \tag{3.3.194}$$

The equation for the vorticity $\boldsymbol{\omega}_0 = \omega_0(x, y)\mathbf{k}$ is

$$\omega_0(x, y) = -\Delta\psi_0(x, y), \tag{3.3.195}$$

where Δ is the Laplacian. Further, taking the curl of the momentum equation (3.3.193b) shows that

$$\operatorname{curl} \operatorname{curl} \boldsymbol{\omega}_0 = 0 \quad \text{or} \quad \Delta\omega = 0. \tag{3.3.196}$$

Thus, the vorticity field is a harmonic function and can be represented, in general, outside the circular cylinder $r = \frac{1}{2}$ by a series with unknown coefficients. The velocity field must be symmetric with respect to the x-axis and the vorticity field antisymmetric. Thus, the general form is the familiar solution of Laplace's equation in cylindrical coordinates

$$\omega_0(r, \theta) = \sum_{n=0}^{\infty} (a_n r^n + b_n r^{-n}) \sin n\theta. \tag{3.3.197}$$

Equation (3.3.195) for the stream function ψ_0 becomes

$$\frac{\partial^2\psi_0}{\partial r^2} + \frac{1}{r}\frac{\partial\psi_0}{\partial r} + \frac{1}{r^2}\frac{\partial\psi_0}{\partial\theta^2} = -\sum_{n=0}^{\infty}(a_n r^n + b_n r^{-n})\sin n\theta, \qquad (3.3.198)$$

and the boundary condition of no slip

$$\psi_0\left(\frac{1}{2},\theta\right) = \frac{\partial\psi_0}{\partial r}\left(\frac{1}{2},\theta\right) = 0 \qquad (3.3.199)$$

can now be found. Thus, let

$$\psi_0(r,\theta) = \sum_{n=1}^{\infty}\Psi^{(n)}(r)\sin n\theta, \qquad (3.3.200)$$

so that we have

$$L^{(n)}\Psi^{(n)} \equiv \frac{d^2\Psi^{(n)}}{dr^2} + \frac{1}{r}\frac{d\Psi^{(n)}}{dr} - \frac{n^2}{r^2}\Psi^{(n)} = -a_n r^n - b_n r^{-n}. \qquad (3.3.201)$$

It can be verified that

$$L^{(n)}r^m = (m^2 - n^2)r^{m-2}, \qquad (3.3.202)$$

so that by choosing $m = n + 2$, we have

$$L^{(n)}r^{n+2} = (4n+4)r^n, \qquad (3.3.203)$$

which is good for all $n \neq -1$. Further, for $n = -1$, we obtain

$$L^{(n)}r\log r = 2/r. \qquad (3.3.204)$$

Thus, introducing new constants, the general solution of (3.3.201) is

$$\Psi^{(1)} = A_1 r^3 + B_1 r\log r + C_1 r + \frac{D_1}{r}, \qquad (3.3.205a)$$

$$\Psi^{(n)} = A_n r^{2+n} + B_n r^{2-n} + C_n r^n + \frac{D_n}{r^n}, \quad n = 2, 3, \ldots, \qquad (3.3.205b)$$

and

$$\frac{d\Psi^{(1)}}{dr} = 3A_1 r^2 + B_1(\log r + 1) + C_1 - \frac{D_1}{r^2}, \qquad (3.3.206a)$$

$$\frac{d\Psi^{(n)}}{dr} = (2+n)A_n r^{1+n} + (2-n)B_n r^{1-n} + nC_n r^{n-1} - n\frac{D_n}{r^{n+1}}, \quad n = 2, 3, \ldots. \qquad (3.3.206b)$$

By applying the boundary condition at the body surface $r = \frac{1}{2}$, (3.3.199), we obtain two relations between the four constants A_n, B_n, C_n, and D_n. Further determination of the solution must come from the boundary conditions at infinity, which would read

$$q_x = \sin\theta\frac{\partial\psi}{\partial r}(r,\theta) + \frac{\cos\theta}{r}\frac{\partial\psi}{\partial\theta}(r,\theta) = 1 \quad \text{as} \quad r \to \infty, \qquad (3.3.207)$$

$$q_y = -\cos\theta \frac{\partial\psi}{\partial r}(r,\theta) + \frac{\sin\theta}{r}\frac{\partial\psi}{\partial\theta}(r,\theta) = 0 \quad \text{as} \quad r \to \infty, \tag{3.3.208}$$

or

$$\frac{\partial\psi}{\partial r}(r,\theta) \to \sin\theta \quad \text{as} \quad r \to \infty; \qquad \frac{1}{r}\frac{\partial\psi}{\partial\theta}(r,\theta) \to \cos\theta \quad \text{as} \quad r \to \infty. \tag{3.3.209}$$

If the condition (3.3.209) is imposed to fix $C_1 = 1$, $A_1 = B_1 = A_n = C_n = 0$, then the two boundary conditions at the wall cannot be satisfied. Thus, this condition has to be given up and replaced by a condition of matching at infinity. The inner expansion is not uniform at infinity. The situation is possibly a little clearer for the corresponding problem for a sphere where, although the first term of the inner expansion can satisfy the conditions at infinity, the second cannot and becomes larger than the first at some distance from the origin. In general, only one more constant B_1 is needed, so that we can choose

$$B_1 \neq 0, \quad A_1 = 0, \quad A_n = 0, \quad C_n = 0, \quad n = 2, 3, \ldots \tag{3.3.210}$$

and obtain the weakest possible divergence of the solution at infinity. This has to be verified by matching. Thus, from the boundary condition at the surface, (3.3.199), applied to (3.3.205) and (3.3.206), we obtain

$$\Psi^{(1)}\left(\frac{1}{2}\right) = 0 = \frac{B_1}{2}\log\frac{1}{2} + \frac{C_1}{2} + 2D_1, \tag{3.3.211a}$$

$$\Psi^{(n)}\left(\frac{1}{2}\right) = 0 = B_n 2^{n-2} + 2^n D_n, \quad n = 2, 3, \ldots, \tag{3.3.211b}$$

$$\frac{d\Psi^{(1)}}{dr}\left(\frac{1}{2}\right) = 0 = B_1\left(\log\frac{1}{2} + 1\right) + C_1 - 4D_1, \tag{3.3.211c}$$

$$\frac{d\Psi^{(n)}}{dr}\left(\frac{1}{2}\right) = 0 = (2-n)B_n 2^{n-1} - nD_n 2^{n+1}, \quad n = 2, 3, \ldots. \tag{3.3.211d}$$

Equations (3.3.211b) and (3.3.211d) imply

$$B_n = D_n = 0, \quad n = 2, 3, \ldots, \tag{3.3.212}$$

whereas (3.3.211a) and (3.3.211c) give two relations between the three constants B_1, C_1, and D_1. Thus, the first term of the inner expansion becomes

$$\psi_0(x, r) = [B_1 r \log r + C_1 r + (D_1/r)]\sin\theta \tag{3.3.213}$$

and

$$q_x = \alpha_0(\text{Re})\{B_1 \log r + C_1 + B_1 \sin^2\theta + (D_1/r^2)\cos 2\theta\} + \alpha_1(\text{Re})q_{1_x} + \cdots. \tag{3.3.214}$$

Outer (Oseen) expansion

In order to construct the outer expansions, a suitable outer variable has to be chosen. It was a basic idea of Kaplun to use the characteristic length ν/U for defining the expansion in the far field. Thus, in these units the body radius is very small and approaches zero in the limit. It can then be anticipated that the first term of the outer expansion is the undisturbed stream, since the body of infinitesimal size has no arresting power. Compare this procedure with that in the model example in Sec. 2.5.2. The formalities involve a limit process with $\tilde{x}, \tilde{y}$ fixed, Re $\to$ 0, where

$$\tilde{x} = (\mathrm{Re})x, \quad \tilde{y} = (\mathrm{Re})y \tag{3.3.215}$$

since x, y are based on the diameter. In these units, the body surface itself is

$$\tilde{r} = \sqrt{\tilde{x}^2 + \tilde{y}^2} = \frac{1}{2}\mathrm{Re}, \quad \mathrm{Re} \to 0. \tag{3.3.216}$$

If the Navier-Stokes equations (3.3.191) are written in these units, the parameter Re disappears. The pressure is again based on ρU^2:

$$\widetilde{\mathrm{div}}\, \mathbf{q} = 0, \tag{3.3.217a}$$

$$\mathbf{q} \cdot \widetilde{\mathrm{grad}}\mathbf{q} + \widetilde{\mathrm{grad}} p = -\widetilde{\mathrm{curl}}\, \omega, \quad \omega = \widetilde{\mathrm{curl}}\, \mathbf{q}, \tag{3.3.217b}$$

where ($\tilde{\ }$) means space derivatives with respect to ($\tilde{x}, \tilde{y}$). The form of the outer expansion is thus assumed to be a perturbation about the uniform free stream, and we have

$$q_x(x, y; \mathrm{Re}) = 1 + \beta(\mathrm{Re})u(\tilde{x}, \tilde{y}) + \beta_1(\mathrm{Re})u_1(\tilde{x}, \tilde{y}) + \dots, \tag{3.3.218a}$$

$$q_y(x, y; \mathrm{Re}) = \beta(\mathrm{Re})v(\tilde{x}, \tilde{y}) + \dots, \tag{3.3.218b}$$

$$p(x, y; \mathrm{Re}) = \beta(\mathrm{Re})\tilde{p}(\tilde{x}, \tilde{y}) + \dots. \tag{3.3.218c}$$

From this expansion, it is clear that the first approximation equation is linearized about the free stream. The transport operator is

$$\mathbf{q} \cdot \widetilde{\mathrm{grad}} = \frac{\partial}{\partial \tilde{x}} + \beta(\mathrm{Re}) \left(u \frac{\partial}{\partial \tilde{x}} + v \frac{\partial}{\partial \tilde{y}} \right) + \dots. \tag{3.3.219}$$

Thus, we have

$$\frac{\partial u}{\partial \tilde{x}} + \frac{\partial v}{\partial \tilde{y}} = 0, \tag{3.3.220a}$$

$$\frac{\partial u}{\partial \tilde{x}} + \frac{\partial \tilde{p}}{\partial \tilde{x}} = \frac{\partial^2 u}{\partial \tilde{x}^2} + \frac{\partial^2 u}{\partial \tilde{y}^2}, \tag{3.3.220b}$$

$$\frac{\partial v}{\partial \tilde{x}} + \frac{\partial \tilde{p}}{\partial \tilde{y}} = \frac{\partial^2 v}{\partial \tilde{x}^2} + \frac{\partial^2 v}{\partial \tilde{y}^2}. \tag{3.3.220c}$$

These are the equations proposed by Oseen as a model for high Reynolds number flow, but they appear here as part of an actual approximation scheme for low Re.

Matching

The matching of the two expansions can now be discussed. We introduce the matching variables

$$x_\eta = \eta(\text{Re})x, \quad y_\eta = \eta(\text{Re})y, \tag{3.3.221}$$

for a class of function $\eta(\text{Re})$ belonging to $\text{Re} \ll \eta(\text{Re}) \ll 1$ and consider the limit as $\text{Re} \to 0$ with x_η and y_η fixed. Therefore, in this limit, we know that $r = (r_\eta/\eta) \to \infty$, and $\widetilde{r} = (\text{Re}/\eta)r_\eta \to 0$. We express the inner and outer expansions (3.3.214) and (3.3.218a) for q_x in terms of r_η and require

$$\lim_{\substack{\epsilon\to 0\\ r_\eta \text{ fixed}}} \left\{\alpha_0(\text{Re})\left[B_1 \log \frac{r_\eta}{\eta(\text{Re})} + C_1 + B_1 \sin^2\theta + D_1 \frac{\eta^2(\text{Re})}{r_\eta^2} \cos 2\theta\right]\right.$$

$$\left. +\alpha_1(\text{Re})q_x + \ldots - 1 - \beta(\text{Re})u\left(\frac{\text{Re}}{\eta}x_\eta, \frac{\text{Re}}{\eta}y_\eta\right) - \ldots\right\} = 0. \tag{3.3.222}$$

It is clear from (3.3.222) that a solution of the Oseen equation (3.3.220) must be found, in which

$$u(\widetilde{x}, \widetilde{y}) \to a \log \widetilde{r} + \ldots \quad \text{as } \widetilde{r} \to 0, \tag{3.3.223a}$$

$$u(\widetilde{x}, \widetilde{y}) \to a \log \frac{\text{Re}r_\eta}{\eta} + \ldots \tag{3.3.223b}$$

in the matching domain. Once this solution is used, the dominant remaining terms are

$$-\alpha_0(\text{Re})B_1 \log \eta(\text{Re}) + \ldots - 1 + \beta(\text{Re})a \log \eta(\text{Re}) + \beta(\text{Re})a \log\left(\frac{1}{\text{Re}}\right) + \ldots.$$

Matching is accomplished by choosing

$$\alpha_0(\text{Re}) = \beta(\text{Re}), \quad \beta_1 = a, \tag{3.3.224}$$

and

$$\beta(\text{Re}) = \frac{1}{\log(1/\text{Re})}, \quad a = 1. \tag{3.3.225}$$

Thus, (3.3.223) provides the necessary boundary condition for the solution of the first outer approximation. The stream function, pressure, and other velocity component can also be considered and matched. In this way, the complete first approximation to the flow near the body is found (see (3.3.211a) and (3.3.211c)):

$$B_1 = 1, \quad C_1 = -\frac{1}{2}\log\frac{1}{2} - \frac{1}{2}, \quad D_1 = \frac{1}{8}. \tag{3.3.226}$$

The continuation of this procedure enables the various higher approximations to be carried out.

Note that in (3.3.191b), the Navier-Stokes equations expressed in inner variables, the nonlinear terms are $O(\text{Re})$ and hence transcendentally small compared to

$\beta(\text{Re}) = 1/[\log(1/\text{Re})]$ as Re $\to$ 0. In particular, when successive terms of the inner expansions are constructed, each satisfies the same Stokes equations (3.3.193). The nonlinear effects appear only explicitly in the outer equation and outer expansion. Thus, the nonlinearity indicates the existence of terms,

$$\beta_1(\text{Re}) = \beta^2(\text{Re}) = \log^2\left(\frac{1}{\text{Re}}\right), \tag{3.3.227}$$

so that $(u_1, v_1, \widetilde{p}_1)$ satisfy nonhomogeneous Oseen equations. Of course, terms of intermediate order satisfying the homogeneous Oseen equations may appear between (u, u_1) to complete the matching. For the incompressible case, it turns out that the outer expansion includes the inner expansion and that a uniformly valid solution is found from the outer expansion with a boundary condition satisfied on $\widetilde{r} = \text{Re}/2$. Such a result cannot be expected in the more general compressible case.

A much more sophisticated version of this problem and the general problem of low Re flow appears in [3.12], [3.14], and [3.17].

Independently of the above work, similar results were also obtained in [3.35].

3.3.5 *Potential Induced by a Point Source of Current in the Interior of a Biological Cell*

Certain boundary-value problems become singular, in the perturbation sense, because the solution fails to exist for a limiting value of a parameter. For example, if a heat source is turned on and maintained inside a finite conducting body that is imperfectly insulated at its surface, a steady temperature will be reached. If the insulation is made more and more perfect, the body will heat up more quickly and in the limiting case of perfect insulation, a steady state is never reached.

An analogous problem that occurs in electrophysiology is discussed in this section. In certain experiments, in order to measure passive electrical properties, a microelectrode is used to introduce a point source of current into a cell, and the potential is measured at another point. The analysis of this experiment depends on the theoretical treatment sketched in the following.

Problem formulation

The model for the cell is a finite body of characteristic dimension a enclosed by a membrane of thickness δ, surrounded by a perfectly conducting external medium (constant potential). The more general case of finite external conductivity can be worked out by similar methods and appears in [3.32]. The geometry and coordinate system are shown in Figure 3.3.5.

The conductivities of the cell interior and membrane are σ_i and σ_m (mhos/cm), respectively. The membrane thickness and conductivity are considered to approach zero individually in such a way that the ratio σ_m/δ, the surface conductivity, remains finite. For a typical cell used in physiological experiments, $\delta = 10^{-6}$cm and $a = 10^{-3}$ to 5×10^{-2}cm. The membrane is also assumed to have a surface

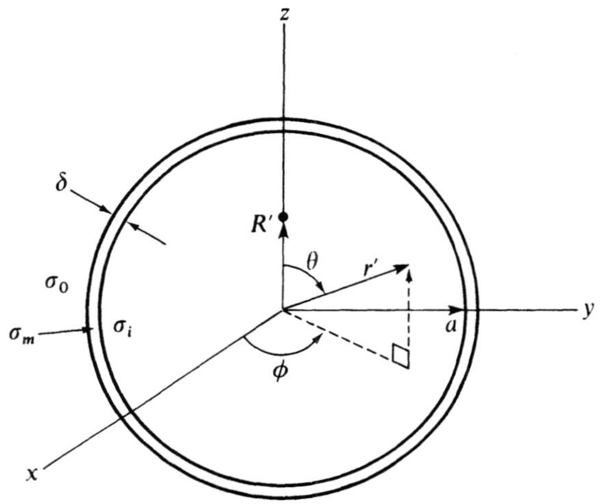

FIGURE 3.3.5. Coordinate System for Spherical Cell

capacitance $c_m \sim 1\mu$ farad/cm^2. Typical values are $\sigma_m = 3 \times 10^{-10}$ mhos/cm, $\sigma_i = 7 \times 10^{-3}$ mhos/cm, and the basic dimensionless small parameter of the problem is

$$\epsilon = \frac{\sigma_m a}{\delta \sigma_i} < 10^{-3}.$$

Let ( )′ temporarily denote quantities with physical dimensions, and assume that a point source of current (4π amps) at $\mathbf{r}' = \mathbf{R}'$ is turned on in a quiescent system. The current density $\mathbf{J}'$ amps/cm^2 is then given by

$$\operatorname{div}' \cdot \mathbf{J}' = 4\pi\delta(\mathbf{r}' - \mathbf{R}')H(t'), \tag{3.3.228a}$$

where δ is the Dirac delta function respresenting a point source in three dimensions, and H is the Heaviside step function:

$$H(t') = \begin{cases} 0 & t' < 0 \\ 1 & t' > 0. \end{cases}$$

Ohm's law is

$$\mathbf{J}' = -\sigma_i \operatorname{grad}' V', \qquad V' = \text{ potential (volts)}.$$

Thus, V' obeys the Laplace equation

$$\Delta' V' = -\frac{4\pi}{\sigma_i}\delta(\mathbf{r}' - \mathbf{R}')H(t'). \tag{3.3.228b}$$

The boundary condition that results when the membrane is considered as a discontinuity (n' = outward normal) is given by

$$-\sigma_i \frac{\partial V'}{\partial n'} = \frac{\sigma_m}{\delta} V' + C_m \frac{\partial V'}{\partial t'}. \tag{3.3.229}$$

This boundary condition balances the current flowing into the membrane under Ohm's law with the current flowing across the membrane and the accumulation of charge on the membrane. A detailed derivation appears in [3.32]. If suitable dimensionless variables (V, r, t) are introduced by

$$\begin{aligned} V &= a\sigma_i V', \\ r &= r'/a, \\ t &= \frac{\sigma_m}{C_m \delta} t', \end{aligned} \tag{3.3.230}$$

then the equation inside the cell is again the Laplace equation

$$\Delta V = -4\pi\delta(\mathbf{r} - \mathbf{R})H(t), \tag{3.3.231}$$

and on the cell boundary

$$-\frac{\partial V}{\partial n} = \epsilon\left(V + \frac{\partial V}{\partial t}\right). \tag{3.3.232}$$

This problem has the character discussed earlier. For $\epsilon = 0$, no steady state exists. We will now discuss the asymptotic behavior of $V(\mathbf{r}, t; \epsilon)$ as $\epsilon \to 0$.

Outer (long time) expansion

The first expansion we consider is that associated with the limit $\epsilon \to 0$, $\mathbf{r}$, t fixed in the coordinates as chosen. The limit problem with $\epsilon = 0$ has no solution. This corresponds physically to the fact that a very large potential develops for small ϵ. Hence, we try

$$V(\mathbf{r}, t; \epsilon) = \alpha_0(\epsilon)V_0(\mathbf{r}, t) + \alpha_1(\epsilon)V_1(\mathbf{r}, t) + \ldots, \tag{3.3.233}$$

where now $\alpha_0(\epsilon) \to \infty$ as $\epsilon \to 0$, but the $\alpha_j(\epsilon)$ still form an asymptotic sequence. If $\alpha_1(\epsilon) = 1$, then we have

$$\Delta V_0 = 0. \tag{3.3.234}$$

$$\Delta V_1 = -4\pi\delta(\mathbf{r} - \mathbf{R})H(t), \tag{3.3.235a}$$

$$\Delta V_{2,3} = 0, \tag{3.3.235b}$$

with the following boundary conditions:

$$\frac{\partial V_0}{\partial n} = 0 \quad \text{on the cell surface} \tag{3.3.236a}$$

$$-\frac{\partial V_1}{\partial n} = \begin{cases} 0 & \text{if } \alpha_0(\epsilon) \ll 1/\epsilon, \\ V_0 + \partial V_0/\partial t & \text{if } \alpha_0(\epsilon) = O_s\left(\frac{1}{\epsilon}\right). \end{cases} \tag{3.3.236b}$$

It is clear that the second case in (3.3.236b) is the only one to have a solution, so we choose $\alpha_0 = \epsilon^{-1}$ for simplicity, and V_1 satisfies

$$-\frac{\partial V_1}{\partial n} = V_0 + \frac{\partial V_0}{\partial t} \quad \text{on the cell surface.} \tag{3.3.237}$$

Thus, $\alpha_2 = \epsilon, \alpha_3 = \epsilon^2$, etc., and $-\partial V_k/\partial n = V_{k-1} + \partial V_{k-1}/\partial t, k = 2, 3, \ldots$ on the cell surface.

The solution for V_0 is thus uniform inside the cell:

$$V_0 = f_0(t). \tag{3.3.238}$$

As is typical in singular boundary-value problems, this is all the information that can be obtained from a study of V_0. In order to find out more about $f_0(t)$, it is necessary to use a solvability condition derived from the equation and boundary condition for V_1. Integration of (3.3.235) and the use of Gauss's theorem gives (d^3r = infinitesimal volume element)

$$\iiint_{\text{cell volume}} \Delta V_1 d^3r = -4\pi = \iint_{\text{cell surface}} \frac{\partial V_1}{\partial n} dS. \tag{3.3.239}$$

Using the boundary condition (3.3.237), this gives

$$\iint_{\text{cell surface}} \left(\frac{df_0}{dt}(t) + f_0(t)\right) dS = 4\pi,$$

i.e.,

$$\frac{df_0}{dt} + f_0 = 4\frac{\pi}{A}, \tag{3.3.240}$$

where A is the surface area of the cell membrane. The solution for $f_0(t)$ is

$$V_0 = f_0 = \frac{4\pi}{A} + a_0 e^{-t}. \tag{3.3.241}$$

Thus, if we assume that the potential is initially zero, we have

$$V_0(\mathbf{r}, t) = \frac{4\pi}{A}(1 - e^{-t}). \tag{3.3.242}$$

This result shows that the cell builds up to a large uniform potential independent of cell shape but dependent on cell surface area A.

Next, we consider the problem for V_1 using the result for V_0 to write the boundary condition (3.3.237) in the form

$$\frac{\partial V_1}{\partial n} = -\frac{4\pi}{A}. \tag{3.3.243}$$

A corresponding solvability condition derived from the problem for V_2 exists for the potential V_1:

$$\iiint_{\text{cell volume}} \Delta V_2 d^3r = 0 = \iint_{\text{cell surface}} \frac{\partial V_2}{\partial n} dS.$$

Thus,

$$\iint_{\text{cell surface}} \left(V_1 + \frac{\partial V_1}{\partial t} \right) dS = 0. \tag{3.3.244}$$

We can effectively split V_1 into a steady-state part, which is a characteristic function G_1 for the domain, and a transient part. Let

$$V_1(\mathbf{r}, t) = G_1(\mathbf{r}) + f_1(\mathbf{r}, t), \tag{3.3.245}$$

where

$$\Delta G_1 = -\, 4\pi\delta(\mathbf{r} - \mathbf{R}) \tag{3.3.246a}$$

$$-\frac{\partial G_1}{\partial n} = \frac{4\pi}{A}; \tag{3.3.246b}$$

$$\iint_{\text{cell surface}} G_1 dS = 0. \tag{3.3.246c}$$

The condition (3.3.246c) serves to define the arbitrary constant that would exist for G_1 otherwise. Correspondingly, we have the problem for f_1:

$$\Delta f_1 = 0,$$

$$\frac{\partial f_1}{\partial n} = 0, \tag{3.3.247}$$

$$\iint_{\text{cell surface}} \left(f_1 + \frac{\partial f_1}{\partial t} \right) dS = 0.$$

Again, we see that, because of the equations and boundary conditions, $f_1 = f_1(t)$, and it follows that

$$f_1 = a_1 e^{-t}. \tag{3.3.248}$$

Now the representation for V_1 is

$$V_1 = G_1(\mathbf{r}, \mathbf{R}) + a_1 e^{-t}, \tag{3.3.249}$$

and a new difficulty comes to light. The initial condition $V_1 = 0$ cannot be satisfied so that this particular limit process expansion ($\epsilon \to 0, t$ fixed) is not initially valid.

Initially valid expansion

In order to construct an initially valid expansion, we must take into account that there is another important short time scale in the problem, and that $\partial V/\partial t$ in the boundary condition (3.3.232) can be large. However, since the time is short, the potential has not yet had time to reach a large value. A consistent expansion that keeps the time derivative term in the boundary condition is

$$V(\mathbf{r}, t; \epsilon) = v_1(\mathbf{r}, t^*) + \epsilon v_2(\mathbf{r}, t^*) + \ldots, \tag{3.3.250}$$

where $t^* = t/\epsilon$. The limit process associated with this is, of course, $\epsilon \to 0$, $\mathbf{r}$, t^* fixed. The following sequence of problems results:

$$\Delta v_1 = -4\pi\delta(\mathbf{r} - \mathbf{R})H(t^*), \tag{3.3.251a}$$

$$-\frac{\partial v_1}{\partial n} = \frac{\partial v_1}{\partial t^*} \quad \text{on the cell surface,} \tag{3.3.251b}$$

$$v_1 = 0 \quad \text{at } t = 0 \quad \text{on the cell surface.} \tag{3.3.251c}$$

$$\Delta v_2 = 0. \tag{3.3.252a}$$

$$-\frac{\partial v_2}{\partial n} = \frac{\partial v_2}{\partial t^*} + v_1 \quad \text{on the cell surface,} \tag{3.3.252b}$$

$$v_2 = 0 \quad \text{at } t = 0 \quad \text{on the cell surface.} \tag{3.3.252c}$$

Some indication of the general form that the solution must have can be obtained by integration over the cell volume. From (3.3.251), we have for $t^* > 0$

$$\iiint_{\text{cell volume}} \Delta v_1 d^3r = -4\pi = \iint_{\text{cell surface}} \frac{\partial v_1}{\partial n} dS$$
$$= -\frac{\partial}{\partial t^*} \iint_{\text{cell surface}} v_1 dS. \tag{3.3.253}$$

Thus, using the initial condition, we find

$$\iint_{\text{cell surface}} v_1 dS = 4\pi t^*. \tag{3.3.254}$$

Part of v_1 must increase linearly with t^*. The following decomposition of v_1 is suggested:

$$v_1(\mathbf{r}, t^*) = u_1(\mathbf{r}, t^*) + h(\mathbf{r})t^*, \tag{3.3.255}$$

where $u_1(\mathbf{r}, t^*)$ is a potential that does not grow with time. The boundary condition (3.3.251b) becomes

$$\frac{\partial u_1}{\partial n} + t^* \frac{\partial h}{\partial n} = -\frac{\partial u_1}{\partial t^*} - h \quad \text{on the cell surface.} \tag{3.3.256}$$

This implies that

$$\frac{\partial h}{\partial n} = 0 \quad \text{on the cell surface} \tag{3.3.257}$$

since u_1 does not grow with time. Since $\Delta h = 0$ inside the cell, it follows that

$$h(\mathbf{r}) = \text{const.} \tag{3.3.258}$$

This constant can be evaluated from the integral condition (3.3.254)

$$h(\mathbf{r}) = \frac{4\pi}{A}, \qquad A = \text{cell surface area.} \tag{3.3.259}$$

Thus, we have

$$v_1(\mathbf{r}, t^*) = u_1(\mathbf{r}, t^*) + \frac{4\pi}{A} t^*. \tag{3.3.260}$$

Now, u_1 satisfies the problem

$$\Delta u_1 = -4\pi\delta(\mathbf{r}-\mathbf{R})H(t^*), \tag{3.3.261a}$$

$$\frac{\partial u_1}{\partial n} = -\frac{\partial u_1}{\partial t^*} - \frac{4\pi}{A} \quad \text{on the cell surface,} \tag{3.3.261b}$$

$$u_1 = 0 \quad \text{at } t^* = 0 \quad \text{on the cell surface.} \tag{3.3.261c}$$

It is clear that as u_1 approaches a steady state, it will tend to G_1, the characteristic function for the cell, defined by (3.3.246). To get some idea how this approach might take place, let us assume that the geometry is such that the characteristic function can be represented by a separation-of-variables type of expansion:

$$G_1(\mathbf{r}, \mathbf{R}) = \sum_k c_k f_k(\rho_1)\psi_k(\rho_2, \rho_3). \tag{3.3.262}$$

Here, $\rho_1(\mathbf{r}) = \rho_c = $ const. defines the cell surface. Thus ρ_1 is a coordinate normal to the surface; ρ_2, ρ_3 are coordinates in the surface. The function $f_k(\rho_1)$ satisfies an equation of the form

$$\frac{d}{d\rho_1}\frac{K_2^2}{K_1}\frac{df_k}{d\rho_1} - \lambda_k^2 K_1 f_k = 0, \qquad K_{1,2}(\rho_1) > 0, \tag{3.3.263}$$

where λ_k^2 is a separation constant. Typically, $\rho_1 = 0$ is a singular point inside the cell and, depending on the type of expansion, a delta function may appear on the right-hand side of (3.3.263). In any case, the energy integral

$$\int_0^{\rho_c} f_k \left\{ \frac{d}{d\rho_1}\left(\frac{K_2^2}{K_1}\frac{df_k}{d\rho_1}\right) - \lambda_k^2 K_1 f_k \right\} d\rho_1 = 0$$

implies that $f_k(df_k/d\rho_1) > 0$ at the cell surface, $((K_2^2/K_1)f_k(df_k/d\rho_1) \to 0$ as $\rho_1 \to 0)$ or

$$\frac{df_k}{d\rho_1}(\rho_c) = \mu_k^2 f_k(\rho_c) \quad \text{on the cell surface.} \tag{3.3.264}$$

Now we can try to expand u_1 in terms of the same functions

$$u_1(\mathbf{r}, t) = G_1(\mathbf{r}, \mathbf{R}) + \sum_k a_k(t^*) f_k(\rho_1)\psi_k(\rho_2, \rho_3), \tag{3.3.265}$$

where $a_k(0) = -c_k$ so that the singularity at $\mathbf{r} = \mathbf{R}$ is removed just at $t^* = 0$. But for all $t^* > 0$, it remains. The boundary condition (3.3.261b) then becomes for $t^* > 0$

$$\sum_k a_k(t^*)\frac{df_k}{d\rho_1}(\rho_c)\psi_k(\rho_2, \rho_3) = -\sum_k \frac{da_k}{dt^*} f_k(\rho_c)\psi_k(\rho_2, \rho)3) \tag{3.3.266}$$

or using (3.3.264)

$$\frac{da_k}{dt^*} + \mu_k^2 a_k = 0, \tag{3.3.267}$$

where $\mu_k^2 = \frac{f_k'}{f_k}$ at the cell surface. Therefore,

$$a_k(t^*) = -c_k e^{-\mu_k^2 t^*}. \tag{3.3.268}$$

The preceding calculations can be regarded as symbolic, but they are verified in detail for the explicit case of a sphere. In summary, for $t^* > 0$,

$$v_1 = G_1(\mathbf{r}, \mathbf{R}) - \sum_k c_k e^{-\mu_k^2 t^*} f_k(\rho_1)\psi_k(\rho_2, \rho_3) + \frac{4\pi}{A} t^*. \tag{3.3.269}$$

In a similar way, the form of the short-time correction potential v_2 can also be found. In the calculation of v_1 only the term corresponding to membrane capacitance remains in the boundary condition. Now, for v_2, the effect of membrane resistance appears, since (3.3.252b) implies that

$$-\iint_{\text{cell surface}} \frac{\partial v_2}{\partial n} dS = 0 = \frac{\partial}{\partial t^*}\iint_{\text{cell surface}} v_2 dS - \iint_{\text{cell surface}} u_1 dS + 4\pi t^*. \tag{3.3.270}$$

A suitable decomposition for v_2 is

$$v_2(\mathbf{r}, t^*) = h_2(\mathbf{r})\frac{t^{*2}}{2} + u_2(\mathbf{r}, t^*), \tag{3.3.271}$$

where u_2 does not grow with t^*.

Then the boundary condition (3.3.252b) is

$$-\frac{\partial h_2}{\partial n}\frac{t^{*2}}{2} - \frac{\partial u_2}{\partial n} = h_2(\mathbf{r})t^* + u_1 + \frac{4\pi}{A}t^* + \frac{\partial u_2}{\partial t^*}.$$

Again, $\partial h_2/\partial n = 0$, $h_2 = \text{const.} = -4\pi/A$. The resulting problem for u_2 is

$$\Delta u_2 = 0, \tag{3.3.272a}$$

$$-\frac{\partial u_2}{\partial n} = \frac{\partial u_2}{\partial t^*} + \sum_k c_k(1 - e^{-\mu_k^2 t^*}) f_k(\rho_k)\psi_m(\rho_2, \rho_3), \tag{3.3.272b}$$

$$u_2 = 0 \quad \text{at } t^* = 0. \tag{3.3.272c}$$

Thus, a representation for u_2 is sought:

$$u_2 = \sum_k b_k(t^*) f_k(\rho_1)\psi_k(\rho_2, \rho_3). \tag{3.3.273}$$

The boundary condition (3.3.272b) then gives

$$\frac{db_k}{dt^*} + \mu_k^2 b_k = -c_k(1 - e^{-\mu_k^2 t^*}), \quad b_k(0) = 0 \tag{3.3.274}$$

so that

$$b_k(t^*) = -\frac{c_k}{\mu_k^2}(1 - e^{-\mu_k^2 t^*}) + c_k t^* e^{-\mu_k^2 t^*}. \tag{3.3.275}$$

Matching

Now we can discuss the matching between the long-time and short-time expansions in terms of $t_\eta = t/\eta(\epsilon)$, where η belongs to an appropriate subclass of $\epsilon \ll \eta \ll 1$.

Thus, for t_η fixed $t^* = (\eta t_\eta/\epsilon) \to \infty$ and $t = \eta t_\eta \to 0$. We have, for the long-time expansion,

$$V_0 = \frac{4\pi}{A}(1 - e^{-\eta t_\eta}) = \frac{4\pi}{A}\left(\eta t_\eta - \frac{\eta^2 t_\eta^2}{2!} + \ldots\right),$$

$$V_1 = G_1(\mathbf{r}, \mathbf{R}) + a_1 e^{-\eta t_\eta} = G_1(\mathbf{r}, \mathbf{R}) + a_1(1 - \eta t_\eta + \ldots),$$

and for the short-time expansion, neglecting transcendentally small terms,

$$v_1 = G_1(\mathbf{r}, \mathbf{R}) + \frac{4\pi}{A}\frac{\eta t_\eta}{\epsilon}$$

$$-v_2 = \sum_k \frac{c_k}{\mu_k^2} f_k(\rho_1)\psi_k(\rho_2, \rho_3) + \frac{4\pi}{A}\frac{\eta^2 t_\eta^2}{2\epsilon^2}.$$

Comparing $(1/\epsilon)V_0 + V_1$ with $v_1 + \epsilon v_2$, we see that the term $(4\pi/A)\eta t_\eta$ matches, that G_1 matches, and that we must choose

$$a_1 = 0.$$

The term $(4\pi/A)$ $(\eta^2 t_\eta^2/2)$ also matches, and neglected terms are $O(\eta^3, \epsilon^2)$. Therefore, the overlap domain for the matching to $O(\epsilon)$ is $\epsilon|\log\epsilon| \ll \eta \ll \epsilon^{1/3}$.

The voltage response at a typical point, as indicated by this theory, is given in Figure 3.3.6. For further discussion, see also [3.30].

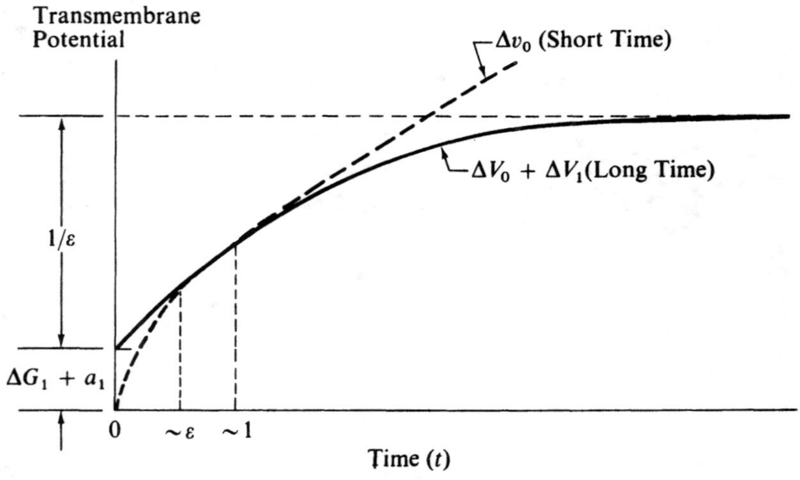

FIGURE 3.3.6. Matching of Short-Time and Long-Time Expansions

3.3.6 Green's Function; Infinite Cylindrical Cell

A number of problems in biology require the solution of Laplace's equation with a boundary condition that describes the properties of the membrane surrounding a biological cell, separating the interior from the exterior, and buffering the internal environment from external disturbance. The membrane serves as an electrical buffer because its resistivity is much greater than the resistivity of the cell interior. The membrane boundary condition, therefore, contains a small parameter ϵ, the ratio of the internal resistance to the membrane resistance in appropriate units (see Sec. 3.3.5).

Problem formulation

Here we consider a problem that arises when the electrical properties of very long cylindrical cells are investigated by the application of current to the interior of the cell from a microelectrode, a glass micropipette filled with conducting salt solution. The potential in the interior of the cell obeys Laplace's equation. The boundary condition is that the normal derivative of the potential at the inside surface of the membrane (proportional to the normal component of current) is proportional to the potential difference across the membrane. If the microelectrode is considered a point source of current, the solution to the problem is the Green's function for the electric potential in a cylinder with a membrane boundary condition.

The same method applies with other boundary conditions or source distributions in a part of the cell near the origin. There are also, of course, analogies with other problems for the Laplace equation, for example, steady heat conduction or incompressible flow. A more detailed discussion of this problem appears in [3.31] and in references therein. Another approach by classical analysis appears in [3.29].

The problem for determining the potential $V(x, r, \theta; \epsilon)$ may be written, in cylindrical coordinates,

$$\frac{1}{r}(rV_r)_r + \frac{1}{r^2}V_{\theta\theta} + V_{xx} = -\frac{1}{r}\delta(x)\delta(r-R)\delta(\theta), \tag{3.3.276a}$$

$$V_r(x, 1, \theta; \epsilon) + \epsilon V(x, 1, \theta; \epsilon) = 0, \tag{3.3.276b}$$

$$V(\pm\infty, r, \theta) = 0. \tag{3.3.276c}$$

When ϵ is small, the boundary condition at $r = 1$ in (3.3.276b) implies that the current flow will be predominantly in the axial direction, i.e., only a small fraction of the local current, $O(\epsilon)$, crosses the membrane in an axial distance of $O(1)$. We are tempted to try to find an expansion in the small parameter ϵ, in which the leading term is the potential for $\epsilon = 0$. Denoting this term by $V_1(x, r, \theta)$, we see from (3.3.276a, b) that V_1 satisfies

$$\frac{1}{r}(rV_{1_r})_r + \frac{1}{r^2}V_{1_{\theta\theta}} + V_{1_{xx}} = -\frac{1}{r}\delta(x)\delta(r-R)\delta(\theta)$$

$$V_{1_r}(x, 1, \theta) = 0. \tag{3.3.277}$$

The boundary condition at $r = 1$ implies that no current crosses the membrane; all the current is confined to the interior of the cell. Consequently, V_1 must contain a part that is linearly decreasing with increasing $|x|$. This would lead to a potential $V_1 \to -\infty$ as $|x| \to \infty$, making it impossible to satisfy the boundary condition $V = 0$ at $|x| = \infty$. To avoid this divergence, any expansion that contains V_1 can be valid only over a limited range of x, designated the *near field*, which contains the source point. At large distances from the source, we must look for another, far-field, expansion.

We expect that as $\epsilon \to 0$, the region of validity of any near-field expansion of which V_1 is a part becomes larger. If there is a linearly decaying potential over a large distance, and the potential approaches zero as $|x| \to \infty$, then the potential at $x = 0$ must be very large, i.e., $V(0, r, \theta) \to \infty$ as $\epsilon \to 0$. Clearly, V_1 must be $O(1)$ and cannot be the leading term in the expansion. The leading term can be found by matching to the far-field solution, and therefore we first solve the far-field problem.

Far-field expansion

In the far field, a long distance from the source, current flow is predominantly in the axial direction. Since only a small fraction of the current within the cell, at any value of x, leaks out of the cylinder in an axial distance of $O(1)$, the variation in the x direction will be slow. We therefore, for convenience in ordering the far-field expansion, write the far-field potential in terms of a new "slow" variable $\widetilde{x}$. Denoting the potential in the far field by W, we write the following expansion:

$$V = W(\widetilde{x}, r, \theta; \epsilon) = \zeta_0(\epsilon)W_0(\widetilde{x}, r, \theta) + \zeta_1(\epsilon)W_1(\widetilde{x}, r, \theta) + \dots, \tag{3.3.278}$$

where the slow variable is defined by

$$\widetilde{x} = \sigma(\epsilon)x, \tag{3.3.279}$$

for a $\sigma(\epsilon) \ll 1$ to be defined. Thus,

$$\Delta V = 0 = \zeta_0\sigma^2 W_{0_{\widetilde{x}\widetilde{x}}} + \zeta_1\sigma^2 W_{1_{\widetilde{x}\widetilde{x}}} + \dots + \zeta_0\Delta_t W_0 + \zeta_1\Delta_t W_1 + \dots,$$

where $\Delta_t = (1/r)(\partial/\partial r)(r\partial/\partial r) + (1/r^2)(\partial^2/\partial\theta^2)$ is the transverse Laplacian, and on the boundary, $r = 1$, we have

$$W_r + \epsilon W = 0 = \zeta_0 W_{0_r} + \zeta_1 W_{1_r} + \dots + \epsilon\zeta_0 W_0 + \epsilon\zeta_1 W_1 + \dots.$$

The corresponding approximating sequence of problems is

$$\begin{gathered} \Delta_t W_0 = 0 \\ W_{0_r}(\widetilde{x}, 1, \theta) = 0 \\ W_0(\pm\infty, \cdot, \theta) = 0 \\ \Delta_t W_1 = -W_{0_{\widetilde{x}\widetilde{x}}} \end{gathered} \tag{3.3.280}$$

$$W_{1_r}(\tilde{x}, 1, \theta) = -W_0(\tilde{x}, 1, \theta) \tag{3.3.281}$$

$$W_1(\pm\infty, r, \theta) = 0,$$

where we have set

$$\epsilon\zeta_0 = \zeta_1$$

to obtain the surface boundary condition in (3.3.281).

Writing an expansion for $\sigma(\epsilon)$ in the form

$$\sigma(\epsilon) = \sigma_0(\epsilon) + \sigma_1(\epsilon) + \sigma_2(\epsilon) + \ldots,$$

where the $\sigma_i(\epsilon)$ are an asymptotic sequence, we further set

$$\zeta_0\sigma_0^2 = \zeta_1$$

to obtain (3.3.281). Thus

$$\sigma_0 = \sqrt{\epsilon}. \tag{3.3.282}$$

We could take $\sigma = \sigma_0$ with $\sigma_1 = \sigma_2 = \ldots = 0$ and still obtain a sequence of problems of increasing order in ϵ. It will be seen later, however, that we would not be able to maintain uniform validity of the asymptotic expansion for W at large $\tilde{x}$. Assuming $\sigma(\epsilon)$ to have the more general form makes it possible to obtain a uniform expansion. Continuing this procedure with $\epsilon\zeta_1 = \zeta_2$, $\sigma_1 = \alpha_1\epsilon\sigma_0$, etc., we find

$$\Delta_t W_2 = -(W_1 + 2\alpha_1 W_0)_{\tilde{x}\tilde{x}}$$

$$W_{2_r}(\tilde{x}, 1, \theta) = -W_1(\tilde{x}, 1, \theta) \tag{3.3.283}$$

$$W_2(\pm\infty, r, \theta) = 0$$

and

$$\Delta_t W_3 = -[W_2 + 2\alpha_1 W_1 + (\alpha_1^2 + 2\alpha_2)W_0]_{\tilde{x}\tilde{x}}$$

$$W_{3_r}(\tilde{x}, 1, \theta) = -W_2(\tilde{x}, 1, \theta) \tag{3.3.284}$$

$$W_3(\pm\infty, r, \theta) = 0,$$

where the α_i are unknown constants. Thus, the far-field expansion of the potential is taken in the form

$$W(\tilde{x}, r, \theta; \epsilon) = \zeta_0(\epsilon)[W_0(\tilde{x}, r, \theta) + \epsilon W_1(\tilde{x}, r, \theta) + \epsilon^2 W_2(\tilde{x}, r, \theta) + \ldots], \tag{3.3.285}$$

where the axial coordinate variable is

$$\tilde{x} = \sqrt{\epsilon}(1 + \alpha_1\epsilon + \alpha_2\epsilon^2 + \ldots)x. \tag{3.3.286}$$

So far, $\zeta_0(\epsilon)$, the order of the leading term in the W expansion, is unknown. It will be determined by matching to the near field. The constants $\alpha_1, \alpha_2, \ldots$ in the expansion of $\tilde{x}$, which couple different orders of W in the sequence of problems, will be determined by requiring uniform validity of the W expansion for large values of $\tilde{x}$.

We now solve the sequence of problems. The solution to the first problem is independent of r and θ. Thus, we have

$$W_0(\tilde{x}, r, \theta) = F(\tilde{x}), \tag{3.3.287}$$

where $F(\tilde{x})$ is an as yet arbitrary function of $\tilde{x}$. We must go to the second problem to determine its functional form.

From (3.3.281) and (3.3.287), we obtain

$$\frac{1}{r}(rW_{1_r}) + \frac{1}{r^2}W_{1_{\theta\theta}} = -F''(\tilde{x}),$$

$$W_{1_r}(\tilde{x}, 1, \theta) = -F(\tilde{x}), \tag{3.3.288}$$

$$W_1(\pm\infty, r, \theta) = 0.$$

Since the inhomogeneous term in the equation and the boundary condition at $r = 1$ are both independent of θ, clearly, W_1 is independent of θ. Examining (3.3.283)–(3.3.284), the same reasoning then implies that W_2, W_3, ... are all independent of θ.

Integrating (3.3.288), we obtain, for the solution that is bounded at $r = 0$,

$$W_1(\tilde{x}, r) = -\frac{r^2}{4}F''(\tilde{x}) + G(\tilde{x}), \tag{3.3.289}$$

where $G(\tilde{x})$ is an arbitrary function of $\tilde{x}$ that cannot be determined until we go to the next problem for W_2.

Substituting the result (3.3.289) in the $r = 1$ boundary condition yields

$$F'' - 2F = 0, \tag{3.3.290}$$

$$W_0(\tilde{x}) = F(\tilde{x}) = Ae^{-\sqrt{2}|\tilde{x}|}, \tag{3.3.291}$$

where A is a constant to be determined by matching to the near field.

Continuing in the same way, we find

$$\frac{1}{r}(rW_{2_r}) = (r^2 - 4\alpha_1)F - G'' \tag{3.3.292}$$

$$W_{2_r}(\tilde{x}, 1) = F/2 - G$$

$$W_2(\pm\infty, r) = 0$$

and then

$$W_2(\tilde{x}, r) = \frac{r^4}{16}F - r^2\left(\alpha_1 F + \frac{1}{4}G''\right) + H(\tilde{x}). \tag{3.3.293}$$

Substituting the expression for W_2 and F in the $r = 1$ boundary condition yields

$$G'' - 2G = -4A\left(\alpha_1 + \frac{1}{8}\right)e^{-\sqrt{2}|\tilde{x}|}. \tag{3.3.294}$$

The right-hand side is a homogeneous solution of the equation. Therefore, the particular solution contains a term proportional to $\tilde{x}$ times $\exp(-\sqrt{2}|\tilde{x}|)$. If such a term appears in G, then for $|\tilde{x}| = O(\epsilon^{-1})$, the expansion will not be valid uniformly in $\tilde{x}$. To avoid this, we require the right-hand side of (3.3.294) to vanish; this occurs if

$$\alpha_1 = -\frac{1}{8}. \tag{3.3.295}$$

It is now clear why we could not assume the simple relation $\tilde{x} = \sqrt{\epsilon}x$ but required the more general form. The freedom to choose $\alpha_1, \alpha_2, \ldots$ allows us to force all of the $\tilde{x}$ dependence of W into $\exp(-\sqrt{2}\tilde{x})$, eliminating nonuniformities in the expansion.

The solution to (3.3.294) is thus

$$G(\tilde{x}) = Be^{-\sqrt{2}|\tilde{x}|}, \tag{3.3.296}$$

where

$$\tilde{x} = \sqrt{\epsilon}\left(1 - \frac{\epsilon}{8} + \ldots\right)x. \tag{3.3.297}$$

The constant B will be determined by matching to the near field.

Thus, W_1, the second term in the far-field expansion (see (3.2.289)), is

$$W_1(\tilde{x}, r) = \left(-\frac{1}{2}Ar^2 + B\right)e^{-\sqrt{2}|\tilde{x}|}. \tag{3.3.298}$$

We continue the same procedure and calculate $\alpha_2 = 5\epsilon^2/384$.

The order of the expansion $\zeta_0(\epsilon)$ and the three constants (A, B, C) are to be found by matching to the near-field expansion. We introduce the matching variable

$$x_\eta = x\eta(\epsilon) \tag{3.3.299}$$

for a class of $\eta(\epsilon)$ contained in $\sqrt{\epsilon} \ll \eta(\epsilon) \ll 1$, and we take the limit $\epsilon \to 0$ with x_η fixed. Under this limit, the near-field coordinate $x = x_\eta/\sigma \to \infty$ and the far-field coordinate

$$\tilde{x} = \frac{\sqrt{\epsilon}}{\eta}x_\sigma\left(1 - \frac{\epsilon}{8} + \frac{5}{384}\epsilon^2 - \ldots\right) \to 0.$$

Because of the simple way $\tilde{x}$ enters the expansion, it is appropriate to express the far field in terms of x as $\tilde{x} \to 0$ and to compare directly with the near field as $x \to \infty$.

The far field has the expansion

$$W\left[x\sqrt{\epsilon}\left(1 - \frac{\epsilon}{8} + \frac{5\epsilon^2}{384} - \ldots\right), r; \epsilon\right]$$

$$= \zeta_0(\epsilon)\Bigg\{A - \epsilon^{1/2}A\sqrt{2}|x| + \epsilon\left[A\left(x^2 - \frac{r^2}{2}\right) + B\right]$$
$$+ \epsilon^{3/2}\sqrt{2}|x|\left[A\left(\frac{1}{8} - \frac{x^2}{3} + \frac{r^2}{2}\right) - B\right]$$
$$+ \epsilon^2\left[A\left(-\frac{x^2}{4} + \frac{x^4}{6} - \frac{r^2x^2}{2} + \frac{r^2}{8} + \frac{r^4}{16}\right) + B\left(x^2 - \frac{r^2}{2}\right) + C\right]$$
$$+ \epsilon^{5/2}\sqrt{2}|x|\left[A\left(-\frac{5}{384} + \frac{x^2}{8} + \frac{x^2r^2}{6} - \frac{3r^2}{16} - \frac{r^4}{16} - \frac{x^4}{30}\right)\right.$$
$$\left. + B\left(\frac{1}{8} - \frac{x^2}{3} + \frac{r^2}{2}\right) - C\right] + O(\epsilon^3)\Bigg\}. \tag{3.3.300}$$

Near-field expansion

Next, we consider the sequence of near-field problems implied by the far-field behavior.

In the vicinity of the point source, the potential is a rather complex function of position, and there is no simple mathematical representation in terms of elementary functions as there is in the far field. The potential has a singularity at the source point; the current diverges from this point, half going toward $x = +\infty$ and half toward $x = -\infty$. Close to the source, the lines of current flow are diverging outward, equally in all directions. Those lines directed toward the membrane must curve to avoid the membrane as, again, only a small fraction of the local current leaves the cylinder. As the current flows down the cylinder, the lines become predominantly in the axial direction, and the potential joins smoothly onto the far-field potential.

In terms of the asymptotic expansions representing the near and far fields, this behavior requires that the near-field expansion increase in powers of $\sqrt{\epsilon}$ so it can join the expansion (3.3.300) of the far field. Furthermore, in accordance with the earlier arguments, which concluded that the $O(1)$ term in the near field has a linear dependence on $|x|$ as $|x| \to \infty$, we see that the second term in (3.3.300) must be $O(1)$ in order to match the near field. Consequently,

$$\zeta_0(\epsilon) = \frac{1}{\sqrt{\epsilon}}. \tag{3.3.301}$$

The near-field expansion must be of the form

$$V(x, r, \theta; \epsilon) = \epsilon^{-1/2}V_0(x, r, \theta) + V_1(x, r, \theta)$$
$$+ \epsilon^{1/2}V_2(x, r, \theta) + \epsilon V_3(x, r, \theta) + \dots. \tag{3.3.302}$$

Thus, we obtain the following sequence of near-field problems:

$$\Delta V_0 = 0$$

$$V_{0_r}(x, 1, \theta) = 0; \quad V_0(x, r, \theta) \to A \text{ as } |x| \to \infty, \tag{3.3.303}$$

$$\Delta V_1 = -\frac{1}{r}\delta(x)\delta(r-R)\delta(\theta)$$

$$V_{1,r}(x, 1, \theta) = 0$$

$$V_1(x, r, \theta) \to -A\sqrt{2}|x| \text{ as } |x| \to \infty, \tag{3.3.304}$$

$$\Delta V_2 = 0$$

$$V_{2,r}(x, 1, \theta) = -V_0(x, 1, \theta)$$

$$V_2(x, r, \theta) \to A\left(x^2 - \frac{r^2}{2}\right) + B, \text{ as } |x| \to \infty, \tag{3.3.305}$$

$$\Delta V_3 = 0$$

$$V_{3,r}(x, 1, \theta) = -V_1(x, 1, \theta)$$

$$V_3(x, r, \theta) \to \sqrt{2}|x|\left[A\left(\frac{1}{8} - \frac{x^2}{3} + \frac{r^2}{2}\right) - B\right] \text{ as } |x| \to \infty, \tag{3.3.306}$$

$$\Delta V_4 = 0$$

$$V_{4,r}(x, 1, \theta) = -V_2(x, 1, \theta);$$

$$V_4(x, r, \theta) \to A\left(-\frac{x^2}{4} + \frac{x^4}{6} - \frac{r^2x^2}{2} + \frac{r^2}{8} + \frac{r^4}{16}\right)$$

$$+B\left(x^2 - \frac{r^2}{2}\right) + C \text{ as } |x| \to \infty, \tag{3.3.307}$$

$$\Delta V_5 = 0$$

$$V_{5,r}(x, 1, \theta) = -V_3(x, 1, \theta)$$

$$V_{5,r}(x, r, \theta) \to \sqrt{2}|x|\left[A\left(-\frac{5}{384} + \frac{x^2}{8} + \frac{x^2r^2}{6} - \frac{r^4}{16} - \frac{x^4}{30}\right)\right.$$

$$\left. +B\left(\frac{1}{8} - \frac{x^2}{3} + \frac{r^2}{2}\right) - C\right] \text{ as } |x| \to \infty. \tag{3.3.308}$$

The delta function source appears in the V_1 problem, consistent with the linear decrease with x as $|x| \to \infty$. All other orders of the potential are source-free.

Each even(odd) order problem (except for the first two) is coupled to the preceding even(odd) order problem via the boundary condition on the $x = 1$ surface. The physical interpretation of this coupling is that the current crossing the membrane in the nth problem is proportional to the membrane potential in the $(n-2)$nd problem. The even order problems are coupled to the odd order problems by their asymptotic behavior as $|x| \to \infty$, i.e., the constants $A, B, C, \ldots$, appear in both even and odd order problems.

It should be noted that the $V_1, V_3, \ldots$ terms alone are sufficient to satisfy (3.3.276) at small x. It is only from considerations of behavior for large $|x|$, required of the far-field potential, that we conclude that $V_0, V_2, \ldots$ are even necessary. These terms are thus known as *switchback* terms.

By direct substitution of the $|x| \to \infty$ asymptotic forms of V_0, V_2, and V_4 in the respective equations and boundary conditions (3.3.303, 305, 307), it is seen that the $|x| \to \infty$ forms are the solutions valid for all x. This part of the near field is thus completely contained in the far field:

$$V_0 = A, \tag{3.3.309}$$

$$V_2 = A\left(x^2 - \frac{r^2}{2}\right) + B, \tag{3.3.310}$$

$$V_4 = A\left(-\frac{x^2}{4} + \frac{x^4}{6} - \frac{r^2x^2}{2} + \frac{r^2}{8} + \frac{r^4}{16}\right) + B\left(x^2 - \frac{r^2}{2}\right) + C. \tag{3.3.311}$$

Now we evaluate the constant A. Integrating (3.3.304) over the large volume of the cylinder between $-x$ and x, $|x| \to \infty$, and using the divergence theorem, we obtain

$$\begin{aligned} -1 &= \lim_{|x|\to\infty} \int_0^{2\pi} d\theta \int_0^1 r dr \int_{-x}^{x} dx \Delta V_1 \\ &= \lim_{|x|\to\infty} \int_0^{2\pi} d\theta \int_0^1 r dr [V_{1_x}(x, r, \theta) - V_{1_x}(-x, r, \theta)] \\ &= -2\pi A/\sqrt{2}, \end{aligned}$$

where in accordance with the $r = 1$ boundary condition, the integral over the surface of the cylinder is zero, leaving only the integral over the disks at $\pm x$. The last equality follows from substitution of the asymptotic behavior of V_1, as $|x| \to \infty$.

The problem (3.3.304) for V_1 is now definite. In order to solve the problem, it is convenient to decompose the near-field potential V_1 into two terms:

$$V_1(x, r, \theta) = \Phi_1(x, r, \theta) - \frac{|x|}{2\pi}. \tag{3.3.312}$$

We obtain the following problem for Φ_1:

$$\Delta\Phi_1 = -\delta(x)\left[\frac{1}{r}\delta(r-R)\delta(\theta) - \frac{1}{\pi}\right],$$

$$\Phi_{1_r}(x, 1, \theta) = 0, \tag{3.3.313}$$

$$\Phi_1(\pm\infty, r, \theta) = 0.$$

The right-hand side of (3.3.312) consists of a unit point source at $(0, R, 0)$ plus a uniform distribution of sinks in the $x = 0$ plane. The net current source for Φ_1 is zero, i.e., all the current that enters the cylinder at the point $(0, R, 0)$ is removed uniformly in the cross section $(0, r, \theta)$. Unlike the problem for V_1, which contains unit current flowing outward as $|x| \to \infty$, the problem for Φ_1 contains no current flow as $|x| \to \infty$.

The boundary-value problem may be solved by Fourier transformation in the θ and x coordinates. Defining the double Fourier transform of Φ_1 by

$$\begin{aligned} \psi_1^{(n)}(k, r) &= \textstyle\int_0^{2\pi} d\theta e^{-in\theta} \int_{-\infty}^{\infty} dx \cos(kx)\Phi_1(x, r, \theta), \\ \Phi_1(x, r, \theta) &= \textstyle\frac{1}{2\pi^2}\int_0^{\infty} dk \cos(kx) \sum_{n=-\infty}^{\infty} e^{in\theta}\psi_1^{(n)}(k, r), \end{aligned} \tag{3.3.314}$$

noting that Φ_1 is even in x and θ, we see that the problem (3.3.313) becomes, in Fourier transform space,

$$\frac{1}{r}(r\psi_{1_r}^{(n)})_r - \left(k^2 + \frac{n^2}{r^2}\right)\psi_1^{(n)} = -\frac{1}{r}\delta(r-R) + 2\delta_{0n},$$

$$\psi_{1_r}^{(n)}(k, 1, \theta) = 0,$$

where

$$\delta_{0n} = \frac{i}{2n\pi}\left[(-1)^n - 1\right].$$

The solution is

$$\begin{aligned} \psi_1^{(n)}(k, r) = &-\frac{2\delta_{0n}}{k^2} - I_n(kr)\frac{K_n'(k)}{I_n'(k)} I_n(kR) \\ &+ \begin{cases} K_n(kR)I_n(kr), & 0 \le r \le R, \\ K_n(kr)I_n(kR), & R \le r \le 1. \end{cases} \end{aligned} \tag{3.3.315}$$

Taking the inverse transform, we obtain

$$\begin{aligned} V_1(x, r, \theta) = &-\frac{|x|}{2\pi} + \frac{1}{4\pi}(x^2 + r^2 + R^2 - 2rR\cos\theta)^{-1/2}) \\ &- \frac{1}{2\pi^2}\sum_{n=-\infty}^{\infty} e^{in\theta}\int_0^{\infty} dk \cos(kx)\left[\frac{K_n'(k)}{I_n'(k)} I_n(kR)I_n(kr)\right. \\ &\left. + 2\frac{\delta_{0n}}{k^2}\right]. \end{aligned} \tag{3.3.316}$$

The integral over k can be replaced by an equivalent sum by considering the integral in (3.3.316) as a portion of a contour integral, so that

$$V_1(x, r, \theta) = -\frac{|x|}{2\pi} - \frac{1}{2\pi}\sum_{n=-\infty}^{\infty} e^{in\theta} \sum_{s=1}^{\infty} e^{-\lambda_{ns}|x|} \times \frac{J_n(\lambda_{ns}R)J_n(\lambda_{ns}r)}{\lambda_{ns}\left(\frac{n^2}{\lambda_{ns}^2} - 1\right) J_n^2(\lambda_{ns})}, \tag{3.3.317}$$

where λ_{ns} is the sth zero of $J_n'(\lambda)$ excluding the one at $\lambda = 0$. We can see that as $|x| \to \infty$, $V_1 \to -|x|/2\pi$ plus terms that are exponentially small in (3.3.317).

We now turn to the V_3 problem and evaluate the constant B. Integrating the Laplacian in (3.3.306) over the volume of a large cylinder extending from $-x$ to x and using the divergence theorem, we have

$$\begin{aligned} 0 &= \lim_{|x|\to\infty} \int_{-x}^{x} dx \int_0^1 r dr \int_0^{2\pi} d\theta \,\Delta V_3 \\ &= \lim_{|x|\to\infty} \left[\int_{-x}^{x} dx \int_0^{2\pi} d\theta \, V_{3r}(x, 1, \theta) + 2\int_0^1 r dr \int_0^{2\pi} d\theta \, V_{3_x}(x, r, \theta)\right]. \end{aligned} \tag{3.3.318}$$

Using the boundary condition and the transform for V_1, we see that the first integral becomes

$$\begin{aligned} &-\lim_{|x|\to\infty} \int_{-x}^{x} dx \int_0^{2\pi} d\theta \, V_1(x, 1, \theta) \\ &= x^2 - \frac{1}{\pi}\int_{-\infty}^{\infty} dx \int_0^{\infty} dk \cos(kx)\psi_1^{(0)}(k, 1) \\ &= x^2 - \psi_1^{(0)}(0, 1). \end{aligned} \tag{3.3.319}$$

From (3.3.315), we obtain, using the Wronskian of I_n and K_n and the power series expansion of $I_n(k)$,

$$\psi_1^{(0)}(0, 1) = \lim_{k\to 0}\left(-\frac{2}{k^2} + \frac{I_0(kR)}{kI_1(k)}\right) = \frac{1}{2}\left(R^2 - \frac{1}{2}\right). \tag{3.3.320}$$

Using the asymptotic form for large $|x|$ for V_3 from (3.3.306), we see that the second integral in (3.3.318) becomes

$$\begin{aligned} &\int_0^1 r dr \int_0^{2\pi} d\theta 2\sqrt{2}\left[A\left(\frac{1}{8} - x^2 + \frac{r^2}{2}\right) - B\right] \\ &= 2\pi\sqrt{2}\left[A\left(\frac{3}{8} - x^2\right) - B\right]. \end{aligned} \tag{3.3.321}$$

Thus, from (3.3.318),

$$B = \frac{\sqrt{2}}{4\pi}\left(\frac{5}{8} - \frac{R^2}{2}\right). \tag{3.3.322}$$

As a consequence of (3.3.322), W_1 and V_2 depend on R, the distance from the source to the axis of the cylinder, whereas lower-order terms do not.

Having evaluated A and B, we have now obtained the near field and far field up to terms of $O(\epsilon^{1/2})$, i.e., we have obtained V_0, V_1, V_2, W_0, and W_1. These terms represent that part of the potential that is numerically significant in a physiological experiment: all higher-order terms are too small to detect anywhere in a cylindrical cell. In [3.31], the calculations are carried out further to find the constant C.

The leading terms in the far-field expansion and in the near-field expansion are each of order $\epsilon^{-1/2}$. In the near field, the leading term is a constant. Thus, near the point source, the interior of the cylinder is raised to a large, constant potential, relative to the zero potential at infinity. The physical basis for the large potential is that the membrane permits only a small fraction of the current to leave the cylinder per unit length. Consequently, most of the current flows a long distance before getting out, and a large potential drop is required to force this current down the cylinder. The existence of this large constant potential, and its magnitude of $O(\epsilon^{-1/2})$, could only be deduced from considerations of the far field.

The leading term in the far field decays as $\exp(-\sqrt{2\epsilon}|x|)$. Consequently, to lowest order, $1/e$ of the current leaves the cylinder in a distance of $1/\sqrt{2\epsilon}$. The corresponding potential required to drive a current this distance is of $O(\epsilon^{-1/2})$, which is the physical basis for the order of the large potential in the near field. The precise numerical values of the leading terms for V and W were determined by requiring in the limit $|x| \to \infty$, $\tilde{x} \to 0$, that the two terms be identical to the lowest order in ϵ. In the far field, i.e., $\tilde{x} = x\sqrt{\epsilon}(1 - \epsilon/8 + \ldots) \to \infty$, the potential is seen to approach zero exponentially.

The leading term in the far-field expansion is independent of r and θ. Thus, to the lowest order, the far-field current is distributed uniformly over the circular cross section of the cylinder. The leading term is the known result of one-dimensional cable theory, equation 14 in [3.39]. The high-order terms are all independent of the polar angle. They do, however, depend on the radial coordinate, r. The dependence is in the form of a polynomial in r^2, the degree of the polynomial increasing by one in each successive term. We also see that the higher-order terms also depend on R, the radial distance between the source and the axis of the cylinder. The potential is seen to be symmetric with respect to an interchange of r and R. This must be so because the potential is the Green's function (with source at $x = 0, \theta = 0$) for the cylindrical problem.

The solution obtained here is close to the solution derived by Barcilon, Cole, and Eisenberg [3.2], using multiple scaling.

The result of the multiple scale analysis differs from our result only because [3.2] contains a sign error and a secular term $V^{(3)}$ which has not been removed. If these errors are corrected, and the infinite sum over Bessel functions is written in closed form, the multiple scale result, the expansion of the exact solution, and the present results are identical.

3.3.7 Whispering Gallery Modes

The propagation of sound waves with little attenuation inside a curved surface is called the *whispering gallery* effect. Rayleigh commented on this effect in [3.36] and said that the effect is easily observed in the dome of St. Paul's, London. He noted that the explanation of the Astronomer Royal (Airy), that rays from one pole converge toward the opposite pole, was not correct since the effect can be felt all around the circumference. A simple ray explanation was offered, but a more complete theory based on solutions of the wave equation was given in [3.37]. The idea is to examine the normal modes and identify those that can support waves along the surface; these might be excited by a source of disturbance. Rayleigh then studies the planar case and shows that the desired effect can be seen in high-frequency modes, whose wavelength is considerably shorter than the characteristic radius of the circle. Analytically, the result depends on the asymptotic behavior of Bessel functions of nearly equal argument and order.

A similar result is worked out here for the interior of a spherical dome by using boundary-layer ideas. Only axisymmetric modes are considered. The wave equation of acoustics for the velocity potential is

$$\frac{\partial^2\Phi}{\partial r^2} + \frac{2}{r}\frac{\partial\Phi}{\partial r} + \frac{1}{r^2}\frac{\partial^2\Phi}{\partial\theta^2} + \frac{\cot\theta}{r^2}\frac{\partial\Phi}{\partial\theta} - \frac{1}{c^2}\frac{\partial^2\Phi}{\partial t^2} = 0. \tag{3.3.323}$$

The velocity and density perturbations are, as usual,

$$\mathbf{q} = \text{grad}\,\Phi, \tag{3.3.324a}$$

$$\frac{\rho-\rho_0}{\rho_0} = -\frac{1}{c^2}\frac{\partial\Phi}{\partial t}, \tag{3.3.324b}$$

where ρ_0 is the ambient density, $c = \sqrt{\gamma p_0/\rho_0}$ is the speed of sound, p_0 is the ambient pressure, and γ is the ratio of specific heats.

The coordinates are the usual spherical polar coordinates

$$x = r\cos\theta,\quad y = r\sin\theta\cos\psi,\quad z = r\sin\theta\sin\psi, \tag{3.3.325}$$

where θ is the pole angle, and ψ is the azimuth angle. The radius of the sphere is taken to be a. The boundary condition at the wall, $r = a$, is that the radial velocity is zero

$$\frac{\partial\Phi}{\partial r}(a,\theta,t) = 0, \tag{3.3.326}$$

and the solution should be regular at $r = 0$.

A point source at $\theta = 0$ would produce an axisymmetric wave field $\left(\frac{\partial}{\partial\psi} \equiv 0\right)$.

Since these modes propagate almost unattenuated along the θ direction but are confined to a thin region close to $r = a$, we can assume the following form for the modes using complex notation

$$\Phi(r, \theta, t) = e^{i(ka\theta - \omega t)} \left\{ \phi(r^*) + \cdots \right\}, \tag{3.3.327}$$

where $(\omega/k) \cong c$ so that the waves travel in the $+\theta$ direction as approximately one-dimensional. We have introduced the boundary-layer coordinate r^*

$$r^* = \frac{r - a}{\epsilon a}, \qquad r = a(1 + \epsilon r^*) \tag{3.3.328}$$

since the disturbance is confined to a thin region close to the wall. Here ϵ is a small parameter to be related to the wave number k. Note that

$$\frac{\partial \Phi}{\partial r} = \frac{1}{\epsilon a} e^{i(ka\theta - \omega t)} \frac{\partial \phi}{\partial r^*} \tag{3.3.329}$$

so that the wall boundary condition becomes

$$\frac{\partial \phi}{\partial r^*}(0) = 0. \tag{3.3.330}$$

From (3.3.323), we get the equation for $\phi(r^*)$:

$$\frac{1}{\epsilon^2 a^2} \frac{d^2\phi}{dr^{*2}} + \frac{2}{a(1 + \epsilon r^*)} \frac{1}{\epsilon a} \frac{d\phi}{dr^*} + \frac{(-k^2 a^2)}{a^2(1 + \epsilon r^*)^2} \phi$$
$$+ \frac{\cot \theta}{a^2(1 + \epsilon r^*)^2} ika\phi + \frac{\omega^2}{c^2} \phi = 0. \tag{3.3.331}$$

In order to obtain a limiting equation as $\epsilon \to 0$, $(k, \omega) \to \infty$, which can satisfy the radial boundary conditions, it is necessary to allow (ω/k) to vary from c as

$$\frac{\omega}{c} = k(1 + \epsilon\Omega + \cdots), \tag{3.3.332}$$

where the parameter Ω is $O(1)$ and is later identified with an eigenvalue.

Then the dominant terms of (3.3.331) are

$$\frac{1}{\epsilon^2} \frac{d^2\phi}{dr^{*2}} + 2\epsilon k^2 a^2 (r^* + \Omega)\phi = 0, \tag{3.3.333}$$

and a balance is achieved when

$$2\epsilon^3 k^2 a^2 = 1 \qquad \text{or} \qquad \epsilon = \frac{1}{2^{1/3}(ka)^{2/3}}. \tag{3.3.334}$$

A modal dispersion relation results:

$$\frac{\omega}{c} = k\left(1 + \frac{\Omega}{2^{1/3}(ka)^{2/3}} + \cdots\right), \tag{3.3.335}$$

so that the phase velocity of longer waves is reduced. Equation (3.3.333) becomes

$$\frac{d^2\phi}{dr^{*2}} + (r^* + \Omega)\phi = 0, \qquad -\infty < r^* \leq 0. \tag{3.3.336}$$

Equation (3.3.336) describes the radial dependence of the propagating mode. The regularity condition at the origin is replaced by the condition that the modes die out as $r^* \to -\infty$. The solution of (3.3.336), which dies out, is

$$\phi(r^*) = Ai(-(r^* + \Omega)), \tag{3.3.337}$$

where Ai is the Airy function. The second solution, Bi, does not die out as $r^* \to -\infty$. A graph of the Airy functions appears in Figure 3.3.7.

The asymptotic forms of Ai are

$$Ai(-(r^* + \Omega)) \cong \frac{1}{2\sqrt{\pi}} \frac{1}{(-(r^* + \Omega))^{1/4}} e^{-\frac{2}{3}(-(r^*+\Omega))^{3/2}} \text{ as } r^* \to -\infty, \tag{3.3.338a}$$

$$Ai(-(r^* + \Omega)) \cong \frac{1}{\sqrt{\pi}(r^* + \Omega)^{1/4}} \sin\left(\frac{2}{3}(r^* + \Omega)^{3/2} + \frac{\pi}{4}\right) \text{ as } r^* \to +\infty. \tag{3.3.338b}$$

The boundary condition (3.3.330) is thus

$$Ai'(-\Omega) = 0 \tag{3.3.339}$$

so that the spectrum of radial solutions gives values

$$\Omega = \Omega_j, \qquad j = 1, 2, \ldots, \tag{3.3.340}$$

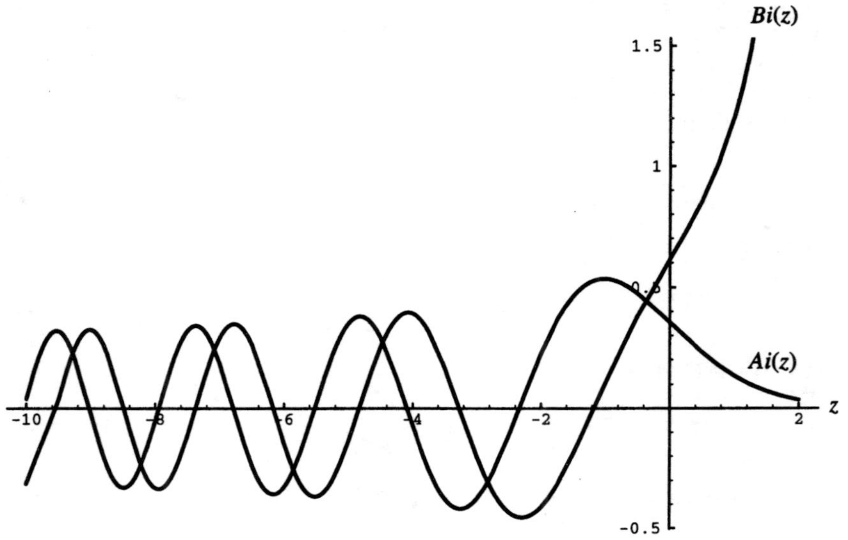

FIGURE 3.3.7. Airy Functions Ai and B_i

where $Ai'(-\Omega_j) = 0$, $\Omega_1 = 1.02\ldots$, $\Omega_2 = 3.25\ldots$. The higher values are easily obtained from the asymptotic formula. The most important mode is the lowest, Ω_1, for which there is only decay and no oscillations, and this is also the one with the largest excitation if the disturbance itself is not oscillating in r^*. These modes are orthogonal in $-\infty < r^* \leq 0$.

The general modes of this finite system are discrete so that the possible wave numbers (and frequencies) are also restricted. In the planar case, this is a condition of periodicity. In this case, we can note that another set of modes could run in the $(-\theta)$ direction. This other set has the form

$$\Phi = e^{i(-ka\theta-\omega t)}\phi(r^*) + \cdots, \tag{3.3.341}$$

where ϕ is the Airy function. But, by symmetry, these modes propagating in $(-\theta)$ must be the same as our original modes emanating from $\theta = \pi$, that is,

$$e^{i(ka(\pi-\theta)-\omega t)} = e^{-i(ka\theta+\omega t)}. \tag{3.3.342}$$

This says that the wave numbers must be restricted so that $ka = 2n$, where n is a large integer or $n = ka/2$. The wavelengths $\lambda = \pi a/n$ are much less than the radius.

An approach to the theory of the whispering gallery along the lines presented here appears in [3.10]. Hamet also considers a mode concentrated on a great circle. Other references, such as [3.26], which summarizes earlier work, connect the whispering gallery effect with the treatment of caustics where ray theory breaks down, but details are not given for this case.

Rayleigh remarks that the same type of effect can be expected to occur for earthquake waves traveling on the surface of Earth. He also notes that the exact shape of the convex surface should not matter; the general phenomenon should still exist.

3.3.8 Boundary Layers in Highly Anisotropic Elastic Materials

Composite materials with very strong fibers embedded in a matrix can be studied in the continuum linear elasticity theory of transversely isotropic materials. The occurrence of boundary layers and "singular" fibers is then manifest. The work here follows the ideas in [3.7] and [3.33] by A.C. Pipkin et al. See the references in these papers for further bibliography.

Only one example is considered here. A cantilever beam is constructed with nearly inextensible fibers along its length. It is loaded by a shearing stress $Ts(Y/H)$ at its free end. See Figure 3.3.8.

The near inextensibility of the fibers is approximated by a modulus of elasticity for straining in the x direction, E, which is much larger than the shear modulus, G.

The boundary conditions for this problem are shown in Figure 3.3.8; we have an applied shear stress T_{XY} and zero normal stress T_{XX} at the end $x = L$, zero

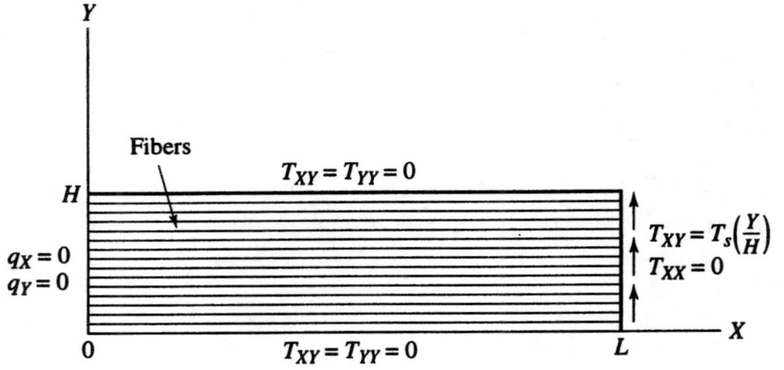

FIGURE 3.3.8. Cantilever Beam in Physical Coordinates

shear and normal stresses on the upper and lower surfaces and built-in conditions at $X = 0$, zero deflections $q_X = q_Y = 0$.

The basic equations are taken to be those of (generalized) plane stress. The transversely isotropic stress-strain relations are

$$E\epsilon_{XX} = T_{XX} - \nu T_{YY}, \tag{3.3.343a}$$

$$E'\epsilon_{YY} = T_{YY} - \nu' T_{XX}, \tag{3.3.343b}$$

$$2G\epsilon_{XY} = T_{XY}, \tag{3.3.343c}$$

where the strains $\epsilon_{XX}, \epsilon_{XY}, \epsilon_{YY}$ are expressed in terms of displacements (q_X, q_Y) in the X, Y directions in the usual way:

$$\epsilon_{XX} = \frac{\partial q_X}{\partial X}, \qquad \epsilon_{XY} = \frac{1}{2}\left(\frac{\partial q_X}{\partial Y} + \frac{\partial q_Y}{\partial X}\right), \qquad \epsilon_{YY} = \frac{\partial q_Y}{\partial Y}. \tag{3.3.344}$$

The modulus E for tension in x is such that $E \gg G$. In (3.3.343b), E' is the tension modulus for the cross-fiber direction; at first, it is of order G but later it can be large also. Also, as a consequence of transverse isotropy, the Poisson ratios ν, ν' are related to the tensile moduli by

$$\frac{\nu}{E} = \frac{\nu'}{E'}. \tag{3.3.345}$$

(For example, see the discussion in pages 107–108 of [3.25]).

The basic equilibrium equations stating that the divergence of the stress tensor is zero are

$$\frac{\partial T_{XX}}{\partial X} + \frac{\partial T_{XY}}{\partial Y} = 0, \tag{3.3.346a}$$

$$\frac{\partial T_{XY}}{\partial X} + \frac{\partial T_{YY}}{\partial Y} = 0. \tag{3.3.346b}$$

A perturbation procedure will be based on the small parameter $\epsilon \downarrow 0$, where

$$\epsilon^2 = \frac{G}{E}, \tag{3.3.347}$$

but first the problem will be put into a suitable dimensionless form. T is the order of magnitude of the applied stress ($s = O(1)$) so that all stresses are scaled with T, and L, H are used to scale X, Y, respectively. Thus,

$$T_{XX}(X, Y) = T\sigma_{xx}(x, y), \quad T_{YY}(X, Y) = T\sigma_{yy}(x, y)$$

$$T_{XY}(x, y) = T\sigma_{xy}(x, y), \tag{3.3.348}$$

where $x = X/L$ and $y = Y/H$. The corresponding displacements are written

$$q_X(X, Y) = \frac{TL}{G} u(x, y), \quad q_Y(X, Y) = \frac{TH}{G} v(x, y). \tag{3.3.349}$$

Then the equilibrium equations are

$$\delta \frac{\partial \sigma_{xx}}{\partial x} + \frac{\partial \sigma_{xy}}{\partial y} = 0, \quad \delta \frac{\partial \sigma_{xy}}{\partial x} + \frac{\partial \sigma_{yy}}{\partial y} = 0. \tag{3.3.350}$$

Here, $\delta = H/L$ is the fineness ratio of the beam.

From (3.3.343) and (3.3.344), the stresses can be expressed in terms of displacement gradients

$$\sigma_{xx} = \frac{1}{\gamma^2 - \nu^2\epsilon^2}\left\{\frac{\gamma^2}{\epsilon^2}\frac{\partial u}{\partial x} + \nu\frac{\partial v}{\partial y}\right\}, \tag{3.3.351a}$$

$$\sigma_{yy} = \frac{1}{\gamma^2 - \nu^2\epsilon^2}\left\{\frac{\partial v}{\partial y} + \nu\frac{\partial u}{\partial x}\right\}, \tag{3.3.351b}$$

$$\sigma_{xy} = \frac{1}{\delta}\frac{\partial u}{\partial y} + \delta\frac{\partial v}{\partial x}, \tag{3.3.351c}$$

where the parameter γ is defined by

$$\gamma^2 = \frac{G}{E'}.$$

Replacing the stresses in the equilibrium equations by displacements from (3.3.351), we have a pair of second-order p.d.e.'s for the displacement field

$$\gamma^2\delta^2\frac{\partial^2 u}{\partial x^2} + \epsilon^2\left(\gamma^2 - \nu^2\epsilon^2\right)\frac{\partial^2 u}{\partial y^2} + \epsilon^2\delta^2\left\{\nu + \gamma^2 - \nu^2\epsilon^2\right\}\frac{\partial^2 v}{\partial x \partial y} = 0, \tag{3.3.352a}$$

$$\delta^2\left(\gamma^2 - \nu^2\epsilon^2\right)\frac{\partial^2 v}{\partial x^2} + \frac{\partial^2 v}{\partial y^2} + (\nu + \gamma^2 - \nu^2\epsilon^2)\frac{\partial^2 u}{\partial x \partial y} = 0. \quad (3.3.352b)$$

The beam is now on the unit square and the boundary conditions can be expressed, also in terms of displacements. See Figure 3.3.9.

The limit of inextensibility has $\epsilon \to 0$. In this case, an "outer" limit $\epsilon \to 0$, $(x, y$ fixed), leads to the following expansion for the displacement field:

$$u(x, y; \epsilon) = u_0(x, y) + \epsilon^2 u_1(x, y) + \cdots, \quad (3.3.353a)$$

$$v(x, y; \epsilon) = v_0(x, y) + \epsilon^2 v_1(x, y) + \cdots. \quad (3.3.353b)$$

It follows from (3.3.352) that

$$\frac{\partial^2 u_0}{\partial x^2} = 0, \quad (3.3.354a)$$

$$\delta^2\gamma^2\frac{\partial^2 v_0}{\partial x^2} + \frac{\partial^2 v_0}{\partial y^2} + (\nu + \gamma^2)\frac{\partial^2 u_0}{\partial x \partial y} = 0. \quad (3.3.354b)$$

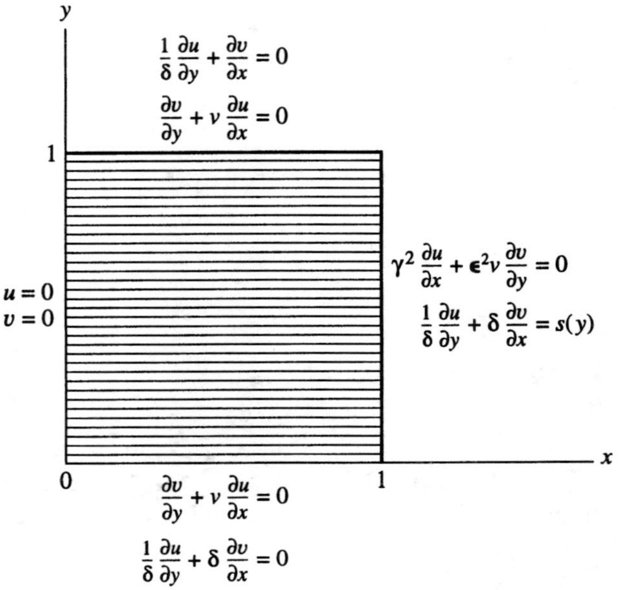

FIGURE 3.3.9. Beam and Boundary Conditions in Scaled Variables

The boundary conditions at $x = 0$ and $x = 1$ give

$$u_0(0, y) = 0, \qquad \frac{\partial u_0}{\partial x}(1, y) = 0,$$

so that

$$u_0(x, y) = 0. \tag{3.3.355}$$

The inextensibility of the fibers in the limit prevents any stretching in the x direction. However y displacements and stresses can be calculated from the vertical force balance (3.3.354b), which now reads

$$\delta^2 y^2 \frac{\partial^2 v_0}{\partial x^2} + \frac{\partial^2 v_0}{\partial y^2} = 0. \tag{3.3.356}$$

The complete set of physical boundary conditions for v_0 is

$$v_0(0, y) = 0, \qquad \text{built-in end} \tag{3.3.357a}$$

$$\left.\begin{aligned} &\frac{\partial v_0}{\partial y}(x, 1) = \frac{\partial v_0}{\partial y}(x, 0) = 0 \qquad (3.3.357b) \\ &\frac{\partial v_0}{\partial x}(x, 1) = \frac{\partial v_0}{\partial x}(x, 0) = 0 \qquad (3.3.357c) \end{aligned}\right\} \text{ zero stresses at } y = 0, 1$$

$$\delta \frac{\partial v_0}{\partial x}(1, y) = s(y), \qquad \text{applied stress.} \tag{3.3.357d}$$

Equation (3.3.357c) implies $v_0(x, 1) = v_0(x, 0) = 0$, but it is necessary to give up this boundary condition in order to find a suitable solution to the Laplace equation for v_0. In order to match to a layer, it is necessary that $\partial v_0/\partial y$ be given correctly. This implies the existence of boundary layers near $y = 0$ and $y = 1$.

The special case $s(y) = s_0 =$ constant is the only case that will be discussed in detail here. A representation of the solution can be given in terms of y eigenfunctions satisfying (3.3.357b). The basic solutions are

$$\begin{aligned} &v_0 = x, && n = 0 \\ &v_0 = \sinh \kappa_n x \cos n\pi y, && n = 1, 2, \cdots \qquad \kappa_n = \frac{n\pi}{\gamma\delta}. \end{aligned} \tag{3.3.358}$$

These solutions also have $v_0 = 0$ at $x = 0$.

Thus, in general

$$v_0(x, y) = a_0 x + \sum_{n=1}^{\infty} a_n \sinh \kappa_n x \cos n\pi y, \tag{3.3.359}$$

but for this special case the $a_n = 0$ for $n = 1, 2, \ldots$.

The shearing stress in the interior (away from $y = 0, 1$) is then given by

$$\sigma_{xy} = \delta \frac{\partial v_0}{\partial x} = s_0. \tag{3.3.360}$$

Next, we consider the stress and displacement boundary layer near $y = 0,\ 0 < x < 1$, which arises because of the necessary neglect of the shear stress boundary condition.

A suitable boundary-layer coordinate is

$$y^* = \frac{y}{\epsilon}, \tag{3.3.361}$$

and the expansion is associated with the limit $\epsilon \downarrow 0,\ (x, y^*)$ fixed. The proposed boundary-layer expansion for the displacements is

$$u(x, y; \epsilon) = \epsilon u^*(x, y^*) + \cdots, \tag{3.3.362a}$$

$$v(x, y; \epsilon) = v^*(x, y^*) + \cdots. \tag{3.3.362b}$$

The assumed orders of magnitude are verified here.

In addition to boundary conditions at the edges of the region, asymptotic matching will be used. Thus, in simple form, as $y^* \to \infty$

$$u^*(x, y^*) \to 0 \tag{3.3.363a}$$

$$v^*(x, y^*) \to v_0(x, 0+) = \frac{s_0}{\delta} x. \tag{3.3.363b}$$

The basic equilibrium equations (3.3.352) now read

$$O(\epsilon): \qquad \gamma^2\delta^2 \frac{\partial^2 u^*}{\partial x^2} + \gamma^2 \frac{\partial^2 u^*}{\partial y^{*2}} + (\nu + \gamma^2)\frac{\partial^2 v^*}{\partial x \partial y^*} = 0, \tag{3.3.364a}$$

$$O(1/\epsilon^2): \qquad \frac{\partial^2 v^*}{\partial y^{*2}} = 0. \tag{3.3.364b}$$

Using the matching, we see that

$$v^*(x, y^*) = v^*(x) = \frac{s_0}{\delta} x. \tag{3.3.365}$$

Thus, the x displacement u^* satisfies the (modified) Laplace equation in the boundary layer

$$\delta^2 \frac{\partial^2 u^*}{\partial x^2} + \frac{\partial^2 u^*}{\partial y^{*2}} = 0. \tag{3.3.366}$$

The boundary and matching conditions for $u^*(x, y^*)$ appear in Figure 3.3.10.

The boundary condition on the shear stress at $x = 1$ has $(\partial u^*/\partial y^*) = 0$, and this implies $u^*(1, y^*) = 0$ from the matching. This is the boundary condition that must be given up to define a problem for the boundary-layer equation (3.3.366). Then, an end boundary layer (not discussed here) can be fitted in to bring $u^*(1, y^*)$ to zero.

An eigenfunction expansion satisfying $u^*(0, y^*) = 0,\ \frac{\partial u^*}{\partial x}(1, y^*) = 0$ for the solution to (3.3.366) is

$$u^*(x, y^*) = \sum_{m=0}^{\infty} b_m^* e^{-(m+\frac{1}{2})\pi\delta y^*} \sin\left(m + \frac{1}{2}\right)\pi x, \tag{3.3.367a}$$

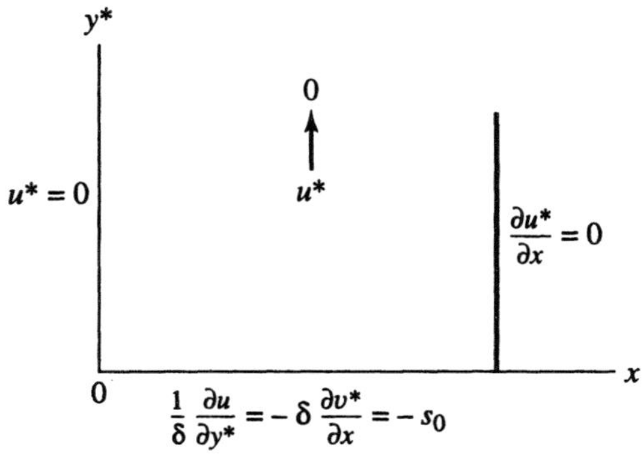

FIGURE 3.3.10. Boundary Conditions for $u^*(x, y^*)$

$$\frac{\partial u^*}{\partial y^*}(x, y^*) = -\delta \sum_{m=0}^{\infty} b_m^* \left(m + \frac{1}{2}\right) \pi e^{-(m+\frac{1}{2})\pi\delta y^*} \sin\left(m + \frac{1}{2}\right)\pi x. \tag{3.3.367b}$$

The boundary condition at $y^* = 0$ shows

$$\sum_{m=0}^{\infty} b_m^* \left(m + \frac{1}{2}\right) \pi \sin\left(m + \frac{1}{2}\right)\pi x = s_0, \qquad 0 < x < 1, \tag{3.3.368}$$

so that

$$b_m^* = \frac{2s_0}{\left(m + \frac{1}{2}\right)^2 \pi^2}. \tag{3.3.369}$$

The shearing stress in the thin boundary layer is then

$$\begin{aligned} \sigma_{xy} &= \frac{1}{\delta}\frac{\partial u^*}{\partial y^*} + \delta \frac{\partial v^*}{\partial x} \\ &= -2s_0 \sum_{m=0}^{\infty} \frac{1}{\left(m + \frac{1}{2}\right)\pi} e^{-\left(m+\frac{1}{2}\right)\pi\delta y^*} \sin\left(m + \frac{1}{2}\right)\pi x + s_0. \end{aligned} \tag{3.3.370}$$

This shows how the shear stress relaxes from its value s_0 in the interior to zero at $y^* = 0$. For the longitudinal stress, we have

$$\sigma_{xx} = \frac{1}{\gamma^2}\left\{\frac{\gamma^2}{\epsilon^2}\epsilon\frac{\partial u^*}{\partial x} + \frac{\nu}{\epsilon}\frac{\partial v^*}{\partial y^*}\right\} = \frac{1}{\epsilon}\frac{\partial u^*}{\partial x}.$$

The representation (3.3.367a) shows

$$\sigma_{xx} = \frac{2s_0}{\epsilon} \sum_{m=0}^{\infty} \frac{1}{\left(m+\frac{1}{2}\right)\pi} e^{-\left(m+\frac{1}{2}\right)\pi\delta y^*} \cos\left(m+\frac{1}{2}\right)\pi x + s_0. \quad (3.3.371)$$

Then the small $O(\epsilon)$ extension produces a very large $O\left(\epsilon^{-1}\right)$ tensile stress in the thin layer next to $y^* = 0$. The resulting (scaled) tension force F_x is finite:

$$F_x = \int_0^{y_l} \sigma_{xx} dy = \frac{2s_0}{\epsilon} \sum_{m=0}^{\infty} \frac{\cos\left(m+\frac{1}{2}\right)\pi x}{\left(m+\frac{1}{2}\right)\pi} \int_0^{y_l} e^{-\left(m+\frac{1}{2}\right)\pi\delta\frac{y}{\epsilon}} dy,$$

where y_l is a value that is $O(1)$. Thus,

$$= \frac{2s_0}{\delta} \sum_{m=0}^{\infty} \frac{\cos\left(m+\frac{1}{2}\right)\pi x}{\left(m+\frac{1}{2}\right)^2 \pi^2} + \text{T.S.T.}, \quad (3.3.372)$$

which is just

$$= \frac{s_0}{\delta}(1-x). \quad (3.3.373)$$

The fibers at $y = 0, 1$ are thus singular fibers from the point of view of the outer expansion. These carry a delta function $\delta(y^*)$ and $\delta(y^{**})$, $y^{**} = \frac{y-1}{\epsilon}$ stress and finite force, from the point of view of the outer (inextensible) expansion. This distribution of tensile stress is at sharp variance with the distribution of the classical isotropic beam theory.

If (3.3.373) is written in physical units, the tension force (per width) is

$$F_X = \frac{TL}{H}\left(1 - \frac{X}{L}\right) = \frac{T}{H}(L - X). \quad (3.3.374)$$

The couple produced by singular forces $\pm F_X$ at $T = H$ just balances that produced by the uniform shear T at (X, L).

If the shear applied at $x = 1$ is not uniform, $T_{XY} = Ts(y)$, then the a_n, $n = 1 \cdots$ in (3.3.359) are not zero. There is another end layer in the outer expansion that decays like

$$\exp[-(1-x)\pi/\gamma\delta].$$

When $\gamma\delta < 1$, this effectively dies out away from the end $x = 1$, and the solution is as already described. This layer can be incorporated into the solution and the situation near the corner becomes more complicated; a full discussion is not given here but appears in the references.

Another view of the limits can be seen from the equation for the Airy stress function Ω. The equilibrium equations (3.3.350) are satisfied identically by using $\Omega(x, y)$ such that

$$\sigma_{xx} = \frac{\partial^2\Omega}{\partial y^2}, \qquad \sigma_{yy} = \delta^2 \frac{\partial^2\Omega}{\partial x^2}, \qquad \sigma_{xy} = -\delta \frac{\partial^2\Omega}{\partial x \partial y}. \quad (3.3.375)$$

Since the strains are

$$\epsilon_{xx} = \frac{T}{G}\frac{\partial u}{\partial x}, \qquad \epsilon_{yy} = \frac{T}{G}\frac{\partial v}{\partial y}, \qquad 2\epsilon_{xy} = \frac{T}{G}\left\{\frac{1}{\delta}\frac{\partial u}{\partial x} + \delta\frac{\partial v}{\partial y}\right\}, \tag{3.3.376}$$

the strain compatibility equation is

$$\delta^2\frac{\partial^2\epsilon_{yy}}{\partial x^2} + \frac{\partial^2\epsilon_{xx}}{\partial y^2} = 2\delta\frac{\partial^2\epsilon_{xy}}{\partial x\partial y}.$$

Expressing strains in terms of stresses by the scaled versions of (3.3.343) results in

$$\frac{\delta^2}{E'}\frac{\partial^2}{\partial x^2}\left(\sigma_{yy} - \nu'\sigma_{xx}\right) + \frac{1}{E}\frac{\partial^2}{\partial y^2}\left(\sigma_{xx} - \nu\sigma_{yy}\right) - \frac{\delta}{G}\frac{\partial^2\sigma_{xy}}{\partial y^2} = 0.$$

Then, in terms of the stress function $\Omega(x, y)$,

$$\gamma^2\delta^4\frac{\partial^4\Omega}{\partial x^4} + (\delta^2 - \epsilon^2(\nu\delta^2 + \nu))\frac{\partial^4\Omega}{\partial x^2\partial y^2} + \epsilon^2\frac{\partial^4\Omega}{\partial x^4} = 0. \tag{3.3.377}$$

This equation is of biharmonic type. Under the outer limit $\epsilon \downarrow 0$, (x, y) fixed, the first term of an expansion of the stress function would satisfy

$$\frac{\partial^2}{\partial x^2}\left(\gamma^2\delta^2\frac{\partial^2\Omega_0}{\partial x^2} + \frac{\partial^2\Omega_0}{\partial y^2}\right) = 0. \tag{3.3.378}$$

The limit equation has real characteristics along $y = \text{const.}$, which is where the singularities appear in the outer solution. If the further limit $\gamma\delta \to 0$ is taken, the first term $\Omega_{0,0}$ of an expansion of Ω_0 would satisfy

$$\frac{\partial^4\Omega_{0,0}}{\partial x^2\partial y^2} = 0, \tag{3.3.379}$$

so that the lines $x = \text{constant}$ are also characteristics. The exponentially decaying end layer collapses in this limit to another singular line.

3.3.9 Limit Process Expansions and Homogenization

In the simplest problems of homogenization, a theory is made to estimate an effective transport coefficient of an inhomogeneous material for use in predicting the behavior of such a material in bulk. Such problems have been considered in various contexts for a long time, at least 150 years. The earliest results seem to have been those of Mossotti and Clausius; their formula will be given later. Many papers have been written on this subject. An excellent review article is that of Landauer [3.22], which summarizes many references; only a few will be given here.

The inhomogeneities are considered compact. An asymptotic theory is constructed based on the volume fraction of inclusion, f_I, being small. An inner expansion is used near each inclusion, based on the length scale of the inclusion, and an outer expansion is set up based on the length scale of the inhomogeneities. Limit processes are associated with each asymptotic expansion, and the expansions can be matched asymptotically to define boundary-value problems and solutions uniquely.

The method is applied here to a single example, a problem first considered by Rayleigh [3.38]. In one context, the problem is to find the effective thermal conductivity of a cubical array of spheres (conductivity k_I) in a matrix of conductivity k_M. Rayleigh used some ideas very close to matching, but not asymptotics. His result for the effective conductivity k_{eff} is of the form

$$\frac{k_{eff}}{k_M} = \text{terms in } f_I,\ f_I^2,\ f_I^3,\ f_I^4,\ f_I^{13/3}\ \cdots. \tag{3.3.380}$$

One aim of this paper is to give some explanation of the occurrence of $f_I^{13/3}$ in the sequence above.

Many papers have also been written about Rayleigh's problem and its simple extensions. The definitive paper based on a numerical method is that of Acrivos and Sangani, A.S. [3.1]. Earlier references appear there.

Basic solutions

Some basic singular solutions of the heat equation in infinite space, which are useful later, are tabulated here. The heat flux (cal/cm^2 sec) is given by

$$\mathbf{q} = -k \nabla T, \qquad k = \text{thermal conductivity}, \qquad T = \text{temperature}.$$

(i) Heat source at the origin

$$-\nabla \cdot \left(\frac{\mathbf{q}}{k}\right) = \nabla^2 T_s = -\delta(x)\delta(y)\delta(z) \tag{3.3.381}$$

$$T_s = \frac{1}{4\pi R}, \qquad R = \sqrt{x^2 + y^2 + z^2}. \tag{3.3.382}$$

(ii) Heat dipole

$$\nabla^2 T_d = -\delta'(x)\delta(y)\delta(z) \tag{3.3.383}$$

$$T_d = \frac{\partial T_s}{\partial x} = \frac{-x}{4\pi(x^2+y^2+z^2)^{3/2}} = -\frac{\cos\theta}{4\pi R^2}, \qquad \theta = \text{pole angle}.$$

See Figure 3.3.11.

(iii) Heat quadrupole

$$\nabla^2 T_q = -\delta''(x)\delta(y)\delta(z) \tag{3.3.384}$$

$$T_q = \frac{\partial T_d}{\partial x} = \frac{1}{4\pi}\left(\frac{2x^2 - (y^2+z^2)}{(x^2+y^2+z^2)^{5/2}}\right) = \frac{1}{4\pi}\,\frac{3\cos^2\theta - 1}{R^3}$$

$$T_q = \frac{1}{4\pi}\left(\frac{2P_2(\cos\theta)}{R^3}\right), \qquad P_2 = \text{Legendre polynomial}. \tag{3.3.385}$$

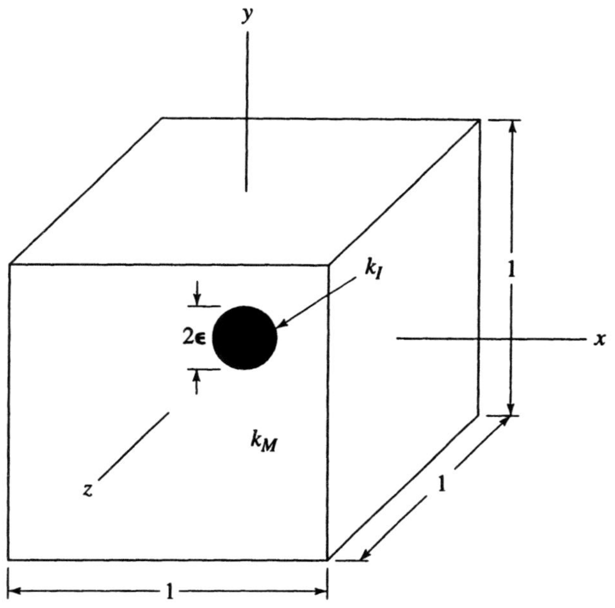

FIGURE 3.3.11. Unit Cube

(iv) General axisymmetric n-pole

$$T_{2^n} = \frac{(-)^n}{4\pi} \frac{n!P_n(\cos\theta)}{R^{n+1}}. \tag{3.3.386}$$

(v) Zonal harmonic with fourfold symmetry in ϕ, the azimuth angle; antisymmetric in x

$$T_z = \frac{1}{4\pi} \frac{5!P_5(\cos\theta)}{R^6} + \ell_5 \frac{\cos 4\phi P_5^4(\cos\theta)}{R^6}, \qquad \ell_5 = \text{const.} \tag{3.3.387}$$

Spherical polar coordinates are used:

$$x = R\cos\theta, \quad y = R\sin\theta\cos\phi, \quad z = R\sin\theta\sin\phi.$$

Canonical Problem I

Consider a sphere of radius a at the origin in an infinite medium; the sphere has conductivity k_I in a medium with conductivity k_M in a uniform temperature gradient at infinity

$$T \to -Gx \qquad \text{as } R \to \infty. \tag{3.3.388}$$

The boundary condition at the interface between the matrix and inclusion is continuity of temperature and heat flux

$$T(a^+, \theta) = T(a^-, \theta), \qquad k_M \frac{\partial T}{\partial R}(a^+, \theta) = k_I \frac{\partial T}{\partial R}(a^-, \theta). \tag{3.3.389}$$

The (well-known) solution outside the sphere looks like a dipole at the origin (or $P_1(\cos\theta)$). The solution is

$$T = \begin{cases} -Gx + G\left(\dfrac{\lambda - 1}{\lambda + 2}\right)\dfrac{a^3}{R^2}\cos\theta, & R > a \\ -\dfrac{3G}{\lambda + 2}x, & R < a. \end{cases} \tag{3.3.390}$$

Array of spheres: asymptotic theory

Consider an array of spheres of (dimensionless) radius ϵ, conductivity k_I, located on the integer points in an infinite matrix. There is an impressed temperature gradient

$$T = -Gx, \tag{3.3.391}$$

which is kept fixed as ϵ varies. The approximation of small volume fraction is connected with the limit $\epsilon \to 0$. We wish to find the effective conductivity for this array. The unit box was shown in Figure 3.3.11.

It is natural to describe the temperature field with the help of two asymptotic expansions of limit process type. In the first, or outer, expansion the observer is at a fixed (x, y, z) point in the unit box and the sphere shrinks to zero. In this case, the first term is (3.3.391). In the second, or inner, expansion, the observer stays close to the sphere as $\epsilon \to 0$; coordinates

$$x^* = \frac{x}{\epsilon},\ y^* = \frac{y}{\epsilon},\ z^* = \frac{z}{\epsilon}$$

are fixed in the limit. In the first instance, the sphere behaves as if it were in an infinite medium. These asymptotic expansions match and enable an effective conductivity to be found. These expansions have, at first, the form

$$T(x, y, z; \epsilon) = -Gx + \alpha_1(\epsilon)T_1(x, y, z) + \alpha_2(\epsilon)T_2(x, y, z) + \cdots \quad \text{outer,} \tag{3.3.392a}$$

$$T(x, y, z; \epsilon) = \beta_1(\epsilon)T_1^*(x^*, y^*, z^*) + \beta_2(\epsilon)T_2^*(x^*, y^*, z^*) + \cdots \quad \text{inner.} \tag{3.3.392b}$$

Since the scaling is uniform, all T_j, T_j^* satisfy Laplace's equation. The gauge functions $\alpha_j(\epsilon)$, $\beta_j(\epsilon)$ are to be found. The first matching goes from outer to inner expansion and can be carried out by writing the outer expansion (first term) in inner coordinates. The idea is that the behavior of the outer expansion as $(x, y, z) \to 0$ must match that of the inner expansion as $(x^*, y^*, z^*) \to \infty$. We have

$$-Gx = -G\epsilon x^* \Leftrightarrow \beta_1(\epsilon)T_1^*(x^*, y^*, z^*) \qquad \text{as } R^* \to \infty. \tag{3.3.393}$$

The symbol ($\Leftrightarrow$) denotes matching. Thus,

$$\beta_1(\epsilon) = \epsilon, \qquad T_1^* \to -Gx^* \text{ as } R^* \to \infty. \tag{3.3.394}$$

Equation (3.3.394) provides the boundary condition at infinity, and the problem to be solved is once again Canonical Problem I. The solution is

$$T_1^*(R^*, \theta) = \begin{cases} -Gx^* + G\frac{\lambda-1}{\lambda+2}\frac{\cos\theta}{R^{*2}}, & R^* > 1 \\ -G\frac{3}{\lambda+2}x^*, & R^* < 1. \end{cases} \tag{3.3.395}$$

The next match goes from inner to outer: T_1^* to T_1. Since the uniform gradient is already matched, we have

$$\epsilon(T_1^* + Gx^*) = \epsilon^3 G\frac{\lambda-1}{\lambda+2}\frac{\cos\theta}{R^2} \Leftrightarrow \alpha_1(\epsilon)T_1(R,\theta) \text{ as } R \to 0. \tag{3.3.396}$$

Thus,

$$\alpha_1(\epsilon) = \epsilon^3, \qquad T_1 \to G\frac{\lambda-1}{\lambda+2}\frac{\cos\theta}{R^2} \qquad \text{as } R \to 0. \tag{3.3.397}$$

T_1 is the temperature distribution in the unit box due to a negative heat dipole at the origin. The boundary conditions can be considered to be periodic. Thus, T_1 satisfies

$$\nabla^2 T_1 = \frac{\partial^2 T_1}{\partial x^2} + \frac{\partial^2 T_1}{\partial y^2} + \frac{\partial^2 T_1}{\partial z^2} = 4\pi\kappa G\delta'(x)\delta(y)\delta(z), \tag{3.3.398}$$

where

$$\kappa = \frac{\lambda-1}{\lambda+2}.$$

Rayleigh showed how to obtain the heat flux from Green's theorem,

$$\iiint (U\,\nabla^2 V - V\,\nabla^2 U)dx\,dy\,dz = \iint \left(U\frac{\partial V}{\partial n} - V\frac{\partial U}{\partial n}\right)dA,$$

applied to the unit box. Let

$$U = x = \cos\theta, \qquad V = T_1(x,y,z).$$

Then

$$4\pi\kappa G\iiint x\delta'(x)\delta(y)\delta(z)dx\,dy\,dz = \iint x\frac{\partial T_1}{\partial x}\left(-\frac{1}{2}, y, z\right)dy\,dz$$

or

$$\iint \frac{\partial T_1}{\partial x}\left(-\frac{1}{2}, y, z\right)dy\,dz = -4\pi\kappa G = -\frac{Q_1}{k_M},$$

where

$$Q_1 = 4\pi\kappa G k_M$$

is the scaled flux. Defining

$$k_{eff} = \frac{\text{heat flux}}{\text{temperature gradient}},$$

we have the Clausius-Mossotti result

$$\frac{k_{eff}}{k_M} = 1 + \epsilon^3(4\pi\kappa) = 1 + 3\kappa f_I + \cdots. \tag{3.3.399}$$

In order to carry the approximation further, we need a representation for $T_1(x, y, z)$. Because of periodicity, a Fourier representation in (y, z) is useful. Let

$$T_1(x, y, z) = \sum\sum_{j,k=-\infty}^{\infty} \tilde{T}_{jk}(x) e^{2\pi i(jy+kz)}. \tag{3.3.400}$$

Thus, from (3.3.398),

$$\frac{d^2\tilde{T}_{jk}}{dx^2} - \Lambda^2 \tilde{T}_{jk} = 4\pi\kappa G\delta'(x), \quad \Lambda^2 = (2\pi)^2(j^2 + k^2).$$

We have $[\tilde{T}_{jk}]_{x=0} = 4\pi\kappa G$, $[\frac{d\tilde{T}_{jk}}{dx}]_{x=0} = 0$, where [ ] is the jump. Using these conditions and periodicity, the desired representation is

$$T_1(x, y, z) = -2\pi\kappa G \sum\sum_{j,k=-\infty}^{\infty} e^{2\pi i(jy+kz)} \frac{1}{\sinh\frac{\Lambda}{2}} \begin{Bmatrix} \sinh\Lambda\left(x + \frac{1}{2}\right) & x < 0 \\ \sinh\Lambda\left(x - \frac{1}{2}\right) & x > 0 \end{Bmatrix}, \tag{3.3.401}$$

$$\frac{\partial T_1}{\partial x}(x, y, z) = -2\pi\kappa G \sum\sum_{j,k=-\infty}^{\infty} e^{2\pi i(jy+kz)} \frac{\Lambda}{\sinh\frac{\Lambda}{2}} \begin{Bmatrix} \cosh\Lambda\left(x + \frac{1}{2}\right) & x < 0 \\ \cosh\Lambda\left(x - \frac{1}{2}\right) & x > 0 \end{Bmatrix}. \tag{3.3.402}$$

The heat flux comes from, e.g.,

$$\iint_{-\frac{1}{2}}^{\frac{1}{2}} \frac{\partial T_1}{\partial x}(x, y, z)\,dy\,dz \equiv \left\langle \frac{\partial T_1}{\partial x} \right\rangle = \text{Fourier mean } (\Lambda \to 0) = -4\pi\kappa G, \tag{3.3.403}$$

as before.

The next matching proceeds from outer T_1 to inner T_2^*. The dipole singularity in (3.3.401) comes from the high-frequency terms; T_1 is odd in x and thus has an expansion near the origin

$$T_1(x, y, z) = \kappa G \frac{\cos\theta}{R^2} - G_1 x + O(x^3, xy^2, xz^2). \tag{3.3.404}$$

The dipole in the box produces a new gradient G_1 that must be matched:

$$\epsilon^3\left(T_1 - \kappa G \frac{\cos\theta}{R^2}\right) = \epsilon^3(-G_1 x + \cdots)$$

$$= -\epsilon^4 G_1 x^* \Leftrightarrow \beta_2(\epsilon) T_2^*(x^*, y^*, z^*), \quad R^* \to \infty$$

or

$$\beta_2(\epsilon) = \epsilon^4, \qquad T_2^* \to -G_1 x^*, \qquad R^* \to \infty. \tag{3.3.405}$$

The solution for T_2^* is again given by the Canonical Problem I, and to obtain T_2^* replace G by G_1 in T_1^*.

The next matching proceeds exactly as before from T_2^* to T_2 to give $\alpha_2 = \epsilon^6$ and a dipole singularity for T_2. It is clear that this procedure can be carried out indefinitely to produce the outer expansion and flux

$$T = -Gx + \sum_{n=1} \epsilon^{3n} T_n(x, y, z), \tag{3.3.406}$$

where

$$T_n = \kappa G_{n-1} \frac{\cos\theta}{R^2} - G_n x + \cdots, \qquad R \to 0, \qquad G_0 = G. \tag{3.3.407}$$

$$\frac{k_{eff}}{k_M} = 1 + 4\pi\kappa\epsilon^3 \sum_{n=0} \epsilon^{3n} \frac{G_n}{G_0}. \tag{3.3.408}$$

It remains to calculate the various G_n and to show how terms of different orders may appear in the expansion.

Calculation of G_1, G_2

Let $T_1^\dagger(x, y, z)$ denote $T_1(x, y, z)$ with the dipole singularity removed. Then

$$T_1^\dagger = T_1 - \kappa G \frac{\cos\theta}{R^2} = x \frac{\partial T_1^\dagger}{\partial x}(0, 0, 0) + O(x^3). \tag{3.3.409}$$

The singularity can be removed from the representation (3.3.401) by subtracting a representation of the temperature field of a dipole at the origin, T_{c_1}, which has periodicity in (y, z) but dies out in x. But then all the reflected dipoles in the plane $x = 0$ have to be added back in, except the one at $x = 0$, T_{c_2}. With reasoning similar to that used before, we find

$$T_{c_1} = 2\pi\kappa G \sum\sum_{j,k=-\infty}^{\infty} \operatorname{sgn}(x)\ e^{-\Lambda|x|} e^{2\pi i(jy+kz)}, \tag{3.3.410}$$

$$T_{c_2} = \kappa G \sum\sum_{j,k=-\infty}^{\infty}{}'' \frac{x}{(x^2 + (y-j)^2 + (z-k)^2)^{3/2}}, \tag{3.3.411}$$

where $\sum\sum''$ denotes not both $j, k = 0$. Thus,

$$T_1^\dagger(x, y, z) = -2\pi\kappa G \sum\sum_{j,k=-\infty}^{\infty} e^{2\pi i(jy+kz)} \begin{Bmatrix} \dfrac{\sinh \Lambda\left(x + \frac{1}{2}\right)}{\sinh \frac{\Lambda}{2}} - e^{\Lambda x} & x < 0 \\ \dfrac{\sinh \Lambda\left(x - \frac{1}{2}\right)}{\sinh \frac{\Lambda}{2}} + e^{-\Lambda x} & x > 0 \end{Bmatrix}$$

$$+\kappa G \sum\sum_{j,k=-\infty}^{\infty}{}' \frac{x}{(x^2+(y-j)^2+(z-k)^2)^{3/2}}, \tag{3.3.412}$$

$$\frac{\partial T_1^\dagger}{\partial x}(0,0,0) = -2\pi\kappa G \sum\sum_{j,k=-\infty}^{\infty} \Lambda\left(\coth\frac{\Lambda}{2}-1\right) + \kappa G \sum\sum_{j,k=-\infty}^{\infty}{}' \frac{1}{(j^2+k^2)^{3/2}}. \tag{3.3.413}$$

Then

$$G_1 = -\frac{\partial T_1^\dagger}{\partial x}(0,0,0)$$

or

$$\frac{G_1}{G} = \kappa(S_{11}-S_{12}), \tag{3.3.414}$$

where

$$S_{11} = 4\pi \sum\sum_{j,k=-\infty}^{\infty} \frac{2\pi\sqrt{j^2+k^2}\,e^{-2\pi\sqrt{j^2+k^2}}}{1-e^{-2\pi\sqrt{j^2+k^2}}}$$

$$S_{12} = 4\pi \sum\sum_{j,k=-\infty}^{\infty}{}' \frac{1}{(j^2+k^2)^{3/2}}.$$

S_{11} converges rapidly so that only about 10 terms are needed to achieve six-figure accuracy. S_{12} converges slowly, and about 10,000 terms are used plus an estimate of the remainder by an integral.

The numerical results are

$$\frac{S_{11}}{4\pi} = 1.0522060\cdots$$

$$\frac{S_{12}}{4\pi} = .7188727\cdots$$

so that

$$\frac{S_{11}-S_{12}}{4\pi} = .333333\cdots, \qquad \frac{G_1}{G} = 4\pi\kappa(.333333\cdots), \tag{3.3.415}$$

a suggestive result. In [3.38], Rayleigh represents $T_1^\dagger$ directly as a sum of dipoles located on all the integer points except $(0, 0, 0)$. The aim is to calculate the temperature gradient at the origin due to this array of dipoles. In our notation,

$$T_1^\dagger = \kappa G \sum\sum\sum_{i,j,k=-\infty}^{\infty}{}''' \frac{x-i}{((x-i)^2+(y-j)^2+(z-k)^2)^{3/2}}, \tag{3.3.416}$$

where $\sum\sum\sum'''$ means not all $(i, j, k) = 0$

$$\frac{\partial T_1^\dagger}{\partial x}(0,0,0) = \kappa G \sum\sum\sum_{j,k=-\infty}^{\infty}{}''' \frac{j^2+k^2-2i^2}{(i^2+j^2+k^2)^{5/2}}. \tag{3.3.417}$$

This sum is not absolutely convergent; its value depends on the way the sum is taken. Rayleigh pointed out that if the sum is taken over a cube $|i|, |j|, |k| \le N$ then the value of the sum is zero due to pairwise cancellation. However, Rayleigh states "the medium is infinite in all directions but is more infinite in the x direction." Thus, the sum is carried out first to infinity in the x direction and then (y, z) go to infinity.

Since the sum over the cube N vanishes, we have

$$\frac{\partial T_1^\dagger}{\partial x}(0,0,0) = \kappa G \sum_{j,k=-N}^{N}\sum \left(\sum_{i=N}^{\infty} + \sum_{i=-N}^{-\infty}\right)\left(\frac{j^2+k^2-2i^2}{(i^2+j^2+k^2)^{5/2}}\right), \quad N \to \infty. \tag{3.3.418}$$

Approximating the sum by an integral gives

$$\frac{\partial T_1^\dagger}{\partial x}(0,0,0) = 2\kappa G \iint_{-N}^{N} dy\, dz \int_{N}^{\infty} \frac{y^2+z^2-2x^2}{(x^2+y^2+z^2)^{5/2}}\, dx, \qquad N \to \infty. \tag{3.3.419}$$

This integral is independent of N; letting $x = NX$, $y = NY$, $z = NZ$ gives

$$\begin{aligned}\frac{\partial T_1^\dagger}{\partial x}(0,0,0) &= 2\kappa G \iint_{-1}^{+1} dY\, dZ \int_{1}^{\infty} (Y^2+Z^2-2X^2)(X^2+Y^2+Z^2)^{-5/2} dX,\\ &= -2\kappa G \iint_{-1}^{+1} \frac{dY\, dZ}{(1+Y^2+Z^2)^{3/2}} = -4\kappa G \int_{-1}^{+1} \frac{dY}{(1+Y^2)\sqrt{2+Y^2}}\\ &= -8\kappa G \tan^{-1}\frac{1}{\sqrt{3}} = -\kappa G \frac{4\pi}{3}.\end{aligned} \tag{3.3.420}$$

Thus, according to this calculation,

$$\frac{G_1}{G} = \frac{4\pi}{3}\kappa. \tag{3.3.421}$$

This result agrees with (3.3.415), makes it more precise, and shows that Rayleigh's summability is correct.

Therefore, from this first series of terms we have the expression for the effective conductivity

$$\frac{k_{eff}}{k_M} = 1 + 4\pi\kappa\epsilon^3 \sum_{n=0}^{\infty} \left(\frac{4\pi\kappa}{3}\right)^n + O(?). \tag{3.3.422}$$

Next, we consider how other terms can intervene in the series (3.3.422).

Modified expansion

Different terms in the expansion for k_{eff} come from matching the next-order terms in the expansion for $T_1^\dagger$ near the origin. These are cubic: x^3, $x^2(y^2+z^2)$, and can be expressed in terms of the regular spherical harmonic expansion

$$T_1^\dagger(x, y, z) = -G_1 x + \frac{\alpha}{3!} R^3 P_3(\cos\theta) + \cdots$$
$$= -G_1 x + \alpha \left(\frac{x^3}{6} - \frac{x}{4}(y^2 + z^2) \right),$$

where

$$\alpha = \frac{\partial^3 T_1^\dagger}{\partial x^3}(0, 0, 0).$$

The previous representation for $T_1^\dagger$, (3.3.412), can be used to find a formula for α:

$$\frac{\partial^3 T_1^\dagger}{\partial x^3}(0, 0, 0) = -2\pi\kappa G \sum\sum_{j,k=-\infty}^{\infty} \Lambda^3 \left(\coth\frac{\Lambda}{2} - 1 \right)$$

$$-9\kappa G \sum\sum_{j,k=-\infty}^{\infty}{}'' \frac{1}{(j^2 + k^2)^{5/2}}$$

$$\alpha = -\kappa G(S_{21} + S_{22}),$$

$$S_{21} = 4\pi \sum\sum_{-\infty}^{\infty} \Lambda^3 \frac{e^{-\Lambda}}{1 - e^{-\Lambda}} = 28.79, \tag{3.3.423}$$

$$S_{22} = 9 \sum\sum_{-\infty}^{\infty}{}'' \frac{1}{(j^2 + k^2)^{5/2}} = 45.82.$$

Let

$$\sigma_2 = \left(\frac{S_{21} + S_{22}}{6} \right) = 12.43. \tag{3.3.424}$$

Then

$$T_1^\dagger = -G_1 x - \kappa G \sigma_2 R^3 P_3(\cos\theta) + O(R^5) \quad \text{as} \quad R \to 0. \tag{3.3.425}$$

Using this, matching from the outer expansion to the inner can be carried out. A term T_{21}^* with appropriate order has to be introduced into (3.3.392b). The previous expansion contains terms $\epsilon T_1^* + \epsilon^4 T_2^* + \epsilon^7 T_3^* + \cdots$. We have

$$\epsilon^3(T_1^\dagger + G_1 x) = \epsilon^3(-\kappa G \sigma_2 R^3 P_3(\cos\theta)) = -\epsilon^6((\kappa G \sigma_2) R^{*^3} P_3(\cos\theta))$$

$$\Leftrightarrow \epsilon^6 T_{21}^*(R^*, \theta) \quad \text{as } R^* \to \infty. \tag{3.3.426}$$

The term T_{21}^* is of order intermediate to T_2^* and T_3^*. The boundary condition at infinity for heat conduction past the unit sphere is

$$T_{21}^* \to -\kappa G \sigma_2 R^{*^3} P_3(\cos\theta), \qquad R^* \to \infty. \tag{3.3.427}$$

This defines a new canonical problem for which the solution can be written

$$T_{21}^* = \begin{cases} \left(-\kappa G\sigma_2 R^{*^3} + \frac{A}{R^{*^4}}\right) P_3(\cos\theta), & R^* > 1 \\ B R^{*^3} P_3(\cos\theta), & R^* < 1. \end{cases}$$

A, B are found from the conditions of continuous temperature and flux on $R^* = 1$, as before

$$T_{21}^* = \begin{cases} \kappa G\sigma_2 \left(-R^{*^3} + \frac{\lambda-1}{\lambda+\frac{4}{3}} \frac{1}{R^{*^4}}\right) P_3(\cos\theta), & R^* > 1 \\ -\kappa G\sigma_2 \frac{\frac{7}{3}}{\lambda+\frac{4}{3}} R^{*^3} P_3(\cos\theta), & R^* < 1. \end{cases} \tag{3.3.428}$$

The next matching of this term from inner to outer introduces a term T_{31} of appropriate order in the outer expansion

$$\begin{aligned} &\epsilon^6 \left(T_{21}^* + \kappa G\sigma_2 R^{*^3} P_3(\cos\theta)\right) \\ &= \epsilon^6 \left(\kappa G\sigma_2 \frac{\lambda - 1}{\lambda + \frac{4}{3}} \frac{P_3(\cos\theta)}{R^{*^4}}\right) = \epsilon^{10} \left(\kappa G\sigma_2 \frac{\lambda - 1}{\lambda + \frac{4}{3}} \frac{P_3(\cos\theta)}{R^4}\right) \\ &\Leftrightarrow \epsilon^{10} T_{31}(R,\theta) \qquad \text{as } R \to 0. \end{aligned} \tag{3.3.429}$$

Since the outer expansion has terms $\epsilon^3 T_1 + \epsilon^6 T_2 + \epsilon^9 T_3 + \epsilon^{12} T_4$, this term is intermediate to T_3, T_4. The singularity of T_{31} at the origin is an octupole

$$T_{31}(R,\theta) \to G\left(\kappa\kappa_{4/3}\sigma_2\right) \frac{P_3(\cos\theta)}{R^4}, \tag{3.3.430}$$

where $\kappa_{4/3} = \frac{\lambda-1}{\lambda+\frac{4}{3}}$. The equation to be satisfied by T_{31} in the unit box with periodic boundary conditions is

$$\nabla^2 T_{31} = \frac{4\pi}{3!} \left(\kappa\kappa_{4/3} G\sigma_2\right) \delta'''(x)\delta(y)\delta(z). \tag{3.3.431}$$

T_{31} has no other singularity so that the next term in its expansion around the origin is linear. Also, since there is no dipole singularity, T_{31} does not contribute to the flux. It follows from (3.3.395) that

$$T_{31} = \frac{1}{6} \kappa_{4/3}\sigma_2 \frac{\partial^2 T_1}{\partial x^2},$$

and $T_{31}^\dagger$, which is T_{31} with the octupole singularity removed, is given by

$$T_{31}^\dagger = \frac{1}{6} \kappa_{4/3}\sigma_2 \frac{\partial^2 T_1^\dagger}{\partial x^2}. \tag{3.3.432}$$

Thus, as $(x, y, z) \to 0$

$$T_{31}^\dagger = x \frac{\partial T_{31}^\dagger}{\partial x}(0,0,0) + \cdots = \frac{1}{6} \kappa_{4/3}\sigma_2 \frac{\partial^3 T_1^\dagger}{\partial x^3}(0,0,0) = -G_{31} x. \tag{3.3.433}$$

The expression for $\frac{\partial^3 T_1^\dagger}{\partial x^3}(0,0,0)$ appears in (3.3.423) so that

$$G_{31} = \kappa\kappa_{4/3}\sigma_2^2 G. \tag{3.3.434}$$

Now this linear gradient term in T_{31} of the outer expansion must be matched to a term of the appropriate order in the inner expansion. We have

$$\epsilon^{10}T_{31}^\dagger = \epsilon^{10}(-G_{31}x + \cdots) = \epsilon^{11}(-G_{31}x^*) \Leftrightarrow \epsilon^{11}T_{41}^*(R^*, \theta) \text{ as } R^* \to \infty. \tag{3.3.435}$$

The order is between T_4^* and T_5^*. The solution of the inner problem for T_{41}^* is again given by Canonical Problem I (3.3.410),

$$T_{41}^* = -G_{31}x^* + \kappa G_{31}\frac{\cos\theta}{R^{*2}}, \qquad R^* > 1. \tag{3.3.436}$$

Matching this term to the outer expansion produces a temperature field T_{41} with a dipole singularity, which then contributes to the flux. We have

$$\begin{aligned}\epsilon^{11}(T_{41}^* + G_{31}x^*) &= \epsilon^{11}\kappa G_{31}\frac{\cos\theta}{R^{*2}}\\ &= \epsilon^{13}\kappa G_{31}\frac{\cos\theta}{R^2} \Leftrightarrow \epsilon^{13}T_{41}(R,\theta) \text{ as } R \to 0.\end{aligned} \tag{3.3.437}$$

The problem for T_{41} in the unit box is

$$\nabla^2 T_{41} = 4\pi\kappa G_{31}\delta'(x)\delta(y)\delta(z). \tag{3.3.438}$$

This adds a term to the heat flux $4\pi\kappa G_{31}$ so that (3.3.408) is modified to

$$\frac{k_{eff}}{k_M} = 1 + 4\pi\kappa\epsilon^3\sum_{n=0}^{4}\left(\frac{4\pi\epsilon^3}{3}\kappa\right)^n + 4\pi\kappa^2\kappa_{4/3}\sigma_2^2\epsilon^{13} + O(\epsilon^{16}), \tag{3.3.439}$$

where $\kappa = \frac{\lambda-1}{\lambda+2}$, $\kappa_{4/3} = \frac{\lambda-1}{\lambda+\frac{4}{3}}$, $\lambda = \frac{k_I}{k_M}$, $\sigma_2 = \text{constant} = 12.43$.

This shows how the term of volume fraction $f_I^{13/3}$ can appear in Rayleigh's formula. The terms with fourfold symmetry (3.3.387) do not yet appear in the expansion. In summary, the inner and outer expansions have the form

$$\text{Outer: } T = -Gx + \epsilon^3 T_1 + \epsilon^6 T_2 + \epsilon^9 T_3 + \epsilon^{10}T_{31} + \epsilon^{12}T_4 + \epsilon^{13}T_{41} + \epsilon^{15}T_5 + \cdots, \tag{3.3.440}$$

$$\text{Inner: } T = \epsilon T_1^* + \epsilon^4 T_2^* + \epsilon^6 T_{21}^* + \epsilon^7 T_3^* + \epsilon^{10}T_4^* + \epsilon^{11}T_{41}^* + \cdots. \tag{3.3.441}$$

Comments

The summability difficulty experienced by Rayleigh due essentially to the slow decay of the dipole field occurs in a similar way in other problems. A classical example is a calculation of the added mass due to the unsteady motion of a sphere in incompressible potential flow. The added mass can be calculated directly from the pressure distribution on the surface of the sphere or more simply by energy considerations (see [3.20]). However, if one attempts to calculate the added mass from momentum considerations, the volume integral representing the momentum of the flow is not absolutely convergent, analogous to Rayleigh's sums. This is noted in a footnote on page 33 in Landau and Lifschitz [3.21] and is cited by Peierls in chapter 7 of [3.28]. Another example of Peierls, in chapter 2, deals with momentum of phonons. The value of the integral depends on the manner in which the boundary surface goes to infinity. The problem of added mass is discussed in Theodorsen [3.40]. It is shown there that the correct momentum in the flow is calculated only for the volume inside a surface, which tends to infinity if the dimension in the direction of motion of a body is much longer than the transverse dimension. This is shown by considering a variety of limiting shapes for the bounding surface. O'Brien in [3.27] discusses another approach to justify Rayleigh's summability, based on ideas of Batchelor.

A more complete discussion of this problem appears in [3.4]. It is to be expected that the method outlined here can be extended to other more interesting cases with small volume fraction. In general, terms of such high order would not be studied. However, different physics, different shapes of inclusions, different arrays, and random distributions all offer problems that can be attacked by the method sketched out here. The core of large volume fraction, of particles close together or touching, demands a different approach based on local boundary layers.

Problems

1. Consider steady heat conduction in a cylindrical rod, $0 \le X \le L, 0 \le R \le a$, with the following boundary conditions of temperature prescribed on all surfaces:

$$\text{at } X = 0: \quad T(0, R) = T^* F(R/a), \tag{3.3.442a}$$

$$\text{at } X = L: \quad T(L, R) = T^* G(R/a), \tag{3.3.442b}$$

$$\text{at } R = a: \quad T(X, a) = T^* H(X/L). \tag{3.3.442c}$$

   a. Construct asymptotic expansions of the solution for $T/T^* = \theta(x, r; \epsilon)$, where $x = X/L, r = R/a$ are fixed, and $\epsilon = a/L \to 0$.
   b. Construct suitable boundary-layer solutions for the ends, and show how they match to the expansion valid away from the ends. Does the solution constructed here represent one-dimensional heat conduction?

2. Consider a plane sound wave of frequency ω, wavelength λ incident on a sphere of radius a. Construct matched inner and outer expansions for the case $a/\lambda \ll L$. Compute the first two terms in the inner expansion.

   The acoustic velocity potential satisfies the wave equation

$$\frac{\partial^2\phi}{\partial x^2} + \frac{\partial^2\phi}{\partial y^2} + \frac{\partial^2\phi}{\partial z^2} - \frac{1}{c^2}\frac{\partial^2\phi}{\partial t^2} = 0, \quad c = \text{sound speed}. \tag{3.3.443}$$

   The incoming plane wave is represented by (in complex notation)

$$\phi = A \exp\left[i\omega\left(t - \frac{x}{c}\right)\right]. \tag{3.3.444}$$

   The boundary condition at the surface of the rigid sphere is $\partial\phi/\partial r = 0$.

3. Consider the self-induced motion of a radially deforming slender body of revolution in an incompressible fluid at rest. Assume that the body always remains neutrally buoyant by requiring the total displaced volume to be a constant. Thus, the velocity is always horizontal. Assume also, as in Sec. 3.3.3, that the body sheds no vortices and that the flow is everywhere irrotational.

If we fix our coordinate system in the body, we have a problem similar to the one discussed in Sec. 3.3.3, except now the velocity at upstream infinity (in the coordinate system fixed to the moving body) is an unknown $q(t;\epsilon)$. It is easy to show that in this coordinate system the Bernoulli equation [see (3.3.123)] is

$$C_p = -\phi_t + x\frac{dq}{dt} + \frac{1}{2}[q^2 - \phi_x^2 - \phi_r^2], \tag{3.3.445}$$

where we have set $\delta = 1$ for simplicity.

Consider the axial equation of motion in dimensionless form

$$m\frac{dq}{dt} = C_D(t), \tag{3.3.446}$$

where m is a dimensionless mass

$$m = 2\int_0^1 F^2(x,t)dx \tag{3.3.447}$$

and $C_D(t)$ is defined by (3.3.124) with C_p as given earlier.

a. By paralleling the discussion in Sec. 3.3.3, show that the inner expansion for the velocity potential is of the form

$$\phi = q(t;\epsilon)x + \epsilon^2\log\epsilon A_1(x,t) + \epsilon^2[A_1(x,t)\log r^* + B_1(x,t)] + O(\epsilon^4\log\epsilon), \tag{3.3.448}$$

where

$$A_1 = \frac{1}{2}G_t(x,t) \tag{3.3.449a}$$

and

$$B_1 = -\frac{1}{4}\int_0^1 \frac{\partial^2 G}{\partial\xi\partial t}(\xi,t)\operatorname{sgn}(x-\xi)\log 2|x-\xi|d\xi. \tag{3.3.449b}$$

So far, the magnitude of q is assumed small ($q \ll 1$) but unknown.

b. Show that using the expression for ϕ to calculate C_p gives

$$C_p = -\frac{1}{2}(\epsilon^2\log\epsilon)G_{tt} - \frac{\epsilon^2}{2}\left(\frac{1}{2}G_{tt}\log G + 2T_{1_t} + \frac{1}{4G}G_t^2\right) + O(\epsilon^4\log\epsilon) \tag{3.3.450}$$

independently of q.

c. Show also that the axial force coefficient now becomes

$$C_D(t) = \epsilon^2\log\epsilon C_{D_0}(t) + \epsilon^2 C_{D_1}(t) + \ldots, \tag{3.3.451}$$

where

$$C_{D_0}(t) = -\int_0^1 \frac{\partial}{\partial t}(G_t G_x)dx, \tag{3.3.452a}$$

$$C_{D_1}(t) = -\int_0^1 \frac{\partial}{\partial t}\left[\frac{1}{2}G_t G_x \log G + 2T_1 G_x\right]dx. \tag{3.3.452b}$$

Thus, applying the axial equation of motion fixes the order of magnitude of q; in fact, we can expand

$$q(t;\epsilon) = \epsilon^2 \log \epsilon q_0(t) + \epsilon^2 q_1(t) + \ldots \tag{3.3.453}$$

and we have

$$m\frac{dq_0}{dt} = C_{D_0}(t), \tag{3.3.454a}$$

$$m\frac{dq_1}{dt} = C_{D_1}(t). \tag{3.3.454b}$$

d. Assume again that G is periodic in t with period λ. It then follows that $C_{D_0}(t)$ is also periodic in t with the same period and *a zero average value*, i.e., the leading term in the axial equation of motion would be of the form

$$m\frac{dq_0}{dt} = \sum_{n=1}^{\infty} \alpha_n \sin\left(\frac{2n\pi t}{\lambda} + \beta_n\right), \tag{3.3.455}$$

where α_n and β_n are known constants once the body deformation is specified. Integrating the axial equation of motion with $q_0(0) = 0$ gives

$$q_0(t) = \frac{1}{m}\sum_{n=1}^{\infty}\left[-\frac{\lambda\alpha_n}{2n\pi}\cos\left(\frac{2n\pi t}{\lambda} + \beta_n\right) + \lambda\alpha_n \cos\beta_n/2n\pi\right]. \tag{3.3.456}$$

Thus, the body will acquire the constant average velocity

$$\langle q\rangle = (\epsilon^2 \log\epsilon)\frac{\lambda}{2m\pi}\sum_{n=1}^{\infty}\frac{\alpha_n}{n}\cos\beta_n + \ldots. \tag{3.3.457}$$

For example, if we assume $G(x,t)$ to have the simple form $G(x,t) = \sin\pi x + a\sin 2\pi x \sin\omega t$ with $|a| < \frac{1}{2}$, it is easy to see that $G(x,t) > 0$ for all t and that $\int_0^1 G(x,t)dx = 2/\pi$. Hence, the volume of the body remains constant. Show that in this case

$$q_0(t) = \frac{2\pi}{3}a\omega(1 - \cos\omega t). \tag{3.3.458}$$

Therefore, the average velocity of motion is

$$\langle q\rangle = \frac{2\pi}{3}a\omega\epsilon^2 \log\epsilon + \ldots. \tag{3.3.459}$$

If we keep in mind that in a perfect fluid any initial velocity imparted to a rigid body is preserved, we conclude that radial deformations of the surface without vortex shedding do not provide a satisfactory mechanism for propulsion. This is because viscous forces would tend to reduce the already small value of $\langle q \rangle$ even further.

Fortunately, aquatic animals derive their propulsion by undulatory motions of their spines as well as vortex shedding from fins (when applicable). The interested reader can refer to [3.3] and [3.24].

References

3.1. A. Acrivos and A.S. Sangani, "The effective conductivity of a periodic array of spheres," *Proc. R. Soc. London A*, **386**, 1983, pp. 263–275.

3.2. V. Barcilon, J.D. Cole, and R.S. Eisenberg, "A singular perturbation analysis of induced electric fields in cells," *SIAM J. Appl. Math.*, **21**, 1971, pp. 339–353.

3.3. S. Childress, *Mechanics of Swimming and Flying*, Cambridge University Press, Cambridge, 1981.

3.4. J.D. Cole, "Limit process expansions and homogenization," *SIAM J. Appl. Math.*, **55**, 1995, pp. 410–424.

3.5. J.D. Cole, "On a quasilinear parabolic equation occurring in aerodynamics," *Q. Appl. Math.*, **9**, 1951, pp. 225–236.

3.6. R. Courant and D. Hilbert, *Methods of Mathematical Physics*, Vol. 1, Interscience Publishers, Inc., New York,1953.

3.7. G.C. Everstine and A.C. Pipkin, "Boundary layers in fiber-reinforced materials," *J. Appl. Mech.*, **40**, 1973, pp. 518–522.

3.8. J. Grasman and B.J. Matkowsky, "A variational approach to singularly perturbed boundary value problems for ordinary and partial differential equations with turning points," *SIAM J. Appl. Math.*, **32**, 1977, pp. 588–597.

3.9. R.P. Gregory and F.Y.M. Wan, "Correct asymptotic theories for the axisymmetric deformation of thin and moderately thick cylindrical shells," *Int. J. Solids and Structures*, **30**, 1993, pp. 1957–1981.

3.10. J.-F. Hamet, "Some acoustic phenomena related to curved surfaces," Ph.D. Thesis, University of California, Los Angeles, 1971.

3.11. E. Hopf, "The partial differential equation $u_t + uu_x = \mu u_{xx}$," *Comm. Pure Appl. Math.*, **3**, 1950, pp. 201–230.

3.12. S. Kaplun, "Low Reynolds number flow past a circular cylinder," *J. Math. Mech.*, **6**, 1957, pp. 595–603.

3.13. S. Kaplun, "The role of coordinate systems in boundary layer theory," *Zeitshrift für Angewandte Mathematik und Physik*, **2**, 1954, pp. 111–135.

3.14. S. Kaplun and P.A. Lagerstrom, "Asymptotic expansions of Navier-Stokes solutions for small Reynolds numbers," *J. Math. Mech.*, **6**, 1957, pp. 515–593.

3.15. J. Kevorkian, *Partial Differential Equations: Analytical Solution Techniques*, Chapman & Hall, New York, London 1990, 1993.

3.16. P.A. Lagerstrom, Laminar Flow Theory, *High Speed Aerodynamics and Jet Propulsion*, F.K. Moore, Ed., Vol. 4, Princeton University Press, Princeton, NJ, 1964, pp. 20–285.

3.17. P.A. Lagerstrom, "Note on the preceding two papers," *J. Math. Mech.*, **6**, 1957, pp. 605–606.

3.18. P.A. Lagerstrom and J.D. Cole, "Examples illustrating expansion procedures for the Navier-Stokes equations," *J. Rat. Mech. Anal.*, **4**, 1955, pp. 817–882.
3.19. P.A. Lagerstrom, L.N. Howard, and C.-S. Liu, *Fluid Mechanics and Singular Perturbations, A Collection of Papers by Saul Kaplun*, Academic Press, New York, 1967.
3.20. H. Lamb, *Hydrodynamics*, Dover, New York, 1945.
3.21. L. Landau and E. Lifschitz, *Fluid Mechanics*, Pergamon Press, New York, 1959.
3.22. R. Landauer, *Electrical conductivity in inhomogeneous media*, A.I.P. Conf. Proceedings #40, American Institute of Physics, New York, 1977.
3.23. A. Libai and J.G. Simmonds, *The Nonlinear Theory of Elastic Shells: One Spatial Dimension*, Academic Press, Boston, 1988.
3.24. M.J. Lighthill, *Mathematical Biofluiddynamics*, Society for Industrial and Applied Mathematics, Philadelphia, 1975.
3.25. A.E.H. Love, *A Treatise on the Mathematical Theory of Elasticity*, 4th ed., Cambridge University Press, 1927.
3.26. D. Ludwig, "Uniform asymptotic expansions for wave propagation and diffraction problems," *SIAM Rev.*, **12**, 1970, pp. 325–331.
3.27. R. O'Brien, "A method for the calculation of the effective transport properties of suspensions of interacting particles," *J. Fluid Mech.*, **91**, 1979, pp. 17–39.
3.28. R. Peierls, *More Surprises in Theoretical Physics*, Princeton University Press, Princeton, NJ, 1991.
3.29. A. Peskoff, "Green's function for Laplace's equation in an infinite cylindrical cell," *J. Math. Phys.*, **15**, 1974, pp. 2112–2120.
3.30. A. Peskoff and R.S. Eisenberg, "The time-dependent potential in a spherical cell using matched asymptotic expansions," *J. Math. Bio.*, **2**, 1975, pp. 277–300.
3.31. A. Peskoff, R.S. Eisenberg, and J.D. Cole, "Matched asymptotic expansions of the Green's function for the electric potential in an infinite cylindrical cell," *SIAM J. Appl. Math.*, **30**, 1976, pp. 222–239.
3.32. A. Peskoff, R.S. Eisenberg, and J.D. Cole, *Potential Induced by a Point Source of Current in the Interior of a Spherical Cell*, University of California at Los Angeles, Rept. U.C.L.A.-ENG-7259, 1972.
3.33. A.C. Pipkin, *Stress Channeling and Boundary Layers in Strongly Anisotropic Solids, Continuum Theory of the Mechanics of Fibre-Reinforced Composites*, A.J.M. Spencer, Ed., CISM Courses and Lectures No. 282, Springer-Verlag, Wien-New York, 1984, pp. 123–145.
3.34. L. Prandtl, Über Flüsigkeiten bei sehr kleiner Reibung. Verh. III, International Math. Kongress, Heidelberg, 1905, Teubner, Leipzig, pp. 484–491.
3.35. I. Proudman and J.R.A. Pearson, "Expansions at small Reynolds number for the flow past a sphere and a circular cylinder," *J. Fluid Mech.*, **2**, Part 3, 1957, pp. 237–262.
3.36. Lord Rayleigh, *Theory of Sound*, Dover, New York, v2, 1945, pp. 126–129.
3.37. Lord Rayleigh, "The problem of the whispering gallery," *Philos. Mag.*, **20**, 1910, pp. 1001–1004.
3.38. Lord Rayleigh, "On the influence of obstacles arranged in rectangular order upon the properties of the medium," *Philos. Mag.*, **34**, 1892, pp. 481–502.
3.39. R.D. Taylor, Cable theory, *Physical Techniques in Biological Research*, W.L. Nastuk, Ed., Vol. VIB., Academic Press, New York, 1963, pp. 219–262.
3.40. T. Theodorsen, *Impulse and Momentum in an Infinite Fluid*, Memorial Volume in Honor of the 60th Birthday of Th. von Karman, Caltech, Pasadena, CA, 1941.

3.41. S. Timoshenko and S. Woinowsky-Krieger, *Theory of Plates and Shells*, 2nd ed., McGraw-Hill Book Co., New York, 1959.
3.42. F.Y.M. Wan and H.J. Weinitschke, "On shells of revolution with the Love-Kirchhoff hypothesis," *J. Eng. Math.*, **22**, 1988, pp. 285–334.
3.43. H. Weyl, "On the differential equations of the simplest boundary layer problems," *Ann. Math.*, **43**, 1942, pp. 381–407.
3.44. G.B. Whitham, *Linear and Nonlinear Waves*, John Wiley and Sons, New York, 1974.

4

The Method of Multiple Scales for Ordinary Differential Equations

Various physical problems are characterized by the presence of a small disturbance which, because it is active over a long time, has a non-negligible cumulative effect. For example, the effect of a small damping force over many periods of oscillation is to produce a decay in the amplitude of an oscillator. A more interesting example having the same physical and mathematical features is that of the motion of a satellite around Earth. Here the dominant force is a spherically symmetric gravitational field. If this were the only force acting on the satellite, the motion would be periodic (for sufficiently low energies). The presence of a thin atmosphere, a slightly nonspherical Earth, a small moon, a distant sun, and so on, all produce small but cumulative effects which, after a sufficient number of orbits, drastically alter the nature of the motion.

It is the aim of this chapter to discuss the method of multiple scales, one of the two principal methods for accounting for small cumulative perturbations over a long time; the other approach is discussed in Chapter 5. The main effort here, as in the preceding chapter, is the exposition of various aspects of the method by means of a series of examples.

A central feature of the method is the nonexistence, for long times, of a limit process expansion of the type used so extensively in previous chapters. As a result, one is led to represent the solution at the outset in the form of a general asymptotic expansion. The nature of such an expansion was discussed in Section 2.1 (see (2.1.52)). This is in contrast to the situation encountered in Chapter 2, where a general asymptotic expansion arose at the last stage of computation when one combined an inner and outer expansion to define the composite solution. Because limit process expansions are not applicable, successive terms in the solution cannot be calculated by the repeated application of limits and, more importantly, rules must be established for the calculation of these terms. Viewed in this light, the method of multiple scales is a generalization of a method proposed by the astronomer Lindstedt for the calculation of periodic solutions. Thus, it is appropriate to begin this chapter with a brief review of Lindstedt's method.

4.1 Method of Strained Coordinates for Periodic Solutions

In his famous treatise on celestial mechanics (see sec. 125 of [4.28]), Poincaré credits the basic idea for this method to Lindstedt, A.. Perhaps due to the inaccessibility of Lindstedt's 1882 paper, some subsequent authors have referred to this as Poincaré's method. Actually, the basic idea was used even earlier, in 1847, by Stokes [4.31] in his study of periodic solutions for water waves (see Sec. 4.1.3). Strictly speaking, one should therefore refer to this as Stokes' method. This has not been the case, and many authors have called it the PLK method (P for Poincaré, L for Lighthill, who introduced a more general version in 1949, and K for Kuo, who applied it inappropriately to viscous flow problems in 1953). To minimize confusion, we will adhere to Van Dyke's nomenclature of the "method of strained coordinates" and refer the reader to [4.33] for an extensive discussion of applications in fluid mechanics. Some of these applications are considered in Chapter 6.

4.1.1 The Weakly Nonlinear Oscillator

Consider the weakly nonlinear oscillator with no damping, modeled in dimensionless variables by

$$\frac{d^2y}{dt^2} + y + \epsilon y^3 = 0, \quad 0 < \epsilon \ll 1, \tag{4.1.1a}$$

$$y(0; \epsilon) = 0, \tag{4.1.1b}$$

$$\frac{dy}{dt}(0; \epsilon) = v > 0. \tag{4.1.1c}$$

We will see that all the solutions of (4.1.1a) are periodic for $\epsilon \geq 0$. Hence, regardless of the initial values of y and dy/dt, the solution will at some later time pass through $y = 0$. Since (4.1.1a) is autonomous, there is no loss of generality in choosing the origin of time when $y = 0$.

We studied the regular expansion of the solution in Sec. 1.3 (see (1.3.17)–(1.3.19)), where we saw that this expansion fails to be uniformly valid if $t = O(\epsilon^{-1})$. The nonuniformity is exhibited by a term proportional to $\epsilon t \cos t$ in the second term of the expansion. Such a term is referred to as *mixed-secular* in the astronomy literature. Here *mixed* indicates the presence of the product of a linear and trigonometric function of time. *Secular* (derived from the Latin *saeculum* for century) was first used in astronomical applications, where ϵ is quite small and ϵt becomes significant only if t is on the order of a century. Actually, the solution of (4.1.1) is bounded. In fact, it happens to be periodic for any v, as we will see.

We can solve (4.1.1) exactly since it describes a conservative system with the energy integral

$$\frac{1}{2}\left(\frac{dy}{dt}\right)^2 + \frac{1}{2}y^2 + \frac{\epsilon y^4}{4} = \text{const.} = \frac{v^2}{2} \tag{4.1.2}$$

obtained by multiplying (4.1.1a) by dy/dt and noting that the result is integrable.

Since the potential energy $y^2/2 + \epsilon y^4/4$ is (for $\epsilon > 0$) a concave function for all y, the solution for any energy level $v^2/2$ describes periodic oscillations in the interval $-y_m \leq y \leq y_m$, where

$$y_m = \left[\frac{-1 + (1 + 2\epsilon v^2)^{1/2}}{\epsilon}\right]^{1/2} = v\left[1 - \frac{\epsilon v^2}{4} + O(\epsilon^2)\right] \tag{4.1.3}$$

is obtained by solving the quadratic equation that results from (4.1.2) when $dy/dt = 0$.

One can proceed further and calculate the formal solution by integrating (4.1.2) once more as follows:

$$t = \pm \int_0^y \frac{ds}{\sqrt{v^2 - s^2 - \epsilon s^4/2}}, \tag{4.1.4}$$

where the upper sign is to be used when dy/dt is positive and the lower sign when dy/dt is negative.

The solution can be expressed as an elliptic integral of the first kind by setting[1]

$$s = -y_m \cos\psi. \tag{4.1.5}$$

We calculate

$$(1 + 2\epsilon v^2)^{1/4} t = \pm \int_{\pi/2}^{\cos^{-1}(-y/y_m)} \frac{d\psi}{\sqrt{1 - k^2 \sin^2\psi}}, \tag{4.1.6}$$

where

$$k^2 = \frac{-1 + \sqrt{1 + 2\epsilon v^2}}{2\sqrt{1 + 2\epsilon v^2}} = \frac{\epsilon v^2}{2} + O(\epsilon^2). \tag{4.1.7}$$

In particular, since the potential energy is an even function of y, the value of t when $y = y_m$ equals one-fourth of the period P, and (4.1.6) gives

$$P(\epsilon) = \frac{4K(k^2)}{(1 + 2\epsilon v^2)^{1/4}}, \tag{4.1.8}$$

where $K(k^2)$ is the complete elliptic integral of the first kind defined by

$$K(k^2) = \int_0^{\pi/2} \frac{d\psi}{\sqrt{1 - k^2 \sin^2\psi}}. \tag{4.1.9}$$

---

[1] The reader will find an extensive discussion of elliptic functions and the various definitions we use here in [4.3].

Since $k^2 = O(\epsilon)$, we can use standard tables to derive the following expansion for the period in powers of ϵ:

$$P = 2\pi[1 - \frac{3}{8}\epsilon v^2 + O(\epsilon^2)]. \tag{4.1.10}$$

We see that for $\epsilon \neq 0$, the period is amplitude dependent. In contrast, for the linear case, $P = 2\pi$ for any value of v.

The result (4.1.10) also follows from (4.1.4) after some algebra if we set the upper limit equal to y_m and expand the resulting definite integral for $P/4$ in powers of ϵ.

Finally, we can invert (4.1.6) and express the result using elliptic functions

$$y(t;\epsilon) = y_m cn[(1 + 2\epsilon v^2)^{1/4} t + K(k^2), k]. \tag{4.1.11}$$

This defines a periodic function of time. A more explicit form of (4.1.11) can be obtained by expressing y as a Fourier series:

$$y(t;\epsilon) = \sum_{n=1}^{\infty} b_n(\epsilon) \sin \frac{2n\pi t}{P(\epsilon)}, \tag{4.1.12}$$

where

$$b_n(\epsilon) = \frac{4}{P} \int_0^{P/2} y(t;\epsilon) \sin \frac{2n\pi t}{P} dt, \tag{4.1.13}$$

and the coefficients can be calculated, in principle, using (4.1.11) for $y(t;\epsilon)$. Either form of the exact solution is cumbersome, particularly if one is only interested in the case $\epsilon \ll 1$.

We note from (4.1.12) that the solution actually depends on the variable $t^+ = 2\pi t/P$ instead of t, and that if $\epsilon \ll 1$, t^+ will have the form

$$t^+ = (1 + \epsilon\omega_1 + \epsilon^2\omega_2 + \ldots)t, \tag{4.1.14}$$

where the ω_i are constants independent of ϵ. Equation (4.1.10) shows, for example, that $\omega_1 = (3/8)v^2$. It is now clear why the regular expansion used in Chapter 1 failed (see (1.3.19)). A term such as

$$\sin \frac{2n\pi t}{P} = \sin n(1 + \epsilon\omega_1 + \epsilon^2\omega_2 + \ldots)t, \tag{4.1.15}$$

occurring in the exact solution, would, under the limit process $\epsilon \to 0$, t fixed, have the expansion

$$\sin \frac{2n\pi t}{P} = \sin nt + n\epsilon\omega_1 t \cos nt + O(\epsilon^2). \tag{4.1.16}$$

Thus, the mixed-secular terms encountered in solving (4.1.1) (see (1.3.19))

$$y = v \sin t + \epsilon v^3 \left(\frac{3t}{8} \cos t - \frac{9}{32} \sin t - \frac{1}{32} \sin 3t \right) + \ldots \tag{4.1.17}$$

are strictly due to the nonuniform representation of trigonometric functions of the variable t^+.

The remedy is easily discerned; we need a limit process expansion in which t^+ is held fixed as $\epsilon \to 0$.

Thus, to accommodate for trigonometric terms with arguments involving t^+, and to allow the Fourier coefficients to have expansions in terms of ϵ, we seek a solution in the form

$$y(t;\epsilon) = f_0(t^+) + \epsilon f_1(t^+) + \epsilon^2 f_2(t^+) + \ldots, \tag{4.1.18}$$

where t^+ is defined by (4.1.14). The ω_i are unknown constants to be determined by the requirement that the f_i be periodic functions of t^+. This is the essential idea of the method of strained coordinates.

Because (4.1.18) is a limit process expansion, it is convenient to first write (4.1.1) in terms of t^+ as follows:

$$(1 + \epsilon\omega_1 + \epsilon^2\omega_2 + \ldots)^2 \frac{d^2 y}{dt^{+2}} + y + \epsilon y^3 = 0; \tag{4.1.19}$$

$$y(0;\epsilon) = 0; \tag{4.1.20}$$

$$(1 + \epsilon\omega_1 + \epsilon^2\omega_2 + \ldots) \frac{dy}{dt^+}(0;\epsilon) = v. \tag{4.1.21}$$

Substituting (4.1.18) into the above then gives the following sequence of initial value problems for the f_i:

$$L(f_0) \equiv \frac{d^2 f_0}{dt^{+2}} + f_0 = 0; \quad f_0(0) = 0, \quad \frac{df_0}{dt^+}(0) = v; \tag{4.1.22}$$

$$L(f_1) = -2\omega_1 \frac{d^2 f_0}{dt^{+2}} - f_0^3; \quad f_1(0) = 0, \quad \frac{df_1}{dt^+}(0) = -\omega_1 v; \tag{4.1.23}$$

$$L(f_2) = -(\omega_1^2 + 2\omega_2) f_0 - 2\omega_1 \frac{d^2 f_1}{dt^{+2}} - 3 f_0^2 f_1;$$

$$f_2(0) = 0; \quad \frac{df_2}{dt^+}(0) = (\omega_1^2 - \omega_2)v; \tag{4.1.24}$$

etc.

We solve (4.1.22) immediately:

$$f_0(t^+) = v \sin t^+. \tag{4.1.25}$$

Using this in (4.1.23) gives

$$L(f_1) = \left(2\omega_1 v - \frac{3}{4}v^3\right) \sin t^+ + \frac{v^3}{4} \sin 3t^+. \tag{4.1.26}$$

We know that a forcing term that is a homogeneous solution of $L(f_1) = 0$, such as the first term on the right-hand side of (4.1.26), will have a mixed-secular term as a response. Such a response becomes unbounded as $t \to \infty$ and is certainly not periodic. Therefore, to ensure the periodicity of f_1, we must set

$$2\omega_1 v - \frac{3}{4}v^3 = 0, \tag{4.1.27}$$

which can only hold (since $v \neq 0$) if

$$\omega_1 = \frac{3}{8}v^2. \tag{4.1.28}$$

Thus, we have recovered the earlier result quite efficiently. Moreover, what remains of (4.1.26) can now be solved subject to the appropriate initial conditions in the form

$$f_1(t^+) = -\frac{9}{32}v^3 \sin t^+ - \frac{v^3}{32}\sin 3t^+. \tag{4.1.29}$$

The procedure can be continued indefinitely. Removal of homogeneous solutions (terms proportional to $\sin t^+$ or $\cos t^+$) from the right-hand side of the $L(f_i)$ defines the ω_i. A feature typical of weakly nonlinear oscillations will be recognized, namely, that higher harmonics of $\sin t^+$ occur to higher orders. In fact, we can deduce for this example that $f_n(t)$ will only involve the $(n+1)$-functions $\sin t^+, \sin 3t^+, \ldots, \sin(2n+1)t^+$. The foregoing procedure also applies if $\epsilon < 0$ as long as $|\epsilon| \ll 1$, because all solutions are periodic for $v = O(1)$.

In Problem 6, we will show that the strained coordinate expansion of a periodic function is uniformly valid over an interval $I : 0 \leq t \leq T(\epsilon) = O(\epsilon^{-1})$.

4.1.2 Rayleigh's Equation

In the preceding example for $\epsilon > 0$, all the solutions of (4.1.1) were periodic regardless of the value of v. We now consider Rayleigh's equation

$$\frac{d^2y}{dt^2} + y + \epsilon\left[-\frac{dy}{dt} + \frac{1}{3}\left(\frac{dy}{dt}\right)^3\right] = 0, \quad 0 < \epsilon \ll 1, \tag{4.1.30}$$

which has only one periodic solution, called a *limit cycle*, corresponding to one particular initial value of y when $dy/dt = 0$.

Equation (4.1.30) is related to the van der Pol equation through the transformation $w = dy/dt$; differentiating (4.1.30) and setting $dy/dt = w$ gives the van der Pol equation for w.

Although one can prove the existence of a limit cycle rigorously, we will instead use a heuristic argument to indicate that such a solution exists. Consider (4.1.30) in the phase-plane of y and dy/dt. If $\epsilon = 0$, the integral curves are circles. For any positive ϵ, the oscillator is subject to an additional "force," $\epsilon[dy/dt - \frac{1}{3}(dy/dt)^3]$. If dy/dt is small, i.e., if the motion starts near the origin of the phase-plane, the term dy/dt is more important than $-\frac{1}{3}(dy/dt)^3$. Hence the net effect of the bracketed term in (4.1.30) is a negative damping, leading to an increase in amplitude. But this cannot go on indefinitely, since eventually the term $-\frac{1}{3}(dy/dt)^3$ would dominate and produce a decay in the amplitude. Similarly, if the motion were initiated with a large value of v, the tendency would be for the amplitude to decrease until a balance was struck between the two opposing forces in the bracketed term. Therefore, it is reasonable to expect that for certain special initial conditions there exists a closed trajectory in the phase plane, i.e., a periodic solution.

We will use the method of strained coordinates to exhibit this periodic solution $y(t; \epsilon)$. Since the initial amplitude that corresponds to the limit cycle solution is unknown, we assume that

$$y(0; \epsilon) = a(\epsilon) = a_0 + \epsilon a_1 + \epsilon^2 a_2, \tag{4.1.31a}$$

$$\frac{dy}{dt}(0; \epsilon) = 0, \tag{4.1.31b}$$

where the unknown constants a_i are to be determined. Note here again that a periodic solution will always pass through $dy/dt = 0$, and we set the origin of the time scale to be zero when this occurs.

We develop $y(t; \epsilon)$ in the form

$$y(t; \epsilon) = f_0(t^+) + \epsilon f_1(t^+) + \epsilon^2 f_2(t^+) + \ldots \tag{4.1.32}$$

with

$$t^+ = (1 + \epsilon\omega_1 + \epsilon^2\omega_2 + \ldots)t. \tag{4.1.33}$$

The equations and initial conditions governing the f_i are

$$L(f_0) \equiv \frac{d^2 f_0}{dt^{+2}} + f_0 = 0, \quad f_0(0) = a_0, \quad \frac{df_0}{dt^+}(0) = 0; \tag{4.1.34}$$

$$L(f_1) = -2\omega_1 \frac{d^2 f_0}{dt^{+2}} + \frac{df_0}{dt^+} - \frac{1}{3}\left(\frac{df_0}{dt^+}\right)^3, \quad f_1(0) = a_1, \quad \frac{df_1}{dt^+}(0) = 0; \tag{4.1.35}$$

$$\begin{aligned} L(f_2) = &- 2\omega_1 \frac{d^2 f_1}{dt^{+2}} - (2\omega_2 + \omega_1^2)\frac{d^2 f_0}{dt^{+2}} + \frac{df_1}{dt^+} + \omega_1 \frac{df_0}{dt^+} \\ &- \left(\frac{df_0}{dt^+}\right)^2 \left[\frac{df_1}{dt^+} + \omega_1 \frac{df_0}{dt^+}\right], \quad f_2(0) = a_2, \quad \frac{df_2}{dt^+}(0) = 0. \end{aligned} \tag{4.1.36}$$

Clearly

$$f_0(t^+) = a_0 \cos t^+, \tag{4.1.37}$$

and using this result in (4.1.35) gives

$$L(f_1) = 2\omega_1 a_0 \cos t^+ + \left(\frac{a_0^3}{4} - a_0\right) \sin t^+ - \frac{a_0^3}{12} \sin 3t^+. \tag{4.1.38}$$

Periodicity requires that

$$2\omega_1 a_0 = 0 \tag{4.1.39}$$

$$\frac{a_0^3}{4} - a_0 = 0. \tag{4.1.40}$$

We discard the trivial solution $a_0 = 0$, ω_1 arbitrary, and set

$$\omega_1 = 0, \tag{4.1.41}$$

$$a_0 = 2. \tag{4.1.42}$$

This determines y to $O(1)$ and t^+ to $O(\epsilon)$ and gives the limit cycle amplitude to be 2.

Next, we calculate the solution for f_1 from (4.1.38) with only $-\frac{2}{3}\sin 3t^+$ remaining on the right-hand side. This gives

$$f_1(t^+) = -\frac{1}{4}\sin t^+ + a_1 \cos t^+ + \frac{1}{12}\sin 3t^+. \tag{4.1.43}$$

Using the results calculated so far, we can evaluate the right-hand side of (4.1.36) and find

$$L(f_2) = \left(4\omega_2 + \frac{1}{4}\right)\cos t^+ + 2a_1 \sin t^+ \\ - \frac{1}{2}\cos 3t^+ - a_1 \sin 3t^+ + \frac{1}{4}\cos 5t^+. \tag{4.1.44}$$

Thus, in order to have f_2 periodic, we must set

$$a_1 = 0, \tag{4.1.45a}$$

$$\omega_2 = -\frac{1}{16}, \tag{4.1.45b}$$

and this procedure can be continued indefinitely. We note that once $f_n(t^+)$ is completely determined, i.e., when a_n is evaluated, we also have evaluated ω_{n+1}.

In this example, the method of strained coordinates determines both the appropriate initial conditions for a periodic solution and the corresponding period.

The basic assumption for the applicability of the method is that the exact solution depends to all orders on *one strained coordinate only*. This is certainly true for a periodic solution. The reader is cautioned that for nonperiodic solutions, particularly when applied to partial differential equations, the method might superficially appear to work but could give incorrect results. Examples of this are cited in [4.33]. Also, as will be pointed out in Problem 7 of Sec. 6.2, the method fails to higher orders for the problem of supersonic thin airfoil theory as analyzed in Lighthill's 1949 study.

4.1.3 The Korteweg–de Vries Equation, Small-Amplitude Periodic Waves

In Chapter 6, we will show that the leading approximation for shallow water waves obeys the Korteweg–de Vries equation that can be written in the following dimensionless form for the flow speed $u(x, t)$:

$$u_t + \left(1 + \frac{3}{2}u\right)u_x + \frac{\delta^2}{6}u_{xxx} = 0. \tag{4.1.46}$$

Here δ is the ratio of the undisturbed water depth to a characteristic wavelength for the flow.

We assume small disturbances and look for a solution where $u = O(\epsilon)$, where $0 < \epsilon \ll 1$ measures the amplitude of the disturbance.

This problem is somewhat simpler than the one Stokes studied, but it serves to illustrate the essential ideas.

Linearized problem, dispersion relation

The basic assumption regarding the structure of the solution is that it is a periodic traveling wave close to the periodic traveling wave that one finds when the nonlinear term $(3/2)uu_x$ in (4.1.46) is ignored. We consider first the linearized problem

$$w_t + w_x + \frac{\delta^2}{6}w_{xxx} = 0, \tag{4.1.47}$$

and look for a traveling wave solution, i.e.,

$$w = W(kx - \omega t), \tag{4.1.48}$$

where k is a fixed constant and ω will be determined in terms of k to make W a periodic function of its argument $\theta \equiv kx - \omega t$. Because $w =$ constant solves (4.1.47) we normalize the solution by requiring $W(\theta)$ to have zero average over one period in θ.

Substituting (4.1.48) into (4.1.47) gives the ordinary differential equation

$$-\omega W' + kW' + \frac{\delta^2 k^3}{6}W''' = 0, \tag{4.1.49}$$

where $' \equiv d/d\theta$. Integrating once gives

$$\frac{\delta^2 k^3}{6}W'' + (k - \omega)W = c_1 = \text{constant}. \tag{4.1.50}$$

Periodic solutions are found only if the constant $6(k - \omega)/\delta^2 k^3$ is positive and we set

$$\frac{6(k - \omega)}{\delta^2 k^3} = \lambda^2 = \text{const}. \tag{4.1.51}$$

The solution of (4.1.50) is then given by

$$W(\theta) = \rho_1 \sin(\lambda\theta + \phi_1) + \frac{c_1}{k - \omega}, \tag{4.1.52}$$

where ρ_1 and ϕ_1 are two more constants. First, set $c_1 = 0$ in order that $W(\theta)$ have zero average. Then we use (4.1.51) to express ω in terms of k, and we write the solution (4.1.52) as

$$W(\theta) = \rho_1 \sin\left[\lambda k(x-t) + (\lambda k)^3 \frac{\delta^2}{6} t + \phi_1\right]. \tag{4.1.53}$$

It is clear from this result that λ and k always occur in the combination λk. Therefore, with no loss of generality, we may set $\lambda = 1$ as this merely rescales k, a constant we may choose at will. With $\lambda = 1$ the expression linking ω to k, called the *dispersion relation*, becomes

$$\omega = k - \frac{\delta^2}{6} k^3. \tag{4.1.54a}$$

The result (4.1.53) (with $\lambda = 1$) describes a sinusoidal wave with wavelength $2\pi/k$ propagating without changing shape to the right with *phase speed*

$$c_p = \frac{\omega}{k} = 1 - \delta^2 k^2/6. \tag{4.1.54b}$$

Weakly nonlinear problem

For $0 < \epsilon \ll 1$, we look for a solution that has the form

$$u(x, t; \epsilon) = \epsilon U(\theta^+; \epsilon) = \epsilon u_1(\theta^+) + \epsilon^2 u_2(\theta^+) + \epsilon^3 u_3(\theta^+) + \dots, \tag{4.1.55}$$

where

$$\theta^+ = kx - \Omega(\epsilon)t, \tag{4.1.56a}$$

$$\Omega(\epsilon) = \omega_0 + \epsilon\omega_1 + \epsilon^2\omega_2 + \dots. \tag{4.1.56b}$$

As in the previous examples, we expect to determine the u_i and the ω_i by requiring the solution U to be periodic in θ^+ with zero average.

Substituting the expansions for u and Ω into (4.1.46) yields the following sequence of equations governing u_1, u_2, and u_3:

$$(k-\omega_0)u_1' + \frac{\delta^2}{6}k^3 u_1''' = 0, \tag{4.1.57a}$$

$$(k-\omega_0)u_2' + \frac{\delta^2}{6}k^3 u_2''' = \omega_1 u_1' - \frac{3}{4}k(u_1^2)', \tag{4.1.57b}$$

$$(k-\omega_0)u_3' + \frac{\delta^2}{6}k^3 u_3''' = \omega_1 u_2' + \omega_2 u_1' - \frac{3}{2}k(u_1 u_2)'. \tag{4.1.57c}$$

It is computationally more convenient to write the solution for u_1 in the form (after setting the first constant of integration equal to zero)

$$u_1(\theta^+) = A_1 \sin\theta^+ + B_1 \cos\theta^+. \tag{4.1.58}$$

Of course, we also require that ω_0 satisfy the dispersion relation (4.1.54a).

Substituting the expression for u_1 into the right-hand side of (4.1.57b), decomposing u_1^2 into its harmonics, using the dispersion relation for ω_0, and integrating once gives

$$\frac{\delta^2 k^3}{6}(u_2'' + u_2) = \omega_1(A_1 \sin\theta^+ + B_1 \cos\theta^+) + c_2 - \frac{3}{8}k(A_1^2 + B_1^2)$$
$$+ \frac{3k}{8}(A_1^2 - B_1^2)\cos 2\theta^+ - \frac{3}{4}kA_1B_1 \sin 2\theta^+. \qquad (4.1.59)$$

To avoid mixed-secular terms in the solution for u_2, we set

$$\omega_1 = 0 \qquad (4.1.60a)$$

to eliminate the $\sin\theta^+$ and $\cos\theta^+$ terms on the right-hand side of (4.1.59).

We set

$$c_2 = \frac{3k}{8}(A_1^2 + B_1^2) \qquad (4.1.60b)$$

so that u_2 has zero average. Solving what remains of (4.1.59) gives

$$u_2 = A_2 \sin\theta^+ + B_2 \cos\theta^+ - \frac{3(A_1^2 - B_1^2)}{4\delta^2 k^2}\cos 2\theta^+$$
$$+ \frac{3A_1B_1}{2\delta^2 k^2}\sin 2\theta^+, \qquad (4.1.61)$$

where A_2 and B_2 are arbitrary constants.

We now use the expression we have found for u_1 and u_2 to reduce (4.1.57c) to the form

$$\frac{\delta^2 k^3}{6}(u_3'' + u_3) = c_3 - \frac{3k}{4}(A_1A_2 + B_1B_2)$$
$$+ A_1\left[\omega_2 - \frac{9}{16}\frac{(A_1^2 + B_1^2)}{\delta^2 k^2}\right]\sin\theta^+$$
$$+ B_1\left[\omega_2 - \frac{9(A_1^2 + B_1^2)}{16\delta^2 k^2}\right]\cos\theta^+$$
$$\frac{A_1B_2 + B_1A_2}{4}\sin 2\theta^+ + \frac{B_1B_2 - A_1A_2}{4}\cos 2\theta^+$$
$$+ \frac{3A_1(3B_1^2 - B_1^2)}{8\delta^2 k^2}\sin 3\theta^+$$
$$+ \frac{3B_1(B_1^2 - 3A_1^2)}{8\delta^2 k^2}\cos 3\theta^+. \qquad (4.1.62)$$

Once again, by choosing

$$c_3 = \frac{3k}{4}(A_1A_2 + B_1B_2), \qquad (4.1.63a)$$

$$\omega_2 = \frac{9}{16}\frac{(A_1^2 + B_1^2)}{\delta^2 k^2}, \qquad (4.1.63b)$$

we ensure that u_3 is periodic in θ^+ with zero average.

This procedure can be continued indefinitely, and we note the following features:

(i) The assumption that the solution is a uniform traveling wave reduces the partial differential equation in x and t to an ordinary differential equation in θ^+.

(ii) The periodicity condition applied to the leading approximation determines the linear dispersion relation. The periodicity condition to higher orders gives the remaining ω_i.

(iii) The $O(1)$ constants $(A_1, B_1), (A_2, B_2), \ldots$ may be chosen arbitrarily.

(iv) The ω_i for $i > 0$ depend on the (A_j, B_j) for $j \leq i - 1$.

(v) The final result corresponds to a *very special* class of initial conditions obtained by setting $t = 0$ in the expression calculated for u.

Problems

1. Consider the weakly nonlinear wave equation

$$u_{tt} - u_{xx} + u + \epsilon u^3 = 0. \tag{4.1.64}$$

For $\epsilon = 0$, this equation has the special periodic solution

$$u = \rho_0 \sin(\theta + \phi_0), \quad \theta = kx - \sqrt{1 + k^2}t \tag{4.1.65}$$

for arbitrary constants ρ_0, ϕ_0, and k.

a. Calculate the periodic solution for $0 < \epsilon \ll 1$ in the form

$$u(x, t; \epsilon) = \rho_0 \sin(\theta^+ + \phi_0) + \epsilon u_1(\theta^+) + \epsilon^2 u_2(\theta^+) + \ldots, \tag{4.1.66}$$

where

$$\theta^+ = kx - \sqrt{1 + k^2}\Omega(\epsilon)t, \quad \Omega(\epsilon) = 1 + \epsilon\omega_1 + \epsilon^2\omega_2 + \ldots. \tag{4.1.67}$$

Show that

$$u_1 = -\frac{\rho_0^3}{32} \sin 3(\theta^+ + \phi_0), \quad \omega_1 = \frac{3\rho_0^2}{8(1 + k^2)}. \tag{4.1.68}$$

b. Suppose the leading approximation has *two* waves with different wavelengths in the form

$$u_0 = \rho_0 \sin(\theta_k^+ + \phi_0) + \alpha_0 \sin(\theta_\ell^+ + \beta_0), \tag{4.1.69}$$

where

$$\theta_k^+ = kx - \sqrt{1 + k^2}\Omega_k(\epsilon)t; \tag{4.1.70a}$$

$$\theta_\ell^+ = \ell x - \sqrt{1 + \ell^2}\Omega_\ell(\epsilon)t; \tag{4.1.70b}$$

and $k \neq \ell$. Does the method of strained coordinates apply?

2. Calculate the solution to $O(\epsilon)$ of

$$\frac{d^2y}{dt^2} + y + \epsilon y|y| = 0 \tag{4.1.71}$$

with $y(0; \epsilon) = 0$, $\dot{y}(0; \epsilon) = v$.

Hint: $\sin t^+ |\sin t^+|$ is an odd periodic function of t^+ and can therefore be expanded in a Fourier sine series over the interval $0 \leq t^+ \leq \pi$.

3. Calculate the limit cycle to $O(\epsilon)$ for

$$\frac{d^2y}{dt^2} + y + \epsilon\left[-\frac{dy}{dt} + \frac{dy}{dt}\left|\frac{dy}{dt}\right|\right] = 0. \tag{4.1.72}$$

4. Consider Mathieu's equation

$$\frac{d^2y}{dt^2} + [\delta(\epsilon) + \epsilon\cos t]y = 0, \quad 0 < \epsilon \ll 1. \tag{4.1.73}$$

It can be shown that for appropriate values of $\delta(\epsilon)$ this equation has periodic solutions with period 2π or 4π. Expand y and δ in the form

$$y(t;\epsilon) = y_0(t) + \epsilon y_1(t) + \ldots, \tag{4.1.74a}$$

$$\delta(\epsilon) = \delta_0 + \epsilon\delta_1 + \epsilon^2\delta_2 + \ldots \tag{4.1.74b}$$

to show that a necessary condition for periodic solutions of period 2π or 4π is $\delta_0 = n^2/4, n = 0, 1, 2, \ldots$. For the cases $n = 0, 1, 2$, calculate δ_1 and $y_1(t)$.

5. Show that Duffing's equation

$$\frac{d^2y}{dt^2} + y + \epsilon y^3 = \epsilon\cos\Omega(\epsilon)t \tag{4.1.75}$$

has periodic solutions with frequency Ω for appropriate initial conditions. Carry out the calculations in detail to $O(\epsilon)$ for the cases

$$\Omega(\epsilon) = 1 + \epsilon\omega_1 + \epsilon^2\omega_2 + \ldots \tag{4.1.76a}$$

and

$$\Omega(\epsilon) = 3 + \epsilon\mu_1 + \epsilon^2\mu_2 + \ldots. \tag{4.1.76b}$$

Observe the occurrence of subharmonics (i.e., a response with frequency equal to a fraction of the impressed frequency) in the second case.

6. Consider the periodic function $f(t;\epsilon)$ having a Fourier series expansion

$$f(t;\epsilon) = \frac{a_0(\epsilon)}{2} + \sum_{n=1}^{\infty}[a_n(\epsilon)\cos n\omega(\epsilon)t + b_n(\epsilon)\sin n\omega(\epsilon)t]. \tag{4.1.77}$$

We assume that the $a_n(\epsilon)$, $b_n(\epsilon)$, and $\omega(\epsilon)$ can be expanded asymptotically for $\epsilon \to 0$ in the form:

$$a_n(\epsilon) \sim \sum_{i=0}^{\infty} a_{ni}\epsilon^i, \tag{4.1.78a}$$

$$b_n(\epsilon) \sim \sum_{i=0}^{\infty} b_{ni}\epsilon^i, \tag{4.1.78b}$$

$$\omega(\epsilon) \sim 1 + \sum_{i=1}^{\infty} \omega_i\epsilon^i, \tag{4.1.78c}$$

with the a_{ni}, b_{ni}, and ω_i constants independent of ϵ.

Clearly, the asymptotic expansion of $f(t;\epsilon)$ to any order ϵ^N by the method of strained coordinates is given by

$$f(t;\epsilon) = \sum_{i=0}^{N} f_i(t_N^+)\epsilon^i + O(\epsilon^{N+1}), \tag{4.1.79}$$

where

$$t_N^+ = \left[1 + \sum_{j=1}^{N+1} \omega_j \epsilon^j + O(\epsilon^{N+2})\right] t; \tag{4.1.80}$$

$$f_i(t_N^+) = \frac{a_{0i}}{2} + \sum_{n=1}^{\infty} (a_{ni} \cos nt_N^+ + b_{ni} \sin nt_N^+). \tag{4.1.81}$$

Show that

$$\frac{f(t;\epsilon) - \sum_{i=0}^{N} f_i(t_N^+)\epsilon^i}{\epsilon^N} = O(\epsilon) + O(\epsilon^2 t) \quad \text{as } \epsilon \to 0. \tag{4.1.82}$$

Thus, the strained coordinate expansion is uniformly valid in the interval I: $0 \leq t \leq T(\epsilon)$, where $T = O(\epsilon^{-1})$.

4.2 Two Scale Expansions for the Weakly Nonlinear Autonomous Oscillator

Hints of the idea of multiple scales appear in the book [4.21] by Krylov and Bogoliubov. However, the main thrust in their book and most of the subsequent Russian literature is on averaging techniques, to be described in Chapter 5. One exception is the paper [4.22] by Kuzmak, which appears to be the first example where an asymptotic expansion depending explicitly on two time scales in proposed. Kuzmak's method concerns perturbed strictly nonlinear oscillators and is discussed in Sec. 4.4.

Independently of Kuzmak's work, and a short time later, there appeared three independent studies, [4.4], [4.19], and [4.26], on the use of multiple-scale expansions. The thesis [4.19], which evolved from ideas suggested by J.D. Cole, was reported in abbreviated form in [4.5] and published in full in [4.18]. The basic idea of multiple scale expansions reappears under various names and guises in a number of subsequent papers, but to our knowledge, the original references are [4.4], [4.19], [4.22], and [4.26].

4.2.1 The Linear Oscillator with Small Linear Damping

An elementary example illustrating the basic ideas of the method of multiple scales is that of a linear oscillator with small linear damping. This example was formulated in Sec. 2.1.1 (see 2.1.4)). Simplifying the notation, we have

$$\frac{d^2y}{dt^2} + 2\epsilon\frac{dy}{dt} + y = 0, \tag{4.2.1}$$

$$y(0;\epsilon) = 0, \tag{4.2.2a}$$

$$\frac{dy}{dt}(0;\epsilon) = 1, \tag{4.2.2b}$$

where ϵ is the ratio of the two time scales T_1, T_2:

$$\epsilon = \frac{T_1}{T_2} = \frac{B}{2\sqrt{KM}}. \tag{4.2.3a}$$

Here T_2 is the damping time

$$T_2 = \frac{2M}{B}, \tag{4.2.3b}$$

which is long if B is small, and $2\pi T_1$ is the period of oscillation for $B = 0$:

$$T_1 = \sqrt{\frac{M}{K}}. \tag{4.2.3c}$$

We assume that T_1 is small compared to T_2.

Exact solution, nonexistence of outer limit

The physical phenomena described by (4.2.1) occur over these two time scales as can be seen clearly if the exact solution

$$y = \frac{e^{-\epsilon t}}{\sqrt{1-\epsilon^2}}\sin\sqrt{1-\epsilon^2}\,t \tag{4.2.4}$$

is written with dimensional variables Y and T:

$$\frac{Y}{A} = \frac{e^{-T/T_2}}{\sqrt{1-(T_1/T_2)^2}}\sin\sqrt{1-\left(\frac{T_1}{T_2}\right)^2}\left(\frac{T}{T_1}\right). \tag{4.2.5}$$

For $\epsilon \ll 1$ the "period" of damped oscillations in approximately $2\pi T_1$ and the *damping time*, which we may define as the time it takes for the damping to have an $O(1)$ effect on the solution, is T_2.

An expansion of this solution was derived in Sec. 2.1 in the form (see (2.1.6)):

$$y(t;\epsilon) = \sin t - \epsilon t \sin t + O(\epsilon^2) + O(\epsilon^2 t) + O(\epsilon^2 t^2). \tag{4.2.6}$$

This expansion is associated with the limit process $\epsilon \to 0$, t fixed and is only initially valid ($0 \le t \le T_0 = O(1)$) due to the presence of the $\epsilon t \sin t$ term. In this example, the first mixed-secular term we encounter to $O(\epsilon)$ is due to the nonuniform representation for large times of the $e^{-\epsilon t}$ term in the exact solution. To $O(\epsilon^2)$, the expansion of $e^{-\epsilon t}$ contributes a term proportional to $\epsilon^2 t^2 \sin t$; a mixed-secular term proportional to $\epsilon^2 t \cos t$ will also occur to $O(\epsilon^2)$ from the nonuniform representation of the $\sin\sqrt{1-\epsilon^2}t$ term in (4.2.4).

It is also evident that mutually contradictory requirements arise if we attempt to represent both $e^{-\epsilon t}$ and $\sin\sqrt{1-\epsilon^2}t$ uniformly for t in the large interval I: $0 \leq t \leq T(\epsilon) = O(\epsilon^{-1})$. In particular, the only uniformly valid representation for $e^{-\epsilon t}$ in this interval is $e^{-\epsilon t}$ itself. Therefore, we need the limit process $\epsilon \to 0$, $\tilde{t} = \epsilon t$ fixed $\neq 0$ in order to represent $e^{-\epsilon t}$ uniformly in I. However, this limit process *does not exist* for $\sin\sqrt{1-\epsilon^2}t = \sin\sqrt{1-\epsilon^2}\tilde{t}/\epsilon$, as the argument of the sine function tends to infinity as $\epsilon \to 0$ with $\tilde{t}$ = fixed $\neq 0$. Another way of saying this is that the decaying oscillatory function defined by (4.2.4) does not have an outer limit.

On the other hand, as pointed out in Sec. 4.1, we need the limit process $\epsilon \to 0$, $t^+ = (1 - (\epsilon^2/2) + \dots)t$ fixed $\neq \infty$, i.e., an expansion in terms of the strained coordinate t^+, in order to uniformly represent $\sin\sqrt{1-\epsilon^2}t$ over I. In this case $e^{-\epsilon t}$ is expressed as $e^{-\epsilon t^+(1+\epsilon^2/2+\dots)}$ and leads to essentially the same nonuniformity in I as the initially valid expansion (4.2.6).

General asymptotic expansion, two scale expansion

Any asymptotic expansion of (4.2.4) must *simultaneously* depict both the decaying and oscillatory behaviors of the solution in order to be uniformly valid in I. It is clear that a limit process expansion will not do, and we broaden our scope and look for a general asymptotic expansion (see (2.1.53)) where each term depends on t and ϵ. In fact, if we avoid expanding $e^{-\epsilon t}$ and simply develop the argument $\sqrt{1-\epsilon^2}t$ without further expanding the sine function, we find the following general asymptotic expansion for the function defined in (4.2.4) in a form that is uniformly valid in I:

$$y(t;\epsilon) = \sum_{n=0}^{N}(\alpha_n e^{-\epsilon t}\sin\Omega_N(\epsilon)t)\epsilon^n + O(\epsilon^{N+1}). \tag{4.2.7}$$

Here α_n follows from the expansion of the factor $(1-\epsilon^2)^{-1/2}$ in (4.2.4), whereas Ω_n results from expanding the frequency $(1-\epsilon^2)^{1/2}$

$$\alpha_{2n} = \frac{1}{2^n n!}\Pi_{k=1}^{n+1}|2k-3|;\ n = 0, 1, \dots, \tag{4.2.8a}$$

$$\alpha_{2n+1} = 0;\ \ n = 0, 1, 2, \dots. \tag{4.2.8b}$$

$$\Omega_0(\epsilon) = 1, \tag{4.2.9a}$$

$$\Omega_{2n-1}(\epsilon) = 1 - \sum_{j=1}^{n}\frac{1}{2^j j!}\Pi_{k=1}^{j}|2k-3|\epsilon^{2j} n = 1, 2, \dots, \tag{4.2.9b}$$

$$\Omega_{2n} = \Omega_{2n-1};\ \ n = 1, 2, \dots. \tag{4.2.9c}$$

In particular, the uniformly valid approximations to order $1, \epsilon, \epsilon^2, \epsilon^3 \ldots$, denoted respectively as $y^{(0)}, y^{(1)}, y^{(2)}, y^{(3)}, \ldots$, are

$$y^{(0)} = e^{-\epsilon t} \sin t, \tag{4.2.10a}$$

$$y^{(1)} = e^{-\epsilon t} \sin \left(1 - \frac{\epsilon^2}{2}\right) t, \tag{4.2.10b}$$

$$y^{(2)} = \left(1 + \frac{\epsilon^2}{2}\right) e^{-\epsilon t} \sin \left(1 - \frac{\epsilon^2}{2}\right) t, \tag{4.2.10c}$$

$$y^{(3)} = \left(1 + \frac{\epsilon^2}{2}\right) e^{-\epsilon t} \sin \left(1 - \frac{\epsilon^2}{2} - \frac{3\epsilon^4}{8}\right) t. \tag{4.2.10d}$$

As pointed out in Problem 6 of Sec. 4.1, we need to account for the frequency to $O(\epsilon^{N+1})$ in order to have uniform validity in I to $O(\epsilon^N)$. In this example, the frequency expansion proceeds in even powers of ϵ.

We also see that the general asymptotic expansion (4.2.7) to any $O(\epsilon^N)$ may be expressed *uniquely* in the form of a series of functions of the *fast scale* $t_N^+ = \Omega_N(\epsilon)t$ and the *slow scale* $\tilde{t} = \epsilon t$. In fact, the exact expression (4.2.4) is itself a unique function of t_∞^+ and $\tilde{t}$ where $t_\infty^+ = \Omega_\infty(\epsilon)t = \sqrt{1-\epsilon^2}t$.

We have

$$y(t; \epsilon) = F(t_\infty^+, \tilde{t}; \epsilon), \tag{4.2.11}$$

where

$$F(t_\infty^+, \tilde{t}; \epsilon) = \frac{e^{-\tilde{t}}}{\sqrt{1-\epsilon^2}} \sin t_\infty^+, \tag{4.2.12}$$

and for any given integer N, F has a *unique two-scale* expansion correct to $O(\epsilon^N)$ uniformly in I of the form

$$F = \sum_{n=0}^{N} F_n(t_N^+, \tilde{t})\epsilon^n + O(\epsilon^{N+1}), \tag{4.2.13}$$

where for N even we have

$$F_0 = e^{-\tilde{t}} \sin t_N^+, \tag{4.2.14a}$$

$$F_1 = 0, \tag{4.2.14b}$$

$$F_2 = \frac{e^{-\tilde{t}}}{2} \sin t_N^+, \tag{4.2.14c}$$

$$F_3 = 0, \tag{4.2.14d}$$

$$\vdots$$

$$F_N = \alpha_N e^{-\tilde{t}} \sin t_N^+, \tag{4.2.14e}$$

$$t_N^+ = \Omega_N(\epsilon)t. \tag{4.2.14f}$$

Two-scale expansion of the differential equation

Guided by the above results, let us attempt to reconstruct the expansion (4.2.13) from the governing differential equation and initial conditions without direct knowledge of the exact solution. The ideas we develop will later be applied to nonlinear problems where the exact solution is not available.

The fundamental assumption is that solutions have a general asymptotic expansion that is uniformly valid in the interval $I\colon 0 \leq t \leq T(\epsilon) = O(\epsilon^{-1})$, and each term in this general asymptotic expansion can be uniquely expressed as a function of t^+, $\tilde{t}$ and a power of ϵ as in (4.2.13).

This assumption, which we will not attempt to justify in each case, enables us to calculate uniformly valid expansions in a wide variety of problems that will be discussed in this and subsequent chapters.

Let us first examine the implications of this assumption. Suppose we encounter a mixed-secular term of the form $\epsilon t_2^+ \sin t_2^+$ in the $O(\epsilon)$ contribution to the expansion, where $t_2^+ = (1 + \epsilon\omega_1 + \epsilon^2\omega_2)t$. Such a term is not consistent with a uniquely defined two scale expansion because we may express this term as follows:

$$\epsilon t_2^+ \sin t_2^+ = \tilde{t} \sin t_2^+ + \epsilon\omega_1\tilde{t} \sin t_2^+ + \epsilon^2\omega_2\tilde{t} \sin t_2^+$$

and redistribute its contribution to various orders. A term need not become unbounded as $t \to \infty$ to be inconsistent. For example, the exponentially decaying term $\epsilon\tilde{t}e^{-\tilde{t}} \sin t_2^+$ is also unacceptable because we can always relabel $\epsilon\tilde{t}$ in terms of t_2^+ and $\tilde{t}$ and change its nominal order, i.e.,

$$\epsilon\tilde{t} = \epsilon^2 t_2^+ - \epsilon^2\omega_1\tilde{t} - \epsilon^3\omega_2\tilde{t} + \dots .$$

We note, however, that it is not possible to uniquely allocate an $O(\epsilon)$ contribution in the frequency in terms of a t^+ and $\tilde{t}$ contribution. To see this, consider a term of the form

$$g = \sin(1 + \epsilon\omega_1 + \epsilon^2\omega_2 + \epsilon^3\omega_3)t$$

for given nonzero constants $\omega_1, \omega_2, \omega_3$. With t_3^+ defined as $(1 + \epsilon\omega_1 + \epsilon^2\omega_2 + \epsilon^3\omega_3)t$, such a term would simply be denoted $g = \sin t_3^+$. However, if we choose another fast scale, say, $t_3^* = (1 + \epsilon^2\omega_2 + \epsilon^3\omega_3)t$ instead of t_3^+, then we find $g = \cos\omega_1\tilde{t} \sin t_3^* + \sin\omega_1\tilde{t} \cos t_3^*$. In fact, there are an infinite number of possible choices of fast scale corresponding to different choices of the $O(\epsilon)$ term in the expansion of the frequency Ω; each of these choices results in a different two-scale representation for g. To avoid this ambiguity, we will henceforth set $\omega_1 = 0$ in the definition of t^+ and account for it via the $\tilde{t}$-dependence of the solution. In Sec. 4.2.5, we will prove that our procedure cannot determine ω_1; however, the choice of ω_1 is irrelevant because ω_1 *cancels out identically* when the two-scale expansion is expressed in terms of t and ϵ.

For notational simplicity, we shall henceforth omit the subscript N in t_N^+, as the number of terms retained in the expansion of the frequency will be clear from the context. We look for a two-scale expansion of the solution of (4.2.1)–(4.2.2) in the form

$$y(t;\epsilon) = F(t^+, \tilde{t}; \epsilon) = \sum_{n=0}^{N} F_n(t^+, \tilde{t})\epsilon^n + O(\epsilon^{N+1}). \tag{4.2.15}$$

Here

$$t^+ = (1 + \epsilon^2\omega_2 + \epsilon^3\omega_3 + \ldots)t, \tag{4.2.16}$$

where $\omega_2, \omega_3, \ldots$ are unknown constants, and $\tilde{t} = \epsilon t$. The chain rule gives

$$\frac{dy}{dt} = (1 + \epsilon^2\omega_2 + \epsilon^3\omega_3 + \ldots)\frac{\partial F}{\partial t^+} + \epsilon\frac{\partial F}{\partial \tilde{t}}, \tag{4.2.17a}$$

$$\frac{d^2y}{dt^2} = (1 + \epsilon^2\omega_2 + \epsilon^3\omega_3 + \ldots)^2\frac{\partial^2 F}{\partial t^{+2}} + 2\epsilon(1 + \epsilon^2\omega_2 + \epsilon^2\omega_3 + \ldots)\frac{\partial^2 F}{\partial t^+\partial\tilde{t}}$$

$$+\epsilon^2\frac{\partial^2 F}{\partial\tilde{t}^2}, \tag{4.2.17b}$$

and using the expansion for F we find to $O(\epsilon^2)$

$$\frac{dy}{dt} = \frac{\partial F_0}{\partial t^+} + \epsilon\left(\frac{\partial F_1}{\partial t^+} + \frac{\partial F_0}{\partial\tilde{t}}\right) + \epsilon^2\left(\frac{\partial F_2}{\partial t^+} + \frac{\partial F_1}{\partial\tilde{t}} + \omega_2\frac{\partial F_0}{\partial t^+}\right) + O(\epsilon^3), \tag{4.2.18a}$$

$$\frac{d^2y}{dt^2} = \frac{\partial^2 F_0}{\partial t^{+^2}} + \epsilon\left(\frac{\partial^2 F_1}{\partial t^{+^2}} + 2\frac{\partial^2 F_0}{\partial t^+\partial\tilde{t}}\right) + \epsilon^2\left(\frac{\partial^2 F_2}{\partial t^{+^2}} + 2\frac{\partial^2 F_1}{\partial t^+\partial\tilde{t}}\right.$$

$$\left.+2\omega_2\frac{\partial^2 F_0}{\partial t^{+^2}} + \frac{\partial^2 F_0}{\partial\tilde{t}^2}\right) + O(\epsilon^3). \tag{4.2.18b}$$

Thus, the sequence of equations that results from (4.2.1) is

$$L(F_0) \equiv \frac{\partial^2 F_0}{\partial t^{+2}} + F_0 = 0, \tag{4.2.19a}$$

$$L(F_1) = -2\frac{\partial^2 F_0}{\partial t^+\partial\tilde{t}} - 2\frac{\partial F_0}{\partial t^+}, \tag{4.2.19b}$$

$$L(F_2) = -2\omega_2\frac{\partial^2 F_0}{\partial t^{+2}} - \frac{\partial^2 F_0}{\partial\tilde{t}^2} - 2\frac{\partial^2 F_1}{\partial t^+\partial\tilde{t}} - 2\frac{\partial F_0}{\partial\tilde{t}} - 2\frac{\partial F_1}{\partial t^+}. \tag{4.2.19c}$$

The first of these is the equation for the free oscillations, while the remainder have the appearance of forced linear oscillations. However, since $F_0 = F_0(t^+, \tilde{t})$, the free linear oscillations that are the solutions to (4.2.19a) have the possibility of being slowly modulated. Thus, we have

$$F_0(t^+, \tilde{t}) = A_0(\tilde{t})\cos t^+ + B_0(\tilde{t})\sin t^+. \tag{4.2.20}$$

According to (4.2.15) and (4.2.18a), the initial conditions $y(0;\epsilon) = 0$, $dy(0;\epsilon)/dt = 1$ become

$$F_0(0,0) = 0, \quad \frac{\partial F_0}{\partial t^+}(0,0) = 1, \tag{4.2.21a}$$

$$F_1(0,0) = 0, \quad \frac{\partial F_1}{\partial t^+}(0,0) = -\frac{\partial F_0}{\partial \tilde{t}}(0,0), \tag{4.2.21b}$$

$$F_2(0,0) = 0, \quad \frac{\partial F_2}{\partial t^+}(0,0) = -\frac{\partial F_1}{\partial \tilde{t}}(0,0) - \omega_2 \frac{\partial F_0}{\partial t^+}(0,0). \tag{4.2.21c}$$

Equation (4.2.21a) yields initial conditions for A_0 and B_0:

$$A_0(0) = 0, \quad B_0(0) = 1. \tag{4.2.22}$$

Nothing more can be found out about $A_0(\tilde{t})$ and $B_0(\tilde{t})$ without considering F_1. This is directly analogous to the situation encountered in Sec. 4.1 for the method of strained coordinates.

Substituting for F_0 into the right-hand side of (4.2.19b) gives

$$L(F_1) = 2\left[\frac{dA_0}{d\tilde{t}} + A_0\right]\sin t^+ - 2\left[\frac{dB_0}{d\tilde{t}} + B_0\right]\cos t^+. \tag{4.2.23}$$

The bracketed terms on the right-hand side of (4.2.23) are functions of $\tilde{t}$ only. Therefore, the particular solutions corresponding to these terms would be functions of $\tilde{t}$ multiplied by the mixed-secular terms $t^+ \sin t^+$ or $t^+ \cos t^+$. Such terms cannot be permitted to occur in the solution because, as discussed earlier, they are inconsistent with a unique F_1.

Therefore, we must eliminate all homogeneous solutions of $L(F_1) = 0$ from the right-hand side of (4.2.23), and this gives the two first-order ordinary differential equations for A_0 and B_0:

$$\frac{dA_0}{d\tilde{t}} + A_0 = 0, \tag{4.2.24a}$$

$$\frac{dB_0}{d\tilde{t}} + B_0 = 0. \tag{4.2.24b}$$

Taking account of the initial conditions, (4.2.22), we find that

$$A_0(\tilde{t}) = 0, \quad B_0(\tilde{t}) = e^{-\tilde{t}}. \tag{4.2.25}$$

The uniformly valid expansion to $O(1)$ is

$$y(t;\epsilon) = e^{-\tilde{t}}\sin t^+ + O(\epsilon), \tag{4.2.26}$$

where $t^+ = (1 + O(\epsilon^2))t$, and this agrees with the exact result (see (4.2.10a)).

Thus far, we have determined the first two terms in the expansion (4.2.15):

$$F_0(t^+,\tilde{t}) = e^{-\tilde{t}}\sin t^+, \quad F_1(t^+,\tilde{t}) = A_1(\tilde{t})\cos t^+ + B_1(\tilde{t})\sin t^+, \tag{4.2.27}$$

and (4.2.19c) defining F_2 becomes

$$L(F_2) = \left[2\left(\frac{dA_1}{d\tilde{t}} + A_1\right) + (2\omega_2 + 1)e^{-\tilde{t}}\right]\sin t^+ - 2\left[\frac{dB_1}{d\tilde{t}} + B_1\right]\cos t^+. \tag{4.2.28}$$

Now, $A_1(\tilde{t})$, $B_1(\tilde{t})$, and the frequency shift ω_2 are to be found from similar considerations applied to (4.2.28).

First, repeating the argument that homogeneous solutions of $L(F_2) = 0$ cannot be permitted, we must set the bracketed terms in (4.2.28) equal to zero. Solving the resulting equations for A_1 and B_1 subject to the initial conditions $A_1(0) = B_1(0) = 0$ (which follow from (4.2.21b)), we find

$$A_1(\tilde{t}) = -(\omega_2 + \frac{1}{2})\tilde{t}e^{-\tilde{t}}, \tag{4.2.29a}$$

$$B_1(\tilde{t}) = 0. \tag{4.2.29b}$$

This means that ϵF_1 would contain a term proportional to $\epsilon\tilde{t}e^{-\tilde{t}}\cos t^+$. Again, such a term cannot be consistent because, as pointed out earlier, it can also be written as $\epsilon^2 e^{-\tilde{t}}t^+\cos t^+ + O(\epsilon^3)$ and shift to $O(\epsilon^2)$ in the expansion. One could also have required that $|F_2/F_1|$ be bounded for large $\tilde{t}$ to disallow such a term. Therefore, we must set

$$\omega_2 = -\frac{1}{2}, \tag{4.2.30}$$

and we find the following uniformly valid result in I to $O(\epsilon)$:

$$y(t, \epsilon) = e^{-\tilde{t}}\sin\left[1 - \frac{\epsilon^2}{2} + O(\epsilon^3)\right]t + O(\epsilon^2) \tag{4.2.31}$$

in agreement with (4.2.10b).

All the necessary reasoning has now been explained to carry out the solution to any order and, in fact, to solve a wide variety of weakly nonlinear problems of the form

$$\frac{d^2y}{dt^2} + y + \epsilon f\left(y, \frac{dy}{dt}\right) = 0. \tag{4.2.32}$$

4.2.2 *Oscillator with Small Cubic Damping*

In suitable dimensionless variables (see (2.1.3)), an oscillator with cubic damping can be represented by

$$\frac{d^2y}{dt^2} + y + \epsilon\left(\frac{dy}{dt}\right)^3 = 0, \quad 0 < \epsilon \ll 1. \tag{4.2.33}$$

It is sufficient to consider the special initial conditions

$$y(0; \epsilon) = 1, \tag{4.2.34a}$$

$$\frac{dy}{dt}(0;\epsilon) = 0 \tag{4.2.34b}$$

since the problem is autonomous and the solution is oscillatory.

Using the expansion (4.2.15) with L, t^+, and $\tilde{t}$ as defined before, we calculate

$$L(F_0) = 0, \tag{4.2.35a}$$

$$L(F_1) = -2\frac{\partial^2 F_0}{\partial t^+ \partial \tilde{t}} - \left(\frac{\partial F_0}{\partial t^+}\right)^3, \tag{4.2.35b}$$

$$\begin{aligned} L(F_2) = {} & -2\frac{\partial^2 F_1}{\partial t^+ \partial \tilde{t}} - 2\omega_2 \frac{\partial^2 F_0}{\partial t^{+2}} - \frac{\partial^2 F_0}{\partial \tilde{t}^2} \\ & - 3\left(\frac{\partial F_0}{\partial t^+}\right)^2 \left(\frac{\partial F_1}{\partial t^+}\right) - 3\left(\frac{\partial F_0}{\partial t^+}\right)^2 \left(\frac{\partial F_0}{\partial \tilde{t}}\right). \end{aligned} \tag{4.2.35c}$$

The basic solution is again

$$F_0(t^+, \tilde{t}) = A_0(\tilde{t}) \cos t^+ + B_0(\tilde{t}) \sin t^+. \tag{4.2.36}$$

Now, using well-known identities to express products of trigonometric functions in terms of their various harmonics, we find the following equation for F_1:

$$\begin{aligned} L(F_1) = {} & 2\left[\frac{dA_0}{d\tilde{t}} + \frac{3}{8}A_0(A_0^2 + B_0^2)\right] \sin t^+ - 2\left[\frac{dB_0}{d\tilde{t}} \right. \\ & \left. + \frac{3}{8}B_0(A_0^2 + B_0^2)\right] \cos t^+ \\ & - \frac{A_0}{4}(A_0^2 - 3B_0^2) \sin 3t^+ - \frac{B_0}{4}(B_0^2 - 3A_0^2) \cos 3t^+. \end{aligned} \tag{4.2.37}$$

Removal of mixed-secular terms in t^+ requires that

$$\frac{dA_0}{d\tilde{t}} + \frac{3}{8}A_0(A_0^2 + B_0^2) = 0, \tag{4.2.38a}$$

$$\frac{dB_0}{d\tilde{t}} + \frac{3}{8}B_0(A_0^2 + B_0^2) = 0, \tag{4.2.38b}$$

which are two coupled nonlinear equations for A_0 and B_0.

The cynical reader might say that we have gained little by replacing a nonlinear second-order equation (4.2.33) by the two nonlinear first-order equations (4.2.38) for the slowly varying functions appearing in the first approximation. Actually, the situation is quite a bit better, as the system (4.2.38) is integrable for any initial values of A_0 and B_0, while (4.2.33) is not integrable at all. We will see in Sec. 4.2.5 that this is true for the general weakly nonlinear oscillator (4.2.32).

To solve (4.2.38), we introduce the slowly varying amplitude $\rho_0(\tilde{t})$ and phase shift $\phi_0(\tilde{t})$ by

$$A_0 = \rho_0 \cos \phi_0, \tag{4.2.39a}$$

$$B_0 = \rho_0 \sin \phi_0 \tag{4.2.39b}$$

and transform the system to[1]

$$\frac{d\rho_0}{d\tilde{t}} + \frac{3}{8}\rho_0^3 = 0, \tag{4.2.40a}$$

$$\frac{d\phi_0}{d\tilde{t}} = 0. \tag{4.2.40b}$$

For our choice of initial conditions $\rho_0(0) = 1, \phi_0(0) = 0$ and integrating (4.2.40) gives

$$A_0 = \rho_0 = \frac{1}{\sqrt{1 + 3\tilde{t}/4}}, \tag{4.2.41a}$$

$$B_0 = \phi_0 = 0. \tag{4.2.41b}$$

Thus, the uniformly valid solution to $O(1)$ is

$$F_0 = \frac{1}{\sqrt{1 + 3\tilde{t}/4}} \cos[1 + O(\epsilon^2)]t. \tag{4.2.42}$$

We see that because of the relatively weaker damping than in the linear case, the solution now decays algebraically.

We can also solve what remains of (4.2.37) for F_1 in the form

$$F_1(t^+, \tilde{t}) = A_1(\tilde{t}) \cos t^+ + B_1(\tilde{t}) \sin t^+ + \frac{\sin 3t^+}{4(3\tilde{t} + 4)^{3/2}}. \tag{4.2.43}$$

It is left as an exercise for the reader to show that

$$A_1(\tilde{t}) = 0, \quad \omega_2 = 0, \tag{4.2.44a}$$

$$B_1(\tilde{t}) = \frac{1}{(3\tilde{t} + 4)^{1/2}} \left[\frac{3}{8(3\tilde{t} + 4)} + \frac{15}{32} \right]. \tag{4.2.44b}$$

4.2.3 *Rayleigh's Equation*

In Sec. 4.1.2, we studied the periodic solutions of Rayleigh's equation

$$\frac{d^2y}{dt^2} + y + \epsilon \left[-\frac{dy}{dt} + \frac{1}{3}\left(\frac{dy}{dt}\right)^3 \right] = 0. \tag{4.2.45}$$

Here we consider the solution for an arbitrary initial displacement a

$$y(0; \epsilon) = a, \tag{4.2.46a}$$

$$\frac{dy}{dt}(0; \epsilon) = 0. \tag{4.2.46b}$$

---

[1] Had we written the solution (4.2.35) in the equivalent form $F_0 = \rho_0 \cos(t^+ - \phi_0)$, we would have directly obtained (4.2.40).

The two-scale expansion (4.2.15) is again used to calculate the following equations for F_0, F_1, and F_2:

$$L(F_0) = 0, \tag{4.2.47a}$$

$$L(F_1) = -2\frac{\partial^2 F_0}{\partial t^+ \partial \tilde{t}} + \frac{\partial F_0}{\partial t^+} - \frac{1}{3}\left(\frac{\partial F_0}{\partial t^+}\right)^3, \tag{4.2.47b}$$

$$L(F_2) = -2\omega_2 \frac{\partial^2 F_0}{\partial t^{+2}} - 2\frac{\partial^2 F_1}{\partial t^+ \partial \tilde{t}} - \frac{\partial^2 F_0}{\partial \tilde{t}^2} + \frac{\partial F_1}{\partial t^+} - \left(\frac{\partial F_0}{\partial t^+}\right)^2 \left[\frac{\partial F_1}{\partial t^+} + \frac{\partial F_0}{\partial \tilde{t}}\right]. \tag{4.2.47c}$$

The solution is similar to the preceding case. We use (4.2.36) for F_0. Now, the linear-damping term and the factor $\frac{1}{3}$ modify (4.2.37) slightly, and (4.2.47b) becomes

$$\begin{aligned} L(F_1) &= \left[2\frac{dA_0}{d\tilde{t}} - A_0 + \frac{A_0}{4}(A_0^2 + B_0^2)\right] \sin t^+ \\ &\quad - \left[2\frac{dB_0}{d\tilde{t}} - B_0 + \frac{B_0}{4}(A_0^2 + B_0^2)\right] \cos t^+ \\ &\quad - \frac{A_0}{12}(A_0^2 - 3B_0^2) \sin 3t^+ - \frac{B_0}{12}(B_0^2 - 3A_0^2) \cos 3t^+. \end{aligned} \tag{4.2.48}$$

The slowly varying functions A_0 and B_0, obtained by setting the bracketed terms on the right-hand side of (4.2.48) equal to zero, are

$$A_0(\tilde{t}) = \frac{2\lambda}{\sqrt{1 - ke^{-\tilde{t}}}}, \tag{4.2.49a}$$

$$B_0(\tilde{t}) = \frac{2\sqrt{1-\lambda^2}}{\sqrt{1 - ke^{-\tilde{t}}}}, \tag{4.2.49b}$$

where λ and k are integration constants.

This result clearly shows the approach to the limit cycle of amplitude 2, found in Sec. 4.1.2, independent now of the initial condition. The approach is exponential over the slow time $\tilde{t}$. For the initial conditions of (4.2.46), we have $\lambda = 1$, $k = (a^2 - 4)/a^2$ and the following uniformly valid solution in I to $O(1)$:

$$y(t;\epsilon) = \frac{2a}{\sqrt{a^2 - (a^2-4)e^{-\tilde{t}}}} \cos[1 + O(\epsilon^2)]t + O(\epsilon). \tag{4.2.50}$$

In particular, if the solution starts on the limit cycle ($a = 2$), it remains there.

The reader can verify that proceeding further defines the solution as follows:

$$F_1 = B_1(\tilde{t}) \sin t^+ + \frac{a^3}{12[a^2 - (a^2-4)e^{-\tilde{t}}]} \sin 3t^+, \tag{4.2.51}$$

where

$$B_1(\tilde{t}) = \frac{A_0}{8} \log\left(\frac{A_0}{a}\right) + \frac{A_0}{64}(A_0^2 + 5a^2 - 32), \tag{4.2.52a}$$

$$\omega_2 = -\frac{1}{16}. \tag{4.2.52b}$$

4.2.4 Scaling

Unbounded oscillations

So far, the three examples of the type characterized by (4.2.32) that we have considered have described bounded oscillations. The question arises whether the two-scale expansion also applies to situations where the solution becomes unbounded.

For example, if we let ϵ be negative in the linear damping case (4.2.1), the method does give the correct result. However, this linear case is exceptional. For example, if we consider (4.2.33), where we have a small cubic damping, and let ϵ be negative, the result given by (4.2.41a) predicts that the amplitude becomes infinite at the finite time $\tilde{t} = \frac{4}{3}$, i.e., $t = -4/3\epsilon$. This is obviously incorrect as the solution y tends to infinity only as $t \to \infty$. Our expansion procedure tacitly assumes that $y(t;\epsilon) = O(1)$ in I, and this is indeed correct for the solution of (4.2.1). However, if f is such that $y(t;\epsilon) \gg 1$ in I, we must rescale y to correctly order the various terms in (4.2.32). Such a scaling does not affect the relative orders of the three terms in (4.2.1) because it is linear. However, as the following examples indicate, the correct y scaling is crucial if f is nonlinear.

Consider first the case of negative cubic damping ((4.2.33) with $\epsilon < 0$). Let

$$\tilde{y} = (-\epsilon)^{\alpha} y \tag{4.2.53}$$

for some positive α. Inserting this into (4.2.33) gives

$$(-\epsilon)^{-\alpha}\frac{d^2\tilde{y}}{dt^2} + (-\epsilon)^{-\alpha}\tilde{y} - (-\epsilon)^{1-3\alpha}\left(\frac{d\tilde{y}}{dt}\right)^3 = 0. \tag{4.2.54}$$

As y gets large with $\epsilon < 0$, the last term in (3.2.54) becomes larger and larger as we increase α until for $\alpha = \frac{1}{2}$ its order of magnitude equals that of the linear terms. We see that when y does become large, we must solve the full equation

$$\frac{d^2\tilde{y}}{dt^2} + \tilde{y} - \left(\frac{d\tilde{y}}{dt}\right)^3 = 0, \tag{4.2.55}$$

and we no longer have a perturbation problem. Thus, the two-scale solution, (4.2.51a), for the case $\epsilon < 0$ is only initially valid. One needs the exact solution of (4.2.55) to describe y for times greater than $O(1)$.

A similar example is the oscillator with a "soft" spring and negative damping:

$$\frac{d^2y}{dt^2} + y - \epsilon\left(\frac{dy}{dt} + y^3\right) = 0, \quad \epsilon > 0, \tag{4.2.56}$$

$$y(0;\epsilon) = 1, \tag{4.2.57a}$$

$$\frac{dy}{dt}(0; \epsilon) = 0. \tag{4.2.57b}$$

It is easy to establish the qualitative behavior of the solution by noting that (4.2.56) implies

$$\frac{dE}{dt} = \epsilon \left(\frac{dy}{dt}\right)^2 > 0 \tag{4.2.58}$$

with

$$E = \frac{1}{2}\left[\left(\frac{dy}{dt}\right)^2 + y^2 - \frac{\epsilon y^4}{2}\right]. \tag{4.2.59}$$

Here E is the energy of the undamped oscillator. Since $dE/dt > 0$, we expect the motion to start as being oscillatory with slowly increasing amplitude. Because after each cycle of oscillation the energy is slightly higher than its initial value, we expect the motion to "climb out" of the potential well of the function $V(y) = \frac{1}{2}y^2 - \epsilon y^4/4$ and "escape" to either $+\infty$ or $-\infty$ when E exceeds $1/4\epsilon$. This is sketched in Figure 4.2.1 for the case when $y \to -\infty$.

Of course, whether y tends to $+\infty$ or $-\infty$ depends very much on the cumulative effect of the negative damping term.

If we attempt to solve (4.2.56) using a two-scale expansion

$$y = F_0(t^+, \tilde{t}) + \epsilon F_1(t^+, \tilde{t}) + \ldots,$$

we obtain the nonsensical result that

$$F_0 = \rho_0(\tilde{t}) \cos[t + \phi(\tilde{t})],$$

with

$$\rho_0 = e^{\tilde{t}/2},$$

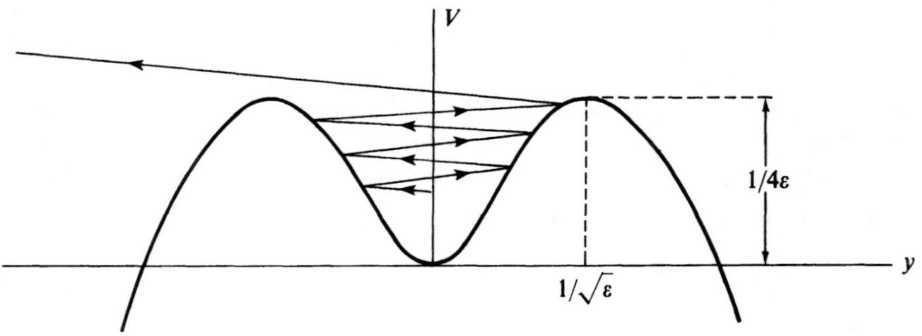

FIGURE 4.2.1. Escape to $y = -\infty$

$$\phi_0 = \frac{3}{8}(1 - e^{\tilde{t}}),$$

which predicts that the solution remains oscillatory with a monotonically increasing amplitude and phase.

Here again, the assumption that the perturbation terms remain $O(\epsilon)$ is incorrect. The correct perturbation problem to be solved is obtained by rescaling y. If we set

$$\tilde{y} = \epsilon^{\alpha} y$$

with $\alpha > 0$, (4.2.56) transforms to

$$\epsilon^{-\alpha}\frac{d^2\tilde{y}}{dt^2} + \epsilon^{-\alpha}\tilde{y} - \epsilon^{1-\alpha}\frac{d\tilde{y}}{dt} - \epsilon^{1-3\alpha}\tilde{y}^3 = 0.$$

Clearly, we must set $-\alpha = 1 - 3\alpha$, i.e., $\alpha = \frac{1}{2}$, to obtain the richest equation when $\epsilon \to 0$. With $\tilde{y}$ so chosen, (4.2.56) becomes

$$\frac{d^2\tilde{y}}{dt^2} + \tilde{y} - \tilde{y}^3 - \epsilon\frac{d\tilde{y}}{dt} = 0. \tag{4.2.60}$$

Thus, the negative damping perturbs the *strictly* nonlinear oscillator, and we expect this to provide a gradual transition from oscillatory motion to escape as pictured in Figure 4.2.1. In Sec. 4.4, we will study the general problem of bounded solutions for a perturbed strictly nonlinear oscillator, and the preescape phase of the solution of (4.2.60) will be worked out there.

Bounded oscillations, f depends on ϵ

It is not only for unbounded oscillations that one must worry about the appropriate scale of the dependent variable. Consider, for example, the problem

$$\frac{d^2y}{dt^2} + y + \epsilon\left(\frac{dy}{dt}\right)^3 + 3\nu\epsilon^2\frac{dy}{dt} = 0, \quad \epsilon > 0, \tag{4.2.61}$$

$$y(0;\epsilon) = 1, \quad \frac{dy}{dt}(0;\epsilon) = 0, \tag{4.2.62}$$

where ν is a positive constant independent of ϵ. Clearly, with $\epsilon > 0$ the motion is damped since both perturbation terms oppose the motion.

If we assume the usual expansion for y in the form

$$y(t,\epsilon) = F_0(t^+, \tilde{t}) + \epsilon F_1(t^+, \tilde{t}) + \ldots, \tag{4.2.63}$$

we find that F_0 is identical with that in (4.2.42) since the term $3\epsilon^2\nu(dy/dt)$ does not affect the determination of F_0. Proceeding further, we find that F_1 is of the form

$$F_1 = A_1(\tilde{t})\cos t^+ + B_1(\tilde{t})\sin t^+ + \frac{1}{4(3\tilde{t} + 4)^{3/2}}\sin 3t^+, \tag{4.2.64}$$

where as before

$$t^+ = [1 + O(\epsilon^3)]t, \tag{4.2.65}$$

$$B_1 = \left[\frac{3}{8(3\tilde{t}+4)^{3/2}} + \frac{15}{32(3\tilde{t}+4)^{1/2}} \right], \tag{4.2.66}$$

and the term multiplied by ν now means that A_1 is not zero but is given by

$$A_1(\tilde{t}) = -\frac{\nu}{(3\tilde{t}+4)^{3/2}} \left(\frac{9}{2}\tilde{t}^2 + 12\tilde{t} \right). \tag{4.2.67}$$

We see that as $\tilde{t} \to \infty$, $A_1 \to -(9\nu/2(3)^{3/2})\tilde{t}^{1/2}$, which means that F_1 becomes unbounded in this limit in contradiction to the damped behavior of the solution. Also, the presence of the quadratic and linear terms in $\tilde{t}$ in the denominator of A_1 implies that there is no consistent way to order this term. Nevertheless, our result is uniformly valid to $O(\epsilon)$ for all t in the large interval $0 \le t \le T(\epsilon) = O(\epsilon^{-1})$.

The inconsistency in F_1 for this example is directly due to our assumption $\epsilon^2(dy/dt) \ll \epsilon(dy/dt)^3$. Initially, when $(dy/dt) = O(1)$, this assumption is correct, but for longer times, as $(dy/dt) \to 0$, the term $3\nu\epsilon^2(dy/dt)$ will be more important than the term $\epsilon(dy/dt)^3$.

Therefore, to determine an ordering of the perturbation that remains consistent for longer times, we rescale y as follows:

$$y^* = \frac{y}{\epsilon^\beta}. \tag{4.2.68}$$

Inserting this into (4.2.61) shows that $\beta = \frac{1}{2}$ and that the correct perturbation problem is defined by

$$\frac{d^2y^*}{dt^2} + y^* + \epsilon^2 \left[\left(\frac{dy^*}{dt} \right)^3 + 3\nu \frac{dy^*}{dt} \right] = 0, \tag{4.2.69}$$

$$y^*(0;\epsilon) = a = \frac{1}{\epsilon^{1/2}}, \tag{4.2.70a}$$

$$\frac{dy^*}{dt}(0;\epsilon) = 0. \tag{4.2.70b}$$

Now, we can construct a two-scale expansion of the form[1]

$$y^*(t;\epsilon) = F_0(t^+, \tilde{\tilde{t}}) + \epsilon^2 F_1(t^+, \tilde{\tilde{t}}) + \dots, \tag{4.2.71}$$

where

$$t^+ = [1 + O(\epsilon^4)]t,$$

$$\tilde{\tilde{t}} = \epsilon^2 t.$$

We calculate

$$F_0 = \rho_0(\tilde{\tilde{t}}) \cos[t^+ + \phi_0(\tilde{\tilde{t}})] \tag{4.2.72}$$

---

[1] Strictly speaking, we should also allow the F_i to depend on $\epsilon^{1/2}$ since this occurs in the initial condition. However, as we will see, disregarding the order of a leads to no difficulties.

and

$$L(F_1) = -2\frac{\partial^2 F_0}{\partial t^+ \partial \tilde{\tilde{t}}} - \left(\frac{\partial F_0}{\partial t^+}\right)^3 - 3\nu \frac{\partial F_0}{\partial t^+}$$
$$= \left(2\frac{d\rho_0}{d\tilde{\tilde{t}}} + \frac{3}{4}\rho_0^3 + 3\nu\rho_0\right)\sin(t^+ + \phi_0)$$
$$+ 2\rho_0 \frac{d\phi_0}{d\tilde{\tilde{t}}}\cos(t^+ + \phi_0) - \frac{\rho_0^3}{4}\sin 3(t^+ + \phi_0). \quad (4.2.73)$$

Eliminating mixed-secular terms determines ρ_0 and ϕ_0 as follows:

$$\rho_0(\tilde{\tilde{t}}) = \frac{2\nu^{1/2}}{[(1 + 4\nu/a^2)e^{3\nu\tilde{\tilde{t}}} - 1]^{1/2}}, \quad (4.2.74)$$

$$\phi_0(\tilde{\tilde{t}}) = 0, \quad (4.2.75)$$

and this shows that the amplitude decays with time.

First, we note that

$$\lim_{\nu \to 0} \epsilon^{1/2}\rho_0 = \frac{2}{\sqrt{4 + 3\tilde{t}}}, \quad \tilde{t} = \epsilon t,$$

as found earlier. Moreover, if we take the limit as $\epsilon \to 0$, $\tilde{t} = \epsilon t$ fixed, the result in (4.2.74) gives the same nonuniform expansion (4.2.63).

4.2.5 General Theory

Guided by the examples worked out thus far, we now consider the two-scale expansion for the general second-order equation

$$\frac{d^2y}{dt^2} + y + \epsilon f\left(y, \frac{dy}{dt}\right) = 0 \quad (4.2.76)$$

with initial conditions

$$y(0; \epsilon) = a, \quad (4.2.77)$$

$$\frac{dy}{dt}(0; \epsilon) = b. \quad (4.2.78)$$

Even though there is no essential difficulty in formally working out the general theory for the case where f also involves ϵ, we do not consider this case to avoid the complications discussed in the last example of the preceding section. Also, we restrict attention to perturbation functions f for which the solution of (4.2.76) is bounded for any choice of a and b.

We seek a solution for Equation (4.2.76) in the form

$$y(t; \epsilon) = F(t^+, \tilde{t}, \epsilon) = \sum_{n=0}^{N} F_n(t^+, \tilde{t})\epsilon^n + O(\epsilon^{N+1}), \quad (4.2.79)$$

where (*Note*: we do not assume $\omega_1 = 0$)

$$t^+ = \left[\sum_{n=0}^{N} \omega_n \epsilon^n + O(\epsilon^{N+1})\right] t, \quad \omega_0 = 1, \tag{4.2.80}$$

$$\tilde{t} = \epsilon t. \tag{4.2.81}$$

We then calculate the following expansions for the various terms appearing in (4.2.76):

$$\frac{dy}{dt} = \frac{\partial F_0}{\partial t^+} + \sum_{n=1}^{N} \left(\frac{\partial F_n}{\partial t^+} + \frac{\partial F_{n-1}}{\partial \tilde{t}} + \sum_{k=1}^{n} \omega_k \frac{\partial F_{n-k}}{\partial t^+} \right) \epsilon^n + O(\epsilon^{N+1}) \tag{4.2.82}$$

$$\begin{aligned}\frac{d^2 y}{dt^2} &= \frac{\partial^2 F_0}{\partial t^{+2}} + \epsilon \left(\frac{\partial^2 F_1}{\partial t^{+2}} + 2\omega_1 \frac{\partial^2 F_0}{\partial t^{+2}} + 2\frac{\partial^2 F_0}{\partial t^+ \partial \tilde{t}} \right) \\ &+ \epsilon^2 \left\{ \sum_{n=0}^{N} \left[\sum_{r=0}^{n+2} \left(\sum_{k=0}^{r} \omega_k \omega_{r-k} \right) \frac{\partial^2 F_{n-r-2}}{\partial t^{+2}} \right.\right. \\ &\left.\left. + 2\sum_{k=0}^{n+1} \omega_k \frac{\partial^2 F_{n-k+1}}{\partial t^+ \partial \tilde{t}} + \frac{\partial^2 F_n}{\partial \tilde{t}^2} \right] \epsilon^n \right\} + O(\epsilon^{N+1}). \end{aligned} \tag{4.2.83}$$

Since the arguments of the perturbation function f are expanded in powers of ϵ, f itself can be expanded in the form

$$f\left(y, \frac{dy}{dt}\right) = f_0 + \epsilon f_1 + O(\epsilon^2), \tag{4.2.84}$$

where

$$f_0 = f\left(F_0, \frac{\partial F_0}{\partial t^+}\right), \tag{4.2.85}$$

$$f_1 = F_1 \frac{\partial f}{\partial y}\left(F_0, \frac{\partial F_0}{\partial t^+}\right) + \left(\frac{\partial F_1}{\partial t^+} + \frac{\partial F_0}{\partial \tilde{t}} + \omega_1 \frac{\partial F_0}{\partial t^+}\right) \frac{\partial f}{\partial \dot{y}}\left(F_0, \frac{\partial F_0}{\partial t^+}\right). \tag{4.2.86}$$

Substituting the above into (4.2.76) gives

$$L(F_0) \equiv \frac{\partial^2 F_0}{\partial t^{+2}} + F_0 = 0, \tag{4.2.87a}$$

$$L(F_1) = -2\frac{\partial^2 F_0}{\partial t^+ \partial \tilde{t}} - 2\omega_1 \frac{\partial^2 F_0}{\partial t^{+2}} - f_0, \tag{4.2.87b}$$

$$\begin{aligned} L(F_2) = &- 2\frac{\partial^2 F_1}{\partial t^+ \partial \tilde{t}} - (2\omega_2 + \omega_1^2)\frac{\partial^2 F_0}{\partial t^{+2}} - \frac{\partial^2 F_0}{\partial \tilde{t}^2} \\ &- 2\omega_1 \left(\frac{\partial^2 F_1}{\partial t^{+2}} + \frac{\partial^2 F_0}{\partial t^+ \partial \tilde{t}} \right) - f_1. \end{aligned} \tag{4.2.87c}$$

The initial conditions, when expanded, give

$$F_0(0,0) = a; \quad F_n(0,0) = 0, \quad n \neq 0, \tag{4.2.88a}$$

$$\frac{\partial F_0}{\partial t^+}(0,0) = b; \quad \frac{\partial F_n}{\partial t^+} = -\left(\frac{\partial F_{n-1}}{\partial \tilde{t}}(0,0)\right.$$

$$\left. + \sum_{k=1}^{n} \omega_k \frac{\partial F_{n-k}}{\partial t^+}(0,0)\right), \quad n \neq 0. \tag{4.2.88b}$$

The solution of the sequence of equations (4.2.87) was worked out in [4.19] explicitly to $O(\epsilon)$. However, the differential equations governing the slowly varying parameters were only worked out explicitly for the solution to $O(1)$. In [4.27], Morrison gives the details of this calculation explicitly to $O(\epsilon)$, and we now present his approach.

We write the solution of (4.2.87a) in the form

$$F_0 = \alpha_0(\tilde{t}) \cos \psi, \tag{4.2.89a}$$

where

$$\psi = t^+ - \phi_0(\tilde{t}) \tag{4.2.89b}$$

and α_0 and ϕ_0 are the slowly varying amplitude and phase of the $O(1)$ solution. The initial conditions imply that

$$\alpha_0(0) = (a^2 + b^2)^{1/2}, \tag{4.2.90a}$$

$$\phi_0(0) = \tan^{-1} \frac{b}{a}. \tag{4.2.90b}$$

If we substitute this result into the right-hand side of (4.2.87b), we find

$$\begin{aligned} L(F_1) = &2\alpha_0' \sin \psi - 2\alpha_0(\phi_0' - \omega_1) \cos \psi \\ &- f(\alpha_0 \cos \psi, -\alpha_0 \sin \psi), \end{aligned} \tag{4.2.91}$$

where a prime denotes $d/d\tilde{t}$.

Now, since the arguments of f depend on $\sin \psi$ and $\cos \psi$, f is a periodic function of ψ with period 2π. If we wish, we can express f in a Fourier series with respect to the variable ψ and the coefficients of this Fourier series *will be functions only of* α_0. This is the approach adopted in [4.19], but it proves to be inconvenient in the calculations to higher order. Instead, Morrison [4.27] proposes to use the Fourier series idea only to isolate the first harmonics of periodic functions, as these first harmonics contribute inconsistent mixed-secular terms in t^+ in the solution to $O(1)$. The first harmonics of f are

$$\frac{1}{\pi}\int_0^{2\pi} f(\alpha_0 \cos \psi, -\alpha_0 \sin \psi) \sin \psi d\psi \equiv 2P_1(\alpha_0), \tag{4.2.92a}$$

$$\frac{1}{\pi}\int_0^{2\pi} f(\alpha_0 \cos \psi, -\alpha_0 \sin \psi) \cos \psi d\psi \equiv 2Q_1(\alpha_0). \tag{4.2.92b}$$

In principle, for a given perturbation function f, one can calculate the two functions P_1 and Q_1. Then, in order to remove terms multiplying $\sin\psi$ and $\cos\psi$ in (4.2.91), we must set

$$\alpha_0' = P_1(\alpha_0), \tag{4.2.93a}$$

$$-\alpha_0(\phi_0' - \omega_1) = Q_1(\alpha_0). \tag{4.2.93b}$$

The above differential equations, subject to the initial conditions given by (4.2.90), define α_0 and ϕ_0 In fact, we can evaluate $\tilde{t}$ in terms of α_0 by quadrature from (4.2.93a) in the form

$$\tilde{t} = \int_{\alpha_0(0)}^{\alpha_0} \frac{ds}{P_1(s)} = K[(\alpha_0, \alpha_0(0)], \tag{4.2.94}$$

which, when inverted, gives

$$\alpha_0 = I[\tilde{t}, \alpha_0(0)]. \tag{4.2.95}$$

If we now denote $Q_1(\alpha_0)/\alpha_0$ by $J[\tilde{t}, \alpha_0(0)]$, we can also compute ϕ_0 by quadrature in the form

$$\phi_0 = \phi_0(0) + \omega_1\tilde{t} - \int_0^{\tilde{t}} J[s, \alpha_0(0)]ds. \tag{4.2.96}$$

It is important to note that the system of two first-order nonlinear equations (4.2.92) will *always* uncouple and can be solved by quadrature [see the remarks following equations (4.2.38)].

Let us now consider the role of the unknown coefficient ω_1 in the solution. If we substitute the result we have for ϕ_0 into (4.2.89a) and write out t^+ and $\tilde{t}$ in terms of ϵ and t according to the definitions (4.2.80) and (4.2.81), we find

$$F_0 = \alpha_0 \cos\left(t + \epsilon^2\omega_2 t + \epsilon^3\omega_3 t + \ldots - \phi_0(0) + \int_0^{\tilde{t}} J ds\right). \tag{4.2.97}$$

We see that the $\epsilon\omega_1 t$ term in t^+ cancels the $\omega_1\tilde{t}$ term in ϕ_0 and *the final result for ψ is independent of ω_1*. In fact, we will see that to all orders the solution can be expressed as a function of ψ and $\tilde{t}$. Thus, our earlier claim based on intuition that ω_1 may be set equal to zero [see equation (4.2.16)] is justified, and henceforth we will adopt this definition of t^+.

The right-hand side of (4.2.91), now free of first harmonics, becomes

$$L(F_1) = 2[P_1(\alpha_0)\sin\psi + Q_1(\alpha_0)\cos\psi] - f_0(\alpha_0, \psi), \tag{4.2.98}$$

where we have used the notation

$$f_0(\alpha_0, \psi) = f(\alpha_0\cos\psi, -\alpha_0\sin\psi). \tag{4.2.99}$$

We now seek the general solution of (4.2.98) in the usual way as

$$F_1 = F_1^{(H)} + F_1^{(P)}, \tag{4.2.100}$$

where $F_1^{(H)}$ is the homogeneous solution

$$F_1^{(H)}(t^+, \tilde{t}) = A_1(\tilde{t}) \cos \psi + B_1(\tilde{t}) \sin \psi \tag{4.2.101a}$$

and $F_1^{(P)}$ is a particular solution assumed in the form

$$F_1^{(P)} = \lambda_1(\alpha_0, \psi) \cos \psi + \alpha_0 \mu_1(\alpha_0, \psi) \sin \psi. \tag{4.2.101b}$$

Moreover, since we have already included arbitrary $\cos \psi$, $\sin \psi$ terms in the homogeneous solution, we require these to be absent from $F_1^{(P)}$, i.e., we set

$$\int_0^{2\pi} \lambda_1(\alpha_0, \psi) d\psi = \int_0^{2\pi} \mu_1(\alpha_0, \psi) d\psi = 0. \tag{4.2.102}$$

Using variation of parameters, we find that λ_1 and μ_1 must satisfy

$$\frac{\partial \lambda_1}{\partial \psi} = f_0(\alpha_0, \psi) \sin \psi - P_1(\alpha_0); \tag{4.2.103}$$

$$\alpha_0 \frac{\partial \mu_1}{\partial \psi} = -f_0(\alpha_0, \psi) \cos \psi + Q_1(\alpha_0). \tag{4.2.104}$$

Note that the right-hand sides of (4.2.103)–(4.2.104) are 2π-periodic functions of ψ, and according to the definitions of P_1 and Q_1 these periodic functions *have zero average value*. Therefore, the quadrature of (4.2.103)–(4.2.104) defines λ_1 and μ_1 as periodic functions of ψ. Again, for a given perturbation function f, one can calculate λ_1 and μ_1 explicitly by quadrature so F_1 only involves the two unknown functions of $\tilde{t}$, A_1 and B_1. These will be defined next by requiring F_2 to be consistent.

The right-hand side of (4.2.87c), with $\omega_1 = 0$, is simply

$$L(F_2) = -f_1 - \left(\frac{\partial^2 F_0}{\partial \tilde{t}^2} + 2 \frac{\partial^2 F_1}{\partial t^+ \partial \tilde{t}} \right) - 2\omega_2 \frac{\partial^2 F_0}{\partial t^{+2}}. \tag{4.2.105}$$

We will now isolate the first harmonics of the terms appearing on the right-hand side. Consider first f_1. Using the definition (4.2.86), we need to compute $\partial F_1/\partial t^+$ and $\partial F_0/\partial \tilde{t}$, and these are

$$\begin{aligned} \frac{\partial F_1}{\partial t^+} = \frac{\partial F_1}{\partial \psi} &= [B_1(\tilde{t}) + \alpha_0(\tilde{t})\mu_1(\alpha_0, \psi) - P_1(\alpha_0)] \cos \psi \\ &\quad - [A_1(\tilde{t}) + \lambda_1(\alpha_0, \psi) - Q_1(\alpha_0)] \sin \psi, \end{aligned} \tag{4.2.106a}$$

$$\begin{aligned} \frac{\partial F_0}{\partial \tilde{t}} &= \alpha_0'(\tilde{t}) \cos \psi + \alpha_0(\tilde{t})\phi_0'(\tilde{t}) \sin \psi \\ &= P_1(\alpha_0) \cos \psi - Q_1(\alpha_0) \sin \psi. \end{aligned} \tag{4.2.106b}$$

Therefore,

$$\frac{\partial F_1}{\partial t^+} + \frac{\partial F_0}{\partial \tilde{t}} = (B_1 + \alpha_0 \mu_1) \cos \psi - (A_1 + \lambda_1) \sin \psi. \tag{4.2.107a}$$

Substituting this and the expression

$$F_1 = (A_1 + \lambda_1) \cos \psi + (B_1 + \alpha_0 \mu_1) \sin \psi \tag{4.2.107b}$$

into (4.2.86) gives

$$f_1 = (A_1 + \lambda_1)\left[\frac{\partial f}{\partial y}\left(F_0, \frac{\partial F_0}{\partial t^+}\right)\cos\psi - \frac{\partial f}{\partial \dot{y}}\left(F_0, \frac{\partial F_0}{\partial t^+}\right)\sin\psi\right]$$
$$+ (B_1 + \alpha_0\mu_1)\left[\frac{\partial f}{\partial y}\left(F_0, \frac{\partial F_0}{\partial t^+}\right)\sin\psi + \frac{\partial f}{\partial \dot{y}}\left(F_0, \frac{\partial F_0}{\partial t^+}\right)\cos\psi\right]. \quad (4.2.108)$$

But, according to the definition (4.2.99), for f_0 we have

$$\frac{\partial f}{\partial y}\left(F_0, \frac{\partial F_0}{\partial t^+}\right)\cos\psi - \frac{\partial f}{\partial \dot{y}}\left(F_0, \frac{\partial F_0}{\partial t^+}\right)\sin\psi = \frac{\partial f_0}{\partial \alpha_0}, \quad (4.2.109a)$$

$$\frac{\partial f}{\partial y}\left(F_0, \frac{\partial F_0}{\partial t^+}\right)(-\alpha_0\sin\psi) - \frac{\partial f}{\partial \dot{y}}\left(F_0, \frac{\partial F_0}{\partial t^+}\right)(\alpha_0\cos\psi) = \frac{\partial f_0}{\partial \psi}. \quad (4.2.109b)$$

Therefore,

$$f_1 = (A_1 + \lambda_1)\frac{\partial f_0}{\partial \alpha_0} - \left(\frac{B_1}{\alpha_0} + \mu_1\right)\frac{\partial f_0}{\partial \psi}. \quad (4.2.110)$$

Consider now the expressions for one-half times the coefficients of $\sin\psi$ and $\cos\psi$ in the Fourier expansion for f_1. These are

$$\frac{1}{2\pi}\int_0^{2\pi} f_1 \sin\psi d\psi = \frac{1}{2\pi}\int_0^{2\pi}\left[(A_1 + \lambda_1)\frac{\partial f_0}{\partial \alpha_0}\sin\psi - \left(\frac{B_1}{\alpha_0} + \mu_1\right)\frac{\partial f_0}{\partial \psi}\sin\psi\right]d\psi \quad (4.2.111a)$$

and

$$\frac{1}{2\pi}\int_0^{2\pi} f_1 \cos\psi d\psi = \frac{1}{2\pi}\int_0^{2\pi}\left[(A_1 + \lambda_1)\frac{\partial f_0}{\partial \alpha_0}\cos\psi - \left(\frac{B_1}{\alpha_0} + \mu_1\right)\frac{\partial f_0}{\partial \psi}\cos\psi\right]d\psi. \quad (4.2.111b)$$

Noting that A_1, B_1, and α_0 are functions only of $\tilde{t}$, and that λ_1, μ_1, $\partial f_0/\partial\alpha_0$, and $\partial f_0/\partial\psi$ are known functions of α_0 and ψ, we write (4.2.111) as

$$\frac{1}{2\pi}\int_0^{2\pi} f_1 \sin\psi d\psi = \frac{A_1}{2\pi}\int_0^{2\pi}\frac{\partial f_0}{\partial \alpha_0}\sin\psi d\psi - \frac{B_1}{2\pi\alpha_0}\int_0^{2\pi}\frac{\partial f_0}{\partial \psi}\sin\psi d\psi + P_2(\alpha_0), \quad (4.2.112a)$$

$$\frac{1}{2\pi}\int_0^{2\pi} f_1 \cos\psi d\psi = \frac{A_1}{2\pi}\int_0^{2\pi}\frac{\partial f_0}{\partial \alpha_0}\cos\psi d\psi - \frac{B_1}{2\pi\alpha_0}\int_0^{2\pi}\frac{\partial f_0}{\partial \psi}\cos\psi d\psi + Q_2(\alpha_0), \quad (4.2.112b)$$

where P_2 and Q_2 are the following known functions of α_0:

$$P_2(\alpha_0) = \frac{1}{2\pi}\int_0^{2\pi} \lambda_1(\alpha_0,\psi)\frac{\partial f_0}{\partial \alpha_0}(\alpha_0,\psi)\sin\psi\, d\psi - \frac{1}{2\pi}\int_0^{2\pi} \mu_1(\alpha_0,\psi)\frac{\partial f_0}{\partial \alpha_0}(\alpha_0,\psi)\sin\psi\, d\psi, \quad (4.2.113a)$$

$$Q_2(\alpha_0) = \frac{1}{2\pi}\int_0^{2\pi} \lambda_1(\alpha_0,\psi)\frac{\partial f_0}{\partial \alpha_0}(\alpha_0,\psi)\cos\psi\, d\psi - \frac{1}{2\pi}\int_0^{2\pi} \mu_1(\alpha_0,\psi)\frac{\partial f_0}{\partial \psi}(\alpha_0,\psi)\cos\psi\, d\psi. \quad (4.2.113b)$$

We can further simplify (4.2.112) by noting, according to (4.2.92), that

$$\frac{dP_1}{d\alpha_0} = \frac{1}{2\pi}\int_0^{2\pi} \frac{\partial f_0}{\partial \alpha_0}(\alpha_0,\psi)\sin\psi\, d\psi \quad (4.2.114a)$$

and

$$\frac{dQ_1}{d\alpha_0} = \frac{1}{2\pi}\int_0^{2\pi} \frac{\partial f_0}{\partial \alpha_0}(\alpha_0,\psi)\cos\psi\, d\psi. \quad (4.2.114b)$$

Moreover, integration by parts and use of (4.2.91) gives

$$\frac{1}{2\pi}\int_0^{2\pi} \frac{\partial f_0}{\partial \psi}(\alpha_0,\psi)\sin\psi\, d\psi = -Q_1(\alpha_0), \quad (4.2.115a)$$

$$\frac{1}{2\pi}\int_0^{2\pi} \frac{\partial f_0}{\partial \psi}(\alpha_0,\psi)\cos\psi\, d\psi = P_1(\alpha_0). \quad (4.2.115b)$$

Thus, (4.2.112) simplify to

$$\frac{1}{2\pi}\int_0^{2\pi} f_1 \sin\psi\, d\psi = P_2(\alpha_0) + A_1(\tilde{t})\frac{dP_1}{d\alpha_0} + \frac{B_1(\tilde{t})}{\alpha_0}Q_1(\alpha_0), \quad (4.2.116a)$$

$$\frac{1}{2\pi}\int_0^{2\pi} f_1 \cos\psi\, d\psi = Q_2(\alpha_0) + A_1(\tilde{t})\frac{dQ_1}{d\alpha_0} - \frac{B_1(\tilde{t})}{\alpha_0}P_1(\alpha_0). \quad (4.2.116b)$$

Next, we consider the contributions from the two terms $(\partial^2 F_0/\partial \tilde{t}^2) + 2(\partial^2 F_1/\partial t^+ \partial \tilde{t})$. Denoting

$$g(\psi,\tilde{t}) \equiv \frac{\partial F_0}{\partial \tilde{t}} + 2\frac{\partial F_1}{\partial t^+}, \quad (4.2.117)$$

we have

$$\begin{aligned}\frac{\partial^2 F_0}{\partial \tilde{t}^2} + 2\frac{\partial^2 F_1}{\partial t^+ \partial \tilde{t}} &= \frac{\partial}{\partial \tilde{t}}[g(\psi,\tilde{t})] \\ &= \frac{\partial g}{\partial \tilde{t}} + \frac{\partial g}{\partial \psi}\frac{\partial \psi}{\partial \tilde{t}} \\ &= \frac{\partial g}{\partial \tilde{t}} + \frac{Q_1}{\alpha_0}\frac{\partial g}{\partial \psi}. \end{aligned} \quad (4.2.118)$$

Thus, one-half times the coefficients of the first harmonics of $(\partial^2 F_0/\partial\tilde{t}^2) + 2(\partial^2 F_1/\partial t^+ \partial\tilde{t})$ is given by

$$\frac{1}{2\pi}\int_0^{2\pi} \frac{\partial}{\partial\tilde{t}}[g(\psi,\tilde{t})]\sin\psi d\psi = \frac{1}{2\pi}\int_0^{2\pi}\frac{\partial g}{\partial\tilde{t}}\sin\psi d\psi + \frac{Q_1}{2\pi\alpha_0}\int_0^{2\pi}\frac{\partial g}{\partial\psi}\sin\psi d\psi, \quad (4.2.119a)$$

$$\frac{1}{2\pi}\int_0^{2\pi} \frac{\partial}{\partial\tilde{t}}[g(\psi,\tilde{t})]\cos\psi d\psi = \frac{1}{2\pi}\int_0^{2\pi}\frac{\partial g}{\partial\tilde{t}}\cos\psi d\psi + \frac{Q_1}{2\pi\alpha_0}\int_0^{2\pi}\frac{\partial g}{\partial\psi}\cos\psi d\psi. \quad (4.2.119b)$$

We integrate the second terms in (4.2.119) by parts and note that the derivative with respect to $\tilde{t}$ can be moved outside the integrals of the first terms. Thus,

$$\frac{1}{2\pi}\int_0^{2\pi} \frac{\partial}{\partial\tilde{t}}[g(\psi,\tilde{t})]\sin\psi d\psi = \frac{1}{2\pi}\frac{d}{d\tilde{t}}\int_0^{2\pi} g(\psi,\tilde{t})\sin\psi d\psi - \frac{Q_1}{2\pi\alpha_0}\int_0^{2\pi} g(\psi,\tilde{t})\cos\psi d\psi, \quad (4.2.120a)$$

$$\frac{1}{2\pi}\int_0^{2\pi} \frac{\partial}{\partial\tilde{t}}[g(\psi,\tilde{t})]\cos\psi d\psi = \frac{1}{2\pi\alpha_0}\frac{d}{d\tilde{t}}\int_0^{2\pi} g(\psi,\tilde{t})\cos\psi d\psi + \frac{Q_1}{2\pi\alpha_0}\int_0^{2\pi} g(\psi,\tilde{t})\sin\psi d\psi. \quad (4.2.120b)$$

Using (4.2.106), we compute

$$g = 2(B_1\cos\psi - A_1\sin\psi) - H(\alpha_0,\psi), \quad (4.2.121a)$$

where H denotes

$$H(\alpha_0,\psi) \equiv (P_1 - 2\alpha_0\mu_1)\cos\psi + (2\lambda_1 - Q_1)\sin\psi. \quad (4.2.121b)$$

We wish to show that H has no contribution in (4.2.120), i.e., that

$$\int_0^{2\pi} H(\alpha_0,\psi)\sin\psi d\psi = \int_0^{2\pi} H(\alpha_0,\psi)\cos\psi d\psi = 0. \quad (4.2.122)$$

Consider the first part of (4.2.122). Using the definition for H, we have

$$\int_0^{2\pi} H(\alpha_0,\psi)\sin\psi d\psi = -\alpha_0\int_0^{2\pi}\mu_1\sin 2\psi d\psi - \int_0^{2\pi}\lambda_1\cos 2\psi d\psi - \pi Q_1, \quad (4.2.123)$$

where we have used trigonometric identities for $\sin^2\psi$, $\cos^2\psi$, and $\sin\psi\cos\psi$, and we have noted that λ_1 has a zero average value [see (4.2.102)]. If we now use integration by parts and (4.2.103)–(4.2.104) to evaluate $\partial\mu_1/\partial\psi$ and $\partial\lambda_1/\partial\psi$, the right-hand side of (4.2.123) vanishes identically. A similar calculation confirms the second part of (4.2.122). Thus, (4.2.120) reduce to

$$\frac{1}{2\pi}\int_0^{2\pi}\left(\frac{\partial^2 F_0}{\partial\tilde{t}^2}+2\frac{\partial^2 F_1}{\partial t^+\partial\tilde{t}}\right)\sin\psi d\psi=-\left(\frac{dA_1}{d\tilde{t}}+\frac{Q_1 B_1}{\alpha_0}\right), \tag{4.2.124a}$$

$$\frac{1}{2\pi}\int_0^{2\pi}\left(\frac{\partial^2 F_0}{\partial\tilde{t}^2}+2\frac{\partial^2 F_1}{\partial t^+\partial\tilde{t}}\right)\cos\psi d\psi=\frac{dB_1}{d\tilde{t}}-\frac{Q_1 A_1}{\alpha_0}. \tag{4.2.124b}$$

Finally, we have

$$\frac{\partial^2 F_0}{\partial t^{+2}}=-F_0=-\alpha_0\sin\psi. \tag{4.2.125}$$

Collecting the coefficients of the $\sin\psi$ terms on the right-hand side of (4.2.105) and setting these equal to zero gives

$$\frac{dA_1}{d\tilde{t}}-A_1\frac{dP_1}{d\alpha_0}=P_2(\alpha_0). \tag{4.2.126a}$$

Similarly, canceling the coefficients of the $\cos\psi$ terms gives

$$\frac{dB_1}{d\tilde{t}}-\frac{P_1}{\alpha_0}B_1=A_1\left(\frac{Q_1}{\alpha_0}-\frac{dQ_1}{d\alpha_0}\right)-Q_2(\alpha_0)+\omega_2\alpha_0. \tag{4.2.126b}$$

These equations, when solved subject to the appropriate initial conditions, define $A_1(\tilde{t})$ and $B_1(\tilde{t})$. We determine ω_2, as illustrated by the examples in the preceding section, by requiring B_1 to be consistent.

We note first that these differential equations are explicit in the sense that all the terms appearing can be computed a priori for a given perturbation function f. Furthermore, the equations are linear and uncoupled since one can solve (4.2.126a) first for $A_1(\tilde{t})$ and then use the result in (4.2.126b).

We conclude this section by writing the explicit solution for y and dy/dt, assuming that the quadratures in (4.2.93), (4.2.103), (4.2.104), and (4.2.126) have been carried out:

$$\begin{aligned} y=&\{\alpha_0(\tilde{t})+\epsilon[A_1(\tilde{t})+\lambda_1(\alpha_0,\psi)]+O(\epsilon^2)\}\\ &\times\cos\left\{\psi-\epsilon\left[\mu_1(\alpha_0,\psi)+\frac{B_1(\tilde{t})}{\alpha_0(\tilde{t})}\right]+O(\epsilon^2)\right\}+O(\epsilon^2), \end{aligned} \tag{4.2.127a}$$

$$\frac{dy}{dt}=-\{\alpha_0+\epsilon[A_1+\lambda_1]+O(\epsilon^2)\}\sin\left\{\psi-\epsilon\left[\mu_1+\frac{B_1}{\alpha_0}\right]+O(\epsilon^2)\right\}+O(\epsilon^2), \tag{4.2.127b}$$

where

$$\psi=(1+\epsilon^2\omega_2+O(\epsilon^3))t+\phi_0(\tilde{t}). \tag{4.2.127c}$$

It is shown in [4.27] (and in more detail in section 3.7.1 of the predecessor of this book) that (4.1.127) is in agreement with the result found by the method of averaging. The method of averaging is discussed in a more general setting in Chapter 5.

4.2.6 Applicability of the Two-Scale Method to Boundary-Layer Problems

It is interesting to note that a two-scale expansion can also be used to solve problems of boundary-layer type as discussed in Chapter 2. For example, consider the initial-value problem for the damped linear oscillator with small mass discussed in Sec. 2.1. Rewriting (2.1.13) and (2.1.14) in terms of the inner variable $t^* = t/\epsilon$ gives [see also (2.1.39)–(2.1.40)]:

$$\frac{d^2y}{dt^{*2}} + \frac{dy}{dt^*} + \epsilon y = 0. \tag{4.2.128}$$

$$y(0;\epsilon) = 0, \tag{4.2.129a}$$

$$\frac{dy}{dt^*}(0;\epsilon) = 1. \tag{4.2.129b}$$

Thus, in the terminology of this chapter, t^* is the fast time and $t = \epsilon t^*$ is the slow time.

We seek a two-scale expansion in the form

$$y(t^*;\epsilon) = F_0(t^*, t) + \epsilon F_1(t^*, t) + \dots. \tag{4.2.130}$$

The equations for F_0 and F_1 are easily calculated as

$$\frac{\partial^2 F_0}{\partial t^{*2}} + \frac{\partial F_0}{\partial t^*} \equiv M(F_0) = 0, \tag{4.2.131a}$$

$$M(F_1) = -2\frac{\partial^2 F_0}{\partial t^* \partial t} - \frac{\partial F_0}{\partial t} - F_0. \tag{4.2.131b}$$

Solving (4.2.131a) gives

$$F_0 = A_0(t)e^{-t^*} + B_0(t). \tag{4.2.132}$$

The initial conditions (4.2.129) imply that $A_0(0) + B_0(0) = 0$ and $-A_0(0) = 1$. Thus $B_0(0) = 1$. Substituting (4.2.132) into the right-hand side of (4.2.131a) gives

$$M(F_1) = \left(\frac{dA_0}{dt} - A_0\right)e^{-t^*} - \left(\frac{dB_0}{dt} + B_0\right). \tag{4.2.133}$$

The solution of (4.2.133) is easily calculated as

$$F_1 = A_1(t)e^{-t^*} + B_1(t) - \left(\frac{dA_0}{dt} - A_0\right)t^* e^{-t^*} - \left(\frac{dB_0}{dt} + B_0\right)t^*. \tag{4.2.134}$$

Clearly, the terms proportional to $t^* e^{-t^*}$ and t^* are inconsistent because if we relabel these in terms of t we obtain $(t/\epsilon)e^{-t/\epsilon}$ and t/ϵ, which would change their order. So we must set

$$\frac{dA_0}{dt} - A_0 = 0, \tag{4.2.135a}$$

$$\frac{dB_0}{dt} + B_0 = 0. \tag{4.2.135b}$$

Solving these with $A_0(0) = -1$, $B_0(0) = 1$ gives

$$A_0(t) = -e^t, \tag{4.2.136a}$$

$$B_0(t) = e^{-t}. \tag{4.2.136b}$$

Thus, the uniformly valid solution to $O(1)$ that we have obtained is

$$F_0 = e^{-t} - e^{-t^*+t}. \tag{4.2.137}$$

In Chapter 2, we calculated the uniformly valid solution to $O(1)$ for this problem from the exact solution (2.1.15) and by matching an outer and inner limit and obtaining the composite solution in the form [see (2.1.51)]

$$y = e^{-t} - e^{-t^*} + O(\epsilon), \tag{4.2.138}$$

and there appears to be a discrepancy between the two results. However, this discrepancy is superficial because the difference between (4.2.137) and (4.2.138) is $e^{-t^*}(1 - e^t)$ and is transcendentally small in the outer region. In the inner region, it is $O(\epsilon)$ and is taken into account in the second term in the inner expansion. Actually, if we compare (4.2.137) with the exact solution (2.1.15) we see that the exponent $(-t^* - t)$ corresponds to the first *two* terms in the expansion of the second exponential in (2.1.15) (see (2.1.16)). Thus, although (4.2.137) and (4.2.138) are asymptotically equivalent to $O(1)$, the former is, in some sense, more accurate. Of course, if we wish to ensure uniformity near $t = \infty$, we must again introduce a strained slow variable [see (2.1.52)] to represent adequately the behavior of the first exponential in (2.1.15) as $t \to \infty$.

The above ideas also carry over to boundary-value problems. For example, see Problem 5 for a boundary-value problem over the unit interval with a boundary layer near the origin.

In spite of the remarkable efficiency and accuracy of the two-scale expansion method for this example, one must keep the following disadvantages in mind.

In general, the choice of the fast and slow scales are not given a priori, nor does one know the locations of boundary layers, corner layers, etc. These questions can be systematically addressed, as we saw in Chapter 2, by constructing and attempting to match appropriate limit process expansions. It is only when the structure of the composite expansion is known that one can hope to apply a multiple-scale expansion.

Often, limit process expansions correspond to definite physical approximations and are intrinsically important. Constructing the composite solution directly may not be necessary.

At any rate, in this book we will not pursue the question of which method should be used if both are applicable. Whenever a problem does have a solution derivable by matching limit process expansions, we will use this approach. If the problem arises in a form where there is a small cumulative perturbation *but no outer limit*, we will use the multiple-scale method.

Problems

1. Carry out the details of the solution to $O(\epsilon)$ for the examples of Sections 4.2.2 and 4.2.3.
2. Carry out the solution to $O(\epsilon)$ for equations of the type (4.2.32) with the following perturbation functions f:
   a. $f = cy^3 + dy/dt$, $c =$ arbitrary constant independent of ϵ,
   b. $f = (dy/dt)|dy/dt|^n$, $n =$ positive integer,
   c. $f = c(d/dt)(y^3) - (dy/dt)^2$, $c =$ positive constant.
3. Comment on the solution of Problem 3c when c is negative.
4. Can one use the two-scale expansion method to solve

$$\frac{d^2y}{dt^2} - y + \epsilon y^3 = 0, \quad \epsilon > 0, \tag{4.2.139}$$

   for which all solutions are periodic, by perturbing about the $\epsilon = 0$ solution? Show that if the appropriate scaling for y is introduced we no longer have a perturbation problem.
5. Consider the following boundary-value problem for $y^*(x^*, \epsilon)$:

$$\frac{d^2y^*}{dx^{*2}} + \frac{dy^*}{dx^*} - \epsilon y^{*^2} = 0, \quad 0 < \epsilon \ll 1. \tag{4.2.140}$$

$$y^*(0; \epsilon) = 0, \tag{4.2.141a}$$

$$y^*\left(\frac{1}{\epsilon}; \epsilon\right) = 1. \tag{4.2.141b}$$

   Clearly this is the same problem as

$$\epsilon \frac{d^2y}{dx^2} + \frac{dy}{dx} - y^2 = 0. \tag{4.2.142}$$

$$y(0; \epsilon) = 0, \quad y(1; \epsilon) = 1, \quad x = \epsilon x^* \tag{4.2.143}$$

   for $y(x; \epsilon) \equiv y^*(x/\epsilon; \epsilon)$, which can be solved by matching an inner and outer expansion. Do this, and compare your result with the solution by the two-scale (x^*, x) method. What happens if the sign of the y^2 term is positive?

4.3 Multiple-Scale Expansions for General Weakly Nonlinear Oscillators

A large variety of physical problems may be expressed, under certain approximations and in suitable variables, in the form of one or more perturbed linear oscillators. Unlike the class of problems discussed in Sec. 4.2, we will encounter in this section examples where the unperturbed frequency as well as the small nonlinear perturbations depend on t (or a suitable independent variable). We will see that these problems may be solved by straightforward individually tailored extensions of the ideas of two-scale expansions. Later, in Sec. 4.5 we will formulate a systematic and generally applicable multiple-scale procedure that starts from a system of first-order equations in a certain standard form, and we will point out how all the examples discussed here may be expressed in this form.

4.3.1 Forced Motion Near Resonance

In this section, forced and free motions of a weakly nonlinear oscillator are considered. All slow variations are considered to depend on $\tilde{t} = \epsilon t$. It is shown how a consistent two-scale expansion can include many previous cases. The general initial-value problem will be considered in order to get some idea of how the solution approaches its final state in the cases of forced motion. This kind of analysis can take the place of a stability analysis in showing which final states are accessible, that is, stable, and which are not.

In variables with physical units, the initial-value problem for the oscillator can be written (forced Duffing equation)

$$M\frac{d^2Y}{dT^2} + B\frac{dY}{dT} + KY + JY^3 = F\cos\Omega T, \tag{4.3.1}$$

$$Y(0) = D, \qquad \frac{dY}{dT}(0) = 0. \tag{4.3.2}$$

The problem can be expressed in dimensionless variables (y, t). Let $t = \Omega_N T$, $\Omega_N = \sqrt{K/M}$ = natural frequency of free linear oscillations, $y = Y/A$, and A = a characteristic amplitude of motion, to be made more precise later. The following parameters then appear:

$$\epsilon = \frac{JA^2}{K} = \frac{\text{nonlinear part of spring force}}{\text{spring force}},$$

$$\epsilon f = \frac{F}{KA} = \frac{\text{weak driving force}}{\text{spring force}}.$$

Near resonance, this weak driving force, $f = O(1)$, is large enough to cause an $O(1)$ displacement. We set $(\Omega/\Omega_N) = 1 + \epsilon\omega$, $\omega = O(1)$. Thus, the driver frequency is close to resonance of the linear system.

We also set $(B/\sqrt{KM}) = \epsilon\beta$, $\beta = O(1)$; this is the ratio of the weak damping to half the critical linear damping. Thus, the problem reads

$$\frac{d^2y}{dt^2} + \epsilon\beta\frac{dy}{dt} + y + \epsilon y^3 = \epsilon f\cos(1+\epsilon\omega)t = \epsilon f\cos(t+\omega\tilde{t}). \tag{4.3.3}$$

$$y(0;\epsilon) = \delta \equiv \frac{D}{A}; \quad \frac{dy}{dt}(0;\epsilon) = 0. \tag{4.3.4}$$

The original general problem depends on seven physical constants (M, B, K, J, D, F, Ω) and by dimensional analysis the dimensionless version should depend on four parameters (ϵ, β, f, ω). The extra parameter δ appears because the length A has been introduced for convenience. The perturbation expansion is expressed in terms of ϵ so that the resulting solution to be studied still depends in general on three parameters, and a wide variety of phenomena can occur.

The two-scale expansion has the form

$$y = F_0(t,\tilde{t}) + \epsilon F_1(t,\tilde{t}) + \dots. \tag{4.3.5}$$

A strained variable of the form $t^+ = t(1+\epsilon^2\omega_2+\dots)$ is not necessary to the order considered here. In the same way as in Sec. 4.2, we derive the following equations for F_0 and F_1:

$$L(F_0) \equiv \frac{\partial^2 F_0}{\partial t^2} + F_0 = 0. \tag{4.3.6}$$

$$F_0(0,0) = \delta, \quad \frac{\partial F_0}{\partial t}(0,0) = 0. \tag{4.3.7}$$

$$L(F_1) = -F_0^3 + f\cos(t+\omega\tilde{t}) - 2\frac{\partial^2 F_0}{\partial t\,\partial\tilde{t}} - \beta\frac{\partial F_0}{\partial t}. \tag{4.3.8}$$

$$F_1(0,0) = 0, \quad \frac{\partial F_1}{\partial t}(0,0) = -\frac{\partial F_0}{\partial \tilde{t}}(0,0). \tag{4.3.9}$$

Let the solution be represented in terms of a slowly varying amplitude R and phase ν relative to the driver

$$F_0(t,\tilde{t}) = R(\tilde{t})\cos(t+\omega\tilde{t}-\nu(\tilde{t})). \tag{4.3.10}$$

Referring to the initial conditions (4.3.7), we can choose

$$R(0) = \delta, \quad \nu(0) = 0. \tag{4.3.11}$$

Equations for the slowly varying amplitude and phase are obtained in the usual way from the condition that mixed-secular terms do not appear in the solution F_1. The equation (4.3.8) for F_1 now reads

$$L(F_1) = -R^3\cos^3(t+\omega\tilde{t}-\nu) + f\cos(t+\omega\tilde{t}) + 2\frac{dR}{d\tilde{t}}\sin(t+\omega\tilde{t}-\nu)$$
$$+\left(\omega - \frac{d\nu}{d\tilde{t}}\right)R\cos(t+\omega\tilde{t}-\nu) + \beta R\sin(t+\omega\tilde{t}-\nu). \tag{4.3.12}$$

The coefficients of $\cos(t + \omega\tilde{t} - \nu)$, $\sin(t + \omega\tilde{t} - \nu)$ must thus both vanish in (4.3.12). To find these equations, use $\cos^3 t = \frac{3}{4}\cos t + \frac{1}{4}\cos 3t$ and $\cos(t + \omega\tilde{t}) = \cos(t + \omega\tilde{t} - \nu)\cos\nu - \sin(t + \omega\tilde{t} - \nu)\sin\nu$. Then, the basic system to be studied is

$$\frac{dR}{d\tilde{t}} + \frac{\beta}{2}R = \frac{1}{2}f\sin\nu, \tag{4.3.13a}$$

$$R\frac{d\nu}{d\tilde{t}} - \left(\omega - \frac{3}{8}R^2\right)R = \frac{1}{2}f\cos\nu. \tag{4.3.13b}$$

Now, we will consider a series of special cases to obtain some simple results and to show how some previous results are contained in this formalism.

Free undamped motion

For this case, the force $f = 0$ and the damping $\beta = 0$. The system (4.3.13) reduces to

$$\frac{dR}{d\tilde{t}} = 0, \quad R(0) = \delta, \tag{4.3.14a}$$

$$\frac{d\nu}{d\tilde{t}} = \omega - \frac{3}{8}R^2, \quad \nu(0) = 0. \tag{4.3.14b}$$

The solution has $R = \delta = \text{const}$. That is, the amplitude of the motion is preserved and the phase ν is

$$\nu = \omega\tilde{t} - \frac{3}{8}\delta^2\tilde{t}. \tag{4.3.15}$$

Returning to the expansion (4.3.5), we find

$$y = \delta\cos(t + \frac{3}{8}\delta^2\tilde{t}) + \dots. \tag{4.3.16}$$

The driver frequency ω drops out, as it must since it is undefined for a free-motion problem. The amplitude A in this case can be identified with the initial displacement, $\delta = 1$. Thus, $y = \cos(t + \frac{3}{8}\tilde{t})$. This frequency shift is exactly that obtained by the method of strained coordinates (see (4.1.10), (4.1.28)).

Free damped motion, $f = 0$

Now, (4.3.13) simplifies to

$$\frac{dR}{d\tilde{t}} + \frac{\beta}{2}R = 0, \quad R(0) = \delta, \tag{4.3.17a}$$

$$\frac{d\nu}{d\tilde{t}} = \omega - \frac{3}{8}R^2, \quad \nu(0) = 0. \tag{4.3.17b}$$

The solution shows

$$R(\tilde{t}) = \delta e^{-(\beta/2)\tilde{t}} \tag{4.3.18}$$

so that the decay is exactly the same as in the linear case. Then

$$\nu(\tilde{t}) = \omega\tilde{t} - \frac{3}{8\beta}\{1 - e^{-\beta\tilde{t}}\}. \tag{4.3.19}$$

The frequency ω again drops out, $\delta = 1$ as above, and

$$y = e^{-(\beta/2)\tilde{t}} \cos\left(t + \frac{3}{8\beta}(1 - e^{-\beta\tilde{t}})\right) + \dots. \tag{4.3.20}$$

The phase lag goes from zero to $(-(3/8\beta))$ as $\tilde{t} \to \infty$. As $\beta \to 0$ we recover the result in (4.3.16).

Forced linear motion

To achieve this case in the previous framework, it is necessary to drop out the term $\frac{3}{8}R^3$, which comes from the spring nonlinearity in (4.3.13). Since in the original equation $J = 0$, ϵ can instead be identified with the weak driving force. That is, let $\delta = 1$, $A = D$, and $f = 1$, and our basic system for (4.3.13) for slowly varying amplitude and phase becomes

$$\frac{dR}{d\tilde{t}} + \frac{\beta}{2}R = \frac{1}{2}\sin\nu, \qquad R(0) = 0, \tag{4.3.21a}$$

$$R\frac{d\nu}{d\tilde{t}} - \omega R = \frac{1}{2}\cos\nu, \qquad \nu(0) = 0. \tag{4.3.21b}$$

As $\tilde{t} \to \infty$, the slow variations approach a steady state for $\beta > 0$ given by $\beta R = \sin\nu$; $2\omega R = -\cos\nu$ or

$$R(\infty) = \frac{1}{\sqrt{\beta^2 + 4\omega^2}}, \qquad \nu(\infty) = \tan^{-1}\left(-\frac{\beta}{2\omega}\right). \tag{4.3.22}$$

This shows the typical resonance amplification of the forced linear oscillator (see Figure 4.3.1). The phase lag ν, as $\tilde{t} \to \infty$, varies from $0(\omega \to -\infty)$ to $\pi(\omega \to \infty)$ and is exactly $\pi/2$ at the linear resonance value $\omega = 0$. As damping gets relatively smaller, the change in phase takes place in a narrower range of frequency around $\omega = 0$. The approach of the system (4.3.21) to the steady state is most easily expressed in terms of

$$\begin{aligned} A(\tilde{t}) &= R(\tilde{t})\cos\nu(\tilde{t}) \\ B(\tilde{t}) &= R(\tilde{t})\sin\nu(\tilde{t}). \end{aligned} \tag{4.3.23}$$

The original solution (4.3.10) reads

$$F_0(t, \tilde{t}) = A(\tilde{t})\cos(t + \omega\tilde{t}) + B(\tilde{t})\sin(t + \omega\tilde{t}) \tag{4.3.24}$$

in these variables. We easily obtain from (4.3.21) the following system:

$$\frac{dA}{d\tilde{t}} + \frac{\beta}{2}A + \omega B = 0, \tag{4.3.25a}$$

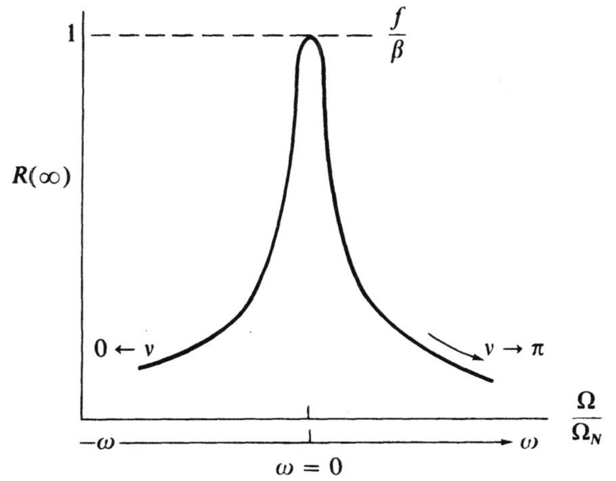

FIGURE 4.3.1. Asymptotic Form of Linear Resonance Curve ($\tilde{t} \to \infty$)

$$\frac{dB}{d\tilde{t}} + \frac{\beta}{2} B - \omega A = \frac{1}{2}, \tag{4.3.25b}$$

for which the steady state corresponding to (4.3.22) is

$$A(\infty) = -\frac{2\omega}{\beta^2 + 4\omega^2}, \quad B(\infty) = \frac{\beta}{\beta^2 + 4\omega^2}. \tag{4.3.26}$$

The difference from the steady state A^*, B^* where

$$A(\tilde{t}) = A^*(\tilde{t}) - \frac{2\omega}{\beta^2 + 4\omega^2}, \quad B(\tilde{t}) = B^*(\tilde{t}) + \frac{\beta}{\beta^2 + 4\omega^2}$$

satisfies

$$\frac{dA^*}{d\tilde{t}} + \frac{\beta}{2} A^* + \omega B^* = 0, \tag{4.3.27a}$$

$$\frac{dB^*}{d\tilde{t}} + \frac{\beta}{2} B^* - \omega A^* = 0. \tag{4.3.27b}$$

The form now suggests that exponential damping can be factored out

$$A^* = \overline{A}(\tilde{t})e^{-(\beta/2)\tilde{t}}, \quad B^* = \overline{B}e^{-(\beta/2)\tilde{t}} \tag{4.3.28}$$

so that

$$\frac{d\overline{A}}{d\tilde{t}} + \omega\overline{B} = 0, \tag{4.3.29a}$$

$$\frac{d\overline{B}}{d\tilde{t}} - \omega\overline{A} = 0, \tag{4.3.29b}$$

which has the simple solution

$$\begin{aligned} \overline{A} &= a\cos\omega\tilde{t} + b\sin\omega\tilde{t}, \\ \overline{B} &= a\sin\omega\tilde{t} - b\cos\omega\tilde{t}. \end{aligned} \tag{4.3.30}$$

Thus

$$\begin{aligned} A(\tilde{t}) &= e^{-(\beta/2)\tilde{t}}(a\cos\omega\tilde{t} + b\sin\omega\tilde{t}) - \frac{2\omega}{\beta^2 + 4\omega}, \\ B(\tilde{t}) &= e^{-(\beta/2)\tilde{t}}(a\sin\omega\tilde{t} - b\cos\omega\tilde{t}) + \frac{\beta}{\beta^2 + 4\omega^2}. \end{aligned} \tag{4.3.31}$$

The initial conditions corresponding to (4.3.21) are $B(0) = 0$, $A(0) = 1$ so that the solution (4.3.31) becomes

$$\begin{aligned} A(\tilde{t}) &= e^{-(\beta/2)\tilde{t}}\left[\left(1 + \frac{2\omega}{\beta^2 + 4\omega^2}\right)\cos\omega\tilde{t} + \frac{\beta\sin\omega\tilde{t}}{\beta^2 + 4\omega^2}\right] - \frac{2\omega}{\beta^2 + 4\omega^2}, \\ B(\tilde{t}) &= e^{-(\beta/2)\tilde{t}}\left[\left(1 + \frac{2\omega}{\beta^2 + 4\omega^2}\right)\sin\omega\tilde{t} - \frac{\beta\cos\omega\tilde{t}}{\beta^2 + 4\omega^2}\right] + \frac{\beta}{\beta^2 + 4\omega^2}. \end{aligned} \tag{4.3.32}$$

The approach to the steady state in the form of oscillatory decay from the given initial state is now clear. This type of behavior disappears as the damping $\beta \to 0$

$$\begin{aligned} A(\tilde{t}) &= \left(1 + \frac{1}{2\omega}\right)\cos\omega\tilde{t} - \frac{1}{2\omega}, \\ B(\tilde{t}) &= \left(1 + \frac{1}{2\omega}\right)\sin\omega\tilde{t}. \end{aligned} \tag{4.3.33}$$

The solution never reaches the steady state, but a slow beating oscillation about the steady state (in $\tilde{t}$) occurs. Finally, if $\beta = 0$ and resonance is approached ($\omega \to 0$), we find

$$\begin{aligned} A(\tilde{t}) &\to 1, \\ B(\tilde{t}) &\to \frac{\tilde{t}}{2}. \end{aligned} \tag{4.3.34}$$

The beat period approaches infinity so that linear growth in $\tilde{t}$ is the ultimate dominating result.

Forced nonlinear motion

According to the basic system (4.3.13) in the general case, the slow variation of amplitude and phase can approach a steady state

$$R(\infty) = \rho, \quad \nu(\infty) = \alpha \tag{4.3.35}$$

given by

$$\begin{aligned} \beta\rho &= f \sin\alpha, \\ -2\omega\rho + \tfrac{3}{4}\rho^3 &= f \cos\alpha. \end{aligned} \tag{4.3.36}$$

These steady states are singular points of the system (4.3.13). The nature of the singular points and the structure of the motion can be discussed qualitatively for very small damping by considering at first $\beta = 0$. Then there are two possible branches to discuss that give different relations between frequency ω and amplitude ρ:

Branch (i)

$$\alpha = 0,$$

$$\omega(\rho) = \frac{3}{8}\rho^2 - \frac{f}{2\rho}. \tag{4.3.37}$$

This branch, in phase with the driver, lies above the free resonance curve $\omega = \frac{3}{8}\rho^2$ (see Figure 4.3.2).

Branch (ii)

$$\alpha = \pi,$$

$$\omega(\rho) = \frac{3}{8}\rho^2 + \frac{f}{2\rho}. \tag{4.3.38}$$

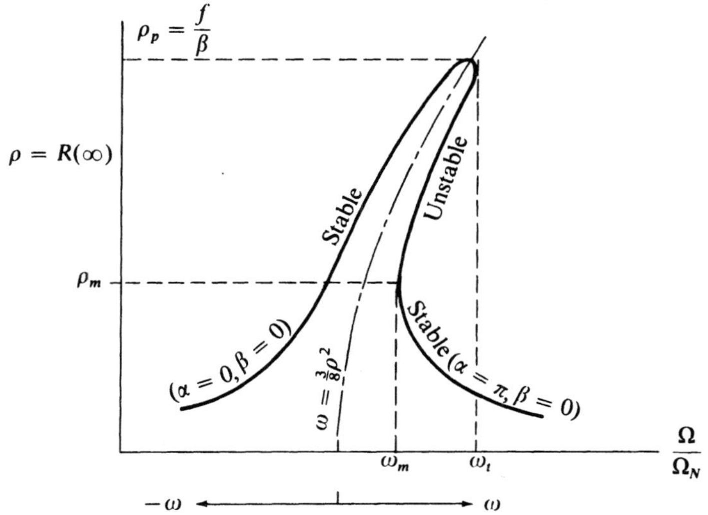

FIGURE 4.3.2. Nonlinear Resonance Diagram f, β Fixed

On this branch, there is a minimum frequency with corresponding amplitude

$$\omega_m = \frac{3^{4/3}}{2^{7/3}} f^{2/3}, \quad \rho_m = \left(\frac{2}{3} f\right)^{1/3}. \tag{4.3.39}$$

This branch lies below the corresponding free branch of the nonlinear motion. If the damping β is finite, then these two branches are joined and the phase varies continuously.

The complete resonance curve is given by

$$\beta^2\rho^2 + \left(2\omega\rho - \frac{3}{4}\rho^3\right)^2 = f^2. \tag{4.3.40}$$

Thus, it can be shown that the peak amplitude occurs exactly on the free "resonance" curve $\omega = \frac{3}{8}\rho^2$ and has the value

$$\rho_p = \frac{f}{\beta}. \tag{4.3.41}$$

The peak amplitude ρ_p of the nonlinear case is thus exactly the same as that of the linear case if the driver frequency is adjusted properly.

The question of whether all points on branches (i) or (ii) can be reached from an arbitrary initial state can be answered by considering the nature of the singular points. This can be done, at first again, for $\beta = 0$. For branch (ii) $\omega > \omega_m, \alpha = \pi$ let

$$\left\{\begin{array}{l} R(\tilde{t}) = \rho + r(\tilde{t}) \\ \nu(\tilde{t}) = \alpha + \theta(\tilde{t}) \end{array}\right\}, \quad \omega = \frac{3}{8}\rho^2 + \frac{f}{2\rho} > \omega_m. \tag{4.3.42}$$

Then linearization of (4.3.13) about the singular point gives

$$\begin{aligned} \frac{dr}{d\tilde{t}} + \frac{f}{2}\theta &= 0, \\ \rho\frac{d\theta}{d\tilde{t}} - \omega(\rho)r + \frac{9}{8}\rho^2 r &= 0. \end{aligned} \tag{4.3.43}$$

For solutions of the form $r = ae^{\lambda\tilde{t}}, \theta = be^{\lambda\tilde{t}}$, we find

$$\lambda^2 = \frac{f}{2}\left(\frac{3}{4}\rho - \frac{f}{2\rho^2}\right). \tag{4.3.44}$$

Thus, on the upper part of this branch ($\rho > \rho_m = \left(\frac{2}{3}f\right)^{1/3}$) there are two real roots $\lambda_{1,2}$ of opposite sign. The singularity is a saddle point. Only one exceptional path runs into this saddle point so that this branch is accessible only from a very special set of initial conditions. Therefore, this branch is labeled unstable in Figure 4.3.2. This singular point is analogous to the unstable equilibrium of a pendulum standing vertically. Therefore, the addition of finite damping cannot change the nature of this singular point. But if $\rho < \rho_m$, $\lambda_{1,2}$ are complex conjugates and the singular point is a center with closed paths around $r = \theta = 0$, the motion in this case is beating with a period depending on $\lambda_{1,2}$. But now the addition of damping

$\beta > 0$ causes the beats to die out and the steady state is approached; the branch is stable.

Similarly, the entire branch (i) is stable and which branch is approached depends, in this view, on the initial conditions. The entire course of all solutions can be studied in the (A, B) phase plane with (R, ν) as polar coordinates. This in fact is one of the main advantages of the present approach to the original nonautonomous system. For R large, the basic system (4.3.13) shows

$$\frac{dR}{d\nu} \to \frac{4}{3}\frac{\beta}{R}, \qquad \frac{dR}{d\tilde{t}} \to -\frac{\beta}{2}R \quad \text{or} \quad R = R_0 e^{-(\beta/2)\tilde{t}}. \tag{4.3.45}$$

The motion spirals inward for $\beta > 0$. In accordance with the previous discussion, all paths run into one singular point for $\omega < \omega_m$ (see Figure (4.3.2)), corresponding to the stable branch with frequencies less than $\frac{3}{8}\rho^2$. For $\omega > \omega_t$ all paths run into the singular point corresponding to the stable part of the branch with frequencies $\omega > \frac{3}{8}\rho^2$. For $\omega_m < \omega < \omega_t$ there is a saddle point corresponding to the unstable branch. The paths through the saddle are separatrices that divide the A, B plane into those initial conditions that run into the different singular points corresponding to the stable branches. A detailed picture depends on the numerical values of (f, β, ω).

4.3.2 *Oscillator with Slowly Varying Frequency*

A natural extension of the two-scale method is to those problems that contain a slowly varying function explicitly. However, the simple example of this section shows that some thought must be given to the proper choice of variables. The classical example is the motion of a pendulum under slow variations in its length. In this context, of course, *slow* means over a time scale much longer than the natural period. In the version corresponding to small amplitudes, the following equation would apply:

$$\frac{d^2y}{dt^2} + \mu^2(\tilde{t})y = 0, \tag{4.3.46}$$

where $\tilde{t} = \epsilon t$ is the slow variable, $\mu^2 > 0$, and $\mu = O(1)$. Arbitrary initial conditions can be chosen, for example, as

$$y(0;\epsilon) = a, \qquad \frac{dy}{dt}(0;\epsilon) = b. \tag{4.3.47}$$

Another problem leading to (4.3.46) is the motion of a charged particle in a magnetic field almost homogeneous in space and varying slowly in time. The equations of motion of a particle of mass m in the xy plane, with a magnetic field $B(t)$ in the z direction, are

$$m\frac{d^2x}{dt^2} = qB\frac{dy}{dt} + qE_x, \qquad m\frac{d^2y}{dt^2} = -qB\frac{dx}{dt} + qE_y, \tag{4.3.48}$$

where q is the charge.

The Maxwell equation

$$\operatorname{curl} E = -\frac{\partial B}{\partial t} \tag{4.3.49}$$

has the local solution

$$E_x = -\frac{y}{2}\frac{dB}{dt} + \dots, \qquad E_y = \frac{x}{2}\frac{dB}{dt} + \dots, \tag{4.3.50}$$

which can be used in (4.3.48) near the origin. Letting

$$u = x + iy, \tag{4.3.51}$$

we find that the system (4.3.48) becomes

$$\frac{d^2u}{dt^2} + i\omega\frac{du}{dt} + \frac{i}{2}\frac{d\omega}{dt}u = 0, \tag{4.3.52}$$

where

$$\omega(t) = \frac{qB(t)}{m}$$

is the cyclotron frequency. For $B = \text{const.}$, the particle motion is a circular orbit about the origin with this frequency. Now, if we introduce the amplitude $\phi(t)$ by

$$u = x + iy = \phi(t)\exp\left[-(i/2)\int_0^t \omega(s)ds\right], \tag{4.3.53}$$

$\phi(t)$ satisfies

$$\frac{d^2\phi}{dt^2} + \frac{\omega^2(t)}{4}\phi = 0. \tag{4.3.54}$$

An attempt to apply a two-scale expansion directly to (4.3.46) with $\mu(\tilde{t})t$ and $\tilde{t}$ as the fast and slow variables fails, as can be quickly verified by the reader. The reason for this failure is in the assumption that μt is the appropriate fast variable for, even though the instantaneous frequency is μ, the phase of the oscillation is not μt but rather the integral of μ with respect to t since μ is *not* constant.

This can be explicitly demonstrated by transforming the independent variable in (4.3.46) from t to t^+ in such a way that the oscillations will have a constant frequency with respect to t^+. If we let

$$t^+ = f(t;\epsilon), \quad f(0;\epsilon) = 0 \tag{4.3.55}$$

and perform the exact transformation (4.3.55) on (4.3.46), we find

$$\frac{d^2y}{dt^{+2}} + \frac{d^2f/dt^2}{(df/dt)^2}\frac{dy}{dt^+} + \frac{\mu^2(\tilde{t})}{(df/dt)^2}y = 0. \tag{4.3.56}$$

$$y(0;\epsilon) = a, \qquad \frac{dy}{dt^+}(0;\epsilon) = \frac{b}{df(0;\epsilon)/dt}. \tag{4.3.57}$$

In order for the oscillations to have constant frequency on the t^+ scale, we must set df/dt proportional to μ. For convenience, we take

$$\frac{df}{dt} = \mu.$$

Now df/dt is the instantaneous frequency on the t scale, and the appropriate fast variable t^+ is

$$t^+ = \int_0^t \mu(\epsilon s)ds = \frac{1}{\epsilon}\int_0^{\tilde{t}} \mu(\tilde{s})d\tilde{s} \equiv f(t;\epsilon). \tag{4.3.58}$$

Since $d^2f/dt^2 = \epsilon d\mu/d\tilde{t}$, the transformed equation for $y^+(t^+;\epsilon)$, where $y^+(f(t;\epsilon);\epsilon) = y(t;\epsilon)$, is

$$\frac{d^2y^+}{dt^{+2}} + \epsilon g(\tilde{t})\frac{dy^+}{dt^+} + y^+ = 0, \quad g(\tilde{t}) = \frac{d\mu/d\tilde{t}}{\mu^2}. \tag{4.3.59}$$

$$y^+(0;\epsilon) = a, \quad \frac{dy^+}{dt^+}(0;\epsilon) = \frac{b}{\mu(0)}. \tag{4.3.60}$$

In these variables, the equation has a small, slowly varying linear damping. Note that the definition (4.3.58) for the fast variable t^+ is a generalization of the usual definition of t^+.

Once the appropriate t^+ is established, we may construct a two-scale expansion of the solution based on either (4.3.46) for $y(t;\epsilon)$ or (4.3.59) for $y^+(t^+;\epsilon)$; the end result is the same.

Using (4.3.59), we expand

$$y(t;\epsilon) = F(t^+,\tilde{t};\epsilon) = F_0(t^+,\tilde{t}) + \epsilon F_1(t^+,\tilde{t}) + \dots \tag{4.3.61}$$

and compute

$$\frac{dy}{dt} = \mu\frac{\partial F}{\partial t^+} + \epsilon\frac{\partial F}{\partial \tilde{t}} = \mu\frac{\partial F_0}{\partial t^+} + \epsilon\left(\mu\frac{\partial F_1}{\partial t^+} + \frac{\partial F_0}{\partial \tilde{t}}\right) + \dots. \tag{4.3.62}$$

$$\begin{aligned}\frac{d^2y}{dt^2} &= \mu^2\frac{\partial^2 F}{\partial t^{+2}} + 2\epsilon\mu\frac{\partial^2 F}{\partial t^+\partial\tilde{t}} + \epsilon\mu'\frac{\partial F}{\partial t^+} + \epsilon^2\frac{\partial^2 F}{\partial \tilde{t}^2} \\ &= \mu^2\frac{\partial^2 F_0}{\partial t^{+2}} + \epsilon\left(\mu^2\frac{\partial^2 F_1}{\partial t^{+2}} + 2\mu\frac{\partial^2 F_0}{\partial t^+\partial\tilde{t}} + \mu'\frac{\partial F_0}{\partial t^+}\right) + \dots.\end{aligned} \tag{4.3.63}$$

Thus, the equations for F_0 and F_1 are

$$\mu^2 L(F_0) \equiv \mu^2\left(\frac{\partial^2 F_0}{\partial t^{+2}} + F_0\right) = 0. \tag{4.3.64}$$

$$\mu^2 L(F_1) = -2\mu\frac{\partial^2 F_0}{\partial t^+\partial\tilde{t}} - \mu'\frac{\partial F_0}{\partial t^+}. \tag{4.3.65}$$

Using

$$F_0(t^+,\tilde{t}) = A_0(\tilde{t})\cos t^+ + B_0(\tilde{t})\sin t^+ \tag{4.3.66}$$

to describe the basic oscillation, we see that (4.3.65) becomes ($' = d/d\tilde{t}$)

$$L(F_1) = \left(A_0 \frac{\mu'(\tilde{t})}{\mu^2(\tilde{t})} + \frac{2}{\mu(\tilde{t})} A_0'\right) \sin t^+ - \left(B_0 \frac{\mu'(\tilde{t})}{\mu^2(\tilde{t})} + \frac{2}{\mu(\tilde{t})} B_0'\right) \cos t^+. \tag{4.3.67}$$

Consistency of the expansion requires that we set the terms in parentheses equal to zero, and we find

$$A_0(\tilde{t}) = A_0(0)\sqrt{\frac{\mu(0)}{\mu(\tilde{t})}}, \quad B_0(\tilde{t}) = B_0(0)\sqrt{\frac{\mu(0)}{\mu(\tilde{t})}}. \tag{4.3.68}$$

The general solution of the initial value problem is thus

$$F_0(t^+, \tilde{t}) = \sqrt{\frac{\mu(0)}{\mu(\tilde{t})}} \left\{ a \cos t^+ + \frac{b}{\mu(0)} \sin t^+ \right\}. \tag{4.3.69}$$

We can use the above explicit solution to verify the well-known result that the action $J \equiv (1/2\mu)[(dy/dt)^2 + \mu^2(\tilde{t})y^2]$ is an *adiabatic invariant* to $O(1)$.

First, we note that there are many possible invariants to $O(1)$, since (4.3.46) with $\epsilon = 0$ is integrable. For example, if we denote the nonconstant instantaneous energy by E:

$$E \equiv \frac{1}{2}\left[\left(\frac{dy}{dt}\right)^2 + \mu^2 y^2\right], \tag{4.3.70}$$

then E, $E\mu$, etc. are all invariants to $O(1)$ because along a solution dE/dt, $d(E\mu)/dt$, etc. are all $O(\epsilon)$. We reserve the term *adiabatic* (i.e., slowly varying) to an invariant whose *derivative is purely oscillatory with zero average on the fast scale*. As a consequence of this property, we expect that assuming an adiabatic invariant to be constant will, in some sense, introduce no cumulative errors as $t \to \infty$. A more precise definition of an adiabatic invariant is given in Sec. 5.1.1. To exhibit the oscillatory nature of $J = E/\mu$, we calculate

$$\frac{dJ}{dt} = \frac{1}{\mu}\frac{dE}{dt} - \frac{\epsilon}{\mu^2} E \frac{d\mu}{d\tilde{t}} = \frac{\epsilon}{2}\frac{d\mu}{d\tilde{t}}\left[y^2 - \frac{1}{\mu^2}\left(\frac{dy}{dt}\right)^2\right]. \tag{4.3.71}$$

Inserting the solution (4.3.69) to $O(1)$ in the above gives

$$\frac{dJ}{dt} = \epsilon\mu(0)\left[a^2 - \frac{b^2}{\mu(0)^2}\right]\frac{\mu'}{2\mu}\cos 2t^+ + \epsilon ab\frac{\mu'}{\mu}\sin 2t^+ + O(\epsilon^2), \tag{4.3.72}$$

which is oscillatory on the scale of t^+ with zero average value. On the other hand, an invariant like E, for example, is less nice because

$$\frac{dE}{dt} = \epsilon\mu\mu' y^2 \tag{4.3.73}$$

and inserting the solution to $O(1)$ shows that

$$\frac{dE}{dt} = \frac{\epsilon\mu(0)}{2}\mu' \left\{ \left[a^2 + \frac{b^2}{\mu(0)^2} \right] + \frac{2ab}{\mu(0)} \sin 2t^+ \right.$$
$$\left. + \left[a^2 - \frac{b^2}{\mu(0)^2} \right] \cos 2t^+ \right\} + O(\epsilon^2). \qquad (4.3.74)$$

Thus, dE/dt has a nonoscillatory component to $O(\epsilon)$ equal to

$$\frac{\epsilon\mu(0)}{2}\mu' \left(a^2 + \frac{b^2}{\mu(0)^2} \right)$$

and is not adiabatic.

4.3.3 Sturm-Liouville Equation; Differential Equation with a Large Parameter

The classical problem of the approximate solution of a differential equation with a large parameter (see (4.3.75) later) falls naturally into the discussion of this chapter. The usual asymptotic expansion valid away from turning points turns out not to be a limit-process expansion but, rather, one of the two-scale type. However, near a turning point, the local behavior dominates, so that a limit-process expansion, valid locally, can be constructed. These two expansions, however, can be matched by the same procedure used for purely limit-process expansions. This extension of our previous ideas should prove useful for many similar problems.

For the consideration of asymptotic distribution of eigenvalues and eigenfunctions and for various other reasons, it is often necessary to obtain the asymptotic behavior of the solutions to the general self-adjoint second-order equation

$$\frac{d}{d\tilde{x}}\left(p(\tilde{x}) \frac{dy}{d\tilde{x}} \right) + [\lambda q(\tilde{x}) - r(\tilde{x})]y = 0 \qquad (4.3.75)$$

as $\lambda \to \infty$. The choice of $\tilde{x}$ for independent variable is for consistency of notation and will become evident later.

A standard method is the transformation of (4.3.75) to an equation of canonical type by the introduction of

$$y(\tilde{x}) = f(\tilde{x})w(z), \quad z = g(\tilde{x}).$$

Over an interval of $\tilde{x}$, where p, q individually have one sign (say positive), (4.3.75) is transformed to

$$\frac{d^2w}{dz^2} + \lambda w = \phi(z)w \qquad (4.3.76)$$

by a suitable choice of (f, g). For large λ, the right-hand side of (4.3.76) makes a small contribution that can be estimated by iteration (see [4.11]). Since for $\lambda \to \infty$ the solutions of (4.3.76) have the form of slowly varying oscillations, it is natural to expect a two-scale expansion to apply to this part of the problem. Besides having a certain unity with what has been discussed earlier, the two-scale method has the advantage that higher approximations are more easily calculated.

When the original equation (4.3.75) has a simple turning point ($q(\tilde{x}) = 0$), the extension of the previous method uses a comparison equation,

$$\frac{d^2w}{dz^2} + \lambda z w = \psi(z)w,$$

which gives the results of the WKBJ... method. The procedure here is different. A local expansion valid near the turning point is constructed and matched to the expansions valid away from the turning point.

The method used here is similar to that of Sec. 4.3.2. The equation is transformed to the form for an oscillator of constant frequency and small damping. A fast variable is $x = \sqrt{\lambda}\tilde{x}$, and $\tilde{x}$ itself is a slow variable. Thus, (4.3.75) can be written

$$\frac{d^2y}{dx^2} + \epsilon \frac{dp/d\tilde{x}}{p(\tilde{x})}\frac{dy}{dx} + \left\{\frac{q(\tilde{x})}{p(\tilde{x})} - \epsilon^2 \frac{r(\tilde{x})}{p(\tilde{x})}\right\} y = 0, \qquad (4.3.77)$$

where $\epsilon = 1/\sqrt{\lambda} \ll 1$. Consider, first, (4.3.77) over an interval $(0 < \tilde{x} < 1)$, where $p, q > 0$, and consider also that $\lambda > 0$, as would be typical for an eigenvalue problem. Then, introduce a new fast variable x^+,

$$x^+ = \psi(x),$$

in order to bring (4.3.77) to the desired form. We have

$$\psi'^2(x)\frac{d^2y}{dx^{+2}} + \psi''(x)\frac{dy}{dx^+} + \epsilon\frac{dp/d\tilde{x}}{p(\tilde{x})}\psi'(x)\frac{dy}{dx^+} + \left\{\frac{q(\tilde{x})}{p(\tilde{x})} - \epsilon^2\frac{r(\tilde{x})}{p(\tilde{x})}\right\} y = 0, \qquad (4.3.78)$$

where primes denote derivatives with respect to x. It is clear that $\psi'(x)$ should be chosen so that

$$\psi'(x) = \sqrt{\frac{q(\tilde{x})}{p(\tilde{x})}} = O(1), \qquad (4.3.79)$$

and the relationship of the new fast variable x^+ to the slow variable is

$$x^+ = \int_0^x \sqrt{\frac{q(\epsilon z)}{p(\epsilon z)}}\,dz = \frac{1}{\epsilon}\int_0^{\tilde{x}} \sqrt{\frac{q(\xi)}{p(\xi)}}\,d\xi. \qquad (4.3.80)$$

With this choice, we have

$$\psi''(x) = \frac{\epsilon}{2}\sqrt{\frac{p}{q}}\left[\frac{1}{p}\frac{dq}{d\tilde{x}} - \frac{q}{p^2}\frac{dp}{d\tilde{x}}\right], \qquad (4.3.81)$$

and (4.3.77) is

$$\frac{d^2y}{dx^{+2}} + \epsilon f(\tilde{x})\frac{dy}{dx^+} + \left\{1 - \epsilon^2\frac{r(\tilde{x})}{p^2(\tilde{x})q(\tilde{x})}\right\} y = 0, \qquad (4.3.82)$$

where

$$f(\tilde{x}) = \frac{1}{2}\frac{p^{1/2}}{q^{3/2}}\frac{dq}{d\tilde{x}} + \frac{1}{2}\frac{dp/d\tilde{x}}{p^{1/2}q^{1/2}}. \tag{4.3.83}$$

Now, for the two-scale expansion, we assume

$$y(x^+;\epsilon) = F_0(x^+,\tilde{x}) + \epsilon F_1(x^+,\tilde{x}) + O(\epsilon^2), \tag{4.3.84a}$$

$$\frac{dy}{dx^+} = \frac{\partial F_0}{\partial x^+} + \frac{d\tilde{x}}{dx^+}\frac{\partial F_0}{\partial \tilde{x}} + \epsilon\frac{\partial F_1}{\partial x^+} + O(\epsilon^2), \tag{4.3.84b}$$

$$\frac{d^2y}{dx^{+2}} = \frac{\partial^2 F_0}{\partial x^{+2}} + 2\frac{d\tilde{x}}{dx^+}\frac{\partial^2 F_0}{\partial x^+\partial\tilde{x}} + \epsilon\frac{\partial^2 F_1}{\partial x^{+2}} + O(\epsilon^2). \tag{4.3.84c}$$

Note, from (4.3.80), that

$$\frac{d\tilde{x}}{dx^+} = \epsilon\sqrt{\frac{p(\tilde{x})}{q(\tilde{x})}}, \tag{4.3.85}$$

and $d^2\tilde{x}/dx^{+2} = O(\epsilon^2)$. It follows from (4.3.82) that the first two approximate equations are

$$L(F_0) \equiv \frac{\partial^2 F_0}{\partial x^{+2}} + F_0 = 0, \tag{4.3.86}$$

$$L(F_1) = -2\sqrt{\frac{p}{q}}\frac{\partial^2 F_0}{\partial x^+\partial\tilde{x}} - f(\tilde{x})\frac{\partial F_0}{\partial x^+}. \tag{4.3.87}$$

Thus, we have

$$F_0(x^+,\tilde{x}) = A_0(\tilde{x})\cos x^+ + B_0(\tilde{x})\sin x^+. \tag{4.3.88}$$

Using the same argument as before, namely, that fast growth (x^+ scale) is not permitted, we obtain differential equations for A_0, B_0 from the right-hand side of (4.3.87). We have

$$\begin{aligned} L(F_1) = &-2\sqrt{\frac{p}{q}}\left\{-\frac{dA_0}{d\tilde{x}}\sin x^+ + \frac{dB_0}{d\tilde{x}}\cos x^+\right\} \\ &- f(\tilde{x})\{-A_0\sin x^+ + B_0\cos x^+\}. \end{aligned} \tag{4.3.89}$$

Here A_0, B_0 satisfy the same differential equation (using the definition of f in (4.3.83)):

$$2\frac{dA_0}{d\tilde{x}} + \frac{1}{2}\left(\frac{1}{q}\frac{dq}{d\tilde{x}} + \frac{1}{p}\frac{dp}{d\tilde{x}}\right)A_0 = 0. \tag{4.3.90}$$

The general solution of (4.3.90) is, thus,

$$A_0(\tilde{x}) = a_0[p(\tilde{x})q(\tilde{x})]^{-1/4}, \tag{4.3.91}$$

and the first approximation is the same as that usually found,

$$y(x^+;\epsilon) = \frac{a_0 \cos x^+ + b_0 \sin x^+}{[p(\tilde{x})q(\tilde{x})]^{1/4}} + \epsilon F_1(x^+, \tilde{x}) + \dots, \tag{4.3.92}$$

where

$$F_1 = A_1(\tilde{x}) \cos x^+ + B_1(\tilde{x}) \sin x^+$$

and

$$x^+ = \frac{1}{\epsilon} \int_0^{\tilde{x}} \sqrt{\frac{q(\xi)}{p(\xi)}}\, d\xi.$$

The same formulation can also be applied to an interval in which $p > 0, q < 0$, but $\lambda > 0$, and the sine and cosine are replaced by exponential functions. In such a region, we have

$$y(x^+;\epsilon) = \frac{c_0 e^{-x^+} + d_0 e^{x^+}}{[-p(\tilde{x})q(\tilde{x})]^{1/4}} + \epsilon F_1(x^+, \tilde{x}) + \dots, \tag{4.3.93}$$

where

$$x^+ = \frac{1}{\epsilon} \int_0^{\tilde{x}} \sqrt{\frac{-q(\xi)}{p(\xi)}}\, d\xi.$$

Next, we consider the behavior of the original equation (4.3.75) near a simple turning point (say $\tilde{x} = 0$, where $q(0) = 0$), and we assume that q, p, and r have the following behavior:

$$q(\tilde{x}) = \alpha\tilde{x} + \dots, \qquad p(\tilde{x}) = \beta + \beta_1\tilde{x} + \dots, \qquad r(\tilde{x}) = \rho + \dots. \tag{4.3.94}$$

The idea is to construct a limit-process expansion valid near $\tilde{x} = 0$. Introduce

$$x^* = \frac{\tilde{x}}{\delta(\epsilon)}, \tag{4.3.95}$$

and consider an expansion procedure in which x^* is fixed and $\delta, \epsilon \to 0$. The form of the expansion is

$$y(x^+;\epsilon) = \sigma(\epsilon)g(x^*) + \sigma_1(\epsilon)g_1(x^*) + \dots \tag{4.3.96}$$

so that (4.3.75) becomes

$$\epsilon^2 \frac{p(\delta x^*)}{\delta^2} \frac{d^2 g}{dx^{*2}} + \epsilon^2 \frac{p'(\delta x^*)}{\delta} \frac{dg}{dx^*} + [q(\delta x^*) - \epsilon^2 r(\delta x^*)]g + \dots = 0. \tag{4.3.97}$$

The dominant terms are

$$\frac{\epsilon^2}{\delta^2} \beta \frac{d^2 g}{dx^{*2}} + \delta(\epsilon)\alpha x^* g = 0, \tag{4.3.98}$$

so that both terms are of the same order if $\epsilon^2/\delta^2 = \delta$ or

$$\delta(\epsilon) = \epsilon^{2/3}. \tag{4.3.99}$$

Thus, the basic turning-point equation is obtained for $g(x^*)$:

$$\frac{d^2g}{dx^{*2}} + k^2x^*g = 0, \quad k^2 = \frac{\alpha}{\beta}. \tag{4.3.100}$$

The problem is thus reduced to knowing the properties of the solution of (4.3.100), and, since these solutions are expressed in terms of Airy functions or ordinary Bessel functions, further progress toward matching can be made. The general solution of (4.3.100) can be written

$$g(x^*) = \sqrt{x^*}\left\{CJ_{-1/3}\left(\frac{2}{3}kx^{*3/2}\right) + DJ_{1/3}\left(\frac{2}{3}kx^{*3/2}\right)\right\}, \tag{4.3.101}$$

with the corresponding analytic continuation to $x^* < 0$. The function $g(x^*)$ is well behaved at $x^* = 0$, the D term in (4.3.101) varies like x^*, and the C term is constant. (No branches!)

The two-scale expansion (4.3.92) is valid in some region $\tilde{x} > \tilde{x}_0$ excluding the turning point. For the matching of (4.3.92) and (4.3.96), we introduce the matching variable

$$x_\eta = \frac{\tilde{x}}{\eta(\epsilon)}, \tag{4.3.102}$$

where $\eta(\epsilon)$ belongs to an appropriate subclass of $\epsilon^{2/3} \ll \eta \ll 1$. Thus, we have

$$\tilde{x} = \eta x_\eta \to 0, \quad x^* = \frac{\eta x_\eta}{\epsilon^{2/3}} \to \infty, \tag{4.3.103}$$

and

$$x^+ = \frac{1}{\epsilon}\int_0^{\eta x_\eta}\sqrt{\frac{q(\xi)}{p(\xi)}}\,d\xi \to \frac{1}{\epsilon}\sqrt{\frac{\alpha}{\beta}}\int_0^{\eta x_\eta}\sqrt{\xi}\,d\xi + \dots,$$

$$x^+ \to \frac{2}{3}\sqrt{\frac{\alpha}{\beta}}\frac{(\eta x_\eta)^{3/2}}{\epsilon} + \dots. \tag{4.3.104}$$

The two-scale expansion should contain all limit-process expansions valid in restricted x neighborhoods and so be able to be matched to (4.3.101). Now the behavior of (4.3.101) for large x^* is necessary for the matching, and the well-known asymptotic expression of $J_\nu(z)$ for large z can be used:

$$J_\nu(z) = \sqrt{\frac{2}{\pi z}}\cos\left(z - \frac{\nu\pi}{2} - \frac{\pi}{4}\right) + O(z^{-1}). \tag{4.3.105}$$

Thus, in terms of the matching variable x_η, we have the following.

Transition

$$y(x^+;\epsilon) = \sigma(\epsilon)\left\{\sqrt{\frac{3}{\pi k}}\frac{\epsilon^{1/6}}{(\eta x_\eta)^{1/4}}\right\}\left\{C\cos\left(\frac{2}{3}k\frac{(\eta x_\eta)^{3/2}}{\epsilon} - \frac{\pi}{12}\right)\right.$$
$$\left. + D\cos\left(\frac{2}{3}k\frac{(\eta x_\eta)^{3/2}}{\epsilon} - \frac{5\pi}{32}\right)\right\} + \dots .$$

Two-scale

$$y(x^+;\epsilon) = \frac{a_0\cos\left(\frac{2}{3}\sqrt{\alpha/\beta}(\eta x_\eta)^{3/2}/\epsilon\right) + b_0\sin\left(\frac{2}{3}\sqrt{\alpha/\beta}(\eta x_\eta)^{3/2}/\epsilon\right)}{(\beta\alpha)^{1/4}(\eta x_\eta)^{1/4}} + \dots .$$

Since $k = \sqrt{\alpha/\beta}$, it is seen that the dominant terms in these two expansions are matched if

$$\sigma(\epsilon) = \epsilon^{-1/6} \tag{4.3.106}$$

and, further, if

$$\sqrt{\frac{3}{\pi}}\beta^{1/4}\left\{C\cos\frac{\pi}{12} + D\cos\frac{5\pi}{32}\right\} = \frac{a_0}{\beta^{1/4}}, \tag{4.3.107}$$

$$\sqrt{\frac{3}{\pi}}\beta^{1/4}\left\{C\sin\frac{\pi}{12} + D\sin\frac{5\pi}{12}\right\} = \frac{b_0}{\beta^{1/4}}. \tag{4.3.108}$$

Equations (4.3.107) and (4.3.108) provide the basic relations for the constants in the solution. A uniformly valid first approximation including the transition point could be written by adding (4.3.101) and (4.3.92) and subtracting the common part. Further, the same procedure can be applied as $x^* \to -\infty$, so that ultimately the relationships between c_0, d_0 of (4.3.93) and a_0, b_0, which provide the analytic continuation, are found.

When an eigenvalue problem is being considered, the asymptotic formulas are used, and the set of ϵ is determined by consideration of the homogeneous boundary conditions.

4.3.4 Two-Scale Expansions in Satellite Motion

We are concerned with a class of motions that remain in a bounded region surrounding a gravitational center. These are called *satellite motions* and are distinguished by the fact that the dominant force is a spherically symmetric Newtonian gravitation perturbed by small effects. To account adequately for the cumulative effect of these small terms, we use a two-scale expansion.

The intimate connection between a weakly nonlinear oscillator and an almost Keplerian orbit was first noticed by Laplace [4.23], and it is a generalization of this idea that we will use to cast satellite problems into the same mathematical mold as the previous examples discussed in this chapter.

Satellite equations in local orbital plane

To fix ideas, consider a planar motion that can be conveniently defined using the polar coordinates (R, θ) as follows:

$$m\left[\frac{d^2R}{dT^2} - R\left(\frac{d\theta}{dT}\right)^2\right] = -\frac{GMm}{R^2} + F_0 f_1, \tag{4.3.109a}$$

$$m\left(R\frac{d^2\theta}{dT^2} + 2\frac{dR}{dT}\frac{d\theta}{dT}\right) = F_0 f_2. \tag{4.3.109b}$$

Here m is the mass of the satellite, and we recognize the left-hand sides of (4.3.109) as the mass times the acceleration expressed in radial (R) and tangential (θ) components. The term GmM/R^2 is the dominant gravitational attraction due to the planet of mass M located at $R = 0$, and G is the universal gravitational constant. The terms multiplied by the characteristic force F_0, which is intended to be small compared to the central gravitational force, are the radial and tangential perturbations. In general, f_1 and f_2 may depend on both coordinates and their derivatives.

We introduce dimensionless variables by choosing some characteristic length L (say the initial value of R) and normalize the time by the characteristic time $(L^3/GM)^{1/2}$ and obtain the following equations in terms of $r = R/L$, θ, and $t = T(L^3/GM)^{-1/2}$.

$$\ddot{r} - r\dot{\theta}^2 = -\frac{1}{r^2} + \epsilon f_1(r, \theta, \dot{r}, \dot{\theta}), \tag{4.3.110a}$$

$$r\ddot{\theta} + 2\dot{r}\dot{\theta} = \epsilon f_2(r, \theta, \dot{r}, \dot{\theta}), \tag{4.3.110b}$$

where a dot denotes d/dt and ϵ is the ratio of the small perturbation force F_0 to the gravitational force

$$\epsilon = \frac{F_0}{(GmM/L^2)} \ll 1. \tag{4.3.111}$$

The unperturbed orbit, $\epsilon = 0$

The general solution of (4.3.110) for $\epsilon = 0$ can be found by quadrature once we note that this system has the two integrals

$$r^2\dot{\theta} = l = \text{const.} = \text{ angular momentum}, \tag{4.3.112a}$$

$$\frac{1}{2}(\dot{r}^2 + r^2\dot{\theta}^2) - \frac{1}{r} = E = \text{const.} = \text{ energy}. \tag{4.3.112b}$$

For details see sections 3.7 and 3.8 of [2.5].

Here, we are only interested in the case of bounded orbits ($E < 0$), and this class of solutions is most conveniently expressed by the equation of the orbit

$$r = \frac{a(1-e^2)}{1 + e\cos(\theta - \omega)} \tag{4.3.113a}$$

and Kepler's equation for the time history

$$t - \tau = a^{3/2}(\Phi - e \sin \Phi). \tag{4.3.113b}$$

The four constants of integration are the semimajor axis a, the eccentricity e, the argument of pericenter ω, and the time of passage through pericenter τ. The relations between a, e and l, E are

$$l = \sqrt{a(1 - e^2)}, \tag{4.3.114a}$$

$$E = -\frac{1}{2a}, \tag{4.3.114b}$$

and the orbit is sketched in Figure 4.3.3.

The time history of the particle in its orbit (4.3.113b) is compactly written using the so-called *eccentric anomaly* Φ, which is an angle related to θ according to

$$\sin \Phi = \frac{\sqrt{1 - e^2} \sin(\theta - \omega)}{1 + e \cos(\theta - \omega)}. \tag{4.3.115}$$

It has a simple geometrical interpretation derived by enclosing the ellipse inside a circle of radius a and drawing a line normal to the major axis of the ellipse from the particle to the circle. If we denote the intersection of this normal with the circle by Q, then Φ is the angle between the pericenter and Q, as shown in Figure 4.3.4.

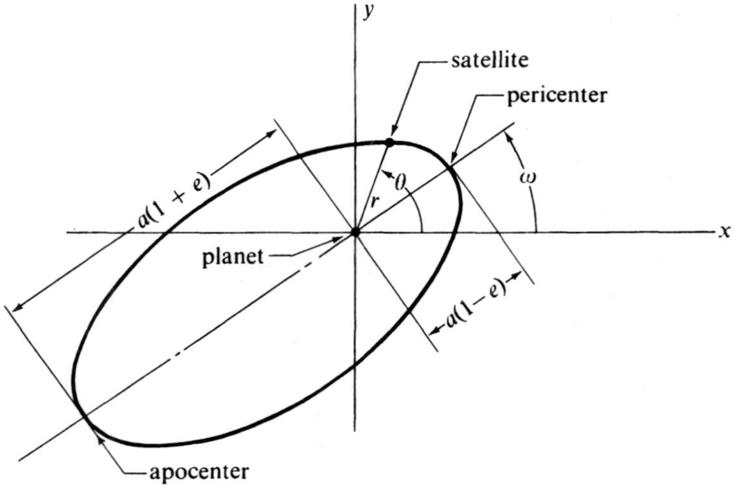

FIGURE 4.3.3. Kepler Ellipse

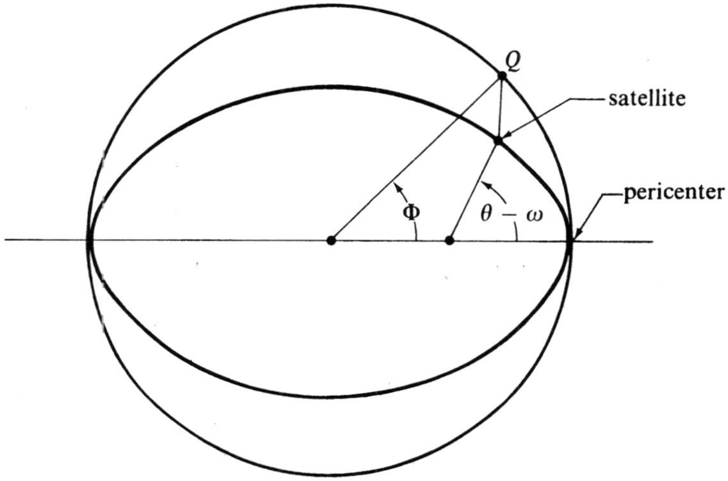

FIGURE 4.3.4. Eccentric Anomaly

The perturbed problem in standard form

The principal result in the foregoing is the structure of (4.3.113a). We note that $u \equiv 1/r$ is a harmonic function of θ. This observation led Laplace to propose the transformation of variables from r and θ in terms of t to u and t in terms of θ. If we perform this transformation on (4.3.110), we find $(' = d/d\theta)$

$$u'' + u - u^4 t'^2 = -\epsilon u^2 t'^2 \left(f_1 + \frac{u'}{u} f_2 \right), \tag{4.3.116a}$$

$$(u^2 t')' = -\epsilon u^3 t'^3 f_2. \tag{4.3.116b}$$

Thus, if $\epsilon = 0$, $u^2 t' = 1/l^2 =$ const., and we recover the solution (4.3.113). With $\epsilon \neq 0$, u obeys the equation of a weakly nonlinear oscillator, and this type of problem is well adapted for solution by the two-scale method.

Actually, we can also handle three-dimensional orbits in the same way by referring the motion to a local orbital plane determined by the instantaneous displacement and velocity vectors. This choice of variables was proposed in [4.32] and the derivation of equations of motion for arbitrary three-dimensional perturbations is given in [4.18] in a form suitable for applying the two-scale expansion technique. Examples of satellite problems using the multiple-scale method can be found in [4.7]–[4.9], [4.12], and [4.17]. In what follows, we restrict attention to two simple planar examples.

Decay of orbit due to drag

Consider the very idealized model of a spherical (hence nonlifting) satellite in orbit around a Newtonian gravitational center and perturbed by a thin, constant-density atmosphere. Actually, a more realistic model of atmospheric density variation with altitude can also be used at the cost of some algebraic complexity (see Problem 8). However, the main qualitative features of the solution are present in this simple model.

Since the drag force acts in the direction opposite to the velocity, $F_0 f_1$ and $F_0 f_2$ in (4.3.109) are given by

$$F_0 f_1 = -D \sin \gamma, \quad F_0 f_2 = -D \cos \gamma, \tag{4.3.117}$$

where D is the magnitude of the drag force and γ is the flight-path angle as shown in Figure 4.3.5.

Now D is given by

$$D = \frac{1}{2} \rho V^2 S C_D, \tag{4.3.118}$$

where ρ is the constant atmospheric density, S is the cross-sectional area of the satellite, and C_D is the drag coefficient, which we also assume to be constant. The

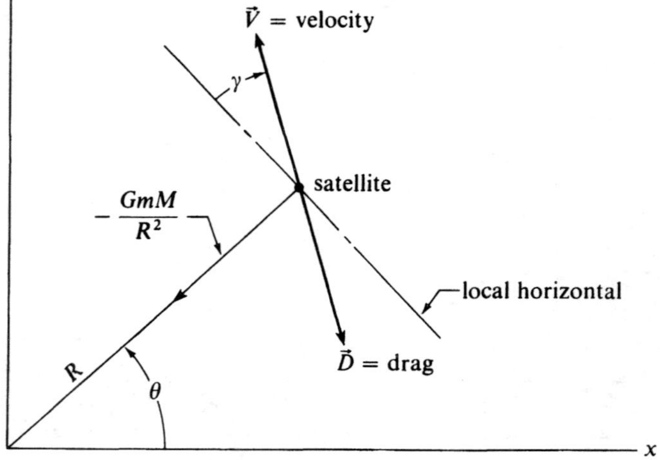

FIGURE 4.3.5. Drag on Satellite

magnitude of the velocity in polar coordinates is

$$V = \left[\left(\frac{dR}{dT} \right)^2 + R^2 \left(\frac{d\theta}{dT} \right)^2 \right]^{1/2}, \tag{4.3.119}$$

and the flight-path angle γ is simply

$$\gamma = \tan^{-1} \frac{dR/dT}{Rd\theta/dT}. \tag{4.3.120}$$

If we use the dimensionless variables of (4.3.110) with R normalized by R_0, its initial value, we obtain

$$\ddot{r} - r\dot{\theta}^2 = -\frac{1}{r^2} - \epsilon\dot{r}(\dot{r}^2 + r^2\dot{\theta}^2)^{1/2}, \tag{4.3.121a}$$

$$r\ddot{\theta} + 2\dot{r}\dot{\theta} = -\epsilon r\dot{\theta}(\dot{r}^2 + r^2\dot{\theta}^2)^{1/2}, \tag{4.3.121b}$$

where the small parameter

$$\epsilon = \frac{C_D \rho S R_0}{2m} \tag{4.3.122}$$

is the ratio of the drag force to the centrifugal force acting on the satellite.

Transforming (4.3.121) to $u(\theta; \epsilon)$ and $t(\theta; \epsilon)$ then gives [see (4.3.116)]

$$u'' + u - u^4 t'^2 = 0, \tag{4.3.123a}$$

$$(u^2 t')' = \epsilon t'(u'^2 + u^2)^{1/2}. \tag{4.3.123b}$$

The initial conditions we adopt correspond to the satellite being at pericenter at $t = 0$. Also, with no loss of generality we may choose the argument of pericenter $\omega = 0$. Hence, the "initial" ($\theta = 0$) conditions are

$$u(0; \epsilon) = 1, \tag{4.3.124a}$$

$$t(0; \epsilon) = 0, \tag{4.3.124b}$$

$$u'(0; \epsilon) = 0, \tag{4.3.124c}$$

$$t'(0; \epsilon) = \sigma. \tag{4.3.124d}$$

Since that constant σ is the reciprocal angular velocity initially, it should be chosen in the interval $2^{-1/2} < \sigma < 1$; see (4.3.126).

If $\epsilon = 0$, the above initial conditions define a unique Keplerian ellipse with constant elements a, e, ω, and τ. With $\epsilon \neq 0$ the motion is more complicated. However, to $O(1)$ it will still be in the form of a Keplerian orbit [see (4.3.130)], but with slowly varying elements. Therefore, it is convenient to express the initial conditions (4.3.124) in terms of equivalent conditions on the initial values of a, e, ω, and τ.

First, we note that since the motion starts from the pericenter, Kepler's equation (4.3.113b) implies that

$$\tau(0) = 0. \tag{4.3.125a}$$

Moreover, we have chosen

$$\omega(0) = 0. \tag{4.3.125b}$$

Since the pericenter distance is $a(1-e)$, (4.3.124a) gives the following condition:

$$a(0)[1 - e(0)] = 1. \tag{4.3.125c}$$

Finally, using the fact that $\ell(0) = \sqrt{a(0)(1 - e(0)^2)} = r^2(0)/t'(0)$ and (4.3.125c) gives

$$\sigma = \frac{1}{\sqrt{1 + e(0)}}. \tag{4.3.125d}$$

Solving (4.3.125) gives

$$a(0) = \frac{\sigma^2}{2\sigma^2 - 1} \equiv a_0 > 0, \tag{4.3.126a}$$

$$e(0) = \frac{1 - \sigma^2}{\sigma^2} \equiv e_0 < 1, \tag{4.3.126b}$$

$$\tau(0) = 0, \tag{4.3.126c}$$

$$\omega(0) = 0. \tag{4.3.126d}$$

Note as $\sigma \downarrow 2^{-1/2}$, $a_0 \to \infty$ (i.e., $E \uparrow 0$ and the motion becomes unbounded). As $\sigma \uparrow 1$, $e_0 \downarrow 0$ (i.e., the initial orbit becomes circular).

We observe that the differential equations (4.3.123) only involve t' and that the time t does not occur explicitly. Hence, we develop u and t' in two-scale form as follows:

$$u(\theta;\epsilon) = u_0(\theta, \tilde{\theta}) + \epsilon u_1(\theta, \tilde{\theta}) + \dots, \tag{4.3.127a}$$

$$t'(\theta;\epsilon) = v_0(\theta, \tilde{\theta}) + \epsilon v_1(\theta, \tilde{\theta}) + \dots, \tag{4.3.127b}$$

where, as usual, $\tilde{\theta} = \epsilon\theta$. Once the expansion (4.3.127b) for t' is known, t can be calculated by quadrature.

Substituting (4.3.127) and the corresponding formulas for the derivatives into (4.3.123) gives the following equations for u_0, v_0 and u_1, v_1:

$$\frac{\partial^2 u_0}{\partial \theta^2} + u_0 = u_0^4 v_0^2, \tag{4.3.128a}$$

$$u_0^2 \frac{\partial v_0}{\partial \theta} + 2u_0 \frac{\partial u_0}{\partial \theta} v_0 = 0. \tag{4.3.128b}$$

$$\frac{\partial^2 u_1}{\partial \theta^2} + u_1 = -2\frac{\partial^2 u_0}{\partial \theta \partial \tilde{\theta}} + 2u_0^4 v_0 v_1 + 4u_0^3 u_1 v_0^2, \tag{4.3.129a}$$

$$u_0^2\left(\frac{\partial v_1}{\partial \theta} + \frac{\partial v_0}{\partial \tilde{\theta}}\right) + 2u_0 u_1 \frac{\partial v_0}{\partial \theta} + 2u_1 \frac{\partial u_0}{\partial \theta} v_0$$
$$+2u_0\left[\frac{\partial u_0}{\partial \theta} v_1 + v_0\left(\frac{\partial u_1}{\partial \theta} + \frac{\partial u_0}{\partial \tilde{\theta}}\right)\right]$$
$$= v_0\left[\left(\frac{\partial u_0}{\partial \theta}\right)^2 + u_0^2\right]^{1/2}. \tag{4.3.129b}$$

The solution of (4.3.128) is most conveniently expressed in terms of the three slowly varying Keplerian elements a, e, and ω as follows. The fourth element τ will only arise after the quadrature of v_0. [*Note*: $l^2 = a(1 - e^2)$.]

$$u_0(\theta, \tilde{\theta}) = p^2[1 + e\cos(\theta - \omega)], \tag{4.3.130a}$$

$$v_0(\theta, \tilde{\theta}) = p^{-3}[1 + e\cos(\theta - \omega)]^{-2}, \tag{4.3.130b}$$

where p is the slowly varying reciprocal angular momentum $p = 1/l = 1/\sqrt{a(1-e^2)}$, and $a(\tilde{\theta})$, $e(\tilde{\theta})$, $\omega(\tilde{\theta})$ have initial values a_0, e_0, 0 defined in (4.3.126).

For simplicity, we shall henceforth neglect terms of order e_0^2. This approximation will be justified by the fact that e is a monotone decreasing function of $\tilde{\theta}$.

Equation (4.3.129b) can be integrated once with respect to θ if we make use of the solution (4.3.130). We find

$$u_0^2 v_1 + 2p\frac{u_1}{u_0} + \frac{dp}{d\tilde{\theta}}\theta - \frac{1}{p}[\theta - e\sin(\theta - \omega)] = g_1(\tilde{\theta}) + O(e^2), \tag{4.3.131}$$

where g_1 is an unknown function of $\tilde{\theta}$. Using (4.3.131) in (4.3.129a) leads to

$$\frac{\partial^2 u_1}{\partial \theta^2} + u_1 = \left(-2e + 2p^2\frac{de}{d\tilde{\theta}} + 4pe\frac{dp}{d\tilde{\theta}}\right)\sin(\theta - \omega)$$
$$- 2ep^2\frac{d\omega}{d\tilde{\theta}}\cos(\theta - \omega) + 2\left(1 - p\frac{dp}{d\tilde{\theta}}\right)\theta$$
$$+ 2pg_1 + O(e^2). \tag{4.3.132}$$

Unless u_1 is bounded, the assumed expansion for u would be inconsistent. Therefore, we must set

$$1 - p\frac{dp}{d\tilde{\theta}} = 0, \tag{4.3.133a}$$

$$-e + p^2\frac{de}{d\tilde{\theta}} + 2pe\frac{dp}{d\tilde{\theta}} = 0, \tag{4.3.133b}$$

$$p^2 e \frac{d\omega}{d\tilde{\theta}} = 0, \tag{4.3.133c}$$

and these can easily be solved as follows:

$$a = \frac{1}{1+2\tilde{\theta}} + O(e_0^2), \tag{4.3.134a}$$

$$e = \frac{e_0}{\sqrt{1+2\tilde{\theta}}} + O(e_0^2), \tag{4.3.134b}$$

$$\omega = 0. \tag{4.3.134c}$$

This shows that to $O(1)$ the effect of drag leaves the pericenter fixed but produces an algebraic decay in the orbital semimajor axis and an algebraic decrease of the eccentricity. Thus, an initially elliptic orbit tends to spiral in and become more and more circular. Moreover, the behavior of e justifies expanding the solution in powers of e.

To determine the time history, i.e., to calculate $\tau(\tilde{\theta})$, is not a straightforward quadrature of v_0. We must keep in mind that a long periodic term (i.e., one depending on $\sin\tilde{\theta}$ or $\cos\tilde{\theta}$) in v_1 will, upon quadrature, drop by one order and contribute to $O(1)$ in t. But, in order to determine all the long periodic terms in v_1, we must study the solutions for v_2 and u_2 and require these to be bounded. Thus, we conclude that to determine $t(\theta)$ to $O(\epsilon^n)$ we must know v_{n+1} completely, and this requires examining the equations for u_{n+2} and v_{n+2}. This question is posed in Problem 6.

The above feature is typical of the solution for the time history in all satellite problems where an exact integral of motion (e.g., an energy integral) is not available. If, however, we consider a problem where there is such an exact integral, one can use this to derive the missing long periodic terms to any order. This question is discussed in [4.10].

Close lunar satellite in the restricted three-body problem

Consider the motion of three gravitational mass points (denoted *Earth*, *moon*, *satellite*) in the limit when one of the masses (satellite) is very much smaller than that of the other two. Clearly, in this limit the two large bodies will describe a Keplerian orbit about their common center of mass, while the satellite will move in this gravitational field, without influencing the motion of the two large bodies. This is the *restricted three-body problem*. If the two large bodies (called *primaries*) describe a circular orbit about their common mass center, the satellite obeys the following set of equations for motion in the orbital plane of the primaries:

$$\frac{d^2\overline{x}}{dt^2} = -(1-\mu)\frac{(\overline{x}-\xi_1)}{\overline{r}_1^3} - \mu\frac{(\overline{x}-\xi_2)}{\overline{r}_2^3}, \tag{4.3.135a}$$

$$\frac{d^2\overline{y}}{dt^2} = -(1-\mu)\frac{(\overline{y}-\eta_1)}{\overline{r}_1^3} - \mu\frac{(\overline{y}-\eta_2)}{\overline{r}_2^3}. \tag{4.3.135b}$$

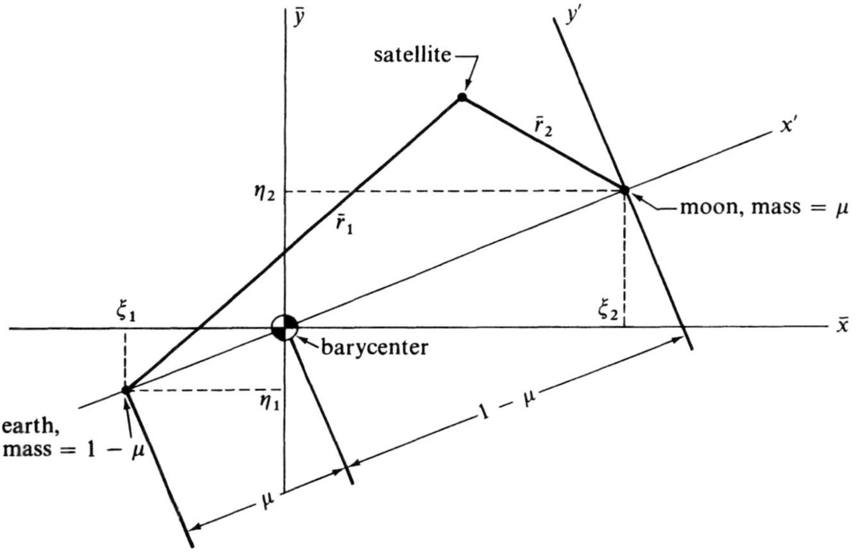

FIGURE 4.3.6. Motion in Barycentric Inertial Frame

The geometry is sketched in Figure 4.3.6.

The variables are normalized by choosing the Earth-moon distance as the unit of length and the reciprocal angular velocity of the Earth-moon orbit for the unit of time. The origin is located at the barycenter, and the dimensionless masses μ and $1-\mu$ are given by

$$\mu = \frac{m_2}{m_1+m_2}, \quad 1-\mu = \frac{m_1}{m_1+m_2}. \tag{4.3.136}$$

For counterclockwise circular motion with unit dimensionless angular velocity, the coordinates of the Earth and moon orbits are defined by

$$\xi_1 = -\mu \cos t, \tag{4.3.137a}$$

$$\eta_1 = -\mu \sin t, \tag{4.3.137b}$$

$$\xi_2 = (1-\mu)\cos t, \tag{4.3.137c}$$

$$\eta_2 = (1-\mu)\sin t. \tag{4.3.137d}$$

The distances $\bar{r}_1$ and $\bar{r}_2$ between the satellite and the two primaries are

$$\bar{r}_1^2 = (\bar{x}-\xi_1)^2 + (\bar{y}-\eta_1)^2, \tag{4.3.138a}$$

$$\overline{r}_2^2 = (\overline{x} - \xi_2)^2 + (\overline{y} - \eta_2)^2. \tag{4.3.138b}$$

For the purpose of studying the motion of a lunar satellite, it is more convenient to refer the motion to the x', y' coordinates centered at the moon and rotating with the system; see Figure 4.3.6. Since

$$\overline{x} = (x' - 1 - \mu)\cos t - y' \sin t, \tag{4.3.139a}$$

$$\overline{y} = (x' + 1 - \mu)\sin t + y' \cos t, \tag{4.3.139b}$$

(4.3.135) transform to

$$\frac{d^2x'}{dt^2} = -(1-\mu)\frac{(x'+1)}{r_1'^3} - \mu\frac{x'}{r'^3} + 2\frac{dy'}{dt} + x' + 1 - \mu, \tag{4.3.140a}$$

$$\frac{d^2y'}{dt^2} = -(1-\mu)\frac{y'}{r_1'^3} - \mu\frac{y'}{r'^3} - 2\frac{dx'}{dt} + y', \tag{4.3.140b}$$

where

$$r_1'^2 = (x'+1)^2 + y'^2, \quad r'^2 = x'^2 + y'^2.$$

We note that in this rotating frame the equations are autonomous. However, in addition to the gravitational forces, we now have the Coriolis force $(2(dy'/dt), -2(dx'/dt))$ and the centrifugal force $(x' + 1 - \mu, y')$. We could also have written (4.3.140) directly by taking proper account of the fictitious forces introduced by referring the motion to the rotating frame. It is easily seen (by multiplying (4.3.140a) by dx'/dt, (4.3.140b) by dy'/dt and adding) that this system has the exact integral (Jacobi integral)

$$\frac{1}{2}\left[\left(\frac{dx'}{dt}\right)^2 + \left(\frac{dy'}{dt}\right)^2\right] - \frac{(1-\mu)}{r_1'} - \frac{\mu}{r_2'}$$
$$-\frac{1}{2}\left[(x'+1-\mu)^2 + y'^2\right] = -\frac{C}{2} = \text{constant}. \tag{4.3.141}$$

For further discussion of this result and various other aspects of the restricted three-body problem, the reader is referred to [2.16].

We wish to study the solution of (4.3.140) for $\mu \ll 1$ for the case of motion close to the moon. (Actually, for the Earth-moon system, $\mu \approx 0.012$.) It is clear that for motion close to the moon it is inappropriate to use the x', y', t variables. In terms of these variables, the moon's attraction will nominally occur only to order μ when, in fact, for arbitrarily small μ there exists some neighborhood of the origin where the lunar gravitation dominates.

We therefore introduce rescaled variables

$$x^* = \frac{x'}{\delta(\mu)}; \quad y^* = \frac{y'}{\delta(\mu)}; \quad t^* = \frac{t}{\gamma(\mu)}, \tag{4.3.142}$$

where δ and γ are to be determined by an order-of-magnitude analysis. We anticipate that for close satellites the period may also need to be rescaled, as it may be significantly smaller than the Earth-moon period that was used in normalizing t.

Substituting the above into (4.3.140a) gives

$$\frac{\delta}{\gamma^2}\frac{d^2x^*}{dt^{*2}} = -(1-\mu)\frac{(1+\delta x^*)}{[1+2\delta x^*+\delta^2(x^{*2}+y^{*2})]^{3/2}} - \frac{\mu}{\delta^2}\frac{x^*}{[x^{*2}+y^{*2}]^{3/2}} + \frac{\delta}{\gamma}\frac{dy^*}{dt^*} + \delta x^* + 1 - \mu. \quad (4.3.143)$$

Developing the right-hand side and multiplying by γ^2/δ gives

$$\frac{d^2x^*}{dt^{*2}} = -\frac{\mu\gamma^2}{\delta^3}\frac{x^*}{[x^{*2}+y^{*2}]^{3/2}} + 2\gamma\frac{dy^*}{dt^*} + 3\gamma^2x^* + O(\mu\gamma^2) + O(\delta\gamma^2). \quad (4.3.144)$$

The fundamental criterion for a lunar satellite is that the lunar gravitation be dominant. This requires that we set $\mu\gamma^2/\delta^3 = 1$. With this premise, the richest equations result by setting $\gamma = 1$, $\delta = \mu^{1/3}$ and we have, in the limit as $\mu \to 0$:

$$\frac{d^2x^*}{dt^2} = -\frac{x^*}{[x^{*2}+y^2]^{3/2}} + 2\frac{dy^*}{dt} + 3x^*, \quad (4.3.145a)$$

$$\frac{d^2y^*}{dt^2} = -\frac{y^*}{[x^{*2}+y^{*2}]^{3/2}} - 2\frac{dx^*}{dt}. \quad (4.3.145b)$$

These equations were first derived by G.W. Hill using physical arguments (see [2.16] for a comprehensive account). They are not significantly simpler than the exact set (4.3.135) even though now the net effect of Earth's gravity plus centrifugal force is simply accounted for by the term $3x^*$ in (4.3.145a).

In deriving (4.3.145) it was assumed that the lunar gravitation was comparable to the Coriolis force and to the net effect of Earth's gravity plus centrifugal forces. This would be true for orbits as far away as $O(\mu^{1/3})$ and having periods comparable to the Earth-moon period. Such orbits are *not* perturbed Kepler ellipses and will not be considered here. Again, we refer the reader to [2.16] for a discussion of the solutions of Hill's equations. Rather, we are interested in closer satellites for which the lunar gravity dominates over the Coriolis force.

In this case, any $\gamma(\mu) \ll 1$ is appropriate. The smaller γ the closer the orbit is to the moon. Setting $\gamma(\mu) = \mu^\alpha \equiv \epsilon, \alpha > 0$, we have the inner variables

$$x^{**} = \frac{x'}{\mu^{(1+2\alpha)/3}}, \quad y^{**} = \frac{y'}{\mu^{(1+2\alpha)/3}}, \quad t^{**} = \frac{t}{\mu^\alpha}, \quad (4.3.146)$$

and the following equations of motion to order ϵ^2:

$$\frac{d^2x^{**}}{dt^{**2}} = -\frac{x^{**}}{r^{**3}} + 2\epsilon\frac{dy^{**}}{dt^{**}} + 3\epsilon^2x^{**}, \quad (4.3.147a)$$

$$\frac{d^2y^{**}}{dt^{**2}} = -\frac{y^{**}}{r^{**3}} - 2\epsilon\frac{dx^{**}}{dt^{**}}, \quad (4.3.147b)$$

where

$$r^{**2} = x^{**2} + y^{**2}.$$

Transforming the above to the standard form, we set

$$x^{**} = r^{**} \cos\theta, \quad y^{**} = r^{**} \sin\theta, \tag{4.3.148}$$

and

$$u \equiv \frac{1}{r^{**}} = u(\theta;\epsilon), \quad p \equiv u^{*2} \frac{dt^{**}}{d\theta} = p(\theta;\epsilon). \tag{4.3.149}$$

Thus, p is the reciprocal angular momentum, and knowing p and u one can calculate the time history by quadrature.

Equations (4.3.147) become ($' = d/d\theta$):

$$\begin{aligned} u'' + u =& p^2 - 2\epsilon\left(\frac{p}{u} + \frac{p}{u^3}u'^2\right) - \frac{3}{2}\epsilon^2\frac{p^2}{u^3} \\ &+ \frac{3}{2}\epsilon^2\frac{p^2}{u^4}u'(\sin 2\theta - u\cos 2\theta) \end{aligned} \tag{4.3.150a}$$

$$p' = -2\epsilon u'\frac{p^2}{u^3} + \frac{3}{2}\epsilon^2\frac{p^3}{u^4}\sin 2\theta. \tag{4.3.150b}$$

We now assume a two-scale expansion for u and p in the form

$$u(\theta;\epsilon) = u_0(\theta,\tilde{\theta}) + \epsilon u_1(\theta,\tilde{\theta}) + \epsilon^2 u_2(\theta,\tilde{\theta}) + \ldots, \tag{4.3.151a}$$

$$p(\theta;\epsilon) = p_0(\theta,\tilde{\theta}) + \epsilon p_1(\theta,\tilde{\theta}) + \epsilon^2 p_2(\theta,\tilde{\theta}) + \ldots, \tag{4.3.151b}$$

where the slow variable $\tilde{\theta}$ is now taken in the form

$$\tilde{\theta} = \epsilon\theta(1 + \epsilon\alpha_1 + \epsilon^2\alpha_2 + \ldots). \tag{4.3.152}$$

The reason for expanding the slow variable here instead of the usual straining of the fast variable is that θ occurs explicitly in the problem, and it is inconvenient to transform it. The final result is, of course, the same for either choice of strained variables.

We will only carry out the solution to $O(1)$; the higher-order calculations are left as an exercise (Problem 9). Substituting the expansions (4.3.151) into (4.3.150) gives

$$\frac{\partial^2 u_0}{\partial\theta^2} + u_0 = p_0^2, \tag{4.3.153a}$$

$$\frac{\partial p_0}{\partial\theta} = 0. \tag{4.3.153b}$$

To order ϵ, we find

$$\frac{\partial^2 u_1}{\partial\theta^2} + u_1 = -2\frac{\partial u_0}{\partial\theta\partial\tilde{\theta}} + 2p_0\left[\frac{1}{u_0} + \frac{1}{u_0^3}\left(\frac{\partial u_0}{\partial\theta}\right)^2\right] + 2p_0p_1, \tag{4.3.154a}$$

$$\frac{\partial p_1}{\partial\theta} + \frac{\partial p_0}{\partial\tilde{\theta}} = -2\frac{\partial u_0}{\partial\theta}\frac{p_0^2}{u_0^3}. \tag{4.3.154b}$$

For reference to Problem 9, we also list the equations governing the terms of order ϵ^2:

$$\frac{\partial^2 u_2}{\partial\theta^2} + u_2 = -2\frac{\partial^2 u_1}{\partial\theta\partial\tilde{\theta}} - \frac{\partial^2 u_0}{\partial\theta^2} - 2\alpha_1\frac{\partial^2 u_0}{\partial\theta\partial\tilde{\theta}}$$
$$+ p_1^2 + 2p_0p_2 - 2\left[\frac{p_1}{u_0} - \frac{p_0u_1}{u_0^2} + \frac{p_1}{u_0^3}\left(\frac{\partial u_0}{\partial\theta}\right)^2\right.$$
$$\left. - 3\frac{p_0}{u_0^4}u_1\left(\frac{\partial u_0}{\partial\theta}\right)^2 - 2\frac{p_0}{u_0^3}\frac{\partial u_0}{\partial\theta}\left(\frac{\partial u_0}{\partial\tilde{\theta}} + \frac{\partial u_1}{\partial\theta}\right)\right]$$
$$- \frac{3}{2}\frac{p_0^2}{u_0^3} + \frac{3}{2}\frac{p_0^2}{u_0^4}\left(\frac{\partial u_0}{\partial\theta}\sin 2\theta - u_0\cos 2\theta\right), \qquad (4.3.155a)$$

$$\frac{\partial p_2}{\partial\theta} + \frac{\partial p_1}{\partial\tilde{\theta}} + \alpha_1\frac{\partial p_0}{\partial\tilde{\theta}} = -2\frac{p_0^2}{u_0^3}\left(\frac{\partial u_0}{\partial\tilde{\theta}} + \frac{\partial u_1}{\partial\theta}\right)$$
$$- \frac{4p_0p_1}{u_0^3}\frac{\partial u_0}{\partial\theta} + 6\frac{p_0^2u_1}{u_0^4}\frac{\partial u_0}{\partial\theta} + \frac{3}{2}\frac{p_0^3}{u_0^4}\sin 2\theta. \qquad (4.3.155b)$$

The solution of the $O(1)$ terms is (see (4.3.131a)

$$p_0 = p_0(\tilde{\theta}), \qquad (4.3.156a)$$

$$u_0 = p_0^2(\tilde{\theta})\{1 + e(\tilde{\theta})\cos[\theta - \omega(\tilde{\theta})]\}, \qquad (4.3.156b)$$

and the slowly varying reciprocal angular momentum p_0, the eccentricity e, and the argument of pericenter ω are to be determined by consistency of the solution to $O(\epsilon)$.

Substituting the above into (4.3.154) gives

$$\frac{\partial^2 u_1}{\partial\theta^2} + u_1 = -2p_0^2 e\frac{d\omega}{d\tilde{\theta}}\cos(\theta - \omega) - \frac{2}{p_0[1 + e\cos(\theta - \omega)]}$$
$$- \frac{2e^2}{p_0}\frac{\sin^2(\theta - \omega)}{[1 + e\cos(\theta - \omega)]^3} + 2p_0p_1$$
$$+ 4p_0\frac{dp_0}{d\tilde{\theta}}e\sin(\theta - \omega) + 2p_0^2\frac{de}{d\tilde{\theta}}\sin(\theta - \omega), \qquad (4.3.157a)$$

$$\frac{\partial p_1}{\partial\theta} = -\frac{dp_0}{d\tilde{\theta}} + \frac{2e}{p_0^2}\frac{\sin(\theta - \omega)}{[1 + e\cos(\theta - \omega)]^3}. \qquad (4.3.157b)$$

Since $dp_0/d\tilde{\theta}$ in (4.3.157b) depends only on $\tilde{\theta}$, integrating this term will give rise to an inconsistent secular term proportional to θ in p_1. We must therefore set $dp_0/d\tilde{\theta} = 0$, i.e., $p_0 =$ constant. Then, integrating (4.3.157b) gives

$$p_1 = \frac{1}{p_0^2[1 + e\cos(\theta - \omega)]^2} + \tilde{p}_1(\tilde{\theta}), \qquad (4.3.158)$$

where $\tilde{p}_1$ is an unknown function of $\tilde{\theta}$.

We now substitute the above expression for p_1 into (4.3.157a) and can explicitly integrate the result. In addition to the secular terms $\psi \sin\psi$ and $\psi\cos\psi$ (with $\psi = \theta - \omega$), we encounter a term of the form $\sin\psi \cdot \sin^{-1}((e+\cos\psi)/(1+e\cos\psi))$. The arc sine consists of a secular part equal to $-\psi$ plus a periodic part. In order to separate the periodic and secular terms in the solution, we introduce the function

$$Z(\psi, e) \equiv -\sin^{-1}\frac{e+\cos\psi}{1+e\cos\psi} - \psi, \tag{4.3.159}$$

which is strictly periodic. In fact, it is easy to show that Z has the Fourier series

$$Z = -\frac{\pi}{2} + 2\sum_{n=1}^{\infty}\frac{1}{n}\left(\frac{1-\sqrt{1-e^2}}{-e}\right)^n \sin n\psi. \tag{4.3.160}$$

Removing the secular terms in u_1 then requires that

$$\frac{d\omega}{d\tilde{\theta}} = -\frac{1}{p_0^3(1-e^2)^{3/2}}; \quad \omega = -\frac{\tilde{\theta}}{p_0^3(1-e^2)^{3/2}}, \tag{4.3.161a}$$

$$\frac{de}{d\tilde{\theta}} = 0; \quad e = \text{const.}, \tag{4.3.161b}$$

and the solution for u_1 can be calculated in the form

$$\begin{aligned} u_1 =& 2p_0\tilde{p}_1 + \frac{1}{p_0(1-e^2)} + \left(A_1(\tilde{\theta}) - \frac{eZ}{p_0(1-e^2)^{3/2}}\right)\sin(\theta-\omega) \\ &+ B_1(\tilde{\theta})\cos(\theta-\omega) - \frac{1}{p_0[1+e\cos(\theta-\omega)]}, \end{aligned} \tag{4.3.162}$$

where the unknown functions A_1 and B_1 are to be found from the solution to $O(\epsilon^2)$.

The principal result to $O(1)$ is that the Keplerian ellipse rotates within the x', y' coordinate system at a clockwise rate exactly equal to the counterclockwise rate of rotation of this coordinate system relative to the barycenter. (See Problem 9.) Thus, the orbit to first order preserves its orientation in space. This well-known result could have, of course, been derived more directly by noting that an orientation-preserving ellipse is a solution of the differential equations to $O(\epsilon)$.

The problem of a close lunar satellite is studied in [4.9] to $O(\epsilon)$ for arbitrary orbital inclination and eccentricity. In this general case, the solution requires use of *two slow variables* $\epsilon\theta$ and $\epsilon^2\theta$, and the calculations are quite involved. The motion is also complicated, as the orbital elements perform oscillatory or secular variations on both time scales.

4.3.5 Coupled Linear Oscillators

Various physical problems lead to systems of coupled linear oscillator equations. For example, in Sec. 1.3.2 we saw that if one expresses the solution of a perturbed linear wave equation in a series of the unperturbed eigenfunctions—a Fourier sine series for the case discussed—then the modal amplitudes obey a system of coupled linear oscillator equations (see (1.3.52), (1.3.53)).

Let us study the general problem for the case of small damping. We have the dimensionless vector equation

$$M\frac{d^2\mathbf{x}}{dt^2} + \epsilon B\frac{d\mathbf{x}}{dt} + K\mathbf{x} = 0. \tag{4.3.163}$$

Here M, B, K are considered to be positive definite symmetric matrices, $\mathbf{x}$ an n-dimensional vector $\mathbf{x} = (x_1, x_2, \ldots, x_n)$. The two-scale formalism can be used to great advantage to discuss the general properties of the system represented by (4.3.163).

One physical system that leads to (4.3.163) is the system of masses, linear springs, and linear dampers sketched in Figure 4.3.7. Every mass is coupled to every other mass with linear springs and dampers. The physical masses are M_j, spring constants K_{ij}, damping coefficients B_{ij}, the deflection from equilibrium is $X_j(T)$, T is physical time, and $i, j = 1, \ldots, n$. Dimensionless variables can be introduced in terms of a characteristic mass M_c and a characteristic spring constant K_c by letting $t = (K_c/M_c)^{1/2}T$ and $\mathbf{x} = \mathbf{X}/A_c$, where A_c is the characteristic amplitude. Then, in (4.3.163)

$$M = \frac{1}{M_c}\begin{pmatrix} M_1 & 0 & \ldots & 0 \\ 0 & M_2 & 0 & \ldots \\ \vdots & 0 & \ddots & \\ 0 & \vdots & & M_n \end{pmatrix} = (m_{ij}), \tag{4.3.164a}$$

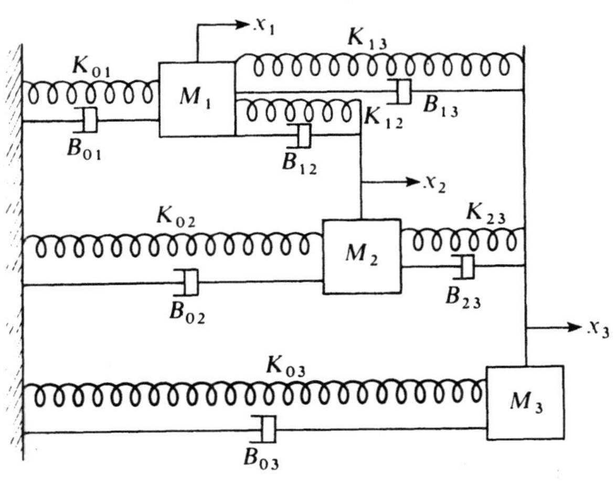

FIGURE 4.3.7. Coupled System, Many Degrees of Freedom

$$B = \frac{1}{B_c}\begin{pmatrix} B_{01}+B_{12}+B_{13}+\ldots & -B_{12} & -B_{13}\ldots \\ -B_{12} & B_{02}+B_{12}+B_{32}+\ldots & \\ -B_{1n} & -B_{2n} & \ldots \end{pmatrix} \equiv (b_{ij}), \tag{4.3.164b}$$

$$K = \frac{1}{K_c}\begin{pmatrix} K_{01}+K_{12}+K_{13}+\ldots & -K_{12} & -K_{13}\ldots \\ -K_{12} & K_{02}+K_{12}+K_{32}+\ldots & \\ -K_{1n} & -K_{2n} & \ldots \end{pmatrix} \equiv (k_{ij}), \tag{4.3.164c}$$

and the small parameter ϵ is

$$\epsilon = \frac{B_c}{(K_c M_c)^{1/2}}. \tag{4.3.165}$$

For example, average values can be used for the characteristic values. In order for an analysis based on small ϵ to give good results, all the system constants should not deviate too much from their average values.

The motion of this model system for small damping should be close in some sense to the undamped case. The modes of free vibration of the undamped system are very convenient for representing the solution to damped initial-value problems or forced motions. Hence, it is reasonable to try to represent the motion of the damped system in terms of the undamped free modes as in the example of Sec. 1.3.2. This can always be done since the eigenvectors, which give the normal mode shapes, form a basis of the n-dimensional space. We expect, in analogy with the one-dimensional case of Sec. 4.2.1, that the physical effect of the damping is to make the free motion die out and to shift the frequencies of vibration from those of the undamped case. The analysis serves to make these intuitive ideas more precise and to clarify the behavior of linear systems with many degrees of freedom. It should be noted, however, that the analysis applies to general symmetric positive definite matrices (M, B, K). We are interested in the general solution of (4.3.163) or equivalently in the solution to the general initial-value problem.

First, we summarize some properties of the modes of undamped free vibration derived from the problem with $\epsilon = 0$:

$$M\frac{d^2\mathbf{z}}{dt^2} + K\mathbf{z} = 0, \quad \mathbf{z} = (z_1, z_2, \ldots, z_n). \tag{4.3.166}$$

The free vibrations are of the form

$$\mathbf{z}(t) = \boldsymbol{\xi}\cos\omega t, \quad \boldsymbol{\xi} = (\xi_1, \ldots, \xi_n) \tag{4.3.167}$$

so that

$$(K - M\omega^2)\boldsymbol{\xi} = 0 \tag{4.3.168}$$

is the system of n linear equations for the ξ_j. The characteristic equation for $\omega^{(j)}$, the natural frequencies (dimensionless) of free vibration, is

$$\det(K - M\omega^2) = 0. \tag{4.3.169}$$

We assume that the n real roots are distinct and order them: $\omega^{(1)}, \omega^{(2)}, \ldots$. Thus, there is one eigenvector $\boldsymbol{\xi}^{(j)}$ corresponding to each frequency such that

$$(K - M\omega^{(j)^2})\boldsymbol{\xi}^{(j)} = 0, \quad j = 1, \ldots, n. \tag{4.3.170}$$

These eigenvectors represent the free mode shapes and can be chosen to be an orthonormal set with respect to the weight function M

$$\boldsymbol{\xi}^{(i)} M \boldsymbol{\xi}^{(j)} = \delta_{ij}. \tag{4.3.171}$$

To study the damped problem, we represent the motion in terms of the undamped modes

$$\mathbf{x}(t) = \sum_{i=1}^{n} \alpha_i(t)\boldsymbol{\xi}^{(i)}, \tag{4.3.172}$$

where the $\alpha_i(t)$ are the modal amplitudes to be found. Then (4.3.163) becomes

$$\sum_{i=1}^{n} M\boldsymbol{\xi}^{(i)} \frac{d^2\alpha_i}{dt^2} + K\boldsymbol{\xi}^{(i)}\alpha_i = -\epsilon \sum_{i=1}^{n} B^{(i)}\boldsymbol{\xi}^{(i)} \frac{d\alpha_i}{dt}. \tag{4.3.173}$$

Using (4.3.170), multiplying by $\boldsymbol{\xi}^{(j)}$, and using (4.3.172), we find (see (1.3.53))

$$\frac{d^2\alpha_j}{dt^2} + \omega^{(j)^2}\alpha_j = -\epsilon \sum_{i=1}^{n} \beta_{ij} \frac{d\alpha_i}{dt}, \tag{4.3.174}$$

where β_{ij} is the quadratic form

$$\beta_{ij} = \boldsymbol{\xi}^{(i)} B \boldsymbol{\xi}^{(j)} \tag{4.3.175}$$

and is thus symmetric. The β_{ij} represent the coupling between the modes due to damping.

The solution of (4.3.174) is now expressed in the form of a two-scale expansion

$$\alpha_j(t;\epsilon) = F_{j0}(t^+, \tilde{t}) + \epsilon F_{j1}(t^+, \tilde{t}) + \epsilon^2 F_{j2}(t^+, \tilde{t}) + \ldots, \tag{4.3.176}$$

where $t^+ = t(1 + \epsilon^2\sigma + \ldots)$ is the fast time and $\tilde{t} = \epsilon t$ is the slow time.

Repeating the calculation of Sec. 4.2.1, we find

$$L(F_{j0}) \equiv \frac{\partial^2 F_{j0}}{\partial t^{+2}} + \omega^{(j)^2} F_{j0} = 0. \tag{4.3.177}$$

$$L(F_{j1}) = -2\frac{\partial^2 F_{j0}}{\partial t^+ \partial \tilde{t}} - \sum_{p=1}^{n} \beta_{jp} \frac{\partial F_{p0}}{\partial t^+}. \tag{4.3.178}$$

$$\begin{aligned} L(F_{j2}) = &-2\sigma \frac{\partial^2 F_{j0}}{\partial t^{+2}} - \frac{\partial^2 F_{j0}}{\partial \tilde{t}^2} - 2\frac{\partial^2 F_{j1}}{\partial t^+ \partial \tilde{t}} \\ &- \sum_{p=1}^{n} \beta_{pj} \left(\frac{\partial F_{p0}}{\partial \tilde{t}} + \frac{\partial F_{p1}}{\partial t^+} \right). \end{aligned} \tag{4.3.179}$$

The general solution of (4.3.177) is

$$F_{j0} = A_{j0}(\tilde{t}) \cos \omega^{(j)} t^{+} + B_{j0}(\tilde{t}) \sin \omega^{(j)} t^{+}. \tag{4.3.180}$$

Thus, (4.3.178) becomes

$$\begin{aligned} L(F_{j1}) =& 2\omega^{(j)} \frac{dA_{j0}}{d\tilde{t}} \sin \omega^{(j)} t^{+} - 2\omega^{(j)} \frac{dB_{j0}}{d\tilde{t}} \cos \omega^{(j)} t^{+} \\ & - \sum_{p=1}^{n} \beta_{jp}(-A_{p0}\omega^{(p)} \sin \omega^{(p)} t^{+} + B_{p0}\omega^{(p)} \cos \omega^{(p)} t^{+}). \end{aligned} \tag{4.3.181}$$

In order to eliminate mixed secular terms in F_{j1}, all driving terms with frequency $\omega^{(j)}$ on the right-hand side of (4.3.181) must vanish. Thus,

$$S(A_{j0}) \equiv 2\frac{dA_{j0}}{d\tilde{t}} + \beta_{jj} A_{j0} = 0, \tag{4.3.182a}$$

$$S(B_{j0}) = 0. \tag{4.3.182b}$$

The result thus far shows that to first order the modes remain uncoupled, each oscillating with its natural frequency, each mode having a damping coefficient $\frac{1}{2}\beta_{jj}$, i.e.,

$$A_{j0} = a_{j0}e^{-(1/2)\beta_{jj}\tilde{t}}, \quad B_{j0} = b_{j0}e^{-(1/2)\beta_{jj}\tilde{t}}. \tag{4.3.183}$$

This coefficient comes from the diagonal elements of the general damping matrix (4.3.175) and is always positive.

In order to calculate the first approximation to the frequency shift, it is necessary to study the equation for F_{j2}. The solution for F_{j1} is now

$$\begin{aligned} F_{j1} =& A_{j1}(\tilde{t}) \cos \omega^{(j)} t^{+} + B_{j1}(\tilde{t}) \sin \omega^{(j)} t^{+} \\ & + \sum_{\substack{p=1 \\ p \neq j}}^{n} \beta_{pj} \frac{\omega^{(p)}}{\omega^{(j)^2} - \omega^{(p)^2}} [A_{p0} \sin \omega^{(p)} t^{+} - B_{p0} \cos \omega^{(p)} t^{+}]. \end{aligned} \tag{4.3.184}$$

This shows that the $O(\epsilon)$ terms of the jth mode contain all the frequencies of the whole system and that the mode is no longer "pure." It now follows from (4.3.151) that

$$\begin{aligned} L(F_{j2}) =& 2\sigma\omega^{(j)^2}(A_{j0} \cos \omega^{(j)} t^{+} + B_{j0} \sin \omega^{(j)} t^{+}) - \frac{d^2 A_{j0}}{d\tilde{t}^2} \cos \omega^{(j)} t^{+} \\ & - \frac{d^2 B_{j0}}{d\tilde{t}^2} \sin \omega^{(j)} t^{+} - 2\frac{dA_{j1}}{dt^{+}} \omega^{(j)} \sin \omega^{(j)} t^{+} - 2\frac{dB_{j1}}{dt^{+}} \omega^{(j)} \cos \omega^{(j)} t^{+} \\ & - 2\sum_{\substack{p=1 \\ p \neq j}}^{n} \beta_{pj} \frac{\omega^{(p)^2}}{\omega^{(j)^2} - \omega^{(p)^2}} \left[\frac{dA_{p0}}{d\tilde{t}} \cos \omega^{(p)} t^{+} + \frac{dB_{p0}}{d\tilde{t}} \sin \omega^{(p)} t^{+} \right] \\ & - \sum_{q=1}^{n} \beta_{qj} \left[\left(\frac{dA_{q0}}{d\tilde{t}} \cos \omega^{(q)} t^{+} + \frac{dB_{q0}}{d\tilde{t}} \sin \omega^{(q)} t^{+} \right. \right. \end{aligned}$$

$$- \omega^{(q)} A_{q1} \sin \omega^{(q)} t^+ + \omega^{(q)} B_{q1} \cos \omega^{(q)} t^+ \Big)$$

$$+ \sum_{\substack{p=1 \\ p \neq q}}^{n} \beta_{pq} \frac{\omega^{(p)^2}}{\omega^{(q)^2} - \omega^{(p)^2}} (A_{p0} \cos \omega^{(p)} t^+ + B_{p0} \sin \omega^{(p)} t^+ \Big] . \quad (4.3.185)$$

Again, to avoid mixed-secular terms, the coefficients of $\cos \omega^{(j)} t^+$, $\sin \omega^{(j)} t^+$ in the right-hand side of (4.3.185) must be set equal to zero. This gives, for $\cos \omega^{(j)} t^+$,

$$\omega^{(j)} S(B_{j1}) = \alpha_{j0} e^{-(1/2)\beta_{jj}\tilde{t}} \left[2\sigma \omega^{(q)^2} + \frac{1}{4} \beta_{jj}^2 - \sum_{\substack{p=1 \\ p \neq j}}^{n} \beta_{pj}^2 \frac{\omega^{(j)^2}}{\omega^{(p)^2} - \omega^{(j)^2}} \right] \quad (4.3.186)$$

when (4.3.183) is used.

An equivalent equation comes from the coefficient of $\sin \omega^{(j)} t^+$. As in Sec. 4.2.1, the right-hand side of (4.3.186) must be set equal to zero, so that inconsistent terms of the form $\tilde{t} \exp(-\beta_{jj}\tilde{t}/2)$ do not appear. Thus, the formula for the "shift" σ is

$$\sigma_j = -\frac{1}{8} \frac{\beta_{jj}^2}{\omega^{(j)^2}} + \sum_{\substack{p=1 \\ p \neq j}}^{n} \frac{\beta_{pj}^2}{\omega^{(p)^2} - \omega^{(j)^2}} \quad \text{for the } j\text{th mode.} \quad (4.3.187)$$

From this, a better approximation to the period of F_{j0} results. The frequency shift now depends on all damping coefficients as well as the natural frequencies of the undamped system. These results show how useful the two-scale method is in elucidating the behavior of a coupled system with small damping.

The old master Lord Rayleigh gave a succinct discussion of this problem in the course of his general discussion of vibrating systems, section 102 of [4.29]. The principal result of this section, the pure damping of each mode and the formula for the damping coefficient (4.3.183), was given by Rayleigh. He also presented the equivalent of (4.3.184) and remarked that the $O(\epsilon)$ modes excited in (4.3.184) are all in phase with each other but "that phase differs by a quarter period from the phase of F_{j0}" (in our notation). This is manifested in (4.3.184). Lastly, he gave in implicit form the recipe for the frequency shift (4.3.187). Earlier in [4.29] Rayleigh formulates conditions under which the three matrices M, B, K can be simultaneously diagonalized. For this case, which does not occur often in practical cases, each damped mode acts separately.

4.3.6 Two Weakly Nonlinear Oscillators

Consider the system

$$\frac{d^2x}{dt^2} + a^2 x = \epsilon y^2, \quad (4.3.188a)$$

$$\frac{d^2y}{dt^2} + b^2 y = 2\epsilon xy, \quad (4.3.188b)$$

where a and b are arbitrary constants and $0 < \epsilon \ll 1$.

This system, which models the motion of a star in a galaxy, was originally studied in [4.6] using a perturbation procedure based on the Hamiltonian structure of (4.3.188). We will discuss this approach in Chapter 5. Here and in Sec. 4.5 we study the solution using different multiple-scale expansions.

(a) *Necessary conditions for bounded solutions*

It is instructive to examine the energy integral ($\dot{} = d/dt$)

$$\frac{1}{2}(\dot{x}^2 + \dot{y}^2) + \frac{1}{2}(a^2x^2 + b^2y^2) - \epsilon xy^2 = E = \text{ constant.} \tag{4.3.189}$$

that is available for the system (4.3.188) in order to establish the conditions under which solutions are always bounded. We take $\epsilon > 0$ and construct the zero-velocity curves that one obtains by setting $\dot{x} = \dot{y} = 0$ in (4.3.189) and considering the one-parameter family

$$V(x, y) \equiv \frac{1}{2}(a^2x^2 + b^2y^2) - \epsilon xy^2 = E \tag{4.3.190}$$

corresponding to different values of E.

For any set of initial conditions: $x(0; \epsilon), y(0; \epsilon), \dot{x}(0; \epsilon), \dot{y}(0; \epsilon)$, one calculates E from (4.3.189). For this value of E, the curve $V(x, y) = E$ determines a "barrier" in the xy plane that cannot be crossed for motion evolving from these initial conditions because $V > E$ on the other side of the zero-velocity curve and, according to (4.3.189), this corresponds to an imaginary velocity. The exceptional case $V_x = V_y = 0$ on a zero-velocity curve corresponds to an equilibrium solution of the system (4.3.188). It then follows that motion resulting from a given set of initial conditions is bounded if the corresponding zero-velocity curve is closed and the initial point lies inside this closed curve.

To study (4.3.190), it is more convenient to rescale x and y and to consider the one-parameter family

$$\xi^2 + \eta^2 - \gamma\xi\eta^2 = C, \tag{4.3.191}$$

where

$$\xi = ax, \quad \eta = by, \quad \gamma = \frac{2\epsilon}{ab^2} > 0, \quad C = 2E. \tag{4.3.192}$$

We have the equilibrium point ($V_\xi = V_\eta = 0$) at (0, 0), which is a center, and the points $(\gamma^{-1}, \pm 2^{1/2}\gamma^{-1})$, which are saddles. Thus, motion with small initial values of x, $\dot{x}$, y, and $\dot{y}$ is stable, while motion with $x(0; \epsilon) \approx b^2/2\epsilon$, $\dot{x}(0; \epsilon) \approx 0$, $y(0; \epsilon) \approx \pm ab/2^{1/2}\epsilon$, $\dot{y}(0; \epsilon) \approx 0$ is unstable.

The zero-velocity curves defined by (4.3.191) can be easily calculated and are shown in Figure 4.3.8. For $0 \leq C < \gamma^{-2}$ one branch of the family consists of the nested set of closed curves surrounding the origin. As $C \to \gamma^{-2}$, these tend to the limiting curve bounded by the parabola $\eta^2 = \gamma^{-2}(1 + \gamma\xi)$ on the left and the vertical line $\xi = \gamma^{-1}$ on the right. The other branch for $0 < C < \gamma^{-2}$ consists of

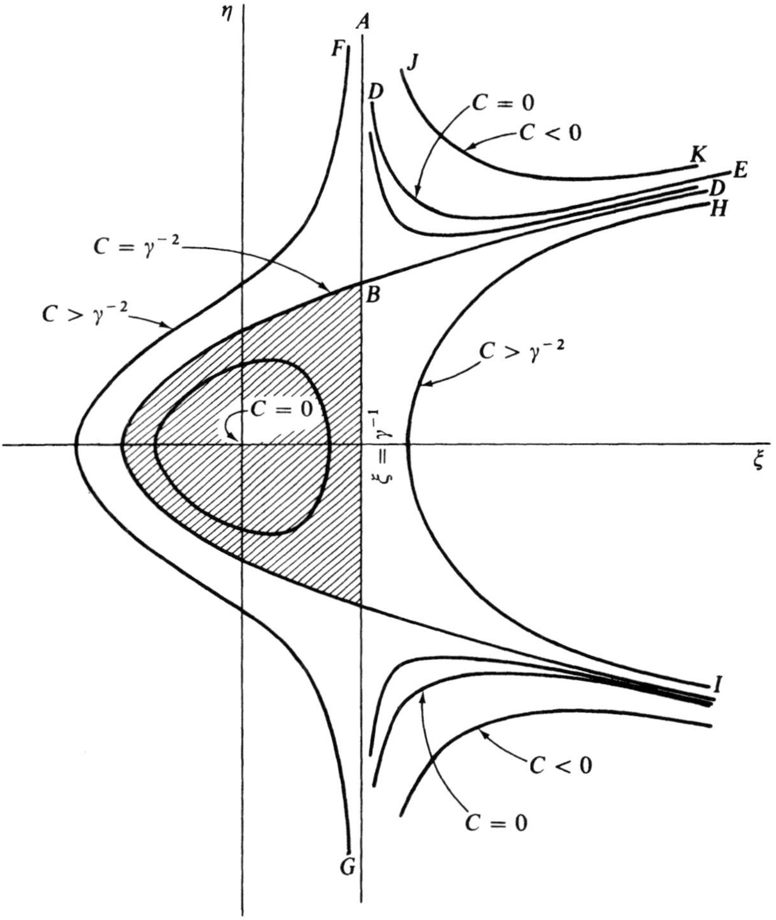

FIGURE 4.3.8. Zero-Velocity Curves, Equation (4.3.191)

the set of curves (and their images for $\eta < 0$) evolving from D-E, when $C = 0$, to the limiting curve A-B-D as $C \to \gamma^{-2}$.

Thus, if $0 < C < \gamma^{-2}$, motion is bounded if the initial values of ξ and η lie inside the shaded region. Translating this to the original variables gives bounded motion as long as

$$0 \leq E < \frac{a^2 b^4}{8\epsilon^2} \tag{4.3.193}$$

and

$$|y(0;\epsilon)| < \frac{a\sqrt{b^2+2\epsilon x(0;\epsilon)}}{2\epsilon}, \quad |x(0;\epsilon)| < \frac{b^2}{2\epsilon}. \tag{4.3.194}$$

For $C > \gamma^{-2}$, we have the two branches represented by the curves F-G and H-I in Figure 4.3.8. Finally, if $C < 0$, the zero-velocity curves generate the family represented by the curve J-K and its mirror image. In either of these cases, bounded solutions are not assured. In the remainder of this section, we will only consider initial values that are $O(1)$ as $\epsilon \to 0$ and values of a and b that are bounded away from zero. Thus, the conditions (4.3.193) and (4.3.194) are trivially satisfied and the solutions will be bounded for all times.

(b) *Multiple-scale expansion*

For the case of n-coupled linear oscillators discussed in the previous section, we were able to express the solution for each mode in terms of its associated fast time $t_j^+ = (1+\epsilon^2\sigma_j + \dots)t$ and the slow time $\tilde{t} = \epsilon t$. Since σ_j is different for each mode, the complete solution actually involves $n+1$ scales: $t_1^+, t_2^+, \dots, t_n^+, \tilde{t}$.

It is also possible to construct an expansion in terms of a hierarchy of successively slower scales: $t_0 = t, t_1 = \epsilon t, t_2 = \epsilon^2 t, \dots, t_N = \epsilon^N t, \dots$, and we will illustrate this more systematic approach next for the system (4.3.188). The idea of using N successively slower times was proposed independently in different contexts including the present one. The original references are [4.9], [4.14], [4.24], and [4.30].

We assume that the solution can be represented in the form

$$x(t;\epsilon) = x_0(t_0, t_1, \dots) + \epsilon x_1(t_0, t_1, \dots) + \epsilon^2 x_2(t_0, t_1, \dots) + \dots, \tag{4.3.195a}$$

$$y(t;\epsilon) = y_0(t_0, t_1, \dots) + \epsilon y_1(t_0, t_1, \dots) + \epsilon^2 y_2(t_0, t_1, \dots) + \dots. \tag{4.3.195b}$$

We then compute

$$\frac{dx}{dt} = \frac{\partial x_0}{\partial t_0} + \epsilon\left(\frac{\partial x_1}{\partial t_0} + \frac{\partial x_0}{\partial t_1}\right) + \epsilon^2\left(\frac{\partial x_2}{\partial t_0} + \frac{\partial x_1}{\partial t_1} + \frac{\partial x_0}{\partial t_2}\right) + \dots, \tag{4.3.196a}$$

$$\begin{aligned}\frac{d^2x}{dt^2} &= \frac{\partial^2 x_0}{\partial t_0^2} + \epsilon\left(\frac{\partial^2 x_1}{\partial t_0^2} + 2\frac{\partial^2 x_0}{\partial t_0 \partial t_1}\right)\\ &\quad + \epsilon^2\left(\frac{\partial^2 x_2}{\partial t_0^2} + 2\frac{\partial^2 x_1}{\partial t_0 \partial t_1} + 2\frac{\partial^2 x_0}{\partial t_0 \partial t_2} + \frac{\partial^2 x_0}{\partial t_1^2}\right) + \dots,\end{aligned} \tag{4.3.196b}$$

and similar expressions for dy/dt and d^2y/dt^2. Note that to order ϵ the procedure gives the same formal results as if t_2 was not involved. Hence, to determine the dependence of x_0 on t_2, we need to consider the terms of order ϵ^2, etc.

Substituting the above expansions into (4.3.188) leads to the following sequence of pairs of equations for the x_i and y_i:

$$L_1(x_0) \equiv \frac{\partial^2 x_0}{\partial t_0^2} + a^2 x_0 = 0, \tag{4.3.197a}$$

$$L_2(y_0) \equiv \frac{\partial^2 y_0}{\partial t_0^2} + b^2 y_0 = 0. \tag{4.3.197b}$$

$$L_1(x_1) = y_0^2 - 2\frac{\partial^2 x_0}{\partial t_0 \partial t_1}, \tag{4.3.198a}$$

$$L_2(y_1) = 2x_0 y_0 - 2\frac{\partial^2 y_0}{\partial t_0 \partial t_1}. \tag{4.3.198b}$$

$$L_1(x_2) = 2y_0 y_1 - 2\frac{\partial^2 x_1}{\partial t_0 \partial t_1} - 2\frac{\partial^2 x_0}{\partial t_0 \partial t_2} - \frac{\partial^2 x_0}{\partial t_1^2}, \tag{4.3.199a}$$

$$L_2(y_2) = 2(x_0 y_1 + x_1 y_0) - 2\frac{\partial^2 y_1}{\partial t_0 \partial t_1} - 2\frac{\partial^2 y_0}{\partial t_0 \partial t_2} - \frac{\partial^2 y_0}{\partial t_1^2}. \tag{4.3.199b}$$

Solving (4.3.197) gives

$$x_0(t_0, t_1, t_2, \ldots) = \alpha_0(t_1, t_2, \ldots)\cos\mu_0; \quad \mu_0 = at_0 + \phi_0(t_1, t_2, \ldots), \tag{4.3.200a}$$

$$y_0(t_0, t_1, t_2, \ldots) = \beta_0(t_1, t_2, \ldots)\cos\nu_0; \quad \nu_0 = bt_0 + \psi_0(t_1, t_2, \ldots), \tag{4.3.200b}$$

where α_0 and β_0 are unknown amplitudes and ϕ_0 and ψ_0 are unknown phases, and all four functions may depend on $t_1, t_2, \ldots$ but not on t_0.

Using (4.3.200) to evaluate the right-hand sides of (4.3.198) gives

$$L_1(x_1) = \frac{\beta_0^2}{2}(1 + \cos 2\nu_0) + 2a\left(\frac{\partial \alpha_0}{\partial t_1}\sin\mu_0 + \alpha_0\frac{\partial \phi_0}{\partial t_1}\cos\mu_0\right), \tag{4.3.201a}$$

$$\begin{aligned} L_2(y_1) &= \alpha_0\beta_0[\cos(\mu_0 + \nu_0) + \cos(\mu_0 - \nu_0)] \\ &\quad + 2b\left(\frac{\partial \beta_0}{\partial t_1}\sin\nu_0 + \beta_0\frac{\partial \psi_0}{\partial t_1}\cos\nu_0\right). \end{aligned} \tag{4.3.201b}$$

The trigonometric terms with μ_0 argument in (4.3.201a) and ν_0 argument in (4.3.201b) are homogeneous solutions and will therefore lead to unbounded contributions on the t_0 scale. Removing these inconsistent terms results in $\partial\alpha_0/\partial t_1 = \partial\phi_0/\partial t_1 = \partial\beta_0/\partial t_1 = \partial\psi_0/\partial t_1 = 0$. Henceforth, we will adopt the superscript notation $f^{(n)}$ to denote functions $f^{(n)}(t_n, t_{n+1}, \ldots)$ that do not depend on t_0, t_1, $\ldots$, t_{n-1}. Thus,

$$\alpha_0 = \alpha_0^{(2)}, \quad \beta_0 = \beta_0^{(2)}, \quad \phi_0 = \phi_0^{(2)}, \quad \psi_0 = \psi_0^{(2)}. \tag{4.3.202}$$

What remains of (4.3.201) can now be solved to give

$$x_1 = \alpha_1^{(1)} \cos \mu_1 + \frac{\beta_0^2}{2a^2} + \frac{\beta_0^2}{2(a^2 - 4b^2)} \cos 2\nu_0, \tag{4.3.203a}$$

$$y_1 = \beta_1^{(1)} \cos \nu_1 - \frac{\alpha_0 \beta_0}{a(a+2b)} \cos(\mu_0 + \nu_0) - \frac{\alpha_0 \beta_0}{a(a-2b)} \cos(\mu_0 - \nu_0), \tag{4.3.203b}$$

where

$$\mu_1 = at_0 + \phi_1^{(1)}, \quad \nu_1 = bt_0 + \psi_1^{(1)}. \tag{4.3.203c, d}$$

We note the occurrence of the divisors a, a^2, $(a + 2b)$, and $(a - 2b)$ in our results. By hypothesis, for bounded solutions, a and b are bounded away from zero and positive (see the discussion following (4.3.194)). Hence, $a > 0, a + 2b > 0$, and these divisors are not troublesome. However, if $a = 2b$, our results become singular even though the solution must be bounded. Clearly, this is a reflection of the inadequacy of the assumed expansion for values of $a \approx 2b$, which is called the *first resonance* condition for (4.3.188). We will return to this case soon. For the time being, we assume $a \neq 2b$ and proceed with the calculations for the next order.

Using the known solutions for x_0, y_0, x_1, and y_1 in (4.3.199) gives

$$\begin{aligned} L_1(x_2) =& \beta_0\beta_1[\cos(\nu_1 + \nu_0) + \cos(\nu_1 - \nu_0)] - \frac{\alpha_0\beta_0^2}{a(a+2b)} \cos(2\nu_0 + \mu_0) \\ & - \frac{\alpha_0\beta_0^2}{a(a-2b)} \cos(2\nu_0 - \mu_0) + S, \end{aligned} \tag{4.3.204a}$$

$$\begin{aligned} L_2(y_2) =& \alpha_0\beta_1[\cos(\mu_0 + \nu_1) + \cos(\mu_0 - \nu_1)] - \alpha_0^2\beta_0[\cos(2\mu_0 + \nu_0) \\ & + \cos(2\mu_0 - \nu_0)] + \alpha_1\beta_0[\cos(\nu_0 + \mu_1) + \cos(\nu_0 - \mu_1)] \\ & + \frac{\beta_0^3}{2(a^2 - 4b^2)} \cos 3\nu_0 + T, \end{aligned} \tag{4.3.204b}$$

where S and T denote the following terms, which are solutions of the homogeneous equations $L_1(x_2) = 0$ and $L_2(y_2) = 0$, respectively:

$$\begin{aligned} S = & 2\left(a \frac{\partial}{\partial t_1} \alpha_1 \cos(\phi_1 - \phi_0) + a \frac{\partial \alpha_0}{\partial t_2}\right) \sin \mu_0 \\ & + 2\left(a \frac{\partial}{\partial t_1} \alpha_1 \sin(\phi_1 - \phi_0) + \alpha_0 a \frac{\partial \phi_0}{\partial t_2} - \frac{\alpha_0 \beta_0^2}{a^2 - 4b^2}\right) \cos \mu_0, \end{aligned} \tag{4.3.205a}$$

$$\begin{aligned} T =& 2\left(b \frac{\partial}{\partial t_1} \beta_1 \cos(\psi_1 - \psi_0) + b \frac{\partial \beta_0}{\partial t_2}\right) \sin \nu_0 \\ & + 2\left\{b \frac{\partial}{\partial t_1} \beta_1 \sin(\psi_1 - \psi_0) + \beta_0 \frac{\partial \psi_0}{\partial t_2} b \right. \end{aligned}$$

$$+ \left. \frac{\beta_0[\beta_0^2(3a^2 - 8b^2) - 4a^2\alpha_0^2]}{4a^2(a^2 - 4b^2)} \right\} \cos \nu_0. \tag{4.3.205b}$$

Clearly, we must set $S = T = 0$ in order to ensure that x_2 and y_2 are bounded on the t_0 scale. In order that $S = 0$ identically, the coefficients of the $\sin \mu_0$ and $\cos \mu_0$ terms in (4.3.205a) must vanish identically. Similarly, $T \equiv 0$ requires that the coefficients of $\sin \nu_0$ and $\cos \nu_0$ in (4.3.205b) must each vanish. Now, examining the coefficients of the $\sin \mu_0$ and $\sin \nu_0$ terms in (4.3.205a), we see that α_1 and β_1 will be *unbounded on the t_1 scale* unless we set $\partial\alpha_0/\partial t_2 = 0$, $\partial\beta_0/\partial t_2 = 0$. Therefore, the amplitudes of the $O(1)$ solution do not depend on t_2 either. We next examine the coefficients of the $\cos \mu_0$ and $\cos \nu_0$ term and note that boundedness of α_1 and β_1 on the t_1 scale requires (with α_0 and β_0 independent of t_2) that we also take

$$\frac{\partial \phi_0}{\partial t_2} = \frac{\beta_0^2}{a(a^2 - 4b^2)}, \tag{4.3.206a}$$

$$\frac{\partial \psi_0}{\partial t_2} = -\frac{\beta_0^2(3a^2 - 8b^2) - 4a^2\alpha_0^2}{4a^2(a^2 - 4b^2)b}, \tag{4.3.206b}$$

which we can immediately integrate. The solution to $O(1)$ is now determined up to terms involving t_2 and is summarized here:

$$\alpha_0 = \alpha_0^{(3)}, \tag{4.3.207a}$$

$$\beta_0 = \beta_0^{(3)}, \tag{4.3.207b}$$

$$\phi_0 = \frac{\beta_0^2}{a(a^2 - 4b^2)} t_2 + k_1^{(3)}, \tag{4.3.207c}$$

$$\psi_0 = -\frac{\beta_0^2(3a^2 - 8b^2) - 4a^2\alpha_0^2}{4a^2(a^2 - 4b^2)b} t_2 + k_2^{(3)}, \tag{4.3.207d}$$

where the functions $\alpha_0^{(3)}$, $\beta_0^{(3)}$, $k_1^{(3)}$, and $k_2^{(3)}$ arise after integration with respect to t_2 and therefore depend only on $t_3, t_4, \ldots$. Actually, if we stop our calculations at this stage, we can regard these functions as constants and evaluate them using the initial values of x, y, dx/dt, and dy/dt. We also have a strong suspicion that α_0, β_0 will turn out to be pure constants, while ϕ_0, ψ_0 will depend linearly on the t_2, t_3, etc. This is borne out by the calculations, at least to the next order, which we do not give. (See Problem 12.)

The need for $N > 2$ time scales is now apparent since the phases for the x and y solutions have *different* corrections and could not have been uniformly represented by a single t^+ variable.

Having eliminated the inconsistent terms with respect to t_2 from (4.3.205), these reduce to the statement that α_1, β_1, ϕ_1, and ψ_1 are independent of t_1, i.e.,

$$\alpha_1 = \alpha_1^{(2)}; \quad \beta_1 = \beta_1^{(2)}; \quad \phi_1 = \phi_1^{(2)}; \quad \psi_1 = \psi_1^{(2)}. \tag{4.3.208}$$

Now, we can also integrate what remains of (4.3.204) to calculate x_2 and y_2 in the form

$$\begin{aligned} x_2 =& \alpha_2^{(1)} \cos \mu_2 + \frac{\beta_0\beta_1}{a^2 - 4b^2} \cos(\nu_1 + \nu_0) + \frac{\beta_0\beta_1}{a^2} \cos(\nu_1 - \nu_0) \\ &+ \frac{\alpha_0\beta_0^2}{4ab(a+b)(a+2b)} \cos(2\nu_0 + \mu_0) \\ &+ \frac{\alpha_0\beta_0^2}{4ab(a-b)(a-2b)} \cos(2\nu_0 - \mu_0), \end{aligned} \tag{4.3.209a}$$

$$\begin{aligned} y_2 =& \beta_2^{(1)} \cos \nu_2 - \frac{\alpha_0\beta_0}{a(a+2b)} [\cos(\mu_0 + \nu_1) + \cos(\mu_0 - \nu_1)] \\ &+ \frac{\alpha_0^2\beta_0}{4a^2(a+b)^2} \cos(2\mu_0 + \nu_0) + \frac{\alpha_0^2\beta_0}{4a^2(a-b)^2} \cos(2\mu_0 - \nu_0) \\ &- \frac{\alpha_1\beta_0}{a(a+2b)} \cos(\nu_0 + \mu_1) - \frac{\alpha_1\beta_0}{a(a-2b)} \cos(\nu_0 - \mu_1) \\ &- \frac{\beta_0^3}{16b^2(a^2 - 4b^2)} \cos 3\nu_0, \end{aligned} \tag{4.3.209b}$$

where

$$\mu_2 = at_0 + \phi_2^{(2)}, \quad \nu_2 = bt_0 + \psi_2^{(2)}. \tag{4.3.210}$$

The reason we calculated the solution for x_2 and y_2 is to exhibit the second resonance for this problem associated with the small divisor $a - b$ when $a = b$. We conclude that to each higher order in ϵ the solution will contain a new divisor which can vanish for a certain ratio of a/b.

These higher-order resonances become gradually weaker in the sense that for any given small but nonvanishing value of the small divisor, the corresponding singularity is strongest for the first resonance and decreases by one order in ϵ for each succeeding resonance. Actually, these higher resonances are not interesting because they do not lead to an exchange of energy between modes as is the case for the first resonance. In the next subsection, we will concentrate on the solution for the case $a \approx 2b$.

Finally, we note that if we truncate the procedure at the stage where the form of the $O(\epsilon^N)$ solution is determined, we have already defined the dependence of the $O(1)$ solution on $t_0, t_1, \ldots, t_N$, the dependence of the $O(\epsilon)$ solution on $t_0, t_1, \ldots, t_{N-1}$, etc.

(c) *Solution near the first resonance*

The procedure for handling the case when $a \approx 2b$ in the previous subsection is quite straightforward and merely involves setting $b = (a/2)+s$ (where s is a small constant) explicitly in the equation. It is easy to verify that the richest equations to $O(\epsilon)$ result by choosing $as + s^2 = \epsilon\kappa$, where κ is an arbitrary $O(1)$ constant (i.e., $s = O(\epsilon)$).

We can then rescale t and ϵ so that with no loss of generality we need only study the system

$$\frac{d^2x}{dt^2} + x = \epsilon y^2, \tag{4.3.211a}$$

$$\frac{d^2y}{dt^2} + \frac{1}{4}y = 2\epsilon xy - \epsilon\kappa y. \tag{4.3.211b}$$

We will only study the dependence of the solution on t_0 and t_1, as the calculations to higher order are difficult. Therefore, we ignore the dependence of x and y on t_2, t_3, etc., and consider a development in the form

$$x = x_0(t_0, t_1) + \epsilon x_1(t_0, t_1) + \ldots, \tag{4.3.212a}$$

$$y = y_0(t_0, t_1) + \epsilon y_1(t_0, t_1) + \ldots. \tag{4.3.212b}$$

The differential equations governing x_0, y_0, x_1, and y_1 are then easily derived in the form

$$L_1(x_0) \equiv \frac{\partial^2 x_0}{\partial t_0^2} + x_0 = 0, \tag{4.3.213a}$$

$$L_2(y_0) \equiv \frac{\partial^2 y_0}{\partial t_0^2} + \frac{1}{4}y_0 = 0, \tag{4.3.213b}$$

$$L_1(x_1) = -2\frac{\partial^2 x_0}{\partial t_0 \partial t_1} + y_0^2, \tag{4.3.214a}$$

$$L_2(y_1) = -2\frac{\partial^2 y_0}{\partial t_0 \partial t_1} - 2x_0 y_0 - \kappa y_0. \tag{4.3.214b}$$

The solution of (4.3.213) is

$$x_0 = \alpha_0(t_1)\cos\mu_0, \quad \mu_0 = t_0 + \phi_0(t_1), \tag{4.3.215a}$$

$$y_0 = \beta_0(t_1)\cos\nu_0, \quad \nu_0 = \frac{t_0}{2} + \psi_0(t_1). \tag{4.3.215b}$$

We substitute these into the right-hand sides (4.3.214) and isolate homogeneous solutions to obtain (primes denote d/dt_1):

$$L_1(x_1) = \left[2\alpha_0' - \frac{\beta_0^2}{2}\sin(2\psi_0 - \phi_0)\right]\sin\mu_0 + \left[2\alpha_0\phi_0' \frac{\beta_0^2}{2}\cos(2\psi_0 - \phi_0)\right]\cos\mu_0 + \frac{\beta_0^2}{2}, \tag{4.3.216a}$$

$$\begin{aligned} L_2(y_1) = {} & [\beta_0' + \alpha_0\beta_0\sin(2\psi_0 - \phi_0)]\sin\nu_0 \\ & + [\beta_0\psi_0' + \alpha_0\beta_0\cos(2\psi_0 - \phi_0) - \kappa\beta_0]\cos\nu_0 \\ & + \alpha_0\beta_0\cos(\mu_0 + \nu_0). \end{aligned} \tag{4.3.216b}$$

The bracketed terms in (4.3.216) multiply solutions of the homogeneous equations $L_1(x_1) = 0$ and $L_2(y_1) = 0$ and must be set equal to zero. Note that the terms that produced small divisors in the preceding subsection are now homogeneous solutions and will be eliminated.

The resulting equations governing the four unknowns α_0, β_0, ϕ_0, and ψ_0 are the coupled nonlinear system

$$2\alpha_0' - \frac{\beta_0^2}{2}\sin(2\psi_0 - \phi_0) = 0, \tag{4.3.217a}$$

$$2\alpha_0\phi_0' + \frac{\beta_0^2}{2}\cos(2\psi_0 - \phi_0) = 0, \tag{4.3.217b}$$

$$\beta_0' + \alpha_0\beta_0\sin(2\psi_0 - \phi_0) = 0, \tag{4.3.217c}$$

$$\beta_0\psi_0' + \alpha_0\beta_0\cos(2\psi_0 - \phi_0) - \kappa\beta_0 = 0. \tag{4.3.217d}$$

Despite the forbidding nature of these equations, they can be solved because two integrals exist for the system. The absence of t_1 then reduces the solution to quadrature.

Before tackling this solution, we integrate what is left of (4.3.216) and calculate x_1 and y_1 in the following form free of small divisors:

$$x_1 = \alpha_1(t_1)\cos\mu_1 + \frac{\beta_0^2}{2}, \qquad \mu_1 = t_0 + \phi_1(t_1), \tag{4.3.218a}$$

$$y_1 = \beta_1(t_1)\cos\nu_1 - \frac{16}{9}\alpha_0\beta_0\cos(\mu_0 + \nu_0), \qquad \nu_1 = \frac{t_0}{2} + \psi_1(t_1). \tag{4.3.218b}$$

A first integral for the system (4.3.217) can be derived by multiplying (4.3.217a) by α_0, (4.3.217c) by β_0, and adding the result. We find

$$\alpha_0^2 + \frac{\beta_0^2}{4} = \text{const.} = 2E_0. \tag{4.3.219}$$

It is easy to verify that E_0 is the leading term of the total energy of the system (4.3.211)

$$E = \frac{1}{2}\left[\left(\frac{dx}{dt}\right)^2 + \left(\frac{dy}{dt}\right)^2\right] + \frac{1}{2}\left(x^2 + \frac{y^2}{4}\right) - \epsilon\left(xy^2 + \frac{\kappa y^2}{2}\right) = \text{const.}, \tag{4.3.220}$$

which is an exact integral.

The second integral of the system (4.3.217) is more subtle. One way of obtaining it is to differentiate (4.3.217b) solved for ϕ_0. Letting $\xi = 2\psi_0 - \phi_0$ temporarily, we obtain

$$\phi_0'' = \frac{\beta_0^2\xi'}{4\alpha_0}\sin\xi - \frac{\beta_0\beta_0'}{2\alpha_0}\cos\xi + \frac{\beta_0^2\alpha_0'}{4\alpha_0^2}\cos\xi. \tag{4.3.221}$$

Now, using (4.3.217a, b, and d) to eliminate ϕ_0', β_0', and ψ_0', we obtain

$$\phi_0'' = -\frac{2\alpha_0'}{\alpha_0}(\phi_0' - \kappa), \tag{4.3.222}$$

which integrates to give

$$\phi_0' = \frac{\lambda_0}{\alpha_0^2} + \kappa, \tag{4.3.223}$$

where λ_0 is a constant of integration. It is interesting to note that λ_0 corresponds to the leading term of the so-called "adelphic" integral discussed by Whittaker [4.35]. In Chapter 5, we will study in detail the procedure for calculating such integrals directly using the Hamiltonian formulation of the problem.

Now, to reduce the solution to quadrature, we square and add (4.3.217a, b), then use the energy integral to find

$$4\alpha_0'^2 + 4\alpha_0^2\phi_0'^2 = \frac{\beta_0^4}{2} = 4(2E_0 - \alpha_0^2)^2. \tag{4.3.224}$$

Using (4.3.223) to express ϕ_0' in terms of α_0 then gives

$$\alpha_0'^2 + \frac{\lambda_0^2}{\alpha_0^2} + p^2\alpha_0^2 - \alpha_0^4 = q, \tag{4.3.225}$$

where

$$p^2 = 4E_0 + \kappa^2, \tag{4.3.226a}$$

$$q = 2E_0^2 - 2\kappa\lambda_0. \tag{4.3.226b}$$

Equation (4.3.225) is an integral involving the amplitude of the x oscillator and can be solved by quadrature. Actually, in view of the apparent[1] singularity in (4.3.225) when $\alpha_0 \to 0$, $\lambda_0 \neq 0$, it is more convenient to express this result in terms of the energy E_1 of the x oscillator to $O(1)$. With

$$E_1 = \frac{\alpha_0^2}{2}, \tag{4.3.227}$$

(4.3.225) becomes

$$E_1'^2 - 8E_1(E_0 - E_1)^2 + 4\kappa E_1(\lambda_0 + \kappa E_1) = -\lambda_0^2. \tag{4.3.228}$$

The solution of (4.3.228) can be carried out using elliptic functions. However, it is more instructive to study the qualitative behavior of E_1 using energy arguments. To fix ideas, let $\kappa = 0$ and denote

$$V(E_1) = -8E_1(E_0 - E_1)^2. \tag{4.3.229}$$

[1] The singularity is not worrisome because α_0 is never equal to zero unless $\lambda_0 = 0$, and this limiting case will be considered in our study of (4.3.228).

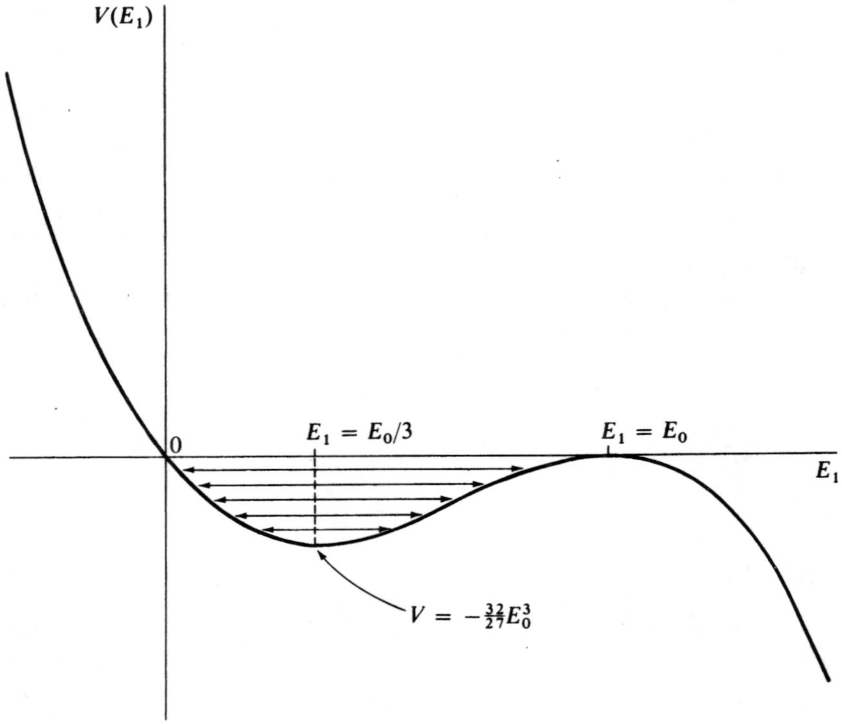

FIGURE 4.3.9. V as a Function of E_1, Equation (4.3.229)

For some fixed E_0, $V(E_1)$ is shown in Figure 4.3.9.

In view of the definition (4.3.227), $E_1 \geq 0$, and from (4.3.219) we must have $E_1 \leq E_0$. Therefore, solutions of (4.3.229) only exist ($E'^2 \geq 0$) and are consistent with the other integrals of motion for values of λ_0 such that

$$0 \leq \lambda_0^2 \leq \frac{32}{27} E_0^3. \tag{4.3.230}$$

For this range of values of λ_0, E_1 oscillates between $E_{1_{\min}}$ and $E_{1_{\max}}$, where $E_{1_{\min}}$ is the smallest root of $V(E_1) = 0$ and $E_{1_{\max}}$ is the second root. We note from Figure 4.3.9 (or the calculation of the roots of the cubic $V(E_1) = 0$) that the third root is larger than E_0 and must be excluded.

The curves in the $E_1 E_1'$ plane are the ovals sketched in Figure 4.3.10 for a fixed E_0. Thus, we have the interesting phenomenon of periodic energy exchange on the t_1 scale between the two oscillators at resonance.

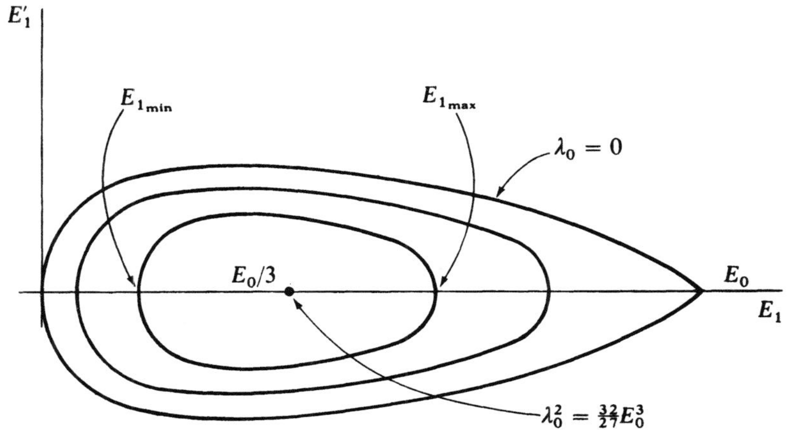

FIGURE 4.3.10. E_1 as a Function of E_1', $\kappa = 0, 0 \leq \lambda_0 \leq \frac{32}{27} E_0^3$, Equation (4.3.228)

The point $E_1 = E_0/3$ is a stable equilibrium point and corresponds to a periodic solution of (4.3.211) and no energy exchange. We can derive this periodic solution directly from (4.3.217) by noting that α_0 and β_0 are constants and ϕ_0 and ψ_0 are linear in t_1 if

$$2\psi_0 - \phi_0 = 0, \tag{4.3.231a}$$

$$\phi_0' = -\frac{\beta_0^2}{4\alpha_0} = \text{const.}, \tag{4.3.231b}$$

$$\psi_0' = \kappa - \alpha = \text{const.} = \frac{\phi_0'}{2}. \tag{4.3.231c}$$

Thus, given some α_0, we calculate the following values for β_0, ϕ_0 and ψ_0:

$$\beta_0 = \sqrt{8\alpha_0(\alpha_0 - \kappa)}, \tag{4.3.232a}$$

$$\phi_0 = 2(\kappa - \alpha_0)t_1 + \text{const.}, \tag{4.3.232b}$$

$$\psi_0 = (\kappa - \alpha_0)t_1 + \text{const.}, \tag{4.3.232c}$$

and the expressions for x and y, which are periodic, at least to $O(1)$, are

$$x = \alpha_0 \cos\{[1 + 2\epsilon(\kappa - \alpha_0) + O(\epsilon^2)]t + \text{const.}\} + O(\epsilon), \tag{4.3.233a}$$

$$y = \sqrt{8\alpha_0(\alpha_0 - \kappa)} \cos\left\{[1 + 2\epsilon(\kappa - \alpha_0) + O(\epsilon^2)]\frac{t}{2} + \text{const.}\right\} + O(\epsilon). \tag{4.3.233b}$$

These results agree with the expressions given in [4.13] to $O(1)$ in amplitude and to $O(\epsilon)$ in frequency. Note that this periodic solution corresponds to very special initial conditions and could also have been derived by using the method of strained coordinates (see Problem 14).

The situation for $\kappa \neq 0$ in (4.3.228) is not much different from the above and is not presented here. Finally, once E_1 (or α_0) is known, we calculate ϕ_0 by integrating (4.3.223). The amplitude β_0 of the y oscillator is obtained from the energy integral (4.3.219). Then ψ_0 can be found directly from (4.3.217a, c).

The interested reader can also consult [4.14], where a numerical verification of these results is presented for some special values of the parameters. In all cases, the agreement is consistent with the order of accuracy of the derived theory.

Problems

1. Consider the problem of beats for a linear oscillator in which the driver frequency ω is close to the natural frequency ω_N:

$$\frac{d^2Y}{dT^2} + \omega_N^2 = F_0 \cos \omega T. \tag{4.3.234}$$

Here F_0 is a constant, and the small parameter is $\epsilon = (\omega_N - \omega)/\omega_N$. For the initial conditions $Y(0) = A$, $dY(0)/dT = 0$ use dimensionless variable $y = Y/A$, $t = \omega_N T$ to express the problem in the form ($\delta = F_0/A\omega_N^2$, $\tilde{t} = \epsilon t$)

$$\frac{d^2y}{dt^2} + y = \delta \cos(t - \tilde{t}) \tag{4.3.235}$$

with $y(0; \epsilon) = 1$, $dy(0; \epsilon)/dt = 0$.

Derive a two-scale expansion of the solution correct to $O(\epsilon)$ and compare this with the exact result.

2. Consider the system

$$\frac{d^2y}{dt^2} + \mu^2 y = 0, \tag{4.3.236a}$$

$$\frac{d\mu}{dt} = \epsilon f(y, \mu) \tag{4.3.236b}$$

analogous to the example discussed in Section 4.3.2 except that now μ is not given explicitly but is a slowly varying dependent variable coupled with y.

Make an exact change of variable from t to t^+ by setting

$$\frac{dt^+}{dt} = \mu. \tag{4.3.237}$$

Now solve the resulting system using a two-scale expansion with t^+ and ϵt^+ as the fast and slow variables. Carry out the solution explicitly for the case $f = y^2$.

3. Generalize (4.3.46) to include weak linear damping and cubic nonlinearity. In dimensional variables, we have

$$M\frac{d^2Y}{dT^2} + B\frac{dY}{dT} + K\mu^2\left(\frac{T}{T_2}\right)Y + JY^3 = 0, \tag{4.3.238}$$

$$Y(0) = D, \quad \frac{dY(0)}{dT} = 0, \tag{4.3.239}$$

where M, B, K, T_2, and D are constants. Thus, T_2 is the time scale over which the frequency varies and $T_1 \equiv (M/K)^{1/2} \ll T_2$. Denote $\epsilon = T_1/T_2$, and choose the dimensionless variables $y = Y/(T_1K/T_2J)^{1/2}, t = T/(M/K)^{1/2}$ to obtain

$$\frac{d^2y}{dt^2} + \mu^2(\tilde{t})y + \epsilon\beta\frac{dy}{dt} + \epsilon y^3 = 0, \tag{4.3.240}$$

$$y(0;\epsilon) = \delta, \quad \frac{dy}{dt}(0;\epsilon) = 0, \tag{4.3.241}$$

where $\tilde{t} = \epsilon t$, $\epsilon\beta = B/(MK)^{1/2}$, $\delta = D/(T_1K/T_2J)^{1/2}$ with $\beta = O(1)$ and $\delta = O(1)$.

Calculate the two-scale expansion of the solution correct to $O(\epsilon)$.

4. We wish to examine whether the subharmonic response to the forced Duffing equation survives in the presence of damping. Study

$$\frac{d^2y}{dt^2} + y + \epsilon\left(y^3 + \beta\frac{dy}{dt}\right) = \epsilon\cos\Omega(\epsilon)t, \tag{4.3.242}$$

$$y(0;\epsilon) = a, \quad \frac{dy}{dt}(0;\epsilon) = b, \tag{4.3.243}$$

$$\Omega = 3 + \epsilon\omega_1 + \ldots, \tag{4.3.244}$$

where $\beta, a, b, \omega_1, \omega_2, \ldots$ are given arbitrary constants independent of ϵ.

Choosing $t^+ = (3 + \epsilon^2\omega_2 + \ldots)t$ and $\tilde{t} = \epsilon t$ as fast and slow variables, construct the solution to $O(1)$ and show that for certain values of a and b the steady-state solution is indeed a subharmonic oscillation.

5. The following equation models a certain resonance behavior in celestial mechanics:

$$\frac{d^2y}{dt^2} + y + 2\epsilon y(1 - 5\cos^2 R) = \epsilon^2 R\cos t, \tag{4.3.245}$$

where $R^2 = y^2 + (dy/dt)^2$. The initial conditions are

$$y(0;\epsilon) = a\cos b, \quad \frac{dy}{dt}(0;\epsilon) = a\sin b, \tag{4.3.246}$$

where a and b are constants independent of ϵ.

a. Construct a two-scale expansion for the solution in the form

$$y = F_0(t, \tilde{t}) + \epsilon F_1(t, \tilde{t}) + \dots \tag{4.3.247}$$

and show that the procedure works routinely as long as $s \equiv 5\cos^2 a - 1 \neq 0$. Carry out the details of the solution to $O(\epsilon)$.

b. Since s appears as a divisor in the solution for F_1, the procedure fails for initial values of a such that $s \approx 0$. Show that for s small it is incorrect to neglect the forcing term $\epsilon^2 R \cos t$ in determining the solution for F_0. Retain this term formally in the equations governing the slowly varying amplitude and phase of F_0, and deduce the structure of the solution when $s \approx 0$.

c. Guided by the results in (b), develop the solution when s is small in terms of appropriate time variables, to order ϵ. In particular, show that the slow variable is now $\epsilon^{3/2}t$, that the expansion should proceed in powers of $\epsilon^{1/2}$, and that we must set $s = \epsilon^{1/2}\bar{s}$ for some fixed $\bar{s}$ as $\epsilon \to 0$.

d. Match the solutions in (a) and (c) in some common overlap *domain in s*, and derive a result that is uniformly valid for all s to $O(\epsilon)$.

6. Calculate the time history $t(\theta; \epsilon)$ for the solution of (4.3.123) to $O(1)$. In particular, derive the expression for $\tau(\tilde{\theta})$ by considering the terms of order ϵ^2 in the expansion.

7. Generalize the system (4.3.117) to include a lift force L

$$L = \frac{1}{2}\rho V^2 S C_L, \quad C_L = \text{constant} \tag{4.3.248}$$

that acts in the direction normal to the velocity vector. In particular, show that to $O(1)$ the effect of lift is a slow motion of the perigee according to

$$\omega(\tilde{\theta}) = -\frac{C_L}{2C_D}\log(1 + 2\tilde{\theta}). \tag{4.3.249}$$

8. Consider the problem of a spherical (nonlifting) satellite in an exponentially varying atmosphere. Assume that the atmospheric density is given by

$$\rho(R) = \rho_0 e^{(R-R_0)/H}, \tag{4.3.250}$$

where H is the scale height. Values for ρ_0 and H can be calculated from the following typical density values: $\rho = 3.65 \times 10^{-15}\text{kg/m}^3$ at 1000 km altitude, and $\rho = 5.604 \times 10^{-7}\text{kg/m}^3$ at 100 km.

Assume that the satellite is initially at apogee at 500 km with $e = 0.01$, and you wish to predict the decay of the orbit down to 100 km. Use appropriate dimensionless variables to model the density in the above altitude range and calculate the slowly varying values of a, e, and ω.

9. Note that a Kepler ellipse with focus at the moon but preserving its orientation relative to the inertial $\bar{x}\,\bar{y}$ frame must rotate with angular velocity -1 (clockwise) relative to the $x'y'$ frame.

a. Show that (4.3.161a) then implies that the orbit to $O(1)$ preserves its orientation in the inertial $\bar{x}\,\bar{y}$ frame.

b. Introduce moon-centered *nonrotating* coordinates, and show that the equations that now correspond to (4.3.147) are Keplerian to $O(\epsilon)$ with $O(\epsilon^2)$ perturbations. This confirms the fact that to $O(\epsilon)$ the solution is an orientation-preserving Kepler ellipse.

c. Continue the solution of (4.3.150) to $O(\epsilon)$ by considering boundedness of u_2 and p_2. Carry out the calculations for small e correct to $O(e)$ and evaluate A, B, $\tilde{p}$, and α_1.

d. Express the limiting form

$$\frac{1}{2}\left[\left(\frac{dx^{**}}{dt^{**}}\right)^2 + \left(\frac{dy^{**}}{dt^{**}}\right)^2\right] - \frac{1}{r^{**}} + \frac{3}{2}\epsilon^2 x^{**^2} = \text{constant} \quad (4.3.251)$$

of the Jacobi integral for a close lunar satellite in terms of u', u, and p to obtain

$$\frac{1}{2p^2}(u'^2 + u^2) - u - \frac{3\epsilon^2}{2u^2}\cos^2\theta = \text{constant}. \quad (4.3.252)$$

Show that this is an exact integral of the system (4.3.150). Use this result to verify that the solution you calculated in part (c) is correct.

10. Apply the results of Sec. 4.3.5 to the simple system of two equal masses (M) with equal springs ($K = K_{01} = K_{12} = K_{02}$) but general damping. Verify that $\beta_{11} < \beta_{22}$ so that the fundamental mode decays more slowly than the second mode. Calculate the approximate period of each mode.

11. Solve the coupled linear system (4.3.174) using the multiple-scale expansion

$$\alpha_j(t;\epsilon) = \alpha_{j0}(t_0, t_1, t_2, + \ldots) + \epsilon\alpha_{j1}(t_0, t_1, t_2, \ldots) + \ldots \quad (4.3.253)$$

and verify that your results agree with those found in Sec. 4.3.5 when (4.3.176) is expressed in the form (4.3.253).

12. For the example of Sec. 4.3.6(b), determine the dependence of α_0, β_0, ϕ_0, and ψ_0 on t_3.

13. Using the results calculated in Sec. 4.3.6(b), verify that the energy integral (4.3.189) is satisfied to $O(\epsilon)$.

14. Rederive the periodic solution (4.3.233) $O(\epsilon)$ using the method of strained coordinates.

4.4 Two-Scale Expansions for Strictly Nonlinear Oscillators

In this section, we generalize the ideas of two-scale expansions to a *strictly* nonlinear second-order equation with solutions that are slowly modulated oscillations. Thus, as $\epsilon \to 0$ the equation remains nonlinear.

The basic technique is due to Kuzmak [4.22], who studies a special form of

$$\ddot{y} + g(y, \tilde{t}) + \epsilon h(y, \dot{y}, \tilde{t}) = 0, \quad (4.4.1)$$

where as usual, $\dot{} = d/dt$, $\tilde{t} = \epsilon t$ and $0 < \epsilon \ll 1$. Here h and g are given functions, analytic with respect to each of their arguments. We assume h to be odd in $\dot{y}$ to model dissipation; Kuzmak assumes h to be proportional to $\dot{y}$. The only other restriction is that for $\epsilon = 0$, the reduced nonlinear oscillator

$$\ddot{y} + g(y, 0) = 0 \tag{4.4.2}$$

has periodic solutions. This condition is satisfied whenever the potential $V(y, 0) = \int_0^y g(s, 0)ds$ is concave in some interval $y_1 < y < y_2$, and we restrict attention to oscillations in this interval.

Kuzmak works out the $O(1)$ solution partially; the equation of the slowly varying phase is not derived. In [4.25], Luke studies nonlinear nearly periodic dispersive waves and extends Kuzmak's results to higher order. Mathematically, the solution for such waves essentially reduces to (4.4.1) with $h = 0$, and Luke states that in this case the phase is constant. Bourland and Haberman [4.2] give a careful analysis of (4.4.1) and derive the equation governing the slowly varying phase. In many applications, including that of sustained resonance, to be discussed in Sec. 5.3, a more general version of (4.4.1) arises. The damping and restoring force terms also depend on n slowly varying quantities ρ_i, $i = 1, \ldots, n$ with the ρ_i governed by n first-order equations of the form $d\rho_i/dt = O(\epsilon)$. This problem is discussed in [4.1].

Here we restrict attention to the simple form (4.4.1) that suffices to illustrate all of the essential features. We develop the two-scale expansion of the solution based primarily on the approach in [4.2]. A specific example is then worked out in detail.

4.4.1 General Theory

Expansion procedure

We assume that the solution of (4.4.1) can be expressed in the following two-scale form:

$$y(t; \epsilon) = Y(t^+, \tilde{t}; \epsilon) = Y_0(t^+, \tilde{t}) + \epsilon Y_1(t^+, \tilde{t}) + O(\epsilon^2), \tag{4.4.3}$$

where $\tilde{t} = \epsilon t$ and

$$\frac{dt^+}{dt} = \omega_0(\tilde{t}) + \epsilon\omega_1(\tilde{t}) + O(\epsilon^2). \tag{4.4.4a}$$

Here $\omega_0(\tilde{t})$ and $\omega_1(\tilde{t})$ are unknowns to be defined by certain consistency requirements on Y_1. Integrating (4.4.4a), we also write

$$t^+ = \frac{\theta(\tilde{t})}{\epsilon} + \phi_0(\tilde{t}) + O(\epsilon); \theta(\tilde{t}) = \int_0^{\tilde{t}} \omega_0(s)ds, \ \phi_0(\tilde{t}) = \phi_0(0) + \int_0^{\tilde{t}} \omega_1(s)ds. \tag{4.4.4b}$$

As usual, we calculate the following expressions for the derivatives ($' \equiv d/d\tilde{t}$):

$$\dot{y} = (\omega_0 + \epsilon\omega_1)\frac{\partial Y}{\partial t^+} + \epsilon\frac{\partial Y}{\partial \tilde{t}} + O(\epsilon^2)$$

$$= \omega_0 \frac{\partial Y_0}{\partial t^+} + \epsilon \left(\omega_0 \frac{\partial Y_1}{\partial t^+} + \omega_1 \frac{\partial Y_0}{\partial t^+} + \frac{\partial Y_0}{\partial \tilde{t}} \right) + O(\epsilon^2), \tag{4.4.5a}$$

$$\ddot{y} = (\omega_0 + \epsilon\omega_1)^2 \frac{\partial^2 Y}{\partial t^{+2}} + 2\epsilon(\omega_0 + \epsilon\omega_1) \frac{\partial^2 Y}{\partial t^+ \partial \tilde{t}} + \epsilon\omega_0' \frac{\partial Y}{\partial t^+} + O(\epsilon^2)$$

$$= \omega_0^2 \frac{\partial^2 Y_0}{\partial t^{+2}} + \epsilon \left(\omega_0^2 \frac{\partial^2 Y_1}{\partial t^{+2}} + 2\omega_0\omega_1 \frac{\partial^2 Y_0}{\partial t^{+2}} + 2\omega_0 \frac{\partial^2 Y_0}{\partial t^+ \partial \tilde{t}} + \omega_0' \frac{\partial Y_0}{\partial t^+} \right)$$
$$+ O(\epsilon^2). \tag{4.4.5b}$$

We also develop h and g as follows:

$$h(y, \dot{y}, \tilde{t}) = h(Y_0, \omega_0 \frac{\partial Y_0}{\partial t^+}, \tilde{t}) + O(\epsilon), \tag{4.4.6a}$$

$$g(y, \tilde{t}) = g(Y_0, \tilde{t}) + \epsilon g_y(Y_0, \tilde{t}) Y_1 + O(\epsilon^2), \tag{4.4.6b}$$

to obtain the following equations governing Y_0 and Y_1:

$$\omega_0^2 \frac{\partial^2 Y_0}{\partial t^{+2}} + g(Y_0, \tilde{t}) = 0, \tag{4.4.7a}$$

$$L(Y_1) \equiv \omega_0^2 \frac{\partial^2 Y_1}{\partial t^{+2}} + g_y(Y_0, \tilde{t})Y_1 = -2\omega_0\omega_1 \frac{\partial^2 Y_0}{\partial t^{+2}}$$
$$- 2\omega_0 \frac{\partial^2 Y_0}{\partial t^+ \partial \tilde{t}} - \omega_0' \frac{\partial Y_0}{\partial t^+} - h\left(Y_0, \omega_0 \frac{\partial Y_0}{\partial t^+}, \tilde{t} \right) \equiv r_1. \tag{4.4.7b}$$

Solution to $O(1)$

The solution of (4.4.7a) can be carried out by quadrature. We multiply it by $(\partial Y_0/\partial t^+)$ and observe that the result can be integrated with respect to t^+ to yield the "energy" integral

$$\frac{\omega_0^2}{2} \left(\frac{\partial Y_0}{\partial t^+} \right)^2 + V(Y_0, \tilde{t}) = E_0(\tilde{t}), \tag{4.4.8a}$$

where $E_0(\tilde{t})$ is the slowly varying energy and V is the potential defined by

$$V(Y_0, \tilde{t}) = \int_0^{Y_0} g(\eta, \tilde{t}) d\eta. \tag{4.4 8b}$$

We next integrate (4.4.8a) and invert the result to express Y_0 in the form

$$Y_0(t^+, \tilde{t}) = f(t^+ + \lambda_0(\tilde{t}), E_0(\tilde{t}), \omega_0(\tilde{t}), \tilde{t}), \tag{4.4.9}$$

where $\lambda_0(\tilde{t})$ is a slowly varying phase shift that arises as an additive integration "constant" to t^+. At this point, it is useful to note that there is no loss of generality setting $\lambda_0 = 0$, as we have already included an arbitrary phase shift $\phi_0(\tilde{t})$ in the definition of t^+. In fact, we see that the two unknowns λ_0 and ϕ_0 appear in the solution only in the combination $\lambda_0 + \phi_0$ so that one or the other of these two constants may be ignored. In [4.2] λ_0 is set equal to zero and ϕ_0 is retained, whereas in [4.25] the converse choice is made.

Since Y_0 is periodic with respect to t^+, the curves in the Y_0, $(\partial Y_0/\partial t^+)$ plane for fixed $\tilde{t}$ (hence with E_0 and ω_0 also fixed) are ovals that are symmetric with respect to the Y_0 axis as sketched in Figure 4.4.1. Note that since $\tilde{t}$ is held fixed, the closed curve in Figure 4.4.1 is not an actual integral curve of (4.4.8a). However, we expect E_0, ω_0, and $\tilde{t}$ to change only by $O(\epsilon)$ after one complete cycle in this "phase plane."

Given initial values for y and $\dot{y}$ at $t = 0$, we have, once $\omega_0(E_0)$ is determined, the initial values $Y_0(\phi_0(0), 0)$ and $(\partial Y_0(\phi_0(0), 0)/\partial t^+)$, which specify a point, say the point marked O, on the oval in Figure 4.4.1. The details of this calculation are discussed at the end of this section.

With no loss of generality, we choose $\phi_0(0)$ so that $t^+ = 0$ when Y_0 first equals $Y_{0_{\min}}$. Note $\phi_0(0) < 0$. Thus, $Y_0(0, -\epsilon\phi_0(0)) = Y_{0_{\min}}$, $(\partial Y_0(0, -\epsilon\phi_0(0))/\partial t^+) = 0$, and Y_0 is an even function of t^+. Moreover, the expression for $f(t^+, E_0, \omega_0, \tilde{t})$ is obtained by inversion from (note that $\lambda_0 \equiv 0$)

$$t^+ = \omega_0(\tilde{t}) \int_{Y_{0_{\min}}(E_0, \tilde{t})}^{Y_0(t^+, \tilde{t})} \frac{d\eta}{\pm\sqrt{2[E_0(\tilde{t}) - V(\eta, \tilde{t})]}}, \tag{4.4.10}$$

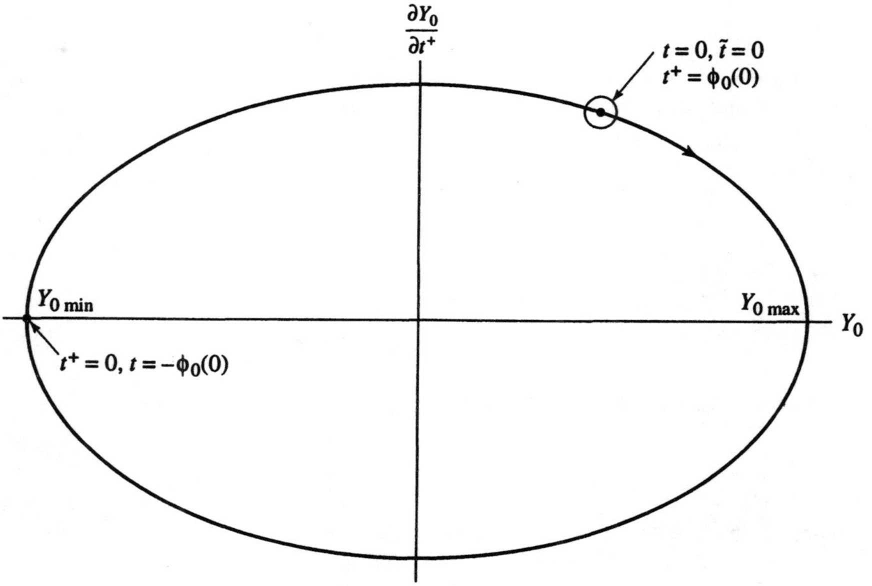

FIGURE 4.4.1. "Phase-Plane" of Y_0, $\frac{\partial Y_0}{\partial t+}$ for Fixed E_0, ω_0, and $\tilde{t}$

where the $\pm$ signs correspond to the signs of $\partial Y_0/\partial t^+$.

The period of oscillation P is then twice the integral from $Y_{0_{\min}}$ to $Y_{0_{\max}}$, i.e.,

$$P(E_0, \omega_0, \tilde{t}) \equiv 2\omega_0 \int_{Y_{0_{\min}}(E_0,\tilde{t})}^{Y_{0_{\max}}(E_0,\tilde{t})} \frac{dY_0}{\sqrt{2[E_0(\tilde{t}) - V(Y_0, \tilde{t})]}}. \tag{4.4.11}$$

At this stage, P is a function of E_0, ω_0, and $\tilde{t}$; once $E_0(\tilde{t})$ and $\omega_0(\tilde{t})$ are defined, we will have P as a function of $\tilde{t}$. In [4.25] Luke points out that unless P is a *constant* we lose uniformity for t large. To see this, let us use the periodicity condition for Y_0

$$Y_0(t^+, \tilde{t}) = Y_0(t^+ + nP(\tilde{t}), \tilde{t}), \tag{4.4.12}$$

where n is an integer. Denoting $t^+ + nP = \xi$ and $Y_0(t^+ + nP, \tilde{t}) = \overline{Y}_0(\xi, \tilde{t})$, we have, upon differentiation of (4.4.12) with respect to t^+,

$$\frac{\partial Y_0}{\partial t^+}(t^+, \tilde{t}) = \frac{\partial \overline{Y}_0}{\partial \xi}(\xi, \tilde{t}). \tag{4.4.13a}$$

Partial differentiation of (4.4.12) with respect to $\tilde{t}$ gives

$$\frac{\partial Y_0}{\partial \tilde{t}}(t^+, \tilde{t}) = \frac{\partial \overline{Y}_0}{\partial \xi}(\xi, \tilde{t}) n \frac{dP}{d\tilde{t}} + \frac{\partial \overline{Y}_0}{\partial \tilde{t}}(\xi, \tilde{t}). \tag{4.4.13b}$$

It then follows that

$$\frac{\partial \overline{Y}_0}{\partial \tilde{t}}(\xi, \tilde{t}) = \frac{\partial Y_0}{\partial \tilde{t}}(t^+ + nP, \tilde{t}) = \frac{\partial Y_0}{\partial \tilde{t}}(t^+, \tilde{t}) - n\frac{dP}{d\tilde{t}}\frac{\partial Y_0}{\partial t^+}(t^+, \tilde{t}). \tag{4.4.14}$$

Thus, for n large, $(\partial Y_0/\partial \tilde{t})$ becomes unbounded unless we set $(dP/d\tilde{t}) = 0$, and we choose $P(E_0, \omega_0(\tilde{t}), \tilde{t}) = P_0 =$ constant. The actual value of the constant P_0 is irrelevant; it may be chosen appropriately for computational convenience. For any fixed constant period P_0, (4.4.11) then gives a relation linking $\omega_0(\tilde{t})$ to $E_0(\tilde{t})$:

$$\omega_0 = \frac{P_0}{2}\left\{\int_{Y_{0_{\min}}(E_0,\tilde{t})}^{Y_{0_{\max}}(E_0,\tilde{t})} \frac{dY_0}{\sqrt{2[E_0(\tilde{t}) - V(Y_0, \tilde{t})]}}\right\}^{-1} \equiv \Omega(E_0, \tilde{t}). \tag{4.4.15}$$

Having fixed $P = P_0$, (4.4.15) allows us to express the solution (4.4.9) in the form

$$Y_0(t^+, \tilde{t}) = f(t^+, E_0(\tilde{t}), \Omega(E_0(\tilde{t}), \tilde{t}), \tilde{t}) \equiv p(t^+, E_0(\tilde{t}), \tilde{t}). \tag{4.4.16}$$

Thus, the solution to $O(1)$ involves the two unknown functions of $\tilde{t}$: (E_0, ϕ_0) or (ω_0, ϕ_0). We need to examine the solution of Y_1 to derive consistency conditions that will determine these two functions.

Solution of Y_1, periodicity conditions

Kuzmak [4.22] points out that $(\partial f/\partial t^+)$ is a solution of $L(Y_1) = 0$. To see this, we take the partial derivative of (4.4.7a), for $Y_0 = f(t^+, E_0, \omega_0, \tilde{t})$, with respect to t^+ holding $\tilde{t}$ (and hence E_0, ω_0) fixed. We find

$$\omega_0^2 \frac{\partial^2}{\partial t^{+2}} \left(\frac{\partial f}{\partial t^+} \right) + g_y(f, \tilde{t}) \frac{\partial f}{\partial t^+} = 0, \tag{4.4.17}$$

and this is just $L\left(\partial f/\partial t^+\right) = 0$. Actually, since t^+ occurs only in the first argument of either f or p, we see that $\frac{\partial p}{\partial t^+}(t^+, E_0(\tilde{t}), \tilde{t})$ also satisfies $L\left(\partial p/\partial t^+\right) = 0$.

Luke [4.25] argues that $(\partial f/\partial E_0)$ is a second linearly independent solution (but that $(\partial p/\partial E_0)$ is not) and uses $(\partial f/\partial t^+)$ and $(\partial f/\partial E_0)$ to construct the general solution for Y_1. Unfortunately, this form of Y_1 is not convenient for the calculation of the equation governing ϕ_0. We will follow the approach used by Bourland and Haberman in [4.2].

In order to construct a second linearly independent solution of $L(Y_1) = 0$ depending on t^+, E_0, and $\tilde{t}$, we begin by writing (4.4.7a) in the form

$$\Omega^2(E_0, \tilde{t}) \frac{\partial^2 p}{\partial t^{+^2}}(t^+, E_0, \tilde{t}) + g(p(t^+, E_0, \tilde{t}), \tilde{t}) = 0, \tag{4.4.18}$$

where we have used (4.4.15) to express ω_0 in terms of E_0 and $\tilde{t}$. Now, taking the partial derivative of (4.4.18) with respect to E_0, holding t^+ and $\tilde{t}$ fixed, gives

$$2\Omega \frac{\partial \Omega}{\partial E_0} \frac{\partial^2 p}{\partial t^{+^2}} + \Omega^2 \frac{\partial^2}{\partial t^{+^2}} \left(\frac{\partial p}{\partial E_0} \right) + g_y\left(p, \tilde{t}\right) \frac{\partial p}{\partial E_0} = 0,$$

i.e.,

$$L\left(\frac{\partial p}{\partial E_0}\right) = -2\Omega \frac{\partial \Omega}{\partial E_0} \frac{\partial^2 p}{\partial t^{+^2}}. \tag{4.4.19}$$

Next, we compute $L(t^+ \frac{\partial p}{\partial t^+})$; we have

$$\begin{aligned} L\left(t^+ \frac{\partial p}{\partial t^+}\right) &= \Omega^2 \frac{\partial^2}{\partial t^{+^2}} \left(t^+ \frac{\partial p}{\partial t^+}\right) + g_y(p, \tilde{t}) t^+ \frac{\partial p}{\partial t^+} \\ &= 2\Omega^2 \frac{\partial^2 p}{\partial t^{+^2}} + t^+ L\left(\frac{\partial p}{\partial t^+}\right) = 2\Omega^2 \frac{\partial^2 p}{\partial t^{+^2}} \end{aligned} \tag{4.4.20}$$

because $L(\partial p/\partial t^+) = 0$. Using (4.4.19) and (4.4.20), we see that

$$L\left(\Omega \frac{\partial p}{\partial E_0} + \frac{\partial \Omega}{\partial E_0} t^+ \frac{\partial p}{\partial t^+}\right) = \Omega L\left(\frac{\partial p}{\partial \tilde{t}_0}\right) + \frac{\partial \Omega}{\partial E_0} L\left(t^+ \frac{\partial p}{\partial t^+}\right) = 0.$$

Therefore,

$$q(t^+, E_0, \tilde{t}) \equiv \Omega(E_0, \tilde{t}) \frac{\partial p}{\partial E_0}(t^+, E_0, \tilde{t}) + \frac{\partial \Omega}{\partial E_0}(E_0, \tilde{t}) t^+ \frac{\partial p}{\partial t^+}(t^+, E_0, \tilde{t}) \tag{4.4.21}$$

is a second homogeneous solution ($L(q) = 0$) that is even in t^+. To ascertain that $(\partial p/\partial t^+)$ and q are linearly independent, we construct the Wronskian

$$W = \frac{\partial p}{\partial t^+} \frac{\partial q}{\partial t^+} - q \frac{\partial^2 p}{\partial t^{+^2}}$$

$$= \frac{\partial p}{\partial t^+}\left(\Omega \frac{\partial^2 p}{\partial E_0 \partial t^+} + \frac{\partial \Omega}{\partial E_0}\frac{\partial p}{\partial t^+} + \frac{\partial \Omega}{\partial E_0} t^+ \frac{\partial^2 p}{\partial t^{+2}}\right)$$

$$- \frac{\partial^2 p}{\partial t^{+2}}\left(\Omega \frac{\partial p}{\partial E_0} + \frac{\partial \Omega}{\partial E_0} t^+ \frac{\partial p}{\partial t^+}\right). \tag{4.4.22}$$

This expression simplifies further. In fact, when (4.4.15) is used for ω_0, (4.4.8a) becomes

$$\frac{1}{2}\Omega^2\left(\frac{\partial p}{\partial t^+}\right)^2 + V(p, \tilde{t}) = E_0. \tag{4.4.23}$$

Taking the partial derivative of this with respect to E_0 gives

$$\Omega \frac{\partial \Omega}{\partial E_0}\left(\frac{\partial p}{\partial t^+}\right)^2 + \Omega^2 \frac{\partial p}{\partial t^+}\frac{\partial^2 p}{\partial t^+ \partial E_0} + g(p, \tilde{t})\frac{\partial p}{\partial E_0} = 1,$$

or, if $\Omega \neq 0$,

$$\Omega \frac{\partial p}{\partial t^+}\frac{\partial^2 p}{\partial t^+ \partial E_0} = \frac{1}{\Omega} - \frac{1}{\Omega} g(p, \tilde{t})\frac{\partial p}{\partial E_0} - \frac{\partial \Omega}{\partial E_0}\left(\frac{\partial p}{\partial t^+}\right)^2. \tag{4.4.24}$$

We use (4.4.24) for the first term on the right-hand side of (4.4.22) and also impose (4.4.18) to find

$$W = \frac{1}{\Omega}. \tag{4.4.25}$$

Thus, $(\partial p/\partial t^+)$ and q are linearly independent solutions of $L = 0$.

Since p is an even periodic function of t^+, $\partial p/\partial t^+$ is odd periodic. If $(\partial \Omega/\partial E_0) = 0$, as in the linear case, the second solution q is also periodic; in general, with $(\partial \Omega/\partial E_0) \neq 0$, q is an even nonperiodic homogeneous solution. Now consider h. Since p is even in t^+, $\partial p/\partial t^+$ is odd in t^+, and h is an odd function of its second argument, we have

$$h\left(p(-t^+, E_0, \tilde{t}),\ \Omega_0 \frac{\partial p}{\partial t^+}(-t^+, \tilde{t}_0, \tilde{t}), \tilde{t}\right)$$

$$= h\left(p(t^+, E_0, \tilde{t}), -\Omega_0 \frac{\partial p}{\partial t^+}(t^+, E_0, \tilde{t}), \tilde{t}\right)$$

$$= -h\left(p(t^+, E_0, \tilde{t}), \Omega_0 \frac{\partial p_0}{\partial t^+}(t^+, E_0, \tilde{t}), \tilde{t}\right). \tag{4.4.26}$$

Hence, $h(p, E_0, \tilde{t})$ is an odd function of t^+.

In view of the above, the right-hand side, r_1, of (4.4.7b) has the following decomposition into even and odd periodic functions of t^+:

$$r_1 = r_{1_{\text{even}}} + r_{1_{\text{odd}}}, \tag{4.4.27}$$

where

$$r_{1_{\text{even}}}(t^+, E_0, \tilde{t}) = -2\Omega\omega_1 \frac{\partial^2 p}{\partial t^{+2}} \tag{4.4.28a}$$

$$r_{1_{\text{odd}}}(t^+, E_0, \tilde{t}) = -2\Omega\left[\frac{\partial^2 p}{\partial t^+ \partial E_0}\frac{dE_0}{d\tilde{t}} + \frac{\partial^2 p}{\partial t^+ \partial \tilde{t}}\right] - \frac{\partial \Omega}{d\tilde{t}}\frac{\partial p}{\partial t^+}$$
$$- h\left(p, \Omega\frac{\partial p}{\partial t^+}, \tilde{t}\right). \tag{4.4.28b}$$

Note that in evaluating $r_{1_{\text{odd}}}$ we have used $\omega_0(\tilde{t}) = \Omega(E_0(\tilde{t}), \tilde{t})$ and $Y_0(t^+, \tilde{t}) = p(t^+, E_0(\tilde{t}), \tilde{t})$. In particular, the bracketed term on the right-hand side of (4.4.28b) is just $(\partial^2 Y_0/\partial t^+ \partial \tilde{t})$.

The particular solution of (4.4.7b) due to $r_{1_{\text{even}}}$ follows from (4.4.20) in the form

$$y_{1p_{\text{even}}}(t^+, E_0, \tilde{t}) = -\frac{\omega_1}{\Omega}t^+\frac{\partial p}{\partial t^+}, \tag{4.4.29a}$$

and we use variation of parameters to compute the odd particular solution in the form

$$y_{1p_{\text{odd}}}(t^+, E_0, \tilde{t}) = \frac{q(t^+, E_0, \tilde{t})}{\Omega(E_0, \tilde{t})}\int_0^{t^+} r_{1_{\text{odd}}}(s, E_0, \tilde{t})\frac{\partial p}{\partial t^+}(s, E_0, \tilde{t})ds$$
$$- \frac{1}{\Omega(E_0, \tilde{t})}\frac{\partial p}{\partial t^+}(t^+, E_0, \tilde{t})\int_0^{t^+} r_{1_{\text{odd}}}(s, E_0, \tilde{t})q(s, E_0, \tilde{t})ds. \tag{4.4.29b}$$

The general solution for Y_1 then has the form

$$Y_1(t^+, \tilde{t}) \equiv y_1(t^+, E_0, \tilde{t}) = A_1(\tilde{t})\frac{\partial p}{\partial t^+}(t^+, E_0, \tilde{t}) + B_1(\tilde{t})q(t^+, E_0, \tilde{t})$$
$$+ y_{1p_{\text{odd}}}(t^+, E_0, \tilde{t}) + y_{1p_{\text{even}}}(t^+, E_0, \tilde{t}). \tag{4.4.30}$$

In order that Y_1 be a periodic function of t^+, its even and odd parts must *individually* be periodic. Consider first the even part of Y_1, i.e., $B_1 q + y_{1p_{\text{even}}}$. Using (4.4.21) for q, (4.4.29a) for $y_{1p_{\text{even}}}$, and noting that $(\partial p/\partial t^+)$ is periodic in t^+, we see that the mixed-secular terms are eliminated by setting

$$B_1(\tilde{t})\frac{\partial \Omega}{\partial E_0}(E_0, \tilde{t}) - \frac{\omega_1(\tilde{t})}{\Omega(E_0, \tilde{t})} = 0, \tag{4.4.31}$$

and (4.4.30) reduces to

$$Y_1(t^+, \tilde{t}) \equiv y_1(t^+, E_0, \tilde{t}) = A_1(\tilde{t})\frac{\partial p}{\partial t^+}(t^+, E_0, \tilde{t})$$
$$+ \frac{\omega_1}{(\partial\Omega/\partial E_0)}\frac{\partial p}{\partial E_0}(t^+, E_0, \tilde{t}) + y_{1p_{\text{odd}}}. \tag{4.4.32}$$

If $(\partial\Omega/\partial E_0) \neq 0$, (4.4.31) determines B_1 once E_0 and ϕ_0 have been calculated (*Note*: $\omega_1 = (d\phi_0/d\tilde{t})$). Thus, (4.4.31) makes no contribution toward the determination of the two unknowns (E_0, ϕ_0) in the $O(1)$ solution unless $(\partial\Omega/\partial E_0) = 0$, in which case we must set $\omega_1 = 0$.

The odd part of y_1 consists of $A_1(\partial p/\partial t^+) + y_{1p_{\text{odd}}}$, and since $(\partial p/\partial t^+)$ is already periodic, we must require $y_{1p_{\text{odd}}}$ to be periodic by itself. We show next that a necessary and sufficient condition for $y_{1P_{\text{odd}}}$ to be periodic is that

$$\int_0^{P_0} r_{1_{\text{odd}}}(t^+, E_0, \tilde{t})\frac{\partial p}{\partial t^+}(t^+, E_0, \tilde{t})dt^+ = 0. \tag{4.4.33}$$

This condition can be simply deduced by noting that $\partial p/\partial t^+$ is the only periodic homogeneous solution. Therefore, $r_{1_{\text{odd}}} + r_{1_{\text{even}}}$ must be orthogonal to this periodic solution. Since $r_{1_{\text{even}}}(\partial p/dt^+)$ is odd, $\int_0^{P_0} r_{1_{\text{even}}}(\partial p/\partial t^+)dt^+ = 0$ automatically, and we are left with (4.4.33). A more explicit derivation of (4.4.33) follows from the periodicity condition

$$y_{1P_{\text{odd}}}(t^+ + P_0, E_0, \tilde{t}) - y_{1P_{\text{odd}}}(t^+, E_0, \tilde{t}) = 0. \tag{4.4.34}$$

A straightforward but tedious calculation (see Problem 1) then shows that (4.4.34) implies (4.4.33), and vice versa.

Let us examine (4.4.33) in detail. Substitution of (4.4.28b) for $r_{1_{\text{odd}}}$ gives

$$2\Omega\left[\int_0^{P_0}\frac{\partial^2 p}{\partial t^+\partial E_0}(t^+, E_0, \tilde{t})\frac{\partial p}{\partial t^+}(t^+, E_0, \tilde{t})\frac{dE_0}{d\tilde{t}}dt^+ \right.$$
$$\left. + \int_0^{P_0}\frac{\partial^2 p}{\partial t^+\partial \tilde{t}}(t^+, E_0, \tilde{t})\frac{\partial p}{\partial t^+}(t^+, E_0, \tilde{t})dt^+\right]$$
$$+\frac{d\Omega}{d\tilde{t}}\int_0^{P_0}\left[\frac{\partial p}{\partial t^+}(t^+, E_0, \tilde{t})\right]^2 dt^+ + \int_0^{P_0} h\left(p, \Omega\frac{\partial p}{\partial t^+}, \tilde{t}\right)\frac{\partial p}{\partial t^+}dt^+ = 0.$$

Combining the first two terms gives

$$\frac{d}{d\tilde{t}}\left\{\Omega(E_0(\tilde{t}), \tilde{t})\int_0^{P_0}\left[\frac{\partial p}{\partial t^+}(t^+, E_0, \tilde{t})\right]^2 dt^+\right\}$$
$$+\int_0^{P_0} h\left(p, \Omega\frac{\partial p}{\partial t^+}, \tilde{t}\right)\frac{\partial p}{\partial t^+}dt^+ = 0. \tag{4.4.35}$$

Let us introduce the following notation for the average action to $O(1)$:

$$J(E_0, \tilde{t}) \equiv \Omega(E_0, \tilde{t})\int_0^{P_0}\left[\frac{\partial p}{\partial t^+}(t^+, E_0, \tilde{t})\right]^2 dt^+. \tag{4.4.36a}$$

Changing the integration variable from t^+ to Y_0, noting that the integrand is even in t^+, and using the energy integral (4.4.8a) gives

$$J(E_0, \tilde{t}) \equiv 2\int_{Y_{0_{\min}}(E_0,\tilde{t})}^{Y_{0_{\max}}(E_0,\tilde{t})}\sqrt{2[E_0(\tilde{t}) - V(Y_0, \tilde{t})}\, dY_0. \tag{4.4.36b}$$

We also denote the dissipation by

$$D(E_0, \tilde{t}) \equiv \int_0^{P_0} h\left(p, \Omega\frac{\partial p}{\partial t^+}, \tilde{t}\right)\frac{\partial p}{\partial t^+}dt^+ \tag{4.4.36c}$$

to express (4.4.35) in the compact form

$$\frac{dJ}{d\tilde{t}} + D = 0. \tag{4.4.37}$$

In general, this is a nonlinear first-order equation for E_0, and its solution defines $E_0(\tilde{t})$ for a given $E_0(0)$.

If h is linear in $\dot{y}$ and does not depend on y, i.e., $h = \tilde{h}(\tilde{t})\dot{y}$, (4.4.37) simplifies further to the linear equation

$$\frac{dJ}{d\tilde{t}} + \tilde{h}(\tilde{t})J = 0. \tag{4.4.38}$$

This has the solution

$$J(E_0, \tilde{t}) = J(E_0(0), 0) \exp\left(-\int_0^{\tilde{t}} \tilde{h}(s)ds\right). \tag{4.4.39}$$

Inverting (4.4.36b) then gives $E_0(\tilde{t})$.

If there is no dissipation ($h = 0$) we find $J =$ constant. This is the generalization (for a nonlinear oscillator) of the adiabatic invariant we computed in Sec. 4.3.2 for the linear oscillator with slowly varying frequency.

Weakly nonlinear problem: $g = \mu^2(\tilde{t})y$

It is instructive to specialize the foregoing results to the case where g is linear in y, i.e., $g = \mu^2(\tilde{t})y$ with $\mu(\tilde{t}) \neq 0$ prescribed. For the time being, we leave h in its general form.

Equation (4.4.7a) becomes

$$\omega_0^2 \frac{\partial^2 Y_0}{\partial t^{+2}} + \mu^2 Y_0 = 0 \tag{4.4.40}$$

with solution

$$f(t^+, E_0, \omega_0, \tilde{t}) = \frac{\sqrt{2E_0}}{\mu} \cos \frac{\mu}{\omega_0} t^+. \tag{4.4.41}$$

Thus, $Y_{0_{\max}} = \sqrt{2E_0}/\mu$ and $Y_{0_{\min}} = -\sqrt{2E_0}/\mu$. If we choose $P_0 = 2\pi$, the equation (4.4.11) for the period becomes

$$\pi = \omega_0 \int_{-\sqrt{2E_0}/\mu}^{\sqrt{2E_0}/\mu} \frac{dY_0}{\sqrt{2E_0 - Y_0^2\mu^2}} = \frac{\omega_0}{\mu}\pi, \tag{4.4.42}$$

and this gives $\omega_0 = \mu$. Thus, (4.4.41) implies

$$p(t^+, E_0, \tilde{t}) = \frac{\sqrt{2E_0}}{\mu} \cos t^+. \tag{4.4.43}$$

We note that the two linearly independent solutions of $L(Y_1) = 0$, i.e., $(\partial p/\partial t^+) = -(\sqrt{2E_0}/\mu) \sin t^+$ and $q = \mu(\partial p/\partial E_0) = (1/\sqrt{2E_0}) \cos t^+$, are both periodic in this case.

Since $(\partial\Omega/\partial E_0) = 0$, (4.4.31) gives $\omega_1(\tilde{t}) = 0$, i.e., $\phi_0(\tilde{t}) = \phi_0(0) =$ constant. The second periodicity condition (4.4.35) gives

$$\frac{d}{d\tilde{t}}\left(\frac{E_0}{\mu}\right) - \frac{1}{2\pi}\int_0^{2\pi} h\left(\frac{2E_0}{\mu}\cos s, -\sqrt{2E_0}\sin s, \tilde{t}\right)\frac{\sqrt{2E_0}}{\mu}\sin s ds = 0. \tag{4.4.44}$$

This equation specifies E_0 for a given h. For example, if $h = \tilde{h}(\tilde{t})\dot{y}$, we find

$$\frac{d}{d\tilde{t}}\left(\frac{E_0}{\mu}\right) + \tilde{h}(\tilde{t})\left(\frac{E_0}{\mu}\right) = 0, \tag{4.4.45}$$

which has the solution

$$\frac{E_0(\tilde{t})}{\mu(\tilde{t})} = \frac{E_0(0)}{\mu(0)}\exp\left(-\int_0^{\tilde{t}} \tilde{h}(s)ds\right). \tag{4.4.46}$$

These are precisely the results one finds by eliminating mixed-secular terms in the usual way. In fact, with $\omega_0 = \mu$, $p = (\sqrt{2E_0}/\mu)\cos t^+$, the equation for Y_1 becomes (see (4.4.7b) and (4.4.28))

$$\mu^2\left(\frac{\partial^2 Y_1}{\partial t^{+2}} + Y_1\right) = r_{1_{\text{even}}} + r_{1_{\text{odd}}}, \tag{4.4.47}$$

where

$$r_{1_{\text{even}}} = 2\omega_1\sqrt{2E_0}\cos t^+, \tag{4.4.48a}$$

$$r_{1_{\text{odd}}} = \left[\sqrt{\frac{2}{E_0}}\frac{dE_0}{d\tilde{t}} - \frac{\sqrt{2E_0}}{\mu}\frac{d\mu}{d\tilde{t}}\right]\sin t^+$$
$$- h\left(\frac{\sqrt{2E_0}}{\mu}\cos t^+, -\sqrt{2E_0}\sin t^+, \tilde{t}\right). \tag{4.4.48b}$$

Removing the $\cos t^+$ term from the right-hand side of (4.4.47) requires that we set $\omega_1 = 0$, as found earlier. Now, in order to remove terms proportional to $\sin t^+$, we isolate the first harmonic in the Fourier expansion of h and find the following contribution that must be set equal to zero in $r_{1_{\text{odd}}}$:

$$\sqrt{\frac{2}{E_0}}\frac{dE_0}{d\tilde{t}} - \frac{\sqrt{2E_0}}{\mu}\frac{d\mu}{d\tilde{t}}$$
$$-\frac{1}{\pi}\int_0^{2\pi} h\left(\frac{\sqrt{2E_0}}{\mu}\cos s. -\sqrt{2E_0}\sin s, \tilde{t}\right)\sin s ds = 0.$$

Multiplying this by $(\sqrt{E_0/2}/\mu)$ gives (4.4.44). Thus, for the weakly nonlinear problem, the two periodicity conditions (4.4.31) and (4.4.35) correspond to the requirement that $\cos t^+$ and $\sin t^+$ terms be absent from the right-hand side of (4.4.47).

$\frac{\partial \Omega}{\partial E_0} \neq 0$; *equation for* ω_1

In general, for the strictly nonlinear problem, $(\partial\Omega/\partial E_0) \neq 0$, and the condition (4.4.31) does not define ω_1 as it involves the new unknown function $B_1(\tilde{t})$. One approach is to examine the periodicity of Y_2 to obtain an equation governing ω_1. It is shown in [4.2] that this equation can also be derived more directly. This derivation is based on the observation made in [4.34] that (4.4.1) implies an *exact* condition for the action.

If we regard $y = Y(t^+, \tilde{t}; \epsilon)$ with $\tilde{t} = \epsilon t$ and t^+ to be defined by (4.4.4) exactly, we find that (4.4.1) becomes

$$(\omega_0 + \epsilon\omega_1)^2 \frac{\partial^2 Y}{\partial t^{+2}} + g(Y, \tilde{t}) + \epsilon \left[\left(\frac{d\omega_0}{d\tilde{t}} + \epsilon \frac{d\omega_1}{d\tilde{t}} \right) \frac{\partial Y}{\partial t^+} \right.$$

$$\left. +2(\omega_0 + \epsilon\omega_1) \frac{\partial^2 Y}{\partial t^+ \partial \tilde{t}} + h(Y, (\omega_0 + \epsilon\omega_1) \frac{\partial Y}{\partial t^+} + \epsilon \frac{\partial Y}{\partial \tilde{t}}, \tilde{t}) \right] + \epsilon^2 \frac{\partial^2 Y}{\partial \tilde{t}^2} = 0, \tag{4.4.49}$$

also exactly.

Let us now multiply (4.4.49) by $(\partial Y/\partial t^+)$ and integrate the result with respect to t^+ from $t^+ = 0$ to $t^+ = P_0$:

$$(\omega_0 + \epsilon\omega_1)^2 \int_0^{P_0} \frac{\partial^2 Y}{\partial t^{+2}} \frac{\partial Y}{\partial t^+} dt^+ + \int_0^{P_0} g(Y, \tilde{t}) \frac{\partial Y}{\partial t^+} dt^+$$

$$+\epsilon \left(\frac{d\omega_0}{d\tilde{t}} + \epsilon \frac{d\omega_1}{d\tilde{t}} \right) \int_0^{P_0} \left(\frac{\partial Y}{\partial t^+} \right)^2 dt^+ + 2\epsilon(\omega_0 + \epsilon\omega_1) \int_0^{P_0} \frac{\partial^2 Y}{\partial t^+ \partial \tilde{t}} \frac{\partial Y}{\partial t^+} dt^+$$

$$+\epsilon \int_0^{P_0} h\left(Y, (\omega_0 + \epsilon\omega_1) \frac{\partial Y}{\partial t^+} + \epsilon \frac{\partial Y}{\partial \tilde{t}}, \tilde{t}\right) \frac{\partial Y}{\partial t^+} dt^+ + \epsilon^2 \int_0^{P_0} \frac{\partial^2 Y}{\partial \tilde{t}^2} \frac{\partial Y}{\partial t^+} dt^+ = 0.$$

The first two terms are

$$\frac{1}{2}(\omega_0 + \epsilon\omega_1)^2 \int_0^{P_0} \frac{\partial}{\partial t^+} \left(\frac{\partial Y}{\partial t^+} \right)^2 dt^+, \quad \int_0^{P_0} \frac{\partial}{\partial t^+} (V(Y, \tilde{t})) dt^+,$$

and they vanish because Y is periodic in t^+ with period P_0. The third and fourth terms combine, and upon dividing out an ϵ we find

$$\frac{d}{d\tilde{t}} \left[(\omega_0 + \epsilon\omega_1) \int_0^{P_0} \left(\frac{\partial Y}{\partial t^+} \right)^2 dt^+ \right]$$

$$+ \int_0^{P_0} h\left(Y, (\omega_0 + \epsilon\omega_1) \frac{\partial Y}{\partial t^+} + \epsilon \frac{\partial Y}{\partial \tilde{t}}, \tilde{t}\right) \frac{\partial Y}{\partial t^+} dt^+$$

$$+\epsilon \int_0^{P_0} \frac{\partial^2 Y}{\partial \tilde{t}^2} \frac{\partial Y}{\partial t^+} dt^+ = 0. \tag{4.4.50}$$

For a solution of (4.4.4) that is periodic in t^+ with period P_0, (4.4.50) is an *exact* result.

Now, if we expand Y as in (4.4.3), the $O(1)$ and $O(\epsilon)$ terms of the expansion of (4.4.50) satisfy

$$\frac{d}{d\tilde{t}}\left[\omega_0 \int_0^{P_0} \left(\frac{\partial Y_0}{\partial t^+}\right)^2 dt^+\right] + \int_0^{P_0} h \frac{\partial Y_0}{\partial t^+} dt^+ = 0, \tag{4.4.51}$$

$$\frac{d}{d\tilde{t}}\left[2\omega_0 \int_0^{P_0} \frac{\partial Y_0}{\partial t^+}\frac{\partial Y_1}{\partial t^+} dt^+ + \omega_1 \int_0^{P_0} \left(\frac{\partial Y_0}{\partial t^+}\right)^2 dt^+\right]$$

$$+ \int_0^{P_0} \left[\frac{\partial h}{\partial y} Y_1 + \frac{\partial h}{\partial \dot{y}}\left(\omega_1 \frac{\partial Y_0}{\partial t^+} + \omega_0 \frac{\partial Y_1}{\partial t^+} + \frac{\partial Y_0}{\partial \tilde{t}}\right)\right] \frac{\partial Y_0}{\partial t^+} dt^+$$

$$+ \int_0^{P_0} h \frac{\partial Y_1}{\partial t^+} dt^+ + \int_0^{P_0} \frac{\partial^2 Y_0}{\partial \tilde{t}^2}\frac{\partial Y_0}{\partial t^+} dt^+ = 0, \tag{4.4.52}$$

where the arguments for h, $(\partial h/\partial y)$, and $(\partial h/\partial \dot{y})$ in (4.4.51)–(4.4.52) are evaluated at $y = Y_0$, $\dot{y} = \omega_0(\partial Y_0/\partial t^+)$. We note that (4.4.51) is just the periodicity condition (4.4.35) derived earlier if we regard $\omega_0(\tilde{t}) = \Omega(E_0(\tilde{t}), \tilde{t})$ and $Y_0(t^+, \tilde{t}) = p(t^+, E_0(\tilde{t}), \tilde{t})$, (see (4.4.15)–(4.4.16)). We also use these conditions for ω_0 and Y_0 together with the expression (4.4.32) for Y_1 in (4.4.52). Parity considerations result in much simplification. We have

$$Y_1 = y_{1_{\text{even}}} + y_{1_{\text{odd}}},$$

where (see (4.4.32))

$$y_{1_{\text{even}}} = \frac{\omega_1}{\partial\Omega/\partial E_0}\frac{\partial p}{\partial E_0},$$

$$y_{1_{\text{odd}}} = A_1 \frac{\partial p}{\partial t^+} + y_{1_{p_{\text{odd}}}}.$$

Therefore, the even and odd parts of $(\partial Y_1/\partial t^+)$ are

$$\left(\frac{\partial Y_1}{\partial t^+}\right)_{\text{even}} = \frac{\partial y_{1_{\text{odd}}}}{\partial t^+},$$

$$\left(\frac{\partial Y_1}{\partial t^+}\right)_{\text{odd}} = \frac{\partial y_{1_{\text{even}}}}{\partial t^+}.$$

Since $(\partial Y_0/\partial t^+)$ is odd, we have

$$\int_0^{P_0} \frac{\partial Y_0}{\partial t^+}\frac{\partial Y_1}{\partial t^+} dt^+ = \int_0^{P_0} \frac{\partial p}{\partial t^+}\frac{\partial y_{1_{\text{even}}}}{\partial t^+} dt^+$$

$$= \frac{\omega_1}{\partial\Omega/\partial E_0}\int_0^{P_0} \frac{\partial p}{\partial t^+}\frac{\partial^2 p}{\partial E_0 \partial t^+} dt^+, \tag{4.4.53a}$$

as the integral involving $(\partial y_{1_{\text{odd}}}/\partial t^+)$ vanishes. Recalling that h is odd in t^+ and noting that $\partial h/\partial y$ is odd whereas $\partial h/\partial \dot{y}$ is even, we have the following simplifications:

$$\int_0^{P_0} \frac{\partial h}{\partial y} Y_1 \frac{\partial Y_0}{\partial t^+} dt^+ = \int_0^{P_0} \frac{\partial h}{\partial y} y_{1_{\text{even}}} \frac{\partial p}{\partial t^+} dt^+$$

$$= \frac{\omega_1}{\partial \Omega/\partial E_0} \int_0^{P_0} \frac{\partial h}{\partial y}\left(p, \Omega \frac{\partial p}{\partial t^+}, \tilde{t}\right) \frac{\partial p}{\partial E_0} \frac{\partial p}{\partial t^+} dt^+, \tag{4.4.53b}$$

$$\omega_0 \int_0^{P_0} \frac{\partial h}{\partial \dot{y}} \frac{\partial Y_1}{\partial t^+} \frac{\partial Y_0}{\partial t^+} dt^+ = \omega_0 \int_0^{P_0} \frac{\partial h}{\partial \dot{y}} \frac{\partial y_{1_{\text{even}}}}{\partial t^+} \frac{\partial Y_0}{\partial t^+} dt^+$$

$$= \frac{\Omega \omega_1}{\partial \Omega/\partial E_0} \int_0^{P_0} \frac{\partial h}{\partial \dot{y}}\left(p, \Omega \frac{\partial p}{\partial t^+}, \tilde{t}\right) \frac{\partial^2 p}{\partial E_0 \partial t^+} \frac{\partial p}{\partial t^+} dt^+, \tag{4.4.53c}$$

$$\int_0^{P_0} \frac{\partial h}{\partial \dot{y}} \frac{\partial Y_0}{\partial \tilde{t}} \frac{\partial Y_0}{\partial t^+} dt^+ = 0, \tag{4.4.53d}$$

$$\int_0^{P_0} h \frac{\partial Y_1}{\partial t^+} dt^+ = \int_0^{P_0} h \frac{\partial y_{1_{\text{even}}}}{\partial t^+} dt^+$$

$$= \frac{\omega_1}{\partial \Omega/\partial E_0} \int_0^{P_0} h\left(p, \Omega \frac{\partial p}{\partial t^+}, \tilde{t}\right) \frac{\partial^2 p}{\partial E_0 \partial t^+} dt^+. \tag{4.4.53e}$$

Finally, since $(\partial^2 Y_0/\partial t^+)$ is even and $(\partial Y_0/\partial t^+)$ is odd, the last term in (4.4.52) vanishes. When the expressions in (4.4.53) are used to simplify (4.4.52), we find

$$\frac{d}{d\tilde{t}}\left[\frac{2\Omega\omega_1}{\partial \Omega/\partial E_0} \int_0^{P_0} \frac{\partial p}{\partial t^+} \frac{\partial^2 p}{\partial E_0 \partial t^+} dt^+ + \omega_1 \int_0^{P_0} \left(\frac{\partial p}{\partial t^+}\right)^2 dt^+\right]$$

$$+ \frac{\omega_1}{\partial \Omega/\partial E_0} \int_0^{P_0} \left\{ h\left(p, \Omega \frac{\partial p}{\partial t^+}, \tilde{t}\right) \frac{\partial^2 p}{\partial E_0 \partial t^+} + \frac{\partial h}{\partial y}\left(p, \Omega \frac{\partial p}{\partial t^+}, \tilde{t}\right) \frac{\partial p}{\partial E_0} \frac{\partial p}{\partial t^+} \right.$$

$$\left. + \frac{\partial h}{\partial \dot{y}}\left(p, \Omega \frac{\partial p}{\partial t^+}, \tilde{t}\right) \frac{\partial p}{\partial t^+}\left[\Omega \frac{\partial^2 p}{\partial E_0 \partial t^+} + \frac{\partial \Omega}{\partial E_0}\left(\frac{\partial p}{\partial t^+}\right)^2\right]\right\} dt^+ = 0. \tag{4.4.54}$$

This is just the linear homogeneous equation

$$\frac{d}{d\tilde{t}}\left[\frac{\omega_1}{\partial \Omega/\partial E_0} \frac{\partial}{\partial E_0}\left(\Omega \int_0^{P_0} \left(\frac{\partial p}{\partial t^+}\right)^2 dt^+\right)\right]$$

$$+ \frac{\omega_1}{\partial \Omega/\partial E_0} \frac{\partial}{\partial E_0} \int_0^{P_0} h\left(p, \Omega \frac{\partial p}{\partial t^+}, \tilde{t}\right) \frac{\partial p}{\partial t^+} dt^+ = 0. \tag{4.4.55}$$

Using the notation (4.4.36) for the action J and dissipation D gives

$$\frac{d}{d\tilde{t}}\left(\frac{J_{E_0}}{\Omega_{E_0}} \omega_1\right) + \frac{D_{E_0}}{\Omega_{E_0}} \omega_1 = 0, \tag{4.4.56}$$

where $\Omega_{E_0} \equiv (\partial\Omega/\partial E_0)$, $J_{E_0} \equiv (\partial J/\partial E_0)$, and $D_{E_0} \equiv (\partial D/\partial E_0)$. We can compute Ω_{E_0} using the expression (4.4.15)

$$\Omega_{E_0} = -\frac{P_0}{2}\left\{\int_{Y_{0_{\min}}(E_0,\tilde{t})}^{Y_{0_{\max}}(E_0,\tilde{t})} \frac{dY_0}{\sqrt{2[E_0(\tilde{t}) - V(Y_0,\tilde{t})]}}\right\}^{-2}$$

$$\cdot \frac{\partial}{\partial E_0}\int_{Y_{0_{\min}}(E_0,\tilde{t})}^{Y_{0_{\max}}(E_0,\tilde{t})} \frac{dY_0}{\sqrt{2[E_0(\tilde{t}) - V(Y_0,\tilde{t})]}}. \tag{4.4.57}$$

Differentiating the expression in (4.4.36b) for J, noting that the integrand vanishes at the upper and lower limits, then using (4.4.15) for Ω gives

$$J_{E_0} = 2\int_{Y_{0_{\min}}(E_0,\tilde{t})}^{Y_{0_{\max}}(E_0,\tilde{t})} \frac{dY_0}{\sqrt{2[E_0(\tilde{t}) - V(Y_0,\tilde{t})]}} = \frac{P_0}{\Omega(E_0,\tilde{t})}. \tag{4.4.58}$$

It follows from (4.4.58) that $\Omega_{E_0} = -J_{E_0E_0}/J_{E_0}^2$, and this is verified by (4.4.57). In computing the partial derivative of the integral in (4.4.57), it is important to keep in mind that the limits $Y_{0_{\min}}$ and $Y_{0_{\max}}$ depend on E_0.

Since $E_0(\tilde{t})$ is defined by the solution of (4.4.37), Ω_{E_0}, J_{E_0}, and D_{E_0} are, in principle, known functions of $\tilde{t}$, and (4.4.55) is a linear equation that defines $\omega_1(\tilde{t})$ in terms of $\omega_1(0)$. In fact, we have

$$\omega_1(\tilde{t})\frac{J_{E_0}}{\Omega_{E_0}} = \frac{\omega_1(0)J_{E_0}(E_0(0),0)}{\Omega_{E_0}(E_0(0),0)}\exp\left[-\int_0^{\tilde{t}}\frac{D_{E_0}(E_0(s),s)}{J_{E_0}(E_0(s),s)}ds\right]. \tag{4.4.59}$$

If V and h do not depend on $\tilde{t}$, we know that $\Omega(E_0)$, $J(E_0)$, and $D(E_0)$ are functions of E_0 only. In this case, $\Omega' = \Omega_{E_0}E_0'$, $J' = J_{E_0}E_0'$, and $D' = D_{E_0}E_0'$, where $' \equiv d/d\tilde{t}$. Therefore, (4.4.56) may be written in the form

$$\left(\frac{\omega_1 J'}{\Omega'}\right)' + \frac{\omega_1 D'}{\Omega'} = 0$$

or

$$J'\left(\frac{\omega_1}{\Omega'}\right)' + \frac{\omega_1}{\Omega'}J'' + \frac{\omega_1 D'}{\Omega'} = 0.$$

But, according to (4.4.37) $J'' = -D'$; hence

$$\left(\frac{\omega_1}{\Omega'}\right)' = 0.$$

Therefore, in this case (ω_1/Ω') is a constant.

A second special case has $h \equiv 0$, i.e., $D \equiv 0$, and (4.4.55) implies that $\omega_1 J_{E_0}/\Omega_{E_0}$ is a constant.

Initial conditions

To complete the solution to $O(1)$, we need to know the initial values $E_0(0)$, $\phi_0(0)$ as well as the initial value $\omega_1(0)$. Suppose we are given general initial conditions to $O(\epsilon)$ for the oscillator (4.4.1):

$$y(0;\epsilon) = \alpha_0 + \epsilon\alpha_1, \tag{4.4.60a}$$

$$\dot{y}(0;\epsilon) = \beta_0 + \epsilon\beta_1, \tag{4.4.60b}$$

where α_0, α_1, β_0, and β_1 are specified constants. Since $t^+ = \phi_0(0)$ and $\tilde{t} = 0$ at $t = 0$, (4.4.60a) and the expansion (4.4.3) for y give

$$Y_0(\phi_0(0), 0) = \alpha_0, \tag{4.4.61a}$$

$$Y_1(\phi_0(0), 0) = \alpha_1. \tag{4.4.61b}$$

Similarly, using the expansion (4.4.5a) for $\dot{y}$, we see that (4.4.60b) gives

$$\omega_0(0)\frac{\partial Y_0}{\partial t^+}(\phi_0(0), 0) = \beta_0, \tag{4.4.62a}$$

$$\omega_0(0)\frac{\partial Y_1}{\partial t^+}(\phi_0(0), 0) + \omega_1(0)\frac{\partial Y_0}{\partial t^+}(\phi_0(0), 0) + \frac{\partial Y_0}{\partial \tilde{t}}(\phi_0(0), 0) = \beta_1. \tag{4.4.62b}$$

If we use the definition (4.4.16) for p, the initial condition (4.4.61a) for Y_0 becomes

$$p(\phi_0(0), E_0(0), 0) = \alpha_0. \tag{4.4.63a}$$

Similarly, (4.4.62a) has the form

$$\Omega(E_0(0), 0)\frac{\partial p}{\partial t^+}(\phi_0(0), E_0(0), 0) = \beta_0. \tag{4.4.63b}$$

These two algebraic equations define the two unknowns $\phi_0(0)$ and $E_0(0)$ in terms of the specified constants α_0, β_0. With $E_0(0)$ known, the solution of (4.4.37) defines $E_0(\tilde{t})$. Using this $E_0(\tilde{t})$ in (4.4.15) for $\Omega(E_0(\tilde{t}), \tilde{t})$ and in (4.4.16) for $p(t^+, E_0(\tilde{t}), \tilde{t})$ specifies $\omega_0(\tilde{t})$ and $Y_0(t^+, \tilde{t})$ completely. In order to complete the solution to $O(1)$, we need to know $\omega_1(0)$ in order to specify $\omega_1(\tilde{t})$ from (4.4.59). To evaluate $\omega_1(0)$, we consider the initial conditions to $O(\epsilon)$.

Using the now known expression for $Y_0(t^+, \tilde{t})$, and using (4.4.32) for Y_1 in (4.4.61b) and (4.4.62b) gives the following pair of linear algebraic equations for $A_1(0)$ and $\omega_1(0)$:

$$A_1(0)\frac{\partial Y_0}{\partial t^+}(\phi_0(0), 0) + \frac{\omega_1(0)}{\Omega_{E_0}(E_0(0), 0)}\frac{\partial p}{\partial E_0}(\phi_0(0), E_0(0), 0)$$
$$= \alpha_1 - y_{1p_{\text{odd}}}(\phi_0(0), E_0(0), 0) \tag{4.4.64a}$$

$$A_1(0)\omega_0(0)\frac{\partial^2 Y_0}{\partial t^{+2}}(\phi_0(0), 0) + \frac{\omega_1(0)}{\Omega_{E_0}(E_0(0), 0)}\left[\omega_0(0)\frac{\partial^2 p}{\partial E_0 \partial t^+}(\phi_0(0), E_0, 0)\right.$$
$$\left. + \Omega_{E_0}(E_0(0), 0)\frac{\partial Y_0}{\partial t^+}(\phi_0(0), 0)\right] = \beta_1 - \frac{\partial Y_0}{\partial \tilde{t}}(\phi_0(0), 0)$$

$$-\omega_0(0)\,\frac{\partial y_{1p_{\text{odd}}}}{\partial t^+}(\phi_0(0), E_0(0), 0). \tag{4.4.64b}$$

Because $E_0(\tilde{t})$ has been defined, both $(\partial p/\partial E_0)$ and $(\partial^2 p/\partial E_0 \partial t^+)$ are known functions of t^+ and $\tilde{t}$, say, $\partial p/\partial E_0 \equiv k(t^+, \tilde{t})$ and $\partial^2 p/\partial E_0 \partial t^+ \equiv \ell(t^+, \tilde{t})$. However, we do not use this notation so as to easily identify the various terms that appear in (4.4.64). In particular, the coefficient determinant for (4.4.64) is

$$K = \frac{\partial Y_0}{\partial t^+}\left(\omega_0 \frac{\partial^2 p}{\partial E_0 \partial t^+} + \Omega_{E_0}\frac{\partial Y_0}{\partial t^+}\right) - \omega_0 \frac{\partial^2 Y_0}{\partial t^{+^2}}\frac{\partial p}{\partial E_0}.$$

Using the expression in (4.4.22) for the Wronskian of $(\partial p/\partial t^+)$ and q evaluated at $t^+ = \phi_0(0)$ and $\tilde{t} = 0$, we find

$$W = \frac{\partial Y_0}{\partial t^+}\left(\omega_0 \frac{\partial^2 p}{\partial E_0 \partial t^+} + \Omega_{E_0}\frac{\partial Y_0}{\partial t^+} + \Omega_{E_0}\phi_0(0)\frac{\partial^2 Y_0}{\partial t^{+^2}}\right)$$
$$- \frac{\partial^2 Y_0}{\partial t^{+^2}}\left(\omega_0 \frac{\partial p}{\partial E_0} + \Omega_{E_0}\phi_0(0)\frac{\partial Y_0}{\partial t^+}\right),$$

where the expressions for K and W are evaluated at $t^+ = \phi_0(0)$ and $\tilde{t} = 0$. After canceling the two terms involving ϕ_0 in W, we see that $K = W$, and we have shown that $W = 1/\Omega$. Therefore $K = 1/\omega_0$, and the system (4.4.64) has a unique solution for $A_1(0)$ and $\omega_1(0)$. Here we are only interested in $\omega_1(0)$. If we wanted to compute y to $O(\epsilon)$, we would derive an equation for $A_1(\tilde{t})$ by examining (4.4.50) to $O(\epsilon^2)$.

Solving (4.4.64) for $\omega_1(0)/\Omega_{E_0}(E_0(0), 0)$ gives

$$\frac{\omega_1(0)}{\Omega_{E_0}(E_0(0), 0)} = \left\{\omega_0\left(\beta_1 \frac{\partial Y_0}{\partial t^+} - \alpha_1\omega_0 \frac{\partial^2 Y_0}{\partial t^{+^2}}\right) - \omega_0 \frac{\partial Y_0}{\partial t^+}\frac{\partial Y_0}{\partial \tilde{t}}\right.$$
$$\left. + \omega_0^2\left(\frac{\partial^2 Y_0}{\partial t^{+^2}} y_{1p_{\text{odd}}} - \frac{\partial Y_0}{\partial t^+}\frac{\partial y_{1p_{\text{odd}}}}{\partial t^+}\right)\right\}_{t^+=\phi_0(0),\tilde{t}=0}. \tag{4.4.65}$$

We now use (4.4.29b) to compute $y_{1p_{\text{odd}}}$, $\partial y_{1p_{\text{odd}}}/\partial t^+$ and find that the term multiplied by ω_0^2 in (4.4.65) simplifies to

$$\omega_0^2\left(\frac{\partial^2 Y_0}{\partial t^{+^2}} y_{1p_{\text{odd}}} - \frac{\partial Y_0}{\partial t^+}\frac{\partial y_{1p_{\text{odd}}}}{\partial t^+}\right)_{t^+=\phi_0(0),\tilde{t}=0} =$$

$$-\left\{\int_0^{t^+} r_{1_{\text{odd}}}(s, E_0, \tilde{t})\frac{\partial Y_0}{\partial t^+}(s, \tilde{t})ds\right\}_{t^+=\phi_0(0),\ \tilde{t}=0}.$$

In the expression (4.4.28b) for $r_{1_{\text{odd}}}$, we set $\Omega = \omega_0$ and

$$\frac{\partial^2 p}{\partial t^+ \partial E_0}\frac{dE_0}{d\tilde{t}} + \frac{\partial^2 p}{\partial t^+ \partial \tilde{t}} = \frac{\partial^2 Y_0}{\partial t^+ \partial \tilde{t}}$$

and find

$$\omega_0^2 \left(\frac{\partial^2 Y_0}{\partial t^{+^2}} y_{1p_{\text{odd}}} - \frac{\partial Y_0}{\partial t^+} \frac{\partial y_{1p_{\text{odd}}}}{\partial t^+} \right)_{t^+=\phi_0(0), \tilde{t}=0} =$$

$$\left\{ \frac{\partial}{\partial \tilde{t}} \left[\omega_0(\tilde{t}) \int_0^{t^+} \left(\frac{\partial Y_0}{\partial t^+}(s, \tilde{t}) \right)^2 ds \right] \right.$$

$$\left. + \int_0^{t^+} \frac{\partial Y_0}{\partial t^+}(s, \tilde{t}) h(Y_0(s, \tilde{t}), \frac{\partial Y_0}{\partial t^+}(s, \tilde{t}), \tilde{t}) ds \right\}_{t^+=\phi_0(0), \tilde{t}=0}$$

Therefore, (4.4.65) has the explicit form

$$\begin{aligned} \frac{\omega_1(0)}{\Omega_{E_0}(E_0(0), 0)} = &\left\{ \frac{\partial}{\partial \tilde{t}} \left[\omega_0(\tilde{t}) \int_0^{t^+} \left(\frac{\partial Y_0}{\partial t^+}(s, \tilde{t}) \right)^2 ds \right] \right. \\ &+ \int_0^{t^+} \frac{\partial Y_0}{\partial t^+}(s, \tilde{t}) h(Y_0(s, \tilde{t}), \frac{\partial Y_0}{\partial t^+}(s, \tilde{t}), \tilde{t}) ds \\ &- \omega_0(\tilde{t}) \frac{\partial Y_0}{\partial t^+}(t^+, \tilde{t}) \frac{\partial Y_0}{\partial \tilde{t}}(t^+, \tilde{t}) + \omega_0(\tilde{t}) \left[\beta_1 \frac{\partial Y_0}{\partial t^+}(t^+, \tilde{t}) \right. \\ &\left.\left. - \omega_0(\tilde{t}) \alpha_1 \frac{\partial^2 Y_0}{\partial t^{+^2}}(t^+, \tilde{t}) \right] \right\}_{t^+=\phi_0(0), \tilde{t}=0} \end{aligned} \tag{4.4.66}$$

As pointed out in [4.2], this expression simplifies further if we use the energy equation (4.4.8a). Taking the total derivative of this expression with respect to $\tilde{t}$ gives

$$\omega_0 \omega_0' \left(\frac{\partial Y_0}{\partial t^+} \right)^2 + \omega_0^2 \frac{\partial Y_0}{\partial t^+} \frac{\partial^2 Y_0}{\partial t^+ \partial \tilde{t}} + \frac{\partial V}{\partial y}(Y_0, \tilde{t}) \frac{\partial Y_0}{\partial \tilde{t}} + \frac{\partial V}{\partial \tilde{t}}(Y_0, \tilde{t}) = E_0', \tag{4.4.67}$$

where $' = d/d\tilde{t}$. We now use (4.4.7a) to set $V_y = g = -\omega_0^2(\partial^2 Y_0/\partial t^{+^2})$, divide out an ω_0, and write the result as

$$\omega_0' \left(\frac{\partial Y_0}{\partial t^+} \right)^2 + 2\omega_0 \frac{\partial Y_0}{\partial t^+} \frac{\partial^2 Y_0}{\partial t^+ \partial \tilde{t}} - \omega_0 \left(\frac{\partial Y_0}{\partial t^+} \frac{\partial^2 Y_0}{\partial t^+ \partial \tilde{t}} + \frac{\partial Y_0}{\partial \tilde{t}} \frac{\partial^2 Y_0}{\partial t^{+^2}} \right)$$

$$= \frac{1}{\omega_0} E_0' - \frac{1}{\omega_0} \frac{\partial V}{\partial \tilde{t}}.$$

But this is

$$\frac{\partial}{\partial \tilde{t}} \left[\omega_0 \left(\frac{\partial Y_0}{\partial t^+} \right)^2 \right] - \omega_0 \frac{\partial}{\partial t^+} \left(\frac{\partial Y_0}{\partial t^+} \frac{\partial Y_0}{\partial \tilde{t}} \right) = \frac{1}{\omega_0} E_0' - \frac{1}{\omega_0} \frac{\partial V}{\partial \tilde{t}}.$$

Integrating the above with respect to t^+ from 0 to t^+ gives (*Note*: $(\partial Y_0/\partial t^+)$ vanishes at $t^+ = 0$)

$$\frac{\partial}{\partial \tilde{t}}\left[\int_0^{t^+} \omega_0(\tilde{t})\left(\frac{\partial Y_0}{\partial t^+}(s,\tilde{t})\right)^2 ds\right] - \omega_0(\tilde{t})\frac{\partial Y_0}{\partial t^+}(t^+,\tilde{t})\frac{\partial Y_0}{\partial \tilde{t}}(t^+,\tilde{t})$$

$$= \frac{t^+}{\omega_0(\tilde{t})}E_0'(\tilde{t}) - \frac{1}{\omega_0(\tilde{t})}\int_0^{t^+}\frac{\partial V}{\partial \tilde{t}}(Y_0(s,\tilde{t}),\tilde{t})ds. \qquad (4.4.68)$$

We now evaluate (4.4.68) at $t^+ = P_0$ and use (4.4.35) to find the following expression for E_0' (Note again that $(\partial Y_0/\partial t^+)$ vanishes at $t^+ = P_0$.):

$$E_0'(\tilde{t}) = -\frac{\omega_0(\tilde{t})}{P_0}\int_0^{P_0} h\left(Y_0, \frac{\partial Y_0}{\partial t^+}, \tilde{t}\right)\frac{\partial Y_0}{\partial t^+}dt^+ + \frac{1}{P_0}\int_0^{P_0}\frac{\partial V}{\partial \tilde{t}}(Y_0,\tilde{t})dt^+. \qquad (4.4.69)$$

Substituting (4.4.69) for $E_0'(\tilde{t})$ into (4.4.68) and evaluating the result at $t^+ = \phi_0(0)$, $\tilde{t} = 0$ gives

$$\left\{\frac{\partial}{\partial \tilde{t}}\left[\int_0^{t^+}\omega_0(\tilde{t})\left(\frac{\partial Y_0}{\partial t^+}(s,\tilde{t})\right)^2 ds\right] - \omega_0(\tilde{t})\frac{\partial Y_0}{\partial t^+}(t^+,\tilde{t})\frac{\partial Y_0}{\partial \tilde{t}}(t^+,\tilde{t})\right\}_{t^+=\phi_0(0),\tilde{t}=0} =$$

$$\frac{\phi_0(0)}{\omega_0(0)P_0}\int_0^{P_0}\frac{\partial V}{\partial \tilde{t}}(Y_0,0)dt^+$$

$$-\frac{\phi_0(0)}{P_0}\int_0^{P_0} h(Y_0, \frac{\partial Y_0}{\partial t^+}, 0)\frac{\partial Y_0}{\partial t^+}dt^+ - \frac{1}{\omega_0(0)}\int_0^{\phi_0(0)}\frac{\partial V}{\partial \tilde{t}}(Y_0,0)ds. \qquad (4.4.70)$$

We use the expression given by (4.4.70) for the sum of the first plus third terms on the right-hand side of (4.4.66) to obtain the final result

$$\frac{\omega_1(0)}{\Omega_{E_0}(E_0(0),0)} = C_1(\alpha_0,\beta_0) + C_2(\alpha_0,\beta_0)\alpha_1 + C_3(\alpha_0,\beta_0)\beta_1, \qquad (4.4.71)$$

where we have introduced the notation

$$C_1(\alpha_0,\beta_0) = \frac{\phi_0(0)}{P_0}\int_0^{P_0}\left[\frac{1}{\omega_0(0)}\frac{\partial V}{\partial \tilde{t}}(Y_0(s,0),0) - h(Y_0,(s,0), \frac{\partial Y_0}{\partial t^+}(s,0),0)\frac{\partial Y_0}{\partial t^+}(s,0)\right]ds$$

$$+\int_0^{\phi_0(0)}\left[h(Y_0(s,0), \frac{\partial Y_0}{\partial t^+}(s,0))\frac{\partial Y_0}{\partial t^+}(s,0) - \frac{1}{\omega_0(0)}\frac{\partial V}{\partial \tilde{t}}(Y_0(s,0),0)\right]ds, \qquad (4.4.72a)$$

$$C_2(\alpha_0, \beta_0) = -\omega_0^2(0) \frac{\partial^2 Y_0}{\partial t^{+2}} (\phi_0(0), 0), \tag{4.4.72b}$$

$$C_3(\alpha_0, \beta_0) = \omega_0(0) \frac{\partial Y_0}{\partial t^+} (\phi_0(0), 0). \tag{4.4.72c}$$

As pointed out earlier, and indicated by the arguments of C_1, C_2, and C_3, these constants involve functions that are completely defined once α_0 and β_0 are specified. Therefore, if α_1 and β_1 are also prescribed arbitrarily, (4.4.71) shows that $\omega_1(0)$ does not vanish in general; it only vanishes for the one-parameter family of values of α_1, β_1, for which

$$C_1 + C_2\alpha_1 + C_3\beta_1 = 0. \tag{4.4.73}$$

With $\omega_1(0) \neq 0$, (4.4.59) gives $\omega_1(\tilde{t}) \neq 0$, i.e., the phase shift of the $O(1)$ oscillations is not constant. A special case for which $\omega_1 \equiv 0$ corresponds to $h \equiv 0$, $V_t = 0$, and $\alpha_1 = \beta_1 = 0$.

In [4.1], initial conditions for which (4.4.73) holds, and hence $\omega_1 \equiv 0$, are denoted as *synchronized* initial conditions as the solution is significantly simplified. It is also pointed out there that for any *numerically* prescribed set of values for ϵ, $y(0; \epsilon)$, and $\dot{y}(0; \epsilon)$, it is always possible to choose α_0, β_0, α_1, and β_1 consistent with the initial data and such that (4.4.73) is satisfied. Thus, any pair of initial values $y(0; \epsilon)$ and $\dot{y}(0; \epsilon)$ can be regarded as synchronized, and henceforce we need not dwell on the variation of the phase shift.

4.4.2 An Example

We consider the problem discussed in [4.22] and generalize this to include a small damping term that is linear in $\dot{y}$ and slowly varying:

$$\ddot{y} + \epsilon\tilde{h}(\tilde{t})\dot{y} + a(\tilde{t})y + b(\tilde{t})y^3 = 0, \tag{4.4.74}$$

where $\tilde{h}$, a, and b are given functions.

The energy integral (4.4.8a) is

$$\frac{\omega_0^2}{2}\left(\frac{\partial Y_0}{\partial t^+}\right)^2 + V(Y_0, \tilde{t}) = E_0(\tilde{t}), \tag{4.4.75}$$

where

$$V(Y_0, \tilde{t}) = \frac{a(\tilde{t})}{2} Y_0^2 + \frac{b(\tilde{t})}{4} Y_0^4. \tag{4.4.76}$$

Examining V for the different possible combinations of the signs of a and b will determine the cases for which (4.4.75) admits periodic solutions.

(i) If $a > 0, b > 0$, Y_0 is periodic for *any* positive E_0, (and (4.4.75) shows that E_0 cannot be negative) because as seen in Figure 4.4.2a, V is concave.

(ii) If $a < 0, b > 0$, V is "W-shaped" as seen in Figure 4.4.2b, and we have two families of periodic solutions centered about $Y_0 = \pm\sqrt{-a/b}$ for negative values of E_0. When E_0 becomes positive, the two periodic solutions coalesce to one centered about $Y_0 = 0$.

(iii) If $a > 0$, $b < 0$, V is "M-shaped" and is given by the reflection of Figure 4.4.2b about the Y_0 axis. Therefore, we have periodic solutions around $Y_0 = 0$ as long as $0 < E_0 < a^2/(-4b)$.

(iv) Finally, if $a < 0$, $b < 0$, V has the opposite sign as in case (i) and is convex. Therefore, none of the solutions is periodic.

Here, we will restrict attention to those cases that admit periodic solutions centered around $Y_0 = 0$. There are three possibilities, all for $E_0 > 0$, and the following values of $Y_{0_{\max}}$, $Y_{0_{\min}}$ are obtained by setting $(\partial Y_0/\partial t^+) = 0$ in (4.4.75) and solving the resulting quadratic equation

$$Y_0^4 + 2(a/b)Y_0^2 - 4E_0/b = 0 \tag{4.4.77}$$

for Y_0^2.

(i) $a > 0, b > 0, E_0 > 0$

$$Y_{0_{\max}} = \sqrt{-\frac{a}{b} + \sqrt{\left(\frac{a}{b}\right)^2 + \frac{4E_0}{b}}} = -Y_{0_{\min}}; \tag{4.4.78a}$$

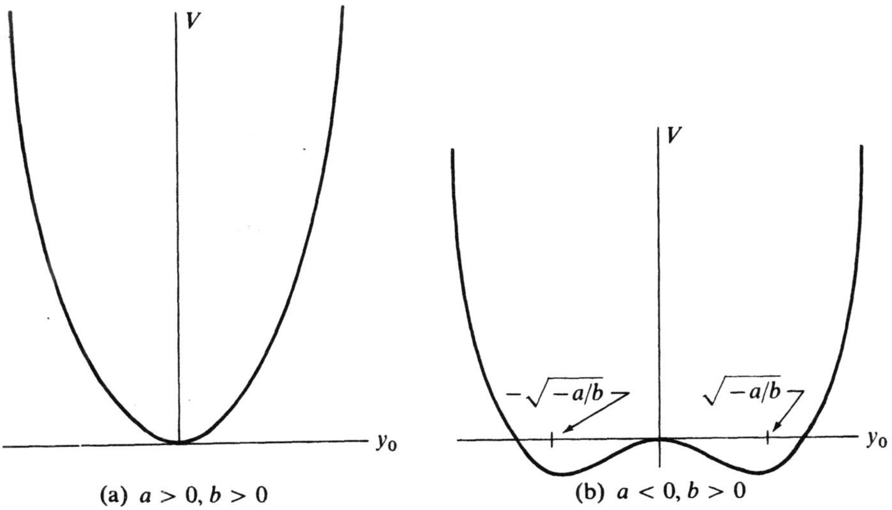

FIGURE 4.4.2. $V(y_0, a, b)$ for Different Signs of a, b

(ii) $a < 0, b > 0, E_0 > 0$

$$Y_{0_{\max}} = \sqrt{-\frac{a}{b} + \sqrt{\left(\frac{a}{b}\right)^2 + \frac{4E_0}{b}}} = -Y_{0_{\min}}; \tag{4.4.78b}$$

(iii) $a > 0, b < 0, 0 < E_0 < a^2/(-4b)$

$$Y_{0_{\max}} = \sqrt{-\frac{a}{b} - \sqrt{\left(\frac{a}{b}\right)^2 + \frac{4E_0}{b}}} = -Y_{0_{\min}}. \tag{4.4.78c}$$

We normalize (4.4.10) by introducing the variable of integration $\xi = \eta / Y_{0_{\max}}$ to obtain

$$t^+ = \omega_0(\tilde{t}) Y_{0_{\max}} \int_{-1}^{Y_0/Y_{0_{\max}}} \frac{d\xi}{\sqrt{2[E_0 - aY_{0_{\max}}^2 \xi^2/2 - bY_{0_{\max}}^4 \xi^4/4]}}. \tag{4.4.79}$$

It is convenient to introduce the notation

$$\nu \equiv -b(\tilde{t}) Y_{0_{\max}}^4 / 4E_0, \tag{4.4.80a}$$

which implies (using the expression given by (4.4.77) for $Y_{0_{\max}}^4$)

$$1 + \nu = 1 - \frac{b}{4E_0}\left[\frac{4E_0}{b} - 2\frac{a}{b} Y_{0_{\max}}^2\right] = \frac{a}{2E_0} Y_{0_{\max}}^2. \tag{4.4.80b}$$

Thus, (4.4.79) takes the form

$$t^+ = \frac{\omega_0(\tilde{t})}{\sqrt{2E_0}} Y_{0_{\max}} \int_{-1}^{Y_0/Y_{0_{\max}}} \frac{d\xi}{\sqrt{(1-\xi^2)(1-\nu\xi^2)}}. \tag{4.4.81}$$

Let us first apply the condition (4.4.11) that ensures the period is independent of $\tilde{t}$; here it is convenient to choose $P_0 = 4$. We find

$$4 = 2\omega_0 \int_{-A_0}^{A_0} \frac{dY_0}{\sqrt{2[E_0 - a(\tilde{t})Y_0^2/2 - b(\tilde{t})Y_0^4/4]}}, \tag{4.4.82}$$

where we have set $Y_{0_{\max}} = A_0$. Notice that (4.4.78) gives a relation linking A_0 to E_0 and the two known functions $a(\tilde{t})$, $b(\tilde{t})$. Introducing the change of variable $\xi = Y_0/A_0$ as in (4.4.79), we find

$$4 = \frac{2\omega_0 A_0}{\sqrt{2E_0}} \int_{-1}^{1} \frac{d\xi}{\sqrt{(1-\xi^2)(1-\nu\xi^2)}}. \tag{4.4.83}$$

The integral in (4.4.83) is just $2K(\nu)$, where K is the complete elliptic integral of the first kind (see (4.1.9))

$$K(\nu) = \int_0^1 \frac{d\xi}{\sqrt{(1-\xi^2)(1-\nu\xi^2)}}. \tag{4.4.84}$$

Therefore, (4.4.83) reduces to

$$\omega_0 = \frac{\sqrt{2E_0}}{K(\nu)A_0}. \tag{4.4.85}$$

It follows from the definition of ν and A_0 that the right-hand side of (4.4.85) is a known function of E_0 and the given functions $a(\tilde{t})$ and $b(\tilde{t})$. Thus, (4.4.85) defines $\Omega(E_0, \tilde{t})$ of (4.4.15). Henceforth, we will use A_0 instead of E_0 in our calculations.

In preparation for inverting (4.4.81), we isolate the integral over $(-1, 0)$ and use (4.4.84)–(4.4.85) to obtain

$$t^+ = 1 + \frac{1}{K(\nu)} \int_0^{Y_0/A_0} \frac{d\xi}{\sqrt{(1-\xi^2)(1-\nu\xi^2)}}. \tag{4.4.86}$$

The inverse is then expressed in terms of the elliptic sine function in the form

$$Y_0 = A_0(\tilde{t})\mathrm{sn}\left[K(\nu)(t^+ - 1), \nu\right]. \tag{4.4.87}$$

It is convenient to express all unknowns in terms of ν. First, we eliminate E_0 from (4.4.80a) and (4.4.80b) to obtain

$$A_0 = \sqrt{\frac{-2a\nu}{b(1+\nu)}}. \tag{4.4.88}$$

Next, we use the definition (4.4.80a) to eliminate E_0 from (4.4.85) to find

$$\omega_0 = \frac{1}{K(\nu)}\sqrt{\frac{a}{1+\nu}}. \tag{4.4.89}$$

Notice that the expressions inside the radicals in (4.4.88) and (4.4.89) are always positive. In particular, for case (i), where $a > 0, b > 0$, we have $-1 < \nu < 0$. For case (ii), where $a < 0, b > 0$, we have $\nu < -1$. Finally, for case (iii), where $a > 0, b < 0$, we have $0 < \nu < 1$.

At this point, our two unknowns A_0 and ω_0 are specified in terms of ν. We obtain a third relation to close the system from the periodicity condition (4.4.39). We have

$$J = \omega_0 \int_0^4 \left(\frac{\partial Y_0}{\partial t^+}\right)^2 dt^+ = 2\omega_0 \int_0^2 \left(\frac{\partial Y_0}{\partial t^+}\right)^2 dt^+,$$

and using the identity $\frac{d}{du}\mathrm{sn}u = \mathrm{cn}u\mathrm{dn}u$ gives

$$\begin{aligned} J &= 2\omega_0 A_0^2 K^2 \int_0^2 \mathrm{cn}^2[K(t^+ - 1), \nu]\mathrm{dn}^2[K(t^+ - 1), \nu]dt^+ \\ &= 2\omega_0 A_0^2 K \int_{-K}^{K} \mathrm{cn}^2(u, \nu)\mathrm{dn}^2(u, \nu)du \\ &= 4\omega_0 A_0^2 K \int_0^K \mathrm{cn}^2(u, \nu)\mathrm{dn}^2(u, \nu)du. \end{aligned} \tag{4.4.90}$$

Thus, (4.4.39) becomes

$$\omega_0 A_0^2 K \int_0^K \mathrm{cn}^2(u,\nu)\mathrm{dn}^2(u,\nu)du = \frac{J_0}{4}\exp\left(-\int_0^{\tilde{t}} \tilde{h}(s)ds\right), \tag{4.4.91}$$

where J_0 is a constant that depends on the initial conditions. According to equation (361.03) of [4.3],

$$L(\nu) \equiv \int_0^K \mathrm{cn}^2(u,\nu)\mathrm{dn}^2(u,\nu)du = \frac{1}{3\nu}\left[(1+\nu)E(\nu) - (1-\nu)K(\nu)\right], \tag{4.4.92}$$

where $E(\nu)$ is the complete elliptic integral of the second kind

$$E(\nu) = \int_0^1 \sqrt{\frac{1-\nu\xi^2}{1-\xi^2}}\,d\xi. \tag{4.4.93}$$

Using (4.4.92) in (4.4.91) defines ω_0 as a function of ν in the form

$$\omega_0 = \frac{(J_0/4)}{A_0^2 K(\nu)L(\nu)}\exp\left(-\int_0^{\tilde{t}} \tilde{h}(s)ds\right). \tag{4.4.94}$$

We eliminate A_0 from (4.4.88) and (4.4.94) to obtain the following equation for ν:

$$\frac{4L^2(\nu)\nu^2}{(1+\nu)^3} = \frac{J_0^2 b^2(\tilde{t})}{16a^3(\tilde{t})}\exp\left(-2\int_0^{\tilde{t}} \tilde{h}(s)ds\right) \equiv c(\tilde{t}), \tag{4.4.95}$$

where the right-hand side, denoted by $c(\tilde{t})$, is a known function of $\tilde{t}$. Therefore, (4.4.95) can be solved for $\nu(\tilde{t})$, and once $\nu(\tilde{t})$ is known, (4.4.89) gives $\omega_0(\tilde{t})$ and (4.4.88) gives $A_0(\tilde{t})$.

A graph of the solution for ν as a function of c, taken from [4.22], is shown in Figure 4.4.3. Note that the sign of c is the same as the sign of a.

(i) If $a > 0, b > 0$, the solution for ν lies in the fourth quadrant and exists for all positive c.
(ii) If $a < 0, b > 0$, the curve for ν lies in the third quadrant and exists for $-\infty < c < -4/9$.
(iii) If $a > 0, b < 0$, the curve for ν lies in the first quadrant. The solution exists only in the range $0 < c < 2/9$. An enlargement of cases (i) and (iii) is shown as an inset in the lower left-hand corner of Figure 4.4.3.

As pointed out earlier for case (ii), we also have the two families of periodic solutions centered about $Y_0 = \pm\sqrt{-a/b}$. The calculations for this case are essentially similar to those given above, and the results are summarized in [4.22].

Escape from a potential well

As a special case, consider the problem of escape from a potential well mentioned in Sec. 4.2.4. In terms of appropriately scaled variables, we have the oscillator

$$\ddot{y} + y - y^3 - \epsilon\dot{y} = 0 \tag{4.4.96}$$

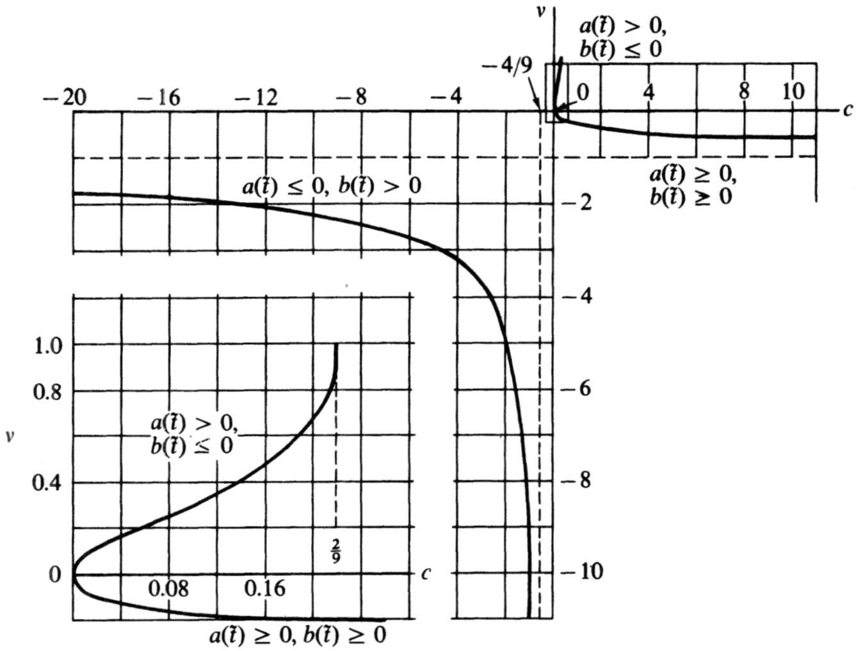

FIGURE 4.4.3. Numerical Results for ν vs. c

with the "M-shaped" potential of case (iii) for constant values: $a = 1$ and $b = -1$. To simplify calculations, we choose the initial conditions

$$y(0;\epsilon) = -1/2 = \alpha_0; \quad \dot{y}(0;\epsilon) = 0 = \beta_0. \tag{4.4.97}$$

Thus, motion starts from $Y_{0_{\min}}$ and we have $\phi_0(0) = 0$. As a consequence, (4.4.72) shows that $C_1 = 0$. The initial conditions (4.4.97) also imply $\alpha_1 = \beta_1 = 0$, and our initial conditions are synchronized (because (4.4.71) gives $\omega_1(0) = 0$, and (4.4.59) gives $\omega_1(\tilde{t}) = 0$).

Since a and b are constants, $c(\tilde{t})$ depends only on the damping, and (4.4.95) gives

$$c(\tilde{t}) = c(0)e^{2\tilde{t}}; \quad c(0) = \frac{4L^2(\nu(0))\nu^2(0)}{(1+\nu(0))^3}. \tag{4.4.98}$$

Thus, c is a monotone increasing function of $\tilde{t}$, and escape occurs when $c(\tilde{t})$ reaches the upper limit 2/9 for periodic motion. If we denote the time elapsed at escape by $\tau(\epsilon)$, we have

$$\tau(\epsilon) = \frac{1}{2\epsilon}\log\frac{2}{9c(0)} = \frac{1}{\epsilon}\log\frac{(1+\nu(0))^{3/2}}{3\sqrt{2}L(\nu(0))\nu(0)}, \tag{4.4.99}$$

and we note that $\tau = O(\epsilon^{-1})$.

The initial values appearing in (4.4.99) can be computed in sequence as follows. The initial energy $E_0(0)$ is obtained by evaluating (4.4.8a) initially. Since (4.4.62a) implies $\omega_0(0)\frac{\partial Y_0}{\partial t^+}(0,0) = 0$ for $\beta_0 = 0$, we have

$$E_0(0) = V(Y_0(0,0),0) = \frac{1}{2}\alpha_0^2 - \frac{1}{4}\alpha_0^4 = \frac{7}{64}. \tag{4.4.100}$$

Using (4.4.88) and (4.4.80a) gives

$$A_0(0) = \frac{1}{2}; \quad \nu(0) = \frac{1}{7}. \tag{4.4.101}$$

To compute $L(\nu(0))$, we use the series expansions for $K(\nu)$ and $E(\nu)$ for $\nu \to 0$

$$K(\nu) = \frac{\pi}{2}\left[1 + \left(\frac{1}{2}\right)^2\nu + \left(\frac{3}{2\cdot 4}\right)\nu^2 + \left(\frac{3\cdot 5}{2\cdot 4\cdot 6}\right)\nu^3 + \dots\right], \tag{4.4.102a}$$

$$E(\nu) = \frac{\pi}{2}\left[1 - \left(\frac{1}{2}\right)^2\nu - \left(\frac{3}{2\cdot 4}\right)\frac{\nu^2}{3} - \left(\frac{3\cdot 5}{2\cdot 4\cdot 6}\right)\frac{\nu^3}{5} - \dots\right], \tag{4.4.102b}$$

to find

$$L(\nu(0)) = 0.771111. \tag{4.4.103}$$

Therefore, (4.4.99) reduces to the simple expression

$$\tau(\epsilon) = \frac{0.9609}{\epsilon}. \tag{4.4.104}$$

It is shown in [4.20] that this result agrees with numerically computed expressions of the escape time for successively smaller values of ϵ; as expected, the error in τ decreases as $\epsilon \to 0$. The reader may consult [4.20] for other examples and for a discussion of approximations that may be used when Y_0 cannot be expressed in terms of elliptic functions.

Problems

1. By direct substitution and using periodicity (with $P_0 = 1$) as appropriate, show that

$$\Omega[y_{1p_{\text{odd}}}(t^+ + 1, E_0, \tilde{t}) - y_{1p_{\text{odd}}}(t^+, E_0, \tilde{t})] = R, \tag{4.4.105}$$

where

$$R = -q(t^+, E_0, \tilde{t})\int_0^1 r_{1_{\text{odd}}}(s, E_0, \tilde{t})\frac{\partial p}{\partial t^+}(s, E_0, \tilde{t})ds$$

$$+\frac{\partial\Omega}{\partial E_0}(E_0,\tilde{t})\frac{\partial p}{\partial t^+}(t^+,E_0,\tilde{t})\left[\int_0^1 r_{1_{\text{odd}}}(s,E_0,\tilde{t})\frac{\partial p}{\partial t^+}(s,E_0,\tilde{t})ds\right.$$
$$\left.-\int_{t^+}^{t^++1} r_{1_{\text{odd}}}(s,E_0,\tilde{t})s\frac{\partial p}{\partial t^+}(s,E_0,\tilde{t})ds\right]. \qquad (4.4.106)$$

Simplify the last term on the right-hand side of (4.4.106) using integration by parts to find

$$R=-q\int_0^1 r_{1_{\text{odd}}}\frac{\partial p}{\partial t^+}ds+\frac{\partial\Omega}{\partial E_0}\frac{\partial p}{\partial t^+}\left[\int_0^{t^+} r_{1_{\text{odd}}}\frac{\partial p}{\partial t^+}ds\right.$$
$$-\int_0^{1+t^+} r_{1_{\text{odd}}}\frac{\partial p}{\partial t^+}ds-t^+\int_0^{1+t^+} r_{1_{\text{odd}}}\frac{\partial p}{\partial t^+}ds$$
$$\left.+t^+\int_0^{t^+} r_{1_{\text{odd}}}\frac{\partial p}{\partial t^+}ds+\int_{t^+}^{1+t^+}\left(\int_0^s r_{1_{\text{odd}}}\frac{\partial p}{\partial t^+}ds'\right)ds\right]. \qquad (4.4.107)$$

Argue that a necessary and sufficient condition for $R=0$ is (4.4.33).

2. Consider the problem of a pendulum with slowly varying length undergoing large amplitude oscillations. In suitable dimensionless variables the governing equation is

$$\frac{d^2y}{dt^2}+a^2(\tilde{t})\sin y=0. \qquad (4.4.108)$$

In a sense, the potential in (4.4.74) is an approximation to the one for (4.4.108) where only two terms of the Taylor expansion for $\sin y$ are retained.

Apply the technique in this section to derive an approximate solution to $O(1)$ in the oscillatory regime.

3. In many problems, it is not possible to express the first approximation Y_0 in terms of special functions, yet Y_0 is even periodic with period 2π; therefore, it can be developed in a Fourier cosine series. Assume $\omega_1=0$ and truncate the series after N terms, i.e.,

$$y_0(t^+,\tilde{t})=\sum_{n=0}^{N}a_n(\tilde{t})\cos nt^+. \qquad (4.4.109)$$

Show by substitution into (4.4.7a) and neglecting harmonics higher than N that

$$-\omega^2(\tilde{t})a_n(\tilde{t})n^2+g_n[a_0(\tilde{t}),a_1(\tilde{t}),\ldots,a_N(\tilde{t}),\tilde{t}]=0 \qquad (4.4.110)$$

for each $n=0,1,2,\ldots,N$, where

$$g_n(a_0,a_1,\ldots,a_N,\tilde{t})=\frac{2}{\pi}\int_0^{\pi}g\left(\sum_{n=0}^{N}a_n\cos nt^+,\tilde{t}\right)\cos nt^+dt^+. \qquad (4.4.111)$$

The above defines a system of $N+1$ equations for the $N+2$ unknowns, $\omega, a_0, \ldots, a_N$. Derive the additional equation needed to close the system by substituting the assumed truncated Fourier series into the periodicity condition (4.4.35) to obtain

$$\frac{d}{d\tilde{t}}\left[\frac{\omega\pi}{2}\sum_{n=0}^{N} n^2 a_n^2\right]$$
$$+\,\omega\int_0^{\pi} h\left[\sum_{n=0}^{N} a_n \cos nt^+, \tilde{t}\right]\left[\sum_{n=0}^{N} na_n \sin nt^+\right]^2 dt^+ = 0. \quad (4.4.112)$$

Thus, the functions $\omega, a_0, \ldots, a_N$ can be computed in principle.

Specialize your results to the problem of Sec. 4.4.2 with $a > 0$ and arbitrary b. Take $N = 1$, and compute ω, a_0, and a_1 and compare your results with the "exact" solution given in Sec. 4.4.2.

4.5 Multiple-Scale Expansions for Systems of First-Order Equations in Standard Form

In this section, we consider the multiple-scale expansion of a general system of first-order equations. Our discussion is based on the expository paper [4.15] to which we refer the reader for a number of physical examples. All the problems discussed so far in this chapter are special cases (when expressed in terms of appropriate dependent variables) of the following system of $M + N$ first-order equations in "standard-form":

$$\frac{dp_m}{dt} = \epsilon F_m(p_i, q_i, \tilde{t}; \epsilon), \; m = 1, 2, \ldots, M, \quad (4.5.1a)$$

$$\frac{dq_n}{dt} = \underline{\omega}_n(p_i, \tilde{t}) + \epsilon G_n(p_i, q_i, \tilde{t}; \epsilon), \; n = 1, \ldots, N. \quad (4.5.1b)$$

Here $0 < \epsilon \ll 1$ and $\tilde{t} = \epsilon t$, as usual. The subscript i indicates that all M of the p_i (or all N of the q_i) are present in the argument. The two requirements for a standard form system are: (i) that the F_m and G_n are $O(1)$ as $\epsilon \to 0$ for each $m = 1, \ldots, M$ and each $n = 1, \ldots, N$ and (ii) that the F_m and G_n are periodic with respect to each of the q_i with period 2π.

A given function, say, $F_m(p_i, q_i, \tilde{t}; \epsilon)$, that is 2π-periodic in each of the q_i can be uniquely decomposed into an *average* term $\underline{F}_m(p_i, \tilde{t}; \epsilon)$ and an *oscillatory* term $\underset{\wedge}{F}_m(p_i, q_i, \tilde{t}; \epsilon)$ with zero average

$$F_m(p_i, q_i, \tilde{t}; \epsilon) = \underline{F}_m(p_i, \tilde{t}; \epsilon) + \underset{\wedge}{F}_m(p_i, q_i, \tilde{t}; \epsilon), \quad (4.5.2)$$

where

$$\underline{F}_m(p_i, \tilde{t}; \epsilon) = \frac{1}{(2\pi)^N}\int_0^{2\pi}\cdots\int_0^{2\pi} F_m(p_i, q_i, \tilde{t}; \epsilon)\,dq_1 \ldots dq_N. \quad (4.5.3)$$

Thus, $\underset{\wedge}{F}_m = F_m - \underline{F}_m$ and

$$\int_0^{2\pi} \cdots \int_0^{2\pi} \underset{\wedge}{F}_m(p_i, q_i, \tilde{t}; \epsilon)\, dq_1 \ldots dq_N = 0.$$

Equivalently, we may expand F_m in a multiple Fourier series to find the average term (4.5.3) and to also compute the oscillatory term $\underset{\wedge}{F}_m$ explicitly (see (4.5.79)–(4.5.80)). Henceforth, we indicate an average function by an underbar and an oscillatory function having zero average with respect to all the q_i by an underhat.

After decomposing the F_m and G_n in (4.5.1) into average and oscillatory parts, we expand these in powers of ϵ and retain terms up to $O(\epsilon^2)$ to write (4.5.1) in the form

$$\frac{dp_m}{dt} = \epsilon \underline{f}_m(p_i, \tilde{t}) + \epsilon \underset{\wedge}{f}_m(p_i, q_i, \tilde{t}) + \epsilon^2 \underline{k}_m(p_i, \tilde{t}) + \epsilon^2 \underset{\wedge}{k}_m(p_i, q_i, \tilde{t}) + O(\epsilon^3), \quad m = 1, 2, \ldots, M, \tag{4.5.4a}$$

$$\frac{dq_n}{dt} = \underline{\omega}_n(p_i, \tilde{t}) + \epsilon \underline{g}_n(p_i, \tilde{t}) + \epsilon \underset{\wedge}{g}_n(p_i, q_i, \tilde{t}) + \epsilon^2 \underline{\ell}_n(p_i, \tilde{t}) + \epsilon^2 \underset{\wedge}{\ell}_n(p_i, q_i, \tilde{t}) + O(\epsilon^3), \quad n = 1, 2, \ldots, N. \tag{4.5.4b}$$

Notice that for $\epsilon = 0$ the p_m are constants $p_m^{(0)}$. Therefore, $q_n = \underline{\omega}_n(p_i^{(0)}, 0)t + q_n^{(0)}$, where the $q_n^{(0)}$ are constants. Thus, for ϵ small, the p_i are slowly varying functions, whereas the q_i are fast variables with respect to t.

The following three special cases of (4.5.4) are of interest:

(i) All the $\underline{f}_i = 0$.
(ii) All the $\underline{\omega}_i$ are independent of the p_i.
(iii) None of the functions on the right-hand side of (4.5.1) depend on $\tilde{t}$.

Of course, it is always possible to have case (iii) by regarding $\tilde{t} = p_{m+1}$ and augmenting the order of the system (4.5.1) to $M + N + 1$ by adding the equation $dp_{m+1}/dt = \epsilon$. However, when the terms on the right-hand side of (4.5.1) actually depend on the slow time $\tilde{t}$, it is more convenient to exhibit this explicitly.

In Chapter 5, we discuss the solution of the system (4.5.1) by the method of averaging; there we also consider the fourth special case, where this system is Hamiltonian. In this special case, we have $M = N$, and there exists a Hamiltonian $h(p_i, q_i, \tilde{t}; \epsilon)$ that is 2π-periodic in each of the q_i such that (4.5.1) is in the form

$$\frac{dp_m}{dt} = -\frac{\partial h}{\partial q_m} \tag{4.5.5a}$$

$$\frac{dq_m}{dt} = \frac{\partial h}{\partial p_m}, \tag{4.5.5b}$$

$m = 1, \ldots, M$.

4.5.1 Transformation to Standard Form

Consider the following system of weakly coupled nearly linear oscillators with slowly varying parameters:

$$\ddot{x}_k + \gamma_k^2(\rho_i, \tilde{t}; \epsilon)x_k = \epsilon\xi_k(x_i, \dot{x}_i, \rho_i, \theta_i, \tilde{t}; \epsilon), \; k = 1, \ldots, K, \tag{4.5.6a}$$

$$\dot{\rho}_r = \epsilon\eta_r(x_i, \dot{x}_i, \rho_i, \theta_i, \tilde{t}; \epsilon), \; r = 1, \ldots, R, \tag{4.5.6b}$$

$$\dot{\theta}_s = \lambda_s(\rho_i, \tilde{t}; \epsilon) + \epsilon\zeta_s(x_i, \dot{x}_i, \rho_i, \theta_i, \tilde{t}; \epsilon), \; s = 1, \ldots, S. \tag{4.5.6c}$$

Here the ξ_k, η_r and ζ_s are 2π-periodic functions of the θ_i. We also assume that all the functions that depend on ϵ have asymptotic expansions in powers of ϵ, e.g.,

$$\gamma_k(\rho_i, \tilde{t}; \epsilon) = \gamma_k^{(0)}(\rho_i, \tilde{t}) + \epsilon\gamma_k^{(1)}(\rho_i, \tilde{t}) + O(\epsilon^2).$$

Note that (4.5.6) includes as a special case each of the examples we have discussed in Secs. 4.1–4.3; the transformation to standard form of the strictly nonlinear oscillator problem in Sec. 4.4 is discussed in Chapter 5. For example, to identify Duffing's equation (4.3.3) with (4.5.6), we set $\theta_1 = t$, $x_1 = y$ and obtain the system

$$\frac{d^2x_1}{dt^2} + x_1 = \epsilon[-\beta\dot{x}_1 - x_1^3 + f\cos(\theta_1 + \omega\tilde{t})] \equiv \epsilon\xi_1(x_1, \dot{x}_1, \theta_1, \tilde{t}), \tag{4.5.7a}$$

$$\dot{\theta}_1 = 1, \tag{4.5.7b}$$

where $K = 1$, $R = 1$, $\gamma_1 = 1$, and $\rho_1 = \eta_1 = \zeta_1 = 0$.

The planar satellite problem in the form (4.3.116) can also be identified with (4.5.6) by setting $u - u^2t' \to x_1$, $u^2t' \to \rho_1$, $t \to \theta_1$, and $\theta \to t$. There are other, more conventional, choices of variables in terms of which the equations governing satellite motion are in standard form. For example, see the discussion in Sec. 3.3 of [4.15] for one such choice.

We now show that the system (4.5.6) can be transformed to the standard form (4.5.4). Part of (4.5.6) is already in standard form, and we only need to transform (4.5.6a). Since for $\xi_k = 0$ (4.5.6a) reduces to K linear decoupled oscillators with slowly varying frequencies γ_k, we refer to the discussion in Sec. 4.3.2 for a single oscillator. We saw there that the choice of the action as a dependent variable resulted in an equation in standard form (see (4.3.72)). To define a point in the phase plane of each oscillator uniquely, we need to specify the action as well as the angle of the vector joining the origin to this point. Thus, we introduce the "action and angle" variables

$$p_k = \frac{\dot{x}_k^2 + \gamma_k^2 x_k^2}{2\gamma_k}, \tag{4.5.8a}$$

$$q_k = \tan^{-1}\left(\frac{\gamma_k x_k}{\dot{x}_k}\right) \tag{4.5.8b}$$

for each $k = 1, \ldots, K$. Henceforth, we will omit the reminder $k = 1, \ldots, K$, etc. The significance of action and angle variables is fully explored in Chapter 5;

for now we should regard (4.5.8) as one of several possible choices for p_k and q_k that lead to the desired standard form. The inverse transformation is

$$x_k = \left(\frac{2p_k}{\gamma_k}\right)^{1/2} \sin q_k, \tag{4.5.9a}$$

$$\dot{x}_k = (2\gamma_k p_k)^{1/2} \cos q_k. \tag{4.5.9b}$$

The K second-order equations (4.5.6a) are to be transformed to $2K$ first-order equations for the $\dot{p}_i$ and $\dot{q}_i$. The first set of the K conditions governing the transformed system follows by differentiating (4.5.9a) with respect to t and setting the result equal to (4.5.9b):

$$\left(\frac{2p_k}{\gamma_k}\right)^{1/2} \dot{q}_k \cos q_k - \frac{\dot{\gamma}_k p_k - \dot{p}_k \gamma_k}{(2\gamma_k^3 p_k)^{1/2}} \sin q_k = (2\gamma_k p_k)^{1/2} \cos q_k. \tag{4.5.10a}$$

The second set of K conditions is obtained by substituting the time derivative of (4.5.9b) for $\ddot{x}_k$ and (4.5.9a) for x_k into the left-hand side of (4.5.6a). We find

$$-(2\gamma_k p_k)^{1/2} \dot{q}_k \sin q_k + \frac{\gamma_k \dot{p}_k + \dot{\gamma}_k p_k}{(2\gamma_k p_k)^{1/2}} \cos q_k + (2\gamma_k^3 p_k)^{1/2} \sin q_k = \epsilon \xi_k. \tag{4.5.10b}$$

We solve the linear system (4.5.10) for $\dot{p}_k$ and $\dot{q}_k$ and use

$$\dot{\gamma}_k = \epsilon \sum_{r=1}^{R} \frac{\partial \gamma_k}{\partial \rho_r} \eta_r + \epsilon \frac{\partial \gamma_k}{\partial \tilde{t}}$$

to obtain

$$\dot{p}_k = \epsilon \left\{ \xi_k \left(\frac{2p_k}{\gamma_k}\right)^{1/2} \cos q_k - \frac{p_k}{\gamma_k} \left[\sum_{r=1}^{R} \frac{\partial \gamma_k}{\partial \rho_r} \eta_r + \frac{\partial \gamma_k}{\partial \tilde{t}} \right] \cos 2q_k \right\}, \tag{4.5.11a}$$

$$\dot{q}_k = \gamma_k + \epsilon \left\{ -\frac{\xi_k}{(2\gamma_k p_k)^{1/2}} \sin q_k + \frac{1}{2\gamma_k} \left[\sum_{r=1}^{R} \frac{\partial \gamma_k}{\partial \rho_r} \eta_r + \frac{\partial \gamma_k}{\partial \tilde{t}} \right] \sin 2q_k \right\}. \tag{4.5.11b}$$

In the above expressions and in (4.5.6b)–(4.5.6c), the arguments of the ξ_k, η_r, and ρ_s functions are evaluated using (4.5.9a) for the x_i and (4.5.9b) for the $\dot{x}_i$.

Equations (4.5.11), (4.5.6b), and (4.5.6c) are in the standard form (4.5.1) with the following notation. For each $k = 1, \ldots, K$ we identify the p_k and q_k in (4.5.11) with the respective p_k and q_k in (4.5.1). We also identify the following expressions in (4.5.1) with their counterparts in (4.5.6):

$$p_{K+r} = \mu_r, \; r = 1, \ldots, R, \tag{4.5.12a}$$

$$q_{K+s} = \theta_s, \; s = 1, \ldots, S, \tag{4.5.12b}$$

$$\epsilon F_k = \text{right-hand side of (4.5.11a)}, \; k = 1, \ldots, K, \tag{4.5.12c}$$

$$F_{K+r} = \eta_r, \; r = 1, \ldots, R, \tag{4.5.12d}$$

$$\omega_k = \gamma_k, \; k = 1, \ldots, K, \tag{4.5.12e}$$

$$\omega_{K+s} = \lambda_s, \; s = 1, \ldots, S, \tag{4.5.12f}$$

$$\epsilon G_k = \text{right-hand side of (4.5.11b)}, \; k = 1, \ldots, K, \tag{4.5.12g}$$

$$G_{K+s} = \zeta_s, \; s = 1, \ldots, S. \tag{4.5.12h}$$

Thus, the M and N of (4.5.1) are $K + R$ and $K + S$, respectively.

Before proceeding with the solution, we must decompose the right-hand sides of (4.5.11) and (4.5.6b)–(4.5.6c) into averaged and oscillatory terms. We also need to expand the ϵ dependence of the various functions on the right-hand side to derive the explicit form to $O(\epsilon^2)$ in (4.5.4).

4.5.2 *A Model Problem in One Degree of Freedom* ($M = N = 1$)

Before considering the expansion procedure for the general system (4.5.1), it is helpful to discuss in some detail the special case where $M = N = 1$. We also assume that F_1 and G_1 have zero average, involve only one harmonic, and take (dropping subscripts)

$$\frac{dp}{dt} = \epsilon A(p, \tilde{t}) \cos q, \tag{4.5.13a}$$

$$\frac{dq}{dt} = \omega(p, \tilde{t}) + \epsilon B(p, \tilde{t}) \sin q. \tag{4.5.13b}$$

Here A, ω, and B are prescribed well-behaved functions of p and $\tilde{t}$. We will use this example in Chapter 5 to motivate the analysis for the method of averaging.

We assume p and q are functions of the two scales τ and $\tilde{t} = \epsilon t$, where the instantaneous frequency $d\tau/dt$ is an unspecified function $\Omega(\tilde{t})$. Expanding in powers of ϵ, we have

$$p(t; \epsilon) = p^{(0)}(\tau, \tilde{t}) + \epsilon p^{(1)}(\tau, \tilde{t}) + \epsilon^2 p^{(2)}(\tau, \tilde{t}) + O(\epsilon^3), \tag{4.5.14a}$$

$$q(t; \epsilon) = q^{(0)}(\tau, \tilde{t}) + \epsilon q^{(1)}(\tau, \tilde{t}) + \epsilon^2 q^{(2)}(\tau, \tilde{t}) + O(\epsilon^3). \tag{4.5.14b}$$

The time derivative has the form

$$\frac{d}{dt} = \Omega(\tilde{t}) \frac{\partial}{\partial \tau} + \epsilon \frac{\partial}{\partial \tilde{t}}.$$

Therefore, we have

$$\frac{dp}{dt} = \Omega \frac{\partial p^{(0)}}{\partial \tau} + \epsilon \left(\frac{\partial p^{(0)}}{\partial \tilde{t}} + \Omega \frac{\partial p^{(1)}}{\partial \tau} \right) + \epsilon^2 \left(\frac{\partial p^{(1)}}{\partial \tilde{t}} + \Omega \frac{\partial p^{(2)}}{\partial \tau} \right) + \ldots \tag{4.5.15}$$

and a similar expression for dq/dt.

Collecting the contributions to (4.5.13) to $O(1)$ and $O(\epsilon)$ gives

$$O(1): \quad \Omega \frac{\partial p^{(0)}}{\partial \tau} = 0, \tag{4.5.16a}$$

$$\Omega \frac{\partial q^{(0)}}{\partial \tau} = \omega(p^{(0)}, \tilde{t}). \tag{4.5.16b}$$

$$O(\epsilon): \quad \Omega \frac{\partial p^{(1)}}{\partial \tau} = -\frac{\partial p^{(0)}}{\partial \tilde{t}} + A(p^{(0)}, \tilde{t}) \cos q^{(0)}, \tag{4.5.17a}$$

$$\Omega \frac{\partial q^{(1)}}{\partial \tau} = -\frac{\partial q^{(0)}}{\partial \tilde{t}} + B(p^{(0)}, \tilde{t}) \sin q^{(0)} + \omega_p(p^{(0)}, \tilde{t}) p^{(1)}. \tag{4.5.17b}$$

Equation (4.5.16a) states that $p^{(0)}$ does not depend on τ

$$p^{(0)} = p^{(0)}(\tilde{t}), \tag{4.5.18a}$$

and when this is used in (4.5.16b) we find, upon integration with respect to τ, that

$$q^{(0)}(\tau, \tilde{t}) = \frac{\omega(p^{(0)}(\tilde{t}), \tilde{t})}{\Omega(\tilde{t})} \tau + \phi^{(0)}(\tilde{t}). \tag{4.5.18b}$$

At this stage, Ω, $p^{(0)}$, and $\phi^{(0)}$ are unknown functions of $\tilde{t}$.

We now consider (4.5.17a) for $p^{(1)}$. Integrating with respect to τ gives

$$p^{(1)}(\tau, \tilde{t}) = -\frac{\tau}{\Omega(\tilde{t})} \frac{dp^{(0)}}{d\tilde{t}} + \frac{A(p^{(0)}, \tilde{t})}{\omega(p^{(0)}, \tilde{t})} \sin q^{(0)} + \theta^{(1)}(\tilde{t}). \tag{4.5.19}$$

The leading term proportional to τ is inconsistent, and we remove it by setting $(dp^{(0)}/d\tilde{t}) = 0$, i.e.,

$$p^{(0)} = \text{constant}. \tag{4.5.20}$$

Substituting known results into the right-hand side of (4.5.17b) gives

$$\Omega \frac{\partial q^{(1)}}{\partial \tau} = \left[-\frac{\omega_{\tilde{t}}(p^{(0)}, \tilde{t})}{\Omega} + \frac{\frac{d\Omega}{d\tilde{t}} \omega(p^{(0)}, \tilde{t})}{\Omega^2} \right] \tau - \frac{d\phi^{(0)}}{d\tilde{t}} + \omega_p(p^{(0)}, \tilde{t}) \theta^{(1)}(\tilde{t})$$
$$+ \left[B(p^{(0)}, \tilde{t}) + \frac{\omega_p(p_0^{(0)}, \tilde{t}) A(p^{(0)}, \tilde{t})}{\omega(p^{(0)}, \tilde{t})} \right] \sin q^{(0)}. \tag{4.5.21}$$

We see that in order to avoid an inconsistent term proportional to τ^2 in $q^{(1)}$, we must set the bracketed coefficient of τ on the right-hand side of (4.5.21) equal to zero. The vanishing of this expression requires that $\Omega(\tilde{t})$ be a constant multiple of $\omega(p^{(0)}, \tilde{t})$. For simplicity, and with no loss of generality, we choose

$$\Omega(\tilde{t}) = \omega(p^{(0)}, \tilde{t}). \tag{4.5.22}$$

Furthermore, to avoid terms proportional to τ in $q^{(1)}$, we must set

$$\frac{d\phi^{(0)}}{d\tilde{t}} = \omega_p(p^{(0)}, \tilde{t}) \theta^{(1)}(\tilde{t}). \tag{4.5.23a}$$

Thus,

$$\phi^{(0)}(\tilde{t}) = \int_0^{\tilde{t}} \omega_p(p^{(0)}, s) \theta^{(1)}(s) ds + \phi^{(0)}(0). \tag{4.5.23b}$$

This means that if $\omega_p \neq 0$, we must wait until we have determined $\theta^{(1)}$ before we can calculate $\phi^{(0)}$. For the special case $\omega_p = 0$, we have $\phi^{(0)}(\tilde{t}) = \text{constant} = \phi^{(0)}(0)$.

We now integrate what remains of (4.5.21) to obtain

$$q^{(1)}(\tau, \tilde{t}) = -\frac{1}{\omega(p^{(0)}, \tilde{t})}\left[B(p^{(0)}, \tilde{t}) + \frac{\omega_p(p^{(0)}, \tilde{t})}{\omega(p^{(0)}, \tilde{t})}\right]\cos q^{(0)} + \phi^{(1)}(\tilde{t}). \tag{4.5.24}$$

To summarize our results so far, we have the expansions of p and q in the form

$$p = p^{(0)} + \epsilon p^{(1)}(\tau, \tilde{t}) + \epsilon^2 p^{(2)}(\tau, \tilde{t}) + O(\epsilon^3), \tag{4.5.25a}$$

$$q = \psi + \epsilon q^{(1)}(\tau, \tilde{t}) + \epsilon^2 q^{(2)}(\tau, \tilde{t}) + O(\epsilon^3), \tag{4.5.25b}$$

where

$$\psi = \tau + \phi^{(0)}(\tilde{t}), \tag{4.5.26a}$$

$$p^{(1)} = \frac{A(p^{(0)}, \tilde{t})}{\omega(p^{(0)}, \tilde{t})}\sin\psi + \theta^{(1)}(\tilde{t}), \tag{4.5.26b}$$

and $q^{(1)}$ is given by (4.5.24).

Before proceeding further it is useful to relate the solution constants to initial values. If we wish to solve (4.5.13) subject to

$$p(0; \epsilon) = p_0, \tag{4.5.27a}$$

$$q(0; \epsilon) = q_0, \tag{4.5.27b}$$

where p_0 and q_0 are constants independent of ϵ, we conclude that

$$p^{(0)} = p_0, \tag{4.5.28a}$$

$$\phi^{(0)}(0) = q_0, \tag{4.5.28b}$$

$$\theta^{(1)}(0) = -\frac{A(p_0, 0)}{\omega(p_0, 0)}\sin q_0, \tag{4.5.28c}$$

$$\phi^{(1)}(0) = \frac{1}{\omega(p_0, 0)}\left[B(p_0, 0) + \frac{\omega_p(p_0, 0)}{\omega(p_0, 0)}A(p_0, 0)\right]\cos q_0. \tag{4.5.28d}$$

To complete the solution to $O(\epsilon)$, we need to determine the two slowly varying functions $\theta^{(1)}(\tilde{t})$ and $\phi^{(1)}(\tilde{t})$. The equations governing the terms of order ϵ^2 are easily obtained in the form

$$\omega\frac{\partial p^{(2)}}{\partial\tau} = -\frac{\partial p^{(1)}}{\partial\tilde{t}} + A_p p^{(1)}\cos\psi - Aq^{(1)}\sin\psi, \tag{4.5.29a}$$

$$\omega\frac{\partial q^{(2)}}{\partial\tau} = -\frac{\partial q^{(1)}}{\partial\tilde{t}} + B_p p^{(1)}\sin\psi + Bq^{(1)}\cos\psi + \omega_p p^{(2)} + \frac{\omega_{pp}}{2}p^{(1)^2}. \tag{4.5.29b}$$

Here ω, A, B, A_p, B_p, ω_p and ω_{pp} are all evaluated for $p = p_0$. Using known expressions for the terms on the right-hand side of (4.5.29a), we find

$$\omega \frac{\partial p^{(2)}}{\partial \tau} = -\frac{d\theta^{(1)}}{d\tilde{t}} - \left[\left(\frac{A}{\omega} \right)_{\tilde{t}} + A\phi^{(1)} \right] \sin \psi + \theta^{(1)} \left(A_p - A \frac{\omega_p}{\omega} \right) \cos \psi$$
$$+ \frac{1}{2\omega} \left(AA_p + AB + A^2 \frac{\omega_p}{\omega} \right) \sin 2\psi. \tag{4.5.30}$$

To avoid terms proportional to τ in $p^{(2)}$, we set $(d\theta^{(1)}/d\tilde{t}) = 0$, i.e., $\theta^{(1)}$ is given by its initial value (4.5.28c). Integrating (4.5.30) defines $p^{(2)}$ in the form

$$p^{(2)} = \frac{1}{\omega} \left[\left(\frac{A}{\omega} \right)_{\tilde{t}} + A\phi^{(1)} \right] \cos \psi + \frac{\theta^{(1)}}{\omega} \left(A_p - \frac{A\omega_p}{\omega} \right) \sin \psi$$
$$- \frac{A}{4\omega^2} \left(A_p + B + \frac{\omega_p A}{\omega} \right) \cos 2\psi + \theta^{(2)}(\tilde{t}). \tag{4.5.31}$$

Also, since $\theta^{(1)}$ is a constant, (4.5.23b) gives $\phi^{(0)}(\tilde{t})$:

$$\phi^{(0)}(\tilde{t}) = -\frac{A(p_0, 0)}{\omega^{(0)}(p_0, 0)} \sin q_0 \int_0^{\tilde{t}} \omega_p(p_0, s) ds + q_0. \tag{4.5.32}$$

We substitute known expressions into the right-hand side of (4.5.29b) to obtain

$$\omega \frac{\partial q^{(2)}}{\partial \tau} = \left[-\frac{d\phi^{(1)}}{d\tilde{t}} + \frac{B_p A}{2\omega} - \frac{B}{2\omega} \left(B + \frac{\omega_p A}{\omega} \right) + \omega_p \theta^{(2)} \right.$$
$$\left. + \frac{\omega_{pp}}{4\omega^2} A^2 + \frac{\omega_p}{2} \theta^{(1)^2} \right] + \left[B\phi^{(1)} + A\phi^{(1)} \frac{\omega_p}{\omega} + \frac{\omega_p}{\omega} \left(\frac{A}{\omega} \right)_{\tilde{t}} \right.$$
$$\left. + \left(\frac{\omega_p A}{\omega^2} \right)_{\tilde{t}} + \left(\frac{B}{\omega} \right)_{\tilde{t}} \right] \cos \psi + \ldots. \tag{4.5.33}$$

In (4.5.33) we have not written the oscillatory terms proportional to $\sin \psi$ and $\cos 2\psi$ that also arise and are easily computed. To avoid terms linear in τ in $q^{(2)}$, we must set the first bracketed expression on the right-hand side of (4.5.33) equal to zero

$$\frac{d\phi^{(1)}}{d\tilde{t}} = \frac{B_p A}{2\omega} - \frac{B}{2\omega} \left(B + \frac{\omega_p A}{\omega} \right) + \omega_p \theta^{(2)} + \frac{\omega_{pp}}{4\omega^2} A^2 + \frac{\omega_p}{2} \theta^{(1)^2}. \tag{4.5.34}$$

We notice that if $\omega_p \neq 0$ this equation for $\phi^{(1)}$ involves the unknown $\theta^{(2)}(\tilde{t})$. For this reason, we must also consider the equation governing $p^{(3)}$ to derive the condition on $\theta^{(2)}$. This is not surprising, as the expression for $\phi^{(0)}$ in (4.5.23) also involved $\theta^{(1)}$ if $\omega_p \neq 0$. In fact, it is easily seen that, to each order, the equation governing $\phi^{(n)}$ involves $\omega_p \theta^{(n+1)}$.

In order to derive the equation governing $\theta^{(2)}(\tilde{t})$, we only need to isolate the average terms on the right-hand side of the equation for $\partial p^{(3)}/\partial \tau$. Extending our expansion of (4.5.13a) to $O(\epsilon^3)$ gives

$$\omega \frac{\partial p^{(3)}}{\partial \tau} = -\frac{\partial p^{(2)}}{\partial \tilde{t}} - \frac{A}{2} q^{(1)^2} \cos\psi - Aq^{(2)} \sin\psi - A_p p^{(1)} q^{(1)} \sin\psi$$
$$+ A_p p^{(2)} \cos\psi + \frac{A_{pp}}{2} p^{(1)^2} \cos\psi. \tag{4.5.35}$$

Since $q^{(2)}$ appears in (4.5.35) in the product $q^{(2)} \sin\psi$, only the term proportional to $\sin\psi$ in $q^{(2)}$ contributes an average, and we can compute this from (4.5.33). A straightforward calculation gives

$$q^{(2)} = \left[\frac{\phi^{(1)}}{\omega}\left(B + A\frac{\omega_p}{\omega}\right) + \frac{\omega_p}{\omega^2}\left(\frac{A}{\omega}\right)_{\tilde{t}} + \frac{1}{\omega}\left(\frac{\omega_p A}{\omega^2}\right)_{\tilde{t}} + \frac{1}{\omega}\left(\frac{B}{\omega}\right)_{\tilde{t}} \right] \sin\psi$$
$$+ \ldots, \tag{4.5.36}$$

where ... indicates terms proportional to $\cos\psi$, and $\sin 2\psi$. We can now compute the average term on the right-hand side of (4.5.35); setting this equal to zero gives

$$\frac{d\theta^{(2)}}{d\tilde{t}} = \frac{1}{2}\left(\frac{A}{\omega}\right)_{\tilde{t}}\left(\frac{A}{\omega}\right)_p - \frac{A}{2\omega}\left(\frac{\omega_p}{\omega^2}A + \frac{B}{\omega}\right)_{\tilde{t}} \equiv h(\tilde{t}). \tag{4.5.37}$$

We again remind the reader that A, B, ω, A_p, ω_p, and B_p are evaluated at $p = p_0$ in (4.5.36). Thus, h is a known function and we have

$$\theta^{(2)} = \int_0^{\tilde{t}} h(s)ds + \theta^{(2)}(0). \tag{4.5.38}$$

We can now evaluate $\phi^{(1)}(\tilde{t})$ by quadrature from (4.5.34), and this completes the solution to $O(\epsilon)$. This solution is summarized as follows:

$$p = p_0 + \epsilon \left[\frac{A(p_0, \tilde{t})}{\omega(p_0, \tilde{t})} \sin\psi + \theta^{(1)} \right] + O(\epsilon^2), \tag{4.5.39a}$$

$$q = \psi + \epsilon \left\{ -\left[\frac{B(p_0, \tilde{t})}{\omega(p_0, \tilde{t})} - \frac{\omega_p(p_0, \tilde{t})}{\omega^2(p_0, \tilde{t})} A(p_0, \tilde{t}) \right] \cos\psi + \phi^{(1)}(\tilde{t}) \right\}$$
$$+ O(\epsilon^2), \tag{4.5.39b}$$

where

$$\psi = \frac{1}{\epsilon}\int_0^{\tilde{t}} \omega(p_0, s)ds + \theta^{(1)} \int_0^{\tilde{t}} \omega_p(p_0, s)ds + q_0, \tag{4.5.40a}$$

$$\phi^{(1)}(\tilde{t}) = \int_0^{\tilde{t}} \left[\frac{B_p A}{2\omega} - \frac{B}{2\omega}\left(B + \frac{\omega_p}{\omega}A\right) + \omega_p \theta^{(2)}(s) \right.$$
$$\left. + \frac{\omega_{pp}}{2}\theta^{(1)^2} + \frac{\omega_{pp}}{4\omega^2}A^2 \right] ds + \phi^{(1)}(0), \tag{4.5.40b}$$

$$\theta^{(1)} = -\frac{A(p_0, 0)}{\omega(p_0, 0)} \sin q_0, \tag{4.5.40c}$$

$$\phi^{(1)}(0) = \left[\frac{1}{\omega}\left(B + \frac{\omega_p}{\omega} A\right)\right]_{p=p_0, \tilde{t}=0} \cos q_0, \tag{4.5.40d}$$

$$\theta^{(2)} = \frac{1}{2}\int_0^{\tilde{t}} \left[\left(\frac{A}{\omega}\right)_{\tilde{t}} \left(\frac{A}{\omega}\right)_p - \left(\frac{A}{\omega}\right)\left(\frac{\omega_p}{\omega^2} A + \frac{B}{\omega}\right)_{\tilde{t}}\right]_{p=p_0, \tilde{t}=s} ds,$$
$$+ \theta^{(2)}(0) \tag{4.5.40e}$$

$$\theta^{(2)}(0) = \left\{ -\frac{1}{\omega}\left[\left(\frac{A}{\omega}\right)_{\tilde{t}} + A\phi^{(1)}\right] \cos q_0 - \frac{\theta^{(1)}}{\omega}\left(A_p - \frac{A\omega_p}{\omega}\right) \sin q_0 \right.$$
$$\left. + \frac{A}{4\omega^2}\left(A_p + B + \frac{\omega_p}{\omega} A\right) \cos 2q_0 \right\}_{p=p_0, \tilde{t}=0}. \tag{4.5.40f}$$

4.5.3 Multiple-Scale Asymptotic Solution for the General Problem

Guided by the approach for the special case discussed in the previous section, we now proceed to calculate the solution for the general case (4.5.4).

Expansion procedure

We assume the following multiple-scale expansion for the p_m and q_n:

$$p_m(t; \epsilon) = p_m^{(0)}(\tilde{t}) + \epsilon p_m^{(1)}(\tau_i, \tilde{t}) + \epsilon^2 p_m^{(2)}(\tau_i, \tilde{t}) + O(\epsilon^3), \tag{4.5.41a}$$
$$q_n(t; \epsilon) = q_n^{(0)}(\tau_n, \tilde{t}) + \epsilon q_n^{(1)}(\tau_i, \tilde{t}) + \epsilon^2 q_n^{(2)}(\tau_i, \tilde{t}) + O(\epsilon^3). \tag{4.5.41b}$$

In view of the fact that $\dot{p}_m = O(\epsilon)$, it is clear that the leading term $p_m^{(0)}$ in the expansion for p_m must depend only on $\tilde{t}$. The τ_i are the fast times associated with ω_i frequencies, and we assume that these are defined by

$$\frac{d\tau_n}{dt} = \Omega_n(\tilde{t}), \tag{4.5.42}$$

where the Ω_n are to be determined. The assumption in (4.5.41b) that each $q_n^{(0)}$ depends on its own associated fast time τ_n and $\tilde{t}$ only is justified systematically later.

It follows from (4.5.42) that the time derivative is given by

$$\frac{d}{dt} = \sum_{j=1}^{N} \Omega_j \frac{\partial}{\partial \tau_j} + \epsilon \frac{\partial}{\partial \tilde{t}}. \tag{4.5.43}$$

Hence the $\dot{p}_m$ and $\dot{q}_n$ have the following expansions:

$$\dot{p}_m = \epsilon\left(p_m^{(0)'} + \sum_{j=1}^{N} \Omega_j \frac{\partial p_m^{(1)}}{\partial \tau_j}\right) + \epsilon^2\left(\frac{\partial p_m^{(1)}}{\partial \tilde{t}} + \sum_{j=1}^{N} \Omega_j \frac{\partial p_m^{(2)}}{\partial \tau_j}\right)$$

$$+ O(\epsilon^3), \tag{4.5.44a}$$

$$\dot{q}_n = \frac{\partial q_n^{(0)}}{\partial \tau_n}\Omega_n + \epsilon\left(\frac{\partial q_n^{(0)}}{\partial \tilde{t}} + \sum_{k=1}^{N}\frac{\partial q_n^{(1)}}{\partial \tau_k}\Omega_k\right)$$
$$+ \epsilon^2\left(\frac{\partial q_n^{(1)}}{\partial \tilde{t}} + \sum_{k=1}^{N}\Omega_k\frac{\partial q_n^{(2)}}{\partial \tau_k}\right) + O(\epsilon^3), \tag{4.5.44b}$$

where a prime indicates $d/d\tilde{t}$.

A typical averaged function $\underline{f}(p_i, \tilde{t})$ such as $\underline{f}_m$, $\underline{\omega}_n$, or $\underline{g}_n$ has the expansion

$$\underline{f}(p_i, \tilde{t}) = \underline{f}(p_i^{(0)}, \tilde{t}) + \epsilon\sum_{j=1}^{M}\frac{\partial \underline{f}(p_i^{(0)}, \tilde{t})}{\partial p_j}p_j^{(1)}$$
$$+ \epsilon^2\left\{\sum_{j=1}^{M}\frac{\partial \underline{f}(p_i^{(0)}, \tilde{t})}{\partial p_j}p_j^{(2)} + \frac{1}{2}\sum_{j=1}^{M}\sum_{r=1}^{M}\frac{\partial^2 \underline{f}(p_i^{(0)}, \tilde{t})}{\partial p_j \partial p_r}p_j^{(1)}p_r^{(1)}\right\}$$
$$+ O(\epsilon^3), \tag{4.5.45a}$$

and an oscillatory function $\hat{f}(p_i, q_i, \tilde{t})$ such as $\hat{f}_m$ or $\hat{g}_m$ takes the form

$$\hat{f}(p_i, q_i, \tilde{t}) = \hat{f}(p_i^{(0)}, q_i^{(0)}, \tilde{t})$$
$$+ \epsilon\left[\sum_{j=1}^{M}\frac{\partial \hat{f}(p_i^{(0)}, q_i^{(0)}, \tilde{t})}{\partial p_j}p_j^{(1)} + \sum_{k=1}^{N}\frac{\partial \hat{f}(p_i^{(0)}, q_i^{(0)}, \tilde{t})}{\partial q_k}q_k^{(1)}\right]$$
$$+ O(\epsilon^2). \tag{4.5.45b}$$

Using (4.5.44) and (4.5.45) in (4.5.4) yields the following differential equations for the various terms in the expansions of p_m and q_n to $O(\epsilon^2)$:

$$\frac{\partial q_n^{(0)}}{\partial \tau_n}\Omega_n = \underline{\omega}_n(p_i^{(0)}, \tilde{t}). \tag{4.5.46}$$

$$p_m^{(0)\prime} + \sum_{k=1}^{N}\Omega_k\frac{\partial p_m^{(1)}}{\partial \tau_k} = \underline{f}_m(p_i^{(0)}, \tilde{t}) + \hat{f}_m(p_i^{(0)}, q_i^{(0)}, \tilde{t}), \tag{4.5.47a}$$

$$\frac{\partial q_n^{(0)}}{\partial \tilde{t}} + \sum_{k=1}^{N}\Omega_k\frac{\partial q_n^{(1)}}{\partial \tau_k} = \sum_{j=1}^{M}\frac{\partial \underline{\omega}_n(p_i^{(0)}, \tilde{t})}{\partial p_j}p_j^{(1)} + \underline{g}_n(p_i^{(0)}, \tilde{t}) + \hat{g}_n(p_i^{(0)}, q_i^{(0)}, \tilde{t}). \tag{4.5.47b}$$

$$\frac{\partial p_m^{(1)}}{\partial \tilde{t}} + \sum_{j=1}^{M}\Omega_j\frac{\partial p_m^{(2)}}{\partial \tau_j} = \sum_{j=1}^{M}\frac{\partial \underline{f}_m(p_i^{(0)}, \tilde{t})}{\partial p_j}p_j^{(1)} + \sum_{j=1}^{M}\frac{\partial \hat{f}_m(p_i^{(0)}, q_i^{(0)}, \tilde{t})}{\partial p_j}p_j^{(1)}$$

$$+\sum_{k=1}^{M}\frac{\partial \underset{\wedge}{f}_m(p_i^{(0)},q_i^{(0)},\tilde{t})}{\partial q_k}q_k^{(1)}+\underline{k}_m(p_i^{(0)},\tilde{t})+\underset{\wedge}{k}_m(p_i^{(0)},q_i^{(0)},\tilde{t}), \tag{4.5.48a}$$

$$\begin{aligned}\frac{\partial q_n^{(1)}}{\partial \tilde{t}}+\sum_{k=1}^{N}\Omega_k\frac{\partial q_n^{(2)}}{\partial \tau_k}&=\sum_{j=1}^{M}\frac{\partial \underline{\omega}_n(p_i^{(0)},\tilde{t})}{\partial p_j}p_j^{(2)}\\&+\frac{1}{2}\sum_{j=1}^{M}\sum_{r=1}^{M}\frac{\partial^2\underline{\omega}_n(p_i^{(0)},\tilde{t})}{\partial p_j\partial p_r}p_j^{(1)}p_r^{(1)}+\sum_{j=1}^{M}\frac{\partial \underline{g}_n(p_i^{(0)},\tilde{t})}{\partial p_j}p_j^{(1)}\\&+\sum_{j=1}^{M}\frac{\partial \underset{\wedge}{g}_n(p_i^{(0)},q_i^{(0)},\tilde{t})}{\partial p_j}p_j^{(1)}+\sum_{k=1}^{N}\frac{\partial \underset{\wedge}{g}_n(p_i^{(0)},q_i^{(0)},\tilde{t})}{\partial q_k}q_k^{(1)}\\&+\underline{l}_n(p_i^{(0)},\tilde{t})+\underset{\wedge}{l}_n(p_i^{(0)},q_i^{(0)},\tilde{t}).\end{aligned} \tag{4.5.48b}$$

Terms of $O(\epsilon)$

The solution of (4.5.46) for $q_n^{(0)}$ is

$$q_n^{(0)}=a_n(\tilde{t})\tau_n+\phi_n^{(0)}(\tilde{t}), \tag{4.5.49}$$

where a_n denotes

$$a_n(\tilde{t})=\underline{\omega}_n(p_i^{(0)}(\tilde{t}),\tilde{t})/\Omega_n(\tilde{t}) \tag{4.5.50}$$

and at this stage the $\phi_n^{(0)}$ are undetermined functions.

In (4.5.47a), unless we remove the averaged terms by setting

$$p_m^{(0)\prime}=\underline{f}_m(p_i^{(0)},\tilde{t}), \tag{4.5.51}$$

the solution of $p_m^{(1)}$ will involve secular terms proportional to τ_i, rendering the expansion (4.5.41a) invalid for t in the interval $0\le t\le T(\epsilon)=O(\epsilon^{-1})$. Notice that (4.5.51) is a *system* of M first-order nonlinear equations for the $p_i^{(0)}$. This system has the trivial solution $p_i^{(0)}=$ const. when all the $\underline{f}_i$ vanish, a special case of considerable significance for the derivation of explicit results. We will see in Chapter 5 that the $\underline{f}_i$ do, in fact, all vanish for the Hamiltonian problem. The solution of the system (4.5.51) of M first-order equations defines the $p_m^{(0)}$ as functions of $\tilde{t}$ and M constants of integration $\alpha_1,\ldots,\alpha_M$

$$p_m^{(0)}=\xi_m(\tilde{t},\alpha_1,\ldots,\alpha_m). \tag{4.5.52}$$

Now, we can solve what remains of (4.5.47a) to express the $p_m^{(1)}$ in the form

$$p_m^{(1)}(\tau_i,\tilde{t})=\underset{\wedge}{P}_m{}^{(1)}(\psi_i,\tilde{t})+\theta_m^{(1)}(\tilde{t}). \tag{4.5.53}$$

Here the $\underset{\wedge}{P}_m{}^{(1)}$ are oscillatory functions of the ψ_i given by

$$\underset{\wedge}{P}_m{}^{(1)}(\psi_i,\tilde{t})=\left\{\int\underset{\wedge}{f}_m(p_i^{(0)},a_i\Omega_is+\phi_i^{(0)},\tilde{t})ds\right\}_{\Omega_is=\tau_i}, \tag{4.5.54a}$$

$$\psi_i = a_i \tau_i + \phi_i^{(0)} \tag{4.5.54b}$$

and the $\theta_m^{(1)}$ are undetermined at this stage. In the present context, whenever we use the notation $\underset{\wedge}{F}(\psi_i, \tilde{t})$ we understand, as before, that $\underset{\wedge}{F}$ is a 2π-periodic function of the ψ_i having a zero average with respect to all the ψ_i.

Equation (4.5.47b) for the $q_n^{(1)}$ can now be decomposed into secular, average, and oscillatory terms in the form:

$$\begin{aligned} a_n'(\tilde{t})\tau_n &+ \left\{ \phi_n^{(0)\prime} - \sum_{j=1}^{M} \frac{\partial \underline{\omega}_n(p_i^{(0)}, \tilde{t})}{\partial p_j} \theta_j^{(1)}(\tilde{t}) - \underline{g}_n(p_i^{(0)}, \tilde{t}) \right\} \\ &+ \left\{ \sum_{k=1}^{N} \Omega_k \frac{\partial q_n^{(1)}}{\partial \tau_k} - \underset{\wedge}{g_n}(p_i^{(0)}, q_i^{(0)}, \tilde{t}) - \sum_{j=1}^{M} \frac{\partial \underline{\omega}_n(p_i^{(0)}, \tilde{t})}{\partial p_j} \underset{\wedge}{P}_j^{(1)} \right\} \\ &= 0. \end{aligned} \tag{4.5.55}$$

First, we must remove the term $a_n' \tau_n$ by taking $a_n' = 0$, for otherwise $q_n^{(1)}$ will involve *quadratic* terms in the τ_i. We set $a_n = 1$ with no loss of generality. Hence,

$$q_n^{(0)} = \tau_n + \phi_n^{(0)}(\tilde{t}) \equiv \psi_n, \tag{4.5.56a}$$

$$\Omega_n(\tilde{t}) = \underline{\omega}_n(p_i^{(0)}(\tilde{t}), \tilde{t}). \tag{4.5.56b}$$

It is now easy to justify the assumption in (4.5.41b) that each q_n depends on its own τ_n and $\tilde{t}$ only. To fix ideas, consider the case $N = 2$. Had we assumed $q_n^{(0)}(\tau_1, \tau_2, \tilde{t})$, (4.5.46) would have read

$$\Omega_1 \frac{\partial q_n^{(0)}}{\partial \tau_1} + \Omega_2 \frac{\partial q_n^{(0)}}{\partial \tau_2} = \underline{\omega}_n, \quad n = 1, 2. \tag{4.5.57}$$

For each n, the solution of (4.5.57) can be written as

$$q_n^{(0)} = \frac{\underline{\omega}_n}{\Omega_m} \tau_m + \Phi\left(\tau_j - \frac{\Omega_j}{\Omega_m} \tau_m, \tilde{t}\right), \tag{4.5.58}$$

where Φ is an arbitrary function of its two arguments and the integers j and m may be taken as 1 or 2 independently. Note that (4.5.49) corresponds to the special case $j = m = n$. Using the result (4.5.58), it is seen that (4.5.55) will have a nonvanishing secular contribution of the form $[(\omega_m/\Omega_m)' - (\Omega_j/\Omega_m)'\Phi_1]\tau_m$ instead of $a_n'(\tilde{t})\tau_n$. Here Φ_1 is the partial derivative of Φ with respect to its first argument. This secular term will only vanish if $j = m = n$ and $\Omega_n = \omega_n$ as we have assumed.

Next, we remove the averaged terms in (4.5.55) by setting the second group of terms in parentheses equal to zero

$$\phi_n^{(0)\prime} = \underline{g}_n(p_i^{(0)}, \tilde{t}) + \sum_{j=1}^{M} \frac{\partial \underline{\omega}_n(p_i^{(0)}, \tilde{t})}{\partial p_j} \theta^{(1)}(\tilde{t}). \tag{4.5.59}$$

We note, as in the example discussed in Sec. 4.5.2, that if the $\frac{\partial \underline{\omega}_n}{\partial p_i}$ do not vanish for all i and n we must postpone calculating the $\phi_n^{(0)}$ until we have found the $\theta_n^{(1)}$. If, however, the $\underline{\omega}_n$ do not depend on the p_i ($\omega_n = \omega_n(\tilde{t})$), we can calculate the $\phi_n^{(0)}$ at this stage by quadrature.

To complete the formal solution to $O(\epsilon)$, we integrate what remains of (4.5.55), the third group of oscillatory terms, and obtain $q_n^{(1)}$ in the form

$$q_n^{(1)}(\tau_i, \tilde{t}) = \underset{\wedge}{Q_n}^{(1)}(\psi_i, \tilde{t}) + \phi_n^{(1)}(\tilde{t}), \tag{4.5.60a}$$

where

$$\underset{\wedge}{Q_n}^{(1)}(\psi_i, \tilde{t}) = \left\{ \int \left[\sum_{j=1}^{M} \frac{\partial \underline{\omega}_n(p_i^{(0)}, \tilde{t})}{\partial p_j} \underset{\wedge}{P_j}^{(1)}(p_i^{(0)}, \Omega_i s + \phi_i^{(0)}, \tilde{t}) + \underset{\wedge}{g_n}(p_i^{(0)}, \Omega_i s + \phi_i^{(0)}, \tilde{t}) \right] ds \right\}_{\Omega_i s = \tau_i}. \tag{4.5.60b}$$

Determination of the unknown averages $\theta_m^{(1)}(\tilde{t})$ and $\phi_n^{(1)}(\tilde{t})$ involves consistency conditions on the solution to $O(\epsilon^2)$, and this is considered next.

Terms of $O(\epsilon^2)$.

If we use the previously calculated results, (4.5.48a) for the $p_m^{(2)}$ may be split up into average and oscillatory terms as

$$\begin{aligned} &\left\{ \theta_m^{(1)\prime} - \sum_{j=1}^{M} \frac{\partial \underline{f}_m(\xi_i, \tilde{t})}{\partial p_j} \theta_j^{(1)} - \underline{k}_m(\xi_i, \tilde{t}) - \gamma_m^{(1)}(\tilde{t}) \right\} \\ &+ \left\{ \sum_{k=1}^{N} \left[\Omega_k(\tilde{t}) \frac{\partial p_m^{(2)}}{\partial \tau_k} + \frac{\partial \underset{\wedge}{P_m}^{(1)}}{\partial \psi_k} \underline{g}_k(\xi_i, \tilde{t}) - \frac{\partial \underset{\wedge}{f_m}(\xi_i, \psi_i, \tilde{t})}{\partial q_k} \phi_k^{(1)}(\tilde{t}) \right] \right. \\ &- \sum_{j=1}^{M} \left[\frac{\partial \underline{f}_m(\xi_i, \tilde{t})}{\partial p_j} \underset{\wedge}{P_j}^{(1)}(\psi_i, \tilde{t}) + \frac{\partial \underset{\wedge}{f_m}(\xi_i, \psi_i, \tilde{t})}{\partial p_j} \theta_j^{(1)}(\tilde{t}) \right] \\ &\left. + \frac{\partial \underset{\wedge}{P_m}^{(1)}(\psi_i, \tilde{t})}{\partial \tilde{t}} - \underset{\wedge}{\Gamma_m}^{(1)}(\psi_i, \tilde{t}) - \underset{\wedge}{k_m}(\xi_i, \psi_i, \tilde{t}) \right\} = 0. \end{aligned} \tag{4.5.61}$$

Here, we have decomposed

$$\sum_{j=1}^{M} \frac{\partial \underset{\wedge}{f_m}(\xi_i, \psi_i, \tilde{t})}{\partial p_j} \underset{\wedge}{P_j}^{(1)} + \sum_{k=1}^{N} \frac{\partial \underset{\wedge}{f_m}(\xi_i, \psi_i, \tilde{t})}{\partial q_k} \underset{\wedge}{Q_k}^{(1)} = \underset{\wedge}{\Gamma_m}^{(1)}(\psi_i, \tilde{t}) + \gamma_m^{(1)}(\tilde{t}) \tag{4.5.62}$$

into its oscillatory part $\underset{\wedge}{\Gamma_m}^{(1)}$ and its average part $\gamma_m^{(1)}$.

The vanishing of the first group of terms in parentheses in (4.5.61) ensures that $p_m^{(2)}$ is free of secular terms and defines $\theta_m^{(1)}$ as the solution of the linear system

$$\theta_m^{(1)\prime} - \sum_{j=1}^{M} \frac{\partial \underline{f}_m(\xi_i, \tilde{t})}{\partial p_j} \theta_j^{(1)} = \underline{k}_m(\xi_i, \tilde{t}) + \gamma_m^{(1)}(\tilde{t}). \tag{4.5.63}$$

Knowing the $\theta_i^{(1)}$, we can calculate the $\phi_n^{(0)}$ using (4.5.59).

What remains of (4.5.61) has the form

$$\sum_{k=1}^{N} \Omega_k(\tilde{t}) \frac{\partial p_m^{(2)}}{\partial \tau_k} = \underset{\wedge}{\Lambda_m}^{(1)}(\psi_i, \tilde{t}), \tag{4.5.64}$$

and the dependence of $\underset{\wedge}{\Lambda_m}^{(1)}$ on the ψ_i is known explicitly. Therefore, integrating (4.5.64) defines $p_m^{(2)}$ to within an additive function of $\tilde{t}$ as follows:

$$p_m^{(2)} = \underset{\wedge}{P_m}^{(2)}(\psi_i, \tilde{t}) + \theta_m^{(2)}(\tilde{t}), \tag{4.5.65a}$$

$$\underset{\wedge}{P_m}^{(2)} = \left\{ \int \underset{\wedge}{\Lambda_m}^{(1)}(\Omega_i s + \phi_i^{(0)}, \tilde{t}) ds \right\}_{\Omega_j s = \tau_j}. \tag{4.5.65b}$$

Next we subdivide (4.5.48b) into average and oscillatory terms using previously calculated results to obtain

$$\begin{aligned}
&\left\{ \phi_n^{(1)\prime} - \sum_{j=1}^{M} \frac{\partial \underline{\omega}_n(\xi_i, \tilde{t})}{\partial p_j} \theta_j^{(2)}(\tilde{t}) - \frac{1}{2} \sum_{j=1}^{M} \sum_{r=1}^{M} \frac{\partial^2 \underline{\omega}_n(\xi_i, \tilde{t})}{\partial p_j \partial p_r} \theta_j^{(1)}(\tilde{t}) \theta_r^{(1)}(\tilde{t}) \right. \\
&\left. \quad - \sum_{j=1}^{M} \frac{\partial \underline{g}_n(\xi_i, \tilde{t})}{\partial p_j} \theta_j^{(1)}(\tilde{t}) - \delta_n^{(1)}(\tilde{t}) - \underline{l}_n(\xi_i, \tilde{t}) \right\} \\
&+ \left\{ \sum_{k=1}^{N} \left[\Omega_k(\tilde{t}) \frac{\partial q_n^{(2)}}{\partial \tau_k} + \frac{\partial \underset{\wedge}{Q_n}^{(1)}}{\partial \psi_k} \underline{g}_k(\xi_i, \tilde{t}) - \frac{\partial \underset{\wedge}{g_n}(\xi_i, \psi_i, \tilde{t})}{\partial q_k} \phi_k^{(1)}(\tilde{t}) \right] \right. \\
&\quad - \sum_{j=1}^{M} \left[\frac{\partial \underline{g}_n(\xi_i, \tilde{t})}{\partial p_j} \underset{\wedge}{P_j}^{(1)}(\psi_i, \tilde{t}) + \frac{\partial \underset{\wedge}{g_n}(\xi_i, \psi_i, \tilde{t})}{\partial p_j} \theta_j^{(1)}(\tilde{t}) \right] \\
&\left. + \frac{\partial \underset{\wedge}{Q_n}^{(1)}(\psi_i, \tilde{t})}{\partial \tilde{t}} - \underset{\wedge}{\Delta_n}^{(1)}(\psi_i, \tilde{t}) - \underset{\wedge}{l_n}(\xi_i, \psi_i, \tilde{t}) \right\} = 0,
\end{aligned} \tag{4.5.66}$$

where again we have decomposed

$$\frac{1}{2} \sum_{j=1}^{M} \sum_{r=1}^{M} \frac{\partial^2 \underline{\omega}_n(\xi_i, \tilde{t})}{\partial p_j \partial p_r} \underset{\wedge}{P_j}^{(1)} \underset{\wedge}{P_r}^{(1)} + \sum_{j=1}^{M} \frac{\partial \underset{\wedge}{g_n}(\xi_i, \psi_i, \tilde{t})}{\partial p_j} \underset{\wedge}{P_j}^{(1)}$$

$$+\sum_{k=1}^{N} \frac{\partial \underset{\wedge}{g_n}(\xi_i, \psi_i, \tilde{t})}{\partial q_k} \underset{\wedge}{Q_k}^{(1)}$$

$$= \underset{\wedge}{\Delta_n}^{(1)}(\psi_i, \tilde{t}) + \delta_n^{(1)}(\tilde{t}) \tag{4.5.67}$$

into an oscillatory part, $\underset{\wedge}{\Delta_n}^{(1)}$, and an average part, $\delta_n^{(1)}$.

We avoid secular terms in $q_n^{(2)}$ by taking

$$\phi_n^{(1)'} = \sum_{j=1}^{M} \frac{\partial \underline{\omega}_n(\xi_i, \tilde{t})}{\partial p_j} \theta_j^{(2)}(\tilde{t}) + \frac{1}{2} \sum_{j=1}^{M} \sum_{r=1}^{M} \frac{\partial^2 \underline{\omega}_n(\xi_i, \tilde{t})}{\partial p_j \partial p_r} \theta_j^{(1)}(\tilde{t}) \theta_r^{(1)}(\tilde{t})$$
$$+ \sum_{j=1}^{M} \frac{\partial \underline{g}_n(\xi_i, \tilde{t})}{\partial p_j} \theta_j^{(1)}(\tilde{t}) + \delta_n^{(1)}(\tilde{t}) + \underline{l}_n(\xi_i, \tilde{t}). \tag{4.5.68}$$

Again, we note that if the $\frac{\partial \omega_n}{\partial p_i}$ do not vanish for all i and n, we must defer calculation of the $\phi_n^{(1)}$ until we have obtained all the $\theta_i^{(2)}$ by removing average terms from the equations for $\partial p_n^{(3)}/\partial \tau$. The details of this calculation were worked out for the special case $M = N = 1$ in Sec. 4.5.2. For brevity, we do not give the corresponding results for the general case.

The remainder of (4.5.66) takes the form

$$\sum_{k=1}^{N} \Omega_k(\tilde{t}) \frac{\partial q_n^{(2)}}{\partial \tau_k} = \underset{\wedge}{\chi_n}^{(1)}(\psi_i, \tilde{t}), \tag{4.5.69}$$

where the $\underset{\wedge}{\chi_n}^{(1)}$ are known explicitly. Hence, integrating (4.5.69) defines $q_n^{(2)}$ as follows:

$$q_n^{(2)} = \underset{\wedge}{Q_n}^{(2)}(\psi_i, \tilde{t}) + \phi_n^{(2)}(\tilde{t}),$$
$$\underset{\wedge}{Q_n}^{(2)} = \left\{ \int \underset{\wedge}{\chi_n}^{(1)}(\Omega_i s + \phi_i^{(0)}, \tilde{t}) ds \right\}_{\Omega_j s = \tau_i}. \tag{4.5.70}$$

This completes the determination of all oscillatory terms to $O(\epsilon^2)$ and all average terms except the $\phi_n^{(1)}$ to $O(\epsilon)$.

Summary of results

In what follows, we list our results for the solution of (4.5.4) in the order that they can be evaluated. We will refer to this summary list in Chapter 5 when we compare results with those obtained by the method of averaging.

The solution to $O(\epsilon)$ has the form

$$p_m(t; \epsilon) = \xi_m(\tilde{t}, \alpha_i) + \epsilon \left[\underset{\wedge}{P_m}^{(1)}(\tau_i, \tilde{t}) + \theta_m^{(1)}(\tilde{t}) \right] + O(\epsilon^2), \tag{4.5.71a}$$

$$q_n(t; \epsilon) = \tau_n + \eta_n(\tilde{t}, \beta_i) + \epsilon \left[Q_n^{(1)}(\tau_i, \tilde{t}) + \phi_n^{(1)}(\tilde{t}) \right] + O(\epsilon^2). \tag{4.5.71b}$$

The fast times τ_n are given by (see (4.5.42), (4.5.56b))

$$\tau_n = \frac{1}{\epsilon}\int_0^{\tilde{t}} \underline{\omega}_n(\xi_i(s,\alpha_i),s)ds, \tag{4.5.72}$$

where the $p_m^{(0)} = \xi_m(\tilde{t},\alpha_i)$ solve the system (see (4.5.51))

$$\frac{dp_m^{(0)}}{d\tilde{t}} = \underline{f}_m(p^{(0)},\tilde{t}) \tag{4.5.73}$$

in terms of $\tilde{t}$ and the M constants $\alpha_i = \alpha_1, \alpha_2, \ldots, \alpha_M$.

The oscillatory components of the p_m to $O(\epsilon)$ are given by (see (4.5.54a))

$$\underset{\wedge}{P}_m{}^{(1)}(\psi_i,\tilde{t}) = \left\{\int \underset{\wedge}{f}_m(\xi_i, \underline{\omega}_i s + \eta_i, \tilde{t})ds\right\}_{\underline{\omega}_i s=\tau_i}, \tag{4.5.74a}$$

$$\psi_n = \tau_n + \eta_n, \tag{4.5.74b}$$

where the η_i are functions of $\tilde{t}$ that we have not yet evaluated. The oscillatory components of the q_n to $O(\epsilon)$ are given by (see (4.5.60b))

$$\underset{\wedge}{Q}_n{}^{(1)}(\psi_i,\tilde{t}) = \left\{\int\left[\sum_{j=1}^{m}\frac{\partial\underline{\omega}_n(\xi_i,\tilde{t})}{\partial p_j}\underset{\wedge}{P}_j{}^{(1)}(\underline{\omega}_i s+\eta_i,\tilde{t}) \right.\right.$$
$$\left.\left. \underset{\wedge}{g}_n(\xi_i, \underline{\omega}_i s + \eta_i, \tilde{t})\right] ds\right\}_{\omega_i s=\tau_i}. \tag{4.5.75}$$

Having defined the $\underset{\wedge}{P}_i{}^{(1)}$ and $\underset{\wedge}{Q}_i{}^{(1)}$, we compute the average terms, $\gamma_i^{(1)}$, using (see (4.5.62))

$$\gamma_m^{(1)}(\tilde{t}) = \frac{1}{(2\pi)^N}\int_0^{2\pi}\cdots\int_0^{2\pi}\left[\sum_{j=1}^{M}\frac{\partial \underset{\wedge}{f}_m(\xi_i,\psi_i,\tilde{t})}{\partial p_j}P_j^{(1)}(\psi_i,\tilde{t}) + \right.$$
$$\left.\sum_{k=1}^{N}\frac{\partial \underset{\wedge}{f}_m(\xi_i,\psi_i,\tilde{t})}{\partial q_k}Q_k^{(1)}(\psi_i,\tilde{t})\right]d\psi_1\ldots d\psi_N. \tag{4.5.76}$$

We can now compute the $\theta_m^{(1)}$, the average terms of order ϵ in p_m, by solving the linear system (see (4.5.63))

$$\frac{d\theta_m^{(1)}}{d\tilde{t}} - \sum_{j=1}^{M}\frac{\partial \underline{f}_m(\xi_i,\tilde{t})}{\partial p_j}\theta_j^{(1)} = \underline{k}_m(\xi_i,\tilde{t}) + \gamma_m^{(1)}(\tilde{t}). \tag{4.5.77}$$

Once the $\theta_m^{(1)}$ are defined, we can compute $\phi_m^{(0)} = \eta_m(\tilde{t};\beta_i)$, the slowly varying phase shifts, by quadrature using (see (4.5.59))

$$\frac{d\phi_n^{(0)}}{d\tilde{t}} = \underline{g}_n(\xi_i,\tilde{t}) + \sum_{j=1}^{M}\frac{\partial\underline{\omega}_n(\xi_i,\tilde{t})}{\partial p_j}\theta_j^{(1)}(\tilde{t}). \tag{4.5.78}$$

The β_i are N *additive* constants of integration.

To complete the solution (4.5.71) to $O(\epsilon)$, we need to also define the $\phi_n^{(1)}(\tilde{t})$. These functions are obtained by solving (4.5.68), once we have computed the averages $\delta_n^{(1)}(\tilde{t})$ using (4.5.67). If any of the $\underline{\omega}_i$ depend on the p_i, we also need to compute the $\theta_m^{(2)}$ by consistency conditions on the $p_m^{(3)}$.

Resonance

In Section 4.3.6, where we studied the motion of two coupled weakly nonlinear oscillators, we saw that resonance is possible to leading and higher orders for certain pairs of frequency values. The conventional multiple-scale expansion fails at resonance because of the presence of certain zero divisors in the solution. Resonance is also possible for the general problem (4.5.1) and is again exhibited by the occurrence of zero divisors.

The possibility of a zero divisor first arises in the solution of (4.5.48a) for the $p_m^{(1)}$ once the condition (4.5.51) has been imposed. To see this, let us expand the oscillatory term on the right-hand side of (4.5.47a) in a multiple Fourier series

$$\underset{\wedge}{f}_m(p_i^{(0)}, q_i^{(0)}, \tilde{t}) = \sum_{r_1=-\infty}^{\infty} \cdots \sum_{r_N=-\infty}^{\infty} f_{r_1 r_2 \ldots r_N}^{(m)}(p_i^{(0)}, \tilde{t}) e^{i(r_1 q_1 + r_2 q_2 + \ldots + r_N q_N)}. \tag{4.5.79}$$

The Fourier coefficients are given by

$$f_{r_1 r_2 \ldots r_N}^{(m)}(p_i^{(0)}, \tilde{t}) = \frac{1}{(2\pi)^N} \int_0^{2\pi} \cdots \int_0^{2\pi} \underset{\wedge}{f}_m(p_i^{(0)}, s_1, s_2, \ldots, s_N, \tilde{t}) e^{-i(r_1 s_1 + r_2 s_2 + \ldots r_N s_N)} ds_1 \ldots ds_N \tag{4.5.80}$$

and $f_{0\ldots 0}^{(m)} = 0$ because we have isolated the average part $\underline{f}_m$. Suppose now that for certain integer values $r_1 = R_1, r_2 = R_2, \ldots r_N = R_N$ (where not all the R_N vanish), we have the resonance condition

$$\sigma \equiv R_1 \underline{\omega}_1 + R_2 \underline{\omega}_2 + \ldots + R_N \underline{\omega}_N = 0$$

at some time $\tilde{t} = \tilde{t}_0$. Here, the $\underline{\omega}_i$ are evaluated for $p_i^{(0)} = \xi_i(\tilde{t}_0, \alpha_1, \ldots, \alpha_M)$ and $\tilde{t} = \tilde{t}_0$. It then follows that upon integration of the term $f_{R_1 \ldots R_N} e^{i(R_1 q_1 + \ldots + R_N q_N)}$ in (4.5.54a), we introduce the divisor σ and this vanishes at $\tilde{t} = \tilde{t}_0$. The same resonant term in $\underset{\wedge}{f}_m$ produces a σ^2 divisor in (4.5.60b) because $\underset{\wedge}{P}_m$ is integrated once more. Additional resonant terms, corresponding to different frequency combinations, may also occur in $\underset{\wedge}{f}_m$ as well as $\underset{\wedge}{g}_n$ to produce other zero divisors in the solution to $O(\epsilon)$. Further resonances may also arise in the solution to $O(\epsilon^2)$ as in the example discussed in Sec. 4.3.6. In Chapter 5, we discuss the procedure for calculating asymptotic solutions when resonances are present. Thus, the results in this section are restricted to problems for which no resonances are possible in the solution to $O(\epsilon^2)$.

Concluding remarks

The procedure outlined in this section ensures that the asymptotic expansions (4.5.41) for the p_i and q_i are uniformly valid to $O(\epsilon)$ over the time interval $O \leq t \leq T(\epsilon) = O(\epsilon^{-1})$, or $0 \leq \tilde{t} \leq \tilde{T}(\epsilon) = O(1)$. Uniform validity may be lost if $\tilde{T}$ is allowed to become large. For example, in solving the system (4.5.63) for the $\theta_m^{(1)}(\tilde{t})$, we may encounter terms that become unbounded as $\tilde{t} \to \infty$, as for the model problem (4.2.61). The need to rescale the dependent variable becomes evident when the governing equation is cast into standard form (see Problem 3).

One advantage of proceeding from the standard form (4.5.1) instead of (4.5.6) for systems of coupled oscillators is that slower time scales $t_2 = \epsilon^2 t$, $t_3 = \epsilon^3 t$, ... are not needed. The appropriate slowly varying phase shifts are correctly taken into account in terms of the $\phi_n^{(0)}, \phi_n^{(1)}, \ldots$ functions (see Problem 4). This approach is particularly efficient for oscillators with slowly varying frequencies because an expansion based on (4.5.6) requires a careful evaluation of second derivative terms as discussed in [4.16].

The procedure outlined in this section is more direct than the one used in [4.15] where the fast scales τ_n are assumed in the form (see (4.5.42) and (4.61) of [4.15])

$$\frac{d\tau_n}{dt} = \omega_n(p_i^{(0)}(\tilde{t}), \tilde{t}) + \epsilon\nu_n(\tilde{t}) + \epsilon^2\mu_n(\tilde{t}). \tag{4.5.81}$$

The unknowns $\nu_n(\tilde{t})$ and $\mu_n(\tilde{t})$ are determined by requiring the equations governing $(d\phi_n^{(0)}/d\tilde{t})$ and $(d\phi_n^{(1)}/d\tilde{t})$ to be independent of the $\theta_n^{(1)}$ and $\theta_n^{(2)}$, respectively. This "bookkeeping" system allows us to compute the solution of the counterpart of (4.5.59) for the $\phi_n^{(0)}$ prior to knowing the $\theta_n^{(1)}$ (and to compute the solution of the counterpart of (4.5.68) prior to knowing the $\theta_n^{(2)}$). Although the calculations using (4.5.81) may appear to be more efficient, there is no essential advantage in deriving the asymptotic solution for the $p_n(t;\epsilon)$ and $q_n(t;\epsilon)$ to a given order. This solution is, in fact, identical to the one given here. Problem 6 outlines the ideas for the special case of (4.5.13).

Problems

1. Consider the general weakly nonlinear oscillator discussed in Sec. 4.2.5

$$\frac{d^2x}{dt^2} + x + \epsilon f\left(x, \frac{dx}{dt}\right) = 0. \tag{4.5.82}$$

a. Show that in this case (4.5.11) reduces to

$$\dot{p} = -\epsilon\left[f(\sqrt{2p}\sin q, \sqrt{2p}\cos q\right]\sqrt{2p}\cos q, \tag{4.5.83a}$$

$$\dot{q} = 1 + \epsilon\left[f(\sqrt{2p}\sin q, \sqrt{2p}\cos q\right]\frac{\sin q}{\sqrt{2p}}. \tag{4.5.83b}$$

b. Calculate the solution to $O(\epsilon)$ using the approach discussed in Sec. 4.5.3 and show that your results agree with those found in Sec. 4.2.5.

2. Consider the weakly nonlinear oscillator with slowly varying frequency governed by

$$\frac{d^2x}{dt^2} + \gamma^2(\tilde{t})x + \epsilon f\left(x, \frac{dx}{dt}, \tilde{t}\right) = 0. \tag{4.5.84}$$

This generalizes the example studied in Sec. 4.3.2.

a. Show that the standard form in this case is

$$\dot{p} = \epsilon\left[-f\left(\left(\frac{2p}{\gamma}\right)^{1/2}\sin q, (2p\gamma)^{1/2}\cos q\right)\left(\frac{2p}{\gamma}\right)^{1/2}\cos q - \frac{p}{\gamma}\gamma'\cos 2q\right], \tag{4.5.85a}$$

$$\dot{q} = \gamma + \epsilon\left[f\left(\left(\frac{2p}{\gamma}\right)^{1/2}\sin q, (2p\gamma)^{1/2}\cos q\right)\frac{\sin q}{(2p\gamma)^{1/2}} - \frac{\gamma'}{2\gamma}\sin 2q\right], \tag{4.5.85b}$$

where $' = d/d\tilde{t}$.

b. Specialize (4.5.85) to the case where $f = \beta(dx/dt) + x^3$ (see (4.3.240)) and derive the solution correct to $O(\epsilon)$. Verify that your results agree with those calculated in Problem 3 of Sec. 4.3.

3.

a. Show that the oscillator (4.2.61) studied in Sec. 4.2.4 has the standard form

$$\dot{p} = \epsilon\left(-12p^2 + 16p^2\cos 2q - \frac{1}{2}p^2\cos 4q\right) + \epsilon^2(-3\nu p - 3\nu p\cos 2q), \tag{4.5.86a}$$

$$\dot{q} = 1 + \epsilon\left(\frac{1}{2}p\sin 2q + \frac{1}{4}p\sin 4q\right) + \epsilon^2\left(\frac{3}{2}\nu\sin 2q\right) \tag{4.5.86b}$$

with initial conditions $p = 1/2$ and $q = \pi/2$ at $t = 0$ corresponding to (4.2.62).

b. Argue that the nonuniformity as $\tilde{t} \to \infty$ is due to the assumption, inherent in (4.5.86a), that $3\nu\epsilon^2 p \ll 12\epsilon p^2$. Show that the rescaling in terms of which these terms are of the same order is $p^* = p/\epsilon$, which corresponds to the rescaling $y^* = y/\sqrt{\epsilon}$ used in (4.2.68).

c. Calculate the two-scale expansion for the rescaled system for p^* and q that follows from (4.5.86) and compare your results with (4.2.74)–(4.2.75).

4. Consider the pair of coupled weakly nonlinear oscillators (see (4.3.188))

$$\frac{d^2x_1}{dt^2} + \gamma_1^2 x_1 = \epsilon x_2^2, \tag{4.5.87a}$$

$$\frac{d^2x_2}{dt^2} + \gamma_2^2 x_2 = 2\epsilon x_1 x_2, \tag{4.5.87b}$$

where γ_1 and γ_2 are given functions of $\tilde{t}$ such that $\gamma_1 \neq 0$, $\gamma_2 \neq 0$, and $\gamma_1 \neq 2\gamma_2$, $\gamma_1 \neq \gamma_2$.

a. Show that the standard form that results for (4.5.87) is

$$\dot{p}_1 = \epsilon \left\{ \frac{p_2}{\gamma_2} \left(\frac{2p_1}{\gamma_1} \right)^{1/2} \left[\cos q_1 - \frac{1}{2} \cos(q_1 + q_2) - \frac{1}{2} \cos(q_1 - 2q_2) \right] - p_1 \frac{\gamma_1'}{\gamma_1} \cos 2q_1 \right\}, \tag{4.5.88a}$$

$$\dot{p}_2 = \epsilon \left\{ \frac{1}{4} \left(\frac{2p_1}{\gamma_1} \right)^{1/2} \left(\frac{2p_2}{\gamma_2} \right) [\cos(q_1 - 2q_2) + \cos(q_1 + 2q_2)] - p_2 \frac{\gamma_2'}{\gamma_2} \cos 2q_2 \right\}, \tag{4.5.88b}$$

$$\dot{q}_1 = \gamma_1 + \epsilon \left\{ - \frac{p_2}{\gamma_2 (2\gamma_1 p_1)^{1/2}} \left[\sin q_1 - \frac{1}{2} \sin(q_1 + 2q_2) - \frac{1}{2} \sin(q_1 - 2q_2) \right] + \frac{\gamma_1'}{2\gamma_1} \sin 2q_1 \right\}, \tag{4.5.88c}$$

$$\dot{q}_2 = \gamma_2 + \epsilon \left\{ - \frac{1}{\gamma_2} \left(\frac{2p_1}{\gamma_1} \right)^{1/2} \left[\sin q_1 - \frac{1}{2} \sin(q_1 + 2q_2) - \frac{1}{2} \sin(q_1 - 2q_2) \right] + \frac{\gamma_2'}{2\gamma_2} \sin 2q_2 \right\}. \tag{4.5.88d}$$

b. Consider the case $\gamma_1 =$ const., $\gamma_2 =$ const. first and calculate the solution for the p_i and q_i to $O(\epsilon)$. In particular, show that the $p_i^{(0)}$ and $\phi_i^{(0)}$ are constants. Then show that the expressions in (4.5.62) for $m = 1, 2$ have no average part ($\gamma_i^{(1)} = 0$), hence the $\theta_i^{(1)}$ are also constants. Finally, show that the $\underline{\delta_i}^{(1)} \neq 0$; hence the $\phi_i^{(1)}$ have terms that are linear in $\tilde{t}$. Show that the expressions you compute for the slowly varying phase shifts $\phi_1^{(1)}$ and $\phi_2^{(1)}$ agree exactly with the results given in (4.3.207) when the notation is reconciled. Exhibit resonance in the solution to $O(\epsilon)$ if $\gamma_1 = 2\gamma_2$ and in the solution to $O(\epsilon^2)$ if $\gamma_1 = \gamma_2$.

c. For the case where γ_1 and γ_2 are functions of $\tilde{t}$, calculate the solution to $O(\epsilon)$. Note that in this case $\gamma_1 = 2\gamma_2$ may occur at some time $\tilde{t} = \tilde{t}_0$, hence we refer to this as "passage through resonance."

5. The following system, which is a special case of (4.5.6), models the mathematical behavior of a more complicated problem that arises in flight mechanics (see [4.1] and the references cited there)

$$\frac{d^2x}{dt^2} + [\omega^2(\tilde{t}) + p^2]x = 0, \tag{4.5.89a}$$

$$\frac{dp}{dt} = \epsilon \omega^2 x \sin \psi, \tag{4.5.89b}$$

$$\frac{d\psi}{dt} = \sqrt{2}p. \tag{4.5.89c}$$

a. Show that (4.5.81) has the standard form

$$\dot{p}_1 = -\epsilon \frac{p_1 \omega \omega'}{\Omega^2} \cos 2q_1 + \epsilon \frac{\omega^2 p_2 (2p_1^3)^{1/2}}{4\Omega^{5/2}} [\cos(3q_1 + q_2) + \cos(q_1 - q_2) - \cos(3q_1 - q_2) - \cos(q_1 + q_2)], \tag{4.5.90a}$$

$$\dot{p}_2 = \epsilon \frac{\omega^2}{2} \left(\frac{2p_1}{\Omega}\right)^{1/2} [\cos(q_1 - q_2) - \cos(q_1 + q_2)], \tag{4.5.90b}$$

$$\dot{q}_1 = \Omega + \epsilon \frac{\omega \omega'}{2\Omega^2} \sin 2q_1 + \epsilon \frac{\omega^2 p_2 (2p_1)^{1/2}}{8\Omega^{5/2}} [\sin(3q_1 - q_2) + \sin(q_1 + q_2) - \sin(q_1 - q_2) - \sin(3q_1 + q_2)], \tag{4.5.90c}$$

$$\dot{q}_2 = \sqrt{2} p_2, \tag{4.5.90d}$$

where ω is a given function of $\tilde{t}$ and

$$\Omega^2 = \omega^2 + p_2^2, \quad p_2 = p, \quad q_2 = \psi. \tag{4.5.91}$$

b. Show that resonance occurs at $\tilde{t} = \tilde{t}_0$, where $\tilde{t}_0$ satisfies

$$\omega(\tilde{t}_0) = p_2(0). \tag{4.5.92}$$

Assume ω is a monotone increasing function of $\tilde{t}$ with $\omega(0) < p_2(0)$, and calculate the multiple-scale expansion to $O(\epsilon)$ for $0 \le \tilde{t} < \tilde{t}_0$.

6. Calculate the solution of the system (4.5.13) using the fast scale τ defined by

$$\frac{d\tau}{dt} = \omega(p^{(0)}, \tilde{t}) + \epsilon \nu(\tilde{t}) + \epsilon^2 \mu(\tilde{t}), \tag{4.5.93}$$

and the slow scale $\tilde{t} = \epsilon t$. Assume that $p(t;\epsilon)$ and $q(t;\epsilon)$ have the expansions

$$p(t;\epsilon) = p^{(0)} + \epsilon p^{(1)}(\tau, \tilde{t}) + \epsilon^2 p^{(2)}(\tau, \tilde{t}) + O(\epsilon^3), \tag{4.5.94a}$$

$$q(t;\epsilon) = \tau + \phi^{(0)}(\tilde{t}) + \epsilon q^{(1)}(\tau, \tilde{t}) + \epsilon^2 q^{(2)}(\tau, \tilde{t}) + O(\epsilon^3), \tag{4.5.94b}$$

where $p^{(0)}$ = constant as in (4.5.25).

a. Show that the right-hand side of (4.5.21) now has the added term $-\gamma(\tilde{t})$. Thus, by picking

$$\gamma(\tilde{t}) = \omega_p(p^{(0)}, \tilde{t})\theta^{(1)}(\tilde{t}), \tag{4.5.95}$$

we remove the $\theta^{(1)}$ dependence from (4.5.21) and find

$$\phi^{(0)} = \text{constant} = \phi^{(0)}(0). \tag{4.5.96}$$

b. Show that with the above choice, $\psi = \tau + \phi^{(0)}$ is exactly the same function of time as given in (4.5.40a).

c. Carry out the corresponding calculation for the right-hand side of (4.5.33) and determine μ by removing the term $\omega_p \theta^{(2)}$. Show that this choice does not affect the solution for $q(t;\epsilon)$ to $O(\epsilon)$.

References

4.1. D.L. Bosley and J. Kevorkian, "On the asymptotic solution of non-Hamiltonian systems exhibiting sustained resonance," *Stud. Appl. Math.*, **98**, 1995, pp. 83–130.

4.2. F.J. Bourland and R. Haberman, "The modulated phase shift for strongly nonlinear slowly varying, and weakly damped oscillators," *SIAM J. Appl. Math.*, **48**, 1988, pp. 737–748.

4.3. P.F. Byrd and M.D. Friedman, *Handbook of Elliptic Integrals for Engineers and Scientists*, 2nd edition, Spring-Verlag, New York, 1971.

4.4. J. Cochran, A new approach to singular perturbation problems, Ph.D. Thesis, Stanford University, Stanford, CA, 1962.

4.5. J.D. Cole and J. Kevorkian, "Uniformly valid asymptotic approximations for certain non-linear differential equations," *Proc. Internat. Sympos. Non-linear Differential Equations and Non-linear Mechanics.*, Academic Press, New York, 1963, pp. 113–120.

4.6. G. Contopoulos, "A third integral of motion in a galaxy," *Z. Astrophys.*, **49**, 1960, p. 273.

4.7. M.C. Eckstein and Y.Y. Shi, "Asymptotic solutions for orbital resonances due to the general geopotential," *Astron. J.*, **74**, 1969, pp. 551–562.

4.8. M.C. Eckstein, Y.Y. Shi, and J. Kevorkian, "Satellite motion for all inclinations around an oblate planet," *Proceedings of Symposium No. 25, International Astronomical Union*, Academic Press, New York, 1966, pp. 291–332.

4.9. M.C. Eckstein, Y.Y. Shi, and J. Kevorkian, "Satellite motion for arbitrary eccentricity and inclination around the smaller primary in the restricted three-body problem," *Astron. J.*, **71**, 1966, pp. 248–263.

4.10. M.C. Eckstein, Y.Y. Shi, and J. Kevorkian, "Use of the energy integral to evaluate higher-order terms in the time history of satellite motion," *Astron. J.*, **71**, 1966, pp. 301–305.

4.11. A. Erdelyi, *Asymptotic Expansions*, Dover Publications, New York, 1956.

4.12. B. Erdi, "The three-dimensional motion of trojan asteroids," *Celest. Mech.*,**18**, 1978, pp. 141–161.

4.13. G.I. Hori, "Nonlinear coupling of two harmonic oscillations," *Publ. Astron. Soc. Jpn.*, **19**, 1967, pp. 229–241.

4.14. H. Kabakow, A perturbation procedure for nonlinear oscillations, Ph.D. Thesis, California Institute of Technology, Pasadena, CA, 1968.

4.15. J. Kevorkian, "Perturbation techniques for oscillatory systems with slowly varying coefficients," *SIAM Rev.*, **29**, 1987, pp. 391–461.

4.16. J. Kevorkian, "Resonance in weakly nonlinear systems with slowly varying parameters," *Stud. Appl. Math.*, **62**, 1980, pp. 23–67.

4.17. J. Kevorkian, "The planar motion of a trojan asteroid," *Periodic Orbits, Stability, and Resonances*, G.E.O. Giacaglia (Editor), D. Reidel Publishing Company, Dordrecht, 1970, pp. 283–303.

4.18. J. Kevorkian, "The two variable expansion procedure for the approximate solution of certain nonlinear differential equation," *Lectures in Applied Mathematics, Vol. 7, Space Mathematics* (J.B. Rosser, Ed.), American Mathematical Society, 1966, pp. 206–275.

4.19. J. Kevorkian, The uniformly valid asymptotic representation of the solutions of certain non-linear ordinary differential equations, Ph.D. Thesis, California Institute of Technology, Pasadena, CA, 1961.

4.20. J. Kevorkian and Y.P. Li, "Explicit approximations for strictly nonlinear oscillators with slowly varying parameters with applications to free-electron lasers," *Stud. Appl. Math.*, **78**, 1988, pp. 111–165.
4.21. N.M. Krylov and N.N. Bogoliubov, *Introduction to Nonlinear Mechanics*, Princeton University Press, Princeton, 1957.
4.22. G.N. Kuzmak, "Asymptotic solutions of non-linear second order differential equations with variable coefficients," *Prikl. Math. Mech.*, **23**, 1959, pp. 515–526. Also appears in English translation.
4.23. P.-S. Laplace, *Mécanique Céleste*, Translated by Nathaniel Bowditch, Vol. 1, Book II, Chap. 15, p. 517, Hillard, Gray, Little, and Wilkins, Boston, 1829.
4.24. W. Lick, "Two-variable expansions and singular perturbation problems," *SIAM J. Appl. Math.*, **17**, 1969, pp. 815–825.
4.25. J.C. Luke, "A perturbation method for nonlinear dispersive wave problems," *Proc. R. Soc. London, Ser. A*, **292**, 1966, pp. 403–412.
4.26. J.J. Mahoney, "An expansion method for singular perturbation problems," *J. Australian Math. Soc.*, **2**, 1962, pp. 440–463.
4.27. J.A. Morrison, "Comparison of the modified method of averaging and the two variable expansion procedure," *SIAM Rev.*, **8**, 1966, pp. 66–85.
4.28. H. Poincaré, *Les Methodes Nouvelles de la Mécanique Celeste*, Vol. II, Dover, New York, 1957.
4.29. Lord Rayleigh, *Theory of Sound*, Second Edition, Dover, New York, 1945.
4.30. G. Sandri, "A new method of expansion in mathematical physics," *Nuovo Cimento*, **B36**, 1965, pp. 67–93.
4.31. G.G. Stokes, "On the Theory of Oscillatory Waves," *Cambridge Trans.*, **8**, 1847, pp. 441–473.
4.32. R.A. Struble, "A geometrical derivation of the satellite equation," *J. Math. Anal. Appl.*, **1**, 1960, p. 300.
4.33. M. Van Dyke, *Perturbation Methods in Fluid Mechanics*, Annotated Edition, Parabolic Press, Stanford, CA, 1975.
4.34. G.B. Whitham, "Two-timing, variational principles and waves," *J. Fluid Mech.*, **44**, 1970, pp. 373–395.
4.35. E.T. Whittaker, *Analytical Dynamics*, Cambridge University Press, London and New York, 1904.

5

Near-Identity Averaging Transformations: Transient and Sustained Resonance

In this chapter, we study another approach for calculating asymptotic solutions for systems in the standard form (4.5.1)

$$\frac{dp_m}{dt} = \epsilon F_m(p_i, q_i, \tilde{t}; \epsilon), \quad m = 1, 2, \ldots, M, \tag{5.1.1a}$$

$$\frac{dq_n}{dt} = \omega_n(p_i, \tilde{t}) + \epsilon G_n(p_i, q_i, \tilde{t}; \epsilon), \quad n = 1, 2, \ldots, N. \tag{5.1.1b}$$

Recall that $0 < \epsilon \ll 1, \tilde{t} = \epsilon t$, and the subscript i indicates that all M of the p_i or all N of the q_i are present in the argument. We assume that for each $m = 1, \ldots, M$ and each $n = 1, \ldots, N$ the F_m and G_n are $O(1)$ as $\epsilon \to 0$, and that the F_m and G_n are 2π-periodic functions of each of the q_i.

The basic idea is to transform the *dependent* variables p_m and q_n to new variables P_m and Q_n in terms of which the system (5.1.1) is as simple as possible to a given order in ϵ. One then solves this transformed system (exactly, asymptotically, or numerically) and uses this solution in the transformation relations linking the (p_m, q_n) to the (P_m, Q_n) to obtain the solution of the original problem. Since the system (5.1.1) is solvable if $\epsilon \equiv 0$, the transformation of dependent variables may be expressed as an asymptotic expansion that reduces to an *identity* transformation as $\epsilon \to 0$. Also, in attempting to obtain the simplest possible governing equation for the new (P_m, Q_n) variables, it is natural to remove all dependence on the Q_n from the transformed problem. Hence, the near-identity transformation one seeks is an *averaging* transformation with respect to the Q_n. As long as resonances are not present, we will show that this tactic of averaging out all the Q_n is effective for deriving asymptotic solutions and is particularly elegant when the system (5.1.1) is Hamiltonian. These ideas are discussed in Secs. 5.1–5.2. The remainder of this chapter is devoted to the study of solutions in resonance, for which the averaging approach needs to be modified and combined with notions from matched asymptotic expansions.

The technique that we discuss has evolved in scope and applicability from the original "method of averaging" proposed in 1937 by Krylov and Bogoliubov [5.18], or its generalizations given in [5.22], [5.23], and [5.25]. Our presentation throughout this chapter emphasizes the idea of near-identity transformations and is based on the point of view championed more recently, starting with [5.11], [5.14], and [5.24]. Actually, the idea of using near-identity transformations to simplify a system was implemented much earlier by von Zeipel [5.26] in his study of the motion of asteroids. We present a more general version of his technique for Hamiltonian systems in standard form in Secs. 5.2–5.3.

Our discussion of passage through resonance in Sec. 5.4 makes use of the basic ideas of matched asymptotic expansions covered in Chapter 2. The need to introduce an interior-layer expansion during resonance and to match this with pre-and post-resonance expansions was first recognized in [5.16] and further developed in [5.14]. A brief qualitative discussion of sustained resonance is given in Sec. 5.5 with references to more detailed recent studies and applications.

This chapter gives an expanded and updated account of many of the results in the expository paper [5.13]. Recent developments on very slowly varying Hamiltonian systems [5.3], on non-Hamiltonian systems [5.2], and on simultaneous resonances [5.27] are left out for brevity.

5.1 General Systems in Standard Form: Nonresonant Solutions

5.1.1 Linear Oscillator with Slowly Varying Frequency: Adiabatic Invariance

Consider the linear oscillator with slowly varying frequency discussed in Sec. 4.3.2

$$\frac{d^2y}{dt^2} + \omega^2(\tilde{t})y = 0, \tag{5.1.2}$$

where $\tilde{t} = \epsilon t$ and ω is a prescribed positive function.

We have various possible options for transforming (5.1.2) to standard form. One obvious choice is based on the solution of (5.1.2) for $\epsilon = 0$ (i.e., $\omega =$ const.). In this case, we have

$$y = \frac{(2E)^{1/2}}{\omega} \sin q, \quad \dot{y} = (2E)^{1/2} \cos q, \tag{5.1.3}$$

where E is the energy

$$E = \frac{1}{2}(\dot{y}^2 + \omega^2 y^2) = \text{constant} \tag{5.1.4a}$$

and q is the phase

$$q = \omega t + \psi; \quad \psi = \text{constant}. \tag{5.1.4b}$$

With $\epsilon \neq 0$, E is no longer constant, and (5.1.3) is not a solution. However, we may regard (5.1.3) as a *transformation of the dependent variables* from $(y, \dot{y})$ to

(E, q), (see (4.5.9)). The inverse transformation is now given by

$$E = \frac{1}{2}(\dot{y}^2 + \omega^2 y^2), \tag{5.1.5a}$$

$$q = \tan^{-1}(\omega y/\dot{y}). \tag{5.1.5b}$$

To compute the equations governing E and q, we proceed as in Sec. 4.5.1 and obtain the standard form system

$$\dot{E} = \epsilon \frac{\omega'}{\omega} E(1 - \cos 2q), \tag{5.1.6a}$$

$$\dot{q} = \omega + \frac{\epsilon\omega'}{2\omega} \sin 2q, \tag{5.1.6b}$$

where $' \equiv d/d\tilde{t}$.

As expected, E is not conserved, nor is the phase shift ψ defined by (5.1.4b). Based on our calculations in Sec. 4.3.2, we also observe that, although $\dot{E} = O(\epsilon)$, the approximation $E =$ const. $+\, O(\epsilon)$ is not uniformly valid to $O(1)$ over the interval $0 \le t \le T(\epsilon) = O(\epsilon^{-1})$ because of the presence of the average term $\epsilon(\omega'/\omega)E$ in the right-hand side of (5.1.6a).

It is natural to seek a new variable J instead of E such that the equation for $\dot{J}$ contains no average term. If we assume J in the general form

$$J = \Phi(E, \tilde{t}) \tag{5.1.7}$$

for an unspecified function Φ, (5.1.6a) gives

$$\dot{J} = \Phi_E \dot{E} + \epsilon \Phi_{\tilde{t}} = \epsilon \left(\frac{\omega'}{\omega} E \Phi_E + \Phi_{\tilde{t}} \right) - \epsilon \frac{\omega'}{\omega} E \Phi_E \cos 2q. \tag{5.1.8}$$

Therefore, the equation for J will be free of an average term if we choose

$$\frac{\omega'}{\omega} E \Phi_E + \Phi_{\tilde{t}} = 0. \tag{5.1.9}$$

The general solution of this linear equation for Φ is that Φ is an arbitrary function of E/ω, the action.

For simplicity, we choose $J = E/\omega$ and obtain the alternate standard form system

$$\dot{J} = -\frac{\epsilon\omega'}{\omega} J \cos 2q, \tag{5.1.10a}$$

$$\dot{q} = \omega + \epsilon \frac{\omega'}{2\omega} \sin 2q, \tag{5.1.10b}$$

where there is no average term in the equation for $\dot{J}$.

In contrast to the situation for E, the approximation $J =$ constant is asymptotically valid to $O(1)$ as $\epsilon \to 0$, *uniformly* in any interval $0 \le t \le T(\epsilon) = O(\epsilon^{-1})$. The property that $\dot{J}$ is a periodic function of q with zero average over one cycle of q holding $\tilde{t}$ fixed is referred to as *adiabatic* invariance.

More generally, consider the standard form system (5.1.1). Let $\mathcal{A}^{(k)}(p_i, q_i, \tilde{t}; \epsilon)$ be a given function of the individual arguments. If the p_i and q_i evolve according to (5.1.1), the time derivative of $\mathcal{A}^{(k)}$ is given by

$$\frac{d\mathcal{A}^{(k)}}{dt} = \epsilon \sum_{m=1}^{M} \frac{\partial \mathcal{A}^{(k)}}{\partial p_m} F_m + \sum_{n=1}^{N} \frac{\partial \mathcal{A}^{(k)}}{\partial q_n}(\omega_n + \epsilon G_n) + \epsilon \frac{\partial \mathcal{A}^{(k)}}{\partial \tilde{t}}$$

$$\equiv \mathcal{B}^{(k)}(p_i, q_i, \tilde{t}; \epsilon). \tag{5.1.11}$$

The function $\mathcal{A}^{(k)}(p_i, q_i, \tilde{t}; \epsilon)$ is said to be an adiabatic invariant of the system (5.1.10) to $O(\epsilon^k)$ as $\epsilon \to 0$ if

(i) $$\int_0^{2\pi} \cdots \int_0^{2\pi} \mathcal{B}^{(k)}(p_i, q_i, \tilde{t}; \epsilon) dq_1 \ldots dq_n = O(\epsilon^{k+2}) \tag{5.1.12a}$$

and

(ii) $$\mathcal{B}^{(k)}(p_i, q_i, \tilde{t}; \epsilon) = O(\epsilon^{k+1}). \tag{5.1.12b}$$

If $\mathcal{A}^{(k)}$ were an *exact* invariant, then $\mathcal{B}^{(k)} \equiv 0$ and (5.1.12) would be trivially satisfied. Thus, an adiabatic invariant is constant to a given order in ϵ in an asymptotic sense, and more importantly, the requirement (5.1.12a) on the average of $\mathcal{B}^{(k)}$ ensures that the error in assuming $\mathcal{A}^{(k)} =$ const. is $O(\epsilon^{k+1})$ *uniformly* in $0 \le t \le T(\epsilon) = O(\epsilon^{-1})$.

For the particular example (5.1.6) with $E \equiv p$, $J \equiv \mathcal{A}^{(0)}$, we see that $\mathcal{A}^{(0)} = p/\omega(\tilde{t})$ is an adiabatic invariant to $O(1)$ because (5.1.10a) gives

$$\dot{\mathcal{A}}^{(0)} = -\frac{\epsilon\omega'}{\omega^2} p \cos 2q \equiv \mathcal{B}^{(0)}, \tag{5.1.13}$$

and $\mathcal{B}^{(0)}$ (with $k = 0$) satisfies both conditions (5.1.12). For this special case, $\mathcal{A}^{(0)}$ is independent of q.

Improved adiabatic invariant

Having calculated an adiabatic invariant to $O(1)$ for the system (5.1.6), we now consider a procedure to improve this result and calculate an adiabatic invariant to $O(\epsilon)$. The idea is to transform J to a new variable, say J_1, such that the equation for $\dot{J}_1$ has a right-hand side that has zero average over q and is $O(\epsilon^2)$. In this example, we need not transform q to accomplish this.

Since J and J_1 agree to $O(1)$, the transformation occurs to $O(\epsilon)$ only. This so-called *near-identity transformation* is assumed in the form

$$J_1 = J + \epsilon T(J, q, \tilde{t}) \tag{5.1.14}$$

that reduces to the identity transformation $J_1 = J$ as $\epsilon \to 0$. The function T is as yet unknown; we propose to choose T such that J_1 is an adiabatic invariant to $O(\epsilon)$.

Differentiating (5.1.14) with respect to t gives

$$\dot{J}_1 = \dot{J} + \epsilon(T_J \dot{J} + T_q \dot{q}) + \epsilon^2 T_{\tilde{t}}. \tag{5.1.15}$$

Now, we use (5.1.10) to express $\dot{J}$ and $\dot{q}$ as functions of J and q to obtain

$$\dot{J}_1 = \epsilon\left(\omega T_q - \frac{\omega'}{\omega} J\cos 2q\right) + \epsilon^2\left(-\frac{\omega'}{\omega} J T_J \cos 2q + \frac{\omega'}{2\omega} T_q \sin 2q + T_{\tilde{t}}\right). \tag{5.1.16}$$

Once we express J in terms of J_1 (using the inverse of (5.1.14) in (5.1.16)), we will have the equation governing $\dot{J}_1$. We defer this step until we have determined T. First, we must eliminate the term of order ϵ in (5.1.16) because we want $\dot{J}_1 = O(\epsilon^2)$. This requirement defines T by quadrature in the form

$$T(J, q, \tilde{t}) = \frac{\omega'}{2\omega^2} J \sin 2q + \underline{T}(J, \tilde{t}), \tag{5.1.17}$$

where $\underline{T}$ is arbitrary. We now use (5.1.17) to evaluate what remains on the right-hand side of (5.1.16) and obtain

$$\dot{J}_1 = \epsilon^2\left[\frac{1}{2}\left(\frac{\omega'}{\omega^2}\right)' J\sin 2q + \underline{T}_{\tilde{t}} - \frac{\omega'}{\omega} J \underline{T}_J \cos 2q\right]. \tag{5.1.18}$$

In order that $J_1(J, q, \tilde{t})$ in (5.1.14) be an adiabatic invariant to $O(\epsilon)$, the right-hand side of (5.1.18) must have a zero average, i.e., $\underline{T}_{\tilde{t}} = 0$, and we choose $\underline{T} \equiv 0$ for simplicity. The adiabatic invariant to $O(\epsilon)$ for the system (5.1.10) is now given by (5.1.14) with $T = (\omega'/2\omega^2)J\sin 2q$

$$J = \mathcal{A}^{(1)}(J, q, \tilde{t}) \equiv J + \frac{\epsilon\omega'}{2\omega^2} J \sin 2q. \tag{5.1.19}$$

We may now write the governing system in terms of the (J_1, q) variables by inverting (5.1.19) asymptotically to any desired order and substituting the result for J into (5.1.18). Our procedure can also be extended to higher order. In general, one can also introduce a near identity transformation for q so that the resulting system is solvable. We discuss this procedure in Sec. 5.1.3 for the general system (5.1.1). We will also show that the preliminary transformation $E \to J_1$, which leads to a zero average for the right-hand side of (5.1.10a), is not necessary; adiabatic invariants can be computed as long as the averaged system for the p_i is solvable.

5.1.2 The Model Problem of Sec. 4.5.2

To illustrate ideas, we first study in detail the example problem that we discussed in Sec. 4.5.2. Instead of the two-scale expansion we used there, we now apply the method of near-identity averaging transformations.

The problem to be solved is

$$\frac{dp}{dt} = \epsilon A(p, \tilde{t}) \cos q, \tag{5.1.20a}$$

$$\frac{dq}{dt} = \omega(p, \tilde{t}) + \epsilon B(p, \tilde{t}) \sin q, \tag{5.1.20b}$$

where A, B, and $\omega \neq 0$ are given well-behaved functions of p and $\tilde{t} = \epsilon t$. The initial conditions are

$$p(0; \epsilon) = p_0, \tag{5.1.21a}$$

$$q(0; \epsilon) = q_0, \tag{5.1.21b}$$

where p_0 and q_0 are constants independent of ϵ. We assume the near-identity transformation $(p, q) \to (P, Q)$ defined to $O(\epsilon^2)$ by

$$P = p + \epsilon T(p, q, \tilde{t}) + \epsilon^2 U(p, q, \tilde{t}), \tag{5.1.22a}$$

$$Q = q + \epsilon L(p, q, \tilde{t}) + \epsilon^2 M(p, q, \tilde{t}). \tag{5.1.22b}$$

For future reference, we also need the inverse transformation that we express in the form

$$p = P + \epsilon R(P, Q, \tilde{t}) + \epsilon^2 S(P, Q, \tilde{t}) + O(\epsilon^3), \tag{5.1.23a}$$

$$q = Q + \epsilon V(P, Q, \tilde{t}) + \epsilon^2 W(P, Q, \tilde{t}) + O(\epsilon^3). \tag{5.1.23b}$$

To derive the functions R, S, V, and W, we substitute (5.1.23) into (5.1.22). In particular, upon expanding arguments in powers of ϵ, we have

$$T(p, q, \tilde{t}) = T(P, Q, \tilde{t}) + \epsilon T_p(P, Q, \tilde{t})R(P, Q, \tilde{t}) + \epsilon T_q(P, Q, \tilde{t})V(P, Q, \tilde{t}) + O(\epsilon^2) \tag{5.1.24}$$

and a similar expression for $L(p, q, \tilde{t})$. In addition,

$$U(p, q, \tilde{t}) = U(P, Q, \tilde{t}) + O(\epsilon), \tag{5.1.25a}$$

$$M(p, q, \tilde{t}) = M(P, Q, \tilde{t}) + O(\epsilon). \tag{5.1.25b}$$

When the above expansions for T, L, U, and M are substituted into (5.1.22) and terms of equal powers in ϵ are collected, we find R, S, V, and W. These results define the inverse transformation (5.1.23) as follows:

$$p = P - \epsilon T(P, Q, \tilde{t}) + \epsilon^2[-U(P, Q, \tilde{t}) + T_p(P, Q, \tilde{t})T(P, Q, \tilde{t}) + T_q(P, Q, \tilde{t})L(P, Q, \tilde{t})] + O(\epsilon^3) \tag{5.1.26a}$$

$$q = Q - \epsilon L(P, Q, \tilde{t}) + \epsilon^2[-M(P, Q, \tilde{t}) + L_p(P, Q, \tilde{t})T(P, Q, \tilde{t}) + L_q(P, Q, \tilde{t})L(P, Q, \tilde{t})] + O(\epsilon^3). \tag{5.1.26b}$$

We want to derive the equations that govern $(\dot{P}, \dot{Q})$ if (P, Q) are related to (p, q) according to (5.1.22) and $(\dot{p}, \dot{q})$ satisfy (5.1.20). First, we compute $\dot{P}$ and $\dot{Q}$ using (5.1.22) (*Note*: $\dot{} \equiv d/dt$):

$$\dot{P} = \dot{p} + \epsilon(T_p\dot{p} + T_q\dot{q}) + \epsilon^2(U_p\dot{p} + U_q\dot{q} + T_{\tilde{t}}) + O(\epsilon^3), \tag{5.1.27a}$$

$$\dot{Q} = \dot{q} + \epsilon(L_p\dot{p} + L_q\dot{q}) + \epsilon^2(M_p\dot{p} + M_q\dot{q} + L_{\tilde{t}}) + O(\epsilon^3). \tag{5.1.27b}$$

The arguments of T_p, T_q, U_p, U_q, L_q, M_p, and M_q are p, q, and $\tilde{t}$. If we now use (5.1.20) for $\dot{p}$ and $\dot{q}$, we find

$$\dot{P} = \epsilon(A\cos q + \omega T_q) + \epsilon^2(T_p A\cos q + T_q B\sin q + U_q\omega + T_{\tilde{t}})$$

$$+ O(\epsilon^3), \tag{5.1.28a}$$

$$\dot{Q} = \omega + \epsilon(B \sin q + \omega L_q) + \epsilon^2(L_q A \cos q + L_q B \sin q + M_q \omega + L_{\tilde{t}}) + O(\epsilon^3), \tag{5.1.28b}$$

where all the arguments on the right-hand side are still $p, q, \tilde{t}$. In order to derive equations for $\dot{P}$ and $\dot{Q}$, we need to express the arguments of the functions that appear on the right-hand sides of (5.1.28) in terms of P, Q. To do this, we substitute the inverse transformation (5.1.26), expand and collect terms. The result is

$$\dot{P} = \epsilon(A \cos Q + \omega T_q) + \epsilon^2(AL \sin Q - A_p T \cos Q - \omega_p T T_q + T_p A \cos Q + T_q B \sin Q + U_q \omega + T_{\tilde{t}} - \omega T_{qp} T - \omega T_{qq} L + \omega_p T_q T) + O(\epsilon^3), \tag{5.1.29a}$$

$$\dot{Q} = \omega + \epsilon(B \sin Q - \omega_p T + \omega L_q) + \epsilon^2\Big(-\omega_p U + \omega_p T_p T + \omega_p T_q L + \frac{\omega_{pp}}{2} T^2 - BL \cos Q - B_p T \sin Q - \omega L_{qp} T - \omega L_{qq} L - \omega_p T L_q + L_p A \cos Q + L_q B \sin Q + M_q \omega + L_{\tilde{t}}\Big) + O(\epsilon^3) \tag{5.1.29b}$$

Now, all the terms on the right-hand sides of (5.1.29) are evaluated at $p = P$, $q = Q$, and $\tilde{t}$. Equations (5.1.29) govern $\dot{P}$ and $\dot{Q}$ for a *general* near-identity transformation (5.1.22). The idea is to pick T, L, U, and M such that (5.1.29) are as simple as possible.

Consider the term of $O(\epsilon)$ in the $\dot{P}$ equation (5.1.29a). This term can be removed by setting

$$T_q(P, Q, \tilde{t}) = -\frac{A(P, \tilde{t})}{\omega(P, \tilde{t})} \cos Q.$$

Quadrature gives

$$T(p, q, \tilde{t}) = -\frac{A(p, \tilde{t})}{\omega(p, \tilde{t})} \sin q + \underline{T}(p, \tilde{t}), \tag{5.1.30}$$

where $\underline{T}$ is an arbitrary function of $\tilde{t}$, and in (5.1.30) we choose to express T in terms of the $p, q, \tilde{t}$ variables as in (5.1.22a). Next, we use (5.1.30) for T in the term of order ϵ on the right-hand side of the $\dot{Q}$ equation (5.1.29b) and obtain

$$\dot{Q} = \omega(P, \tilde{t}) + \epsilon \left\{ \left[B(P, \tilde{t}) + \frac{\omega_p(P, \tilde{t})}{\omega(P, \tilde{t})} A(P, \tilde{t}) \right] \sin Q + \omega(P, \tilde{t}) L_q(P, Q, \tilde{t}) - \omega_p(P, \tilde{t}) \underline{T}(P, \tilde{t}) \right\} + O(\epsilon^2). \tag{5.1.31}$$

The term of order ϵ on the right-hand side of (5.1.31) will be independent of Q if we set $\omega L_q = -[B + (\omega_p/\omega)A] \sin Q$. Quadrature then defines L in the form

$$L(p, q, \tilde{t}) = \frac{1}{\omega(p, \tilde{t})} \left[B(p, \tilde{t}) + \frac{\omega_p(p, \tilde{t})}{\omega(p, \tilde{t})} A(p, \tilde{t}) \right] \cos q + \underline{L}(p, \tilde{t}), \tag{5.1.32}$$

where $\underline{L}$ is an arbitrary function of p and $\tilde{t}$.

The initial conditions (5.1.21) used in conjunction with (5.1.26), (5.1.30), and (5.1.32) imply that

$$P(0) = p_0, \tag{5.1.33a}$$

$$Q(0) = q_0, \tag{5.1.33b}$$

and

$$\underline{T}(p_0, 0) = \frac{A(p_0, 0)}{\omega(p_0, 0)} \sin q_0, \tag{5.1.34a}$$

$$\underline{L}(p_0, 0) = -\frac{1}{\omega(p_0, 0)} \left[B(p_0, 0) + \frac{\omega_p(p_0, 0)}{\omega(p_0, 0)} A(p_0, 0) \right] \cos q_0. \tag{5.1.34b}$$

So far, we have been able to simplify the $(\dot{P}, \dot{Q})$ equations to the form

$$\dot{P} = O(\epsilon^2), \tag{5.1.35a}$$

$$\dot{Q} = \omega(P, \tilde{t}) - \epsilon\omega_p(P, \tilde{t})\underline{T}(P, \tilde{t}) + O(\epsilon^2), \tag{5.1.35b}$$

where the terms of $O(\epsilon^2)$ are periodic in Q but do not necessarily have a zero average. Consider now the effect of such average terms on the right-hand sides of (5.1.35). First, we conclude that

$$P = p_0 + O(\epsilon) \tag{5.1.36a}$$

because an average term (depending on P and $\tilde{t}$) to $O(\epsilon^2)$ will, upon integration, have an $O(\epsilon)$ contribution to P. Since P is known to $O(1)$ only, we can derive the leading contribution to Q only, and this is

$$Q = \frac{1}{\epsilon} \int_0^{\tilde{t}} \omega(p_0, s) ds. \tag{5.1.36b}$$

The results in (5.1.36) *do not suffice* to define (p, q) to $O(\epsilon)$ as is clear from (5.1.26); we need to define (P, Q) as functions of t to $O(\epsilon)$ and to determine $\underline{T}$ and $\underline{L}$. In fact, if ω depends on p *we will need to compute P correct to order ϵ^2*, say, $P = p_0 + \epsilon^2 P^{(2)}(\tilde{t})$ in order to be able to determine Q to $O(\epsilon)$. The detailed justification of this unpleasant situation will be given presently.

We can simplify the terms of order ϵ^2 in (5.1.29) by choosing U and M so as to remove all Q dependence to this order, and we will choose $\underline{T}$ and $\underline{L}$ so as to remove all the average terms. We now proceed with these calculations.

When known expressions for T and L are used to evaluate the term of order ϵ^2 on the right-hand side of (5.1.29a), we find

$$\dot{P} = \epsilon^2 \left[\omega U_q + \frac{1}{2\omega} \left(A^2 \frac{\omega_p}{\omega} - AA_p - AB \right) \sin 2Q + \underline{T}_p A \cos Q + \underline{T}_{\tilde{t}} \right] + O(\epsilon^3). \tag{5.1.37}$$

The only average term of order ϵ^2 on the right-hand side is $\underline{T}_{\tilde{t}}$, and we remove it by choosing $\underline{T}_{\tilde{t}} = 0$, i.e., $\underline{T} = \underline{T}(P)$. The choice of $\underline{T}$ is arbitrary and, as we will see, does not affect the final solution for p. One simple choice is $\underline{T}$ = constant, and we set this constant equal to the initial value given in (5.1.34a). The remaining oscillatory terms of order ϵ^2 are removed by choosing

$$U(p,q,\tilde{t}) = \frac{1}{4\omega^2}\left(A^2\frac{\omega_p}{\omega} - AA_p - AB\right)\cos 2q + \underline{U}, \tag{5.1.38}$$

where $\underline{U}$ is an arbitrary function of $p, \tilde{t}$.

Consider next (5.1.29b) for $\dot{Q}$. When the known expressions for T and L are used, we find

$$\begin{aligned}\dot{Q} &= \omega(P,\tilde{t}) - \epsilon\omega_p(P,\tilde{t})\underline{T} + \epsilon^2\Bigg\{\bigg(\underline{L}_{\tilde{t}} + \frac{AA_p}{2}\frac{\omega_p}{\omega^2} - \frac{\omega_p^2}{\omega^3}A^2 \\ &\quad - \frac{AB\omega_p}{\omega^2} + \frac{\omega_{pp}}{4\omega^2}A^2 + \underline{T}^2\frac{\omega_{pp}}{2} - \frac{B^2}{2\omega} + \frac{AB_p}{2\omega}\bigg) + \bigg[\omega M_q + A\underline{L}_p\cos Q \\ &\quad + \frac{1}{4\omega^3}\Big(2\omega^2B^2 + 2\omega^2AB_p + 3AA_p\omega\omega_p + AB\omega\omega_p \\ &\quad - 5A^2\omega_p^2 + 3A^2\omega\omega_{pp}\Big)\cos 2Q\bigg]\Bigg\} + O(\epsilon^3).\end{aligned} \tag{5.1.39}$$

We choose $\underline{L}_{\tilde{t}}$ to remove the average terms of order ϵ^2 and choose M to remove the oscillatory terms. Quadrature defines $\underline{L}$ in the form

$$\begin{aligned}\underline{L}(p,\tilde{t}) = \int_0^{\tilde{t}}\bigg[&-\frac{AA_p}{2}\frac{\omega_p}{\omega^2} + \frac{\omega_p^2}{\omega^3}A^2 + \frac{AB\omega_p}{\omega^2} - \frac{\omega_{pp}}{4\omega^2}A^2 - \underline{T}^2\frac{\omega_{pp}}{2} \\ &+ \frac{B^2}{2\omega} - \frac{AB_p}{2\omega}\bigg]ds + \underline{L}(p_0, 0).\end{aligned} \tag{5.1.40}$$

$$\begin{aligned}M(p,q,\tilde{t}) = &-\frac{A\underline{L}_p}{\omega}\sin q - \frac{1}{8\omega^4}\Big(2\omega^2B^2 + 2\omega^2AB_p + 3AA_p\omega\omega_p \\ &+ AB\omega\omega_p - 5A^2\omega_p^2 + 3A^2\omega\omega_{pp}\Big)\sin 2q + \underline{M}(p,\tilde{t}).\end{aligned} \tag{5.1.41}$$

In (5.1.40) the arguments are evaluated at $p = P$ and $\tilde{t} = s$, and $\underline{L}(p_0, 0)$ is the constant given by (5.1.34b). Also, we have ignored the arbitrary function of p that arises in (5.1.40) for simplicity. It is easily seen that this function will cancel out of the final result for $p(t;\epsilon)$.

We have now simplified the $(\dot{P}, \dot{Q})$ equations to read

$$\dot{P} = \epsilon^3[\underline{F}^{(3)}(P,\tilde{t};\epsilon) + \hat{F}^{(3)}(P,Q,\tilde{t};\epsilon)], \tag{5.1.42a}$$

$$\begin{aligned}\dot{Q} = \omega(P,\tilde{t}) - \epsilon\omega_p(P,\tilde{t})\frac{A(p_0,\tilde{t})}{\omega(p_0,\tilde{t})}\sin q_0 + \epsilon^3\Big[\underline{G}^{(3)}(P,\tilde{t};\epsilon) \\ + \hat{G}^{(3)}(P,Q,\tilde{t};\epsilon)\Big].\end{aligned} \tag{5.1.42b}$$

The average and oscillatory terms to $O(\epsilon^3)$ have not been evaluated. The terms $\hat{F}^{(3)}$, $\underline{G}^{(3)}$, and $\hat{G}^{(3)}$ do not contribute to the solution of $p(t;\epsilon)$ and $q(t;\epsilon)$ to $O(\epsilon)$.

However, the average term $\underline{F}^{(3)}$ *does have a contribution to* $q(t;\epsilon)$ *to* $O(\epsilon)$ *if* $\omega_p \neq 0$. To show this, let us first solve (5.1.42) asymptotically to $O(\epsilon)$, uniformly over $0 \le t \le T(\epsilon) = O(\epsilon^{-1})$. For example, we may use a two-scale $(\tau, \tilde{t})$ expansion with $\tau = \frac{1}{\epsilon}\int_0^{\tilde{t}} \omega(p_0, s)ds$. It is easily seen that

$$P(t;\epsilon) = p_0 + \epsilon^2 P^{(2)}(\tilde{t}) + O(\epsilon^3), \tag{5.1.43a}$$

where

$$P^{(2)}(\tilde{t}) = \int_0^{\tilde{t}} \underline{F}^{(3)}(p_0, s; 0)ds. \tag{5.1.43b}$$

If ω depends on p, $\omega(P,\tilde{t})$ has an $O(\epsilon^2)$ contribution involving $P^{(2)}$ when (5.1.43a) is used for P, and we find upon integrating (5.1.42b)

$$\begin{aligned} Q(t;\epsilon) = {} & \frac{1}{\epsilon}\int_0^{\tilde{t}} \omega(p_0, s)ds - \frac{A(p_0,\tilde{t})}{\omega(p_0,\tilde{t})}\sin q_0 \int_0^{\tilde{t}} \omega_p(p_0, s)ds + q_0 \\ & + \epsilon \int_0^{\tilde{t}} \omega_p(p_0, s)P^{(2)}(s)ds + O(\epsilon^2). \end{aligned} \tag{5.1.44}$$

This result shows explicitly that if $\omega_p \neq 0$ we need to know the average term of order ϵ^3 in the $\dot{P}$ equation in order to compute Q correct to $O(\epsilon)$. For the special case $\omega_p = 0$, i.e., $\omega = \omega(\tilde{t})$, we have the simple result

$$P = p_0 + O(\epsilon^2), \tag{5.1.45a}$$

$$Q = \frac{1}{\epsilon}\int_0^{\tilde{t}} \omega(s)ds + q_0 + O(\epsilon^2). \tag{5.1.45b}$$

It is not surprising that if ω depends on p we need to consider the average terms of $O(\epsilon^3)$ in the $\dot{P}$ equation in order to compute $q(t;\epsilon)$ to $O(\epsilon)$. The same essential feature was encountered in the two-scale solution of the problem discussed in Sec. 4.5.2 (see (4.5.35)).

To complete the solution for $p(t;\epsilon)$ and $q(t;\epsilon)$ correct to $O(\epsilon)$, we use the expressions (5.1.43) and (5.1.44) for $P(t;\epsilon)$ and $Q(t;\epsilon)$ and our results for T and L in (5.1.26) to find

$$\begin{aligned} p(t;\epsilon) = p_0 + \epsilon \Bigg[& \frac{A(p_0,\tilde{t})}{\omega(p_0,\tilde{t})} \sin\left(\frac{1}{\epsilon}\int_0^{\tilde{t}} \omega(p_0, s)ds \right. \\ & \left. - \frac{A(p_0,0)}{\omega(p_0,0)}\sin q_0 \int_0^{\tilde{t}} \omega_p(p_0, s)ds + q_0\right) \\ & - \frac{A(p_0,0)}{\omega(p_0,0)}\sin q_0 \Bigg] + O(\epsilon^2), \end{aligned} \tag{5.1.46a}$$

$$q(t;\epsilon) = \frac{1}{\epsilon}\int_0^{\tilde{t}} \omega(p_0, s) - \frac{A(p_0,0)}{\omega(p_0,0)}\sin q_0 \int_0^{\tilde{t}} \omega_0(p_0, s)ds + q_0$$

$$-\epsilon\left\{\left[\frac{B(p_0,\tilde{t})}{\omega(p_0,\tilde{t})}+A(p_0,\tilde{t})\frac{\omega_p(p_0,\tilde{t})}{\omega(p_0,\tilde{t})}\right]\cos\left(\frac{1}{\epsilon}\int_0^{\tilde{t}}\omega(p_0,s)ds\right.\right.$$
$$\left.-\frac{A(p_0,0)}{\omega(p_0,0)}\sin q_0\int_0^{\tilde{t}}\omega_p(p_0,s)ds+q_0\right)$$
$$\left.+\underline{L}(p_0,\tilde{t})-\int_0^{\tilde{t}}\omega_p(p_0,s)P^{(2)}(s)ds\right\}+O(\epsilon^2), \qquad (5.1.46b)$$

where $\underline{L}$ is given by (5.1.40).

For the special case $\omega=\omega(\tilde{t})$, the above simplifies to

$$p(t;\epsilon)=p_0+\epsilon\left[\frac{A(p_0,\tilde{t})}{\omega(\tilde{t})}\sin\left(\frac{1}{\epsilon}\int_0^{\tilde{t}}\omega(s)ds+q_0\right)-\frac{A(p_0,0)}{\omega(0)}\sin q_0\right]$$
$$+O(\epsilon^2), \qquad (5.1.47a)$$

$$q(t;\epsilon)=\frac{1}{\epsilon}\int_0^{\tilde{t}}\omega(s)ds+q_0-\epsilon\left\{\frac{B(p_0,\tilde{t})}{\omega(\tilde{t})}\cos\left(\frac{1}{\epsilon}\int_0^{\tilde{t}}\omega(s)ds+q_0\right)\right.$$
$$+\frac{1}{2}\int_0^{\tilde{t}}\left[\frac{B^2(p_0,s)}{\omega(s)}-\frac{A(p_0,s)B_p(p_0,s)}{\omega(s)}\right]ds$$
$$\left.-\frac{B(p_0,0)}{\omega(0)}\cos q_0\right\}+O(\epsilon^2). \qquad (5.1.47b)$$

Comparing (5.1.46a) with the corresponding expression (4.5.39a) calculated using a two-scale expansion, we see that the results are in exact agreement. The expressions (5.1.46b) and (4.5.39b) for $q(t;\epsilon)$ also agree if we can show that

$$\omega_p\theta^{(2)}=\frac{AA_p}{2}\frac{\omega_p}{\omega^2}-\frac{\omega_p^2A^2}{\omega^3}+\omega_p(p_0,\tilde{t})P^{(2)}(\tilde{t}). \qquad (5.1.48)$$

To do so requires that we compute $P^{(2)}$. The condition (5.1.48) is trivially satisfied if $\omega_p=0$.

It is important to note that the choice of the functions $\underline{T}$ and $\underline{L}$ *does not affect the final result* (5.1.46). In particular, had we let $\underline{T}(p,\tilde{t})$ be arbitrary, we would have computed the expression (5.1.37) for $\dot{P}$. In this expression, we can remove oscillatory terms with zero average by choosing U as in (5.1.38) except for the added term $-(\underline{T}_pA/\omega)\sin Q$. Equation (5.1.37) for $\dot{P}$ now reads

$$\dot{P}=\epsilon^2\underline{T}_{\tilde{t}}+O(\epsilon^2) \qquad (5.1.49a)$$

or

$$\frac{dP}{d\tilde{t}}=\epsilon\frac{\partial\underline{T}}{\partial\tilde{t}}(P,\tilde{t})+O(\epsilon^2). \qquad (5.1.49b)$$

Integrating this expression gives

$$P=p_0+\epsilon\underline{T}(p,\tilde{t}). \qquad (5.1.50)$$

Now we substitute (5.1.50) for P and (5.1.30) for T into (5.1.26a) to compute $p(t;\epsilon)$ to $O(\epsilon)$. It is easily seen that the $\epsilon \underline{T}$ contribution in (5.1.50) *exactly cancels out* the contribution due to $\underline{T}$ in (5.1.30), and the final expression for $p(t;\epsilon)$ is unaffected. A similar observation holds for $\underline{L}$ and for the additive functions $\underline{U}$ and $\underline{M}$ that arise to $O(\epsilon^2)$.

In particular, if we wish to calculate $p(t;\epsilon)$ and $q(t;\epsilon)$ correct to $O(\epsilon)$ only, we need not evaluate $\underline{M}$. The function $\underline{U}$ does not affect the expression for $p(t;\epsilon)$ to $O(\epsilon)$; it only contributes to the term of order ϵ in $q(t;\epsilon)$ if $\omega_p \neq 0$. In this case, we may either select $\underline{U}$ so as to eliminate $\underline{F}^{(3)}$ from (5.1.42a), or we may set $\underline{U} = 0$ but take the average terms of $O(\epsilon^3)$ in (5.1.42a) into account in the quadrature for $P(t;\epsilon)$. Either choice leads to the same $q(t;\epsilon)$.

We conclude our discussion of this example by noting that the expression (5.1.22a) linking P to p and q to $O(\epsilon)$ is an adiabatic invariant to $O(\epsilon)$. This follows from the fact that $\dot{P}$ has a zero average to $O(\epsilon^2)$ (in fact, $\dot{P} = O(\epsilon^3)$), see (5.1.11)–(5.1.12) (see Problem 5).

Computing an explicit solution by this method is significantly more tedious than the method of multiple scales. One advantage of the present approach is that it provides an adiabatic invariant.

5.1.3 General System in Many Degrees of Freedom

As in Sec. 4.5, it is convenient to decompose the F_m and G_n in (5.1.1) into average (underbar) and oscillatory (underhat) parts (see (4.5.2)–(4.5.3)). After we expand the result in powers of ϵ and retain terms up to $O(\epsilon^2)$, we have (see (4.5.4))

$$\begin{aligned}\frac{dp_m}{dt} =& \epsilon \underline{f}_m(p_i,\tilde{t}) + \epsilon \underset{\wedge}{f}_m(p_i,q_i,\tilde{t}) + \epsilon^2 \underline{k}_m(p_i,\tilde{t}) \\ &+ \epsilon^2 \underset{\wedge}{k}_m(p_i,q_i,\tilde{t}) + O(\epsilon^3), \quad m = 1,\ldots,M, \end{aligned} \tag{5.1.51a}$$

$$\begin{aligned}\frac{dq_n}{dt} =& \underline{\omega}_n(p_i,\tilde{t}) + \epsilon \underline{g}_n(p_i,\tilde{t}) + \epsilon \underset{\wedge}{g}_n(p_i,q_i,\tilde{t}) + \epsilon^2 \underline{\ell}_n(p_i,\tilde{t}) \\ &+ \epsilon^2 \underset{\wedge}{\ell}_n(p_i,q_i,\tilde{t}) + O(\epsilon^3), \quad n = 1,\ldots,N. \end{aligned} \tag{5.1.51b}$$

Near-identity transformations

Generalizing the approach we used in the previous section for a pair of scalar equations for p and q, we look for a near-identity transformation of *all* the dependent variables $\{p_i, q_i\}$ which will simplify (5.1.51) by removing the oscillatory terms with zero average from the transformed system to $O(\epsilon^2)$.

We denote the new variables by $\{P_i, Q_i\}$ and take the asymptotic expansion of the near-identity transformation from $\{p_i, q_i\}$ to $\{P_i, Q_i\}$ in the form

$$P_m = p_m + \epsilon T_m(p_i,q_i,\tilde{t}) + \epsilon^2 U_m(p_i,q_i,\tilde{t}) + O(\epsilon^3), \tag{5.1.52a}$$

$$Q_n = q_n + \epsilon L_n(p_i,q_i,\tilde{t}) + \epsilon^2 M_n(p_i,q_i,\tilde{t}) + O(\epsilon^3). \tag{5.1.52b}$$

Henceforth, we omit the reminder $m = 1,\ldots,M$ and $n = 1,\ldots,N$ in expressions such as (5.1.52).

We also need the inverse transformation and express this in the form

$$p_m = P_m + \epsilon R_m(P_i, Q_i, \tilde{t}) + \epsilon^2 S_m(P_i, Q_i, \tilde{t}) + O(\epsilon^3), \quad (5.1.53a)$$

$$q_n = Q_n + \epsilon V_n(P_i, Q_i, \tilde{t}) + \epsilon^2 W_n(P_i, Q_i, \tilde{t}) + O(\epsilon^3). \quad (5.1.53b)$$

Generalizing the calculations that led to (5.1.26) to include M p's and N q's gives

$$R_m = - T_m(P_i, Q_i, \tilde{t}), \quad (5.1.54a)$$

$$S_m = - U_m(P_i, Q_i, \tilde{t}) + \sum_{j=1}^{M} \frac{\partial T_m(P_i, Q_i, \tilde{t})}{\partial p_j} T_j(P_i, Q_i, \tilde{t}) + \sum_{k=1}^{N} \frac{\partial T_m(P_i, Q_i, \tilde{t})}{\partial q_i} L_k(P_i, Q_i, \tilde{t}), \quad (5.1.54b)$$

$$V_n = - L_n(P_i, Q_i, \tilde{t}), \quad (5.1.54c)$$

$$W_n = - M_n(P_i, Q_i, \tilde{t}) + \sum_{j=1}^{M} \frac{\partial L_n(P_i, Q_i, \tilde{t})}{\partial p_j} T_j(P_i, Q_i, \tilde{t}) + \sum_{k=1}^{N} \frac{\partial L_n(P_i, Q_i, \tilde{t})}{\partial q_k} L_k(P_i, Q_i, \tilde{t}). \quad (5.1.54d)$$

Thus, once the functions T_m, U_m, L_n, and M_n have been determined, (5.1.53) and (5.1.54) define the $\{p_i, q_i\}$ variables in terms of the $\{P_i, Q_i\}$ and $\tilde{t}$.

In preparation for transforming (5.1.51) to the $\{P_i, Q_i\}$ variables, we note the following expansions for typical terms appearing in the right-hand sides of this system.

An averaged function $\underline{f}(p_i, \tilde{t})$ such as $\underline{f}_m$, $\underline{\omega}_n$, or $\underline{g}_n$ has the expansion

$$\begin{aligned}\underline{f}(p_i, \tilde{t}) = {} & \underline{f}(P_i, \tilde{t}) - \epsilon \sum_{j=1}^{M} \frac{\partial \underline{f}(P_i, \tilde{t})}{\partial p_j} T_j(P_i, \tilde{t}) \\ & + \epsilon^2 \left\{ \sum_{j=1}^{M} \frac{\partial \underline{f}(P_i, \tilde{t})}{\partial p_j} \left[-U_j(P_i, Q_i, \tilde{t}) + \sum_{l=1}^{M} \frac{\partial T_j(P_i, Q_i, \tilde{t})}{\partial p_l} \cdot T_l(P_i, Q_i, \tilde{t}) + \sum_{k=1}^{N} \frac{\partial T_j(P_i, Q_i, \tilde{t})}{\partial q_k} L_k(P_i, Q_i, \tilde{t}) \right] \right. \\ & \left. + \frac{1}{2} \sum_{j=1}^{M} \sum_{r=1}^{M} \frac{\partial^2 \underline{f}(P_i, \tilde{t})}{\partial p_j \partial p_r} T_j(P_i, Q_i, \tilde{t}) T_r(P_i, Q_i, \tilde{t}) \right\} \\ & + O(\epsilon^3), \end{aligned} \quad (5.1.55a)$$

whereas an oscillatory term $\hat{f}$ such as $\hat{f}_m$ or $\hat{g}_m$ is of the form

$$\hat{f}(p_i, q_i, \tilde{t}) = \hat{f}(P_i, Q_i, \tilde{t}) - \epsilon \left\{ \sum_{j=1}^{M} \frac{\partial \hat{f}(P_i, Q_i, \tilde{t})}{\partial p_j} T_j(P_i, Q_i, \tilde{t}) + \sum_{k=1}^{N} \frac{\partial \hat{f}(P_i, Q_i, \tilde{t})}{\partial q_k} L_k(P_i, Q_i, \tilde{t}) \right\} + O(\epsilon^2). \quad (5.1.55b)$$

To compute expressions for $\dot{P}_m$ and $\dot{Q}_n$ (where $\dot{} = d/dt$), we differentiate the transformation relations (5.1.52) with respect to t:

$$\dot{P}_m = \dot{p}_m + \epsilon \left\{ \sum_{j=1}^{M} \frac{\partial T_m}{\partial p_j} \dot{p}_j + \sum_{k=1}^{N} \frac{\partial T_m}{\partial q_k} \dot{q}_k \right\} + \epsilon^2 \left\{ \sum_{j=1}^{M} \frac{\partial U_m}{\partial p_j} \dot{p}_j + \sum_{k=1}^{N} \frac{\partial U_m}{\partial q_k} \dot{q}_k + \frac{\partial T_m}{\partial \tilde{t}} \right\} + O(\epsilon^3) \quad (5.1.56a)$$

$$\dot{Q}_n = \dot{q}_n + \epsilon \left\{ \sum_{j=1}^{M} \frac{\partial L_n}{\partial p_j} \dot{p}_j + \sum_{k=1}^{N} \frac{\partial L_n}{\partial q_k} \dot{q}_k \right\} + \epsilon^2 \left\{ \sum_{j=1}^{M} \frac{\partial M_n}{\partial p_j} \dot{p}_j + \sum_{k=1}^{N} \frac{\partial M_n}{\partial q_k} \dot{q}_k + \frac{\partial L_n}{\partial \tilde{t}} \right\} + O(\epsilon^3). \quad (5.1.56b)$$

Now using (5.1.51) and the associated expansions (5.1.55) in (5.1.56) gives the transformed system to $O(\epsilon^2)$

$$\begin{aligned} \dot{P}_m = {} & \epsilon \left(\underline{f}_m + \hat{f}_m + \sum_{k=1}^{N} \frac{\partial T_m}{\partial q_k} \underline{\omega}_k \right) \\ & + \epsilon^2 \left\{ \underline{k}_m + \hat{k}_m + \frac{\partial T_m}{\partial \tilde{t}} + \sum_{k=1}^{N} \left[\frac{\partial U_m}{\partial q_k} \underline{\omega}_k + \frac{\partial T_m}{\partial q_k} (\underline{g}_k + \hat{g}_k) - \frac{\partial \hat{f}_m}{\partial q_k} L_k \right] \right. \\ & + \sum_{j=1}^{M} \left[-T_j \left(\frac{\partial \underline{f}_m}{\partial p_j} + \frac{\partial \hat{f}_m}{\partial p_j} \right) + \frac{\partial T_m}{\partial p_j} (\underline{f}_j + \hat{f}_j) \right] \\ & \left. - \sum_{k=1}^{N} \sum_{j=1}^{M} T_j \frac{\partial}{\partial p_j} \left(\frac{\partial T_m}{\partial q_k} \underline{\omega}_k \right) - \sum_{k=1}^{N} \sum_{l=1}^{N} \underline{\omega}_k \frac{\partial^2 T_m}{\partial q_k \partial q_l} L_l \right\} + O(\epsilon^3), \end{aligned} \quad (5.1.57a)$$

$$\dot{Q}_n = \underline{\omega}_n + \epsilon \left(\underline{g}_n + \hat{g}_n + \sum_{k=1}^{N} \frac{\partial L_n}{\partial q_k} \underline{\omega}_k - \sum_{j=1}^{M} \frac{\partial \underline{\omega}_n}{\partial p_j} T_j \right)$$

$$
\begin{aligned}
&+\epsilon^2 \left\{ \underline{l}_n + \underset{\wedge}{l}_n + \frac{\partial L_n}{\partial \tilde{t}} + \sum_{k=1}^{N} \left[\frac{\partial M_n}{\partial q_k} \underline{\omega}_k + \frac{\partial L_n}{\partial q_k} (\underline{g}_k + \underset{\wedge}{g}_k) - \frac{\partial \underset{\wedge}{g}_n}{\partial q_k} L_k \right] \right. \\
&+ \sum_{j=1}^{M} \left[-T_j \left(\frac{\partial \underline{g}_n}{\partial p_j} + \frac{\partial \underset{\wedge}{g}_n}{\partial p_j} \right) + \frac{\partial L_n}{\partial p_j} (\underline{f}_j + \underset{\wedge}{f}_j) - \frac{\partial \underline{\omega}_n}{\partial p_j} U_j \right] \\
&+ \sum_{k=1}^{N} \sum_{j=1}^{M} \left[\frac{\partial \underline{\omega}_n}{\partial p_j} \frac{\partial T_j}{\partial q_k} L_k - T_j \frac{\partial}{\partial p_j} \left(\frac{\partial L_n}{\partial q_k} \underline{\omega}_k \right) \right] - \sum_{k=1}^{N} \sum_{l=1}^{N} \underline{\omega}_k \frac{\partial^2 L_n}{\partial q_k \partial q_l} L_l \\
&+ \left. \sum_{j=1}^{M} \sum_{r=1}^{M} \left[\frac{1}{2} \frac{\partial^2 \underline{\omega}_n}{\partial p_j \partial p_r} T_j T_r + \frac{\partial \underline{\omega}_n}{\partial p_r} \frac{\partial T_r}{\partial p_j} T_j \right] \right\} + O(\epsilon^3), \qquad (5.1.57b)
\end{aligned}
$$

where in the arguments of all the terms appearing on the right-hand sides we set the $p_i = P_i$ and the $q_i = Q_i$.

Removal of oscillatory terms to $O(\epsilon)$

The idea is now to choose the unspecified functions T_m, L_n to $O(\epsilon)$ and U_m, M_n to $O(\epsilon^2)$ such that (5.1.57) are free of oscillatory terms. Thus, what remains, the so-called *averaged* equations, will only involve the P_i and $\tilde{t}$.

Clearly, to $O(\epsilon)$, we must set

$$
\sum_{k=1}^{N} \frac{\partial T_m}{\partial q_k} \underline{\omega}_k = -\underset{\wedge}{f}_m, \qquad (5.1.58)
$$

$$
\sum_{k=1}^{N} \frac{\partial L_n}{\partial q_k} \underline{\omega}_k = -\underset{\wedge}{g}_n + \sum_{j=1}^{M} \frac{\partial \underline{\omega}_n}{\partial p_j} \underset{\wedge}{T}_j. \qquad (5.1.59)
$$

In the last term of (5.1.59), we have anticipated that $T_j = \underset{\wedge}{T}_j + \underline{T}_j$ (see (5.1.60)).

For each m (5.1.58) gives a linear equation that only involves T_m and can therefore be easily integrated in the form

$$
T_m(p_i, q_i, \tilde{t}) = \underset{\wedge}{T}_m(p_i, q_i, \tilde{t}) + \underline{T}_m(p_i, \tilde{t}), \qquad (5.1.60a)
$$

where $\underset{\wedge}{T}_m$ is given by the quadrature

$$
\underset{\wedge}{T}_m(p_i, q_i, \tilde{t}) = -\left\{ \int \underset{\wedge}{f}_m(p_i, \underline{\omega}_i s, \tilde{t}) ds \right\}_{\underline{\omega}_i s = q_i}. \qquad (5.1.60b)
$$

The additive unspecified function $\underline{T}_m$ is to be determined by conditions on the $O(\epsilon^2)$ terms of (5.1.57a) and will be considered later.

Similarly, we can solve (5.1.59) in the form

$$
L_n(p_i, q_i, \tilde{t}) = \underset{\wedge}{L}_n(p_i, q_i, \tilde{t}) + \underline{L}_n(p_i, \tilde{t}), \qquad (5.1.61a)
$$

where

$$\underset{\wedge}{L}_n(p_i, q_i, \tilde{t}) = \left\{ \int \sum_{j=1}^{M} \frac{\partial \underline{\omega}_n}{\partial p_j} \underset{\wedge}{T}_j(p_i, \underline{\omega}_i s, \tilde{t}) - \underset{\wedge}{g}_n(p_i, \underline{\omega}_i s, \tilde{t}) ds \right\}_{\underline{\omega}_i s = q_i} . \tag{5.1.61b}$$

As discussed in Sec. 4.5.3, the integrals appearing in (5.1.60b) and (5.1.61b) consist of terms having divisors of the form

$$r_1\underline{\omega}_1 + r_2\underline{\omega}_2 + \ldots + r_N\omega_N,$$

where the r_i range over the negative and positive integers and zero (but all the r_i do not vanish). If such a typical divisor vanishes for certain nonzero values of the r_i we have a resonance, and our approach breaks down. We exclude this situation in the present development and will consider it in detail in Secs. 5.3–5.5.

Removal of oscillatory terms to $O(\epsilon^2)$

Having removed the oscillatory terms of order ϵ from (5.1.57), the remaining system can be further simplified by using the definitions for T_m and L_n.

In particular, by taking the partial derivative of (5.1.58) with respect to p_j, multiplying the result by T_j, and summing over j, we have

$$\sum_{j=1}^{M}\sum_{k=1}^{N} T_j \frac{\partial}{\partial p_j}\left(\frac{\partial T_m}{\partial q_k}\underline{\omega}_k\right) + \sum_{j=1}^{M} \frac{\partial \underset{\wedge}{f}_m}{\partial p_j} T_j = 0. \tag{5.1.62}$$

This result eliminates two of the $O(\epsilon^2)$ terms appearing on the right-hand side of (5.1.57a). Similarly, if we take the partial derivative of (5.1.58) with respect to q_l, multiply the result by L_l, and sum over l, we conclude that

$$\sum_{k=1}^{N}\sum_{l=1}^{N} \underline{\omega}_k \frac{\partial^2 T_m}{\partial q_k \partial q_l} L_l + \sum_{k=1}^{N} \frac{\partial \underset{\wedge}{f}_m}{\partial q_k} L_k = 0, \tag{5.1.63}$$

and this eliminates two more terms from the right-hand side of (5.1.57a).

Using the condition (5.1.59) on L_n provides the following two additional identities that can be used to simplify the $O(\epsilon^2)$ terms in (5.1.57b):

$$\sum_{j=1}^{M}\sum_{k=1}^{N} T_j \frac{\partial}{\partial p_j}\left(\frac{\partial L_n}{\partial q_k}\underline{\omega}_k\right) + \sum_{j=1}^{M} \frac{\partial \underset{\wedge}{g}_n}{\partial p_j} T_j - \sum_{j=1}^{M}\sum_{r=1}^{M} T_j \frac{\partial}{\partial p_j}\left(\frac{\partial \underline{\omega}_n}{\partial p_r}\underset{\wedge}{T}_r\right) = 0, \tag{5.1.64}$$

$$\sum_{k=1}^{N}\sum_{l=1}^{N} \underline{\omega}_k \frac{\partial^2 L_n}{\partial q_k \partial q_l} L_l + \sum_{l=1}^{N} \frac{\partial \underset{\wedge}{g}_n}{\partial q_l} L_l - \sum_{l=1}^{N}\sum_{j=1}^{M} \frac{\partial \underline{\omega}_n}{\partial p_j}\frac{\partial \underset{\wedge}{T}_j}{\partial q_l} L_l = 0. \tag{5.1.65}$$

Equations (5.1.57) now reduce to

$$\dot{P}_m = \epsilon \underline{f}_m + \epsilon^2 \left\{ \left(\sum_{k=1}^{N} \underline{\omega}_k \frac{\partial U_m}{\partial q_k} + \underset{\wedge}{A}_m + \underset{\wedge}{B}_m \right) + \left[\underline{A}_m + \underline{k}_m \right.\right.$$

$$+ \left. \left. \frac{\partial \underline{T}_m}{\partial \tilde{t}} + \sum_{j=1}^{M} \left(\frac{\partial \underline{T}_m}{\partial p_j} \underline{f}_j - \frac{\partial \underline{f}_m}{\partial p_j} \underline{T}_j \right) \right] \right\} + O(\epsilon^3), \tag{5.1.66a}$$

$$\dot{Q}_n = \underline{\omega}_n + \epsilon \left(\underline{g}_n - \sum_{j=1}^{M} \frac{\partial \underline{\omega}_n}{\partial p_j} \underline{T}_j \right) + \epsilon^2 \left\{ \left(\sum_{k=1}^{N} \underline{\omega}_k \frac{\partial M_n}{\partial q_k} + \underset{\wedge}{C}_n + \underset{\wedge}{D}_n \right) \right.$$

$$+ \left[\underline{C}_n + \underline{l}_n + \frac{\partial \underline{L}_n}{\partial \tilde{t}} + \sum_{j=1}^{M} \left(\frac{\partial \underline{L}_n}{\partial p_j} \underline{f}_j - \frac{\partial \underline{g}_n}{\partial p_j} \underline{T}_j - \frac{\partial \underline{\omega}_n}{\partial p_j} \underline{U}_j \right) \right.$$

$$+ \left. \left. \frac{1}{2} \sum_{j=1}^{M} \sum_{r=1}^{M} \frac{\partial}{\partial p_j} \left(\frac{\partial \underline{\omega}_n}{\partial p_r} \underline{T}_j \underline{T}_r \right) \right] \right\} + O(\epsilon^3), \tag{5.1.66b}$$

where we have introduced the notation

$$A_m = \underset{\wedge}{A}_m + \underline{A}_m = \sum_{j=1}^{M} \frac{\partial \underline{T}_m}{\partial p_j} \underset{\wedge}{f}_j + \sum_{k=1}^{N} \frac{\partial \underset{\wedge}{T}_m}{\partial q_k} \underset{\wedge}{g}_k, \tag{5.1.67a}$$

$$\underset{\wedge}{B}_m = \underset{\wedge}{k}_m + \frac{\partial \underset{\wedge}{T}_m}{\partial \tilde{t}} + \sum_{j=1}^{M} \left(\frac{\partial \underset{\wedge}{T}_m}{\partial p_j} \underline{f}_j + \frac{\partial \underline{T}_m}{\partial p_j} \underset{\wedge}{f}_j - \frac{\partial \underline{f}_m}{\partial p_j} \underset{\wedge}{T}_j \right)$$

$$+ \sum_{k=1}^{N} \frac{\partial \underset{\wedge}{T}_m}{\partial q_k} \underline{g}_k, \tag{5.1.67b}$$

$$C_n = \underset{\wedge}{C}_n + \underline{C}_n = -\frac{1}{2} \sum_{j=1}^{M} \sum_{r=1}^{M} \frac{\partial^2 \underline{\omega}_n}{\partial p_j \partial p_r} \underset{\wedge}{T}_j \underset{\wedge}{T}_r + \sum_{j=1}^{M} \frac{\partial \underset{\wedge}{L}_n}{\partial p_j} \underset{\wedge}{f}_j$$

$$+ \sum_{k=1}^{N} \frac{\partial \underset{\wedge}{L}_n}{\partial q_k} \underset{\wedge}{g}_k, \tag{5.1.67c}$$

$$\underset{\wedge}{D}_n = \underset{\wedge}{l}_n + \frac{\partial \underset{\wedge}{L}_n}{\partial \tilde{t}} + \sum_{j=1}^{M} \left(\frac{\partial \underset{\wedge}{L}_n}{\partial p_j} \underline{f}_j + \frac{\partial \underline{L}_n}{\partial p_j} \underset{\wedge}{f}_j - \frac{\partial \underline{g}_n}{\partial p_j} \underset{\wedge}{T}_j - \frac{\partial \underline{\omega}_n}{\partial p_j} \underset{\wedge}{U}_j \right)$$

$$+ \sum_{k=1}^{N} \frac{\partial \underset{\wedge}{L}_n}{\partial q_k} \underline{g}_k + \sum_{j=1}^{M} \sum_{r=1}^{M} \frac{\partial \underline{\omega}_n}{\partial p_r} \frac{\partial \underline{T}_r}{\partial p_j} \underset{\wedge}{T}_j. \tag{5.1.67d}$$

In (5.1.66b) and (5.1.67d), we have anticipated the form (5.1.69) in decomposing U_m. We eliminate the oscillatory terms of order ϵ^2 in (5.1.66a) by setting

$$\sum_{k=1}^{N} \underline{\omega}_k \frac{\partial U_m}{\partial q_k} + \underset{\wedge}{A}_m + \underset{\wedge}{B}_m = 0. \tag{5.1.68}$$

Solving this defines U_m in the form

$$U_m(p_i, q_i, \tilde{t}) = \underset{\wedge}{U}_m(p_i, q_i, \tilde{t}) + \underline{U}_m(p_i, \tilde{t}), \tag{5.1.69a}$$

where

$$\underset{\wedge}{U}_m(p_i, q_i, \tilde{t}) = -\left\{\int \left[\underset{\wedge}{A}_m(p_i, \underline{\omega}_i s, \tilde{t}) + \underset{\wedge}{B}_m(p_i, \underline{\omega}_i s, \tilde{t})\right] ds\right\}_{\underline{\omega}_i s = q_i} \tag{5.1.69b}$$

and $\underline{U}_m(p_i, \tilde{t})$ is to be chosen by conditions on the $O(\epsilon^3)$ terms of what remains of (5.1.66a).

Similarly, removing oscillatory terms of order ϵ^2 in (5.1.66b) provides the condition

$$\sum_{k=1}^{N} \underline{\omega}_k \frac{\partial M_n}{\partial q_k} + \underset{\wedge}{C}_n + \underset{\wedge}{D}_n = 0, \tag{5.1.70}$$

which defines M_n in the form

$$M_n(p_i, q_i, \tilde{t}) = \underset{\wedge}{M}_n(p_i, q_i, \tilde{t}) + \underline{M}_n(p_i, \tilde{t}), \tag{5.1.71a}$$

where

$$\underset{\wedge}{M}_n(p_i, q_i, \tilde{t}) = -\left\{\int [\underset{\wedge}{C}_n(p_i, \underline{\omega}_i s, \tilde{t}) + \underset{\wedge}{D}_n(p_i, \underline{\omega}_i s, \tilde{t})] ds\right\}_{\underline{\omega}_i s = q_i}. \tag{5.1.71b}$$

The averaged equations, the role of the arbitrary terms in the transformation

The transformed equations for $\dot{P}_m$ and $\dot{Q}_n$ now only depend on the P_i and $\tilde{t}$. These are the "averaged" equations, which are more appropriately written in terms of the $\tilde{t}$ derivative in the form:

$$\frac{dP_m}{d\tilde{t}} = \underline{f}_m + \epsilon\left[\underline{A}_m + \underline{k}_m + \frac{\partial \underline{T}_m}{\partial \tilde{t}} + \sum_{j=1}^{M}\left(\frac{\partial \underline{T}_m}{\partial p_j}\underline{f}_j - \frac{\partial \underline{f}_m}{\partial p_j}\underline{T}_j\right)\right] + O(\epsilon^2), \tag{5.1.72a}$$

$$\begin{aligned}\frac{dQ_n}{d\tilde{t}} = {} & \frac{\underline{\omega}_n}{\epsilon} + \underline{g}_n - \sum_{j=1}^{M} \frac{\partial \underline{\omega}_n}{\partial p_j}\underline{T}_j \\ & + \epsilon\left[\underline{C}_n + \underline{l}_n + \frac{\partial \underline{L}_n}{\partial \tilde{t}} + \sum_{j=1}^{M}\left(\frac{\partial \underline{L}_n}{\partial p_j}\underline{f}_j - \frac{\partial \underline{g}_n}{\partial p_j}\underline{T}_j - \frac{\partial \underline{\omega}_n}{\partial p_j}\underline{U}_j\right)\right. \\ & \left. + \frac{1}{2}\sum_{j=1}^{M}\sum_{r=1}^{M}\frac{\partial}{\partial p_j}\left(\frac{\partial \underline{\omega}_n}{\partial p_r}\underline{T}_j\underline{T}_r\right)\right] + O(\epsilon^2),\end{aligned} \tag{5.1.72b}$$

where we keep in mind that the arguments p_i are set equal to P_i in the terms on the right-hand sides.

Consider now the role of the additive functions $\underline{T}_m$ and $\underline{L}_n$ that arose in the solutions for T_m and L_n and that now remain in the averaged equations (5.1.72). Our objective is to solve these equations in order to define the P_i and Q_i as functions of time and $M+N$ constants of integration. Using this result in (5.1.53) then determines the p_i and q_i as functions of t and $M+N$ constants and completes the solution of the original system (5.1.51).

The choice $\underline{T}_m = 0$, $\underline{L}_n = 0$ at the outset would have considerably simplified the calculations and the resulting averaged equations, without compromising the basic requirement of eliminating oscillatory terms. (It is not necessary to compute a *general* solution of (5.1.58) and (5.1.59); any particular solution accomplishes the averaging.) However, in general, even with the $\underline{T}_m$, $\underline{L}_n$, and $\underline{U}_m$ equal to zero *there still remain uncanceled average* terms of order ϵ that must be taken into account in (5.1.72). In particular, one must first solve the system

$$\frac{dP_m}{d\tilde{t}} = \underline{f}_m + \epsilon(\underline{A}_m + \underline{k}_m), \tag{5.1.73a}$$

either exactly or asymptotically to $O(\epsilon)$, before proceeding to the quadrature for the Q_n from

$$\frac{dQ_n}{d\tilde{t}} = \frac{\omega_n}{\epsilon} + \underline{g}_n + \epsilon(\underline{C}_n + \underline{l}_n). \tag{5.1.73b}$$

If the $O(1)$ problem in (5.1.73a) cannot be *solved exactly* and one has to resort to a numerical solution, the $O(\epsilon)$ terms are burdensome only to the extent of increasing computation time. In this case, the choice $\underline{T}_m = 0$, $\underline{L}_n = 0$ (and $\underline{U}_m = \underline{M}_n = 0$) is quite reasonable.

If, however, the system

$$\frac{dP_m}{d\tilde{t}} = \underline{f}_m(P_i, \tilde{t}) \tag{5.1.74}$$

can be solved exactly, one has the P_i in the form of an M-parameter family of curves:

$$P_m = \xi_m(\tilde{t}, \alpha_1, \ldots, \alpha_M), \tag{5.1.75}$$

where the α_i are M integration constants. In this case, we will show that *one can also solve exactly* the system of equations that results for the $\underline{T}_m$ by requiring the $O(\epsilon)$ terms in (5.1.72a) to vanish. As a result, *the P_i will be given by* (5.1.75) *to* $O(\epsilon)$, and the expression (5.1.53a) will not need to be further expanded to define the p_i as functions of time. This choice of $\underline{T}_m$ introduces a significant simplification in the computation of (5.1.53a) and will be adopted. Of course, the final expressions in (5.1.53) for the p_m and q_n as functions of time cannot depend on the choice of $\underline{T}_m$ and $\underline{L}_n$. This consistency statement is verified at the end of this section and was illustrated for the example discussed in Sec. 5.1.2.

As a first step in the calculation of $\underline{T}_m$, we note that on the solution curve (5.1.75) $\underline{T}_m$ is a function of $\tilde{t}$ and the α_m, and we denote this function by $T_m^*(\tilde{t}, \alpha_i)$, i.e.,

$$\underline{T}_m(\xi_i(\tilde{t}, \alpha_1, \ldots, \alpha_M), \tilde{t}) \equiv T_m^*(\tilde{t}, \alpha_i). \tag{5.1.76a}$$

Therefore,

$$\frac{dT_m^*}{d\tilde{t}} = \sum_{j=1}^{M} \frac{\partial \underline{T}_m(\xi_i, \tilde{t})}{\partial p_j} \frac{\partial \xi_j}{\partial \tilde{t}} + \frac{\partial \underline{T}_m(\xi_i, \tilde{t})}{\partial \tilde{t}}, \tag{5.1.76b}$$

and using the fact that the ξ_j satisfy (5.1.74), we have

$$\frac{dT_m^*}{d\tilde{t}} = \sum_{j=1}^{M} \frac{\partial \underline{T}_m(\xi_i, \tilde{t})}{\partial p_j} \underline{f}_j(\xi_i, \tilde{t}) + \frac{\partial \underline{T}_m(\xi_i, \tilde{t})}{\partial \tilde{t}}. \tag{5.1.76c}$$

If we use (5.1.76c), the requirement that (5.1.72a) be free of $O(\epsilon)$ terms reduces to solving the following linear system of *ordinary differential equations* for the T_m^*:

$$\frac{dT_m^*}{d\tilde{t}} - \sum_{j=1}^{M} \frac{\partial \underline{f}_m(\xi_i, \tilde{t})}{\partial p_j} T_j^* = -\underline{A}_m(\xi_i, \tilde{t}) - \underline{k}_m(\xi_i, \tilde{t}), \tag{5.1.77}$$

where the matrix $(\partial \underline{f}_m / \partial p_j)$ and the column vectors $\underline{A}_m$ and $\underline{k}_m$ are known functions of $\tilde{t}$ and the α_i.

It was pointed out by Kuzmak [4.22] and later by Luke [4.25] for a related second-order problem that (5.1.77) is solvable if one has the solution of (5.1.74) (see the discussion following (4.4.16)). This is not surprising since the homogeneous system (5.1.77) is simply the variational system associated with (5.1.74). In fact, the fundamental matrix for (5.1.77) is the $M \times M$ matrix

$$\{\Psi(\tilde{t}, \alpha_i)\}_{rs} = \frac{\partial \xi_r}{\partial \alpha_s}. \tag{5.1.78}$$

The determinant of $\{\Psi\}_{rs}$ is just the Jacobian of the system (5.1.75), which is nonzero because we have M independent solutions of (5.1.74). To show that each column vector of $\{\Psi\}_{rs}$ solves the homogeneous equation (5.1.77), we take the partial derivative of (5.1.74), written in the form

$$\frac{\partial \xi_r}{\partial \tilde{t}} = \underline{f}_r(\xi_i(\tilde{t}, \alpha_i), \tilde{t}), \tag{5.1.79a}$$

with respect to α_s. The result

$$\frac{\partial}{\partial \tilde{t}} \left(\frac{\partial \xi_r}{\partial \alpha_s} \right) = \sum_{j=1}^{M} \frac{\partial \underline{f}_r(\xi_i, \tilde{t})}{\partial p_j} \frac{\partial \xi_j}{\partial \alpha_s} \tag{5.1.79b}$$

is just the homogeneous equation (5.1.77).

Therefore, a particular solution of (5.1.77) for T_m^* is

$$T_m^*(\tilde{t}, \alpha_i) = \sum_{s=1}^{M} \sum_{j=1}^{M} \Psi_{ms}(\tilde{t}, \alpha_i) \int_0^{\tilde{t}} \Phi_{sj}(\sigma, \alpha_i) J_j(\sigma, \alpha_i) d\sigma, \tag{5.1.80a}$$

where Ψ_{ms} are the components of the fundamental matrix, Φ_{sj} are the components of its inverse, and

$$J_j(\tilde{t}, \alpha_i) = -(\underline{A}_j + \underline{k}_j). \tag{5.1.80b}$$

Moreover, the particular solution (5.1.80) for T_m^* is all one needs in order to eliminate the $O(\epsilon)$ terms in (5.1.72a); the homogeneous solution may be chosen as the trivial zero solution for simplicity.

To calculate $\underline{T}_m(P_i, \tilde{t})$, we first solve the system (5.1.75) for the M constants α_i in terms of P_i and $\tilde{t}$, say,

$$\alpha_m = \rho_m(P_i, \tilde{t}), \tag{5.1.81}$$

and then use (5.1.81) in (5.1.76a) to define $\underline{T}_m$ as

$$\underline{T}_m(P_i, \tilde{t}) = T_m^*(\tilde{t}, \rho_i(P_i, \tilde{t})). \tag{5.1.82}$$

We can also remove the terms of order ϵ in (5.1.72b) by setting

$$\frac{dL_n^*}{d\tilde{t}} = \underline{K}_n, \tag{5.1.83a}$$

where

$$L_n^*(\tilde{t}, \alpha_i) = \underline{L}_n(\xi_i(\tilde{t}, \alpha_1, \ldots, \alpha_M), \tilde{t}), \tag{5.1.83b}$$

$$\frac{dL_n^*}{d\tilde{t}} = \sum_{j=1}^{M} \frac{\partial \underline{L}_n(\xi_i, \tilde{t})}{\partial p_j} \frac{\partial \xi_j}{\partial \tilde{t}} + \frac{\partial \underline{L}_n(\xi_i, \tilde{t})}{\partial \tilde{t}}, \tag{5.1.83c}$$

$$\underline{K}_n = \sum_{j=1}^{M} \left[\frac{\partial \underline{\omega}_n}{\partial p_j} \underline{U}_j + \frac{\partial \underline{g}_n}{\partial p_j} \underline{T}_j \right] - \underline{C}_n - \underline{l}_n - \frac{1}{2} \sum_{j=1}^{M} \sum_{r=1}^{M} \frac{\partial}{\partial p_j} \left(\frac{\partial \underline{\omega}_n}{\partial p_r} \underline{T}_j \underline{T}_r \right). \tag{5.1.83d}$$

Now, all the terms on the right-hand side of (5.1.83a) except the $\underline{U}_j$ are known functions of $\tilde{t}$. If the procedure is to be continued to the next order, the $\underline{U}_j$ will be used to remove the average terms of order ϵ^3 in the equation for the (dP_m/dt). However, if the calculations are to be terminated at this stage, we set the $\underline{U}_j = 0$. But, *we must keep track of the average terms of order* ϵ^2 *in* (5.1.72a), because these terms contribute to the solution of the $q_n(t; \epsilon)$ to $O(\epsilon)$ if the $\underline{\omega}_n$ depend on the p_i. The choice of the $\underline{U}_j$ does not affect the final result. Integrating (5.1.83a) gives

$$L_n^*(\tilde{t}, \alpha_i) = \int_0^{\tilde{t}} K_n(s) ds + L_n^*(0, \alpha_i). \tag{5.1.84}$$

Having picked the $\underline{T}_m$ according to (5.1.77) to remove the terms of order ϵ in (5.1.72a), the P_m satisfy the system

$$\frac{dP_m}{d\tilde{t}} = \underline{f}_m(P_i, \tilde{t}) + \epsilon^2 \underline{F}_m{}^{(3)}(P_i, \tilde{t}) + \epsilon^2 \hat{F}_m{}^{(3)}(P_i, Q_i, \tilde{t}) + O(\epsilon^3). \tag{5.1.85a}$$

We have not computed the expressions for the $\underline{F}_m{}^{(3)}$ and $\hat{F}_m{}^{(3)}$. The asymptotic solution for the P_m correct to $O(\epsilon^2)$ then has the form

$$P_m(\tilde{t}; \epsilon) = \xi_m(\tilde{t}, \alpha_i) + \epsilon^2 P_m^{(2)}(\tilde{t}) + O(\epsilon^3), \tag{5.1.85b}$$

where the $P_m^{(2)}$ are the contributions due to the $\underline{F}_m{}^{(3)}$. To obtain the equation for the $(dQ_n/d\tilde{t})$, we substitute (5.1.84) for the L_n in (5.1.72b), set the $\underline{U}_i = 0$, and expand the $\underline{\omega}_n(P_i, \tilde{t})$ using (5.1.85b) to obtain

$$\frac{dQ_n}{d\tilde{t}} = \frac{\underline{\omega}_n(\xi_i, \tilde{t})}{\epsilon} + \underline{g}_n(\xi_i, \tilde{t}) - \sum_{j=1}^{M} \frac{\partial \underline{\omega}_n}{\partial p_j}(\xi_i, \tilde{t})\underline{T}_j(\xi_i, \tilde{t})$$
$$+ \epsilon \sum_{j=1}^{M} \frac{\partial \underline{\omega}_n}{\partial p_j}(\xi_i, \tilde{t})P_j^{(2)}(\tilde{t}) + O(\epsilon^2). \quad (5.1.85c)$$

Quadrature then defines the $Q_n(\tilde{t}; \epsilon)$ correct to $O(\epsilon)$ in the form

$$Q_n(t; \epsilon) = \frac{1}{\epsilon} \int_0^{\tilde{t}} \underline{\omega}_n(\xi_i(s, \alpha_i), s)ds + \int_0^{\tilde{t}} \Bigg[\underline{g}_n(\xi_i(s, \alpha_i), s)$$
$$- \sum_{j=1}^{M} \frac{\partial \underline{\omega}_n}{\partial p_j}(\xi_i(s, \alpha_i), s)\underline{T}_j(\xi_i(s, \alpha_i), s) \Bigg] ds + \beta_n$$
$$+ \epsilon \int_0^{\tilde{t}} \sum_{j=1}^{M} \frac{\partial \underline{\omega}_n}{\partial p_j}(\xi_i(s, \alpha_i), s)P_j^{(2)}(s)ds + O(\epsilon^2), \quad (5.1.85d)$$

where the β_n are N integration constants. The last term on the right-hand side of (5.1.85d) is the contribution due to the expansion of the $\underline{\omega}_i$. Note again that this term is absent if the ω_i are all independent of the p_i.

Using the expressions given by (5.1.85) for the $P_m(t; \epsilon)$ and $Q_n(t; \epsilon)$ in (5.1.33) defines the solution for the original variables p_m and q_n as functions of time and $M + N$ integration constants. This solution is summarized next.

Summary of results; comparison with results by multiple scales

We have computed the solution for $p(t; \epsilon)$ and $q(t; \epsilon)$ to $O(\epsilon)$ in the *implicit* form (see (5.1.53)–(5.1.54))

$$p_m(t; \epsilon) = P_m(t; \epsilon) - \epsilon[\hat{T}_m(P_i(t; \epsilon), Q_i(t; \epsilon), \tilde{t}) + \underline{T}_m(P_i(t; \epsilon), \tilde{t})] + O(\epsilon^2), \quad (5.1.86a)$$

$$q_n(t; \epsilon) = Q_n(t; \epsilon) - \epsilon[\hat{L}_n(P_i(t; \epsilon), Q_i(t; \epsilon), \tilde{t}) + \underline{L}_n(P_i(t; \epsilon), \tilde{t})] + O(\epsilon^2). \quad (5.1.86b)$$

If we use the expressions derived in (5.1.85) for the $P_m(t; \epsilon)$ and $Q_n(t; \epsilon)$ and retain terms up to $O(\epsilon)$ only, we obtain the following *explicit* formulas:

$$p_n(t; \epsilon) = \xi_m(\tilde{t}, \alpha_i) + \epsilon[P_m^{(1)}(\psi_i, \tilde{t}) + \theta_m^{(1)}(\tilde{t})] + O(\epsilon^2) \quad (5.1.87a)$$
$$q_n(t; \epsilon) = \psi_n + \epsilon[Q_n^{(1)}(\psi_i, \tilde{t}) + \phi_n^{(1)}(\tilde{t})] + O(\epsilon^2). \quad (5.1.87b)$$

In (5.1.87a), we have introduced the following notation in order to facilitate comparison with the results in Sec. 4.5.3. The asymptotic solution of the system (5.1.85a) to $O(\epsilon^2)$ defines the P_m in the form (see (5.1.85b)

$$P_m(\tilde{t}; \epsilon) = \xi_m(\tilde{t}, \alpha_i) + \epsilon^2 P_m^{(2)}(\tilde{t}) + O(\epsilon^3), \quad (5.1.88)$$

involving the M integration constants $\alpha_1, \ldots, \alpha_M$. Comparing (5.1.85a) with (4.5.73) shows that the ξ_m are the *same* functions in both cases. The fast phases ψ_n are given by (see (5.1.85d)

$$\psi_n = \tau_n + \eta_n, \tag{5.1.89}$$

where

$$\tau_n = \frac{1}{\epsilon} \int_0^{\tilde{t}} \underline{\omega}_n(\xi_i(s, \alpha_i), s) ds \tag{5.1.90}$$

and

$$\eta_n(\tilde{t}, \beta_i) = \int_0^{\tilde{t}} \left[\underline{g}_n(\xi_i(s, \alpha_i), s) + \sum_{j=1}^{M} \frac{\partial \underline{\omega}_n(\xi_i(s, \alpha_i), s)}{\partial p_j} \theta_j^{(1)}(s) \right] ds + \beta_n. \tag{5.1.91}$$

In (5.1.87a) and (5.1.91), we have denoted

$$-\underline{T}_m(\xi_i(\tilde{t}, \alpha_i), \tilde{t}) \equiv \theta_m^{(1)}(\tilde{t}). \tag{5.1.92}$$

We have also denoted

$$-\underset{\wedge}{T}_m(\xi_i(\tilde{t}, \alpha_i), \psi_i, \tilde{t}) \equiv \underset{\wedge}{P}_m{}^{(1)}(\psi_i, \tilde{t}), \tag{5.1.93}$$

$$-\underset{\wedge}{L}_n(\xi_i(\tilde{t}, \alpha_i), \psi_i, \tilde{t}) \equiv \underset{\wedge}{Q}_n{}^{(1)}(\psi_i, \tilde{t}), \tag{5.1.94}$$

$$-\underline{L}_n(\xi_i(\tilde{t}, \alpha_i), \tilde{t}) + \int_0^{\tilde{t}} \sum_{j=1}^{M} \frac{\partial \underline{\omega}_n(\xi_i(s, \alpha_i), s)}{\partial p_j} P_j^{(2)}(s) ds \equiv \phi_n^{(1)}(\tilde{t}). \tag{5.1.95}$$

We now show that the $\theta_m^{(1)}$, $\underset{\wedge}{P}_m{}^{(1)}$, $\underset{\wedge}{Q}_n{}^{(1)}$, and $\phi_n^{(1)}$ appearing in (5.1.87) are identical with their respective counterparts (bearing the same notation) appearing in (4.5.71).

Consider first the $\underset{\wedge}{P}_m{}^{(1)}$. According to (5.1.60b), these are given by

$$\underset{\wedge}{P}_m{}^{(1)}(\psi_i, \tilde{t}) = \left\{ \int \underset{\wedge}{f}_m(\xi_i, \underline{\omega}_i s + \eta_i, \tilde{t}) ds \right\}_{\underline{\omega}_i s = \tau_i}, \tag{5.1.96}$$

and this is the same expression as (4.5.74a). Similarly, multiplying (5.1.61b) by -1 and using (5.1.93)–(5.1.94) gives

$$\underset{\wedge}{Q}_n{}^{(1)}(\psi_i, \tilde{t}) = \left\{ \int \left[\sum_{j=1}^{M} \frac{\partial \underline{\omega}_n(\xi_i, \tilde{t})}{\partial p_j} P_j^{(1)}(\underline{\omega}_i s + \eta_i, \tilde{t}) + \underset{\wedge}{g}_n(\xi_i, \underline{\omega}_i s + \eta_i, \tilde{t}) \right] ds \right\}_{\omega_i s = \tau_i}. \tag{5.1.97}$$

This expression is identical with (4.5.75).

According to (5.1.92) and (5.1.76a), we have $\theta_m^{(1)} = -T_m^*$. Therefore, it follows from (5.1.77) that $\theta_m^{(1)}$ solves

$$\frac{d\theta_m^{(1)}}{d\tilde{t}} - \sum_{j=1}^{M} \frac{\partial \underline{f}_m(\xi_i, \tilde{t})}{\partial p_j} \theta_j^{(1)} = \underline{A}_m + \underline{k}_m, \tag{5.1.98}$$

and this agrees with (4.5.77) if we can show that

$$\underline{A}_m = \gamma_m^{(1)}. \tag{5.1.99}$$

Similarly, it follows from (5.1.95) and (5.1.83b) that

$$\frac{d\phi_n^{(1)}}{d\tilde{t}} = -\frac{dL_n^*}{d\tilde{t}} + \sum_{j=1}^{M} \frac{\partial \underline{\omega}_n(\xi_i(\tilde{t}, \alpha_i), \tilde{t})}{\partial p_j} P_j^{(2)}(\tilde{t}), \tag{5.1.100}$$

and using (5.1.83a) for $(dL^*/d\tilde{t})$ with $\underline{U}_j = 0$ gives

$$\frac{d\phi_n^{(1)}}{d\tilde{t}} = \sum_{j=1}^{M} \frac{\partial \underline{g}_n(\xi_i, \tilde{t})}{\partial p_j} \theta_j^{(1)} + \frac{1}{2} \sum_{j=1}^{M} \sum_{r=1}^{M} \frac{\partial}{\partial p_j} \left(\frac{\partial \underline{\omega}_n}{\partial p_r} \underline{T}_j \underline{T}_r \right) +$$

$$\sum_{j=1}^{M} \frac{\partial \underline{\omega}_n(\xi_i(\tilde{t}, \alpha_i), \tilde{t})}{\partial p_j} P_j^{(2)}(\tilde{t}) + \underline{C}_n + \underline{\ell}_n. \tag{5.1.101}$$

Therefore, the two equations (4.5.68) and (5.1.101) defining $\phi_n^{(1)'}$ agree if we can show that

$$\sum_{j=1}^{M} \frac{\partial \underline{\omega}_n(\xi_i, \tilde{t})}{\partial p_j} \theta_j^{(2)}(\tilde{t}) + \delta_n^{(1)} = \underline{C}_n + \sum_{j=1}^{M} \frac{\partial \underline{\omega}_n}{\partial p_j} P_j^{(2)}(\tilde{t})$$

$$+ \frac{1}{2} \sum_{j=1}^{M} \sum_{r=1}^{M} \frac{\partial \underline{\omega}_n}{\partial p_r} \frac{\partial}{\partial p_j} (\underline{T}_j, \underline{T}_r). \tag{5.1.102}$$

Once we have verified the two conditions (5.1.99) and (5.1.102), we will have shown the complete equivalence of our results to $O(\epsilon)$ by the methods of multiple scales and averaging.

The explicit verification of (5.1.99) is given in Sec. 4.4.3 of [5.13] and is omitted for brevity. It is also pointed out in this reference that $\underline{A}_m = \gamma_m^{(1)} = 0$ for the Hamiltonian problem. The verification of (5.1.102) remains open, as we have neither calculated the $\theta_m^{(2)}$ (using multiple scales) nor the $P_m^{(2)}$ (using averaging); we only computed $\theta^{(2)}$ for the example problem discussed in Sec. 4.5.2. If all the $\underline{\omega}_i$ are functions of $\tilde{t}$ only, (5.1.102) reduces to showing $\delta_n^{(1)} = C_n^{(1)}$, and this straightforward proof is analogous to the proof of (5.1.99).

If we proceed no further, (5.1.87a) defines $p_m(t;\epsilon)$ to $O(\epsilon)$ explicitly. Equation (5.1.87b) defines $q(t;\epsilon)$ to $O(\epsilon)$ explicitly if all the $\underline{\omega}_i$ depend only on $\tilde{t}$. If, however, the $\underline{\omega}_i$ also depend on the p_i, we need to know the $P_m^{(2)}(\tilde{t})$ in (5.1.88) in order to define $q(t;\epsilon)$ to $O(\epsilon)$. Once the $P_n^{(2)}(\tilde{t})$ are known (or if $(\partial\omega_i/\partial p_j) = 0$

for all i, j), the expressions in (5.1.87) give the uniformly valid expansions of the p_n and q_n to $O(\epsilon)$ for all t in $0 \leq t \leq T(\epsilon) = O(\epsilon^{-1})$.

To compute the $P_m^{(2)}$, we need to evaluate the average terms of order ϵ^2 in the equation (5.1.72a) for the $(dP_m/d\tilde{t})$. Extension of these results to higher order is conceptually straightforward but impractical in general. One can, however, make further progress for special examples such as the one discussed in Secs. 4.5.2 and 5.1.2, particularly with the help of symbolic manipulation software.

In summary, it is important to bear in mind that in order to determine the average terms to any given order in the solution for the p_m, one must consider the averaged differential equation for (dP_m/dt) to one higher order. The same is true for q_m if all the $\underline{\omega}_i$ are independent of the p_i. Otherwise, we need to derive the averaged differential equation for (dP_m/dt) to *two orders higher*.

We conclude this discussion by reiterating that the choice of the functions $\underline{T}_m$ and $\underline{L}_n$ does not affect the final result for the expansions of the $p_m(t;\epsilon)$ and $q_n(t;\epsilon)$ to $O(\epsilon)$. Thus, for example, the expression (5.1.87a) defining the $p_m(t;\epsilon)$ to $O(\epsilon)$ is the same whether we evaluate the P_m using (5.1.73a) (i.e., if we set the $\underline{T}_m = 0$) or if we use (5.1.74) (i.e., if we define the $\underline{T}_m$ through the solution of (5.1.77)). In the first instance, $R_m = -\underset{\wedge}{T}_m$, but the expression for the P_m in (5.1.53a) contains average terms of order ϵ. In the second instance, precisely the same average terms show up in R_m, and the P_m have no $O(\epsilon)$ average terms. The same is true for $\underline{L}_n$. This fact was illustrated for the example problem of Sec. 5.1.2 and will be highlighted again for a two-dimensional example at the end of this section.

Adiabatic invariants

The principal advantage of the technique we are studying in this section is that it leads in a natural way to M explicit adiabatic invariants as defined by the conditions (5.1.12).

In the equations (5.1.81) linking the α_m to the P_m and $\tilde{t}$, let us use (5.1.52a) to express the P_m in terms of the p_i, q_i, and $\tilde{t}$. Denoting the result by $\mathcal{A}_m$, we obtain

$$\mathcal{A}_m(p_i, q_i, \tilde{t}; \epsilon) \equiv \alpha_m = \rho_m(p_i, \tilde{t}) + \epsilon \sum_{j=1}^{M} \frac{\partial \rho_m(p_i, \tilde{t})}{\partial P_j} T_j(p_i, q_i, \tilde{t}). \quad (5.1.103)$$

To verify that the $\mathcal{A}_m$ are M adiabatic invariants to $O(\epsilon)$, we compute $d\alpha_m/dt$ using (5.1.81) to obtain

$$\frac{d\alpha_m}{dt} = \epsilon \left(\frac{\partial \rho_m}{\partial \tilde{t}} + \sum_{j=1}^{M} \frac{\partial \rho_m}{\partial P_j} \frac{dP_j}{d\tilde{t}} \right). \quad (5.1.104)$$

We have shown that the P_m obey (see (5.1.85a))

$$\frac{dP_m}{d\tilde{t}} = \underline{f}_m(P_i, \tilde{t}) + \epsilon^2[\underline{F}_m{}^{(3)}(P_i, \tilde{t}) + \underset{\wedge}{F}{}^{(3)}(P_i, Q_i, \tilde{t})] + O(\epsilon^3) \quad (5.1.105)$$

but have not computed the $\underline{F}_m{}^{(3)}$ and $\underset{\wedge}{F}{}^{(3)}$ explicitly. Thus, (5.1.104) reduces to

$$\frac{d\alpha_m}{dt} = \epsilon^3 \left[\underline{\lambda}_m(P_i, \tilde{t}) + \underset{\wedge}{\lambda}_m(P_i, Q_i, \tilde{t}) \right] + O(\epsilon^4), \tag{5.1.106}$$

where

$$\underline{\lambda}_m = \sum_{j=1}^{M} \frac{\partial \rho_m}{\partial P_j} \underline{F}_j{}^{(3)}, \quad \underset{\wedge}{\lambda}_m = \sum_{j=1}^{M} \frac{\partial \rho_m}{\partial P_j} \underset{\wedge}{F}_j{}^{(3)}. \tag{5.1.107}$$

Since the $P_m = p_m + O(\epsilon)$ and the $Q_n = q_n + O(\epsilon)$, (5.1.106) states that the $\mathcal{A}_m$ are adiabatic invariants to $O(\epsilon^2)$ if all the averages $\underline{F}_i{}^{(3)}$ vanish. If the $\underline{F}_i{}^{(3)} \neq 0$, the $\mathcal{A}_m$ are adiabatic to $O(\epsilon)$ only.

A model problem with $M = N = 2$

We illustrate the application of our results for the following model problem which is analogous to (4.5.90) but is algebraically somewhat simpler:

$$\dot{p}_1 = -\epsilon\psi(\tilde{t}) p_1^2 p_2 \cos(q_1 - q_2) - \epsilon\theta(\tilde{t}) p_1 p_2^2 \cos(q_1 + q_2) \equiv \epsilon \underset{\wedge}{f}_1, \tag{5.1.108a}$$

$$\dot{p}_2 = \epsilon\psi(\tilde{t}) p_1^2 \cos(q_1 - q_2) - \epsilon\theta(\tilde{t}) p_1 p_2^2 \cos(q_1 + q_2) + \epsilon d(\tilde{t}) p_2 \sin(q_1 - q_2) \equiv \epsilon \underset{\wedge}{f}_2, \tag{5.1.108b}$$

$$\dot{q}_1 = \phi(\tilde{t}) + 2\epsilon\psi(\tilde{t}) p_1 p_2 \sin(q_1 - q_2) + \epsilon\theta(\tilde{t}) p_2^2 \sin(q_1 + q_2) + \epsilon c(\tilde{t})(p_1/p_2) \cos q_2 \equiv \underline{\omega}_1 + \epsilon \underset{\wedge}{g}_1, \tag{5.1.108c}$$

$$\dot{q}_2 = p_2 + \epsilon\psi(\tilde{t}) p_1^2 \sin(q_1 - q_2) + 2\epsilon\theta(\tilde{t}) p_1 p_2 \sin(q_1 + q_2) \equiv \underline{\omega}_2 + \epsilon \underset{\wedge}{g}_2. \tag{5.1.108d}$$

Here $\phi(\tilde{t}) > 0$, $\psi(\tilde{t})$, $c(\tilde{t})$, and $d(\tilde{t})$ are prescribed arbitrarily, and the restriction that ϕ be nonzero is to avoid having a resonance at the outset associated with q_1.

We wish to solve (5.1.108) to $O(\epsilon)$ subject to the initial conditions $p_n(0) = \alpha_n$, $(\alpha_2 > 0)$ and $q_n(0) = \beta_n$. Again, we exclude $\alpha_2 = 0$ to avoid a resonance associated with the critical argument q_2 in (5.1.108c).

We note that if c and d are identically equal to zero, the system is associated with the Hamiltonian (see (4.5.5))

$$h(p_i, q_i, \tilde{t}; \epsilon) = \phi p_1 + p_2^2/2 + \epsilon\psi p_1^2 p_2 \sin(q_1 - q_2) + \epsilon\theta p_1 p_2^2 \sin(q_1 + q_2). \tag{5.1.109}$$

It follows from (5.1.60b) and (5.1.61b) that the oscillatory terms $\underset{\wedge}{T}_m$ and $\underset{\wedge}{L}_n$ in the near-identity transformation (5.1.52) are given by

$$\underset{\wedge}{T}_1(p_i, q_i, \tilde{t}) = \frac{\psi p_1^2 p_2}{\phi - p_2} \sin(q_1 - q_2) + \frac{\theta p_1 p_2^2}{\phi + p_2} \sin(q_1 + q_2), \tag{5.1.110a}$$

$$\underset{\wedge}{T}_2(p_i, q_i, \tilde{t}) = -\frac{\psi p_2 p_1^2}{\phi - p_2} \sin(q_1 - q_2) + \frac{\theta p_1 p_2^2}{\phi + p_2} \sin(q_1 + q_2)$$

$$+\frac{dp_2}{\phi - p_2}\cos(q_1 - q_2), \tag{5.1.110b}$$

$$\underset{\wedge}{L_1}(p_i, q_i, \tilde{t}) = \frac{2\psi p_1 p_2}{\phi - p_2}\cos(q_1 - q_2) + \frac{\theta p_2^2}{\phi + p_2}\cos(q_1 + q_2) - \frac{cp_1}{p_2^2}\sin q_2, \tag{5.1.110c}$$

$$\underset{\wedge}{L_2}(p_i, q_i, \tilde{t}) = \frac{\psi\phi p_1^2}{(\phi - p_2)^2}\cos(q_1 - q_2) + \frac{\theta p_1 p_2(2\phi + p_2)}{(\phi + p_2)^2}\cos(q_1 + q_2) + \frac{dp_2}{(\phi - p_2)^2}\sin(q_1 - q_2). \tag{5.1.110d}$$

We remove average terms of order ϵ^2 from the equations for the $\dot{P}_n$ and $\dot{Q}_n$ to define the $\underline{T}_n$ and $\underline{L}_n$. Thus, integrating (5.1.77) (with $\underline{f}_n = \underline{k}_n = 0$) gives

$$\underline{T}_n(p_i, \tilde{t}) = -\int_0^{\tilde{t}} \underline{A}_n(p_i, s)ds + \underline{T}_n(\alpha_i, 0), \tag{5.1.111}$$

where $\underline{A}_1$ and $\underline{A}_2$ follow from (5.1.67a) in the form

$$\underline{A}_1(p_i, \tilde{t}) = \frac{d\phi\psi p_1^2 p_2}{2(\phi - p_2)^2}, \tag{5.1.112a}$$

$$\underline{A}_2(p_i, \tilde{t}) = -\frac{d\phi\psi p_1^2 p_2}{2(\phi - p_2)^2}. \tag{5.1.112b}$$

We note for future reference that the $\underline{A}_n = 0$ if $d = 0$, and this case includes the Hamiltonian problem.

Integration of (5.1.83a) (with $\underline{U}_n = \underline{g}_n = \underline{\ell}_n = 0$) gives

$$\underline{L}_n(p_i, \tilde{t}) = -\int_0^{\tilde{t}} \underline{C}_n(p_i, s)ds + \underline{L}_n(\alpha_i, 0), \tag{5.1.113}$$

where $\underline{C}_1$ and $\underline{C}_2$ follow from (5.1.67c) in the form

$$\underline{C}_1(p_i, \tilde{t}) = \frac{\psi^2 p_1^2 p_2}{(\phi - p_2)^2}\left[\phi(2p_1 - 3p_2) + p_2(3p_2 - p_1)\right] - \frac{\theta^2 p_2^3}{2(\phi + p_2)^2}\left[\phi(4p_1 + p_2) + p_2(p_2 + 3p_1)\right], \tag{5.1.114a}$$

$$\underline{C}_2(p_i, \tilde{t}) = \frac{\phi\psi^2 p_1^3}{2(\phi - p_2)^3}\left[\phi(p_1 - 4p_2) + p_2(p_1 + 4p_2)\right] - \frac{\theta^2 p_1 p_2^2}{(\phi + p_2)^3}\left[\phi^2(3p_1 + 2p_2) + p_2(p_1 + p_2)(p_2 + 3\phi)\right] + \frac{d^2 p_2(\phi + p_2)}{2(\phi - p_2)^3}. \tag{5.1.114b}$$

The solution for the $P_n(t;\epsilon)$ and $Q_n(t;\epsilon)$ given in (5.1.85) for this example are

$$P_1 = \alpha_1 + \epsilon^2 P_1^{(2)}(\tilde{t}), \tag{5.1.115a}$$

$$P_2 = \alpha_2 + \epsilon^2 P_2^{(2)}(\tilde{t}), \tag{5.1.115b}$$

$$Q_1 = \frac{1}{\epsilon}\int_0^{\tilde{t}} \phi(s)ds + \beta_1, \tag{5.1.115c}$$

$$Q_2 = \alpha_2 t - \int_0^{\tilde{t}} \underline{T}_2(\alpha_i, s)ds + \beta_2 + \epsilon \int_0^{\tilde{t}} P_2^{(2)}(s)ds, \tag{5.1.115d}$$

where we have not computed $P_1^{(2)}$ and $P_2^{(2)}$. For this example, since $\partial\omega_i/\partial p_j = 0$ for all i, j except $i = j = 2$, only $P_2^{(2)}$ contributes to the solution of Q_2 to $O(\epsilon)$.

To compute the solution for the p_i and q_i as functions of time we use (5.1.86). The initial conditions imply

$$\underline{T}_n(\alpha_i, 0) = -\hat{T}_n(\alpha_i, \beta_i, 0), \tag{5.1.116a}$$

$$\underline{L}_n(\alpha_i, 0) = -\hat{L}_n(\alpha_i, \beta_i, 0). \tag{5.1.116b}$$

Thus, (5.1.87) give

$$p_1(t;\epsilon) = \alpha_1 - \epsilon\left[\hat{T}_1(\alpha_i, \tau_i + \eta_i, \tilde{t}) + \underline{T}_1(\alpha_i, \tilde{t})\right] + O(\epsilon^2), \tag{5.1.117a}$$

$$p_2(t;\epsilon) = \alpha_2 - \epsilon\left[\hat{T}_2(\alpha_i, \tau_i + \eta_i, \tilde{t}) + \underline{T}_2(\alpha_i, \tilde{t})\right] + O(\epsilon^2), \tag{5.1.117b}$$

$$q_1(t;\epsilon) = \psi_1 - \epsilon\left[\hat{L}_1(\alpha_i, \tau_i + \eta_i, \tilde{t}) + \underline{L}_1(\alpha_i, \tilde{t})\right] + O(\epsilon^2), \tag{5.1.117c}$$

$$q_2(t;\epsilon) = \psi_2 - \epsilon\left[\hat{L}_2(\alpha_i, \tau_i + \eta_i, \tilde{t}) + \underline{L}_2(\alpha_i, \tilde{t}) - \int_0^{\tilde{t}} P_2^{(2)}(s)ds\right] + O(\epsilon^2), \tag{5.1.117d}$$

where

$$\psi_n = \tau_n + \eta_n, \tag{5.1.118a}$$

$$\tau_1 = \frac{1}{\epsilon}\int_0^{\tilde{t}} \phi(s)ds, \tag{5.1.118b}$$

$$\tau_2 = \alpha_2 t, \tag{5.1.118c}$$

$$\eta_1 = \beta_1, \tag{5.1.118d}$$

$$\eta_2 = -\int_0^{\tilde{t}} \underline{T}_2(\alpha_i, s)ds + \beta_2. \tag{5.1.118e}$$

All the terms in (5.1.117) except for $P_2^{(2)}(\tilde{t})$ have been defined.

This completes the solution to $O(\epsilon)$. Since neither ϕ nor α_2 vanish individually, the only divisors that may vanish in our results are those in powers of $(\phi - \alpha_2)$. Thus, the solution is well defined as long as the resonance condition

$$\phi(\tilde{t}) - \alpha_2 = 0 \tag{5.1.119}$$

does not arise. This condition depends on the functional expression for $\phi(\tilde{t})$ as well as the initial value α_2 of p_2. For example, if $\phi(\tilde{t}) = 2 - e^{-\tilde{t}}$ and $\alpha_2 > 2$ or $\alpha_2 < 1$, the divisors $(\phi - \alpha_2)^m$, $m = 1, 2, 3$ in our results will never vanish. However, for values of α_2 in the interval $1 < \alpha_2 < 2$, we have zero divisors when $\tilde{t} = \tilde{t}_0 \equiv -\log(2 - \alpha_2)$.

According to the discussion following (5.1.103), we have two adiabatic invariants to $O(\epsilon)$

$$\mathcal{A}_n^{(1)}(p_i, q_i, \tilde{t}; \epsilon) = p_m + \epsilon \left[\underset{\wedge}{T}_n(p_i, q_i, \tilde{t}) + \underline{T}_n(p_i, \tilde{t}) \right]. \tag{5.1.120}$$

These are adiabatic to $O(\epsilon)$ only because the $\underline{F}_i{}^{(3)}$ in (5.1.105), which we have not evaluated, may be nonvanishing.

Again, these invariants are only valid as long as the resonance condition (5.1.119) does not hold. We shall explore the case for (5.1.119) and more generally the possibility of having $p_2 \approx \phi(\tilde{t})$ in Secs. 5.3 and 5.4.

To conclude this discussion, note that another point of view would have $\underline{T}_n$ and $\underline{L}_n$ equal to zero in (5.1.117). In fact, the near-identity transformation (5.1.86) would now read:

$$p_n = P_n^* - \epsilon \underset{\wedge}{T}_n(P_i^*, Q_i^*, \tilde{t}) + O(\epsilon^2), \tag{5.1.121a}$$

$$q_n = Q_n^* - \epsilon \underset{\wedge}{L}_n(P_i^*, Q_i^*, \tilde{t}) + O(\epsilon^2). \tag{5.1.121b}$$

In this case, the averaged equations obey (see (5.1.73)):

$$\dot{P}_n^* = \epsilon \underline{A}_n(P_i^*, \tilde{t}) + O(\epsilon^2), \tag{5.1.122a}$$

$$\dot{Q}_n^* = \underline{\omega}_n(P_i^*, \tilde{t}) + \epsilon \underline{C}_n(P_i^*, \tilde{t}) + O(\epsilon^2). \tag{5.1.122b}$$

Solving these to $O(\epsilon)$ gives

$$P_n^* = P_n - \epsilon \underline{T}_n(P_i, \tilde{t}) + O(\epsilon^2), \tag{5.1.123a}$$

$$Q_n^* = Q_n - \epsilon \underline{L}_n(P_i, \tilde{t}) + O(\epsilon^2), \tag{5.1.123b}$$

where the P_n, Q_n, $\underline{T}_n$, and $\underline{L}_n$ appearing in (5.1.123) are precisely those defined earlier. Therefore, the final expressions for the $p_n(t; \epsilon)$ and $q_n(t; \epsilon)$ are identical to $O(\epsilon)$ for both points of view.

A number of misprints appearing in Sec. 4.1 of [5.13] have been corrected in the results given in their section. In particular, the need to keep track of the $\underline{F}_m{}^{(3)}$ in (5.1.105), when the $\underline{\omega}_i$ depend on the p_i, was overlooked in this reference. Also, in equation (4.22b) of [5.13] (which corresponds to (5.1.72b), the term $-\sum_{j=1}^{M} \frac{\partial \omega_n}{\partial p_j} \underline{T}_j$ was incorrectly multiplied by ϵ causing errors in the subsequent expressions for the Q_n.

Problems

1. Consider the general weakly nonlinear oscillator (4.5.82) in the standard form (4.5.83). Calculate the solution correct to $O(\epsilon)$ using the general results derived in Sec. 5.1.3. Verify that these results agree with those found in Sec. 4.2.5.
2. Analyze the oscillator (4.5.85) for the case $f = \beta\dot{x} + x^3$ using the general results in Sec. 5.1.3. Verify that your solution to $O(\epsilon)$ agrees with the two-scale expansion derived in Problem 2 of Sec. 4.5.
3. Reconsider the oscillator (4.5.86) with initial conditions $p = 1/2$, $q = \pi$ at $t = 0$. Calculate the asymptotic solution correct to $O(\epsilon)$ using the general results in Sec. 5.1.3. In particular, identify the steps in your calculations that avoid the nonuniformity that was encountered in (4.2.67).
4. Consider Mathieu's equation

$$\frac{d^2x}{dt^2} + (\delta + \epsilon \cos t)x = 0 \tag{5.1.124}$$

for the case

$$\delta = 1 + \epsilon\delta_1 + \epsilon^2\delta_2 + \ldots, \quad \delta_i = \text{constant.} \tag{5.1.125}$$

Transform (5.1.124) to standard form and solve the resulting system by averaging near-identity transformation. Derive the conditions on δ_1 and δ_2 in order that solutions be stable.

5. Solve the system (5.1.20) asymptotically to $O(\epsilon)$ using the general results of Sec. 5.1.3 and verify that your solution agrees with (5.1.46).
6. According to the discussion following (5.1.103), the system (5.1.20) has the following adiabatic invariant to $O(\epsilon)$:

$$\mathcal{A}^{(1)}(p, q, \tilde{t}) = p + \epsilon T(p, q, \tilde{t}), \tag{5.1.126}$$

where (5.1.30) gives

$$T = -\frac{A(p, \tilde{t})}{\omega(p, \tilde{t})} \sin q + \underline{T}, \tag{5.1.127}$$

and based on the discussion following (5.1.37), $\underline{T}$ is an arbitrary function of p but does not depend on $\tilde{t}$.

Verify by direct differentiation and use of the governing system (5.1.20) that

$$\frac{d\mathcal{A}^{(1)}}{dt} = \mathcal{B}^{(1)}(p, q, \tilde{t}; \epsilon), \tag{5.1.128}$$

where $\mathcal{B}^{(1)} = O(\epsilon^2)$ and $\int_0^{2\pi} \mathcal{B}^{(1)}(p, q, \tilde{t}; \epsilon)dq = O(\epsilon^3)$.

7. Verify by direct differentiation and use of the governing system (5.1.108) that $\mathcal{A}_1^{(1)}$ and $\mathcal{A}_2^{(1)}$ defined by (5.1.120) are adiabatic invariants to $O(\epsilon)$.
8. Consider Keplerian motion in the plane.

a. Assume small eccentricity e, and express the dimensionless solution derived in Sec. 4.3.4 (see (4.3.112)–(4.3.115)) in the series form

$$r = a\left[1 - e\cos M + \frac{e^2}{2}(1 - \cos 2M) + O(e^3)\right], \tag{5.1.129a}$$

$$\theta = \omega + M + 2e\sin M + \frac{5}{4}e^2\sin 2M + O(e^3), \tag{5.1.129b}$$

where $M = a^{-3/2}(t - \tau)$.

b. Now consider the problem of a satellite perturbed by a small drag force as given by (4.3.121). Instead of the variables $r, \dot{r}, \theta, \dot{\theta}$, choose a, e, ω, and τ, and regard these as functions of time. Derive the system of equations correct to $O(e)$ for $\dot{a}, \dot{e}, \dot{\omega}, \dot{\tau}$, and $\dot{M}$ as functions of a, e, ω, τ, and M in the standard form

$$\dot{a} = \epsilon f_1(a, e, \omega, \tau, M), \tag{5.1.130a}$$

$$\dot{e} = \epsilon f_2(a, e, \omega, \tau, M), \tag{5.1.130b}$$

$$\dot{\omega} = \epsilon f_3(a, e, \omega, \tau, M), \tag{5.1.130c}$$

$$\dot{\tau} = \epsilon f_4(a, e, \omega, \tau, M), \tag{5.1.130d}$$

$$\dot{M} = a^{-3/2} + \epsilon g(a, e, \omega, \tau, M), \tag{5.1.130e}$$

where $f_1, \ldots, g$ are 2π-periodic in M.

c. Solve the system (5.1.130) correct to $O(1)$ in ϵ and verify that your results agree with (4.3.134).

5.2 Hamiltonian System in Standard Form; Nonresonant Solutions

In this section, we consider the important special case where $M = N$ and the system (5.1.1) is Hamiltonian. More precisely, we assume that there exists a Hamiltonian function $h(p_i, q_i, \tilde{t}; \epsilon)$ that is 2π-periodic in the q_i such that the system (5.1.1) is of the form:

$$\frac{dp_n}{dt} = -\frac{\partial h}{\partial q_n}, \tag{5.2.1a}$$

$$\frac{dq_n}{dt} = \frac{\partial h}{\partial p_n}. \tag{5.2.1b}$$

In (5.2.1) and henceforth we omit the reminder $n = 1, \ldots, N$.

If we decompose h into its oscillatory and average parts and expand it in powers of ϵ, we have (see (4.5.2)–(4.5.3))

$$\begin{aligned} h(p_i, q_i, \tilde{t}; \epsilon) = {} & \underline{h}_0(p_i, \tilde{t}) + \epsilon \underline{h}_1(p_i, \tilde{t}) + \epsilon \hat{h}_1(p_i, q_i, \tilde{t}) \\ & + \epsilon^2 \underline{h}_2(p_i, \tilde{t}) + \epsilon^2 \hat{h}_2(p_i, q_i, \tilde{t}) + O(\epsilon^3). \end{aligned} \tag{5.2.2}$$

Thus, (5.2.1) becomes

$$\frac{dp_n}{dt} = -\epsilon \frac{\partial \underset{\wedge}{h_1}}{\partial q_n} - \epsilon^2 \frac{\partial \underset{\wedge}{h_2}}{\partial q_n} + O(\epsilon^3), \tag{5.2.3a}$$

$$\frac{dq_n}{dt} = \frac{\partial \underline{h}_0}{\partial p_n} + \epsilon \left(\frac{\partial \underline{h}_1}{\partial p_n} + \frac{\partial \underset{\wedge}{h_1}}{\partial p_n} \right) + \epsilon^2 \left(\frac{\partial \underline{h}_2}{\partial p_n} + \frac{\partial \underset{\wedge}{h_2}}{\partial p_n} \right) + O(\epsilon^3), \tag{5.2.3b}$$

and we identify the following partial derivatives of h with the terms appearing on the right-hand sides of (5.1.51):

$$\begin{gathered} \underline{f}_n = 0, \quad \underset{\wedge}{f}_n = -\frac{\partial \underset{\wedge}{h_1}}{\partial q_n}, \quad \underline{k}_n = 0, \quad \underset{\wedge}{k}_n = -\frac{\partial \underset{\wedge}{h_2}}{\partial q_n}, \\ \underline{\omega}_n = \frac{\partial \underline{h}_0}{\partial p_n}, \quad \underline{g}_n = \frac{\partial \underline{h}_1}{\partial p_n}, \quad \underset{\wedge}{g}_n = \frac{\partial \underset{\wedge}{h_1}}{\partial p_n}, \quad \underline{l}_n = \frac{\partial \underline{h}_2}{\partial p_n}, \quad \underset{\wedge}{l}_n = \frac{\partial \underset{\wedge}{h_2}}{\partial p_n}. \end{gathered} \tag{5.2.4}$$

Note that the requirement that h be periodic in the q_i excludes the occurrence of the averaged terms $\underline{f}_n$, $\underline{k}_n$, in the equation for $\dot{p}_n$ because such terms only result from contributions to h, which are *linear* in the q_i.

5.2.1 Summary of Basic Concepts

A discussion of the method of near-identity averaging transformation for the special case of the Hamiltonian system (5.2.3) requires knowledge of Hamiltonian mechanics, and we digress briefly here to review some of the basic concepts. This review is directed strictly toward the reader who has some familiarity with the subject. Otherwise, we urge first studying a standard text on dynamics, e.g., [5.8], where several chapters are devoted to the material in this section.

Hamiltonian systems

Consider a system of $2N$ first-order differential equations in the form

$$\frac{dp_n}{dt} = -\frac{\partial h}{\partial q_n}, \tag{5.2.5a}$$

$$\frac{dq_n}{dt} = \frac{\partial h}{\partial p_n}. \tag{5.2.5b}$$

We refer to the q_n as the *coordinates* and to the p_n as the *momenta* conjugate to the q_n; the function h is called the *Hamiltonian* and depends on the p_n, q_n, and t. Thus,

$$h = h(p_1, p_2, \ldots, p_N, q_1, q_2, \ldots, q_N, t). \tag{5.2.6}$$

A system of differential equations that is expressible in the special form (5.2.5) is called a *Hamiltonian system*. Many dynamical and other systems can be expressed in Hamiltonian form. Often, one obtains the system (5.2.5) from the N *Lagrange equations*

$$\frac{d}{dt}\left(\frac{\partial L}{\partial \dot{q}_n}\right) - \frac{\partial L}{\partial q_n} = 0 \qquad (5.2.7)$$

defined for a *Lagrangian function*

$$L = L(q_i, \dot{q}_i, t), \qquad (5.2.8)$$

by eliminating the $\dot{q}_n$ in favor of the q_n through a *Legendre* transformation

$$p_n = \frac{\partial L}{\partial \dot{q}_n}, \qquad (5.2.9)$$

and using the following definition of h:

$$h = \sum_{n=1}^{N} p_n \dot{q}_n - L. \qquad (5.2.10)$$

Thus, one first solves the system (5.2.9) for the $\dot{q}_n$ as functions of the p_n, q_n, and t, and then substitutes this in (5.2.10) to obtain $h(p_i, q_i, t)$.

It is important to note that not all dynamical systems can be expressed in Hamiltonian form. For example, for a dissipative system neither a Lagrangian nor a Hamiltonian exists. In this sense, the technique we present in this section is more restricted in applicability than the techniques discussed in Secs. 4.5 and 5.1.

The system (5.2.5) is a direct consequence of the *variational principle* (Hamilton's modified principle)

$$\delta \int_{t_1}^{t_2} \left[\sum_{n=1}^{N} p_n \dot{q}_n - h\right] dt = 0, \qquad (5.2.11)$$

subject to the requirements that t_1, t_2 as well as the values of the q_n, $\dot{q}_n$, and p_n at t_1 and t_2 be fixed.

Canonical transformations

We will now use the variational principle (5.2.11) to generate a *canonical transformation*. A canonical transformation is defined as a change of the variables (p_i, q_i) to a new set (P_i, Q_i) defined explicitly in the form

$$P_n = \Phi_n(p_i, q_i, t), \qquad (5.2.12a)$$

$$Q_n = \Psi_n(p_i, q_i, t), \qquad (5.2.12b)$$

such that the Hamiltonian structure is preserved in terms of the new (P_i, Q_i) variables. That is to say, the equations governing the (dP_n/dt) and (dQ_n/dt) that result from (5.2.5) under the transformation (5.2.12) must be derivable from a Hamiltonian function $H(P_i, Q_i, t)$ in order for (5.2.12) to be canonical:

$$\frac{dP_n}{dt} = -\frac{\partial H}{\partial Q_n}, \qquad (5.2.13a)$$

$$\frac{dQ_n}{dt} = \frac{\partial H}{\partial P_n}. \tag{5.2.13b}$$

In general, an arbitrary transformation of the form (5.2.12) is not canonical.

Since the system (5.2.5) is a consequence of the variational principle (5.2.11), and since a canonical transformation preserves the Hamiltonian form of the differential equations, it must be true that (5.2.13) is also a consequence of the variational principle

$$\delta \int_{t_1}^{t_2} \left[\sum_{n=1}^{N} P_n \dot{Q}_n - H \right] dt = 0, \tag{5.2.14}$$

subject to fixed endpoints (t_1, t_2) and fixed values of the Q_n, $\dot{Q}_n$, and P_n at these endpoints. Subtracting (5.2.11) from (5.2.14) then implies that the difference of the resulting integrands is the differential of some arbitrary function S. We will refer to S as a *generating function*, and it may depend on the $4N + 1$ variables p_i, q_i, P_i, Q_i, and t. However, since a canonical transformation defines $2N$ relations, we consider generating functions depending only on N old and N new variables plus the time. There are four possible choices for S: $S_1(q_i, Q_i, t)$, $S_2(q_i, P_i, t)$, $S_3(p_i, Q_i, t)$, and $S_4(p_i, P_i, t)$.

For example, if we select S_1, we have, as a consequence of subtracting (5.2.14) from (5.2.11),

$$\sum_{n=1}^{N} p_n \dot{q}_n - h(p_i, q_i, t) - \sum_{n=1}^{N} P_n \dot{Q}_n + H(P_i, Q_i, t) = \frac{d}{dt} S_1(q_i, Q_i, t)$$

$$= \sum_{n=1}^{N} \left(\frac{\partial S_1}{\partial q_n} \dot{q}_n + \frac{\partial S_1}{\partial Q_n} \dot{Q}_n \right) + \frac{\partial S_1}{\partial t}. \tag{5.2.15}$$

It then follows that we must set

$$p_n = \frac{\partial S_1}{\partial q_n}, \tag{5.2.16a}$$

$$P_n = -\frac{\partial S_1}{\partial Q_n}, \tag{5.2.16b}$$

$$H = h + \frac{\partial S_1}{\partial t}. \tag{5.2.16c}$$

It is easy to show that corresponding to each of the three remaining choices S_2, S_3, and S_4, there result analogous transformation relations. In the remainder of this section we will be concerned with generating functions of the type S_2 that depend on the old coordinates q_i, the new momenta P_i, and t. The expressions corresponding to (5.2.16) for S_2 take the form

$$p_n = \frac{\partial S_2}{\partial q_n}, \tag{5.2.17a}$$

$$Q_n = \frac{\partial S_2}{\partial P_n}, \tag{5.2.17b}$$

$$H = h + \frac{\partial S_2}{\partial t}. \qquad (5.2.17c)$$

The $2N + 1$ relations (5.2.16) or (5.2.17) define a canonical transformation *implicitly* in the sense that these relations do not correspond directly to the explicit definition (5.2.12). However, such an explicit transformation is easy to derive, in principle. Consider, for example, the relations (5.2.17). In order to obtain (5.2.12a), we solve the N equations (5.2.17a) for the P_n

$$p_n = \frac{\partial S_2}{\partial q_n}(q_i, P_i, t) \Longrightarrow P_n = \Phi_n(p_i, q_i, t). \qquad (5.2.18a)$$

We then use (5.2.18a) in (5.2.17b)

$$Q_n = \frac{\partial S_2}{\partial P_n}(q_i, \Phi_i(p_i, q_i, t), t) \equiv \Psi_n(p_i, q_i, t). \qquad (5.2.18b)$$

Finally, the transformed Hamiltonian is obtained from (5.2.17c) as follows:

$$\begin{aligned} H(P_i, Q_i, t) &\equiv h(\phi_i(P_i, Q_i, t), \psi_i(P_i, Q_i, t), t) \\ &\quad + \frac{\partial S_2}{\partial t}(\psi_i(P_i, Q_i, t), P_i, t), \end{aligned} \qquad (5.2.19)$$

where ϕ_i and ψ_i are the inverse transformations to (5.2.12), i.e.,

$$p_n = \phi_n(P_i, Q_i, t), \qquad (5.2.20a)$$

$$q_n = \psi_n(P_i, Q_i, t). \qquad (5.2.20b)$$

Hamilton-Jacobi equation

The fundamental role of canonical transformations is exhibited by the following observation. If we can find a canonical transformation, generated, for example, by a function $S_2(q_i, P_i, t)$ such that H vanishes identically, then (5.2.13) implies that the P_n and Q_n are all constant. Consequently, the transformation relations (5.2.20) define the solution of the original system (5.2.5) as a function of t and the $2N$ integration constants (P_i, Q_i).

If we ask how one would go about finding such an S_2, we see immediately from (5.2.17c) that S_2 must solve the first-order partial differential equation

$$0 = \frac{\partial S_2}{\partial t} + h\left(\frac{\partial S_2}{\partial q_i}, q_i, t\right). \qquad (5.2.21)$$

This is called the *Hamilton-Jacobi* equation associated with the Hamiltonian h. Solution techniques for this and similar nonlinear first-order equations are discussed in texts on partial differential equations, e.g., chapter 6 of [5.12]. A review of the main results of this theory would cause too much of a digression here, particularly since we will be concerned essentially with time-independent Hamiltonians for which a parallel theory applies.

We now show that one can always formally eliminate the time t from h by regarding t as the $N{+}1$st coordinate Q_{N+1} and introducing the appropriate conjugate momentum P_{N+1}. In fact, consider the following generating function S_2:

$$S_2 = \sum_{n=1}^{N} q_n P_n + t P_{N+1}. \tag{5.2.22}$$

Using the relations (5.2.17), we find

$$p_n = P_N, \ n = 1, \ldots, N, \tag{5.2.23a}$$

$$Q_n = q_n, \ n = 1, \ldots, N, \tag{5.2.23b}$$

$$Q_{N+1} = t, \tag{5.2.23c}$$

$$H = h(P_1, \ldots, P_N, Q_1, \ldots, Q_N, Q_{N+1}) + P_{N+1}. \tag{5.2.23d}$$

Consequently, given a Hamiltonian h depending on the N coordinates q_n, the N momenta p_n, and t, we can formally associate with it a time-independent Hamiltonian H of the $M = N+1$ coordinates $q_1, \ldots, q_N, t$, and $M = N+1$ momenta $p_1, \ldots, p_N, p_{N+1}$.

Thus, there is no loss of generality in studying the time-independent Hamiltonian. For such a Hamiltonian, h, assume that we have found a canonical transformation, $h \to H$, which renders H independent of the Q_n. It then follows from (5.2.13a) that the new momenta P_n are constants, and since the $(\partial H/\partial P_n)$ depend on the P_n only, these also are constants, say,

$$\frac{\partial H}{\partial P_n} = \nu_n = \text{constant}. \tag{5.2.24}$$

Equation (5.2.13b) then implies that the solution for the Q_n is simply

$$Q_n = \nu_n t + Q_n(0), \tag{5.2.25}$$

where the $Q_n(0)$ are also constants. Therefore, in this case the general solution of the system (5.2.5) for the q_n and p_n in terms of t and the $2N$ constants follows immediately from the transformation relations (5.2.20).

We also note that, in general, for any $h(p_i, q_i, t)$

$$\frac{dh}{dt} = \sum_{n=1}^{N} \left(\frac{\partial h}{\partial p_n} \dot{p}_n + \frac{\partial h}{\partial q_n} \dot{q}_n \right) + \frac{\partial h}{\partial t}, \tag{5.2.26a}$$

and using (5.2.5) for the $\dot{p}_n$ and $\dot{q}_n$, we find

$$\frac{dh}{dt} = \frac{\partial h}{\partial t}. \tag{5.2.26b}$$

Now if $(\partial h/\partial t) = 0$, it follows that h is an integral of the system (5.2.5), say,

$$h(p_i, q_i, t) = \alpha_1 = \text{constant}. \tag{5.2.27}$$

If we use the formulas (5.2.17) to characterize the generating function $W(q_i, P_i)$ of a transformation from $h(p_i, q_i)$ to $H(P_i, Q_i)$ such that H is independent of the Q_n, we find that W obeys

$$h\left(\frac{\partial W}{\partial q_1}, \frac{\partial W}{\partial q_2}, \ldots, \frac{\partial W}{\partial q_n}, q_1, q_2, \ldots, q_n \right) = \alpha_1. \tag{5.2.28}$$

This is the time-independent form of the Hamilton-Jacobi equation; it also follows from (5.2.21), for time-independent Hamiltonians, from the substitution $S_2 = W - \alpha_1 t$.

The general solution of (5.2.28) is the so-called *complete integral* in the form

$$W = W(q_i, \alpha_1, \alpha_2, \ldots, \alpha_{N-1}) + \alpha_N \tag{5.2.29}$$

involving the N constants, $\alpha_1, \alpha_2, \ldots, \alpha_N$. The new momenta P_n can be chosen as any N arbitrary functions of the N constants α_n. Note that since (5.2.28) does not involve W explicitly, we only need to find a solution involving the $N - 1$ independent constants $\alpha_1, \ldots, \alpha_{N-1}$; then the Nth constant is additive.

Finding the solution (5.2.29) is particularly simple if the Hamiltonian function h is separable for the given choice of variables in the following sense. Assume a solution of (5.2.28) for W in the form

$$W = \sum_{n=1}^{N} W_n(q_n, \alpha_1, \ldots, \alpha_N), \tag{5.2.30}$$

where each W_n involves only the one coordinate q_n. If substitution of (5.2.30) into (5.2.28) decomposes the latter into a sequence of N ordinary differential equations of the form

$$h_n\left(q_n \frac{\partial W_n}{\partial q_n}, \alpha_1, \ldots, \alpha_N\right) = \alpha_n, \tag{5.2.31}$$

then these can be individually solved by quadrature. Note that each h_n only depends on one q_n and the corresponding $\partial W_n / \partial q_n$. An example of a separable system is the motion of a particle in a central force field and is discussed in [5.8]. See also the next section for a discussion of the planar problem.

Whether a given Hamiltonian is separable in a given set of coordinates is a straightforward question to answer. However, the basic question of whether there exist appropriate coordinates with respect to which a given Hamiltonian system is separable is more difficult (in general not possible) to answer a priori. Since separability of the Hamiltonian implies the explicit solvability of the system (5.2.5), the answer to this question is of fundamental importance. For systems that are close to a solvable one, our goal will be to derive successively more accurate canonical transformations to variables such that the transformed Hamiltonian is independent of the coordinates (hence solvable) to any desired degree of accuracy.

Action and angle variables

In many practical applications, particularly in celestial mechanics, the unperturbed system is periodic and is described most concisely by action and angle variables, which we will consider next.

Assume that we have a time-independent Hamiltonian that is separable in some system of coordinates and momenta $\{q_n, p_n\}$. We call this system *periodic* if the motion in *each* of the 2-dimensional planes (q_n, p_n), $n = 1, \ldots, N$ describes either a simple closed curve (*libration*) or corresponds to the p_n being a periodic function of the q_n (*rotation*). Thus, for the case of libration both q_n and p_n must be periodic functions of t with the same period. For the case of rotation, only p_n is periodic in t with some given period T_n, whereas q_n has a secular component added to a T_n-periodic function, i.e., $\dot{q}_n$ is T_n-periodic in t with a nonzero average value. Note also that the periods involved in each (q_n, p_n) plane need not be the same.

Since the system is assumed to be separable, one can determine a priori what the projected motions in each of the (q_n, p_n) planes are without solving the problem. In fact, using (5.2.30) and (5.2.17a) with $S_2 = W$, and $(\partial/\partial t) = 0$, we have the following relations:

$$p_n = \frac{\partial W_n}{\partial q_n}(q_n, \alpha_1, \ldots, \alpha_N) \tag{5.2.32}$$

linking each pair (p_n, q_n) with the N constants $\alpha_1, \ldots, \alpha_N$. One could choose the new momenta P_n as N independent functions of the α_n. For this choice, (5.2.17), with $S_2 = W$ and $(\partial/\partial t) = 0$, defines a canonical transformation to a new Hamiltonian that does not involve the Q_n.

A particular choice of the P_n in terms of the α_n with some useful properties consists of letting $P_n = J_n$, the "action," defined by

$$J_n = \oint p_n dq_n, \tag{5.2.33}$$

where the integral is taken over one complete cycle in the (p_n, q_n) plane for *fixed values of* $\alpha_1, \ldots, \alpha_N$. Thus, for the case of libration, J_n is the area inside the closed (p_n, q_n) curve, whereas for the case of rotation it is simply the area under the p_n vs. q_n curve for one period. Using (5.2.32), J_n can be computed by

$$J_n = \oint \frac{\partial W_n}{\partial q_n}(q_n, \alpha_1, \ldots, \alpha_N) dq_n, \tag{5.2.34}$$

and this defines each J_n as a function of $\alpha_1, \ldots, \alpha_N$. The J_n as functions of the $\alpha_1, \ldots, \alpha_N$ are independent, and these functions can be inverted to express each α_n in terms of $J_1, \ldots, J_N$. Substituting this result into (5.2.30) for W defines this in the form

$$W = W(q_i, J_i), \tag{5.2.35}$$

which is one of the standard forms [see (5.2.17)] for a generating function. The new coordinates corresponding to the choice (5.2.33) are the "angle" variables w_i defined according to (5.2.17b) by

$$w_n = \frac{\partial W}{\partial J_n}, \tag{5.2.36}$$

where (5.2.35) is to be used in computing the partial derivatives.

The transformed Hamiltonian H is now a function of the J_n only:

$$H = H(J_i). \tag{5.2.37}$$

Hence (5.2.13b) for the angles w_n reduces to

$$\frac{dw_n}{dt} = \frac{\partial H}{\partial J_n} \equiv \nu_n(J_i). \tag{5.2.38}$$

The solution in terms of the action and angle variables is therefore given by

$$J_n = J_n(0), \tag{5.2.39a}$$

$$w_n = \nu_n(J_i)t + w_n(0), \tag{5.2.39b}$$

where the $J_n(0)$ and $w_n(0)$ are initial values.

Now consider the relation (5.2.36), which expresses the w_n in terms of the q_n and J_n. We wish to compute $\Delta_j w_i$, the change in a given w_i that results from (5.2.36) when q_j undergoes one complete cycle holding the remaining q_n and all the J_n fixed. Note that this change in w_i does not necessarily have to correspond to a solution in time; it is merely the change in the value of w_i that is predicted by (5.2.36). We have

$$\Delta_j w_i = \oint dw_i, \tag{5.2.40}$$

where the integral is computed over a complete cycle in w_i holding all the J_n and all the q_n except q_j fixed. Thus,

$$dw_i = \frac{\partial w_i}{\partial q_j} dq_j. \tag{5.2.41}$$

If we use (5.2.36) for w_i in (5.2.41) and substitute this in (5.2.40), we obtain

$$\Delta_j w_i = \oint \frac{\partial^2 W}{\partial q_j \partial J_i} dq_j. \tag{5.2.42}$$

Since the contour is for fixed values of $J_1, \ldots, J_N$, we may take the partial derivative with respect to J_i outside the integral and then use (5.2.32)–(5.2.33) to obtain

$$\Delta_j w_i = \frac{\partial}{\partial J_i} \oint \frac{\partial W}{\partial q_j} dq_j = \frac{\partial}{\partial J_i} \oint p_j dq_j = \frac{\partial J_j}{\partial J_i} = \delta_{ij}, \tag{5.2.43}$$

where δ_{ij} is the Kronecker delta. Thus, w_i changes by a unit amount only if q_i varies through a complete cycle; w_i returns to its original value if any of the other coördinates $q_j \neq q_i$ are varied. If the period associated with one cycle in q_i is T_i, (5.2.39b) and (5.2.43) imply

$$\nu_i = \frac{1}{T_i}. \tag{5.2.44}$$

Thus, we can compute the frequencies, $2\pi/T_i$, for each of the q_i motions directly from (5.2.44) without having to solve the system (5.2.5); all we need is the expression (5.2.37) for H in terms of the J_n.

In many applications, it is more convenient to normalize the action and angle variables so that

$$\Delta_j w_i = 2\pi\delta_{ij}. \tag{5.2.45}$$

This is accomplished by the canonical transformation $\{J_i, w_i\} \to \{J_i^*, w_i^*\}$ given by the rescaling

$$w_n^* = 2\pi w_n, \tag{5.2.46a}$$

$$J_n^* = \frac{1}{2\pi} J_n. \tag{5.2.46b}$$

For future reference, we also note that

$$\Delta_j W = J_j, \tag{5.2.47}$$

and this property of the action variable follows from the definition of $\Delta_j W$ and J_j in (5.2.34)

$$\Delta_j W = \oint dW = \oint \frac{\partial W}{\partial q_j} dq_j = J_j.$$

5.2.2 Transformation to Standard Form

In this section, we study a series of progressively more challenging examples that illustrate how to proceed from the primitive formulation of a physical problem to its transformation to the standard form (5.2.3).

The linear oscillator with slowly varying frequency

We consider as a first example the problem of Sec. 4.3.2

$$\frac{d^2 q}{dt^2} + \mu^2(\epsilon t) q = 0, \tag{5.2.48}$$

where $\mu \neq 0$ is a prescribed function. Equation (5.2.48) follows from (5.2.7) for the Lagrangian

$$L(q, \dot{q}, \tilde{t}) = \frac{1}{2}\dot{q}^2 + \frac{1}{2}\mu^2(\tilde{t}) q^2.$$

We define p using (5.2.9)

$$p = \frac{\partial L}{\partial \dot{q}} = \dot{q},$$

and we obtain the Hamiltonian

$$h(p, q, \tilde{t}) = \frac{1}{2} p^2 + \frac{1}{2}\mu^2(\tilde{t}) q^2 \tag{5.2.49}$$

from (5.2.10). Obviously, the Hamiltonian (5.2.49) in terms of the p, q variables that we have is not in standard form. We note, however, that the unperturbed problem for $\epsilon = 0$, i.e., $\mu =$ constant, is periodic. In fact, the motion for

$$h = \frac{1}{2} p^2 + \frac{1}{2}\mu^2 q^2 = \alpha = \text{const.} \tag{5.2.50}$$

is a libration defined by an ellipse in the p, q plane. The half-axes of this ellipse are $\sqrt{2\alpha}$ and $\sqrt{2\alpha/\mu}$ along the p and q directions, respectively. We can therefore introduce action and angle variables and transform (5.2.49) to the trivial (solvable) standard form (5.2.37). To do so, we consider the Hamilton-Jacobi equation (5.2.28) for the generating function W

$$\frac{1}{2}\left(\frac{\partial W}{\partial q}\right)^2 + \frac{1}{2}\mu^2 q^2 = \alpha. \tag{5.2.51}$$

For our one-degree-of-freedom system, this equation is always solvable, and quadrature gives

$$\begin{aligned} W &= \int^q \sqrt{2\alpha - \mu^2 s^2}\, ds \\ &= \frac{\alpha}{\mu}\left[\sin^{-1}\frac{\mu q}{\sqrt{2\alpha}} + \frac{\mu q}{\sqrt{2\alpha}}\left(1 - \frac{\mu^2 q^2}{2\alpha}\right)^{1/2}\right] \equiv W(q, \alpha). \end{aligned} \tag{5.2.52}$$

If we choose the transformed momentum p to be the energy α,

$$P = \alpha,$$

(5.2.52) defines the generating function $W(q, P)$ of the canonical transformation $(p, q) \to (P, Q)$. Using the relations (5.2.17), we compute this transformation in the implicit form

$$p = \frac{\partial W}{\partial q} = \sqrt{2P - \mu^2 q^2}, \tag{5.2.53a}$$

$$Q = \frac{\partial W}{\partial P} = \int^q \frac{ds}{\sqrt{2P - \mu^2 s^2}} = \frac{1}{\mu}\sin^{-1}\frac{\mu q}{\sqrt{2P}}, \tag{5.2.53b}$$

$$H = h. \tag{5.2.53c}$$

The explicit transformation corresponding to (5.2.12) then follows in the form

$$P = \frac{1}{2}p^2 + \frac{1}{2}\mu^2 q^2, \tag{5.2.54a}$$

$$Q = \frac{1}{\mu}\tan^{-1}\frac{\mu q}{p}, \tag{5.2.54b}$$

with

$$H = P. \tag{5.2.54c}$$

Its inverse (5.2.20) is

$$p = \sqrt{2P}\cos\mu Q, \tag{5.2.55a}$$

$$q = \frac{\sqrt{2P}}{\mu}\sin\mu Q. \tag{5.2.55b}$$

A second choice for the transformed momentum is to have P equal the normalized action (see (5.2.33), (5.2.46b))

$$P = \frac{1}{2\pi}\oint p\,dq. \tag{5.2.56a}$$

In (5.2.56a), the contour is the ellipse in the pq plane defined by (5.2.50) for α constant. Thus, $P = A/2\pi$, where A is the area of this ellipse. Since $A = \pi \cdot \sqrt{2\alpha} \cdot \sqrt{2\alpha}/\mu$, we find

$$P = \alpha/\mu. \tag{5.2.56b}$$

For this choice of P, the generating function W is

$$W = P\left[\sin^{-1}\left(\frac{\mu}{2P}\right)^{1/2} q + \left(\frac{\mu}{2P}\right)^{1/2} q\left(1 - \frac{\mu}{2P}q^2\right)^{1/2}\right]. \tag{5.2.57}$$

The implicit transformation relations that replace (5.2.53) are

$$p = \frac{\partial W}{\partial q} = \sqrt{2\mu P - \mu^2 q^2}, \tag{5.2.58a}$$

$$Q = \frac{\partial W}{\partial P} = \sin^{-1}\left(\frac{\mu}{2P}\right)^{1/2} q, \tag{5.2.58b}$$

$$H = \mu P. \tag{5.2.58c}$$

The explicit transformation corresponding to (5.2.54)–(5.2.55) is now

$$P = \frac{p^2 + \mu^2 q^2}{2\mu}, \quad Q = \tan^{-1}\frac{\mu q}{p} \tag{5.2.59}$$

with inverse

$$p = (2\mu P)^{1/2}\cos Q, \quad q = \left(\frac{2P}{\mu}\right)^{1/2}\sin Q. \tag{5.2.60}$$

We note that this was the transformation used in (4.5.8).

There is no intrinsic distinction between the two choices for P if $\mu = $ constant. The essential advantage for the second choice (5.2.56) only becomes apparent when we consider the case $\mu(\tilde{t})$.

Suppose we use the generating function W derived by examining the Hamilton-Jacobi equation with $\mu = $ constant to analyze the case $\mu = \mu(\tilde{t})$. The function W, in which μ now depends on $\tilde{t}$, *still defines a canonical transformation*. In fact, the relationship linking the old and new variables is the *same* as for the $\mu = $ constant case. The only change occurs in the definition of the new Hamiltonian, as we now need to account for the $(\partial S_2/\partial t)$ term with $S_2 = W$ in (5.2.17c). Thus,

$$H = h + \epsilon\frac{\partial W}{\partial \tilde{t}}. \tag{5.2.61}$$

If we choose $P = \alpha$, the energy, the right-hand side of (5.2.61), when expressed in terms of the (P, Q) variables of (5.2.54), gives

$$H = P + \frac{\epsilon\mu' P}{\mu}\left[\frac{\sin 2\mu Q}{2\mu} - Q\right], \tag{5.2.62}$$

where $' \equiv (d/d\tilde{t})$. For this choice of variables, the transformed Hamiltonian is *not in standard form* because the period in Q depends on $\tilde{t}$. As a result, we find that in the associated equations for $\dot{P}$ and $\dot{Q}$

$$\dot{P} = -\frac{\partial H}{\partial Q} = \frac{\epsilon\mu' P}{\mu}(1 - \cos 2\mu Q), \tag{5.2.63a}$$

$$\dot{Q} = \frac{\partial H}{\partial P} = 1 + \frac{\epsilon\mu'}{\mu}\left(\frac{\sin 2\mu Q}{2\mu} - Q\right), \tag{5.2.63b}$$

the right-hand side of (5.2.63b) has a nonperiodic term. We also note that the energy is not an adiabatic invariant.

In contrast, if we choose P to be the normalized action, α/μ, (5.2.61) gives

$$H = \mu P + \epsilon \frac{\mu'}{\mu} P \sin 2Q. \tag{5.2.64}$$

This Hamiltonian is in standard form. Moreover, the governing equations for $\dot{P}$, $\dot{Q}$ that follow,

$$\dot{P} = -\frac{\partial H}{\partial Q} = -\epsilon \frac{\mu' P}{\mu} \cos 2Q, \tag{5.2.65a}$$

$$\dot{Q} = \frac{\partial H}{\partial P} = \mu + \epsilon \frac{\mu'}{2\mu} \sin 2Q, \tag{5.2.65b}$$

indicate that $P = \alpha/\mu$ is an adiabatic invariant to $O(1)$. We will show next that the adiabatic invariance of the action is a basic property of the action of all one-degree-of-freedom Hamiltonians that are periodic for $\epsilon = 0$.

General periodic Hamiltonian in one degree of freedom

Consider the general Hamiltonian $h(p, q, \tilde{t})$ that describes a periodic motion in the pq plane for $\epsilon = 0$. The associated Hamilton-Jacobi equation for *fixed* $\tilde{t}$ and α is

$$h\left(\frac{\partial W}{\partial q}, q, \tilde{t}\right) = \alpha. \tag{5.2.66}$$

Quadrature defines W in the form

$$W = W(q, \alpha, \tilde{t}). \tag{5.2.67}$$

We now propose to use this W to derive a generating function for a canonical transformation $(p, q) \to (P, Q)$. For the time being, let us assume P to be an unspecified function of α and $\tilde{t}$:

$$P = P(\alpha, \tilde{t}). \tag{5.2.68}$$

Using (5.2.68) to solve for α as a function of P and $\tilde{t}$ and substituting this result in (5.2.67) gives

$$W = W(q, \alpha(P, \tilde{t}), \tilde{t}) \equiv F(q, P, \tilde{t}). \tag{5.2.69}$$

The expression in (5.2.69) is now regarded as a generating function of the type (5.2.17) involving the old coordinate q, the new momentum P, and $\tilde{t}$. The implicit transformation for the P, Q is defined by (see (5.2.17))

$$p = \frac{\partial F}{\partial q}, \tag{5.2.70a}$$

$$Q = \frac{\partial F}{\partial P}. \tag{5.2.70b}$$

The new Hamiltonian $H(P, Q, t)$ that follows from (5.2.17c) is

$$H(P, Q, \tilde{t}; \epsilon) = H_0(P, \tilde{t}) + \epsilon H_1(P, Q, \tilde{t}), \tag{5.2.71}$$

where

$$H_0(P, \tilde{t}) = \alpha(P, \tilde{t}), \tag{5.2.72a}$$

$$H_1(P, Q, \tilde{t}) = \frac{\partial F}{\partial \tilde{t}}(q(P, Q, \tilde{t}), P, \tilde{t}). \tag{5.2.72b}$$

As usual, the notation for H_1 indicates that we take the partial derivative of $F(q, P, \tilde{t})$ with respect to its third argument $\tilde{t}$. Then, in the resulting expression, we replace the first argument by the expression obtained by solving (5.2.70a) for q as a function of P, Q, and $\tilde{t}$.

Hamilton's differential equations associated with H are

$$\frac{dP}{dt} = -\frac{\partial H}{\partial Q} = -\epsilon \frac{\partial H_1}{\partial Q}, \tag{5.2.73a}$$

$$\frac{dQ}{dt} = \frac{\partial H}{\partial P} = \frac{\partial H_0}{\partial P} + \epsilon \frac{\partial H_1}{\partial P}. \tag{5.2.73b}$$

The time derivative of H is (see (5.2.26b))

$$\frac{dH}{dt} = \frac{\partial H}{\partial t} = \epsilon \frac{\partial H_0}{\partial \tilde{t}} + \epsilon^2 \frac{\partial H_1}{\partial \tilde{t}}. \tag{5.2.73c}$$

Note that although H is an invariant to $O(1)$ (i.e., $(dH/dt) = O(\epsilon)$), it is not an adiabatic invariant because $(\partial H_0/\partial \tilde{t}) \neq 0$. On the other hand, if we are able to choose P such that $(\partial H_1/\partial Q)$ has zero average, (5.2.73a) then implies that P is an adiabatic invariant to $O(1)$.

In order that $(\partial H_1/\partial Q)$ have a zero average, we must have

$$\oint \frac{\partial H_1}{\partial Q} dq = 0, \tag{5.2.74}$$

where the contour is the closed curve in the pq plane defined by $h(p, q, \tilde{t}) = \alpha$, with $\tilde{t}$ and α fixed. In view of the definition (5.2.72b) for H_1, we must set

$$\Delta \frac{\partial F}{\partial \tilde{t}} = 0, \tag{5.2.75}$$

where $\Delta(\partial F/\partial \tilde{t})$ denotes the change in $\partial F/\partial \tilde{t}$ after q undergoes a complete cycle holding P fixed. Since $(\partial F/\partial \tilde{t})$ is evaluated holding both q and P fixed, it follows from (5.2.75) that we must require

$$\Delta \frac{\partial F}{\partial \tilde{t}} = \frac{\partial}{\partial \tilde{t}}(\Delta F) = 0. \tag{5.2.76}$$

In general, ΔF will depend on both P and $\tilde{t}$ and $(\partial(\Delta F)/\partial\tilde{t}) \neq 0$, *unless* ΔF *is a function of* P *alone*. The simplest choice is to have $\Delta F = P$, and it follows from (5.2.47) that P must be the action J.

With the choice $P = J$, the generating function is now $F(q, J, \tilde{t})$, where

$$J = \oint p\, dq = J(q, p, \tilde{t}). \tag{5.2.77a}$$

The angle variable w follows from (5.2.70b):

$$w = \frac{\partial F}{\partial J} = w(q, p, \tilde{t}). \tag{5.2.77b}$$

The transformed Hamiltonian has the form (5.2.71) with $P = J$, $Q = w$, and we have shown that H_1 is periodic in w with zero average.

To illustrate ideas, consider the nonlinear oscillator discussed in Sec. 4.4.2

$$\ddot{q} + a(\tilde{t})q + b(\tilde{t})q^3 = 0. \tag{5.2.78}$$

We have set $\tilde{h} = 0$ in (4.4.74) to have a Hamiltonian problem, and we are denoting $y = q$, $p = \dot{q}$.

The Hamiltonian $h(p, q, \tilde{t})$ is given by

$$h(p, q, \tilde{t}) = \frac{p^2}{2} + a(\tilde{t})\frac{q^2}{2} + b(\tilde{t})\frac{q^4}{4} = \alpha. \tag{5.2.79}$$

The Hamilton-Jacobi equation for W gives

$$W(q, J, \tilde{t}) = \int^q \sqrt{2[\alpha(J, \tilde{t}) - a(\tilde{t})\frac{\xi^2}{2} - b(\tilde{t})\frac{\xi^4}{4}]}\, d\xi, \tag{5.2.80}$$

where $\alpha(J, \tilde{t})$ is the solution of

$$J = \oint \sqrt{2[\alpha(J, \tilde{t}) - a(\tilde{t})\frac{q^2}{2} - b(\tilde{t})\frac{q^4}{4}]}\, dq. \tag{5.2.81}$$

The canonical transformation to the action and angle variables (J, w) is defined implicitly by

$$p = \frac{\partial W}{\partial q} = \sqrt{2[\alpha(J, \tilde{t}) - a(\tilde{t})\frac{q^2}{2} - b(\tilde{t})\frac{q^4}{4}]}, \tag{5.2.82a}$$

$$w = \frac{\partial W}{\partial J} = \frac{\partial \alpha}{\partial J}\int^q \frac{d\xi}{\sqrt{2[\alpha(J, \tilde{t}) - a(\tilde{t})\frac{\xi^2}{2} - b(\tilde{t})\frac{\xi^4}{4}]}}. \tag{5.2.82b}$$

The three relations (5.2.81), (5.2.82a), and (5.2.82b) must be solved algebraically to define $\alpha(J, \tilde{t})$ and the explicit canonical transformation $(p, q) \to (J, w)$. Notice that the quadrature (5.2.82b), which is crucial for this algebraic solution, *is precisely the same one* that arises in the direct solution of the problem using a two-scale expansion (see (4.4.79)). Thus, there is no particular advantage, at least to leading order, in transforming (5.2.79) to standard form as opposed to solving it directly by the method discussed in Sec. 4.4. However, once the Hamiltonian has been transformed to standard form, the calculation of higher-order terms is significantly simpler. A discussion of the merits of the various approaches for solving strictly nonlinear perturbed oscillators is given in [4.20]. The situation when the quadrature (5.2.82b) cannot be carried out in terms of elliptic functions is also addressed in this reference. Other examples of transformation to standard form for strictly nonlinear perturbed oscillators can be found in [5.21].

Perturbed oscillator

The ideas in the preceding examples can be easily generalized to include small, possibly time-dependent perturbations. Consider, for example, the general weakly nonlinear oscillator

$$\frac{d^2q}{dt^2} + \gamma^2(\tilde{t})q + \epsilon f(q, t; \epsilon) = 0, \tag{5.2.83}$$

where we assume that f is a 2π-periodic function of t, $\tilde{t} = \epsilon t$, and $\gamma(\tilde{t}) \neq 0$.

Equation (5.2.83) is derivable from the time-dependent Hamiltonian

$$h_1(p, q, t; \epsilon) = \frac{1}{2}[p^2 + \gamma^2(\tilde{t})q^2] + \epsilon g(q, t; \epsilon) = E(t), \tag{5.2.84}$$

where

$$p = \dot{q}, \tag{5.2.85a}$$

$$g = \int_0^q f(y, t; \epsilon)dy. \tag{5.2.85b}$$

Thus, g is also 2π-periodic in t.

We now transform (5.2.84) to a two-degree-of-freedom system by regarding t as a second coordinate and introducing its conjugate momentum (see (5.2.22)–(5.2.23)). The transformed Hamiltonian is formally independent of t (it still depends on $\tilde{t}$) but involves two degrees of freedom

$$h(p_1, p_2, q_1, q_2, \tilde{t}; \epsilon) = \frac{1}{2}(p_1^2 + \mu^2 q_1^2)$$

$$+\epsilon g(q_1, q_2; \epsilon) + p_2 = \alpha_1 = \text{constant}. \tag{5.2.86}$$

We have set $q_1 = q$, $q_2 = t$, $p_1 = p$, $p_2 = \alpha_1 - E$, and α_1 is an arbitrary constant.

The unperturbed ($\epsilon = 0$) Hamiltonian is clearly periodic and separable. Therefore, we introduce the normalized action and angle variables in the (p_1, q_1) plane according to (5.2.59)–(5.2.60):

$$P_1 = \frac{p_1^2 + \gamma^2 q_1^2}{2\gamma}, \qquad Q_1 = \tan^{-1}\frac{\gamma q_1}{p_1}. \tag{5.2.87}$$

Since the motion in the (p_2, q_2) plane is the trivial rotation $p_2 = \text{constant}$, the normalized action and angle variables are simply

$$P_2 = p_2, \qquad Q_2 = q_2. \tag{5.2.88}$$

The transformed Hamiltonian is then found in the standard form (see (5.2.64))

$$H(P_1, P_2, Q_1, Q_2, \tilde{t}; \epsilon) = \gamma P_1 + P_2$$

$$+\epsilon \left\{ g(\sqrt{2P_1/\gamma} \sin Q_1, Q_2; \epsilon) + \frac{\gamma'}{\gamma} P_1 \sin 2Q_1 \right\}. \tag{5.2.89}$$

Coupled oscillators

We refer again to the problem of two coupled weakly nonlinear oscillators discussed in Sec. 4.5 (see (4.5.87)–(4.5.88))

$$\frac{d^2q_1}{dt^2} + \gamma_1^2(\tilde{t})q_1 = \epsilon q_2^2, \tag{5.2.90a}$$

$$\frac{d^2q_2}{dt^2} + \gamma_2^2(\tilde{t})q_2 = 2\epsilon q_1 q_2. \tag{5.2.90b}$$

This system follows from the Hamiltonian

$$h(p_1, p_2, q_1, q_2, \tilde{t}; \epsilon) = \frac{1}{2}(p_1^2 + p_2^2) + \frac{1}{2}(\gamma_1^2 q_1^2 + \gamma_2^2 q_2^2) - \epsilon q_1 q_2^2, \tag{5.2.91}$$

where $p_i = \dot{q}_i$.

Consider the Hamilton-Jacobi equation for the unperturbed problem ($\epsilon = 0$, μ_i = constant)

$$\frac{1}{2}\left(\frac{\partial W}{\partial q_1}\right)^2 + \frac{1}{2}\left(\frac{\partial W}{\partial q_2}\right)^2 + \frac{1}{2}\gamma_1^2 q_1^2 + \frac{1}{2}\gamma_2^2 q_2^2 = \text{constant}. \tag{5.2.92}$$

Obviously W is separable

$$W(q_1, q_2, \alpha_1, \alpha_2) = W_1(q_1, \alpha_1, \alpha_2) + W_2(q_2, \alpha_1, \alpha_2), \tag{5.2.93}$$

and each of the W_n obeys (5.2.51). Therefore, choosing the normalized actions, $P_i = \alpha_i/\gamma_i$, as the new momenta gives (see (5.2.57))

$$W = \sum_{i=1}^{2} P_i \left[\sin^{-1}\left(\frac{\gamma_i}{2P_i}\right)^{1/2} q_i + \left(\frac{\gamma_i}{2P_i}\right)^{1/2} q_i \left(1 - \frac{\gamma_i q_i^2}{2P_i}\right)^{1/2} \right]. \tag{5.2.94}$$

Letting the μ_i be dependent on $\tilde{t}$, we regard (5.2.94) as the generating function of a time-dependent canonical transformation $(p_i, q_i) \to (P_i, Q_i)$. Calculations virtually identical to those for the single oscillator give the explicit transformation (5.2.59)–(5.2.60) for $i = 1, 2$, and the following new Hamiltonian in standard form:

$$H(P_1, P_2, Q_1, Q_2, \tilde{t}; \epsilon) = \gamma_1 P_1 + \gamma_2 P_2 + \epsilon \frac{P_2}{2\gamma_2}\left(\frac{2P_1}{\gamma_1}\right)^{1/2} [\sin(Q_1 - 2Q_2)$$

$$+ \sin(Q_1 + 2Q_2) - 2\sin Q_1] + \frac{\epsilon}{2}\left[\frac{P_1}{\gamma_1}\gamma_1' \sin 2Q_1 + \frac{P_2}{\gamma_2}\gamma_2' \sin 2Q_2\right]. \tag{5.2.95}$$

Charged particle in a spatially slowly varying magnetic field

Consider the motion of a charged particle in a magnetic field that varies slowly in the x direction and that is cylindrically symmetric about the x axis. With appropriate dimensionless units, we can normalize the mass of the particle, its charge, and the speed of light to equal unity. We assume there is no electric field and that the magnetic vector potential $\mathbf{A}$ (which is cylindrically symmetric) has the special form

$$\mathbf{A} = \frac{r}{2} f(\tilde{x})\mathbf{e}_\theta. \tag{5.2.96}$$

Here $\tilde{x} = \epsilon x$, f is an arbitrary nonvanishing function that specifies the slow axial variation of $\mathbf{A}$, and $\mathbf{e}_\theta$ is a unit vector in the tangential direction, i.e., in the direction tangent to the cylinder $r = \sqrt{y^2 + z^2}$ for $r =$ constant.

Using the Cartesian x, y, z coordinates, the Lagrangian is

$$L(x, y, z, \dot{x}, \dot{y}, \dot{z}; \epsilon) = \frac{1}{2}(\dot{x}^2 + \dot{y}^2 + \dot{z}^2) + \frac{f}{2}(y\dot{z} - z\dot{y}), \tag{5.2.97}$$

where $\dot{} \equiv (d/dt)$. The derivation of this result can be found in [5.8]. If we introduce cylindrical polar coordinates (x, r, θ), we have

$$L(x, r, \theta, \dot{x}, \dot{r}, \dot{\theta}; \epsilon) = \frac{1}{2}(\dot{x}^2 + \dot{r}^2 + r^2\dot{\theta}^2) + \frac{f}{2}r^2\dot{\theta}. \tag{5.2.98}$$

The absence of θ from L implies that

$$p_\theta \equiv \frac{\partial L}{\partial \dot{\theta}} = r^2\left[\dot{\theta} + \frac{f(\tilde{x})}{2}\right] = \ell = \text{constant}. \tag{5.2.99}$$

This is an *exact* integral of the motion, and its significance will become clear once we have studied the $\epsilon = 0$ problem.

The equations of motion in Cartesian coordinates follow directly from (5.2.97) using (5.2.7):

$$\ddot{x} = \epsilon f'(y\dot{z} - z\dot{y})/2, \tag{5.2.100a}$$

$$\ddot{y} - f\dot{z} = \epsilon f'\dot{x}z/2, \tag{5.2.100b}$$

$$\ddot{z} + f\dot{y} = -\epsilon f'\dot{x}y/2, \tag{5.2.100c}$$

where $' \equiv (d/d\tilde{x})$. If $\epsilon = 0$, we can solve these in the form

$$x = x(0) + \dot{x}(0)t, \tag{5.2.101a}$$

$$y = y_g + \rho\cos(ft - \phi), \tag{5.2.101b}$$

$$z = z_g - \rho\sin(ft - \phi). \tag{5.2.101c}$$

Here $x(0)$ and $\dot{x}(0)$ are the initial values of x and $\dot{x}$, whereas y_g, z_g, ρ, and ϕ are the remaining four constants of the motion in the yz plane. We see that the motion has two distinct components: a uniform translation in the x direction superposed on a uniform clockwise rotation with constant angular velocity f around the "guiding

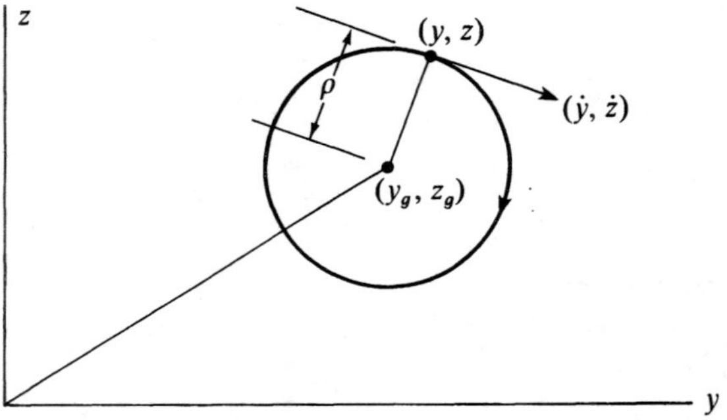

FIGURE 5.2.1. Circular Motion Around the Guiding Center

center" located at $y = y_g$, $z = z_g$. Figure 5.2.1 shows a projection of the motion in the yz plane.

Thus, the composite motion is a spiralling trajectory around the axis $y = y_g$, $z = z_g$. The constants y_g, z_g, ρ, and ϕ are related to the initial values of y, z, $\dot{y}$, and $\dot{z}$ according to

$$\rho = \sqrt{\dot{y}(0)^2 + \dot{z}(0)^2}/f, \tag{5.2.102a}$$

$$\sin\phi = \dot{y}(0)/\rho f; \quad \cos\phi = -\dot{z}(0)/\rho f, \tag{5.2.102b}$$

$$y_g = y(0) + \dot{z}(0)/f, \tag{5.2.102c}$$

$$z_g = z(0) - \dot{y}(0)/f. \tag{5.2.102d}$$

Note that $r_g \equiv \sqrt{y_g^2 + z_g^2} \geq \rho$, and the equality only occurs if the particle passes through the origin.

The integral of motion (5.2.99) takes the form

$$p_\theta = r^2(\dot{\theta} + f/2) = \frac{f}{2}(y_g^2 + z_g^2 - \rho^2) = \text{constant}, \tag{5.2.103}$$

which remains true even if f depends on $\tilde{x}$. Since r_g is constant if f is constant, (5.2.103) reduces, in this case, to the statement that the angular momentum about the guiding center is constant. For the case of slowly varying f, we expect ρ and r_g also to vary slowly with x and (5.2.103) shows that if f is chosen so that the particle's clockwise angular momentum about the guiding center increases then $r_g^2 f$ must decrease and vice versa.

We now consider the transformation of the system (5.2.100) to standard form for the case of variable f. Using the Lagrangian (5.2.97) in Cartesian coordinates, we compute the momenta p_x, p_y, p_z from (5.2.9)

$$p_x = \dot{x}, \tag{5.2.104a}$$

$$p_y = \dot{y} - fz/2; \quad \dot{y} = p_y + fz/2, \tag{5.2.104b}$$

$$p_z = \dot{z} + fy/2' \quad \dot{z} = p_z - fy/2. \tag{5.2.104c}$$

The Hamiltonian then follows from (5.2.10)

$$\overline{h}(p_x, p_y, p_z, x, y, z, \tilde{x}) = \frac{1}{2}\left[p_x^2 + (p_y + fz/2)^2 + (p_z - fy/2)^2\right]$$

$$= \text{constant.} \tag{5.2.105}$$

The constancy of $\overline{h}$, even for variable f, follows from the fact that $\overline{h}$ does not depend on t. In fact, we note that $\overline{h}$ is just the kinetic energy, $\frac{1}{2}(\dot{x}2 + \dot{y}2 + \dot{z}2)$, in this case. Thus, the speed of the particle is an exact integral of motion. This result also follows from (5.2.100). First, we note that the second and third equations imply that $y\dot{z} - z\dot{y} = \ell - (f/2)(y^2 + z^2)$. We then write (5.2.100a) as

$$\ddot{x} = \frac{\epsilon f'}{2}\left[1 - \frac{f}{2}(y^2 + z^2)\right]. \tag{5.2.106}$$

Now, if we multiply (5.2.106) by $\dot{x}$, (5.2.100b) by $\dot{y}$, (5.2.100c) by $\dot{z}$, and add these, we find $\frac{1}{2}(\dot{x}2 + \dot{y}2 + \dot{z}2) = \text{constant}$.

It is convenient to make a preliminary transformation to (p_i, q_i) variables that exhibit some of the features of the unperturbed motion. Since the axial motion is rectilinear, we do not transform x and choose $q_1 = x$. If we choose p_2 and q_2 proportional to $\dot{y}$ and $\dot{z}$, respectively, we will be able to depict the nearly uniform circular motion of the particle around the guiding center in terms of action and angle variables in the (p_2, q_2) plane. Finally, since the coordinates of the guiding center are constant if $\epsilon = 0$, we choose q_3 and p_3 proportional to these.

Noting that

$$\dot{z} = p_z - fy/2, \quad \dot{y} = p_y + fz/2,$$

$$y_g = \frac{1}{f}\left(p_z + \frac{fy}{2}\right), \quad z_g = -\frac{1}{f}\left(p_y - \frac{fz}{2}\right),$$

if $\epsilon = 0$, we seek a canonical transformation $(x, y, z, p_x, p_y, z) \to (p_i, q_i)$ in the form

$$q_1 = x, \quad p_1 = p_x + F, \tag{5.2.107a}$$

$$q_2 = A(p_z - fy/2), \quad p_2 = B(p_y + fz/2), \tag{5.2.107b}$$

$$q_3 = D(p_y - fz/2), \quad p_3 = C(p_z + fy/2), \tag{5.2.107c}$$

where F is unknown and we expect it to be $O(\epsilon)$. The remaining four constants A, B, C, and D are unknown functions of $\tilde{x}$.

In order that (5.2.107) define a canonical transformation, we must have [see (5.2.15)]

$$\mathbf{p} \cdot d\mathbf{q} \equiv \sum_{i=1}^{3} p_i dq_i = p_x dx + p_y dy + p_z d_z + dS, \tag{5.2.108a}$$

where dS is an exact differential. We will use the condition (5.2.108a), rather than an explicit derivation of a generating function, to fix the unknowns A, B, C, D, E, and F. When (5.2.107) is used in the right-hand side of (5.2.108a), we find

$$\begin{aligned}\mathbf{p} \cdot d\mathbf{q} &= p_1 dx + B(p_y + fz/2)d[A(p_z + fy/2) - Afy] \\ &\quad + D(p_z + fy/2)d[C(p_y + fz/2) - Cfz]. \end{aligned} \tag{5.2.108b}$$

Clearly, we must set $A = D$ and $D = C$. Equation (5.2.108b) then simplifies to

$$\begin{aligned}\mathbf{p} \cdot d\mathbf{q} &= [p_1 - \epsilon B(Af)' y p_y - \epsilon A(Bf)' z p_z]dx \\ &\quad - ABf p_y dy - ABf p_z dz + dS. \end{aligned} \tag{5.2.108c}$$

Comparing this result with (5.2.108a) shows that we must set $ABf = -1$ in order to have a canonical transformation. It is convenient to choose $A = -1/\sqrt{f}$, $B = 1/\sqrt{f}$ to define the exact canonical transformation

$$q_1 = x, \quad p_1 = p_x - \epsilon f'(yp_y + zp_z)/2f, \tag{5.2.109a}$$

$$q_2 = -(p_z - fy/2)/\sqrt{f}, \quad p_2 = (p_y + fz/2)/\sqrt{f}, \tag{5.2.109b}$$

$$q_3 = -(p_y - fz/2)/\sqrt{f}, \quad p_3 = (p_z + fy/2)/\sqrt{f}. \tag{5.2.109c}$$

The transformed Hamiltonian becomes ($\tilde{q}_1 = \epsilon q_1 = \epsilon x$)

$$h(p_i, q_i, \tilde{q}_1) = \frac{1}{2}\left[p_1 + \epsilon \frac{f'}{2f}(p_2 p_3 - q_2 q_3)\right]^2 + \frac{f}{2}(q_2^2 + p_2^2) = \text{constant}.$$

Notice that x, p_3, and q_3 only occur in the form ϵx, ϵp_3, and ϵq_3.

We now introduce normalized action and angle variables in the (p_2, q_2) and (p_3, q_3) planes (see (5.2.60) with $\mu = 1$)

$$p_2 = -\sqrt{2J_2^*} \sin w_2^*, \quad q_2 = \sqrt{2J_2^*} \cos w_2^*, \tag{5.2.110a}$$

$$p_3 = \sqrt{2J_3^*} \cos w_3^*, \quad q_3 = \sqrt{2J_3^*} \sin w_3^*. \tag{5.2.110b}$$

The reader can verify by direct calculation that $(p_2 dq_2 + p_3 dq_3 - J_2^* dw_2^* - J_3^* dw_3^*)$ is an exact differential, hence (5.2.110) is canonical. Moreover, (5.2.110a) are orientation preserving in the sense that the direction of increasing w_2^* is along the motion (clockwise) in the (p_2, q_2) plane. The transformed Hamiltonian becomes

$$\begin{aligned} H^*(p_1, J_2^*, J_3^*, q_1, w_2^*, w_3^*, \tilde{q}_1) &= \frac{1}{2}\left[p_1 - \frac{\epsilon f'}{f}\sqrt{J_2^* J_3^*} \sin(w_2^* + w_3^*)\right]^2 \\ &\quad + f J_2^* = \text{constant}. \end{aligned} \tag{5.2.111}$$

Finally, we introduce a new angle variable to denote $w_2^* + w_3^*$ by the canonical transformation

$$w_2 = w_2^* + w_3^*, \quad J_2 = J_2^*, \tag{5.2.112a}$$

$$w_3 = w_3^*, \quad J_3 = J_3^* - J_2^* \tag{5.2.112b}$$

to write the Hamiltonian in the standard form

$$H(p_1, J_2, J_3, q_1, w_2, w_3, \tilde{q}_1) = \frac{1}{2}\left[p_1 - \frac{\epsilon f'}{f}\sqrt{J_2(J_2 + J_3)}\,\sin w_2\right]^2$$

$$+ f J_2 = \text{constant.}$$

The relationship between the J_i, w_i variables and the Cartesian coordinates is summarized in Figure 5.2.2

It follows directly from the result (5.2.112) that, in addition to the exact integrals H and J_3 (which are just h and p_θ as defined earlier), J_2 is an adiabatic invariant to $O(1)$. The momentum p_1 is an invariant to $O(1)$ also, but it is not adiabatic unless f is a periodic function of $\tilde{x}$ with zero average.

Two-body problem with slowly varying mass

As a final example, we consider the planar motion of a particle around a gravitational center with slowly varying mass. The equations of motion in dimensionless form are (see (4.3.110))

$$\ddot{r} - r\dot{\theta}2 = -\frac{k}{r^2}, \tag{5.2.113a}$$

$$r\ddot{\theta} + 2\dot{r}\dot{\theta} = 0. \tag{5.2.113b}$$

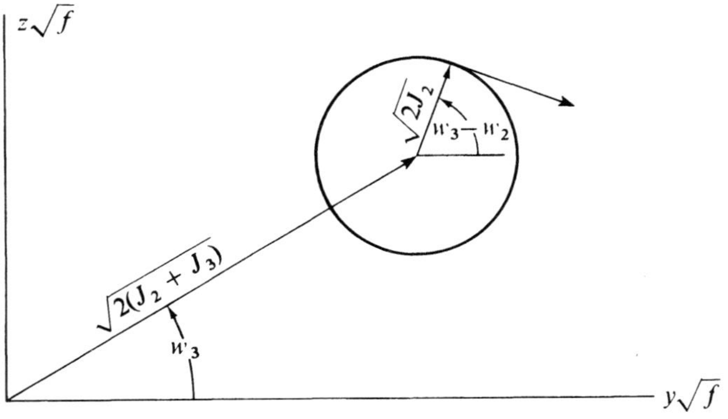

FIGURE 5.2.2. Action and Angle Variables

The mass of the gravitational center is assumed to vary slowly by regarding k to be a prescribed function of $\tilde{t}$ with $k(\tilde{t}) > 0$.

The Lagrangian associated with (5.2.113) is

$$L = \frac{1}{2}(\dot{r}2 + r^2\dot{\theta}2) + \frac{k}{r}, \tag{5.2.114}$$

as can be verified from (5.2.7) with $q_1 = r$ and $q_2 = \theta$. The Hamiltonian obtained from (5.2.10) is

$$\overline{h}(p_r, p_\theta, r, \theta, \tilde{t}) = \frac{1}{2}\left(p_r^2 + \frac{p_\theta^2}{r^2}\right) - \frac{k}{r}, \tag{5.2.115}$$

where (5.2.8) gives the momenta p_r and p_θ as

$$p_r = \frac{\partial L}{\partial \dot{r}} = \dot{r}, \quad p_\theta = \frac{\partial L}{\partial \dot{\theta}} = r^2\dot{\theta}. \tag{5.2.116}$$

Note that the angular momentum is a constant, $p_\theta = \ell$, because $\overline{h}$ does not depend on θ. However, for variable k, the Hamiltonian is not a constant.

We consider first the case k = const. to construct the generating function W. This obeys

$$\frac{1}{2}\left(\frac{\partial W}{\partial r}\right)^2 + \frac{1}{2r^2}\left(\frac{\partial W}{\partial \theta}\right)^2 - \frac{k}{r} = E = \text{constant}. \tag{5.2.117}$$

For periodic (bounded) motion $E < 0$. Assuming that W is additively separable, $W = W_1(r) + W_2(\theta)$ gives

$$\frac{r^2}{2}\left(\frac{\partial W_1}{\partial r}\right)^2 - kr - Er^2 = -\frac{1}{2}\left(\frac{\partial W_2}{\partial \theta}\right)^2 = \text{constant} = -\frac{\ell^2}{2}.$$

Therefore,

$$W = \int^r \sqrt{2E + \frac{2k}{\rho} - \frac{\ell^2}{\rho^2}}\, d\rho + \ell\theta. \tag{5.2.118}$$

This expression for W depends on r, θ, and the two constants E and ℓ. Let us introduce, instead of E and ℓ, the two normalized actions in the (p_r, r) and (p_θ, θ) planes. According to (5.2.33), these are

$$P_r = \frac{1}{2\pi}\oint \sqrt{2E + \frac{2k}{r} - \frac{\ell^2}{r^2}}\, dr, \tag{5.2.119a}$$

$$P_\theta = \frac{1}{2\pi}\oint \ell d\theta. \tag{5.2.119b}$$

In (5.2.119a), the contour is the closed curve $\frac{1}{2}p_r^2 - \frac{2k}{r} + \frac{\ell^2}{r^2} = E$, whereas in (5.2.119b) the contour is the constant $p_\theta = \ell$ over $0 \leq \theta \leq 2\pi$. The evaluation of P_r is best accomplished by transforming the integral into a contour integral in the complex plane to obtain (see ([5.8])

$$P_r = -\ell + \frac{k}{\sqrt{-2E}}. \tag{5.2.120a}$$

The integration of (5.2.119b) gives

$$P_\theta = \ell. \tag{5.2.120b}$$

It is more convenient to choose

$$P_1 = P_r + P_\theta, \quad P_2 = P_\theta \tag{5.2.121}$$

as the new momenta, and the generating function for the canonical transformation $(p_r, p_\theta, r, \theta) \to (P_1, P_2, Q_1, Q_2)$ then follows from (5.2.118)

$$W(r, \theta, P_1, P_2, k) = P_2\theta + \int^r \sqrt{\frac{-k^2}{P_1^2} + \frac{2k}{\rho} - \frac{P_2^2}{\rho^2}}\, d\rho. \tag{5.2.122}$$

The relations between P_1, P_2 and ℓ, E are

$$P_1 = \frac{k}{\sqrt{-2E}}, \quad P_2 = \ell, \tag{5.2.123}$$

and the relations to the semimajor axis a and eccentricity e follow from (4.3.114):

$$P_1 = \sqrt{ka}, \quad P_2 = \sqrt{ka(1-e^2)}. \tag{5.2.124}$$

The implicit formulas defining our canonical transformation are (see (5.2.32), (5.2.36))

$$p_r = \frac{\partial W}{\partial r} = \sqrt{\frac{-k^2}{P_1^2} + \frac{2k}{r} - \frac{P_2^2}{r^2}} \equiv F(P_1, P_2, r, k), \tag{5.2.125a}$$

$$p_\theta = \frac{\partial W}{\partial \theta} = P_2, \tag{5.2.125b}$$

$$Q_1 = \frac{\partial W}{\partial P_1} = \frac{k^2}{P_1^3}\int^r \frac{d\rho}{F(P_1, P_2, \rho, k)}, \tag{5.2.125c}$$

$$Q_2 = \frac{\partial W}{\partial P_2} = \theta - P_2\int^r \frac{d\rho}{\rho^2 F(P_1, P_2, \rho, k)}. \tag{5.2.125d}$$

To derive the explicit canonical transformation relations, we first solve (5.1.125c) for r as a function of P_1, P_2, Q_1. We use this result in (5.1.125d) to derive θ as a function of P_1, P_2, Q_1, Q_2. Equation (5.2.125b) gives p_θ explicitly. Finally, using the previously calculated relation for r in terms of P_1, P_2, Q_1 in (5.2.125a) defines p_r explicitly. For the case k = constant, the transformed Hamiltonian is $H = E$, and using the first equation in (5.2.123) we find

$$H = -\frac{k^2}{2P_1^2}. \tag{5.2.126}$$

The Hamilton differential equations for (5.2.126) are simply

$$\begin{aligned} \dot{P}_1 &= -\frac{\partial H}{\partial Q_1} = 0, \quad \dot{P}_2 = -\frac{\partial H}{\partial Q_2} = 0 \\ \dot{Q}_1 &= \frac{\partial H}{\partial P_1} = \frac{k^2}{P_1^3}, \quad Q_2 = \frac{\partial H}{\partial P_2} = 0. \end{aligned} \tag{5.2.127}$$

Therefore, the solution is

$$P_1 = \frac{k}{\sqrt{-2E}} = \sqrt{ka} = \text{ constant}, \tag{5.2.128a}$$

$$P_2 = \ell = \sqrt{ka(1-e^2)} = \text{ constant}, \tag{5.2.128b}$$

$$Q_1 = \frac{k^2}{P_1^3}(t-\tau) = \frac{k^{1/2}}{a^{3/2}}(t-\tau), \quad \tau = \text{ constant}, \tag{5.2.128c}$$

$$Q_2 = \omega = \text{ constant}. \tag{5.2.128d}$$

In addition to the pair of constants (E, ℓ) or (a, e), we have the constant τ, which is the time of pericenter passage, and the constant ω which is the argument of pericenter (see (4.3.113)). The angle Q_1 is called the *mean anomaly*; its derivative gives the average angular velocity of the particle. Inserting the solution (5.2.128) into the relations linking $(p_r, p_\theta, r, \theta)$ to (P_1, P_2, Q_1, Q_2) defines the former as functions of time and the four constants of motion.

If k is slowly varying, we may still use the generating function (5.2.122) to derive the *same* implicit transformation relations (5.2.125), which in turn lead to the *same* explicit relations linking $(p_r, p_\theta, r, \theta)$ to (P_1, P_2, Q_1, Q_2). However, H now has the added term $\epsilon \partial W/\partial \tilde{t}$.

We compute

$$\frac{\partial W}{\partial \tilde{t}} = k' \int^r \frac{-\frac{k}{P_1^2} - \frac{1}{\rho}}{F(P_1, P_2, \rho, k)} d\rho, \tag{5.2.129a}$$

and this can be evaluated explicitly

$$\frac{\partial W}{\partial \tilde{t}} = \frac{k'}{P_1 k}\sqrt{-k^2r^2 + 2krP_1^2 - P_1^2 P_2} = \frac{k'}{k} r p_r. \tag{5.2.129b}$$

However, we need to express $(\partial W \partial \tilde{t})$ entirely in terms of the new variables. To do so, we first express r in terms of the eccentric anomaly Φ. Using (4.3.113a) and (4.3.115), we find

$$r = a(1 - e\cos\Phi). \tag{5.2.130}$$

Thus,

$$p_r = \dot{r} = ae\dot{\Phi}\sin\Phi. \tag{5.2.131}$$

To compute $\dot{\Phi}$, we differentiate Kepler's equation (4.3.113b), which for our case reads

$$t - \tau = \sqrt{\frac{a^3}{k}}(\Phi - e\sin\Phi), \tag{5.2.132}$$

to find

$$\dot{\Phi} = \sqrt{\frac{k}{a^3}} \frac{1}{1 - e\cos\Phi}. \tag{5.2.133}$$

Note that in deriving (5.2.133) we must treat k as a constant.

Combining the above and using (5.2.124) to express a in terms of P_1 gives

$$\frac{\partial W}{\partial \tilde{t}} = \frac{k'}{k} P_1 e \sin\Phi, \quad e = \sqrt{1 - \frac{P_2^2}{P_1^2}}. \tag{5.2.134}$$

To express $\sin\Phi$ in terms of Q_1, we first combine (5.2.132) and (5.2.128c) to find

$$Q_1 = \Phi - e\cos\Phi. \tag{5.2.135}$$

It then follows that $\sin\Phi$ has the Fourier sine series

$$\sin\Phi = \frac{2}{e} \sum_{n=1}^{\infty} \frac{J_n(ne)}{n} \sin nQ_1,$$

when J_n is the Bessel function of the first kind of order n.

Thus, $H = -(k^2/2P_1^2) + \epsilon(\partial W/\partial\tilde{t})$ may be written in the standard form

$$H(P_i, Q_i, \tilde{t}; \epsilon) = -\frac{k^2(\tilde{t})}{2P_1^2} + \epsilon \frac{2k'}{k} \sum_{n=1}^{\infty} \frac{J_n(ne)}{n} \sin nQ_1, \tag{5.2.136}$$

where $e(P_1, P_2)$ is defined in (5.2.134).

For variable k, the energy is no longer an exact integral. However, in view of the absence of Q_2 in (5.2.136), the angular momentum P_2 remains exactly constant. It also follows from the equations governing $\dot{P}_1$ and $\dot{Q}_2$

$$\dot{P}_1 = -\frac{\partial H}{\partial Q_1} = \epsilon \frac{2k'}{k} \sum_{n=1}^{\infty} J_n(ne)\cos nQ_1, \tag{5.2.137a}$$

$$\begin{aligned}\dot{Q}_2 &= \frac{\partial H}{\partial P_2} \\ &= -\epsilon \frac{2k'}{k} \sum_{n=1}^{\infty} \frac{J_{n-1}(ne) - J_{n+1}(ne)}{2P_1} \frac{P_2}{\sqrt{P_1^2 - P_2^2}} \sin nQ_1,\end{aligned} \tag{5.2.137b}$$

that $P_1 = \sqrt{ka}$ and $Q_2 = \omega$ are adiabatic invariants to $O(1)$. Geometrically, the constancy of P_1 to $O(1)$ means that the semimajor axis $a = k^{-1} + O(\epsilon)$, where the $O(\epsilon)$ terms are oscillatory. Similarly, the constancy of P_2 implies that the eccentricity e is a constant to $O(1)$ with $O(\epsilon)$ oscillations. Finally, the adiabatic invariance of Q_2 means that the argument of pericenter ω is also constant to $O(1)$ with $O(\epsilon)$ oscillations.

For sufficiently small e, we may simplify (5.2.136) by retaining terms in $J_n(ne)$ only up to a certain order in e. This calculation also follows from the asymptotic inversion of (5.2.135) [see (1.2.20)]

$$\Phi = Q_1 + e \sin Q_1 + \frac{e^2}{2} \sin 2Q_1 + O(e^3). \tag{5.2.138}$$

Thus,

$$\sin \Phi = \sin Q_1 + \frac{e}{2} \sin 2Q_1 + \frac{e^2}{8} (\sin 3Q_1 - 3 \sin Q_1) + O(e^2). \tag{5.2.139}$$

Using this result, we obtain the following expansion for H:

$$H(P_i, Q_i, \tilde{t}; \epsilon) = -\frac{k^2(\tilde{t})}{2P_1^2} + \epsilon e \frac{k'}{k} P_1 \left[\sin Q_1 + \frac{e}{2} \sin 2Q_1 \right.$$

$$\left. + \frac{e^2}{8} (\sin 3Q_1 - 3 \sin Q_1) + \dot{O}(e^3) \right]. \tag{5.2.140}$$

We conclude the discussion of the various examples in this section by noting that much physical insight about the solution can be derived without actually solving the governing equations but by merely transforming the Hamiltonian to standard form.

5.2.3 *Near-Identity Averaging Transformations*

In Sec. 5.2.2, we studied several physical examples that can be formulated in terms of a Hamiltonian in the standard form (5.2.2). In Sec. 5.2.2, the notation for the final set of coordinates and momenta varied. Henceforth, we will adopt the notation in (5.2.2), where the (p_i, q_i) represent those variables that result in a standard form.

Our goal is to find a near-identity canonical transformation $(p_i, q_i) \to (P_i, Q_i)$ so that the transformed Hamiltonian in terms of the (P_i, Q_i) variables is free of the Q_i to any desired order in ϵ. We propose to define this canonical transformation in terms of a generating function depending on the q_i, P_i, and $\tilde{t}$. Thus, for a Hamiltonian system it is possible to bypass the system of differential equations (5.2.3) and to deal directly with the Hamiltonian itself in order to calculate the necessary transformations. This is one of several simplifying features associated with Hamiltonian systems.

Generating function

Since to $O(1)$ the generating function must define the identity transformation, we assume it in the following expanded form in powers of ϵ:

$$F(q_i, P_i, \tilde{t}; \epsilon) = \sum_{n=1}^{N} q_n P_n + \epsilon F_1(q_i, P_i, \tilde{t}) + \epsilon^2 F_2(q_i, P_i, \tilde{t}) + O(\epsilon^3). \tag{5.2.141}$$

The transformation $(p_i, q_i) \rightarrow (P_i, Q_i)$ and the new Hamiltonian H are then given implicitly by (see (5.2.17))

$$p_n = \frac{\partial F}{\partial q_n}, \tag{5.2.142a}$$

$$Q_n = \frac{\partial F}{\partial P_n}, \tag{5.2.142b}$$

$$H = h + \epsilon \frac{\partial F}{\partial \tilde{t}}. \tag{5.2.142c}$$

If we denote the explicit transformation as in Sec. 5.1.3 (see (5.1.52))

$$P_n = p_n + \epsilon T_n(p_i, q_i, \tilde{t}) + \epsilon^2 U_n(p_i, q_i, \tilde{t}) + O(\epsilon^3), \tag{5.2.143a}$$

$$Q_n = q_n + \epsilon L_n(p_i, q_i, \tilde{t}) + \epsilon^2 M_n(p_i, q_i, \tilde{t}) + O(\epsilon^3), \tag{5.2.143b}$$

we calculate the following expressions for the T_n, L_n, U_n, and M_n in terms of F_1 and F_2:

$$T_n(p_i, q_i, \tilde{t}) = -\frac{\partial F_1}{\partial q_n}, \tag{5.2.143c}$$

$$L_n(p_i, q_i, \tilde{t}) = \frac{\partial F_1}{\partial P_n}, \tag{5.2.143d}$$

$$U_n(p_i, q_i, \tilde{t}) = -\frac{\partial F_2}{\partial q_n} + \sum_{j=1}^{N} \frac{\partial^2 F_1}{\partial q_n \partial P_j} \frac{\partial F_1}{\partial q_j}, \tag{5.2.143e}$$

$$M_n(p_i, q_i, \tilde{t}) = \frac{\partial F_2}{\partial P_n} - \sum_{j=1}^{N} \frac{\partial^2 F_1}{\partial P_n \partial P_j} \frac{\partial F_1}{\partial q_j}. \tag{5.2.143f}$$

In the right-hand sides of (5.2.143), the various derivatives of F_1 and F_2 are evaluated at $P_i = p_i$.

Similarly, the inverse transformation has the form (see (5.1.53))

$$p_n = P_n + \epsilon R_n(P_i, Q_i, \tilde{t}) + \epsilon^2 S_n(P_i, Q_i, \tilde{t}) + O(\epsilon^3), \tag{5.2.144a}$$

$$q_n = Q_n + \epsilon V_n(P_i, Q_i, \tilde{t}) + \epsilon^2 W_n(P_i, Q_i, \tilde{t}) + O(\epsilon^3), \tag{5.2.144b}$$

where

$$R_n(P_i, Q_i, \tilde{t}) = \frac{\partial F_1}{\partial q_n}, \tag{5.2.144c}$$

$$V_n(P_i, Q_i, \tilde{t}) = -\frac{\partial F_1}{\partial P_n}, \tag{5.2.144d}$$

$$S_n(P_i, Q_i, \tilde{t}) = \frac{\partial F_2}{\partial q_n} - \sum_{j=1}^{N} \frac{\partial^2 F_1}{\partial q_n \partial q_j} \frac{\partial F_1}{\partial P_j}, \tag{5.2.144e}$$

$$W_n(P_i, Q_i, \tilde{t}) = -\frac{\partial F_2}{\partial P_n} + \sum_{j=1}^{N} \frac{\partial^2 F_1}{\partial P_n \partial q_j} \frac{\partial F_1}{\partial P_j}. \tag{5.2.144f}$$

In the right-hand sides of (5.2.144), the various derivatives of F_1 and F_2 are evaluated at $q_i = Q_i$.

Using the formulas in (5.2.144) to express the p_i, q_i in terms of P_i, Q_i in (5.2.142c), we obtain the following expansion for the transformed Hamiltonian H:

$$H(P_i, Q_i, \tilde{t}; \epsilon) = \underline{H}_0(P_i, \tilde{t}) + \epsilon H_1(P_i, Q_i, \tilde{t}) + \epsilon^2 H_2(P_i, Q_i, \tilde{t}) + O(\epsilon^3), \tag{5.2.145}$$

where

$$\underline{H}_0 = \underline{h}_0, \tag{5.2.146a}$$

$$H_1 = \underline{h}_1 + \underset{\wedge}{h}_1 + \sum_{j=1}^{N} \underline{\omega}_j \frac{\partial F_1}{\partial q_j}, \tag{5.2.146b}$$

$$H_2 = \underline{h}_2 + \underset{\wedge}{h}_2 + \frac{\partial F_1}{\partial \tilde{t}} + \sum_{j=1}^{N} \left(\underline{\omega}_j \frac{\partial F_2}{\partial q_j} + \frac{\partial \underset{\wedge}{h}_1}{\partial p_j} \frac{\partial F_1}{\partial q_j} + \frac{\partial \underline{h}_1}{\partial p_j} \frac{\partial F_1}{\partial q_j} - \frac{\partial \underset{\wedge}{h}_1}{\partial q_j} \frac{\partial F_1}{\partial P_j} \right)$$
$$+ \sum_{j=1}^{N} \sum_{k=1}^{N} \left(\frac{1}{2} \frac{\partial \underline{\omega}_j}{\partial p_k} \frac{\partial F_1}{\partial q_j} \frac{\partial F_1}{\partial q_k} - \underline{\omega}_j \frac{\partial^2 F_1}{\partial q_k \partial q_j} \frac{\partial F_1}{\partial P_k} \right). \tag{5.2.146c}$$

Removal of oscillatory terms

Removal of oscillatory terms to $O(\epsilon)$ requires that F_1 obey (see (5.2.146b))

$$\sum_{j=1}^{N} \underline{\omega}_j \frac{\partial F_1}{\partial q_j} + \underset{\wedge}{h}_1 = 0. \tag{5.2.147}$$

This determines F_1 in the form

$$F_1(q_i, P_i, \tilde{t}) = \underset{\wedge}{F}_1(q_i, P_i, \tilde{t}) + \underline{F}_1(P_i, \tilde{t}), \tag{5.2.148a}$$

where

$$\underset{\wedge}{F}_1 = -\left\{ \int \underset{\wedge}{h}_1(P_i, \underline{\omega}_i s, \tilde{t}) ds \right\}_{\underline{\omega}_i s = q_i}, \tag{5.2.148b}$$

and $\underline{F}_1$ is an arbitrary function.

It is easy to see that (5.2.147) is equivalent to the two conditions (5.1.58) and (5.1.59) defining T_m and L_n for the present special case of a Hamiltonian system that satisfies (5.2.4). If we differentiate (5.2.147) with respect to q_n and use the fact that $\partial h_1/\partial q_n = -\underset{\wedge}{f}_n$ (see (5.2.4)) and $\partial F_1/\partial q_n = -T_n$ (see (5.2.143c)), we obtain the condition (5.1.58). Similarly, if we differentiate (5.2.147) with respect to P_n and make use of the identities $\partial \underset{\wedge}{h}_1/\partial p_n = \underset{\wedge}{g}_n$, $\partial F_1/\partial P_n = L_n$, $\partial \underline{\omega}_n/\partial p_j = \partial \underline{\omega}_j/\partial p_n$, we obtain (5.1.59). Thus, for a Hamiltonian system, the two formulas defining the near-identity transformations to $O(\epsilon)$ are merely the q_n and p_n derivatives of the single formula (5.2.147) for the $O(\epsilon)$ term of the generating function.

The expression (5.2.148b) for $\hat{F}_1$ involves divisors of the form $r_1\underline{\omega}_1 + r_2\underline{\omega}_2 + \ldots r_N\underline{\omega}_N$ contributed by terms proportional to $\exp i(r_1q_1 + r_2q_2 + \ldots r_Nq_N)$ in the Fourier expansion of $\hat{h}_1$. Here, the r_i range over the positive and negative integers and zero (but all the r_i do not vanish). If for a certain set of values of the r_i, denoted by $r_i = R_i$ (not all equal to zero), the resonance condition $\sigma \equiv R_1\underline{\omega}_1 + R_2\underline{\omega}_2 + \ldots + R_N\underline{\omega}_N = 0$ occurs at some time $\tilde{t} = \tilde{t}_0$, then the contribution to $\hat{F}_1$ of the term proportional to $\exp i(R_1q_1 + R_2q_2 + \ldots + R_Nq_N)$ in $\hat{h}_1$ will have a zero divisor. This zero divisor invalidates the expansion (5.2.141) of the generating function, thereby invalidating the asymptotic expansions (5.2.143) or (5.2.144). This is the identical situation of resonance, mentioned in Secs. 4.5.3 and 5.1.3, that we exclude in the present analysis. We address this question in Secs. 5.3–5.5.

Consider now the $O(\epsilon^2)$ terms in the Hamiltonian. Using the result (5.2.147) simplifies the expression in (5.2.146c) to the form:

$$H_2 = \underline{h}_2 + \hat{h}_2 + \frac{\partial F_1}{\partial \tilde{t}} + \sum_{j=1}^{N}\left(\underline{\omega}_j\frac{\partial F_2}{\partial q_j} + \frac{\partial \underline{h}_1}{\partial p_j}\frac{\partial \hat{F}_1}{\partial q_j}\right)$$
$$-\sum_{j=1}^{N}\sum_{k=1}^{N}\left(\frac{1}{2}\frac{\partial \underline{\omega}_j}{\partial p_k}\frac{\partial \hat{F}_1}{\partial q_j}\frac{\partial \hat{F}_1}{\partial q_k} + \underline{\omega}_k\frac{\partial^2 \hat{F}_1}{\partial q_k \partial P_j}\frac{\partial \hat{F}_1}{\partial q_j}\right). \tag{5.2.149}$$

In general, the terms under the double summation on the right-hand side of (5.2.149) will have both an average and oscillatory part. Denoting

$$\sum_{j=1}^{N}\sum_{k=1}^{N}\left(\frac{1}{2}\frac{\partial \underline{\omega}_j}{\partial p_k}\frac{\partial \hat{F}_1}{\partial q_j}\frac{\partial \hat{F}_1}{\partial q_k} + \underline{\omega}_k\frac{\partial^2 \hat{F}_1}{\partial q_k \partial P_j}\frac{\partial \hat{F}_1}{\partial q_j}\right) = \underline{Z}(P_i,\tilde{t}) + \hat{Z}(P_i,Q_i,\tilde{t}), \tag{5.2.150}$$

we remove oscillatory terms from H_2 by setting

$$\sum_{j=1}^{N}\underline{\omega}_j\frac{\partial F_2}{\partial q_j} = -\hat{h}_2 - \frac{\partial \hat{F}_1}{\partial \tilde{t}} - \sum_{j=1}^{N}\frac{\partial \hat{h}_1}{\partial p_j}\frac{\partial \hat{F}_1}{\partial q_j} + \hat{Z}. \tag{5.2.151}$$

This determines F_2, and H_2 reduces to

$$H_2(P_i,\tilde{t}) = \underline{h}_2(P_i,\tilde{t}) + \frac{\partial \underline{F}_1(P_i,\tilde{t})}{\partial \tilde{t}} - \underline{Z}(P_i,\tilde{t}). \tag{5.2.152}$$

Since

$$\dot{P}_n = -\frac{\partial H}{\partial Q_n} = O(\epsilon^3), \tag{5.2.153a}$$

and the right-hand side of (5.2.153a) is a periodic function of the Q_i with zero average, setting $P_n = \text{const.}$ only introduces errors of $O(\epsilon^3)$. This means that we can choose $\underline{F}_1$ by quadrature to reduce (5.2.152) to $H_2 = 0$. The Q_n then obey

$$\dot{Q}_n = \frac{\partial H}{\partial P_n} = \frac{\partial \underline{h}_0(P_i, \tilde{t})}{\partial p_n} + \epsilon \frac{\partial \underline{h}_1(P_i, \tilde{t})}{\partial p_n} + O(\epsilon^3). \tag{5.2.153b}$$

It is easy to show that when (5.1.68) and (5.1.70) are specialized to the Hamiltonian case they reduce respectively to the q_n and p_n derivatives of (5.2.151). Moreover, the determination of $\underline{F}_1$ to render $H_2 = 0$ is equivalent to the conditions defining the $\underline{T}_m$ and $\underline{L}_n$ given in (5.1.77) and (5.1.83), respectively. In particular, the $\underline{T}_m = 0$ for the Hamiltonian problem, as follows directly from (5.1.143c), since $\partial F_1/\partial q_n$ has a zero average. This result can also be verified explicitly by noting that $\underline{k}_n = 0$ and that the definition (5.1.67a) implies that the $\underline{A}_n = 0$ in this case. It then follows that (5.1.77) has no inhomogeneous terms, and we can take the trivial solution $T_m^* = 0$.

The solution to $O(\epsilon)$, adiabatic invariants

In view of (5.2.153), the solution (5.1.85) calculated earlier takes the simple form

$$P_n = \alpha_n + O(\epsilon^3), \quad \alpha_n = \text{constant}, \tag{5.2.154a}$$

$$Q_n = \tau_n^{(0)} + \phi_n^{(0)}(\tilde{t}, \alpha_i) + \beta_n + O(\epsilon^2), \quad \beta_n = \text{constant}, \tag{5.2.154b}$$

where

$$\tau_n^{(0)} = \frac{1}{\epsilon} \int_0^{\tilde{t}} \underline{\omega}_n(\alpha_i, \sigma) d\sigma \tag{5.2.155a}$$

and

$$\phi_n^{(0)} = \int_0^{\tilde{t}} \frac{\partial \underline{h}_1(\alpha_i, \sigma)}{\partial p_n} d\sigma. \tag{5.2.155b}$$

Note that the neglected terms in (5.2.154a) for the P_n are $O(\epsilon^3)$ because the terms of order ϵ^3 on the right-hand side of (5.2.153a) for $\dot{P}_n$ have zero average. In contrast, the terms of $O(\epsilon^3)$ on the right-hand side of (5.2.153b) will generally have an average. Therefore, the remainder in (5.2.154b) is $O(\epsilon^2)$.

An important feature of the Hamiltonian structure of the system (5.2.3) is the absence of averaged terms *to all orders* from the equation (5.2.153a) for the $\dot{P}_n$. This is directly due to the assumption that h is a periodic function of the q_i. As a result of this property, we conclude that the expression linking the P_n to the original coordinates, momenta, and $\tilde{t}$ is an *adiabatic invariant to $O(\epsilon^2)$*. Using (5.2.143) in (5.1.52a) defines $\mathcal{A}_n^{(2)}$ in the form

$$\mathcal{A}_n^{(2)}(p_i, q_i, \tilde{t}; \epsilon) \equiv P_n = p_n - \epsilon \frac{\partial F_1(q_i, p_i, \tilde{t})}{\partial q_n} + \epsilon^2 \left\{ -\frac{\partial F_2(q_i, p_i, \tilde{t})}{\partial q_n} + \sum_{j=1}^{N} \frac{\partial^2 F_1(q_i, p_i, \tilde{t})}{\partial q_n \partial P_j} \frac{\partial F_1(q_i, p_i, \tilde{t})}{\partial q_i} \right\}, \tag{5.2.156}$$

and $(d\mathcal{A}_n^{(2)}/dt) = O(\epsilon^3)$ and has a zero average.

Concluding remarks

Our discussion of perturbation solutions for Hamiltonian systems is based on the use of a generating function to define the near-identity transformation. This approach generalizes von Zeipel's basic idea and is rather straightforward in concept. Its drawback is that the generating function involves a *mixture* of the old coordinates and new momenta, thus giving an *implicit* definition of the canonical transformation. The derivation of explicit results becomes progressively more tedious for higher-order calculations. To some extent, this is mitigated by the use of symbolic manipulation software as discussed in [5.3].

An alternate approach that we have not discussed in this book is based on the use of Lie transforms. Using this method, no functions of mixed variables appear, and calculations to higher order are easier to implement using symbolic manipulators. However, there is no significant advantage for calculations up to $O(\epsilon^2)$. A discussion of this method is beyond the scope of this book. The interested reader is referred to Sec. 2.5 of [5.21] and the references cited there for the original sources.

5.2.4 Examples

We illustrate the application of the results in Sec. 5.2.3 for some of the examples we have considered.

The model problems of Sec. 4.5.2

The Hamiltonian system of one degree of freedom

$$\frac{dp}{dt} = \epsilon A(p, \tilde{t}) \cos q = -\frac{\partial h}{\partial q}, \tag{5.2.157a}$$

$$\frac{dq}{dt} = \omega(p, \tilde{t}) - \epsilon \frac{\partial A}{\partial p}(p, \tilde{t}) \sin q = \frac{\partial h}{\partial p} \tag{5.2.157b}$$

is a special case of the model problem (4.5.13) or (5.1.20) if we set $B = -A_p$; it follows from the Hamiltonian

$$h(p, q, \tilde{t}; \epsilon) = \underline{h}_0(p, \tilde{t}) + \epsilon \underset{\wedge}{h}_1(p, q, \tilde{t}), \tag{5.2.158}$$

where

$$\underline{h}_0(p, \tilde{t}) = \int^p \omega(s, \tilde{t}) ds, \quad \underset{\wedge}{h}_1(p, q, \tilde{t}) = -A(p, \tilde{t}) \sin q. \tag{5.2.159}$$

As in our previous calculations, we wish to solve (5.2.157) for $p(t; \epsilon)$ and $q(t; \epsilon)$ subject to the initial conditions (see (4.5.27), (5.1.21))

$$p(0; \epsilon) = p_0, \quad q(0; \epsilon) = q_0. \tag{5.2.160}$$

Equation (5.2.148b) gives

$$F_1(q, P, \tilde{t}) = -\frac{A(P, \tilde{t})}{\omega(P, \tilde{t})} \cos q + \underline{F}_1(P, \tilde{t}), \tag{5.2.161}$$

from which it follows that

$$\frac{\partial F_1}{\partial q} = \frac{A}{\omega} \sin q, \qquad \frac{\partial^2 F_1}{\partial q \partial P} = \left(\frac{A}{\omega}\right)_P \sin q. \tag{5.2.162}$$

Using (5.2.150), we find

$$\underline{Z}(P, \tilde{t}) = \frac{1}{4}\left(\frac{A^2}{\omega}\right)_P, \quad \underset{\wedge}{Z}(P, Q, \tilde{t}) = -\frac{1}{4}\left(\frac{A^2}{\omega}\right)_P \cos 2Q. \tag{5.2.163}$$

Therefore, we remove average terms in H_2 by setting (see (5.2.152))

$$\frac{\partial \underline{F}_1}{\partial \tilde{t}} - \frac{1}{4}\left(\frac{A^2}{\omega}\right)_P = 0, \tag{5.2.164}$$

and we remove oscillatory terms in H_2 by setting (see (5.2.151))

$$\omega(P, \tilde{t}) \frac{\partial F_2}{\partial q} = \left(\frac{A}{\omega}\right)_{\tilde{t}} \cos q - \frac{1}{4}\left(\frac{A^2}{\omega}\right)_P \cos 2q. \tag{5.2.165}$$

Equation (5.2.164) gives

$$\underline{F}_1(P, \tilde{t}) = \frac{1}{4}\int_0^{\tilde{t}} \left[\frac{A^2(P, s)}{\omega(P, s)}\right]_P ds, \tag{5.2.166}$$

where we have set the additive function of P that arises in $\underline{F}_1$ equal to zero with no loss of generality. Equation (5.2.165) gives

$$F_2(q, P, \tilde{t}) = \frac{1}{\omega}\left(\frac{A}{\omega}\right)_{\tilde{t}} \sin q - \frac{1}{8\omega}\left(\frac{A^2}{\omega}\right)_P \sin 2q, \tag{5.2.167}$$

where we have set the additive function $\underline{F}_2(P, \tilde{t}) = 0$ because we are going to terminate the calculations for the generating function F at $O(\epsilon^2)$.

With the above choices for F_1 and F_2, the transformed Hamiltonian is

$$H(P, Q, \tilde{t}; \epsilon) = \int^P \omega(p, \tilde{t}) dp + O(\epsilon^3), \tag{5.2.168}$$

where the terms of $O(\epsilon^3)$ that we have ignored have zero average in Q.

Let us now derive the initial conditions for $P(t; \epsilon)$ and $Q(t; \epsilon)$ that follow from (5.2.160). We use the explicit transformation relations (5.2.143) to find

$$P(0; \epsilon) = p_0 - \epsilon \frac{\partial F_1}{\partial q}(q_0, p_0, 0) + \epsilon^2 \left[-\frac{\partial F_2}{\partial q}(q_0, p_0, 0)\right.$$

$$+ \frac{\partial^2 F_1}{\partial q \partial P}(q_0, p_0, 0) \frac{\partial F_1}{\partial q}(q_0, p_0, 0) \Bigg] + O(\epsilon^3), \quad (5.2.169a)$$

$$Q(0;\epsilon) = q_0 + \epsilon \frac{\partial F_1}{\partial P}(q_0, p_0, 0) + O(\epsilon^2). \quad (5.2.169b)$$

Evaluating the right-hand sides gives

$$P(0;\epsilon) = p_0 + \epsilon\theta^{(1)}(0) + \epsilon^2\theta^{(2)}(0) + O(\epsilon^3), \quad (5.2.170a)$$

$$Q(0;\epsilon) = q_0 + \epsilon\phi^{(1)}(0) + O(\epsilon^2), \quad (5.2.170b)$$

where

$$\theta^{(1)}(0) = -\frac{A(p_0, 0)}{\omega(p_0, 0)} \sin q_0, \quad (5.2.171a)$$

$$\theta^{(2)}(0) = \left[\frac{1}{4}\left(\frac{A^2}{\omega^2}\right)_p - \frac{1}{\omega}\left(\frac{A}{\omega}\right)_{\tilde{t}} \cos q_0 + \frac{1}{4}\frac{\omega_p}{\omega^3} A^2 \cos 2q_0 \right]_{p=p_0,\ \tilde{t}=0} \quad (5.2.171b)$$

$$\phi^{(1)}(0) = -\left[\left(\frac{A}{\omega}\right)_p \cos q_0 \right]_{p=p_0,\ \tilde{t}=0}. \quad (5.2.171c)$$

We note that $\theta^{(1)}(0)$ as defined above agrees identically with the expression (4.5.28c) for the $\theta^{(1)}(0)$ in Sec. 4.5.2. The expressions for $\theta^{(2)}(0)$ and $\phi^{(1)}(0)$ also agree with the corresponding expressions (4.5.40f) and (4.5.28d), respectively, if we set $B = -A_p$ in the latter formulas.

The equations for $\dot{P}$ and $\dot{Q}$ associated with the Hamiltonian (5.2.168) are

$$\dot{P} = O(\epsilon^3), \quad (5.2.172a)$$

$$\dot{Q} = \omega(P, \tilde{t}) + O(\epsilon^3). \quad (5.2.172b)$$

Therefore, since the $O(\epsilon^3)$ terms that we have ignored have zero average,

$$P(t;\epsilon) = P(0;\epsilon) + O(\epsilon^3)$$
$$= p_0 + \epsilon\theta^{(1)}(0) + \epsilon^2\theta^{(2)}(0) + O(\epsilon^3). \quad (5.2.173)$$

Using this expression for P in (5.2.172b) and expanding ω gives

$$\dot{Q} = \omega(p_0, \tilde{t}) + \epsilon\omega_p(p_0, \tilde{t})\theta^{(1)}(0) + \epsilon^2 \Bigg[\frac{1}{2}\omega_{pp}(p_0, \tilde{t})\theta^{(1)^2}(0)$$
$$+ \omega_p(p_0, \tilde{t})\theta^{(2)}(0) \Bigg] + O(\epsilon^3). \quad (5.2.174)$$

Integrating (5.2.174) subject to the initial condition (5.2.169b) gives

$$Q(t;\epsilon) = \psi + \epsilon \Bigg\{ \int_0^{\tilde{t}} \left[\frac{1}{2}\omega_{pp}(p_0, s)\theta^{(1)^2}(0) + \omega_p(p_0, s)\theta^{(2)}(0) \right] ds$$
$$+ \phi^{(1)}(0) \Bigg\} + O(\epsilon^3), \quad (5.2.175a)$$

where

$$\psi = \frac{1}{\epsilon}\int_0^{\tilde{t}} \omega(p_0, s)ds + \theta^{(1)}(0)\int_0^{\tilde{t}} \omega_p(p_0, s)ds + q_0. \qquad (5.2.175b)$$

Note that ψ as given in (5.2.175b) is the same expression as (4.5.40a) defining ψ in Sec. 4.5.2.

We now use (5.2.173) for $P(t; \epsilon)$ and (5.2.175a) for $Q(t; \epsilon)$ in (5.2.144) to obtain the solution for $p(t; \epsilon)$ and $q(t; \epsilon)$ to $O(\epsilon)$. According to (5.2.144), p and q are given by

$$p(t; \epsilon) = P(t; \epsilon) + \epsilon\frac{\partial F_1}{\partial q}(Q(t; \epsilon), P(t; \epsilon), \tilde{t}) + O(\epsilon^2), \quad (5.2.176a)$$

$$q(t; \epsilon) = Q(t; \epsilon) - \epsilon\frac{\partial F_1}{\partial P}(Q(t; \epsilon), P(t; \epsilon), \tilde{t}) + O(\epsilon^2). \quad (5.2.176b)$$

Therefore,

$$p(t; \epsilon) = p_0 + \epsilon\left[\theta^{(1)}(0) + \frac{A(p_0, \tilde{t})}{\omega(p_0, \tilde{t})}\sin\psi\right] + O(\epsilon^2), \qquad (5.2.177a)$$

$$q(t; \epsilon) = \psi + \epsilon\left\{\int_0^{\tilde{t}}\left[\frac{1}{2}\omega_{pp}(p_0, s)\theta^{(1)^2}(0) + \omega_p(p_0, s)\theta^{(2)}(0) - \frac{1}{4}\left(\frac{A^2}{\omega}\right)_{pp}\right]_{p=p_0,\ \tilde{t}=s} ds + \phi^{(1)}(0) + \left(\frac{A}{\omega}\right)_p\Bigg|_{p=p_0}\cos\psi\right\} + O(\epsilon^2). \qquad (5.2.177b)$$

If we set $B = -A_p$ in (4.5.39) and the various defintions of the terms that appear there, we find that the result agrees with (5.2.177) to $O(\epsilon)$.

It is evident that if the basic system (4.5.13) is Hamiltonian, the averaging procedure using a generating function to simplify the Hamiltonian as discussed in Sec. 5.2.3 is somewhat simpler than the two-scale expansion procedure of Sec. 4.5.2, which is simpler than the averaging method of Sec. 5.1.2 where one simplifies the individual differential equations. Of course, if the basic system is not Hamiltonian, the method presented in Sec. 5.2.3 does not apply. These remarks concerning the relative efficiency of the various methods carry over to systems with multiple degrees of freedom.

Again, the fundamental merit of the averaging procedure (for Hamiltonian or non-Hamiltonian systems) is the capability of deriving adiabatic invariants. For the example of the system (5.2.157), we conclude from (5.2.156) that

$$\mathcal{A}^{(2)}(p, q, \tilde{t}) = p - \epsilon\frac{A}{\omega}\sin q + \epsilon^2\left[\frac{1}{4}\left(\frac{A^2}{\omega^2}\right)_p - \frac{1}{\omega}\left(\frac{A}{\omega}\right)_{\tilde{t}}\cos q + \frac{1}{4}\frac{\omega_p}{\omega^3}A^2\cos 2q\right] \qquad (5.2.178)$$

is an adiabatic invariant to $O(\epsilon^2)$.

It is instructive to verify the adiabatic invariance of $\mathcal{A}^{(2)}$ explicitly. We take the total derivative of $\mathcal{A}^{(2)}$ with respect to t to obtain

$$\begin{aligned}\dot{\mathcal{A}}^{(2)} &= \dot{p} - \epsilon\left(\frac{A}{\omega}\right)\dot{q}\cos q - \epsilon\left(\frac{A}{\omega}\right)_p \dot{p}\sin q - \epsilon^2\left(\frac{A}{\omega}\right)_{\tilde{t}}\sin q \\ &+ \frac{\epsilon^3}{4}\left(\frac{A^2}{\omega^2}\right)_{p\tilde{t}} + \frac{\epsilon^2}{\omega}\left(\frac{A}{\omega}\right)_{\tilde{t}}\dot{q}\sin q - \epsilon^2\frac{\omega_p}{2\omega^3}A^2\dot{q}\sin 2q \\ &+ O(\epsilon^3), \end{aligned} \tag{5.2.179}$$

where the terms of order ϵ^3 that we have ignored have zero average. We now substitute (5.2.157a) for $\dot{p}$ and (5.2.157b) for $\dot{q}$ to find that *all terms* of $O(\epsilon)$ and $O(\epsilon^2)$ cancel identically from the right-hand side of (5.2.179). Therefore, since the terms of $O(\epsilon^3)$ that we have ignored have zero average with respect to q, we have proven that $\mathcal{A}^{(2)}$ is an adiabatic invariant to $O(\epsilon^2)$.

We conclude the discussion of this model problem by noting that the transformed variables P, Q calculated in Sec. 5.1.2 *do not correspond* to the P, Q variables in this section for the special case $B = -A_p$. To see this, it suffices to compare the expressions for P derived by each method. When we use the results derived in Sec. 5.1.2 for T and U in (5.1.22a), we find

$$P = p + \epsilon\left[-\frac{A(p,\tilde{t})}{\omega(p,\tilde{t})}\sin q + \frac{A(p_0,0)}{\omega(p_0,0)}\sin q_0\right] + \epsilon^2 A^2\frac{\omega_p}{4\omega^3}\cos 2q + O(\epsilon^3).$$

This is different from the expression (5.2.178), which gives P as a function of p, q and $\tilde{t}$ for a canonical transformation. Thus, even if the basic system is Hamiltonian, the procedure of averaging the individual differential equations does not necessarily result in a canonical transformation. In particular, for the model problem (5.2.157), averaging of the individual differential equations does not ensure that the average term of order ϵ^3 in (5.1.42a) is zero. In fact, $\underline{F}^{(3)} \neq 0$, and one needs to evaluate it to compute the solution for Q to $O(\epsilon)$. In contrast, use of canonical transformations avoids this difficulty. Of couse, if the basic system is not Hamiltonian, the option of using the approach of this section is not available.

The model problem of Sec. 5.1.3

We reconsider the system (5.1.108) for the special case $c = d = 0$ for which the problem is associated with the Hamiltonian (5.1.109)

$$\begin{aligned}h(p_i, q_i, \tilde{t}; \epsilon) &= \phi(\tilde{t})p_1 + \frac{1}{2}p_2^2 + \epsilon\psi(\tilde{t})p_1^2 p_2\sin(q_1 - q_2) \\ &+ \epsilon\theta(\tilde{t})p_1 p_2^2\sin(q_1 + q_2).\end{aligned} \tag{5.2.180}$$

Thus, $\underline{h}_1 = \underline{h}_2 = \hat{h}_2 = 0$, and

$$\underline{h}_0(p_1, p_2, \tilde{t}) = \phi(\tilde{t})p_1 + \frac{1}{2}p_2^2 \tag{5.2.181a}$$

$$\underset{\wedge}{h}_1(p_1, p_2, q_1, q_2, \tilde{t}) = \psi(\tilde{t}) p_1^2 p_2 \sin(q_1 - q_2)$$
$$+ \theta(\tilde{t}) p_1 p_2^2 \sin(q_1 + q_2). \qquad (5.2.181b)$$

Using (5.2.148b), we calculate

$$\underset{\wedge}{F}_1(q_1, q_2, P_1, P_2, \tilde{t}) = \frac{\psi P_1^2 P_2}{\phi - P_2} \cos(q_1 - q_2) + \frac{\theta P_1 P_2^2}{\phi + P_2} \cos(q_1 + q_2). \qquad (5.2.182)$$

We note that the expressions calculated in (5.1.110) for $\underset{\wedge}{T}_1$, $\underset{\wedge}{T}_2$, $\underset{\wedge}{L}_1$, and $\underset{\wedge}{L}_2$ (with $c = d = 0$) follow from (5.2.143c) and (5.2.143d) for this $\underset{\wedge}{F}_1$. We also recall that $\underline{A}_1 = \underline{A}_2 = 0$ for the Hamiltonian problem according to (5.1.112). Thus, $\underline{T}_1 = \underline{T}_2 = 0$.

The expression defining $\underline{Z} + \underset{\wedge}{Z}$ that follows from (5.2.150) is

$$\underline{Z} + \underset{\wedge}{Z} = \frac{1}{2}\left(\frac{\partial F_1}{\partial q_2}\right)^2 + \phi \left(\frac{\partial^2 F_1}{\partial q_1 \partial P_1}\frac{\partial F_1}{\partial q_1} + \frac{\partial^2 F_1}{\partial q_1 \partial P_2}\frac{\partial F_1}{\partial q_2}\right)$$
$$p_2 \left(\frac{\partial^2 F_1}{\partial q_2 \partial P_1}\frac{\partial F_1}{\partial q_1} + \frac{\partial^2 F_1}{\partial q_2 \partial P_2}\frac{\partial F_1}{\partial q_2}\right). \qquad (5.2.183)$$

Using the expression for $\underset{\wedge}{F}_1$ given in (5.2.182), we can compute all the derivatives that occur in $\underline{Z} + \underset{\wedge}{Z}$. (Note that $\underline{F}_1$ does not contribute to this expression since it is independent of q_1 and q_2.) This calculation is straightforward using symbolic manipulation software. We find (using Mathematica)

$$\underline{Z}(P_1, P_2, \tilde{t}) = \frac{\psi^2 P_1^3 P_2}{4(\phi - P_2)^2}\left[2\phi(2P_2 - P_1) + P_2(P_1 - 4P_2)\right]$$
$$+ \frac{\theta^2 P_1 P_2^3}{4(\phi + P_2)^2}\left[2\phi(2P_1 + P_2) + P_2(3P_1 + 2P_2)\right]. \qquad (5.2.184)$$

$$\underset{\wedge}{Z}(P_1, P_2, q_1, q_2, \tilde{t}) = C_{2,0}\cos 2q_1 + C_{2,-2}\cos 2(q_1 - q_2)$$
$$+ C_{0,2}\cos 2q_2 + C_{2,2}\cos 2(q_1 + q_2), \qquad (5.2.185)$$

where

$$C_{2,0}(P_1, P_2, \tilde{t}) = \frac{\psi\theta P_1^2 P_2^2}{2(\phi^2 - P_2^2)}; \quad C_{2,-2}(P_1, P_2, \tilde{t}) = \frac{\psi P_1^3 P_2}{4(\phi - P_2)^2}$$
$$C_{0,2}(P_1, P_2, \tilde{t}) = -\frac{\psi\theta P_1^2 P_2^2}{2(\phi^2 - P_2^2)}; \quad C_{2,2}(P_1, P_2, \tilde{t}) = -\frac{\theta P_1 P_2^3}{4(\phi + P_2)^2}. \qquad (5.2.186)$$

With $\underline{Z}$ known, quadrature gives $\underline{F}_1$ in the form

$$\underline{F}_1(P_1, P_2, \tilde{t}) = \int_0^{\tilde{t}} \underline{Z}(P_1, P_2, s)\,ds. \qquad (5.2.187)$$

We remove the oscillatory term in H_2 by setting

$$\phi \frac{\partial F_2}{\partial q_1} + P_2 \frac{\partial F_2}{\partial q_2} = \hat{Z} - \frac{\partial \overset{\wedge}{F_1}}{\partial \tilde{t}} = C_{1,-1} \cos(q_1 - q_2) + C_{1,1} \cos(q_1 + q_2)$$

$$+C_{2,0} \cos 2q_1 + C_{2,-2} \cos 2(q_1 - q_2) + C_{0,2} \cos 2q_2 + C_{2,2} \cos 2(q_1 + q_2), \tag{5.2.188}$$

where

$$C_{1,-1} = \frac{P_1^2 P_2}{(\phi - P_2)^2} \left[\phi' \psi - \psi'(\phi - P_2)\right], \tag{5.2.189a}$$

$$C_{1,1} = \frac{P_1 P_2^2}{\phi + P_2)^2} \left[\phi' \theta - \theta'(\phi + P_2)\right]. \tag{5.2.189b}$$

It follows from (5.2.188) that

$$F_2(q_1, q_2, P_1, P_2, \tilde{t}) = \frac{C_{1,-1}}{\phi - P_2} \sin(q_1 - q_2) + \frac{C_{1,1}}{\phi + P_2} \cos(q_1 + q_2)$$

$$+ \frac{C_{2,0}}{2\phi} \sin 2q_1 + \frac{C_{2,-2}}{2(\phi - P_2)} \sin 2(q_1 - q_2) + \frac{C_{0,2}}{2P_2} \sin 2q_2$$

$$+ \frac{C_{2,2}}{2(\phi - P_2)} \sin 2(q_1 + q_2). \tag{5.2.190}$$

This completes the averaging to $O(\epsilon^2)$, and we have

$$P_1(t; \epsilon) = P_1(0; \epsilon) + O(\epsilon^3); \qquad P_2(t; \epsilon) = P_2(0; \epsilon) + O(\epsilon^3)$$

$$Q_1(t; \epsilon) = \frac{1}{\epsilon} \int_0^{\tilde{t}} \phi(s) ds + Q_1(0; \epsilon) + O(\epsilon^2) \tag{5.2.191}$$

$$Q_2(t; \epsilon) = P_2(0)t + Q_2(0; \epsilon) + O(\epsilon^2),$$

where the four constants $P_1(0; \epsilon)$, $P_2(0; \epsilon)$, $Q_1(0; \epsilon)$, $Q_2(0; \epsilon)$ follow from evaluating (5.2.143a) and (5.2.143b) at $t = 0$. Since we have defined all the right-hand sides of (5.2.143), we have explicit relations linking $P_1(0; \epsilon) \ldots Q_2(0; \epsilon)$ to the four initial values of p_1, p_2, q_1, and q_2. We do not list these relations for brevity.

Once the P_i and Q_i are defined in (5.2.191), we use these expressions in (5.2.144) to obtain the solution for the p_1 and p_2 correct to $O(\epsilon^2)$ and the solution for q_1 and q_2 correct to $O(\epsilon)$.

We can also exhibit to adiabatic invariants to $O(\epsilon^2)$ using (5.2.156). As F_1 and F_2 have been defined explicitly, the explicit formulas for $\mathcal{A}_1^{(2)}$ and $\mathcal{A}_2^{(2)}$ are easy to compute but are omitted for brevity.

All our results are based on the assumptions that $\phi(\tilde{t}) > 0$, $p_2(0) > 0$, and $\phi(\tilde{t}) - p_2(0) \neq 0$ to avoid resonances, as discussed in Sec. 5.1.3.

Problems

1. Consider the weakly nonlinear oscillator with slowly varying frequency
$$\frac{d^2x}{dt^2} + \gamma^2(\tilde{t})x + \epsilon x^3 = 0, \tag{5.2.192}$$
which is a special case of (4.5.84) with $f = x^3$.
a. Calculate the Hamiltonian for (5.2.192) in standard form.
b. Derive an adiabatic invariant correct to $O(\epsilon^2)$, and solve for $x(t;\epsilon)$ to $O(\epsilon)$.
2. Consider the Hamiltonian (5.2.95) for the case $\gamma_1 = \text{const.}$, $\gamma_2 = \text{const.}$, $\gamma_1 \neq 2\gamma_2$, $\gamma_1 \neq \gamma_2$. Calculate the solution to $O(\epsilon)$ based on the method of this section, and compare your results with those in Sec. 4.3.6b.
3. Now study Problem 2 for the case $\gamma_1(\tilde{t})$ and $\gamma_2(\tilde{t})$ using the method of this section. Show that to $O(\epsilon)$ the two adiabatic invariants corresponding to $n = 1$, 2 in (5.2.156) are
$$\mathcal{A}_1^{(1)} = p_1 + \epsilon\left\{\frac{p_1}{2\gamma_1^2}\gamma_1' \sin 2q_1 + \frac{p_2}{\gamma_2}\left(\frac{p_1}{2\gamma_1}\right)^{1/2}\left[-\frac{2\sin q_1}{\gamma_1} + \frac{\sin(q_1 - 2q_2)}{\gamma_1 - 2\gamma_2} + \frac{\sin(q_1 + 2q_2)}{\gamma_1 + 2\gamma_2}\right]\right\} \tag{5.2.193a}$$
$$\mathcal{A}_2^{(1)} = p_2 + \epsilon\left\{\frac{p_2}{2\gamma_2^2}\gamma_2' \sin 2q_2 + \frac{2p_2}{\gamma_2}\left(\frac{p_1}{2\gamma_1}\right)^{1/2}\left[\frac{\sin(q_1 + 2q_2)}{\gamma_1 + 2\gamma_2} + \frac{\sin(q_1 - 2q_2)}{\gamma_1 - 2\gamma_2}\right]\right\} \tag{5.2.193b}$$
Calculate the solution to $O(\epsilon)$ and compare your results with those you found in Problem 4c of Sec. 4.5.
4. Use the Hamiltonian (5.2.140) to compute the adiabatic invariants to $O(\epsilon)$ for satellite motion with slowly varying k.
5. Consider the weakly nonlinear one-dimensional wave equation in dimensionless variables
$$\overline{u}_{\overline{t}\overline{t}} - \overline{u}_{\overline{x}\overline{x}} + k\overline{u} + \epsilon F(\overline{u}) = 0, \tag{5.2.194}$$
where k is a positive constant and F is a prescribed function of $\overline{u}$ with $F(0) = 0$.
We wish to solve (5.2.194) on the *slowly varying* domain $\overline{x}_1(\tilde{t}) \leq \overline{x} \leq \overline{x}_2(\tilde{t})$, where $\overline{x}_1$ and $\overline{x}_2$ are prescribed functions of $\tilde{t} = \epsilon t$ with $\overline{x}_1(0) = 0, \overline{x}_2(0) = \pi$. The boundary conditions are
$$\overline{u}(\overline{x}_1(\tilde{t}), \overline{t}; \epsilon) = \overline{u}(\overline{x}_2(\tilde{t}), \overline{t}; \epsilon) = 0 \quad \text{for } \overline{t} > 0. \tag{5.2.195}$$
The initial conditions are
$$\overline{u}(\overline{x}, 0; \epsilon) = \overline{u}_0(\overline{x}), \quad \overline{u}_t(\overline{x}, 0; \epsilon) = \overline{v}_0(\overline{x}). \tag{5.2.196}$$

a. Introduce the transformation

$$x = \frac{\bar{x} - \bar{x}_1(\tilde{t})}{\ell(\tilde{t})}\pi, \quad t = \bar{t}, \quad u\left(x, t; \epsilon\right) = \bar{u}(\bar{x}_1(\tilde{t}) + \frac{\ell(\tilde{t})}{\pi}x, t; \epsilon\right), \tag{5.2.197}$$

where $\ell(\tilde{t}) = \bar{x}_2(\tilde{t}) - \bar{x}_1(\tilde{t}) > 0$, and show that the problem to be solved transforms to

$$u_{tt} - \frac{\pi^2}{\ell^2(\tilde{t})}u_{xx} + ku + \epsilon\left[F(u) - \frac{2}{\ell}(x\ell' + \pi\bar{x}_1')u_{xt}\right] + O(\epsilon^2) \tag{5.2.198}$$

with zero boundary conditions on the *fixed* domain $0 \leq x \leq \pi$

$$u(0, t; \epsilon) = u(\pi, t; \epsilon) = 0, \quad t > 0, \tag{5.2.199}$$

and the new initial conditions

$$u(x, 0; \epsilon) = u_0(x) \tag{5.2.200a}$$

$$u_t(x, 0; \epsilon) = v_0(x) + \frac{\epsilon}{\pi}(x\ell'(0) + \pi\bar{x}_1'(0))\frac{du_0}{dx} \equiv v_0(x) + \epsilon v_1(x). \tag{5.2.200b}$$

b. Assume the modal expansion

$$u(x, t; \epsilon) = \sum_{n=1}^{\infty} q_n(t; \epsilon)\sin nx \tag{5.2.201}$$

and show that the $q_n(t; \epsilon)$ satisfy the following infinite set of oscillator equations with slowly varying frequencies and weak nonlinear coupling:

$$\frac{d^2q_n}{dt^2} + \omega_n^2(\tilde{t})q_n + \epsilon\left[f_n(q_i) + \sum_{m=1}^{\infty} C_{nm}(\tilde{t})\frac{dq_m}{dt}\right] = O(\epsilon^2), \tag{5.2.202}$$

where

$$\omega_n^2(\tilde{t}) = k + \frac{n^2\pi^2}{\ell^2(\tilde{t})}, \tag{5.2.203}$$

$$f_n(q_i) = \frac{2}{\pi}\int_0^{\pi} F\left(\sum_{m=1}^{\infty} q_m \sin mx\right)\sin nx dx, \tag{5.2.204}$$

$$C_{nm}(\tilde{t}) = \begin{cases} \ell'/\ell, & n = m \\ \frac{4nm}{\ell(n^2-m^2)}[(-1)^{n+m}\bar{x}_2' - \bar{x}_1'], & n \neq m. \end{cases} \tag{5.2.205}$$

Thus, the initial conditions (5.2.200) give the following initial values for q_n and $\dot{q}_n$:

$$q_n(0; \epsilon) = \frac{2}{\pi}\int_0^{\pi} u_0(\xi)\sin n\xi d\xi, \tag{5.2.206a}$$

$$\dot{q}_n(0; \epsilon) = \frac{2}{\pi}\int_0^{\pi} [v_0(\xi) + \epsilon v_1(\xi)]\sin n\xi d\xi. \tag{5.2.206b}$$

Show also that the f_n are derivable from the potential $g(q_i)$ according to

$$f_n = \frac{\partial g}{\partial q_n}, \tag{5.2.207}$$

where

$$g(q_i) = \frac{2}{\pi}\int_0^\pi G\left(\sum_{m=1}^{\infty} q_m \sin mx\right) dx, \tag{5.2.208a}$$

$$G(u) = \int_0^u F(s)ds. \tag{5.2.208b}$$

c. Introduce the scaled modal amplitudes $q_n^*(t;\epsilon)$ according to

$$q_n(t;\epsilon) = \alpha_n(\tilde{t})q_n^*(t;\epsilon)$$

and show that the system that results from (5.2.202) for the q_n^* has zero diagonal terms for the transformed matrix C_{nm}^* if we choose

$$\alpha_n(\tilde{t}) = \ell^{-1/2}(\tilde{t}). \tag{5.2.209}$$

Show that the q_n^* obey

$$\frac{d^2 q_n^*}{dt^2} + \omega_n^2(\tilde{t})q_n^* + \epsilon\left[\frac{\partial V}{\partial q_n^*} + \sum_{m=1}^{\infty} C_{nm}^*(\tilde{t})\frac{dq_m^*}{dt}\right] = O(\epsilon^2), \tag{5.2.210}$$

where

$$V(q_i^*, \tilde{t}) = \ell(\tilde{t})g\left(\frac{q_i^*}{\sqrt{\ell}}\right), \tag{5.2.211a}$$

$$C_{nm}^* = \begin{cases} 0, & n = m \\ C_{nm}(\tilde{t}), & n \neq m. \end{cases} \tag{5.2.211b}$$

d. Show that (5.2.210) is Hamiltonian to $O(\epsilon)$ and follows from

$$h^*(p_i^*, q_i^*, \tilde{t};\epsilon) = \frac{1}{2}\sum_{n=1}^{\infty}[p_n^{*^2} + \omega_n^2(\tilde{t})q_n^{*^2}]$$

$$+\epsilon V(q_i^*, \tilde{t}) - \frac{\epsilon}{2}\sum_{n=1}^{\infty}\sum_{m=1}^{\infty} C_{nm}^*(\tilde{t})q_m^* p_n^*, \tag{5.2.212}$$

where

$$p_n^* = \dot{q}_n^* + \frac{\epsilon}{2}\sum_{m=1}^{\infty} C_{nm}^*(\tilde{t})q_m^*. \tag{5.2.213}$$

e. Introduce the canonical transformation (see (5.2.94))

$$p_n^* = (2\omega_n p_n)^{1/2}\cos q_n, \quad q_n^* = \left(\frac{2p_n}{\omega_n}\right)\sin q_n \tag{5.2.214}$$

to transform (5.2.212) to the standard form

$$h(p_i, q_i, \tilde{t}; \epsilon) = \sum_{m=1}^{\infty} \omega_n(\tilde{t}) p_m + \epsilon \left\{ \sum_{m=1}^{\infty} \frac{p_m}{2} \frac{\omega'_m}{\omega_m} \sin 2q_m \right.$$

$$- \sum_{k=1}^{\infty} \sum_{m=1}^{\infty} \frac{C^*_{nm}(\tilde{t})}{2} \left(\frac{\omega_k p_k p_m}{\omega_m} \right)^{1/2} \left[\sin(q_m + q_k) + \sin(q_m - q_k) \right]$$

$$\left. + U(p_i, q_i, \tilde{t}) \right\}, \tag{5.2.215a}$$

where

$$U(p_i, q_i, \tilde{t}) = V\left(\left(\frac{2p_i}{\omega_i} \right)^{1/2} \sin q_i, \tilde{t} \right). \tag{5.2.215b}$$

f. Specialize your results to the case $F = u^2$. Thus,

$$G(u) = \frac{u^3}{3}, \tag{5.2.216a}$$

$$g(q_i) = \frac{2}{\pi} \int_0^{\pi} \left[\sum_{m=1}^{\infty} q_m \sin mx \right]^3 dx. \tag{5.2.216b}$$

Show that

$$g(q_i) = \frac{1}{3} \sum_{m=1}^{\infty} a_{mmm} q_m^3 + \sum_{m=1}^{\infty} \sum_{k=1}^{\infty} a_{mkk} q_m q_k^2$$

$$+ \sum_{\substack{m=1 \\ m \neq k \neq \ell}}^{\infty} \sum_{k=1}^{\infty} \sum_{\ell=1}^{\infty} a_{mk\ell} q_m q_k q_\ell, \tag{5.2.217}$$

where

$$a_{mmm} = \frac{4[1 - (-1)^m]}{3m\pi}, \tag{5.2.218a}$$

$$a_{mkk} = \begin{cases} \frac{4k^2[(-1)^m - 1]}{m\pi(m^2 - 4k^2)}, & m \neq 2k \\ 0, & m = 2k \end{cases}, \tag{5.2.218b}$$

$$a_{mk\ell} = \begin{cases} \frac{2}{3\pi} \left[\frac{\ell}{(m+k)^2 - \ell^2} + \frac{m}{(m-k)^2 - \ell^2} \right], & m + k + \ell = \text{odd} \\ 0, & m + k + \ell = \text{even}. \end{cases} \tag{5.2.218c}$$

Equation (5.2.218c) is for $m \neq k \neq \ell$ and corrects an error in (A1.18c) of [5.17] where the nonzero contribution in $a_{mk\ell}$ for odd values of $m + k + \ell$ was overlooked.

5.3 Order Reduction and Global Adiabatic Invariants for Solutions in Resonance

In the results derived in Secs. 4.5.3, 5.1.3, and 5.2.3, we noted the possible occurrence of resonance, i.e., the presence of zero divisors of the form $r_1\underline{\omega}_1 + r_2\underline{\omega}_2 + \ldots + r_N\underline{\omega}_N$ in the solution. Here the r_i range over the positive or negative integers and zero (but all the r_i do not vanish). The $\underline{\omega}_i$ are functions of $\tilde{t}$, which are either prescribed a priori (if the $\underline{\omega}_i$ in (5.1.1b) depend only on $\tilde{t}$) or are defined in terms of $\tilde{t}$ once the p_i are solved to leading order (if the $\underline{\omega}_i$ depend on the p_i and $\tilde{t}$). In this section, we begin our discussion of the solution procedure for the case of resonance. If we exclude the degenerate case where individual $\underline{\omega}_n$ may vanish, the simplest situation that leads to resonance occurs if two angles are in resonance, i.e., if $R_m\underline{\omega}_m - R_n\underline{\omega}_n = 0$ for positive integers R_m and R_n and $\underline{\omega}_m \neq 0$, $\underline{\omega}_n \neq 0$.

5.3.1 The Hamiltonian Problem for a Resonant Pair

The ideas are best introduced using the Hamiltonian problem (5.2.3) for which one need consider only the Hamiltonian function (5.2.2) and canonical transformations of this function rather than the associated system of differential equations.

Since we can always relabel the indices m and n to be 1 and 2, say, by an appropriate canonical transformation, there is no loss of generality in characterizing a resonant pair by the condition $\underline{\omega}_1 - (s/r)\underline{\omega}_2 = 0$, associated with the two coordinates q_1 and q_2 and the given positive, relatively prime integers s and r. Thus, the Hamiltonian $h(p_i, q_i, \tilde{t}; \epsilon)$ must involve a term that depends *only* on the particular combination $(q_1 - (s/r)q_2)$ of coordinates. Removal of this critical term by the averaging near-identity transformation (5.2.143) leads to a zero divisor in $\hat{F}_1$ and the solution. For example, the Hamiltonian (5.2.95) with $\gamma_1 \neq 0$, $\gamma_2 \neq 0$ has a resonance associated with the term involving $\sin(q_1 - 2q_2)$ for any choice of $\gamma_1(\tilde{t})$ and $\gamma_2(\tilde{t})$ for which $\gamma_1(\tilde{t}) - 2\gamma_2(\tilde{t}) = 0$ at some $\tilde{t} = \tilde{t}_0$. In this case, $s = 2$, $r = 1$. We have also pointed out the resonance associated with the term involving $\sin(q_1 - q_2)$ in the Hamiltonian (5.2.180). In this case, the occurrence of resonance depends on the choice of $\phi(\tilde{t})$ and the initial condition $p_2(0)$.

A resonance such as $(\underline{\omega}_1 - (s/r)\underline{\omega}_2) = 0$ can be avoided by isolating the combination $(q_1 - (s/r)q_2)$ and relabeling it as a new variable $\overline{q}_1$ in a transformed system, then removing all the $\overline{q}_i$ *except* $\overline{q}_1$ by an averaging transformation analogous to (5.2.143). The details of the procedure are discussed next for a general case.

Isolating and averaging transformations

The generating function

$$S(q_i, \overline{p}_i) = \left(q_1 - \frac{s}{r}q_2\right)\overline{p}_1 + \sum_{n=2}^{N} q_n\overline{p}_n, \tag{5.3.1}$$

defines the following canonical transformation $\{p_i, q_i\} \longrightarrow \{\overline{p}_i, \overline{q}_i\}$, which *isolates* the critical combination $(q_1 - (s/r)q_2)$:

$$\overline{q}_1 = q_1 - \frac{s}{r} q_2, \tag{5.3.2a}$$

$$\overline{q}_n = q_n, \quad n \neq 1, \tag{5.3.2b}$$

$$\overline{p}_2 = \frac{s}{r} p_1 + p_2, \tag{5.3.2c}$$

$$\overline{p}_n = p_n, \quad n \neq 2. \tag{5.3.2d}$$

The problem can now be defined in terms of the transformed Hamiltonian $\overline{h}$:

$$\overline{h}(\overline{p}_i, \overline{q}_i, \tilde{t}; \epsilon) = \overline{\underline{h}}_0(\overline{p}_i, \tilde{t}) + \epsilon \overline{\underline{h}}_1(\overline{p}_i, \tilde{t}) + \epsilon \overline{\hat{h}}_c(\overline{p}_i, \overline{q}_i, \tilde{t})$$
$$+ \epsilon \overline{\hat{h}}_s(\overline{p}_i, \overline{q}_i, \tilde{t}) + \epsilon^2 \overline{\underline{h}}_2(\overline{p}_i, \tilde{t}) + O(\epsilon^2), \tag{5.3.3}$$

where the $O(\epsilon^2)$ contribution, $\overline{\underline{h}}_2$, only accounts for the averaged terms and the oscillatory terms to this order have been ignored. The right-hand side of (5.3.3) is calculated from the basic Hamiltonian (5.2.2) using the general condition (5.2.19). Since S is time independent, we have

$$\overline{\underline{h}}_j = \underline{h}_j \left(\overline{p}_1, -\frac{s}{r}\overline{p}_1 + \overline{p}_2, \ldots, \overline{p}_N, \tilde{t} \right), \quad j = 0, 1, 2, \tag{5.3.4a}$$

$$\overline{\hat{h}}_c + \overline{\hat{h}}_s = \hat{h}_1 \left(\overline{p}_1, -\frac{s}{r}\overline{p}_1 + \overline{p}_2, \ldots, \overline{p}_N, \overline{q}_1 + \frac{s}{r}\overline{q}_2, \overline{q}_2, \ldots, \overline{q}_N, \tilde{t} \right). \tag{5.3.4b}$$

The $O(\epsilon)$ oscillatory terms in $\overline{h}$ are decomposed into a critical term $\overline{\hat{h}}_c$, which only involves the $\overline{q}_1$ coordinate, and the noncritical term $\overline{\hat{h}}_s$, which must therefore satisfy the condition

$$\int_0^{2\pi} \cdots \int_0^{2\pi} \overline{\hat{h}}_s(\overline{p}_i, \overline{q}_i, \tilde{t}) d\overline{q}_2 \ldots d\overline{q}_N = 0 \tag{5.3.5}$$

of having a zero average with respect to $\overline{q}_2, \overline{q}_3, \ldots, \overline{q}_N$.

We now introduce the generating function for the near-identity averaging transformation to $O(\epsilon)$, analogous to (5.2.141), defined by

$$\overline{F}(\overline{q}_i, P_i, \tilde{t}; \epsilon) = \sum_{n=1}^{N} \overline{q}_n P_n + \epsilon \overline{F}_1(\overline{q}_i, P_i, \tilde{t}) + O(\epsilon^2). \tag{5.3.6}$$

Note that the P_n, Q_n defined by (5.3.6) are not the same as those in (5.2.143).

Proceeding as in Sec. 5.2.3, we pick $\overline{F}_1$ such that the transformed Hamiltonian in terms of the new $\{P_i, Q_i\}$ variables does not involve $Q_2, \ldots, Q_N$. The calculations are similar to those given in Sec. 5.2.3, and we only summarize the results.

The transformation $\overline{F}_1$ obeys (see (5.2.147))

$$\sum_{j=1}^{N} \frac{\partial \overline{\underline{h}}_0}{\partial \overline{p}_j}(P_i, \tilde{t}) \frac{\partial \overline{F}_1(Q_i, P_i, \tilde{t})}{\partial \overline{q}_j} + \overline{\hat{h}}_s(P_i, Q_i, \tilde{t}) = 0. \tag{5.3.7a}$$

Using (5.3.4a) to express $\overline{h}_0$ in terms of $\underline{h}_0$ and denoting $\partial\underline{h}_0/\partial\overline{p}_n = \underline{\omega}_n$ as in (5.2.4) gives

$$\sum_{j=1}^{N} \underline{\sigma}_j(P_i, \tilde{t}) \frac{\partial \overline{F}_1(Q_i, P_i, \tilde{t})}{\partial \overline{q}_j} + \underset{\wedge}{\overline{h}_s}(P_i, Q_i, \tilde{t}) = 0, \tag{5.3.7b}$$

where we have introduced the notation

$$\underline{\sigma}_1(P_i, \tilde{t}) = \underline{\omega}_1(P_i, \tilde{t}) - \frac{s}{r}\underline{\omega}_2(P_i, \tilde{t}), \tag{5.3.8a}$$

$$\underline{\sigma}_n(P_i, \tilde{t}) = \underline{\omega}_n(P_i, \tilde{t}), \quad n \neq 1. \tag{5.3.8b}$$

In view of the property (5.3.5) for $\underset{\wedge}{\overline{h}_s}$, the solution of (5.3.7b) in the form (5.2.148) *does not involve the critical divisor* $\underline{\sigma}_1$. As in (5.2.148), we find

$$\overline{F}_1(\overline{q}_i, P_i, \tilde{t}) = -\left\{\int \underset{\wedge}{\overline{h}_s}(P_i, \sigma_i s, \tilde{t})ds\right\}_{\sigma_i s = \overline{q}_i} + \overline{\underline{F}}_1(P_i, \tilde{t}), \tag{5.3.9}$$

where $\overline{\underline{F}}_1$ is an arbitrary function to be determined later.

The *partially averaged* Hamiltonian is now

$$H(P_i, Q_1, \tilde{t}; \epsilon) = \underline{H}_0(P_i, \tilde{t}) + \epsilon H_1(P_i, Q_1, \tilde{t}) + \epsilon^2 \underline{H}_2(P_i, \tilde{t}) + O(\epsilon^2), \tag{5.3.10}$$

and the neglected $O(\epsilon^2)$ terms have zero average over all the Q_i. The expressions on the right-hand side of (5.3.10) are defined in terms of those in the original Hamiltonian by

$$\underline{H}_j(P_i, \tilde{t}) = \underline{h}_j\left(P_1, -\frac{s}{r}P_1 + P_2, P_3, \ldots, P_N, \tilde{t}\right), \quad j = 0, 1, \tag{5.3.11a}$$

$$H_1(P_i, Q_i, \tilde{t}) = \underline{H}_1(P_i, \tilde{t}) + \underset{\wedge}{H_c}(P_i, Q_1, \tilde{t}), \tag{5.3.11b}$$

$$\underline{H}_2(P_i, \tilde{t}) = \underline{h}_2\left(P_1, -\frac{s}{r}P_1 + P_2, P_3, \ldots, P_N, \tilde{t}\right) + \frac{\partial \overline{\underline{F}}_1}{\partial \tilde{t}}(P_i, \tilde{t}) - \underline{Z}(P_i, \tilde{t}), \tag{5.3.11c}$$

where $\underset{\wedge}{H_c}$ is the critical term

$$\underset{\wedge}{H_c}(P_i, Q_1, \tilde{t}) = \underset{\wedge}{\overline{h}_c}(P_i, Q_1, \tilde{t}). \tag{5.3.12a}$$

In (5.3.11c), $\underline{Z}$ is the average of the expression below (see (5.2.150)):

$$\underline{Z}(P_i, \tilde{t}) = \left\langle \sum_{j=1}^{N}\sum_{k=1}^{N} \left(\frac{1}{2} \frac{\partial^2 \overline{h}_0}{\partial \overline{p}_j \partial \overline{p}_k} \frac{\partial \overline{F}_1}{\partial \overline{q}_j} \frac{\partial \overline{F}_1}{\partial \overline{q}_k} + \frac{\partial \overline{h}_0}{\partial \overline{p}_k} \frac{\partial^2 \overline{F}_1}{\partial \overline{q}_k \partial \overline{p}_j} \frac{\partial \overline{F}_1}{\partial \overline{q}_j} \right) \right\rangle, \tag{5.3.12b}$$

where the $\overline{p}_i$ arguments are set equal to P_i and the $\overline{q}_i = Q_i$ in the terms on the right-hand side.

As in Sec. 5.2.3, we remove the average terms of $O(\epsilon^2)$ from the transformed Hamiltonian by choosing $\overline{F}_1$ such that the right-hand side of (5.3.11c) vanishes. It then follows from (5.3.10) that the P_i and Q_i obey

$$\dot{P}_1 = -\epsilon \frac{\partial \underset{\wedge}{H_c}}{\partial Q_1} + O(\epsilon^2), \tag{5.3.13a}$$

$$\dot{Q}_1 = \underline{\sigma}_1(P_i, \tilde{t}) + \epsilon \left(\frac{\partial \underline{H}_1}{\partial P_1} + \frac{\partial \underset{\wedge}{H_c}}{\partial P_1} \right) + O(\epsilon^2). \tag{5.3.13b}$$

$$\dot{P}_n = O(\epsilon^2), \quad n \neq 1, \tag{5.3.14a}$$

$$\dot{Q}_n = \underline{\sigma}_n(P_i, \tilde{t}) + \epsilon \left(\frac{\partial \underline{H}_1}{\partial P_n} + \frac{\partial \underset{\wedge}{H_c}}{\partial P_n} \right) + O(\epsilon^2), \quad n \neq 1. \tag{5.3.14b}$$

The neglected $O(\epsilon^2)$ terms are all oscillatory with zero average with respect to all the Q_i.

The reduced problem

According to (5.3.14a), $P_2, \ldots, P_N$ are $N - 1$ constants if oscillatory terms of order ϵ^2 are ignored. This means that to $O(\epsilon)$, P_1 and Q_1 obey the *second-order* system (5.3.13) in which $P_2, \ldots, P_N$ are regarded constant. After this "reduced problem" is solved for $P_1(t)$ and $Q_1(t)$, we can compute $Q_2, \ldots, Q_N$ by quadrature from (5.3.14b). Once $P_1(t)$ is known, the condition $\underline{H}_2(P_i, t) = 0$ also defines $\overline{F}_1$ by quadrature using (5.3.11c). Thus, the essential calculation that remains to be carried out is the solution of the reduced problem; we discuss this in Secs. 5.4–5.5.

For future reference, we rewrite (5.3.13) in the compact form (correct to $O(\epsilon)$)

$$\dot{P} = -\frac{\partial \mathcal{H}}{\partial Q} = -\epsilon \frac{\partial \underset{\wedge}{\mathcal{H}_1}}{\partial Q}, \tag{5.3.15a}$$

$$\dot{Q} = \frac{\partial \mathcal{H}}{\partial P} = \underline{\sigma}(P, \tilde{t}) + \epsilon \left(\frac{\partial \underline{\mathcal{H}}_1}{\partial P} + \frac{\partial \underset{\wedge}{\mathcal{H}_1}}{\partial P} \right) \tag{5.3.15b}$$

associated with the *one-dimensional Hamiltonian*:

$$\mathcal{H}(P, Q, \tilde{t}) = \underline{\mathcal{H}}_0(P, \tilde{t}) + \epsilon[\underline{\mathcal{H}}_1(P, \tilde{t}) + \underset{\wedge}{\mathcal{H}_1}(P, Q, \tilde{t})]. \tag{5.3.16}$$

The simplified notation $P_1 = P$, $Q_1 = Q$, $P_2 = \text{const.}, \ldots, P_N = \text{const.}$ has been used, and we denote $\underline{\mathcal{H}}_0 = \underline{H}_0$, $\underline{\sigma} = \underline{\sigma}_1 = \partial \underline{\mathcal{H}}_0 / \partial P$, $\underline{\mathcal{H}}_1 = \underline{H}_1$, $\underset{\wedge}{\mathcal{H}_1} = \underset{\wedge}{H_c}$.

The solution to $O(\epsilon)$, global adiabatic invariants

Combining (5.3.2) with the transformation generated by (5.3.6) defines the $\{p_i, q_i\}$ in terms of the $\{P_i, Q_i\}$ and vice versa as follows to $O(\epsilon)$:

$$p_2 = P_2 - \frac{s}{r} P_1 + \epsilon \left(\frac{\partial \overline{F}_1}{\partial \overline{q}_2} - \frac{s}{r} \frac{\partial \overline{F}_1}{\partial \overline{q}_1} \right), \tag{5.3.17a}$$

$$p_n = P_n + \epsilon \frac{\partial \overline{F}_1}{\partial \overline{q}_n}, \quad n \neq 2, \tag{5.3.17b}$$

$$q_1 = Q_1 + \frac{s}{r} Q_2 - \epsilon \left(\frac{\partial \overline{F}_1}{\partial P_1} + \frac{s}{r} \frac{\partial \overline{F}_1}{\partial P_2} \right), \tag{5.3.17c}$$

$$q_n = Q_n - \epsilon \frac{\partial \overline{F}_1}{\partial P_n}, \quad n \neq 1, \tag{5.3.17d}$$

where $\overline{q}_n = Q_n$ in the arguments of the derivatives of $\overline{F}_1$.

Once the reduced problem has been solved for $P_1(t; \epsilon)$ and $Q_1(t; \epsilon)$ and the quadratures of (5.3.14b) have been carried out, we have all the P_i and Q_i as functions of time and $2N$ constants of integration. Using these results in (5.3.17) then gives the solution for the p_i and q_i to $O(\epsilon)$.

Consider now the inverse relations of (5.3.17) to $O(\epsilon)$:

$$P_2 = p_2 + \frac{s}{r} p_1 - \epsilon \frac{\partial \overline{F}_1}{\partial \overline{q}_2}, \tag{5.3.18a}$$

$$P_n = p_n - \epsilon \frac{\partial \overline{F}_1}{\partial \overline{q}_n}, \quad n \neq 2, \tag{5.3.18b}$$

$$Q_1 = q_1 - \frac{s}{r} q_2 + \epsilon \frac{\partial \overline{F}_1}{\partial P_1}, \tag{5.3.18c}$$

$$Q_n = q_n + \epsilon \frac{\partial \overline{F}_1}{\partial P_n}, \quad n \neq 1, \tag{5.3.18d}$$

where in the arguments of the derivatives of $\overline{F}_1$ we must set $\overline{q}_1 = q_1 - (s/r)q_2$, $\overline{q}_2 = q_2, \ldots, \overline{q}_N = q_N$, and $P_1 = p_1$, $P_2 = p_2 + (s/r)p_1$, $P_3 = p_3, \ldots$, $P_N = p_N$. Since $P_2, P_3, \ldots, P_N$ remain constant to $O(\epsilon)$ through the $\underline{\sigma}_1 = 0$ resonance, we conclude from (5.3.18a,b) that

$$\overline{\mathcal{A}}_2^{(1)}(p_i, q_i, \tilde{t}) \equiv p_2 + \frac{s}{r} p_1 - \epsilon \frac{\partial \overline{F}_1}{\partial \overline{q}_2}(\overline{q}_i, \overline{p}_i, \tilde{t}), \tag{5.3.19a}$$

$$\overline{\mathcal{A}}_n^{(1)}(p_i, q_i, \tilde{t}) \equiv p_n - \epsilon \frac{\partial \overline{F}_1}{\partial \overline{q}_n}(\overline{q}_i, \overline{p}_i, \tilde{t}), \quad n \neq 1, 2 \tag{5.3.19b}$$

are $N - 1$ *global adiabatic invariants* to $O(\epsilon)$ associated with the $\underline{\sigma}_1 = 0$ resonance. The $\overline{q}_i$, P_i arguments of the $(\partial \overline{F}_1/\partial q_n)$ are evaluated as in (5.3.18). We refer to these as *global* adiabatic invariants because their validity is not affected by the $\underline{\sigma}_1 = 0$ resonance.

Of course, the $N-2$ invariants given by (5.3.19b) are the same to $O(\epsilon)$ as those computed in Sec. 5.2.3 (see (5.2.156) with $(\partial F_1/\partial q_n) = (\partial \overline{F}_1/\partial q_n)$, $n \neq 1, 2$) because the $\underline{\sigma}_1 = 0$ divisor does not occur in (5.2.156) for $n \neq 1, 2$. The essential interesting result is (5.3.19a). It shows that even though $\mathcal{A}_1$ and $\mathcal{A}_2$, as given in (5.2.156), are singular for $\underline{\sigma}_1 \to 0$, the expression $\overline{\mathcal{A}}_2^{(1)} =$ const. remains valid to $O(\epsilon)$.

Coupled oscillators

To illustrate the results of this section, let us consider the Hamiltonian system (5.2.95) for a pair of coupled oscillators. In the notation of this section, the Hamiltonian is

$$h(p_1, p_2, q_1, q_2, \tilde{t}; \epsilon) = \underline{h}_0(p_1, p_2, \tilde{t}) + \epsilon \underset{\wedge}{h}_1(p_1, p_2, q_1, q_2, \tilde{t}), \tag{5.3.20}$$

where

$$\underline{h}_0 = \underline{\omega}_1(\tilde{t})p_1 + \underline{\omega}_2(\tilde{t})p_2, \tag{5.3.21a}$$

$$\begin{aligned}\underset{\wedge}{h}_1 &= \frac{p_2}{2\underline{\omega}_2}\left(\frac{2p_1}{\underline{\omega}_1}\right)^{1/2}[\sin(q_1 - 2q_2) + \sin(q_1 + 2q_2) - 2\sin q_1] \\ &\quad + \frac{1}{2}\frac{p_1}{\underline{\omega}_1}\underline{\omega}_1{}'\sin 2q_1 + \frac{1}{2}\frac{p_2}{\underline{\omega}_2}\underline{\omega}_2{}'\sin 2q_2.\end{aligned} \tag{5.3.21b}$$

Thus, $\underline{h}_1 = \underline{h}_2 = \underset{\wedge}{h}_2 = 0$ for this example.

The functions $\underline{\omega}_1(\tilde{t}) > 0$ and $\underline{\omega}_2(\tilde{t}) > 0$ are prescribed arbitrarily except for the requirement that for some fixed $\tilde{t} = \tilde{t}_0$ we have the resonance condition

$$\underline{\omega}_1(\tilde{t}_0) - 2\underline{\omega}_2(\tilde{t}_0) = 0. \tag{5.3.22}$$

This is a special case of a general class of Hamiltonians discussed in [5.14]. The isolating transformation (5.3.2) for this case ($s = 2, r = 1$) is

$$\begin{aligned}\overline{q}_1 &= q_1 - 2q_2, \quad \overline{q}_2 = q_2 \\ \overline{p}_1 &= p_1, \quad \overline{p}_2 = 2p_1 + p_2.\end{aligned} \tag{5.3.23}$$

Therefore, the transformed Hamiltonian $\overline{h}$ is

$$\begin{aligned}\overline{h}(\overline{p}_1, \overline{p}_2, \overline{q}_1, \overline{q}_2, \tilde{t}; \epsilon) &= \underline{\sigma}_1(\tilde{t})\overline{p}_1 + \underline{\omega}_2\overline{p}_2 + \epsilon\left\{\frac{\overline{p}_2 - 2\overline{p}_1}{2\underline{\omega}_2}\left(\frac{2\overline{p}_1}{\underline{\omega}_1}\right)^{1/2}\Big[\sin\overline{q}_1\right. \\ &\quad + \sin(\overline{q}_1 + 4\overline{q}_2) - 2\sin(\overline{q}_1 + 2\overline{q}_2)\Big] \\ &\quad \left. + \frac{1}{2}\left(\frac{\overline{p}_1}{\underline{\omega}_1}\right)\underline{\omega}_1{}'\sin(2\overline{q}_1 + 4\overline{q}_2) + \frac{1}{2}\left(\frac{\overline{p}_2 - 2\overline{p}_1}{\underline{\omega}_2}\right)\underline{\omega}_2{}'\sin 2\overline{q}_2\right\},\end{aligned} \tag{5.3.24}$$

where $\underline{\sigma}_1(\tilde{t}) = \underline{\omega}_1(\tilde{t}) - 2\underline{\omega}_2(\tilde{t})$.

We remove all the oscillatory terms *except* the critical term involving $\sin \overline{q}_1$ by the canonical transformation, $(\overline{p}_i, \overline{q}_i) \to (P_i, Q_i)$, generated by (5.3.6) with $\overline{F}_1$ given by

$$\overline{F}_1(\overline{q}_1, \overline{q}_2, P_1, P_2, \tilde{t}) = \frac{P_2 - 2P_1}{2\underline{\omega}_2}\left(\frac{2P_1}{\underline{\omega}_1}\right)^{1/2}\left[\frac{\cos(\overline{q}_1 + 4\overline{q}_2)}{\underline{\omega}_1 + 2\underline{\omega}_2} - \frac{2\cos(\overline{q}_1 + 2\overline{q}_2)}{\underline{\omega}_1}\right] - \frac{1}{2}\left(\frac{P_1}{\underline{\omega}_1}\right)\underline{\omega}_1{}'\frac{\cos(2\overline{q}_1 + 4\overline{q}_2)}{2\underline{\omega}_1}$$

$$-\frac{1}{2}\left(\frac{P_2 - 2P_1}{\underline{\omega}_2}\right)\underline{\omega}_2{}'\frac{\cos 2\overline{q}_2}{2\underline{\omega}_2} + \underline{\overline{F}}_1(P_1, P_2, \tilde{t}). \tag{5.3.25}$$

For brevity, we omit the straightforward but tedious calculations for $\underline{Z}$ and hence $\underline{F}_1$. The ideas are identical to those discussed in previous examples. (See Problem 1 for a special case.)

The partially avaraged Hamiltonian (5.3.10) has the form

$$H(P_1, P_2, Q_1, Q_2, \tilde{t}; \epsilon) = \underline{\sigma}_1(\tilde{t})P_1 + \underline{\omega}_2(\tilde{t})P_2 - \epsilon A(P_1, P_2, \tilde{t}) \sin Q_1 + O(\epsilon^2), \tag{5.3.26a}$$

where

$$A(P_1, P_2, \tilde{t}) = \frac{P_1 - 2P_2}{2\underline{\omega}_2(\tilde{t})}\left(\frac{2P_1}{\underline{\omega}_1(\tilde{t})}\right)^{1/2}, \tag{5.3.26b}$$

and the terms of $O(\epsilon^2)$ neglected in (5.3.26) have zero average with respect to Q_1 and Q_2.

Thus, the P_i and Q_i satisfy

$$\dot{P}_1 = \epsilon A(P_1, P_2, \tilde{t}) \cos Q_1 + O(\epsilon^2), \tag{5.3.27a}$$

$$\dot{Q}_1 = \sigma_1(\tilde{t}) - \epsilon \frac{\partial A}{\partial P_1} \sin Q_1 + O(\epsilon^2). \tag{5.3.27b}$$

$$\dot{P}_2 = O(\epsilon^2), \tag{5.3.28a}$$

$$\dot{Q}_2 = \underline{\omega}_2(\tilde{t}) - \epsilon \frac{\partial A}{\partial P_2} \sin Q_1 + O(\epsilon^2). \tag{5.3.28b}$$

Equation (5.3.28a) implies that P_2 = constant to $O(\epsilon)$. Therefore, the reduced system consisting of (5.3.27a) and (5.3.27b) decouples from (5.3.28). Once (5.3.27) is solved, we can compute Q_2 from (5.3.28b) by quadrature. The reader will note that the reduced system (5.3.27) with P_2 = constant is a special case (where σ_1 only depends on $\tilde{t}$) of the model problem we have studied in previous sections (see (4.5.13), (5.1.20), (5.2.157)) assuming $\sigma_1 \neq 0$. We will study this problem for the resonance case when $\sigma_1(\tilde{t}_0) = 0$ in Sec. 5.4. Once $P_1(t; \epsilon)$ and $Q_1(t; \epsilon)$ have been defined and we have evaluated the quadrature for $Q_2(t; \epsilon)$, we can use the transformation relations (5.3.17) to derive the solution for the $p_i(t; \epsilon)$ and $q_i(t; \epsilon)$:

The global adiabatic invariant to $O(\epsilon)$ valid through the $\sigma_1 = 0$ resonance is obtained from (5.3.19a) using (5.3.25).

$$\overline{\mathcal{A}}_2^{(1)} = p_2 + 2p_1 + \epsilon \left\{ \frac{p_2}{\underline{\omega}_2} \left(\frac{2p_1}{\omega_1} \right)^{1/2} \left[\frac{2\sin(q_1 + 2q_2)}{\underline{\omega}_1 + 2\underline{\omega}_2} - \frac{2}{\underline{\omega}_1} \sin q_1 \right] + \frac{p_1}{\underline{\omega}_1{}^2} \omega_1' \sin 2q_1 + \frac{p_2}{\underline{\omega}_2{}^2} \omega_2' \sin 2q_2 \right\}. \quad (5.3.29)$$

Direct differentiation and use of the equations for $\dot{p}$ and $\dot{q}$ confirms the validity of this result. The interested reader may also wish to investigate the validity of $\mathcal{A}_2^{(1)}$ by solving for the $p_i(t)$ and $q_i(t)$ numerically and then substituting these results into (5.3.29). It will be seen that this numerically computed expression for $\mathcal{A}_2^{(1)}$ oscillates with an amplitude of $O(\epsilon^2)$ about its initial value right through resonance. A particular example for $\underline{\omega}_1 = 2 - \tilde{t}/2$, $\underline{\omega}_2 = (1/4) + \tilde{t}/2$, $\epsilon = 0.05$, $p_1(0) = 0.25$, $q_1(0) = 0$, $p_2(0) = 0.5$, $q_2(0) = 0$ appears in Figure 5.3.1(a) taken from [5.3]. For this choice of $\underline{\omega}_1$ and $\underline{\omega}_2$, resonance occurs at $\tilde{t} = 1$, and it is seen that the individual actions p_1 and p_2 oscillate with small amplitude about levels that undergo a transition through $\tilde{t} = 1$. (This behavior is discussed in detail in Sec. 5.4.4.) In contrast, the $O(1)$ adiabatic invariant $p_2 + 2p_1$ oscillates with $O(\epsilon)$ amplitude about its initial level, and the adiabatic invariant $\mathcal{A}_2^{(1)}$ given by (5.3.29) oscillates with $O(\epsilon^2)$ amplitude about this level. Other numerical examples are given in [5.17].

In [5.3], passage through resonance is studied for the case where slow variations occur with respect to a variable $t^* = \epsilon^2 t$. In this case, the actions undergo an $O(1)$ transition through resonance, whereas the counterpart of the adiabatic invariant $\mathcal{A}_2^{(1)}$ in (5.3.29) (given by (2.16) of [5.3]) remains constant to $O(\epsilon^2)$. This is shown in Figure 5.3.1(b), which is calculated for the same initial conditions and ω values as those in Figure 5.3.1(a). We do not discuss the details for this case here; a brief summary of the pertinent features is given at the end of Sec. 5.3.2. For a more complete account, refer to [5.3]. The important special case of our results for the case of constant ω_n is explored in Problem 1.

5.3.2 *The Non-Hamiltonian Problem for a Resonant Pair*

When the basic system is non-Hamiltonian but can be expressed in the standard form (5.1.51), the same procedure that we used in Sec. 5.3.1 still applies, except one cannot bypass the individual differential equations and must perform the various transformations on these equations.

Isolating and averaging transformations

We assume as in Sec. 5.3.1 that the critical pair of coordinates have been relabeled q_1 and q_2, that the resonance condition is $r\omega_1 - s\omega_2 = 0$, and that we have isolated

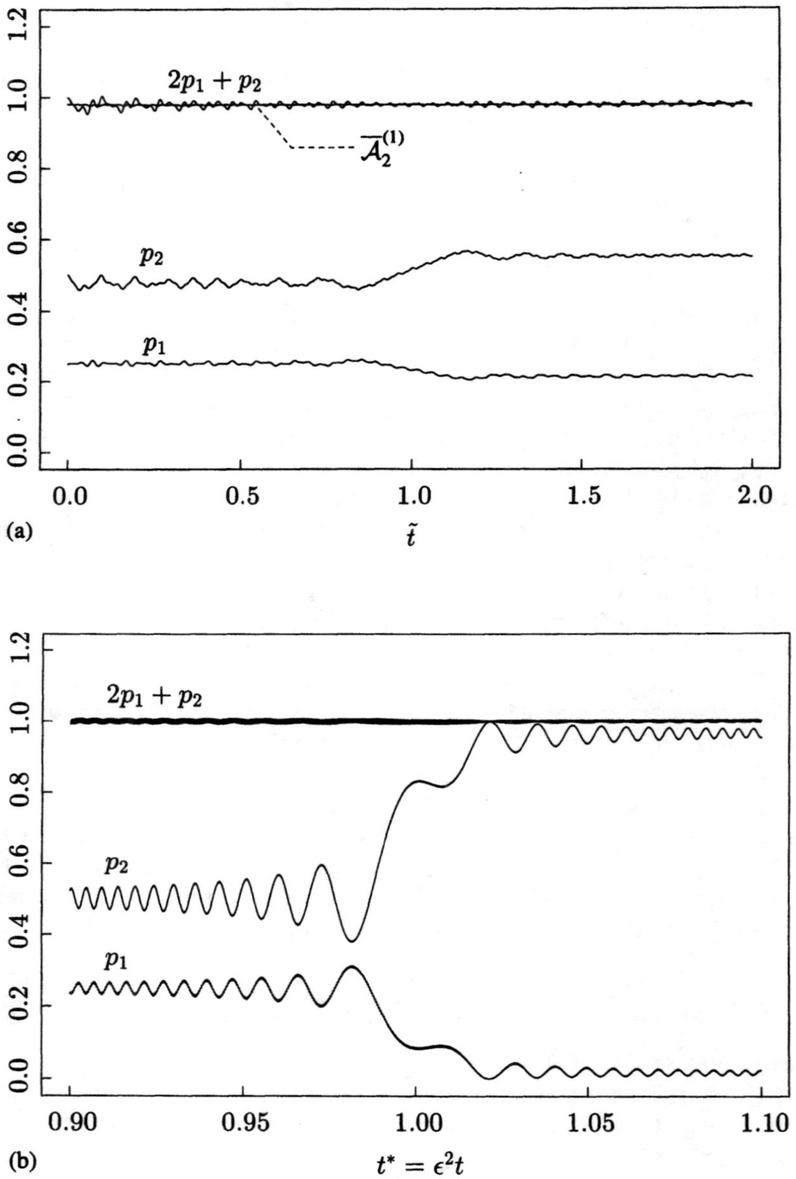

FIGURE 5.3.1. Global Adiabatic Invariant through Resonance for Coupled Oscillators

the critical terms depending only on $rq_1 - sq_2$ in the various functions $\underset{\wedge}{f}_m$ and $\underset{\wedge}{g}_n$. Thus, we write (5.1.51) in the form

$$\dot{p}_m = \epsilon \underline{f}_m(p_i, \tilde{t}) + \epsilon \underset{\wedge}{f}_{m_c}(p_i, rq_1 - sq_2, \tilde{t}) + \epsilon \underset{\wedge}{f}_{m_s}(p_i, q_i, \tilde{t})$$

$$+ \epsilon^2 \underline{k}_m(p_i, \tilde{t}) + O(\epsilon^2), \tag{5.3.30a}$$

$$\dot{q}_n = \underline{\omega}_n(p_i, \tilde{t}) + \epsilon \underline{g}_n(p_i, \tilde{t}) + \epsilon \underset{\wedge}{g}_{n_c}(p_i, rq_1 - sq_2, \tilde{t}) + \epsilon \underset{\wedge}{g}_{n_s}(p_i, q_i, \tilde{t})$$

$$+ \epsilon^2 \underline{\ell}_n(p_i, \tilde{t}) + O(\epsilon^2), \tag{5.3.30b}$$

where the neglected terms of $O(\epsilon^2)$ have zero average with respect to all the q_i. One can use an even simpler isolating transformation than (5.3.2) involving only the critical angles q_1, q_2 without transforming the p_i, as there is no longer a requirement that the transformation be canonical. This is the strategy in [5.3]. However, in order to facilitate the process of specializing our results to the Hamiltonian problem, we use the following generalization of (5.3.2):

$$\overline{q}_1 = q_1 - \frac{s}{r} q_2, \tag{5.3.31a}$$

$$\overline{q}_n = q_n, \quad n = 2, \ldots, N, \tag{5.3.31b}$$

$$\overline{p}_2 = \frac{s}{r} p_1 + p_2, \tag{5.3.31c}$$

$$\overline{p}_m = p_m, \quad m = 1, 3, \ldots, M. \tag{5.3.31d}$$

Using (5.3.31) and (5.3.30) gives the following transformed system that generalizes the system associated with (5.3.3):

$$\dot{p}_m = \epsilon \overline{\underline{f}}_m(\overline{p}_i, \tilde{t}) + \epsilon \underset{\wedge}{\overline{f}}_{m_c}(\overline{p}_i, \overline{q}_1, \tilde{t}) + \epsilon \underset{\wedge}{\overline{f}}_{m_s}(\overline{p}_i, \overline{q}_i, \tilde{t})$$

$$+ \epsilon^2 \overline{\underline{k}}_m(\overline{p}_i, \tilde{t}) + O(\epsilon^2), \tag{5.3.32a}$$

$$\dot{q}_n = \underline{\sigma}_n(\overline{p}_i, \tilde{t}) + \epsilon \overline{\underline{g}}_n(\overline{p}_i, \tilde{t}) + \epsilon \underset{\wedge}{\overline{g}}_{n_c}(\overline{p}_i, \overline{q}_1, \tilde{t})$$

$$+ \epsilon \underset{\wedge}{\overline{g}}_{n_s}(\overline{p}_i, \overline{q}_i, \tilde{t}) + \epsilon^2 \overline{\underline{l}}_n(\overline{p}_i, \tilde{t}) + O(\epsilon^2), \tag{5.3.32b}$$

where again the neglected terms of order ϵ^2 are all oscillatory with zero average over all the $\overline{q}_i$. The right-hand sides of (5.3.32) are related to the expressions in (5.3.30) as follows:

$$\overline{\underline{f}}_2 = (s/r)\underline{f}_1 + \underline{f}_2; \quad \overline{\underline{f}}_m = \underline{f}_m, \quad m \neq 2, \tag{5.3.33a,b}$$

$$\underset{\wedge}{\overline{f}}_{2_c} = (s/r)\underset{\wedge}{f}_{1_c} + \underset{\wedge}{f}_{2_c}; \quad \underset{\wedge}{\overline{f}}_{2_s} = (s/r)\underset{\wedge}{f}_{1_s} + \underset{\wedge}{f}_{2_s}, \tag{5.3.33c,d}$$

$$\underset{\wedge}{\overline{f}}_{m_c} = \underset{\wedge}{f}_{m_c}; \quad \underset{\wedge}{\overline{f}}_{m_s} = \underset{\wedge}{f}_{m_s}, \quad m \neq 2, \tag{5.3.33e,f}$$

$$\overline{\underline{k}}_2 = (s/r)\underline{k}_1 + \underline{k}_2; \quad \overline{\underline{k}}_m = \underline{k}_m, \quad m \neq 2. \tag{5.3.33g,h}$$

$$\underline{\sigma}_1 = \underline{\omega}_1 - (s/r)\underline{\omega}_2; \quad \underline{\sigma}_n = \underline{\omega}_n, \quad n \neq 1, \tag{5.3.34a,b}$$

$$\overline{g}_1 = \underline{g}_1 - (s/r)g_2; \quad \overline{g}_n = \underline{g}_n; \quad n \neq 1, \tag{5.3.34c,d}$$

$$\underset{\wedge}{\overline{g}}_{1_c} = \underset{\wedge}{g}_{1_c} - \frac{s}{r}\underset{\wedge}{g}_{2_c}; \quad \underset{\wedge}{\overline{g}}_{1_s} = \underset{\wedge}{g}_{1_s} - (s/r)\underset{\wedge}{g}_{2_s}, \tag{5.3.34e,f}$$

$$\underset{\wedge}{\overline{g}}_{n_c} = \underset{\wedge}{g}_{n_c}; \quad \underset{\wedge}{\overline{g}}_{n_s} = \underset{\wedge}{g}_{n_s}, \quad n \neq 1, \tag{5.3.34g,h}$$

$$\overline{l}_1 = \underline{l}_1 - (s/r)\underline{l}_2; \quad \overline{l}_n = \underline{l}_n, \quad n \neq 1. \tag{5.3.34i,j}$$

The arguments of the functions appearing on the right-hand sides of (5.3.33)–(5.3.34) are evaluated according to

$$p_2 = -(s/r)\overline{p}_1 + \overline{p}_2, \tag{5.3.35a}$$

$$p_m = \overline{p}_m, \quad m \neq 2, \tag{5.3.35b}$$

$$q_1 = \overline{q}_1 + (s/r)\overline{q}_2, \tag{5.3.35c}$$

$$q_n = \overline{q}_n, \quad n \neq 1. \tag{5.3.35d}$$

As in Sec. 5.1.3, we introduce the near-identity averaging transformations from $\{\overline{p}_i, \overline{q}_i\}$ to $\{P_i, Q_i\}$ defined by

$$P_m = \overline{p}_m + \epsilon\overline{T}_m(\overline{p}_i, \overline{q}_i, \tilde{t}) + O(\epsilon^2), \tag{5.3.36a}$$

$$Q_n = \overline{q}_n + \epsilon\overline{L}_n(\overline{p}_i, \overline{q}_i, \tilde{t}) + O(\epsilon^2), \tag{5.3.36b}$$

with inverse

$$\overline{p}_m = P_m - \epsilon\overline{T}_m(P_i, Q_i, \tilde{t}) + O(\epsilon^2), \tag{5.3.37a}$$

$$\overline{q}_n = Q_n - \epsilon\overline{L}_n(P_i, Q_i, \tilde{t}) + O(\epsilon^2). \tag{5.3.37b}$$

Removing the noncritical terms from the transformed system for $\dot{P}_m$, $\dot{Q}_n$ gives (see (5.1.58) and (5.1.59))

$$\sum_{k=1}^{N} \frac{\partial\overline{T}_m}{\partial\overline{q}_k}\underline{\sigma}_k = -\underset{\wedge}{f}_{m_s}, \tag{5.3.38a}$$

$$\sum_{k=1}^{N} \frac{\partial\overline{L}_n}{\partial\overline{q}_k}\underline{\sigma}_k = -\underset{\wedge}{g}_{n_s} + \sum_{j=1}^{M} \frac{\partial\underline{\sigma}_n}{\partial\overline{p}_j}\underset{\wedge}{\overline{T}}_j, \tag{5.3.38b}$$

and these define $\overline{T}_m$ and $\overline{L}_n$ to within arbitrary additive functions $\underline{\overline{T}}_m$ and $\underline{\overline{L}}_n$ of the P_i and $\tilde{t}$ as in (5.1.60)–(5.1.61). We choose the $\underline{\overline{T}}_m$ and $\underline{\overline{L}}_n$ as in Sec. 5.1.3 to remove averaged terms of order ϵ^2. The results are formally identical to those worked out earlier (see (5.1.77) and (5.1.83)) and are not repeated. The reduced problem in which all the coordinates except Q_1 are absent then becomes

$$\frac{dP_m}{dt} = \epsilon\underline{\overline{f}}_m(P_i, \tilde{t}) + \epsilon\underset{\wedge}{\overline{f}}_{m_c}(P_i, Q_1, \tilde{t}) + O(\epsilon^2), \tag{5.3.39a}$$

$$\frac{dQ_n}{dt} = \underline{\sigma}_n(P_i, \tilde{t}) + \epsilon\underline{\overline{g}}_n(P_i, \tilde{t}) + \epsilon\underset{\wedge}{\overline{g}}_{n_c}(P_i, Q_1, \tilde{t}) + O(\epsilon^2), \tag{5.3.39b}$$

where the neglected $O(\epsilon^2)$ terms have zero average with respect to all the Q_i.

Unlike the Hamiltonian case, $P_2, \ldots, P_N$ *are in general not constant*, and the equations governing P_1 and Q_1 *are not decoupled* from the other equations. We must first solve the $(M+1)$st-order system consisting of (5.3.39a) for $m = 1, \ldots, M$ and (5.3.39b) for $n = 1$. Having determined Q_1 and all the P_i, one can then calculate the $(N-1)$ remaining Q_i from (5.3.39b) by quadrature.

In particular, the occurrence of Q_1 in all the equations (5.3.39a) for $\dot{P}_m$ implies that, in general, no global adiabatic invariants are available for the non-Hamiltonian problem. Special cases, where one or more of the P_n depend only on $\tilde{t}$, can, of course, be constructed for non-Hamiltonian problems. Examples of this and the associated adiabatic invariants can be found in Sec. 3 of [5.17]. We also refer to [5.2] for a discussion of the solution of (5.3.39) for sustained resonance. This solution involves a generalization of the method discussed in Sec. 5.5.

Once the P_i and Q_i have been calculated, the solution for the p_i and q_i follows from the combined transformations given in (5.3.35) and (5.3.37).

The model problem (5.1.108)

To illustrate ideas, let us reconsider the model problem (5.1.108) for situations where the divisor $(\phi - p_2)$ may vanish. The isolating transformation (5.3.31) with $s = r = 1$ gives

$$\frac{d\overline{p}_1}{dt} = -\epsilon\psi\overline{p}_1^2(\overline{p}_2 - \overline{p}_1)\cos\overline{q}_1 - \epsilon\theta\overline{p}_1(\overline{p}_2 - \overline{p}_1)^2\cos(\overline{q}_1 + 2\overline{q}_2), \quad (5.3.40a)$$

$$\frac{d\overline{p}_2}{dt} = -2\epsilon\theta\overline{p}_1(\overline{p}_2 - \overline{p}_1)^2\cos(\overline{q}_1 + 2\overline{q}_2) + \epsilon d(\overline{p}_2 - \overline{p}_1)\sin\overline{q}_1, \quad (5.3.40b)$$

$$\frac{d\overline{q}_1}{dt} = \underline{\sigma}_1 + \epsilon\psi\overline{p}_1(2\overline{p}_2 - 3\overline{p}_1)\sin\overline{q}_1$$

$$+ \epsilon\theta(\overline{p}_2 - 3\overline{p}_1)(\overline{p}_2 - \overline{p}_1)\sin(\overline{q}_1 + 2\overline{q}_2) + \frac{\epsilon c\overline{p}_1}{\overline{p}_2 - \overline{p}_1}\cos\overline{q}_2, \quad (5.3.40c)$$

$$\frac{d\overline{q}_2}{dt} = \underline{\sigma}_2 + \epsilon\psi\overline{p}_1^2\sin\overline{q}_1 + 2\epsilon\theta\overline{p}_1(\overline{p}_2 - \overline{p}_1)\sin(\overline{q}_1 + 2\overline{q}_2), \quad (5.3.40d)$$

where

$$\underline{\sigma}_1 = \phi(\tilde{t}) - \overline{p}_2 + \overline{p}_1, \quad (5.3.40e)$$

$$\underline{\sigma}_2 = \overline{p}_2 - \overline{p}_1, \quad (5.3.40f)$$

and the terms multiplied by ψ are critical. Note that in (5.3.40e) $\underline{\sigma}_1 = \phi - p_2$ in terms of the original variables.

Using (5.3.38) to eliminate the noncritical $O(\epsilon)$ terms gives

$$\underset{\wedge}{\overline{T}}_1 = \frac{\theta\overline{p}_1(\overline{p}_2 - \overline{p}_1)^2}{\underline{\sigma}_1 + 2\underline{\sigma}_2}\sin(\overline{q}_1 + 2\overline{q}_2), \quad (5.3.41a)$$

$$\underset{\wedge}{\overline{T}}_2 = \frac{2\theta\overline{p}_1(\overline{p}_2 - \overline{p}_1)^2}{\underline{\sigma}_1 + 2\underline{\sigma}_2}\sin(\overline{q}_1 + 2\overline{q}_2), \quad (5.3.41b)$$

$$\underset{\wedge}{\overline{L}}_1 = \frac{\theta(\overline{p}_2 - \overline{p}_1)[\phi(\overline{p}_2 - 3\overline{p}_1) + (\overline{p}_2 - \overline{p}_1)(\overline{p}_2 - 2\overline{p}_1)]}{(\underline{\sigma}_1 + 2\underline{\sigma}_2)^2} \cos(\overline{q}_1 + 2\overline{q}_2)$$

$$+ \frac{c\overline{p}_1}{\underline{\sigma}_2{}^2} \sin \overline{q}_2, \tag{5.3.41c}$$

$$\underset{\wedge}{\overline{L}}_2 = \frac{\theta\overline{p}_1(\overline{p}_2 - \overline{p}_1)(2\phi + \overline{p}_2 - \overline{p}_1)}{(\underline{\sigma}_1 + 2\underline{\sigma}_2)^2} \cos(\overline{q}_1 + 2\overline{q}_2). \tag{5.3.41d}$$

Now, since $\underline{\sigma}_1 + 2\underline{\sigma}_2 = \phi + \overline{p}_2 - \overline{p}_1 = \phi + p_2$ and $\underline{\sigma}_2 = \overline{p}_2 - \overline{p}_1 = p_2$, none of the divisors in (5.3.41) vanish as long as $\phi \neq 0$ and $p_2 \neq 0$. We also verify that for the Hamiltonian problem, where $c = d = 0$, the formulas for the $\underset{\wedge}{T}_n$ and $\underset{\wedge}{L}_n$ follow from the generating function

$$\underset{\wedge}{\overline{F}}_1(P_i, \overline{q}_i, \tilde{t}) = \frac{\theta P_1(P_2 - P_1)^2}{\phi + P_2 - P_1} \cos(\overline{q}_1 + 2\overline{q}_2) \tag{5.3.42}$$

when we use $\underset{\wedge}{T}_n = -(\partial \underset{\wedge}{\overline{F}}_1 / \partial \overline{q}_n)$ and $\underset{\wedge}{\overline{L}}_n = (\partial \underset{\wedge}{\overline{F}}_1 / \partial P_n)$ according to (5.2.143c) and (5.2.143d).

As in Sec. 5.2.3, we can remove the average terms of order ϵ^2 from the equations for $\dot{P}_n$ and $\dot{Q}_n$ by appropriate choices of the $\underline{T}_n$ and $\underline{L}_n$. The calculations are similar to those given earlier and are not repeated. We only point out that $\underline{T}_1$ and $\underline{T}_2$ are both constant.

The averaged equations obtained from (5.3.40) by removing the noncritical terms are

$$\dot{P}_1 = -\epsilon\psi P_1^2(P_2 - P_1)\cos Q_1 + O(\epsilon^2), \tag{5.3.43a}$$

$$\dot{P}_2 = \epsilon d(P_2 - P_1)\sin Q_1 + O(\epsilon^2), \tag{5.3.43b}$$

$$\dot{Q}_1 = \phi - P_2 + P_1 + \epsilon\psi P_1(2P_2 - 3P_1)\sin Q_1 + O(\epsilon^2), \tag{5.3.43c}$$

$$\dot{Q}_2 = P_2 - P_1 + \epsilon\psi P_1^2 \sin Q_1 + O(\epsilon^2), \tag{5.3.43d}$$

and all the neglected $O(\epsilon^2)$ terms have zero average with respect to Q_1 and Q_2.

Consider first the case $d \equiv 0$ (which does not necessarily imply that (5.3.40) is Hamiltonian since c need not be identically equal to zero). In this case, $P_2 =$ const. to $O(\epsilon)$ and the reduced problem is of second order, consisting of (5.3.43a) and (5.3.43c). Quadrature defines Q_2 once P_1 and Q_1 have been determined. Furthermore, we have the following global adiabatic invariant to $O(\epsilon)$ (see (5.3.36a) with $\overline{p}_1 = p_1, \overline{p}_2 = p_1 + p_2, \overline{q}_1 + 2\overline{q}_2 = q_1 + q_2, \underline{T}_2 = 0$)

$$\overline{\mathcal{A}}_2^{(1)}(p_i, q_i, \tilde{t}; \epsilon) = p_1 + p_2 + \frac{2\epsilon\theta p_1 p_2^2}{\phi + p_2} \sin(q_1 + q_2). \tag{5.3.44}$$

It is instructive to explicitly verify the adiabatic invariance of $\overline{\mathcal{A}}_2^{(1)}$. Differentiating the expression in (5.3.44) gives the *exact* expression

$$\frac{d\overline{\mathcal{A}}_2^{(1)}}{dt} = \dot{p}_1 + \dot{p}_2 + \frac{2\epsilon\theta p_1 p_2^2}{\phi + p_2}(\dot{q}_1 + \dot{q}_2)\cos(q_1 + q_2)$$

$$+ \frac{2\epsilon^2 p_1 p_2}{(\phi + p_2)^2}[\theta'(\phi + p_2) - \phi' p_2]\sin(q_1 + q_2)$$
$$+ \frac{2\epsilon\theta}{(\phi + p_2)^2}[\dot{p}_1 p_2^2(\phi + p_2)$$
$$+ \dot{p}_2 p_1 p_2(2\phi + p_2)]\sin(q_1 + q_2). \tag{5.3.45}$$

Now, if we use (5.1.108) to express $\dot{p}_1, \dot{p}_2, \dot{q}_1, \dot{q}_2$ in terms of the $p_i, q_i, \tilde{t}$, we find that only terms of order ϵ^2 remain uncanceled on the right-hand side of (5.3.45). Moreover, all these remaining terms have zero average with respect to q_1 and q_2 and are free of the critical divisor $(\phi - p_2)$. This confirms the adiabatic invariance of $\overline{\mathcal{A}}_2^{(1)}$ to $O(\epsilon)$ in a form valid through the resonance.

If $d \not\equiv 0$, the reduced problem is of order three and obeys the system (5.3.43a)–(5.3.43c) for P_1, Q_1, P_2. Equation (5.3.43d) determines Q_2 by quadrature once P_1, Q_1, P_2 have been calculated. This illustrates the possibility of having a reduced problem of order as high as $M + 1$ (in this case $M + 1 = 3$) for a general non-Hamiltonian problem. While the approximate solution of this reduced problem is somewhat more involved now than for the case $P_2 = $ const., the important difference is the nonexistence of a global adiabatic invariant. This is easily seen by noting that there exists no transformation (which is free of zero divisors) of the variables P_n, Q_n to new variables $\overline{P}_n, \overline{Q}_n$ such that the equation for one of the $\overline{P}_n$ is of the form $d\overline{P}_n/dt = O(\epsilon^2)$.

5.3.3 Generalizations

A resonance involving three or more angles

The case of three or more angles q_i in resonance has no essential mathematical distinction from the case of a resonant pair that we considered in Sec. 5.3.2. Consider, for example, the situation where $r_1\omega_1 + r_2\omega_2 + \ldots + r_R\omega_R = 0$ for some $R \le N$, and nonzero integers $r_1, \ldots, r_R$. The isolating transformation generated by (see (5.3.1))

$$S(q_i, p_i) = \left(q_1 + \frac{r_2}{r_1}q_2 + \ldots + \frac{r_R}{r_1}q_R\right)\overline{p}_1 + \sum_{n=2}^{N} q_n\overline{p}_n \tag{5.3.46}$$

defines the following implicit transformation:

$$p_n = \frac{\partial S}{\partial q_1} = \overline{p}_n, \quad n = 1, R+1, \ldots, N, \tag{5.3.47a}$$
$$p_n = \frac{\partial S}{\partial q_n} = \frac{r_n}{r_1}\overline{p}_1 + \overline{p}_n, \quad n = 2, \ldots, R, \tag{5.3.47b}$$
$$\overline{q}_1 = \frac{\partial S}{\partial \overline{p}_1} = q_1 + \frac{r_2}{r_1}q_2 + \ldots + \frac{r_R}{r_1}q_R, \tag{5.3.47c}$$
$$\overline{q}_n = \frac{\partial S}{\partial \overline{p}_n} = q_n, \quad n \ne 1. \tag{5.3.47d}$$

The corresponding explicit isolating transformation (canonical) is then

$$\overline{p}_n = p_n, \quad n = 1, R+1, \ldots, N, \tag{5.3.48a}$$

$$\overline{p}_n = p_n - \frac{r_n}{r_1} p_1, \quad n = 2, \ldots, R, \tag{5.3.48b}$$

$$\overline{q}_1 = q_1 + \frac{r_2}{r_1} q_2 + \ldots + \frac{r_R}{r_1} q_R, \tag{5.3.48c}$$

$$\overline{q}_n = q_n, \quad n \neq 1. \tag{5.3.48d}$$

The following problem of three oscillators having a weak quadratic coupling illustrates ideas. Consider the system

$$\frac{d^2 y_n}{dt^2} + \omega_n^2(\tilde{t}) y_n = \epsilon \kappa_n y_j y_k \quad n = 1, 2, 3, \ j \neq k \neq n, \tag{5.3.49a}$$

where the $\omega_n > 0$ are prescribed functions of $\tilde{t} = \epsilon t$, and the κ_n are positive constants that do not depend on ϵ. We note that the transformation $x_n = (\kappa_j \kappa_k)^{1/2} y_n$ removes the κ_n:

$$\frac{d^2 x_n}{dt^2} + \omega_n^2(\tilde{t}) x_n = \epsilon x_j x_k, \quad n = 1, 2, 3, \ j \neq k \neq n. \tag{5.3.49b}$$

Now, introducing the same canonical transformation (5.2.59)–(5.2.60) as we used for the case of two oscillators, we find the following Hamiltonian in standard form:

$$h(p_i, q_i, \tilde{t}; \epsilon) = \sum_{n=1}^{3} \omega_n(\tilde{t}) p_n + \epsilon \left(\frac{p_1 p_2 p_3}{2\omega_1 \omega_2 \omega_3} \right)^{1/2} [-\sin(q_1 + q_2 - q_3)$$

$$+ \sin(q_1 + q_2 + q_3) - \sin(q_1 - q_2 + q_3) + \sin(q_1 - q_2 - q_3)]$$

$$+\epsilon \sum_{n=1}^{3} \frac{p_n}{2\omega_n} \omega_n' \sin 2q_n. \tag{5.3.50}$$

Because of symmetry, we need only study the resonance $\omega_1 + \omega_2 - \omega_3 = 0$, i.e., $r_1 = r_2 = 1, r_3 = -1, R = 3$. The isolating transformation (5.3.48) gives

$$\overline{p}_1 = p_1, \ \overline{p}_2 = p_2 - p_1, \ \overline{p}_3 = p_3 + p_1, \tag{5.3.51a}$$

$$\overline{q}_1 = q_1 + q_2 - q_3, \ \overline{q}_2 = q_2, \ \overline{q}_3 = q_3, \tag{5.3.51b}$$

and the transformed Hamiltonian becomes

$$\overline{h}(\overline{p}_i, \overline{q}_i, \tilde{t}; \epsilon) = \sigma_1(\tilde{t})\overline{p}_1 + \omega_2(\tilde{t})\overline{p}_2 + \omega_3(\tilde{t})\overline{p}_3$$

$$+\epsilon \left[\frac{\overline{p}_1(\overline{p}_2 + \overline{p}_1)(\overline{p}_3 - \overline{p}_1)}{2\omega_1\omega_2\omega_3} \right]^{1/2} [-\sin \overline{q}_1 + \sin(\overline{q}_1 + 2\overline{q}_3)$$

$$- \sin(\overline{q}_1 - 2\overline{q}_2 + 2\overline{q}_3) + \sin(\overline{q}_1 - 2\overline{q}_2)] + \epsilon \left[\frac{\overline{p}_1}{2\omega_1} \omega_1' + \sin 2(\overline{q}_1 - \overline{q}_2 + \overline{q}_3) \right.$$

$$+\frac{\overline{p}_2+\overline{p}_1}{2\omega_2}\omega_2'\sin 2\overline{q}_2+\frac{\overline{p}_3-\overline{p}_1}{2\omega_3}\omega_3'\sin 2\overline{q}_3\Bigg], \tag{5.3.52}$$

where $\sigma_1 = \omega_1 + \omega_2 - \omega_3$.

The transformation generated by (5.3.6)–(5.3.7) (see Problem 2) leads to the partially averaged Hamiltonian

$$\overline{H}(P_i, Q_i, \tilde{t}; \epsilon) = \sigma_1(\tilde{t})P_1 - \epsilon\left[\frac{P_1(P_2+P_1)(P_3-P_2)}{2\omega_1\omega_2\omega_3}\right]^{1/2}\sin Q_1 + O(\epsilon^2), \tag{5.3.53}$$

where the terms of $O(\epsilon^2)$ that we have ignored have zero average with respect to Q_1, Q_2, Q_3. Thus, P_1 and P_2 are adiabatic invariants to $O(\epsilon)$, and the reduced problem is formally identical to (5.3.27a)–(5.3.27b) with

$$A = \left[\frac{P_1(P_2+P_1)(P_3-P_2)}{2\omega_1\omega_2\omega_3}\right]^{1/2}, \tag{5.3.54}$$

where P_2 and P_3 are regarded as constants.

Simultaneous resonances

In many problems of physical interest, different resonant combinations may occur simultaneously (some examples are listed in [5.27]). The simplest possibility is the simultaneous vanishing of two different resonant pairs, e.g., $m\omega_1$-$n\omega_2$ and $r\omega_2$-$s\omega_3$ for positive integers m, n, r, and s. Clearly, the minimal degrees of freedom that admit two resonant pairs is $N = 3$.

In the case of simultaneous resonances, we need to first isolate the various critical angle combinations before proceeding to average out the remaining non-critical terms. The order of the resulting reduced problem is twice the number of simultaneous resonances.

To fix ideas, consider the following Hamiltonian with three degrees of freedom

$$h(p_i, q_i, \tilde{t}; \epsilon) = \omega_1 p_1 + \omega_2 p_2 + \omega_3 p_3 + \epsilon V_1 \sin(mq_1 - nq_2 + \phi)$$
$$+\epsilon V_2\sin(rq_2 - sq_3 + \psi) + \sum_{i=1}^{3}\sum_{j=1}^{3}\sum_{k=1}^{3} V_{ijk}\sin(iq_1 + jq_2 + kq_3 + \phi_{ijk}). \tag{5.3.55}$$

Here ω_1, ω_2, and ω_3 are given positive functions of $\tilde{t} = \epsilon t$. The (m, n) and (r, s) are positive relatively prime pairs of integers. The V_1, V_2, and V_{ijk} are prescribed functions of p_1, p_2, p_3, and $\tilde{t}$, and the phase shifts ϕ, ψ, ϕ_{ijk} are assumed constant. We also assume that at some time $\tilde{t} = \tilde{t}_0$ we have the two resonance conditions

$$\sigma(\tilde{t}_0) \equiv m\omega_1(\tilde{t}_0) - n\omega_2(\tilde{t}_0) = 0, \tag{5.3.56a}$$
$$\mu(\tilde{t}_0) \equiv r\omega_2(\tilde{t}_0) - s\omega_3(\tilde{t}_0) = 0. \tag{5.3.56b}$$

If we denote the two critical arguments $mq_1 - nq_2 = \overline{q}_1$ and $rq_2 - sq_3 = \overline{q}_2$, the following generating function for an isolating transformation is appropriate (see (5.3.1)):

$$S(q_i, \overline{p}_i) = (mq_1 - nq_2)\overline{p}_1 + (rq_2 - sq_3)\overline{p}_2 + q_3\overline{p}_3. \tag{5.3.57}$$

It then follows that the explicit canonical transformation is

$$\overline{p}_1 = \frac{1}{m} p_1, \quad \overline{p}_2 = \frac{n}{rm} p_1 + \frac{1}{r} p_2, \quad \overline{p}_3 = \frac{ns}{rm} p_1 + \frac{s}{r} p_2 + p_3, \tag{5.3.58a}$$

$$\overline{q}_1 = mq_1 - nq_2, \quad \overline{q}_2 = rq_2 - sq_3, \quad \overline{q}_3 = q_3, \tag{5.3.58b}$$

and the transformed Hamiltonian is

$$\overline{h}(\overline{p}_i, \overline{q}_i, \tilde{t}; \epsilon) = \sigma(\tilde{t})\overline{p}_1 + \mu(\tilde{t})\overline{p}_2 + \omega_3(\tilde{t})\overline{p}_3$$

$$+\epsilon U_1(\overline{p}_i, \tilde{t}) \sin(\overline{q}_1 + \phi) + \epsilon U_2(\overline{p}_i, \tilde{t}) \sin(\overline{q}_2 + \psi)$$

$$+\epsilon \sum_{i=1}^{\infty}\sum_{j=1}^{\infty}\sum_{k=1}^{\infty} U_{ijk}(\overline{p}_i, \tilde{t}) \sin\left[\frac{ir\overline{q}_1 + (in + mj)\overline{q}_2 + r_{ijk}\overline{q}_3}{mr} + \phi_{ijk}\right], \tag{5.3.59}$$

where $U_1 = V_1$, $U_2 = V_2$ and $U_{ijk} = V_{ijk}$, all evaluated for $p_1 = m\overline{p}_1$, $p_2 = -n\overline{p}_1 + r\overline{p}_2$ and $p_3 = -s\overline{p}_2 + \overline{p}_3$. We have also set $r_{ijk} = ins + mjs + mrk$.

We now average out the last term in (5.3.59), thereby making the transformed Hamiltonian independent of $\overline{q}_3$. The reduced problem for the averaged variables P_1, P_2, Q_1, and Q_2 is governed by the *two-degree-of-freedom* Hamiltonian to $O(\epsilon)$

$$\overline{H}(P_i, Q_i, \tilde{t}; \epsilon) = \sigma(\tilde{t})P_1 + \mu(\tilde{t})P_2 + \omega_3(\tilde{t})P_3$$

$$+\epsilon U_1(P_i, \tilde{t}) \sin(Q_1 + \phi) + \epsilon U_2(P_i, \tilde{t}) \sin(Q_2 + \psi) + O(\epsilon^2), \tag{5.3.60}$$

where $P_3 =$ constant to $O(\epsilon)$.

A particular example involving three modes with two simultaneous resonances is explored in Problem 4.

Slow variations on the $t^* = \epsilon^2 t$ scale

For an important class of problems, the coefficients in (5.3.30) depend on $t^* = \epsilon^2 t$ instead of $\tilde{t} = \epsilon t$. Thus, slow variations occur over the longer time interval $0 \le t \le T(\epsilon) = O(\epsilon^{-2})$. We next summarize the necessary modifications of the procedure in Sec. 5.3.2 to handle this case.

The isolating transformation (5.3.35) is still appropriate. However, the averaging transformation (5.3.36) now has the $\overline{T}_m$ and $\overline{L}_n$ depending on t^* instead of $\tilde{t}$. Equations (5.3.38) remain valid, but we cannot remove average terms of order ϵ^2 from (5.3.39). In fact, *we must now also keep track of average terms of $O(\epsilon^3)$* because upon integration such terms contribute to the $O(\epsilon)$ solution.

As it is not possible to remove all higher-order average terms, it is convenient to set the $\overline{T}_m$ and $\overline{L}_n$ equal to zero at the outset. Thus, the equations governing the averaged variables P_m and Q_n take the form

$$\frac{dP_m}{dt} = \epsilon \underline{\overline{f}}_m(P_i, t^*) + \epsilon \overline{f}_{\underset{\wedge}{m_c}}(P_i, Q_1, t^*) + \epsilon^2 \underline{\overline{A}}_m(P_i, t^*)$$

$$+ \epsilon^3 \underline{A}_m{}^*(P_i, t^*) + O(\epsilon^2), \tag{5.3.61a}$$

$$\frac{dQ_n}{dt} = \underline{\sigma}_n(P_i, t^*) + \epsilon\overline{\underline{g}}_n(P_i, t^*) + \epsilon\overline{g}_{\hat{n}_c}(P_i, Q_1, t^*)$$
$$+ \epsilon^2\overline{\underline{C}}_n(P_i, \tilde{t}) + \epsilon^3\underline{C}_n{}^*(P_i, \tilde{t}) + O(\epsilon^2). \quad (5.3.61b)$$

Here the $O(\epsilon^2)$ neglected terms have zero average with respect to all the Q_n; the $\underline{A}_m{}^*$ and $\underline{C}_n{}^*$ are average terms of order ϵ^3 that must be computed explicitly. Therefore, the reduced problem (5.3.61) is considerably more difficult to derive than (5.3.39).

Equations (5.3.61) are somewhat simpler for the special case of a Hamiltonian system because there are no average terms on the right-hand side of (5.3.61a) and the $\overline{f}_{\hat{m}_c}$ are zero for $m > 1$. Therefore, as in Sec. 5.2.3, $P_2, \ldots, P_N$ are adiabatic invariants to $O(\epsilon)$ and the solution for P_1 and Q_1 can be implemented independently of the rest of the system. Details of this derivation as well as a discussion of solutions in resonance can be found in [5.3] and [5.5].

Problems

1. Consider the system (4.3.211) to illustrate the results in this section for the case of constant frequencies in resonance.

   a. Show that the Hamiltonian for this system is

$$h^*(p_1^*, p_2^*, q_1^*, q_2^*; \epsilon) = \frac{1}{2}(p_1^{*^2} + p_2^{*^2}) + \frac{1}{2}\left(q_1^{*^2} + \frac{q_2^{*^2}}{4}\right)$$
$$- \epsilon\left(q_1^* q_2^{*^2} - \frac{\kappa q_2^{*^2}}{2}\right), \quad (5.3.62)$$

   where $q_1^* = x$, $q_2^* = y$, $p_1^* = \dot{x}$, and $q_2^* = \dot{y}$.

   b. Introduce the canonical transformation (see (5.2.59)–(5.2.60))

$$p_1^* = (2p_1)^{1/2}\cos q_1, \quad p_2^* = p_2^{1/2}\cos q_2 \quad (5.3.63a)$$
$$q_1^* = (2p_1)^{1/2}\sin q_1, \quad q_2^* = 2(p_2)^{1/2}\sin q_2 \quad (5.3.63b)$$

   and show that the Hamiltonian in standard form becomes

$$h(p_1, p_2, q_1, q_2; \epsilon) = p_1 + \frac{p_2}{2} + \epsilon\left\{\kappa p_2 - \kappa p_2\cos 2q_2 + \right.$$
$$\left.(2p_1)^{1/2}p_2[-2\sin q_1 + \sin(q_1 - 2q_2) + \sin(q_1 + 2q_2)]\right\}. \quad (5.3.64)$$

   c. Introduce the isolating transformation (5.3.23) to obtain

$$\overline{h}(\overline{p}_1, \overline{p}_2, \overline{q}_1, \overline{q}_2; \epsilon) = \frac{1}{2}\overline{p}_2 + \epsilon\{\kappa(\overline{p}_2 - 2\overline{p}_1)(1 - \cos 2\overline{q}_2)$$
$$+(2\overline{p}_1)^{1/2}(\overline{p}_2 - 2\overline{p}_1)[-2\sin(\overline{q}_1 + 2\overline{q}_2) + \sin\overline{q}_1 + \sin(\overline{q}_1 + 4\overline{q}_2)]\}. \quad (5.3.65)$$

   Thus, aside from the terms involving κ, (5.3.65) is a special case of (5.3.24) for $\underline{\omega}_1 = 1$, $\underline{\omega}_2 = 1/2$.

d. Introduce the generating function (see (5.3.6))

$$\overline{F}(\overline{q}_1, \overline{q}_2, P_1, P_2; \epsilon) = \overline{q}_1 P_1 + \overline{q}_2 P_2 + \epsilon \overline{F}_1(\overline{q}_1, \overline{q}_2, P_1, P_2) \quad (5.3.66)$$

for the canonical transformation $(\overline{q}_i, \overline{p}_i) \rightarrow (Q_i, P_i)$ that averages out all except the $\kappa(\overline{p}_2 - 2\overline{p}_1)$ term and the critical $\sin \overline{q}_1$ term in (5.3.65). Show that the noncritical oscillatory terms of $O(\epsilon)$ are removed by choosing

$$\overline{F}_1(\overline{q}_1, \overline{q}_2, P_1, P_2) = \kappa(P_2 - 2P_1) \sin 2\overline{q}_2 +$$

$$(2P_1)^{1/2}(P_2 - 2P_1) \left[\frac{\cos(\overline{q}_1 + 4\overline{q}_2)}{2} - 2\cos(\overline{q}_1 + 2\overline{q}_1) \right]. \quad (5.3.67)$$

e. Now show that the partially averaged Hamiltonian that results is given *exactly* by

$$\overline{H}(P_1, P_2, Q_1, Q_2; \epsilon) = \frac{P_2}{2} + \epsilon \left[\underline{\overline{H}}_1(P_1, P_2) + \underset{\wedge}{\overline{H}}_1(P_1, P_2, Q_1) \right]$$

$$+\epsilon^2 \left[\underline{\overline{H}}_2(P_1, P_2) + \underset{\wedge}{\overline{H}}_2(P_1, P_2, Q_1, Q_2) \right], \quad (5.3.68)$$

where

$$\underline{\overline{H}}_1 = \kappa(P_2 - 2P_1), \quad (5.3.69a)$$

$$\underset{\wedge}{\overline{H}}_1 = (2P_1)^{1/2}(P_2 - 2P_1) \sin Q_1, \quad (5.3.69b)$$

$$\underline{\overline{H}}_2 = \kappa^2(P_2 - 2P_1) + 7P_1^2 - 8P_1P_2 + \frac{9}{4}P_2^2, \quad (5.3.69c)$$

$$\begin{aligned}\underset{\wedge}{\overline{H}}_2 = &-\left(2P_1^2 - 4P_1P_2 + \frac{3}{2}P_2^2\right)\cos 2Q_2 + \kappa^2(P_2 - 2P_1)\cos 4Q_2 \\ &- (8P_1^2 - 8P_1P_2 + 2P_2^2)\cos(2Q_1 + 4Q_2) + \Big(2P_1^2 - 4P_1P_2 \\ &+ \frac{3}{2}P_2^2\Big)\cos(2Q_1 + 6Q_2) + \left(P_1^2 - \frac{P_2^4}{4}\right)\cos(2Q_1 + 8Q_2) \\ &+ 2\kappa(2P_1)^{1/2}(P_2 - 2P_1)\sin Q_1 \\ &+ 3\kappa\left(\frac{P_1}{2}\right)^{1/2}(2P_1 - P_2)\sin(Q_1 + 2Q_2) \\ &+ 2\kappa(2P_1)^{1/2}(P_2 - 2P_1)\sin(Q_1 + 4Q_2) \\ &+ 3\kappa\left(\frac{P_1}{2}\right)^{1/2}(2P_1 - P_2)\sin(Q_1 + 6Q_2). \end{aligned} \quad (5.3.69d)$$

First, we note that for the case of constant frequencies it is not possible to eliminate average terms from the $O(\epsilon^2)$ Hamiltonian by introducing an additive function of P_1 and P_2 to the expression (5.3.67) for $\overline{F}_1$. Second, because we have introduced an averaging transformation to $O(\epsilon)$ only, the expression for $\hat{\overline{H}}_2$ involves nonresonant terms as well as the critical term proportional to $\sin Q_1$. If we were to continue the averaging to higher order, the nonresonant terms could be eliminated.

f. It follows from (5.3.68) that the P_n and Q_n satisfy

$$\frac{dP_1}{dt} = -\epsilon(2P_1)^{1/2}(2P_2 - P_1)\cos Q_1 - \epsilon^2 \frac{\partial \hat{H}_2}{\partial Q_1}, \tag{5.3.70a}$$

$$\frac{dP_2}{dt} = -\epsilon^2 \frac{\partial \hat{H}_2}{\partial Q_2}, \tag{5.3.70b}$$

$$\frac{dQ_1}{dt} = -2\kappa\epsilon + \epsilon \frac{P_2 - 6P_1}{(2P_1)^{1/2}} \sin Q_1 + \epsilon^2 \frac{\partial \hat{H}_2}{\partial P_1}, \tag{5.3.70c}$$

$$\frac{dQ_2}{dt} = \frac{1}{2} + \epsilon[\kappa + (2P_1)^{1/2}\sin Q_1] + \epsilon^2 \frac{\partial \hat{H}_2}{\partial P_2}. \tag{5.3.70d}$$

Thus, the expression linking P_2 to the (p_i, q_i) is an adiabatic invariant $\mathcal{A}_2^{(1)}(p_i, q_i)$ to $O(\epsilon)$. Derive this expression and verify explicitly that its total derivative with respect to t is $O(\epsilon^2)$ and has a zero average with respect to q_1 and q_2.

g. Since $\overline{H}$ is independent of t, it is an *exact* integral. Since P_2 is adiabatic invariant to $O(\epsilon)$, we conclude from (5.3.68) that $\overline{H}_1(P_1, P_2) + \hat{\overline{H}}_1(P_1, P_2, Q_1)$ is another adiabatic invariant to $O(\epsilon)$. Express P_1, P_2, and Q_1 in terms of the p_n, q_n and verify explicitly that the resulting expression has a total derivative with respect to t that is $O(\epsilon^2)$ and has a zero average with respect to q_1 and q_2. Relate the two adiabatic invariants you found to the two constants E_0 and λ_0 in (4.3.219) and (4.3.223).

h. With P_2 = constant to $O(\epsilon)$, the reduced problem for P_1 and Q_1 satisfies the *strictly nonlinear* perturbed Hamiltonian system, *which is no longer in standard form*

$$\frac{dP_1}{d\tilde{t}} = -(2P_1)^{1/2}(2P_1 - P_2)\cos Q_1 + O(\epsilon), \tag{5.3.71a}$$

$$\frac{dQ_1}{d\tilde{t}} = -2\kappa + \frac{P_2 - 6P_1}{(2P_1)^{1/2}} \sin Q_1 + O(\epsilon). \tag{5.3.71b}$$

Use the approach discussed for the problem (5.2.79) to transform (5.3.71) to standard form. Further examples of the outcome (5.3.71) and its transformation back to standard form can be found in [5.27].

2. Derive the generating function for the partial averaging transformation of (5.3.52) to (5.3.53). Use this result to compute the two adiabatic invariants corresponding to P_2 = const. and P_3 = const.
3. Parallel the derivations of Problem 1 to the case of (5.3.52) with $\omega_1 = 1$, $\omega_2 = 1, \omega_3 = 2$.
4. Specialize the results in Problem 5 of Sec. 5.2 to the case $\overline{x}_1 = 0, \overline{x}_2 = \pi$, $k = 7$. Thus, $C_{nm} \equiv 0$ and $\omega_n^2 = 7 + n^2$ in (5.2.202). Show that for this

special case there are two *simultaneous* resonances involving the three modes $n = 1, 5, 11$. When all nonresonant terms are averaged out, show that, to leading order, these three modes derive their behavior from the following terms in the Hamiltonian for (5.2.202) with $p_n = \dot{q}_n$

$$h^* = \frac{1}{2}(p_1^2 + p_5^2 + p_{11}^2) + \frac{1}{2}(8q_1^2 + 32q_5^2 + 128q_{11}^2) - \frac{\epsilon}{\pi}\left(\frac{8}{105}a_1^2 a_5 + \frac{200}{231}a_5^2 a_{11}\right). \tag{5.3.72}$$

Derive the reduced problem for this case.

5.4 Prescribed Frequency Variations, Transient Resonance

In the previous section, we showed that the system (5.3.30) of order $(M + N)$ can be transformed to the reduced system (5.3.39) of order $(M + 1)$ by near-identity averaging transformation of all angle variables except the critical one exhibiting a resonance. For the special case where $M = N$ and (5.3.30) is Hamiltonian, the reduced problem is of order 2 and is governed by the one-degree-of-freedom Hamiltonian (5.3.16). The transformations that lead to the reduced system are free of zero divisors and remain valid through resonance. Therefore, the resonant behavior of the solution is embodied in the reduced problem. In this section, we discuss the solution of the reduced problem for the important special case where the frequencies depend only on $\tilde{t}$. Thus, the critical divisor, $\sigma_1(\tilde{t}) = \underline{\omega}_1 - (s/r)\underline{\omega}_2$ (see (5.3.8a), (5.3.34)), is a function of $\tilde{t}$ only. The case where σ_1 also depends on P is discussed in Sec. 5.5. We also restrict attention to the Hamiltonian problem (5.3.16) as it contains, in compact form, all the essential features of the more general problem (5.3.39).

It is convenient to develop the right-hand sides of (5.3.15) in Fourier series to derive explicit results. Dropping underbars and underhats for simplicity, we have the Hamiltonian

$$\mathcal{H}(P, Q, \tilde{t}; \epsilon) = \sigma(\tilde{t})P + \epsilon \sum_{r=-\infty}^{\infty} \mathcal{C}_r(P, \tilde{t})e^{irQ}, \tag{5.4.1}$$

where we have set

$$\underline{\mathcal{H}}_0(P, \tilde{t}) = \sigma(\tilde{t})P, \tag{5.4.2a}$$

$$\underline{\mathcal{H}}_1(P, \tilde{t}) = \mathcal{C}_0(P, \tilde{t}), \tag{5.4.2b}$$

$$\mathcal{C}_r(P, \tilde{t}) = \frac{1}{2\pi}\int_0^{2\pi} \mathcal{H}_1(P, Q, \tilde{t})e^{irQ}dQ. \tag{5.4.2c}$$

Equations (5.3.15) then have the explicit series form:

$$\dot{P} = -\epsilon \sum_{r=-\infty}^{\infty} \mathcal{B}_r(P, \tilde{t})e^{irQ}, \tag{5.4.3a}$$

$$\dot{Q} = \sigma(\tilde{t}) + \epsilon \sum_{r=-\infty}^{\infty} \mathcal{D}_r(P, \tilde{t})e^{irQ}, \tag{5.4.3b}$$

where

$$\mathcal{B}_r = ir\mathcal{C}_r, \tag{5.4.4a}$$

$$\mathcal{D}_r = \frac{\partial \mathcal{C}_r}{\partial P}. \tag{5.4.4b}$$

We also assume that $\sigma(\tilde{t})$ has a simple zero at $\tilde{t} = \tilde{t}_0$ and may be expanded in the form

$$\sigma(\tilde{t}) = \mu_2(\tilde{t} - \tilde{t}_0) + \mu_{22}(\tilde{t} - \tilde{t}_0)^2 + O(\tilde{t} - \tilde{t}_0)^3, \tag{5.4.5}$$

where $\mu_2 \neq 0, \mu_{22}, \ldots,$ are prescribed constants. The generalization to a higher-order zero is discussed in Problem 2.

The special case of (5.4.3) with only the first harmonic present is discussed in [5.14]. The model problem of (5.3.27) is also in this form with $\mathcal{C}_1 = Ai/2$; $\mathcal{C}_{-1} = -Ai/2$; $\mathcal{C}_n = 0$; $|n| \neq 1$.

5.4.1 *Solution before Resonance* ($\tilde{t} < \tilde{t}_0$)

To solve the system (5.4.3) away from resonance, we have the choice of using either the approach in Sec. 5.1.3 or the multiple-scale expansion discussed in Sec. 4.5.3. We use the latter, as it is significantly simpler for calculating $P(t;\epsilon)$ and $Q(t;\epsilon)$.

We expand P and Q as functions of the fast time

$$\tau = \frac{1}{\epsilon}\int_0^{\tilde{t}} \sigma(s)ds \tag{5.4.6}$$

and the slow time $\tilde{t} = \epsilon t$

$$P(t;\epsilon) = \tilde{P}^{(0)}(\tilde{t}) + \epsilon P^{(1)}(\tau, \tilde{t}) + \epsilon^2 P^{(2)}(\tau, \tilde{t}) + O(\epsilon^3), \tag{5.4.7a}$$

$$Q(t;\epsilon) = \tau + \tilde{Q}^{(0)}(\tilde{t}) + \epsilon Q^{(1)}(\tau, \tilde{t}) + \epsilon^2 Q^{(2)}(\tau, \tilde{t}) + O(\epsilon^3), \tag{5.4.7b}$$

where we have anticipated the form of the solution to $O(1)$ in terms of $\tilde{P}^{(0)}$, τ, and $Q^{(0)}(\tilde{t})$.

Substituting (5.4.7) into (5.4.3) and evaluating time derivatives by using

$$\frac{d}{dt} = \sigma\frac{\partial}{\partial \tau} + \epsilon\frac{\partial}{\partial \tilde{t}} \tag{5.4.8}$$

gives the following system to $O(\epsilon)$:

$$\sigma\frac{\partial P^{(1)}}{\partial \tau} + \tilde{P}^{(0)\prime} = -\sum_{r=-\infty}^{\infty} \mathcal{B}_r(\tilde{P}^{(0)}, \tilde{t})e^{ir(\tau+\tilde{Q}^{(0)})}, \tag{5.4.9a}$$

$$\sigma\frac{\partial Q^{(1)}}{\partial \tau} + \tilde{Q}^{(0)\prime} = \sum_{r=-\infty}^{\infty} \mathcal{D}_r(\tilde{P}^{(0)}, \tilde{t})e^{ir(\tau+\tilde{Q}^{(0)})}. \tag{5.4.9b}$$

Since $\mathcal{B}_0 = 0$, (see (5.4.4a)), there is no average term on the right-hand side of (5.4.9a). Thus, $\tilde{P}^{(0)\prime} = 0$, i.e., $\tilde{P}^{(0)} = \rho_0$, a constant determined by the initial condition on P. To remove the average term from the right-hand side of (5.4.9b), we set $\tilde{Q}^{(0)\prime} = \mathcal{D}_0(\rho_0, \tilde{t})$, and this determines $\tilde{Q}^{(0)}(\tilde{t})$ in the form

$$\tilde{Q}^{(0)}(\tilde{t}) = \kappa_0 + \int_0^{\tilde{t}} \mathcal{D}_0(\rho_0, s)ds, \quad \kappa_0 = \text{constant}. \tag{5.4.10}$$

Integrating what remains of (5.4.9) with respect to τ defines $P^{(1)}$ and $Q^{(1)}$ as follows:

$$P^{(1)}(\tau, \tilde{t}) = -\frac{1}{\sigma} \sum{}' \frac{\mathcal{B}_r(\rho_0, \tilde{t})}{ir} e^{ir(\tau+\tilde{Q}^{(0)})} + \tilde{P}^{(1)}(\tilde{t}), \tag{5.4.11a}$$

$$Q^{(1)}(\tau, \tilde{t}) = \frac{1}{\sigma} \sum{}' \frac{\mathcal{B}_r(\rho_0, \tilde{t})}{ir} e^{ir(\tau+\tilde{Q}^{(0)})} + \tilde{Q}^{(1)}(\tilde{t}), \tag{5.4.11b}$$

where the notation $\sum'$ indicates $\sum_{r=-\infty,\, r\neq 0}^{\infty}$, and $\tilde{P}^{(1)}$ and $\tilde{Q}^{(1)}$ are functions of $\tilde{t}$ to be determined at the next stage.

In fact, all we need for determining $\tilde{P}^{(1)}$ and $\tilde{Q}^{(1)}$ are the average terms in the $O(\epsilon^2)$ equations. We derive the equations for $P^{(2)}$ and $Q^{(2)}$ using previously obtained results as follows:

$$\sigma \frac{\partial P^{(2)}}{\partial \tau} + \frac{\partial P^{(1)}}{\partial \tilde{t}} = - \sum_{r=-\infty}^{\infty} \left[\frac{\partial \mathcal{B}_r(\rho_0, \tilde{t})}{\partial P} P^{(1)} + \mathcal{B}_r(\rho_0, \tilde{t}) ir Q^{(1)} \right] e^{ir(\tau+\tilde{Q}^{(0)})} + \text{Osc.}, \tag{5.4.12a}$$

$$\sigma \frac{\partial Q^{(2)}}{\partial \tau} + \frac{\partial Q^{(1)}}{\partial \tilde{t}} = \sum_{r=-\infty}^{\infty} \left[\frac{\partial \mathcal{D}_r(\rho_0, \tilde{t})}{\partial P} P^{(1)} + \mathcal{D}_r(\rho_0, \tilde{t}) ir Q^{(1)} \right] e^{ir(\tau+\tilde{Q}^{(0)})} + \text{Osc.}, \tag{5.4.12b}$$

where Osc. indicates terms with zero average with respect to τ.

When the expressions for $P^{(1)}$ and $Q^{(1)}$ given in (5.4.11) are used in (5.4.12), we find that we must choose $\tilde{P}^{(1)}$ and $\tilde{Q}^{(1)}$ as follows in order to remove average terms not depending on τ from the right-hand sides of (5.4.12):

$$\frac{d\tilde{P}^{(1)}}{d\tilde{t}} = \frac{1}{\sigma} \sum{}' \left[\frac{\partial \mathcal{B}_r(\rho_0, \tilde{t})}{\partial P} \frac{\mathcal{B}_{-r}(\rho_0, \tilde{t})}{ir} - \mathcal{D}_r(\rho_0, \tilde{t}) \mathcal{B}_{-r}(\rho_0, \tilde{t}) \right] = 0, \tag{5.4.13a}$$

$$\frac{d\tilde{Q}^{(1)}}{d\tilde{t}} = \tilde{P}^{(1)} \frac{\partial \mathcal{D}_0(\rho_0, \tilde{t})}{\partial P} + \frac{1}{\sigma} \sum{}' \left[\frac{\partial \mathcal{D}_r(\rho_0, \tilde{t})}{\partial P} \frac{\mathcal{B}_{-r}(\rho_0, \tilde{t})}{ir} - \mathcal{D}_r(\rho_0, \tilde{t}) \mathcal{D}_{-r}(\rho_0, \tilde{t}) \right]. \tag{5.4.13b}$$

In deriving (5.4.13), we have used the identity

$$\left\langle\left(\sum' f_r e^{ir\tau}\right)\left(\sum' g_r e^{ir\tau}\right)\right\rangle = \sum' f_r g_{-r} = \sum' f_{-r} g_r$$

for the average of the product of two oscillatory functions.

As discussed in Sec. 5.2.3, the vanishing of the right-hand side of (5.4.13a) is a direct consequence of the Hamiltonian form of (5.4.3). For the general case (5.3.39), this simplification is not available, which means that one needs to go to higher order to compute the solution to $O(\epsilon)$.

We have therefore defined $\tilde{P}^{(1)}$ and $\tilde{Q}^{(1)}$ in the explicit form

$$\tilde{P}^{(1)} = \rho_1 = \text{constant}, \tag{5.4.14a}$$

$$\begin{aligned}\tilde{Q}^{(1)} &= \kappa_1 + \rho_1 \int_0^{\tilde{t}} \left\{ \frac{\partial \mathcal{D}_0(\rho_0, s)}{\partial P} + \frac{1}{\sigma(s)} \right. \\ &\left. \cdot \sum{}' \left[\frac{\partial \mathcal{D}_r(\rho_0, s)}{\partial P} \frac{\mathcal{B}_{-r}(\rho_0, s)}{ir} - \mathcal{D}_r(\rho_0, s)\mathcal{D}_{-r}(\rho_0, s) \right] \right\} ds, \end{aligned} \tag{5.4.14b}$$

where $\kappa_1 = $ constant. This completes the determination of the preresonance expansion (5.4.7) explicitly to $O(\epsilon)$. The constants ρ_0, κ_0, ρ_1, and κ_1 may be chosen so that $P(0;\epsilon)$ and $Q(0;\epsilon)$ satisfy given initial condition to $O(\epsilon)$.

As expected, the σ divisors in (5.4.11) and (5.4.14b) indicate that the terms of order ϵ in the expansions (5.4.7) become singular as $\tilde{t} \to \tilde{t}_0$. In the next section, we derive an expansion that is valid for $\tilde{t} \approx \tilde{t}_0$.

5.4.2 *Solution Near Resonance,* $(\tilde{t} \approx \tilde{t}_0)$

To calculate the solution of (5.4.3) for $\tilde{t} \approx \tilde{t}_0$, we look for an *interior layer* expansion centered around $\tilde{t} = \tilde{t}_0$. As usual, we introduce a rescaled slow time

$$\hat{t} = \frac{\tilde{t} - \tilde{t}_0}{\epsilon^\alpha} = \epsilon^{1-\alpha} t - \frac{\tilde{t}_0}{\epsilon^\alpha}$$

for an appropriate positive constant α to be determined. Since $\tilde{t} - \tilde{t}_0$ is small in this interior layer, we look for a *regular* expansion for P and Q. If we denote $P(t;\epsilon) = \hat{P}(\hat{t};\epsilon)$ and $Q(t;\epsilon) = \hat{Q}(\hat{t};\epsilon)$, we obtain the following equations to leading order from (5.4.3):

$$\epsilon^{1-\alpha}\frac{d\hat{P}}{d\hat{t}} = -\epsilon \sum_{r=-\infty}^{\infty} \mathcal{B}_r(\hat{P}, \tilde{t}_0)e^{ir\hat{Q}} + O(\epsilon^{1+\alpha}), \tag{5.4.15a}$$

$$\epsilon^{1-\alpha}\frac{d\hat{Q}}{d\hat{t}} = \mu_2\epsilon^\alpha \hat{t} + \mu_{22}\epsilon^{2\alpha}\hat{t}^2 + \epsilon \sum_{r=-\infty}^{\infty} \mathcal{D}_r(\hat{P}, \tilde{t}_0)e^{ir\hat{Q}}. \tag{5.4.15b}$$

Upon dividing by $\epsilon^{1-\alpha}$, we see that we must set $\alpha = 1/2$ in order to have a distinguished limit. Thus,

$$\hat{t} = \frac{\tilde{t} - \tilde{t}_0}{\epsilon^{1/2}}, \tag{5.4.16}$$

and (5.4.15) becomes

$$\frac{d\hat{P}}{d\hat{t}} = -\epsilon^{1/2} \sum_{r=-\infty}^{\infty} \mathcal{B}_r(\hat{P}, \tilde{t}_0)e^{ir\hat{Q}} + O(\epsilon), \tag{5.4.17a}$$

$$\frac{d\hat{Q}}{d\hat{t}} = \mu_2\hat{t} + \epsilon^{1/2}\left[\mu_{22}\hat{t}^2 + \sum_{r=-\infty}^{\infty} \mathcal{D}_r(\hat{P}, \tilde{t}_0)e^{ir\hat{Q}}\right] + O(\epsilon). \tag{5.4.17b}$$

We expand $\hat{P}(\hat{t}; \epsilon)$ and $\hat{Q}(\hat{t}; \epsilon)$ in powers of $\epsilon^{1/2}$ as follows:

$$\hat{P}(\hat{t}; \epsilon) = \hat{P}_0(\hat{t}) + \epsilon^{1/2}\hat{P}_{1/2}(\hat{t}) + O(\epsilon), \tag{5.4.18a}$$

$$\hat{Q}(\hat{t}; \epsilon) = \hat{Q}_0(\hat{t}; \epsilon) + \epsilon^{1/2}\hat{Q}_{1/2}(\hat{t}) + O(\epsilon). \tag{5.4.18b}$$

In (5.4.18b), we have allowed $\hat{Q}_0$ to depend on ϵ in anticipation of the fact that Q is $O(\epsilon^{-1})$ at $\tilde{t} = \tilde{t}_0$. Thus, the integration constant in $\hat{Q}_0$ must be $O(\epsilon^{-1})$. An equivalent but more cumbersome alternative is to assume the expansion for $\hat{Q}$ starting with a term of $O(\epsilon^{-1})$.

Substitution of (5.4.18) into (5.4.17) gives

$$\frac{d\hat{P}_0}{d\hat{t}} = 0, \qquad \frac{d\hat{Q}_0}{d\hat{t}} = \mu_2\hat{t}. \tag{5.4.19}$$

$$\frac{d\hat{P}_{1/2}}{d\hat{t}} = -\sum_{r=-\infty}^{\infty} \mathcal{B}_r(\hat{P}_0, \tilde{t}_0)e^{ir\hat{Q}_0}, \tag{5.4.20a}$$

$$\frac{d\hat{Q}_{1/2}}{d\hat{t}} = \mu_{22}\hat{t}^2 + \sum_{r=-\infty}^{\infty} \mathcal{D}_r(\hat{P}_0, \tilde{t}_0)e^{ir\hat{Q}_0}. \tag{5.4.20b}$$

The solution of (5.4.19) is

$$\hat{P}_0 = \hat{\rho}_0 = \text{constant}, \tag{5.4.21a}$$

$$\hat{Q}_0 = \hat{\kappa}_0(\epsilon) + \frac{\mu_2\hat{t}^2}{2}, \qquad \kappa_0 = \text{constant}. \tag{5.4.21b}$$

Substituting these expressions into (5.4.20) and integrating gives

$$\hat{P}_{1/2} = -\sum{}' \mathcal{B}_r^{(0)}e^{ir\hat{\kappa}_0} \int_0^{\hat{t}} e^{ir\mu_2 s^2/2}ds + \hat{\rho}_{1/2}, \tag{5.4.22a}$$

$$\hat{Q}_{1/2} = \frac{\mu_{22}}{3}\hat{t}^3 + \sum{}' \mathcal{D}_r^{(0)}e^{ir\hat{\kappa}_0} \int_0^{\hat{t}} e^{ir\mu_2 s^2/2}ds + \hat{\kappa}_{1/2}, \tag{5.4.22b}$$

where $\hat{\rho}_{1/2}$ and $\hat{\kappa}_{1/2}$ are integration constants and we have introduced the notation

$$\mathcal{B}_r^{(0)} = \mathcal{B}_r(\hat{\rho}_0, \tilde{t}_0), \quad \mathcal{D}_r^{(0)} = \mathcal{D}_r(\hat{\rho}_0, \tilde{t}_0). \tag{5.4.23}$$

Matching the resonance expansion (5.4.18) with the preresonance expansion (5.4.7) will provide the relations linking the unknowns $\hat{\rho}_0$, $\hat{\rho}_{1/2}$, $\hat{\kappa}_0$, and $\hat{\kappa}_{1/2}$ with the given initial values ρ_0 and κ_0. This procedure is discussed next.

5.4.3 Matching of Preresonance and Resonance Expansions

The matching of these two expansions can be carried out using the matching variable

$$t_\eta = \frac{\tilde{t} - \tilde{t}_0}{\eta(\epsilon)} \tag{5.4.24}$$

for $\eta(\epsilon)$ in an appropriate overlap domain contained in $\epsilon^{1/2} \ll \eta \ll 1$. To avoid encumbering the notation unnecessarily, we omit the usual procedure of expressing the outer (preresonance) and interior (resonance) expansions in terms of t_η. Instead, we express the outer in terms of $\hat{t}$ and expand the interior for large *negative* $\hat{t}$. As long as we keep track of all terms ignored in this process, we can implement the calculations and determine the overlap domain.

First, we consider the outer expansion and express τ and $\tilde{t}$ in terms of $\hat{t}$. We have

$$\tau = \frac{\tau_0}{\epsilon} + \frac{\mu_2}{2}\hat{t}^2 + \epsilon^{1/2}\frac{\mu_{22}}{3}\hat{t}^3 + O(\epsilon\hat{t}^4), \tag{5.4.25}$$

$$\tilde{t} = \tilde{t}_0 + \epsilon^{1/2}\hat{t}, \tag{5.4.26}$$

where τ_0 is the constant

$$\tau_0 = \int_0^{\tilde{t}_0} \sigma(s)ds. \tag{5.4.27}$$

When these expressions are used in (5.4.7) and the results are expanded, we obtain

$$P = \rho_0 - \epsilon^{1/2}\frac{1}{\mu_2\hat{t}}\sum{}' \frac{\mathcal{B}_r(\rho_0, \tilde{t}_0)}{ir} e^{ir\psi_0} + O(\epsilon\hat{t}^2), \tag{5.4.28a}$$

$$\begin{aligned} Q = {} & \psi_0 + \epsilon^{1/2}\left(\frac{\mu_{22}\hat{t}^3}{3} + \frac{1}{\mu_2\hat{t}}\sum{}' \frac{\mathcal{D}_r(\rho_0, \tilde{t}_0)}{ir} e^{ir\psi_0}\right) + O(\epsilon\hat{t}^4) \\ & + O(\epsilon \log \epsilon^{1/2}|\hat{t}|), \end{aligned} \tag{5.4.28b}$$

where

$$\psi_0 = \frac{\tau_0}{\epsilon} + \frac{\mu_2\hat{t}^2}{2} + \kappa_0 + \int_0^{\tilde{t}_0} \mathcal{D}_0(\rho_0, s)ds. \tag{5.4.29}$$

As $|\hat{t}|$ is large in the matching domain, we have exhibited the orders of the largest neglected terms in (5.4.28) in terms of ϵ and $\hat{t}$. In particular, truncating the expansion for τ, as in (5.4.25), contributes the $O(\epsilon\hat{t}^2)$ term in (5.4.28a) and the $O(\epsilon\hat{t}^4)$ term in (5.4.28b). The logarithmic error term in (5.4.28b) arises from ignoring the singular part of $\tilde{Q}_1$ in (5.4.14b).

Next, we consider the behavior of the resonance expansion for $\hat{t} \to -\infty$. The crucial calculation concerns the behavior of the complex Fresnel integral

$$w_r(\hat{t}) = \int_0^{\hat{t}} e^{ir\mu_2 s^2/2} ds \tag{5.4.30}$$

as $\hat{t} \to -\infty$. For the matching with the solution after resonance, we also need the behavior of (5.4.30) as $\hat{t} \to \infty$. For $\hat{t} > 0$, we write w_r in the form

$$w_r(\hat{t}) = I - J(\hat{t}), \tag{5.4.31}$$

where I is the constant

$$I = \int_0^{\infty} e^{ir\mu_2 s^2/2} ds = \frac{1}{2}\sqrt{\frac{\pi}{|r\mu_2|}}[1 + i\,\mathrm{sgn}(r\mu_2)] \tag{5.4.32a}$$

and

$$J(\hat{t}) = \int_{\hat{t}}^{\infty} e^{ir\mu_2 s^2/2} ds. \tag{5.4.32b}$$

For $\hat{t} < 0$, we write

$$\begin{aligned} w_r(\hat{t}) &= \int_{-\infty}^{\hat{t}} e^{ir\mu_2 s^2/2} ds - \int_{-\infty}^{0} e^{ir\mu_2 s^2/2} ds \\ &= J(-\hat{t}) - I. \end{aligned} \tag{5.4.33}$$

To compute the asymptotic behavior of J as $|\hat{t}| \to \infty$, we change the variable of integration to $\sigma = r\mu_2 s^2/2$. Integration by parts then shows that

$$J(\hat{t}) = \frac{i e^{ir\mu_2 \hat{t}^2/2}}{r\mu_2 |\hat{t}|} + O(\hat{t}^{-3}) \text{ as } |\hat{t}| \to \infty. \tag{5.4.34}$$

Thus,

$$w_r(\hat{t}) = \frac{\mathrm{sgn}(\hat{t})}{2}\sqrt{\frac{\pi}{|\mu_2 r|}}[1 + i\,\mathrm{sgn}(r\mu_2)] - \frac{i e^{ir\mu_2\hat{t}^2/2}}{r\mu_2\hat{t}} + O(\hat{t}^{-3}) \text{ as } |\hat{t}| \to \infty. \tag{5.4.35}$$

The resonance expansion can now be expressed in the following form for $|\hat{t}| \to \infty$:

$$\begin{aligned} P = \hat{\rho}_0 + \epsilon^{1/2} \Bigg\{ -\sum{}' B_r^{(0)} e^{ir\hat{\kappa}_0} \\ \cdot \left[-\frac{i e^{ir\mu_2\hat{t}^2/2}}{r\mu_2\hat{t}} + \frac{1}{2}\left(\frac{\pi}{|\mu_2 r|}\right)^{1/2} (\mathrm{sgn}\hat{t})(1 + i\,\mathrm{sgn} r\mu_2) \right] + \hat{\rho}_{1/2} \Bigg\} \\ + O(\epsilon^{1/2}\hat{t}^{-3}), \end{aligned} \tag{5.4.36a}$$

$$\begin{aligned} Q = \frac{\mu_2\hat{t}^2}{2} + \hat{\kappa}_0(\epsilon) + \epsilon^{1/2} \Bigg\{ \frac{\mu_{22}\hat{t}^3}{3} + \mathcal{D}_0^{(0)}\hat{t} + \sum{}' \mathcal{D}_r^{(0)} e^{ir\hat{\kappa}_0} \\ \cdot \left[-\frac{i e^{ir\hat{\mu}_2\hat{t}^2/2}}{r\mu_2\hat{t}} + \frac{1}{2}\left(\frac{\pi}{|\mu_2 r|}\right)^{1/2} (\mathrm{sgn}\hat{t})(1 + i\,\mathrm{sgn} r\mu_2) \right] + \hat{\kappa}_{1/2} \Bigg\} \end{aligned}$$

$$+ O(\epsilon^{1/2}\hat{t}^{-3}). \tag{5.4.36b}$$

Here, the $O(\epsilon^{1/2}\hat{t}^{-3})$ remainder is due to ignoring the term of $O(\hat{t}^{-3})$ in (5.4.35).

If we compare (5.4.28) with (5.4.36) (with $\hat{t} < 0$), we see that the $O(1)$ terms match by choosing

$$\hat{\rho}_0 = \rho_0, \tag{5.4.37a}$$

$$\hat{\kappa}_0(\epsilon) = \frac{1}{\epsilon}\int_0^{\tilde{t}_0} \sigma(s)ds + \int_0^{\tilde{t}_0} \mathcal{D}_0(\rho_0, s)ds + \kappa_0. \tag{5.4.37b}$$

As anticipated earlier, we have $\hat{\kappa}_0 = O(\epsilon^{-1})$. In view of (5.4.37a), we see from (5.4.23) that $\mathcal{B}_r^{(0)} = \mathcal{B}_r(\rho_0, \tilde{t}_0)$ and $\mathcal{D}_r^{(0)} = \mathcal{D}_r(\rho_0, \tilde{t}_0)$. Therefore, the terms proportional to $\epsilon^{1/2}\hat{t}^{-1}$ in (5.4.28) that are contributed by the σ divisors match identically with corresponding terms in (5.4.36). These terms are not singular for matching to $O(\epsilon^{1/2})$ but do become singular in a higher-order matching. There are no constant terms of order $\epsilon^{1/2}$ in (5.4.28). Therefore, the constant terms of $O(\epsilon^{1/2})$ in (5.4.36) must vanish. This condition defines the constants $\hat{\rho}_{1/2}$ and $\hat{\kappa}_{1/2}$

$$\hat{\rho}_{1/2} = -\frac{1}{2}\sum{}' \mathcal{B}_r^{(0)} e^{ir\hat{\kappa}_0}\left(\frac{\pi}{|\mu_2 r|}\right)^{1/2}(1 + i\,\text{sgn}\mu_2 r), \tag{5.4.38a}$$

$$\hat{\kappa}_{1/2} = \frac{1}{2}\sum{}' \mathcal{D}_r^{(0)} e^{ir\hat{\kappa}_0}\left(\frac{\pi}{|\mu_2 r|}\right)^{1/2}(1 + i\,\text{sgn}\mu_2 r). \tag{5.4.38b}$$

Finally, we note that the matching of the two expansions is valid to $O(\epsilon^{1/2})$. (We have not considered the $O(\epsilon)$ terms in (5.4.17) or the singular terms in the $O(\epsilon^2)$ preresonance expansion.) To exhibit overlap, we must show that the neglected terms, divided by $\epsilon^{1/2}$, vanish in the limit as $\epsilon \to 0$ with the matching variable t_η of (5.4.24) held fixed for an appropriate overlap subdomain of $\epsilon^{1/2} \ll \eta \ll 1$.

The largest neglected term in (5.4.28), when divided by $\epsilon^{1/2}$, is of order $\epsilon^{1/2}\hat{t}^4$ or $O(\eta^4/\epsilon^{3/2})$ as $\epsilon \to 0$ with t_η fixed. Thus, we must have $\eta \ll \epsilon^{3/8}$, and this restriction implies that the logarithmic term will also vanish. Similarly, the $O(\epsilon^{1/2}\hat{t}^{-3})$ terms that were ignored in (5.4.36), when divided by $\epsilon^{1/2}$, will vanish if $\hat{t}^{-3} \to 0$. In the matching region $\hat{t}^{-3} = O(\epsilon^{3/2}\eta^{-3})$, and this vanishes if $\epsilon^{1/2} \ll \eta$. Thus, the overlap domain for the matching to $O(\epsilon^{1/2})$ is $\epsilon^{1/2} \ll \eta \ll \epsilon^{3/8}$.

5.4.4 Solution after Resonance; Jump in the Action; Uniformly Valid Results

The calculations for $\tilde{t} > \tilde{t}_0$ are analogous to those in Sec. 5.4.1 with two exceptions. First, since σ changes sign at $\tilde{t} = \tilde{t}_0$, we must define the fast time for $\tilde{t} > \tilde{t}_0$ as

$$\tau^+ = \frac{1}{\epsilon}\int_0^{\tilde{t}_0} \sigma(s)ds - \frac{1}{\epsilon}\int_{\tilde{t}_0}^{\tilde{t}} \sigma(s)ds. \tag{5.4.39}$$

This ensures that τ and τ^+ are equal at $\tilde{t}_0$ and that τ^+ is a monotone increasing function of time. Secondly, the fact that the resonance solution now has *uncanceled*

$O(\epsilon^{1/2})$ terms as $\hat{t} \to \infty$ (because of the sign change in the first term of w_r) means that we need a post-resonance expansion in powers of $\epsilon^{1/2}$. The calculations are not repeated, and we summarize the results:

$$P = \rho_0^+ + \epsilon^{1/2}\rho_{1/2}^+ + O(\epsilon), \tag{5.4.40a}$$

$$Q = -\tau^+ + \kappa_0^+ + \int_0^{\tilde{t}_0} \mathcal{D}_0(\rho_0^+, s)ds + \epsilon^{1/2}\kappa_{1/2}^+ + O(\epsilon), \tag{5.4.40b}$$

where $\rho_0^+, \rho_{1/2}^+, \kappa_0^+, \kappa_{1/2}^+$ are integration constants.

These constants are determined by matching with the resonance solution ($\hat{t} \to \infty$), and we find

$$\rho_0^+ = \rho_0, \tag{5.4.41a}$$

$$\kappa_0^+ = \frac{2\tau_0}{\epsilon} + \kappa_0, \tag{5.4.41b}$$

$$\rho_{1/2}^+ = 2\hat{\rho}_{1/2} = -\sum{}' \mathcal{B}_r(\rho_0, \tilde{t}_0)e^{ir\hat{\kappa}_0}\left(\frac{\pi}{|\mu_2 r|}\right)^{1/2}(1 + i\,\mathrm{sgn}\mu_2 r), \tag{5.4.41c}$$

$$\kappa_{1/2}^+ = 2\hat{k}_{1/2} = \sum{}' \mathcal{D}_r(\rho_0, \tilde{t}_0)e^{ir\hat{\kappa}_0}\left(\frac{\pi}{|\mu_2 r|}\right)^{1/2}(1 + i\,\mathrm{sgn}\mu_2 r). \tag{5.4.41d}$$

It is interesting to examine the behavior of the two actions p_1 and p_2 as given to $O(\epsilon^{1/2})$ away from resonance. We use (see (5.3.2c), (5.3.2d), and (5.3.6))

$$p_1 = P_1 + O(\epsilon) = P + O(\epsilon), \tag{5.4.42a}$$

$$p_2 = -\frac{s}{r}P_1 + P_2 + O(\epsilon) = -\frac{s}{r}P + \text{ constant } + O(\epsilon), \tag{5.4.42b}$$

to relate p_1 and p_2 to P_1 and P_2. Recall that P_2 = constant to $O(\epsilon)$, and that we have set $P_1 = P$ in this section.

The solution we calculated in Sec. 5.4.1 for P for $\tilde{t} < t_0$ has $P = \rho_0 + O(\epsilon)$. On the other hand, for $\tilde{t} > \tilde{t}_0$, (5.4.41a) and (5.4.41b) give $P = \rho_0 + 2\epsilon^{1/2}\hat{\rho}_{1/2} + O(\epsilon)$, where $\hat{\rho}_{1/2}$ is defined in (5.4.38a). Thus, to $O(\epsilon^{1/2})$, the two actions remain constant in their respective domains. However, the value of these constants *jumps* by $O(\epsilon^{1/2})$ across resonance. In fact, if we denote the jump by [ ], we have

$$[p_1] = 2\epsilon^{1/2}\hat{\rho}_{1/2} + O(\epsilon), \tag{5.4.43a}$$

$$[p_2] = -\frac{2s}{r}\epsilon^{1/2}\hat{\rho}_{1/2} + O(\epsilon). \tag{5.4.43b}$$

This result was first given in [5.7] (see equation (7)) without details, presumably on the basis of the difference in the asymptotic behavior of the resonance solution at $|\hat{t}| \to \pm\infty$.

A typical numerical solution for p_1 and p_2 (see Figure 5.3.1) shows that for $|\tilde{t} - \tilde{t}_0| \ll \epsilon^{1/2}$, these variables oscillate with $O(\epsilon)$ amplitude about *different* mean values that agree with our results. For $\tilde{t} - \tilde{t}_0 = O(\epsilon^{1/2})$, the numerical results show that p_1 and p_2 make a smooth transition from one level to the other. On the other hand, the adiabatic invariant $\overline{\mathcal{A}}_2^{(1)}$ given in (5.3.19a) exhibits oscillations of amplitude $O(\epsilon)$ only for all $\tilde{t}$. Several other numerical examples are worked out in [5.17]. The more interesting case where the ω_k depend on $t^* = \epsilon^2 t$ is discussed in [5.3]. Here p_1 and p_2 undergo $O(1)$ jumps across resonance. A detailed numerical verification of the matching results and a method for enhancing the accuracy of the solution after resonance is given in [5.1].

We can use our results to construct a uniformly valid (for all $\tilde{t}$) expansion for P and Q to $O(\epsilon^{1/2})$. It is easily seen that this composite expansion is

$$P = \rho_0 - \epsilon^{1/2}\Bigg\{\sum{}' \mathcal{B}_r(\rho_0, \tilde{t}_0)e^{ir\hat{\kappa}_0} \cdot \left[\int_0^{\hat{t}} e^{ir\mu_2 s^2/2}ds + \frac{1}{2}\left(\frac{\pi}{|\mu_2 r|}\right)^{1/2}(1 + i\,\mathrm{sgn}\mu_2 r)\right]\Bigg\} + O(\epsilon), \quad (5.4.44a)$$

$$Q = \tau + \kappa_0 + \int_0^{\tilde{t}_0} \mathcal{D}_0(\rho_0, s)ds + \epsilon^{1/2}\Bigg\{\sum{}' \mathcal{D}_r(\rho_0, \tilde{t}_0)e^{ir\hat{\kappa}_0} \cdot \left[\int_0^{\hat{t}} e^{ir\mu_2 s^2/2}ds + \frac{1}{2}\left(\frac{\pi}{|\mu_2 r|}\right)^{1/2}(1 + i\,\mathrm{sgn}\mu_2 r)\right]\Bigg\} + O(\epsilon). \quad (5.4.44b)$$

This expansion exhibits a typical internal-layer or shock-layer behavior wherein a resonance solution (of short duration on the $\tilde{t}$ scale, hence the term *transient* resonance) smoothly connects the preresonance and post-resonance solutions. These outer solutions differ by $O(\epsilon^{1/2})$ across $\tilde{t} = \tilde{t}_0$ and become singular to $O(\epsilon)$ at $\tilde{t} = \tilde{t}_0$.

Problems

1. Specialize the results of this section to the case (see (5.3.26a))

$$\mathcal{H}(P, Q, \tilde{t}; \epsilon) = \sigma(\tilde{t})P - \epsilon A(P, \tilde{t}) \sin Q, \quad (5.4.45)$$

where σ has the behavior given in (5.4.5) and A is an arbitrary function of P and $\tilde{t}$. In particular, show that

$$\hat{\rho}_0 = \rho_0, \quad \hat{\kappa}_0 = \frac{1}{\epsilon}\int_0^{\tilde{t}} \sigma(s)ds + \kappa_0. \quad (5.4.46)$$

$$\hat{\rho}_{1/2} = -\frac{1}{2}A(\rho_0, \tilde{t}_0)\sqrt{\frac{\pi}{|\mu_2|}}(\mathrm{sgn}(\mu)\sin\alpha - \cos\alpha), \quad (5.4.47a)$$

$$\hat{\kappa}_{1/2} = -\frac{1}{2}\frac{\partial A}{\partial P}(\rho_0, \tilde{t})\sqrt{\frac{\pi}{|\mu_2|}}(\sin\alpha + \mathrm{sgn}(\mu_2)\cos\alpha), \quad (5.4.47b)$$

where

$$\alpha = \frac{\tau_0}{\epsilon} + \kappa_0. \quad (5.4.47c)$$

2. Consider the Hamiltonian (5.4.45) but assume that $\sigma(\tilde{t})$ has the behavior

$$\sigma(\tilde{t}) = \mu_{22}(\tilde{t} - \tilde{t}_0)^2 + O(\tilde{t} - \tilde{t}_0)^3 \text{ as } \tilde{t} \to \tilde{t}_0. \tag{5.4.48}$$

   a. Show that the appropriate resonance variable is now

$$\bar{t} = \frac{\tilde{t} - \tilde{t}_0}{\epsilon^{1/3}} \tag{5.4.49}$$

   and that the resonance expansion must proceed in powers of $\epsilon^{1/3}$

$$\overline{P}(\bar{t};\epsilon) = \sum_{j=0}^{k} \epsilon^{j/3}\overline{P}_{j/3}(\bar{t}) + O(\epsilon^{(k+1)/3}), \tag{5.4.50a}$$

$$\overline{Q}(\bar{t};\epsilon) = \sum_{j=0}^{k} \epsilon^{j/3}\overline{Q}_{j/3}(\bar{t}) + O(\epsilon^{(k+1)/3}). \tag{5.4.50b}$$

   b. Calculate the first two terms of this expansion in the form

$$\overline{P}_0 = \bar{\rho}_0 = \text{constant}, \tag{5.4.51a}$$

$$\overline{Q}_0 = \frac{\mu_{22}}{3}\bar{t}^3 + \bar{\kappa}_0, \quad \bar{\kappa}_0 = \text{constant}. \tag{5.4.51b}$$

$$\overline{P}_{1/3} = A(\bar{\rho}_0, \tilde{t}_0)\int_0^{\bar{t}} \cos\left(\frac{\mu_{22}}{3}s^3 + \bar{\kappa}_0\right) ds + \bar{\rho}_{1/3}, \tag{5.4.52a}$$

$$\overline{Q}_{1/3} = -\frac{\partial A}{\partial P}(\bar{\rho}_0, \tilde{t}_0)\int_0^{\bar{t}} \sin\left(\frac{\mu_{22}}{3} + \bar{\kappa}_0\right) ds + \bar{\kappa}_{1/3}. \tag{5.4.52b}$$

   c. Derive the asymptotic behavior as $|\bar{t}| \to \infty$ of the integral

$$\overline{w} = \int_0^{\bar{t}} e^{i\mu_{22}s^3/3} ds, \tag{5.4.53}$$

   and use this result to carry out the matching to $O(\epsilon^{1/3})$. What is the overlap domain of this matching and what are the values of $\bar{\rho}_0$, $\bar{\kappa}_0$, $\bar{\rho}_{1/3}$, $\bar{\kappa}_{1/3}$ in terms of the initial values ρ_0, κ_0?

3. Consider the Hamiltonian (5.4.45) for the case where the slow time is $t^* = \epsilon^2 t$ instead of $\tilde{t}$. Assume also that $\sigma(t^*)$ has the same behavior as in (5.4.5), i.e.,

$$\sigma(t^*) = \mu_2(t^* - t_0^*) + O((t^* - t_0^*)^2) \text{ as } t^* \to t_0^*. \tag{5.4.54}$$

   Show that the appropriate slow time for the resonance solution is now

$$t_1 = \frac{t^* - t_0^*}{\epsilon} \tag{5.4.55}$$

   and that the leading terms Q_0^* and P_0^* in the resonance expansion obey

$$\frac{dP_0^*}{dt_1} = A(P_0^*, t_0^*)\cos Q_0^*, \tag{5.4.56a}$$

$$\frac{dQ_0^*}{dt_1} = \mu_2 t_1 - \frac{\partial A}{\partial P}(P_0^*, t_0^*)\sin Q_0^*. \tag{5.4.56b}$$

Thus, the problem is strictly nonlinear and nonautonomous.

5.5 Frequencies that Depend on the Actions, Transient or Sustained Resonance

Because the calculations for this case are quite involved if one considers the general reduced problem (5.3.15), we restrict our discussion to the special case where the Hamiltonian involves a single harmonic

$$\mathcal{H}(P, Q, \tilde{t}; \epsilon) = \underline{\mathcal{H}}_0(P, \tilde{t}) - \epsilon A(P, \tilde{t}) \sin Q. \tag{5.5.1}$$

Thus, the equations for P and Q are

$$\frac{dP}{dt} = -\frac{\partial \mathcal{H}}{\partial Q} = \epsilon A(P, \tilde{t}) \cos Q, \tag{5.5.2a}$$

$$\frac{dQ}{dt} = \frac{\partial \mathcal{H}}{\partial P} = \sigma(P, \tilde{t}) - \epsilon A_p(P, \tilde{t}) \sin Q, \tag{5.5.2b}$$

where

$$\sigma(P, \tilde{t}) = \frac{\partial \mathcal{H}_0}{\partial P}. \tag{5.5.3}$$

We wish to solve (5.5.2) for the initial conditions $P(0; \epsilon) = \rho_0$, $Q(0; \epsilon) = \kappa_0$. This is, in fact, the one-degree-of-freedom example we have studied in this and the preceding chapters (see (5.2.157) with $\omega = \sigma$, $p_0 = \rho_0$, $q_0 = \kappa_0$).

The solution prior to resonance ($\sigma \neq 0$) is given by (5.2.177). We note that the expression for q now involves singular terms proportional to σ^{-2} in addition to the σ^{-1} singularities that were present for the case where σ depends only on $\tilde{t}$ discussed in Sec. 5.4. We will omit the lengthy calculations needed to derive the behavior of this solution in the matching region. These calculations are essentially similar to those for the derivation of (5.4.28)–(5.4.29).

The main goal of this section is the study of (5.5.2) near resonance, i.e., for $\tilde{t} \approx \tilde{t}_0$, where $\tilde{t}_0$ is the solution of the algebraic equation

$$\sigma(\rho_0, \tilde{t}_0) = 0. \tag{5.5.4}$$

Note that the resonance condition $\sigma = 0$ gives (5.5.4) to leading order because $P = \rho_0 + O(\epsilon)$.

5.5.1 Transient Resonance

We investigate the behavior of solutions in transient resonance by expanding P and Q as in Sec. 5.4.2 (see (5.4.18)) in terms of the interior-layer variable $\hat{t}$ given by (5.4.16). Thus,

$$P(t; \epsilon) = \rho_0 + \epsilon^{1/2} \hat{P}_{1/2}(\hat{t}) + O(\epsilon), \tag{5.5.5a}$$

$$Q(t; \epsilon) = \hat{Q}_0(\hat{t}) + \epsilon^{1/2} \hat{Q}_{1/2}(\hat{t}) + O(\epsilon), \tag{5.5.5b}$$

where in (5.5.5a) we have anticipated the fact that P equals its initial value, ρ_0, to leading order (see (5.4.19)).

We expand $\sigma(P, \tilde{t})$ around $P = \rho_0, \tilde{t} = \tilde{t}_0$ to obtain

$$\sigma(P, \tilde{t}) = \epsilon^{1/2}(\mu_1 \hat{P}_{1/2} + \mu_2 \hat{t}) + O(\epsilon), \tag{5.5.6}$$

where

$$\mu_1 = \frac{\partial \sigma(\rho_0, \tilde{t}_0)}{\partial P}, \quad \mu_2 = \frac{\partial \sigma(\rho_0, \tilde{t}_0)}{\partial \tilde{t}}. \tag{5.5.7}$$

The equations governing $\hat{P}_{1/2}$ and $\hat{Q}_0$ then follow from (5.5.2)

$$\frac{d\hat{P}_{1/2}}{d\hat{t}} = A(\rho_0, \tilde{t}_0) \cos \hat{Q}_0, \tag{5.5.8a}$$

$$\frac{d\hat{Q}_0}{d\hat{t}} = \mu_1 \hat{P}_{1/2} + \mu_2 \hat{t}. \tag{5.5.8b}$$

The equations for $\hat{P}_1$ and $\hat{Q}_{1/2}$ are easy to derive (see (6.46) of [5.13]); we do not write them down, as we will confine our discussion to (5.5.8).

In contrast to the case $\sigma(\tilde{t})$, where $\mu_1 = 0$, the presence of the term $\mu_1 \hat{P}_{1/2}$ in (5.5.8b) leads to a *nonlinear* problem to leading order. In fact, if we differentiate (5.5.8b) with respect to $\hat{t}$ and use (5.5.8a) to eliminate $\hat{P}_{1/2}$, we find

$$\frac{d^2\hat{Q}_0}{d\hat{t}^2} + \lambda_1 \cos \hat{Q}_0 = \mu_2, \tag{5.5.9}$$

where we have set

$$\lambda_1 = -\mu_1 A(\rho_0, \tilde{t}_0). \tag{5.5.10}$$

This is the equation of motion (in suitable dimensionless variables) for a pendulum with a constant tangential force μ_2; the displacement angle $\hat{Q}_0$ is measured in the counterclockwise sense from the *horizontal*. Equation (5.5.9) and its more general version, to be derived in Sec. 5.5.2, are fundamental for describing resonance for the case where the frequencies depend on the momenta. The solution of (5.5.9), involving two constants of integration, when substituted into (5.5.8a), defines $\hat{P}_{1/2}$ by quadrature in terms of the same two constants. The higher-order terms $\hat{P}_1$ and $\hat{Q}_{1/2}$ obey a *linear system with variable coefficients* that depend on $\hat{Q}_0(\hat{t})$ and $\hat{P}_{1/2}(\hat{t})$.

A qualitative understanding of (5.5.9) is all that is needed to establish the solution behavior in transient resonance and the sense in which this solution matches with the solution (5.2.177). Detailed calculations for a special case are given in [5.15].

We multiply (5.5.9) by $(d\hat{Q}_0/d\hat{t})$ and integrate the result to obtain the energy integral

$$\frac{1}{2}\left(\frac{d\hat{Q}_0}{d\hat{t}}\right)^2 + V(\hat{Q}_0, \lambda_1, \mu_2) = E = \text{ constant}, \tag{5.5.11}$$

where

$$V(\hat{Q}_0, \lambda_1, \mu_2) = \lambda_1 \sin \hat{Q}_0 - \mu_2 \hat{Q}_0. \tag{5.5.12}$$

If the solution of (5.5.9) is known for all initial values of $\hat{Q}_0$ and $(d\hat{Q}_0/d\hat{t})$ for $\lambda_1 > 0$ and $\mu_2 > 0$, we can use symmetry arguments to calculate the solution for other sign combinations of λ_1 and μ_2 by noting that (5.5.9) is invariant under the transformation $\hat{Q}_0 \to -\hat{Q}_0$, $\lambda_1 \to -\lambda_1$, $\mu_2 \to -\mu_2$. Thus, it suffices to consider the case $\lambda_1 > 0$, $\mu_2 > 0$. We also note that the energy integral (5.5.11) is invariant under the transformation $\hat{Q}_0 \to \hat{Q}_0 + 2n\pi$, $E \to E + 2n\pi\mu_2$ for each $n = 0, \pm 1, \pm 2, \ldots$. Therefore, the integral curve for any given value of the constant E in the $(\hat{Q}_0, (d\hat{Q}_0/d\hat{t}))$ plane may be translated to the right (or left) along the $\hat{Q}_0$ axis a distance $2n\pi$, $(n = 1, 2, \ldots)$ to obtain the integral curve for $E - 2n\pi\mu_2$ (or $E + 2n\pi\mu_2$). Thus, it suffices to study the family of integral curves for $-2\pi\mu_2 \leq E \leq 2\pi\mu_2$. We must, however, distinguish the two cases: $\mu_2 > \lambda_1$ and $\mu_2 < \lambda_1$.

If $\mu_2 > \lambda_1$, $(\partial V/\partial \hat{Q}_0)$ does not vanish and there are no equilibrium solutions: $\hat{Q}_0 = (d\hat{Q}_0/d\hat{t}) = 0$. Figure 5.5.1(a) shows V as a function of $\hat{Q}_0$ for three fixed E values: $E = E_1 > 0$, $E = 0$, and $E = -E_1 < 0$; the integral curves for these values of E are sketched in Figure 5.5.1(b), where the arrows indicate the direction of increasing $\hat{t}$.

Recall that for the matching with the preresonance solution we need the asymptotic expansion of $\hat{Q}_0$ as $\hat{t} \to -\infty$. The actual calculation of this expansion is straightforward (see Problem 6 of Sec. 1.2). One of the essential qualitative results of this matching is that it fixes the value of E for a given pair of the initial parameters (ρ_0, κ_0). Once E is known, the solution during resonance proceeds along the corresponding curve in the phase plane. The actual value of E is qualitatively unimportant for the case $\mu_2 > \lambda_1$; all integral curves originate at $\hat{t} \to -\infty$, $\hat{Q}_0 \to \infty$, $(d\hat{Q}_0/d\hat{t}) \to -\infty$ (where they match with the preresonance solution) and eventually reflect about the $\hat{Q}_0$ axis. This implies that transient resonance evolves for this case just as for the case $\sigma(\tilde{t})$ of Sec. 5.4. The details of the matching and the calculation of the solution after resonance are similar to those discussed in Sec. 5.4 and are omitted for brevity.

If $\mu_2 < \lambda_1$, $(\partial V/\partial \hat{Q}_0) = 0$ at the points $\hat{Q}_0 = \cos^{-1}(\mu_2/\lambda_1)$ on the $\hat{Q}_0$ axis. Consider again values of E in the range $-2\pi\mu_2 \leq E \leq 2\pi\mu_2$. As shown in Figure 5.5.2(a), the first equilibrium point on the negative Q_0 axis at $\hat{Q}_0 = \cos^{-1}(\mu_2/\lambda_1) < 0$ is a center, whereas the first positive equilibrium point at $\hat{Q}_0 = \cos^{-1}(\mu_2/\lambda_1) > 0$ is a saddle. The integral curves in the phase plane are sketched in Figure 5.5.2(b) and this pattern is 2π-periodic as discussed earlier.

The integral curve that tends to the saddle as $\hat{t} \to \infty$ corresponds to the constant energy, $E = E_c \equiv (\lambda_1^2 - \mu_2^2)^{1/2} - \mu_2 \cos^{-1}(\mu_2/\lambda_1)$. All the other curves that originate in the preresonance region ($\hat{t} \to -\infty$, $\hat{Q}_0 \to \infty$, $(d\hat{Q}_0/d\hat{t}) \to -\infty$) tend to $\hat{Q}_0 \to \infty$, $(d\hat{Q}_0/d\hat{t}) \to \infty$ as $\hat{t} \to \infty$. Therefore, all curves with $E \neq E_c$ *also describe transient resonance*. This observation follows directly from the fact that as $\hat{t} \to \infty$, the behavior of $\hat{Q}_0$ is the same as that for $\hat{t} \to -\infty$, whereas $(d\hat{Q}_0/d\hat{t})$ evolves with the opposite sign. Moreover, the phase portrait implies that

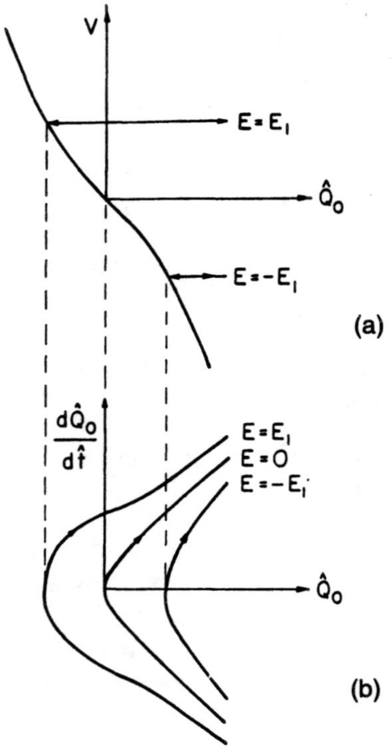

FIGURE 5.5.1. (a) $V(Q_0, \lambda_1, \mu_2)$; (b) Phase Plane for $\mu_2 > \lambda_1$

none of the trajectories inside the separatrix, $E = E_c$, surrounding the center are accessible to solutions originating from the preresonance region.

Suppose that matching with the preresonance solution indicates that $E = E_c$. If the system (5.5.8) were to correctly describe the solution during resonance, we would conclude that as $\hat{t} \to \infty$, $\hat{Q}_0 \to \cos^{-1}(\mu_2/\lambda_1)$ and $(d\hat{Q}_0/d\hat{t}) \to 0$, i.e., that $\hat{P}_{1/2} \to -(\mu_2/\lambda_1)\hat{t}$ according to (5.5.8b). But this implies that the term of order $\epsilon^{1/2}$ in the expansion (5.5.6) for σ is *identically equal* to zero, in contradiction to the assumed orders that led to (5.5.8). In fact, if $E \approx E_c$, $\sigma(P, \tilde{t})$ remains close to zero for a period that is longer than postulated. It is natural to ask whether, in this case, there exists a resonance solution for which $\sigma(P, \tilde{t})$ remains close to zero for a period that is $O(1)$ on the $\tilde{t}$ scale. We investigate this possibility next.

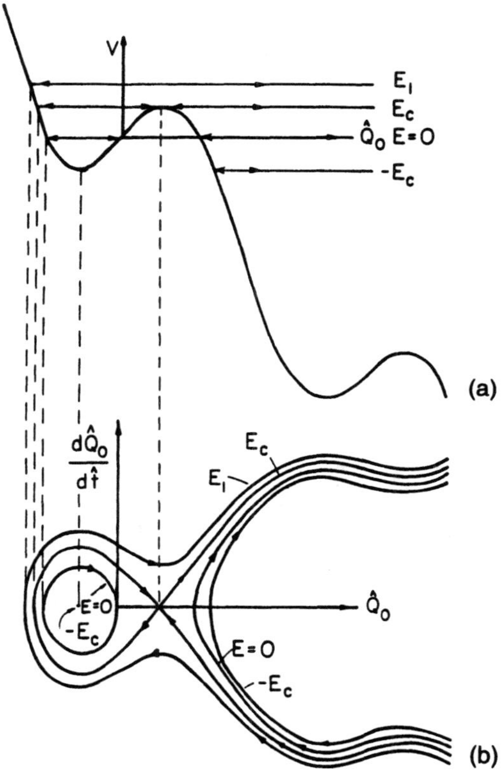

FIGURE 5.5.2. (a) $V(\hat{Q}_0, \lambda_1, \mu_2)$; (b) Phase Plane for $\mu_2 < \lambda_1$

5.5.2 *Sustained Resonance*

If the system (5.5.2) has solutions where $\sigma(P, \tilde{t})$ remains close to zero over the interval I: $(\tilde{t} - \tilde{t}_0) = O(1)$ or longer, it is inconsistent to expand the $\tilde{t}$ dependence of σ as in (5.5.6) in deriving this solution. Moreover, the leading term in the expansion for P is not necesssarily a constant but is given by $P_c(\tilde{t})$, the solution of the algebraic relation,

$$\sigma(P_c, \tilde{t}) = 0. \tag{5.5.13}$$

Thus, we introduce the rescaled variables P^*, Q^* defined by

$$P(t; \epsilon) = P_c(\tilde{t}) + \epsilon^{1/2} P^*(\hat{t}; \epsilon), \tag{5.5.14a}$$

$$Q(t; \epsilon) = Q^*(\hat{t}; \epsilon), \tag{5.5.14b}$$

and obtain the following system to leading order from (5.5.2):

$$\frac{dP^*}{d\hat{t}} = -\frac{dP_c}{d\tilde{t}} + A(P_c, \tilde{t}) \cos Q^* + O(\epsilon^{1/2}), \tag{5.5.15a}$$

$$\frac{dQ^*}{d\hat{t}} = \frac{\partial\sigma(P_c, \tilde{t})}{\partial P} P^* + O(\epsilon^{1/2}). \tag{5.5.15b}$$

We must keep in mind that the rescaling (5.5.14) that results in (5.5.15) only makes sense if (5.5.15) has solutions where $P^* = O(1)$ on the interval I. A solution with this property is said to be in "sustained resonance." Although the terms of $O(\epsilon^{1/2})$ that we have ignored in (5.5.15) are necessary to describe the solution, it is suficient to focus on the leading terms for a qualitative description of sustained resonance.

As in Sec. 5.5.1, we differentiate (5.5.15b) with respect to $\hat{t}$ and use (5.5.15a) to eliminate P^* to find the pendulum equation

$$\frac{d^2Q^*}{d\hat{t}^2} + \lambda_1(\tilde{t}) \cos Q^* = \lambda_2(\tilde{t}), \tag{5.5.16}$$

where

$$\lambda_1(\tilde{t}) = -\frac{\partial\sigma(P_c, \tilde{t})}{\partial P} A(P_c, \tilde{t}), \tag{5.5.17a}$$

$$\lambda_2(\tilde{t}) = -\frac{\partial\sigma(P_c, \tilde{t})}{\partial P} \frac{dP_c}{d\tilde{t}}. \tag{5.5.17b}$$

Now, we have a pendulum with slowly varying length and tangential force. Note also that since $\tilde{t} = \epsilon t = \epsilon^{1/2}\hat{t} + \text{const.}$, we may regard $\hat{t}$ as a fast scale and $\tilde{t}$ as a slow scale with respect to the small parameter $\epsilon^{1/2}$.

In order to demonstrate the existence of sustained resonance, we need to show that (5.5.15) has solutions for which $P^* = O(1)$ over the time interval I. In terms of (5.5.16), this means that we seek solutions where $(dQ^*/d\hat{t}) = O(1)$ over I. We are interested in the case $|\lambda_2(\tilde{t})| < |\lambda_1(\tilde{t})|$ only; otherwise the solution of Sec. 5.5.1 applies. Now, we no longer have the exact integral (5.5.11), as E, λ_1, and λ_2 vary on the $\tilde{t}$ scale. One can, however, describe the *local* behavior of solutions in terms of a slowly varying phase plane. To fix ideas, assume that $E(\tilde{t})$ is a monotone decreasing function of $\tilde{t}$. Then, if we ignore the variation of V with $\tilde{t}$, we may describe the behavior of solutions near the saddle point in Fig. 5.5.2(a) by replacing the horizontal (E = const.) lines by trajectories where E *decreases* with time as shown in Figure 5.5.3(a).

Let us consider a one-parameter family of solutions, all originating from a given $Q^*(0)$ to the right of the saddle point; the initial value of $(dQ^*/d\hat{t})$ varies, thus $E(0)$ is different for each trajectory. We see from Figure 5.5.3(a) that there are two critical values $E_{\min}(0)$ and $E_{\max}(0)$ of the initial energy. For all initial values $E(0)$ such that $E_{\min}(0) < E(0) < E_{\max}(0)$, trajectories are "trapped" in the "potential well" to the left of the saddle. On the other hand, if $E(0) < E_{\max}(0)$ or if $E(0) < E_{\min}(0)$, the trajectories approach the saddle and are reflected without entering the well. The corresponding phase portraits are shown in Figure 5.5.3(b), where the shaded region represents trajectories that are captured in sustained resonance.

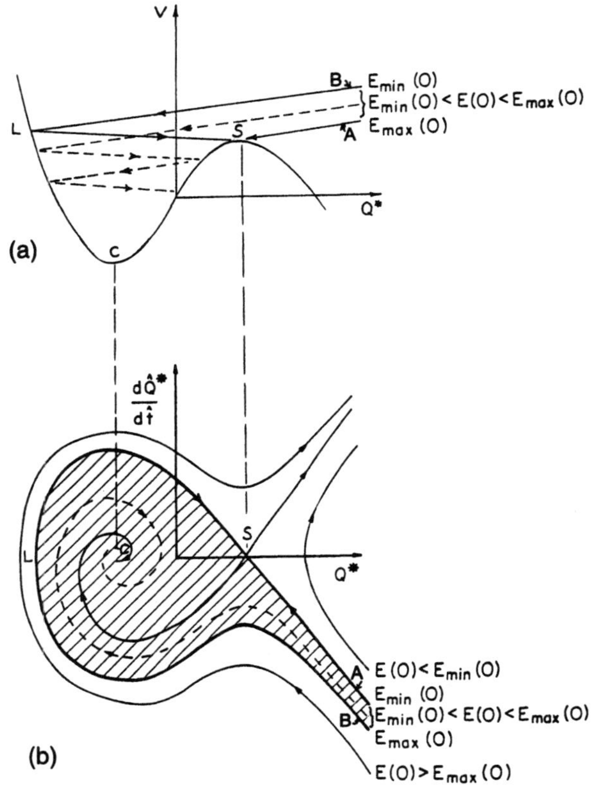

FIGURE 5.5.3. (a) $V(\hat{Q}_0, \lambda_1, \lambda_2)$, $(dE/d\tilde{t}) < 0$, (b) Slowly Varying Phase Plane for $0 < \lambda_2 < \lambda_1$

The above is a simplified characterization of actual solutions where not only does E vary but so does V (and hence the locations of the saddle and center) on the $\tilde{t}$ scale. A numerical description of these features was given in [5.19] for the reduced problem associated with (4.5.81). The conditions necessary for sustained resonance to occur are derived in analytic form in [5.9]–[5.11]. In many applications, one is interested in the converse problem of predicting *escape* from a potential well. This is the case, for example, for the motion of electrons in a free-electron laser. Various aspects are discussed in [5.4] and [5.20]. The behavior of a slowly varying nonlinear oscillator in a double potential well is discussed in [5.6]. Here, the slowly varying potential is W-shaped and the oscillator alternately escapes from one well to be captured in sustained resonance in the neighboring well. A detailed discussion is also given in this reference for the matching procedure necessary to predict the time of escape from one well and capture in the other.

References

5.1. D.L. Bosley, "An improved matching procedure for transient resonance layers in weakly nonlinear oscillatory systems," *SIAM J. Appl. Math.*, **56**, 1996.

5.2. D.L. Bosley and J. Kevorkian, "On the asymptotic solution of non-Hamiltonian systems exhibiting sustained resonance," *Stud. Appl. Math.*, **94**, 1995, pp. 83–130.

5.3. D.L. Bosley and J. Kevorkian, "Adiabatic invariance and transient resonance in very slowly varying Hamiltonian systems," *SIAM J. Appl. Math.*, **52**, 1992, pp. 494–527.

5.4. D.L. Bosley and J. Kevorkian, "Free-electron lasers with very slow wiggler taper," *IEEE J. Quantum Electron.*, **27**, 1991, pp. 1078–1089.

5.5. D.L. Bosley and J. Kevorkian, "Sustained resonance in very slowly varying oscillatory Hamiltonian systems," *SIAM J. Appl. Math.*, **51**, 1991, pp. 439–471.

5.6. F.J. Bourland and R. Haberman, "Separatrix crossing: time-invariant potentials with dissipation," *SIAM J. Appl. Math.*, **50**, 1990, pp. 1716–1744.

5.7. B.V. Chirikov, "The passage of a nonlinear oscillating system through resonance," *Sov. Phys. Dokl*, **4**, 1959, pp. 390–394.

5.8. H. Goldstein, *Classical Mechanics*, 2nd ed., Addison-Wesley, Reading, MA, 1980.

5.9. R. Haberman, "Energy bounds for the slow capture by a center in sustained resonance," *SIAM J. Appl. Math.*, **43**, 1983, pp. 244–256.

5.10. W.L. Kath, "Conditions for sustained resonance II," *SIAM J. Appl. Math.*, **43**, 1983, pp. 579–583.

5.11. W.L. Kath, "Necessary conditions for sustained resonance," *SIAM J. Appl. Math.*, **43**, 1983, pp. 314–324.

5.12. J. Kevorkian, *Partial Differential Equations: Analytical Solution Techniques*, Chapman and Hall, New York, London, 1990, 1993.

5.13. J. Kevorkian, "Perturbation techniques for oscillatory systems with slowly varying coefficients," *SIAM Rev.*, **29**, 1987, pp. 391–461.

5.14. J. Kevorkian, "Adiabatic invariance and passage through resonance for nearly periodic Hamiltonian systems," *Stud. Appl. Math.*, **66**, 1982, pp. 95–119.

5.15. J. Kevorkian, "On a model for re-entry roll resonance," *SIAM J. Appl. Math.*, **26**, 1974, pp. 638–669.

5.16. J. Kevorkian, "Passage through resonance for a one-dimensional oscillator with slowly varying frequency," *SIAM J. Appl. Math.*, **20**, 1971, pp. 364–373. See also Errata in **26**, 1974, p. 686.

5.17. J. Kevorkian and H.K. Li, "Resonant modal interactions and adiabatic invariance for a nonlinear wave equation in a variable domain," *Stud. Appl. Math.*, **71**, 1984, pp. 1–64.

5.18. N.M. Krylov and N.N. Bogoliubov, *Introduction to Nonlinear Mechanics*, Acad. Sci., Ukrain, S.S.R., 1937. Translated by S. Lefschetz, Princeton University Press, Princeton, NJ, 1947.

5.19. L. Lewin and J. Kevorkian, "On the problem of sustained resonance," *SIAM J. Appl. Math.*, **35**, 1978, pp. 738–754.

5.20. Y.P. Li and J. Kevorkian, "The effects of wiggler taper rate and signal field gain rate in free-electron lasers," *IEEE J. Quantum Electron.*, **24**, 1988, pp. 598–608.

5.21. A.J. Lichtenberg and M.A. Lieberman, *Regular and Chaotic Dynamics*, Springer-Verlag, New York, 1992.
5.22. Y.A. Mitropolski, *Problèmes de la Théorie Asymptotique des Oscillations Non Stationnaires*, Gauthier-Villars, Paris, 1966. Translated from the Russian.
5.23. J.A. Morrison, "Generalized method of averaging and the von Zeipel method," *Progress in Astronautics and Aeronautics* **17**, Methods in Astrodynamics and Celestial Mechanics, R.L. Duncombe and V.G. Szebehely, Eds., Academic Press, New York, 1966, pp. 117–138.
5.24. J.C. Neu, "The method of near-identity transformations and its applications," *SIAM J. Appl. Math.*, **38**, 1980, pp. 189–208.
5.25. V.M. Volosov, "Averaging in systems of ordinary differential equations," *Russ. Math. Surveys*, **17**, 1963, pp.1–126.
5.26. H. von Zeipel, "Recherche sur le mouvement des petites planètes," *Ark. Astron. Mat. Fys.*, **11–13**, 1916.
5.27. L. Wang, D.L. Bosley, and J. Kevorkian, "Asymptotic analysis of a class of three-degree-of-freedom Hamiltonian systems near stable equilibria," *Physica D*, **88**, 1995, pp. 87–115.

6

Multiple-Scale Expansions for Partial Differential Equations

Multiple-scale and averaging methods have a broad range of applicability for systems of ordinary differential equations, as discussed in Chapters 4 and 5. In contrast, asymptotic solution techniques for partial differential equations are more recent and may be implemented, in general, only with multiple-scale expansions. Some of the early use of multiple scales concerned problems where the unperturbed state has a simple, usually periodic, structure, and the leading effect of weak nonlinearities is to introduce a slow modulation of the parameters. A number of representative examples are discussed in Sec. 6.1. In Sec. 6.2, we study systems of conservation laws that are perturbed about a uniform state. The ideas are illustrated using examples from shallow water flow, gas dynamics, and other applications. The final section, 6.3, gives a brief introduction to multiple-scale homogenization.

6.1 Nearly Periodic Waves

The problems that we study in this section are characterized by the feature that in the absence of weak nonlinearities solutions are periodic waves. We then solve the perturbed problem by letting the parameters that occur in the leading order periodic wave vary slowly and by including other nearly periodic waves to higher order.

6.1.1 Series Solutions

A model problem

We study the nonlinear wave equation

$$\frac{\partial^2 U}{\partial T^2} - C^2 \frac{\partial^2 U}{\partial X^2} + \frac{U_0}{T_0^2} g\left(\frac{U}{U_0}\right) = 0 \qquad (6.1.1)$$

in dimensional variables X, T, U. Here g is an arbitrary positive nonlinear function, the constant C has dimensions of velocity, and U_0 and T_0 are characteristic values of U and T. Consider the general initial-value problem on $-\infty < x < \infty$

$$U(X, 0) = U_1 \phi\left(\frac{X}{L_0}\right), \quad \frac{\partial U}{\partial T}(X, 0) = \frac{U_1}{T_1}\psi\left(\frac{X}{L_0}\right). \tag{6.1.2}$$

Here U_1 is a characteristic value of the initial data which we will assume small relative to U_0. The characteristic length and time scales for the initial conditions are L_0 and T_1, respectively.

We choose the dimensionless variables

$$x^* = \frac{X}{CT_0}, \quad t^* = \frac{T}{T_0}, \quad u^* = \frac{U}{U_0} \tag{6.1.3}$$

to obtain

$$\frac{\partial^2 u^*}{\partial t^{*2}} - \frac{\partial^2 u^*}{\partial x^{*2}} + g(u^*) = 0. \tag{6.1.4}$$

$$u^*(x^*, 0; \epsilon^*) = \epsilon^* \phi(\lambda x^*), \quad \frac{\partial u^*}{\partial t^*}(x^*, 0; \epsilon^*) = \sigma\epsilon^*\psi(\lambda x^*), \tag{6.1.5}$$

where

$$\epsilon^* = \frac{U_1}{U_0} \ll 1, \quad \sigma = \frac{T_0}{T_1}, \quad \lambda = \frac{CT_0}{L_0}. \tag{6.1.6}$$

Thus, $\epsilon^* \ll 1$ indicates that the initial disturbance (6.1.5) is small. The dimensionless parameters σ and λ are both assumed to be $O(1)$. With no loss of generality, we may regard $g(0) = 0$; this results from replacing u^* by the normalized dependent variable $u^* - g(0)$. Since we are interested in $\epsilon^* \ll 1$, we introduce the rescaled dependent variable $u(x^*, t^*; \epsilon^*) = \epsilon^{*-1} u^*(x^*, t^*; \epsilon^*)$ and expand g to obtain

$$\frac{\partial^2 u}{\partial t^{*2}} - \frac{\partial^2 u}{\partial x^{*2}} + c_1 u + \epsilon^* c_2 u^2 + \epsilon^{*^2} c_3 u^3 = O(\epsilon^{*^3}) \tag{6.1.7}$$

after dividing by ϵ^*. The initial conditions for u become

$$u(x^*, 0; \epsilon^*) = \phi(\lambda x^*), \quad \frac{\partial u}{\partial t^*}(x^*, 0; \epsilon^*) = \sigma\psi(\lambda x^*). \tag{6.1.8}$$

In (6.1.7), we have set

$$c_1 = g'(0), \quad c_2 = \frac{1}{2} g''(0), \quad c_3 = \frac{1}{6} g'''(0). \tag{6.1.9}$$

For bounded solutions, we must have $c_1 \geq 0$. If $c_1 > 0$, we can simplify (6.1.8) further by choosing the rescaled independent variables

$$x = \sqrt{c_1}\, x^*, \quad t = \sqrt{c_1}\, t^* \tag{6.1.10}$$

to obtain

$$u_{tt} - u_{xx} + u + \epsilon^* \frac{c_2}{c_1} u^2 + \epsilon^{*^2} \frac{c_3}{c_1} u^3 = O(\epsilon^{*^3}), \tag{6.1.11}$$

$$u(x, 0; \epsilon^*) = \phi(\lambda x), \quad u_t(x, 0; \epsilon^*) = \frac{\sigma}{\sqrt{c_1}} \psi(\lambda x). \tag{6.1.12}$$

If $g(u)$ is odd, $c_2 = 0$ and we have the following model problem ($\epsilon^{*^2} c_3/c_1 = \epsilon$) to leading order in ϵ:

$$u_{tt} - u_{xx} + u + \epsilon u^3 = 0, \tag{6.1.13}$$

$$u(x, 0; \epsilon) = \phi(\lambda x), \quad u_t(x, 0; \epsilon) = \frac{\sigma}{\sqrt{c_1}} \psi(\lambda x). \tag{6.1.14}$$

Another interesting special case corresponds to $c_1 = c_2 = 0$, $c_3 > 0$. In this case, denoting $\epsilon^{*^2} c_3 = \epsilon$ and dropping asterisks gives

$$u_{tt} - u_{xx} + \epsilon u^3 = 0. \tag{6.1.15}$$

We will show that (6.1.15) has a fundamentally more complicated solution than (6.1.13).

Perturbed traveling waves

Let us first study the case $c_1 > 0$, which leads to (6.1.13). In Sec. 4.1, we used the method of strained coordinates to compute periodic traveling wave solutions for weakly nonlinear partial differential equations. In particular, (6.1.13) has the periodic traveling wave (see Problem 1 of Sec. 4.1)

$$u(x, t; \epsilon) = \rho_0 \sin(\theta^+ + \phi_0) - \epsilon \frac{\rho_0^3}{32} \sin 3(\theta^+ + \phi_0) + O(\epsilon^2), \tag{6.1.16}$$

where

$$\theta^+ = kx - \sqrt{1 + k^2} \left(1 + \frac{3\rho_0^2 \epsilon}{8(1 + k^2)} + O(\epsilon^2) \right) t, \tag{6.1.17}$$

and ρ_0, ϕ_0, and k are arbitrary constants. The dispersion relation for the linear ($\epsilon = 0$) problem is $\omega = \sqrt{1 + k^2}$, and the frequency shift is $\omega_1 = 3\rho_0^2/8(1 + k^2)$. This solution corresponds to very special initial conditions that we can derive a posteriori by setting $t = 0$ in (6.1.16) and its time derivative

$$u(x, 0; \epsilon) = \rho_0 \sin(kx + \phi_0) - \epsilon \frac{\rho_0^3}{32} \sin 3(kx + \phi_0) + O(\epsilon^2), \tag{6.1.18a}$$

$$\begin{aligned} u_t(x, 0; \epsilon) = &- \rho_0 \sqrt{1 + k^2} \cos(kx + \phi_0) \\ &- \frac{3\rho_0^3 \epsilon}{8\sqrt{1 + k^2}} \left[\cos(kx + \phi_0) - \frac{1 + k^2}{4} \cos 3(kx + \phi_0) \right] \\ &+ O(\epsilon^2). \end{aligned} \tag{6.1.18b}$$

If we consider the special case $\phi_0 = 0$ in (6.1.18) and further modify these initial conditions by setting the terms of order ϵ equal to zero,

$$u(x, 0; \epsilon) = \rho_0 \sin kx + O(\epsilon^2), \tag{6.1.19a}$$

$$u_t(x, 0; \epsilon) = -\rho_0 \sqrt{1 + k^2} \cos kx + O(\epsilon^2), \tag{6.1.19b}$$

we expect the solution to $O(\epsilon)$ to be altered. In particular, we cannot expect the solution to depend on the single strained fast phase θ^+.

We assume a two-scale expansion, as in Sec. 4.2, involving the fast scale $t^+ = (1+\epsilon^2\omega_2 + O(\epsilon^3))t$ and the slow scale $\tilde{t} = \epsilon t$

$$u(x,t;\epsilon) = \sum_{m=0}^{M} \epsilon^m u_m(x, t^+, \tilde{t}) + O(\epsilon^{M+1}), \tag{6.1.20}$$

and derive the following equations governing u_0, u_1, and u_2:

$$L(u_0) \equiv \frac{\partial^2 u_0}{\partial t^{+2}} - \frac{\partial^2 u_0}{\partial x^2} + u_0 = 0, \tag{6.1.21a}$$

$$L(u_1) = -2\frac{\partial^2 u_0}{\partial t^+ \partial \tilde{t}} - u_0^3, \tag{6.1.21b}$$

$$L(u_2) = -2\frac{\partial^2 u_1}{\partial t^+ \partial \tilde{t}} - 2\omega_2 \frac{\partial^2 u_0}{\partial t^{+2}} - \frac{\partial^2 u_0}{\partial \tilde{t}^2} - 3u_0^2 u_1. \tag{6.1.21c}$$

The initial conditions (6.1.19) imply that u_0 and u_1 must satisfy

$$u_0(x,0,0) = \rho_0 \sin kx, \qquad \frac{\partial u_0(x,0,0)}{\partial t^+} = -\rho_0\sqrt{1+k^2}\cos kx, \tag{6.1.22}$$

$$u_1(x,0,0) = 0, \qquad \frac{\partial u_1(x,0,0)}{\partial t^+} = -\frac{\partial u_0(x,0,0)}{\partial \tilde{t}}. \tag{6.1.23}$$

Let us introduce the following notation for $n = 1, 2, \ldots$

$$\theta_n^+ = nkx - \sqrt{1+n^2k^2}\,t^+, \qquad \theta_n^- = nkx + \sqrt{1+n^2k^2}\,t^+, \tag{6.1.24}$$

to describe a wave having wave number nk traveling to the right (θ_n^+) or the left (θ_n^-). The solution of (6.1.21a) for the initial conditions (6.1.22) is

$$u_0(x, t^+, \tilde{t}) = \rho(\tilde{t}) \sin[\theta_1^+ + \phi(\tilde{t})], \tag{6.1.25}$$

where

$$\rho(0) = \rho_0, \qquad \phi(0) = 0. \tag{6.1.26}$$

To determine $\rho(\tilde{t})$ and $\phi(\tilde{t})$, we consider (6.1.21b), in which we use (6.1.25) to evaluate the right-hand side

$$L(u_1) = 2\rho'(1+k^2)^{1/2}\cos(\theta_1^+ + \phi) - \left[2\rho(1+k^2)^{1/2}\phi' + \frac{3}{4}\rho^3\right]\sin(\theta_1^+ + \phi) + \frac{\rho^3}{4}\sin 3(\theta_1^+ + \phi), \tag{6.1.27}$$

where $' = d/d\tilde{t}$. The terms multiplied by $\cos(\theta_1^+ + \phi)$ and $\sin(\theta_1^+ + \phi)$ give rise to inconsistent mixed-secular contributions to u_1. We remove these terms by setting their coefficients equal to zero. Solving the resulting equations subject to (6.1.26) gives

$$\rho = \rho_0 = \text{ constant}, \quad \phi = -\frac{3\rho_0^2 \tilde{t}}{8(1+k^2)^{1/2}}. \tag{6.1.28}$$

We note that (6.1.25), with ρ and ϕ defined by (6.1.28), is identical to the leading term of the periodic solution (6.1.16). However, since our initial conditions to $O(\epsilon)$ and higher are not consistent with the periodic solution, we expect $u_1, u_2, \ldots$ to be more complicated.

It follows from (6.1.23) and the result for u_0 that $(\partial u_1/\partial t^+)$ has the initial value

$$\frac{\partial u_1(x, 0, 0)}{\partial t^+} = \frac{3\rho_0^3}{8\sqrt{1+k^2}} \cos kx. \tag{6.1.29}$$

The solution of what remains of (6.1.27) can be expressed in the series form

$$u_1(x, t^+, \tilde{t}) = \sum_{n=1}^{\infty} \{A_n(\tilde{t}) \sin[\theta_n^+ + \alpha_n(\tilde{t})]$$

$$+B_n(\tilde{t}) \sin[\theta_n^- + \beta_n(\tilde{t})]\} - \frac{\rho_0^3}{32} \sin 3(\theta_1^+ + \phi). \tag{6.1.30}$$

The initial conditions imply that all the α_n and β_n are initially equal to zero. Also, all the A_n and B_n vanish initially, except for A_1, A_3, B_1, and B_3. Requiring u_2 to be consistent on the t^+ scale shows that only A_1, A_3, B_1, B_3, α_1, α_3, β_1, and β_3 are needed; all the other coefficients vanish identically. The details are omitted for brevity (see Problem 1). We find

$$u_1(x, t^+, \tilde{t}) = \frac{3\rho_0^3}{16(1+k^2)} \left[-\sin(\theta_1^+ + \phi(\tilde{t})) + \sin(\theta_1^- - 2\phi(\tilde{t}))\right]$$

$$+\frac{\rho_0^3}{64} \left\{\left[1 + 3\left(\frac{1+k^2}{1+9k^2}\right)^{1/2}\right] \sin\left(\theta_3^+ - \frac{3\rho_0^2 \tilde{t}}{4\sqrt{1+9k^2}}\right)\right.$$

$$\left. + \left[1 - 3\left(\frac{1+k^2}{1+9k^2}\right)^{1/2}\right] \sin\left(\theta_3^- + \frac{3\rho_0^2 \tilde{t}}{4\sqrt{1+9k^2}}\right)\right\} - \frac{\rho_0^3}{32} \sin 3(\theta_1^+ + \phi(\tilde{t})). \tag{6.1.31}$$

Consistency of u_2 on the $\tilde{t}$ scale also determines ω_2:

$$\omega_2 = \frac{\rho_0^4(3k^2 - 51)}{256(1+k^2)^2}. \tag{6.1.32}$$

This result shows that u_1 has four waves in addition to the third harmonic term with amplitude $\rho_0^3/32$. First, we have two waves with phase speed $\pm(1+k^2)^{1/2}/k$ associated with the initial condition; these are the $\sin(\theta_1^+ + \phi)$ and $\sin(\theta_1^- - 2\phi)$ terms. In addition, the sine terms involving θ_3^+ and θ_3^- represent waves with phase speed $\pm(1+9k^2)^{1/2}/3k$. Another example illustrating the use of multiple scales to represent nearly periodic waves is outlined in Problem 3.

Equation (6.1.13) also possesses a hierarchy of more complicated solutions corresponding to more general initial conditions. For example, instead of a single wave initially, we can study the solution corresponding to a discrete set of such waves (the case $N = 2$ is considered in Problem 2):

$$u(x,0;\epsilon) = \sum_{i=1}^{N} A_i \sin k_i x, \tag{6.1.33}$$

$$u_t(x,0;\epsilon) = -\sum_{i=1}^{N} A_i\sqrt{1+k_i^2}\cos k_i x. \tag{6.1.34}$$

As one might expect, there will be resonance between these waves for certain ratios of the wave numbers, in exact analogy with the resonance between the normal modes of weakly nonlinear coupled oscillators (see Sec. 4.3.6). Routine extension of the ideas in Sec. 4.3.6 can be used to study resonance cases. We will not go into the details here. The main results for $N = 2$ and $N = 3$ are outlined in parts (c) and (d) of Problem 2.

The most general initial-value problem is, of course,

$$u(x,0;\epsilon) = p(x), \tag{6.1.35}$$

$$u_t(x,0;\epsilon) = q(x) \tag{6.1.36}$$

for arbitrary (well-behaved) functions p and q. In this case u_0 is, in general, not expressible as a discrete set of uniform periodic waves. One approach is to solve (6.1.21a) using Fourier transforms. We find

$$u_0(x,t,\tilde{t}) = \frac{1}{2\sqrt{2\pi}}\int_{-\infty}^{\infty}\left\{\left[P(k,\tilde{t}) - i\frac{Q(k,\tilde{t})}{\sqrt{1+k^2}}\right]\exp[i(kx+\sqrt{1+k^2}t)]\right.$$

$$\left.+\left[P(k,\tilde{t}) + i\frac{Q(k,\tilde{t})}{\sqrt{1+k^2}}\right]\exp[i(kx-\sqrt{1+k^2}t]\right\}dk, \tag{6.1.37}$$

where $P(k,0)$ and $Q(k,0)$ are the Fourier transforms of the initial values $p(x)$ and $q(x)$, respectively. The technical difficulties associated with deriving the equations governing P and Q from consistency arguments on u_1 are formidable in general.

A similar idea is to use Fourier transforms directly on the governing equation (6.1.13). This approach is explored in [6.3] and [6.26]. If the nonlinear terms in the governing equation are in convolution form, the resulting ordinary differential equation for the Fourier transform of u is just the equation for a weakly nonlinear oscillator. This can be easily solved using multiple scales and then inverted. For further details, see [6.3] and Problem 5. This idea is not restricted to (6.1.13), and a number of other examples are discussed in [6.3].

Consider now the special case of (6.1.7) with $c_1 = c_2 = 0$ that results in (6.1.15). Now the linear problem ($\epsilon = 0$) is nondispersive, thus any initial condition of the form

$$u(x,0) = p(kx), \quad u_t(x,0) = -kp'(kx) \tag{6.1.38}$$

gives the traveling wave $u(x,t) = p(k(x-t))$. For simplicity, and to compare with the results for $c_1 > 0$, let us pick the initial conditions:

$$u(x,0;\epsilon) = \rho_0 \sin kx, \quad u_t(x,0;\epsilon) = -\rho_0 k \cos kx. \tag{6.1.39}$$

If we attempt to construct an expansion analogous to (6.1.20), where u_0 has only the primary wave, $u_0 = \rho(\tilde{t}) \sin[k(x-t) + \phi(\tilde{t})]$, we discover immediately that u_0^3 contains the third harmonic $-(\rho^3/4) \sin 3[k(x-t) + \phi(\tilde{t})]$, *which is also a homogeneous solution*. Thus, to eliminate this term we must also include a third harmonic of the primary wave in u_0. But now u_0^3 produces the fifth harmonic, etc. We therefore conclude that the appropriate series form for u_0 should be

$$u_0(x,t^+,\tilde{t}) = \sum_{n=0}^{\infty} \rho_{2n+1}(\tilde{t}) \sin[(2n+1)k(x-t^+) + \phi_{2n+1}(\tilde{t})]. \tag{6.1.40}$$

Thus, even if the initial data consists of a wave with only one harmonic, the perturbed solution for $0 < \epsilon \ll 1$ must involve all the odd higher harmonics of this wave. The calculations of the higher-order terms are now much more complicated, as we have to compute products and powers of infinite series and then decompose these into their harmonics.

A boundary-value problem

An example of a class of problems where the representation of u_0 in infinite series form has limited success is given in [6.16]. Consider the following generalization of (6.1.13):

$$u_{tt} - u_{xx} + u = \epsilon f(u, u_t). \tag{6.1.41}$$

We have the homogeneous boundary conditions

$$u(0,t;\epsilon) = u(\pi,t;\epsilon) = 0, \quad t > 0 \tag{6.1.42}$$

and general initial conditions

$$u(x,0;\epsilon) = p(x), \quad u_t(x,0;\epsilon) = q(x). \tag{6.1.43}$$

We assume the two-scale expansion

$$u(x,t;\epsilon) = u_0(x,t^+,\tilde{t}) + \epsilon u_1(x,t^+,\tilde{t}) + O(\epsilon^2), \tag{6.1.44}$$

where, as usual, $t^+ = (1 + \epsilon^2\omega_2 + \ldots)t$ and $\tilde{t} = \epsilon t$. We find that u_0 obeys

$$L(u_0) \equiv \frac{\partial^2 u_0}{\partial t^{+2}} - \frac{\partial^2 u_0}{\partial x^2} + u_0 = 0, \tag{6.1.45}$$

$$u_0(0,t^+,\tilde{t}) = u_0(\pi,t^+,\tilde{t}) = 0, \quad t^+ > 0, \tag{6.1.46}$$

$$u_0(x,0,0) = p(x), \quad \frac{\partial u_0}{\partial t^+}(x,0,0) = q(x), \tag{6.1.47}$$

and u_1 obeys

$$L(u_1) = -2\frac{\partial^2 u_0}{\partial t^+ \partial \tilde{t}} + f\left(u_0, \frac{\partial u_0}{\partial t^+}\right), \tag{6.1.48}$$

$$u_1(0, t^+, \tilde{t}) = u_1(\pi, t^+, \tilde{t}) = 0, \quad t^+ > 0, \tag{6.1.49}$$

$$u_1(x, 0, 0) = 0, \qquad \frac{\partial u_1(x, 0, 0)}{\partial t^+} = -\frac{\partial u_0(x, 0, 0)}{\partial \tilde{t}}. \tag{6.1.50}$$

We express the solution of (6.1.45)–(6.1.47) using a series of eigenfunctions

$$u_0(x, t^+, \tilde{t}) = \sum_{n=1}^{\infty}[a_n(\tilde{t}) \cos\sqrt{1+n^2}t^+ + b_n(\tilde{t}) \sin\sqrt{1+n^2}t^+] \sin nx. \tag{6.1.51}$$

The initial conditions imply that

$$a_n(0) = \frac{2}{\pi}\int_0^{\pi} p(x) \sin nx\, dx, \tag{6.1.52}$$

$$b_n(0) = \frac{2}{\pi\sqrt{1+n^2}}\int_0^{\pi} q(x) \sin nx\, dx. \tag{6.1.53}$$

We now determine the conditions governing the evolution of the a_n and b_n by considering u_1. In order that u_1 be bounded, the right-hand side of (6.1.48) must be orthogonal to all the homogeneous solutions

$$(\sin\sqrt{1+n^2}t^+) \sin nx, \qquad (\cos\sqrt{1+n^2}t^+) \sin nx. \tag{6.1.54}$$

This is a well-known result for perturbed eigenvalue problems and can be proven directly (e.g., see [6.16]). However, it is more instructive to derive the result by expanding u_1 in a series of eigenfunctions.

Let

$$u_1 = \sum_{n=1}^{\infty} \alpha_n(t^+, \tilde{t}) \sin nx. \tag{6.1.55}$$

Clearly, f must also have such an expansion, i.e.,

$$f\left(u_0, \frac{\partial u_0}{\partial t^+}\right) = \sum_{n=1}^{\infty} f_n(t^+, \tilde{t}) \sin nx, \tag{6.1.56}$$

where

$$f_n(t^+, \tilde{t}) = \frac{2}{\pi}\int_0^{\pi} f\left(u_0(x, t^+, \tilde{t}), \frac{\partial u_0}{\partial t^+}(x, t^+, \tilde{t})\right) \sin nx\, dx. \tag{6.1.57}$$

Therefore, for each n, the α_n are governed by the equation

$$\frac{\partial^2 \alpha_n}{\partial t^{+^2}} + (1+n^2)\alpha_n = 2\frac{da_n}{d\tilde{t}}\sqrt{1+n^2} \sin\sqrt{1+n^2}t^+$$

$$-2\frac{db_n}{d\tilde{t}}\sqrt{1+n^2}\cos\sqrt{1+n^2}t^+ + f_n(t^+,\tilde{t}), \tag{6.1.58}$$

which is free of x.

We seek a solution for α_n by variation of parameters in the form (see the discussion in Sec. 4.2 following (4.2.100))

$$\alpha_n(t^+,\tilde{t}) = P_n(t^+,\tilde{t})\sin\psi + Q_n(t^+,\tilde{t})\cos\psi; \quad \psi = \sqrt{1+n^2}t^+. \tag{6.1.59}$$

Substituting the above into (6.1.58) gives

$$\frac{\partial P_n}{\partial t^+}\cos\psi - \frac{\partial Q_n}{\partial t^+}\sin\psi = \frac{da_n}{d\tilde{t}}\sin\psi - \frac{db_n}{d\tilde{t}}\cos\psi + \frac{f_n(t^+,\tilde{t})}{\sqrt{1+n^2}}, \tag{6.1.60}$$

where we have set

$$\frac{\partial P_n}{\partial t^+}\sin\psi + \frac{\partial Q_n}{\partial t^+}\cos\psi = -\frac{da_n}{d\tilde{t}}\cos\psi - \frac{db_n}{d\tilde{t}}\sin\psi. \tag{6.1.61}$$

Therefore, solving (6.1.60)-(6.1.61), we find

$$P_n = -\left[\frac{db_n}{d\tilde{t}}t^+ - \int_0^{t^+}\frac{f_n\cos\psi}{\sqrt{1+n^2}}dt^+\right], \tag{6.1.62}$$

$$Q_n = -\left[\frac{da_n}{d\tilde{t}}t^+ + \int_0^{t^+}\frac{f_n\sin\psi}{\sqrt{1+n^2}}dt^+\right]. \tag{6.1.63}$$

Note that since f_n depends on t^+ only through $\sin\psi$ and $\cos\psi$, the integrals in (6.1.62)–(6.1.63) can, at worst, grow like t^+ as $t^+ \to \infty$. In order that u_1 be bounded, we must require that

$$\frac{db_n}{d\tilde{t}} - \lim_{t^+\to\infty}\frac{1}{t^+}\int_0^{t^+}\frac{f_n\cos\psi\, dt^+}{\sqrt{1+n^2}} = 0, \tag{6.1.64a}$$

$$\frac{da_n}{d\tilde{t}} + \lim_{t^+\to\infty}\frac{1}{t^+}\int_0^{t^+}\frac{f_n\sin\psi\, dt^+}{\sqrt{1+n^2}} = 0. \tag{6.1.64b}$$

Equations (6.1.64) are a system of infinitely many nonlinear equations for the a_n and b_n to be solved subject to the initial conditions (6.1.52)–(6.1.53).

We were able to derive compact formulas for the a_n and b_n because the boundary conditions (6.1.42) allowed us to expand the solution in a series of eigenfunctions and to essentially eliminate the x dependence from the results. This is not true in general for initial-value problems on the infinite interval, and the corresponding expressions governing the slowly varying coefficients are more difficult to derive.

Having eliminated the mixed-secular terms from u_1, the solution can easily be derived. Of course, it makes little sense to proceed further unless one can solve (6.1.64). In general, this is a formidable task, and one would have to resort to some approximation scheme such as truncating the series after a relatively few terms. Given the notoriously slow convergence of trigonometric series, this does not appear to be a very useful approach either.

A rare example where (6.1.64) were solved *exactly* appears in [6.16] for the case where the perturbation function f is (see Equation (4.2.45))

$$f = u_t - \frac{1}{3}u_t^3. \tag{6.1.65}$$

The calculations are quite involved and will not be given here. In Sec. 6.2, we study a class of problems where we can avoid use of infinite series expansions and the associated difficulties.

6.1.2 Waves Slowly Varying in Space and Time

In view of the behavior of linear dispersive waves in the far field (see Chapter 11 of [6.27] and section 3.8 of [6.17]), it is natural to look for solutions of the nonlinear problem in the form of traveling waves with amplitudes, phases, wave numbers, and frequencies that vary slowly in space and time. The basic ideas for this class of solutions were introduced in [6.28] and are based on a variational approach using a Lagrangian formulation. A unified treatment can be found in [6.27]. A treatment based on multiple-scale expansions is given in [6.22], and we present this point of view here.

Weakly nonlinear problem

We illustrate ideas using (6.1.13) with $0 < \epsilon \ll 1$. We look for a solution that generalizes the traveling wave $u = A\sin(kx - \sqrt{1+k^2}t + \phi)$, which is available for $\epsilon = 0$ with arbitrary constant values of the amplitude A, wave number k, and phase shift ϕ, to allow these parameters to vary slowly with x and t. More precisely, we look for a class of solutions having the structure

$$u(x, t; \epsilon) = F(\theta, \tilde{x}, \tilde{t}; \epsilon) = u_0(\theta, \tilde{x}, \tilde{t}) + \epsilon u_1(\theta, \tilde{x}, \tilde{t}) + O(\epsilon^2), \tag{6.1.66}$$

where F is a periodic function of the fast scale θ and depends also on the slow scales $\tilde{x} = \epsilon x$ and $\tilde{t} = \epsilon t$. The fast variable θ is a function of x, t, and ϵ defined by the pair of equations

$$\theta_x = k(\tilde{x}, \tilde{t}), \tag{6.1.67a}$$

$$-\theta_t = \omega(\tilde{x}, \tilde{t}), \tag{6.1.67b}$$

where k and ω are as yet undetermined functions of $\tilde{x}, \tilde{t}$. This is a direct generalization of the periodic solution (6.1.16) (where k and ω were constants) to a class of solutions with slowly varying parameters.

We note that the definition of θ by (6.1.67) is consistent only if $\theta_{xt} = \theta_{tx}$, i.e.,

$$k_{\tilde{t}}(\tilde{x}, \tilde{t}) = -\omega_{\tilde{x}}(\tilde{x}, \tilde{t}). \tag{6.1.68}$$

This condition represents conservation of waves and is one of the needed conditions linking the various unknowns occurring in the solution. The other conditions will be provided by requiring F to be periodic with respect to θ.

Once k and ω have been determined, we can compute θ by quadrature in the form

$$\theta(x, t; \epsilon) = \frac{1}{\epsilon}\left[\int_0^{\tilde{x}} k(\xi, \tilde{t})d\xi - \int_0^{\tilde{t}} \omega(\tilde{x}, \tau)d\tau - \int_0^{\tilde{x}}\int_0^{\tilde{t}} k_{\tilde{t}}(\xi, \tau)\, d\tau\, d\xi\right]. \tag{6.1.69}$$

It is important to note that although θ can be conveniently expressed in terms of $\tilde{x}$ and $\tilde{t}$, it should be regarded as a function of x, t, and ϵ for the purposes of a multiple-scale expansion. We have assumed that $\theta(0, 0; \epsilon) = 0$, and using (6.1.68) shows that the integrand in the last term may also be set equal to $-\omega_{\tilde{x}}$.

Derivatives of u have the form

$$u_t = -\omega F_\theta + \epsilon F_{\tilde{t}}, \tag{6.1.70a}$$

$$u_x = kF_\theta + \epsilon F_{\tilde{x}}, \tag{6.1.70b}$$

$$u_{tt} = \omega^2 F_{\theta\theta} - 2\epsilon\omega F_{\theta\tilde{t}} - \epsilon\omega_{\tilde{t}} F_\theta + \epsilon^2 F_{\tilde{t}\tilde{t}}, \tag{6.1.70c}$$

$$u_{xx} = k^2 F_{\theta\theta} + 2\epsilon k F_{\theta\tilde{x}} + \epsilon k_{\tilde{x}} F_\theta + \epsilon^2 F_{\tilde{x}\tilde{x}}. \tag{6.1.70d}$$

If we now expand F as in (6.1.66), we obtain the following equations governing u_0, u_1, u_2:

$$L(u_0) \equiv (\omega^2 - k^2)u_{0_{\theta\theta}} + u_0 = 0, \tag{6.1.71a}$$

$$L(u_1) = 2\omega u_{0_{\theta\tilde{t}}} + 2k u_{0_{\theta\tilde{x}}} + (\omega_{\tilde{t}} + k_{\tilde{x}})u_{0_\theta} - u_0^3, \tag{6.1.71b}$$

$$L(u_2) = 2\omega u_{1_{\theta\tilde{t}}} + 2k u_{1_{\theta\tilde{x}}} + (\omega_{\tilde{t}} + k_{\tilde{x}})u_{1_\theta} - u_{0_{\tilde{t}\tilde{t}}} + u_{0_{\tilde{x}\tilde{x}}} - 3u_0^2 u_1. \tag{6.1.71c}$$

Thus, the assumption that the solution depends only on one fast variable (rather than on x and t individually) leads to an ordinary differential operator for each of the u_i.

Periodicity with respect to θ gives us the dispersion relation

$$0 < \omega^2 - k^2 = \text{constant} \tag{6.1.72}$$

and, as in the linear ($\epsilon = 0$) case, we may normalize the period to be 2π by choosing the constant in (6.1.72) to be unity.

The solution of (6.1.71a) is

$$u_0 = \rho(\tilde{x}, \tilde{t}) \sin\psi; \quad \psi = \theta + \phi(\tilde{x}, \tilde{t}), \tag{6.1.73}$$

where the amplitude ρ and phase ϕ are unknowns to be determined by the periodicity requirement on u_1. Using the above expression for u_0, we find that (6.1.71b) reduces to

$$L(u_1) = (2\omega\rho_{\tilde{t}} + 2k\rho_{\tilde{x}} + \rho\omega_{\tilde{t}} + \rho k_{\tilde{x}})\cos\psi - 2\left[\rho\omega\phi_{\tilde{t}} + \rho k\phi_{\tilde{x}} + \frac{3\rho^3}{8}\right]\sin\psi + \frac{\rho^3}{4}\sin 3\psi. \tag{6.1.74}$$

The terms proportional to $\cos\psi$ and $\sin\psi$ produce mixed-secular contributions in u_1 and must be removed. This means that ρ and ϕ obey the following first-order partial differential equations:

$$(\omega\rho^2)_{\tilde{t}} + (k\rho^2)_{\tilde{x}} = 0, \tag{6.1.75a}$$

$$\omega\phi_{\tilde{t}} + k\phi_{\tilde{x}} = -\frac{3}{8}\rho^2. \tag{6.1.75b}$$

The above, together with (6.1.68) and (6.1.72), are a set of four equations for the four unknowns ρ, ϕ, k, ω. We will now consider the solutions of these equations.

First, we use the dispersion relation to eliminate ω from the consistency condition (6.1.68). This reduces to the quasilinear equation

$$k_{\tilde{t}} + \frac{k}{\sqrt{1+k^2}}k_{\tilde{x}} = 0. \tag{6.1.76}$$

We solve this by integrating the characteristic equations

$$\frac{d\tilde{t}}{ds} = 1, \tag{6.1.77a}$$

$$\frac{d\tilde{x}}{ds} = \frac{k}{\sqrt{1+k^2}}, \tag{6.1.77b}$$

$$\frac{dk}{ds} = 0, \tag{6.1.77c}$$

which pass through some initial curve

$$k(\tilde{x}, 0) = K(\tilde{x}). \tag{6.1.78}$$

The solution is conveniently expressed in the parametric form

$$k(\tilde{x}, \tilde{t}) = K(\xi), \tag{6.1.79a}$$

$$\xi = \tilde{x} - \frac{K(\xi)\tilde{t}}{\sqrt{1+K^2(\xi)}}. \tag{6.1.79b}$$

Thus, along the characteristic curve $\xi = \text{const.}$ (which is the straight line with slope $K/\sqrt{1+K^2}$, given by (6.1.79b)), k maintains its initial value. Notice that $K/\sqrt{1+K^2}$ is the group velocity of waves with wave numbers near K according to the linear theory. Thus, the result that holds for large t in the linear problem is now true *everywhere* for this special class of solutions of the weakly nonlinear problem. This is not surprising since at this stage the nonlinear term has not yet played a role, and (6.1.79) just corresponds to a necessary kinematic condition for slowly varying waves that obey the dispersion relation (6.1.72).

Observe also that the solution (6.1.79) becomes meaningless if the characteristics cross for $t > 0$. This means that the initial value K *must be a monotone nondecreasing function of* x, i.e., $K'(\xi) \geq 0$. For any initial value K that produces converging characteristics, i.e., $K' < 0$, (6.1.13) cannot have a solution of the assumed form in those parts of the $\tilde{x}, \tilde{t}$ plane where characteristics cross. In fact, since equation (6.1.13) does not admit shocks, we cannot avoid the difficulty for $K' < 0$ by resorting to a weak solution of (6.1.76).

Once k is defined, ω is given by the dispersion relation. Obviously, $\omega(\tilde{x}, \tilde{t})$ also maintains its initial value

$$\omega(\tilde{x}, 0) = \sqrt{1 + K^2(\tilde{x})} \tag{6.1.80}$$

along each of the characteristics $\xi = \text{constant}$.

To solve (6.1.75a) for ρ, we use the known expressions for k and ω to find

$$\rho_{\tilde{t}} + \frac{k}{\sqrt{1+k^2}}\rho_{\tilde{x}} = -\frac{k_{\tilde{x}}}{2(1+k^2)^{3/2}}\rho. \tag{6.1.81}$$

This is linear and can easily be solved by integrating the characteristic equations. We also express the result in parametric form as follows:

$$\rho = \frac{\rho_0(\xi)}{\left\{1 + \frac{\tilde{t}K'(\xi)}{[1+K^2(\xi)]^{3/2}}\right\}^{1/2}}, \tag{6.1.82}$$

where ξ is the solution of (6.1.79b) and ρ_0 is the initial value of ρ, i.e.,

$$\rho_0(\tilde{x}) = \rho(\tilde{x}, 0). \tag{6.1.83}$$

Now, because of the inhomogeneous term on the right-hand side of (6.1.81), we have the attenuation factor proportional to $\tilde{t}^{-1/2}$, for large $\tilde{t}$, in the amplitude. This result is also very reminiscent of the asymptotic expression for the linear problem and is again to be expected since the nonlinear term has not yet been used; it only contributes to the solution for ϕ.

With ρ known, (6.1.75b) for ϕ becomes

$$\phi_{\tilde{t}} + \frac{k}{\sqrt{1+k^2}}\phi_{\tilde{x}} = -\frac{3}{8}\frac{\rho^2}{\sqrt{1+k^2}}. \tag{6.1.84}$$

This can also be easily solved to give

$$\phi(\tilde{x}, \tilde{t}) = \phi_0(\xi) - \frac{3\rho_0^2(1+K^2)}{8K'}\log\frac{(1+K^2)^{3/2} + K'\tilde{t}}{(1+K^2)^{3/2}}, \tag{6.1.85}$$

where ρ_0 and K are the functions of ξ that we calculated earlier, and $\phi_0(\tilde{x}) = \phi(\tilde{x}, 0)$. Note that for $K' = 0$ we recover the earlier result (see (6.1.28))

$$\phi = \phi_0(\xi) - 3\rho_0^2(\xi)\tilde{t}/8\sqrt{1+K^2} \tag{6.1.86}$$

along each characteristic. For $K' > 0$, ϕ decreases more slowly with $\tilde{t}$. Since $K' < 0$ is not meaningful we do not have to deal with an unpleasant logarithmic singularity in ϕ at finite $\tilde{t}$.

This completes the solution to $O(1)$, and the results to $O(\epsilon)$ can be computed in a similar manner by considering the equation for u_2. (See Problem 4.)

The following limitations must be borne in mind regarding the above class of solutions:

1. As in the case of the strictly periodic solution, our results correspond to very special initial conditions.
2. These initial conditions are in the form of a given wave with spatially slowly varying amplitude, wave number (frequency), and phase, i.e., our results are a formal solution of (6.1.13) in the limit $\epsilon \to 0$ for the initial-value problem

$$u(x, 0; \epsilon) = \rho_0(\tilde{x}) \sin \left[\frac{1}{\epsilon} \int_0^{\tilde{x}} K(\xi) d\xi + \phi(\tilde{x}, 0) \right] + O(\epsilon) \qquad (6.1.87a)$$

$$u_t(x, 0; \epsilon) = -\rho_0(\tilde{x}) \sqrt{1 + K^2(\tilde{x})} \cos \left[\frac{1}{\epsilon} \int_0^{\tilde{x}} K(\xi) d\xi + \phi(\tilde{x}, 0) \right] + O(\epsilon). \qquad (6.1.87b)$$

An example where initial conditions of this form arise naturally is given in Problem 2b.
3. The slowly varying initial wave number K cannot be prescribed arbitrarily; we must have $K' \geq 0$ to avoid multiple-valued solutions.
4. Solutions to more general initial-value problems cannot be calculated from the above by superposition.
5. The asymptotic behavior for $t \to \infty$ of an arbitrary initial-value problem for (6.1.13) is not in the form (6.1.66).

Strictly nonlinear problem

The idea of a nearly periodic traveling wave does generalize to the strictly nonlinear case ($\epsilon = 1$). In fact, we may consider the more general wave equation

$$u_{tt} - u_{xx} + V'(u) = 0 \qquad (6.1.88)$$

for functions $V(u)$ that are concave over some interval $u_1 < u < u_2$. In this case, (6.1.88) has traveling waves that are *exactly* periodic. We derive these by assuming u in the form

$$u = F(\theta), \quad \theta = kx - \omega t \qquad (6.1.89)$$

and find that F obeys the nonlinear oscillator equation

$$\mu^2 \frac{d^2 F}{d\theta^2} + V'(F) = 0, \quad \mu^2 = \omega^2 - k^2 > 0. \qquad (6.1.90)$$

Multiplying by $(dF/d\theta)$ and integrating gives the energy integral

$$\frac{\mu^2}{2} \left(\frac{dF}{d\theta} \right)^2 + V(F) = E = \text{constant}. \qquad (6.1.91)$$

For periodic solutions, (6.1.91) must describe a closed curve in the $F \frac{dF}{d\theta}$ plane for any fixed value of E.

Let $u_{\min}(E)$ and $u_{\max}(E)$ denote the minimum and maximum values achieved by u over one period, i.e., the two roots of $V(u) = E$. Integrating (6.1.91) gives (see (4.4.10))

$$\theta - \eta = \mu \int_{u_{\min}(E)}^{u} \frac{ds}{\pm\sqrt{2[E - V(s)]}}, \tag{6.1.92}$$

where the $\pm$ signs correspond to the signs of $(du/d\theta)$, and the phase shift η is the value of θ when $u = u_{\min}(E)$. The period in θ is then given by

$$P(E, \mu) = 2\mu \int_{u_{\min}(E)}^{u_{\max}(E)} \frac{ds}{\sqrt{2[E - V(s)]}}. \tag{6.1.93}$$

For the linear problem ($V' = F$), this gives $P = 2\pi\mu$, a relation that does not depend on E. Normalizing P to equal 2π gives the dispersion relation (6.1.72), $\mu = 1$. For the nonlinear case, (6.1.92) gives P as a function of E and μ. If we normalize P to again equal 2π, (6.1.93) may be solved for μ in terms of E, or

$$\omega = \Omega(k, E). \tag{6.1.94}$$

To compute F, we invert (6.1.92) to obtain

$$u = F(\theta - \eta, \mu, E). \tag{6.1.95}$$

The details of the calculation are identical to those for the strictly nonlinear oscillator discussed in Sec. 4.4.

Once F is found, we can derive initial conditions $u(x, 0)$ and $u_t(x, 0)$, which generate this traveling wave solution. These are of the form

$$u(x, 0) = p(x, E, \omega, k), \tag{6.1.96a}$$

$$u_t(x, 0) = q(x, E, \omega, k), \tag{6.1.96b}$$

where E and k may be chosen arbitrarily and ω is the function of k and E given by (6.1.94).

The nearly periodic traveling waves generalize the initial conditions (6.1.96) by allowing k and E (hence ω) to vary slowly in x, i.e., to be functions of $\tilde{x} = \epsilon x$. Thus, ϵ is the small parameter that measures this slow variation. In this case, we look for a solution of (6.1.88) that has the form (6.1.66) with k and ω satisfying (6.1.67). We then find that u_0 and u_1 are governed by

$$\mu^2 \frac{\partial^2 u_0}{\partial \theta^2} + V'(u_0) = 0, \tag{6.1.97a}$$

$$\mu^2 \frac{\partial^2 u_1}{\partial \theta^2} + V''(u_0)u_1 = 2\omega \frac{\partial^2 u_0}{\partial \theta \partial \tilde{t}} + 2k \frac{\partial^2 u_0}{\partial \theta \partial \tilde{x}} + (\omega_{\tilde{t}} + k_{\tilde{x}}) \frac{\partial u_0}{\partial \theta}, \tag{6.1.97b}$$

where

$$\mu^2 = \omega^2 - k^2. \tag{6.1.97c}$$

The solution of (6.1.97a) proceeds as for the case of constant k and E. We require the period to be a constant (independent of $\tilde{x}$ and $\tilde{t}$) for exactly the same reason as discussed in Sec. 4.5 (see the discussion following (4.4.11)). If we normalize this constant to equal 2π, we again obtain (6.1.94), which defines ω in terms of k and E.

The consistency condition

$$k_{\tilde{t}} + \omega_{\tilde{x}} = 0 \tag{6.1.98}$$

gives a second relation linking ω to k. We now write the solution of (6.1.97a) in the form

$$u_0 = F(\psi, \mu, E), \quad \psi = \theta - \eta, \tag{6.1.99}$$

where the phase shift η depends on $\tilde{x}$ and $\tilde{t}$. Thus, we need to define $\omega(\tilde{x}, \tilde{t})$, $k(\tilde{x}, \tilde{t})$, $E(\tilde{x}, \tilde{t})$, and $\eta(\tilde{x}, \tilde{t})$ to determine the solution to $O(1)$. Equations (6.1.94) and (6.1.98) give two relations on these four functions. Two more conditions follow by requiring that u_1 be periodic in θ. The details are identical to those in Sec. 4.4 and are not repeated. The reader is also referred to the basic source, [6.22].

6.1.3 Weakly Nonlinear Stability

In this section, we use two simple mathematical models to illustrate the asymptotic behavior of solutions for a class of problems that arise in hydrodynamic stability and other applications. These problems are parabolic and depend on a parameter that determines the stability of a small periodic disturbance with a given wave number. The main goal is to determine the effect of a weak nonlinearity and to derive the asymptotic behavior of solutions. An account of the general theory and a survey of the literature can be found in [6.10]. The linear stability of various problems in fluid mechanics is discussed in [6.9], where some aspects of the weakly nonlinear problem are also discussed. As problems of physical significance are generally rather involved and not solvable analytically, we restrict attention to simple mathematical models that illustrate the basic ideas.

One-dimensional model equation; discrete modes

In [6.24], Matkowsky studies the model nonlinear diffusion equation

$$u_t + \lambda f(u) = u_{xx}, \quad \lambda = \text{ constant}$$

depending on the parameter λ in the interval $0 \leq x \leq \pi$ with zero boundary conditions at $x = 0, \pi$, and a small initial disturbance. To be more specific, let us assume that f is odd in u and keep only two terms in its Taylor expansion. Also, let us rescale variables so that the parameter multiplies the u_{xx} term (as is the case in hydrodynamics). Thus, consider

$$u_t - u + u^3 = \frac{1}{R} u_{xx}, \quad R = \text{ constant } > 0, \quad 0 \leq x \leq \pi, \quad 0 \leq t, \tag{6.1.100}$$

with the homogeneous boundary conditions

$$u(0, t; \epsilon) = u(\pi, t; \epsilon) = 0, \quad t > 0 \tag{6.1.101}$$

and initial condition

$$u(x, 0; \epsilon) = \epsilon h(x). \tag{6.1.102}$$

Here ϵ is a positive small parameter that measures the amplitude of the initial disturbance and the parameter R is $O(1)$. It is convenient to rescale the dependent variable: $u = \epsilon v$, $v = O(1)$ so that the problem reads

$$v_t - v + \epsilon^2 v^3 = \frac{1}{R} v_{xx}. \tag{6.1.103}$$

$$v(0, t; \epsilon) = v(\pi, t; \epsilon) = 0, \quad t > 0. \tag{6.1.104}$$

$$v(x, 0; \epsilon) = h(x). \tag{6.1.105}$$

Consider first the unperturbed linear problem for $\epsilon = 0$. Separation of variables shows that the eigenfunctions are $\sin nx$, and we expand $v(x, t; 0)$ in a Fourier sine series

$$v(x, t; 0) = \sum_{n=1}^{\infty} w_n(t) \sin nx, \tag{6.1.106}$$

where the w_n obey

$$\frac{dw_n}{dt} + \left(\frac{n^2}{R} - 1\right) w_n = 0. \tag{6.1.107}$$

Therefore,

$$w_n(t) = w_n(0)e^{\sigma_n t}, \quad \sigma_n = 1 - \frac{n^2}{R} \tag{6.1.108a}$$

and

$$w_n(0) = \frac{2}{\pi} \int_0^{\pi} h(x) \sin nx dx. \tag{6.1.108b}$$

We notice that if $0 < R < 1$ *all* the w_n decay exponentially. Now let R be a fixed constant larger than one. There is a critical integer, n_R, equal to the first integer that is larger than $\sqrt{R}$, such that the w_n still decay exponentially for all $n \geq n_R$. However, each of the w_n for $n < n_R$ grows exponentially, and the largest growth rate σ_n corresponds to $n = 1$. Thus, we say that w_1 is the *linearly most unstable mode*.

The linear theory is not very interesting. It predicts that either all modes decay exponentially ($R < 1$) or some modes grow exponentially ($R > 1$), thus invalidating the linearization assumption. The question arises whether the weak nonlinearity can stabilize a marginally unstable solution of the linear problem.

To investigate this question, let us assume that R is slightly larger than unity, thereby making w_1 marginally unstable. We set

$$R = 1 + \epsilon^a \alpha, \tag{6.1.109}$$

where α is a positive constant independent of ϵ and a is to be determined. We assume that v has a multiple-scale expansion of the form

$$v(x, t; \epsilon) = v_0(x, t, \bar{t}) + \epsilon^{\gamma} v_1(x, t, \bar{t}) + o(\epsilon^{\gamma}), \tag{6.1.110}$$

where the slow time is $\bar{t} = \epsilon^{\beta} t$. The constants β and γ are also to be determined. We compute

$$v_t = \frac{\partial v_0}{\partial t} + \epsilon^{\beta} \frac{\partial v_0}{\partial \bar{t}} + \epsilon^{\gamma} \frac{\partial v_1}{\partial t} + \ldots, \tag{6.1.111a}$$

$$v_x = \frac{\partial v_0}{\partial x} + \epsilon^{\gamma} \frac{\partial v_1}{\partial x} + \ldots, \tag{6.1.111b}$$

$$v_{xx} = \frac{\partial^2 v_0}{\partial x^2} + \epsilon^{\gamma} \frac{\partial^2 v_1}{\partial x^2} + \ldots. \tag{6.1.111c}$$

Substituting the expansions (6.1.111) as well as the expansion $R^{-1} = 1 - \epsilon^a \alpha + \ldots$ into (6.1.103) gives

$$\frac{\partial v_0}{\partial t} - v_0 - \frac{\partial^2 v_0}{\partial x^2} + \epsilon^{\gamma} \left(\frac{\partial v_1}{\partial t} - v_1 - \frac{\partial^2 v_1}{\partial x^2} \right) + \epsilon^{\beta} \frac{\partial v_0}{\partial \bar{t}}$$

$$+ \epsilon^2 v_0^3 + \epsilon^a \alpha \frac{\partial^2 v_0}{\partial x^2} + \ldots = 0, \tag{6.1.112}$$

where ... denotes terms that tend to zero faster than any of the terms retained.

We see that the richest equation for v_1 requires that we set $\beta = \gamma$ in order to include the variation of v_0 with respect to $\bar{t}$. We also need to set $\gamma = 2$ to include the nonlinear contribution v_0^3. Finally, we must set $a = \gamma$ to account for the departure of R from the neutral value $R = 1$. These three conditions give $\gamma = \beta = a = 2$, i.e., $\bar{t} = \epsilon^2 t$, $R = 1 + \epsilon^2 \alpha$ and the expansion for v has the form

$$v(x, t; \epsilon) = v_0(x, t; \bar{t}) + \epsilon^2 v_1(x, t, \bar{t}) + \epsilon^4 v_2(x, t, \bar{t}) + \ldots. \tag{6.1.113}$$

The equations and initial and boundary conditions governing v_0 and v_1 are then given by

$$L(v_0) \equiv \frac{\partial v_0}{\partial t} - v_0 - \frac{\partial^2 v_0}{\partial x^2} = 0, \tag{6.1.114a}$$

$$v_0(0, t; \bar{t}) = v_0(\pi, t, \bar{t}) = 0, \quad t > 0, \tag{6.1.114b}$$

$$v_0(x, 0, 0) = h(x). \tag{6.1.114c}$$

$$L(v_1) = -\frac{\partial v_0}{\partial \bar{t}} - v_0^3 - \alpha \frac{\partial^2 v_0}{\partial x^2}, \tag{6.1.115a}$$

$$v_1(0, t, \bar{t}) = v_1(\pi, t, \bar{t}) = 0, \quad t > 0, \tag{6.1.115b}$$

$$v_1(x, 0, 0) = 0. \tag{6.1.115c}$$

The solution of (6.1.114a) that satisfies the boundary conditions at the two ends is

$$v_0(x, t, \bar{t}) = \sum_{n=1}^{\infty} B_n(\bar{t}) e^{(1-n^2)t} \sin nx. \tag{6.1.116}$$

The initial condition (6.1.114c) gives

$$B_n(0) = \frac{2}{\pi} \int_0^{\pi} h(x) \sin nx \, dx. \tag{6.1.117}$$

To determine the $B_n(\bar{t})$, we require the solution for v_1 to be consistent. Although it is a straightforward matter to keep track of all the B_n in the right-hand side of (6.1.115a) and to derive an evolution equation for each, we note that all except B_1 are multiplied by the decaying exponential $e^{(1-n^2)t}$ in (6.1.116) and are thus irrelevant after a short time. To highlight this fact, let us just keep explicit track of B_1 and B_2 (which is multiplied by e^{-3t} in the solution).

After some algebra, we find that (6.1.115a) reduces to

$$\begin{aligned} L(v_1) = &-\left(\frac{dB_1}{d\bar{t}} + \frac{3}{4}B_1^3 - \alpha B_1\right)\sin x \\ &-\left(\frac{dB_2}{d\bar{t}} + \frac{3}{2}B_1^2 B_2 - 4\alpha B_2\right)e^{-3t}\sin 2x \\ &+ \text{H.S.} + \text{H.T.} \end{aligned} \tag{6.1.118}$$

Here H.S. denotes homogeneous solutions of the form $F_n(\bar{t})e^{(1-n^2)t} \sin nx$, and H.T. denotes harmless terms of the form $G_n(\bar{t})e^{-rt} \sin nx$, where $0 < r \neq n^2 - 1$. Each of the F_n for $n \geq 2$ is a linear homogeneous differential relation for B_n. The H.T. contribute the perfectly consistent contribution $G_n(e^{-rt} \sin nx)/(n^2 - r - 1)$ to v_1, whereas each of the H.S. contributes the *inconsistent* contribution $tF_n(\bar{t})e^{(1-n^2)t} \sin nx$. Therefore, we must set equal to zero the coefficients of $\sin x$, $e^{-3t} \sin 2x$, $e^{-8t} \sin 3x, \ldots$. This gives the following equations for $B_1, B_2, \ldots$:

$$\frac{dB_1}{d\bar{t}} = -\frac{3}{4}B_1^3 + \alpha B_1. \tag{6.1.119}$$

$$\frac{dB_2}{d\bar{t}} = 4\alpha B_2 - \frac{3}{2}B_1^2 B_2. \tag{6.1.120}$$

The amplitude equation (6.1.119) is sometimes referred to as a *Landau equation* (see [6.9] and [6.10] for further historical remarks).

We see from (6.1.119) that $B_1 = 0$ is an unstable equilibrium point, whereas $B_1 = \pm 2\sqrt{\alpha/3}$ are stable points. Thus, all solutions with $B_1(0) > 0$ tend to $B_1 = 2\sqrt{\alpha/3}$ and all solutions with $B_1(0) < 0$ tend to $B_1 = -2\sqrt{\alpha/3}$. The actual solution of (6.1.119) is

$$B_1(\bar{t}) = \frac{2B_1(0)\sqrt{\alpha/3}}{\left[B_1^2(0) + \left(\frac{4\alpha}{3} - B_1^2(0)\right)e^{-2\alpha\bar{t}}\right]^{1/2}}. \tag{6.1.121}$$

The limiting behavior for B_2 as $\bar{t} \to \infty$ is now easily derived. Because $\frac{3}{2}B_1^2 \to -2\alpha$, we see that as $\bar{t} \to \infty$, B_2 obeys $(dB_2/d\bar{t}) = 2\alpha B_2 + \ldots$, i.e., $B_2 \sim$

$e^{2\alpha\bar{t}} + \ldots$. This exponential growth on the $\bar{t}$ scale is dominated by the exponential decay term e^{-3t} that multiplies B_2. Therefore, $B_2(\bar{t})e^{-3t}$ decays exponentially fast, as do $B_3 e^{-8t}$, etc.

In summary, the marginally unstable ($R - 1 = O(\epsilon^2)$) linear solution that we have studied is equilibrated by the weak nonlinearity and tends to the steady state

$$v_0 = \pm 2\sqrt{\alpha/3} \sin x, \tag{6.1.122}$$

where the plus or minus signs correspond respectively to the signs of $B_1(0)$.

It is important to keep in mind that the existence of the steady state (6.1.122) depends critically on the choice of boundary conditions. For example, if we replace (6.1.104) by the periodic boundary conditions

$$v(0, t; \epsilon) = v(\pi, t; \epsilon), \quad v_x(0, t; \epsilon) = v_x(\pi, t; \epsilon) \tag{6.1.123}$$

for $t > 0$, we find that the appropriate expansion to replace (6.1.106) is

$$v(x, t; 0) = \frac{A_0(0)}{2} e^t + \sum_{n=1}^{\infty} \left(A_n(0) e^{\left(1 - \frac{4n^2}{R}\right)t} \cos 2nx \right.$$

$$\left. + B_n(0) e^{\left(1 - \frac{4n^2}{R}\right)t} \sin 2nx \right). \tag{6.1.124}$$

Thus the linear theory predicts that the $n = 0$ mode grows exponentially for any $0 < R = O(1)$. Therefore, a weakly nonlinear stability analysis based on perturbing (6.1.124) is not possible.

A second important restriction is the requirement $R = 1 + \epsilon^2\alpha$ corresponding to the linearly most unstable mode $n = 1$. Because for a general $h(x)$ all the initial modal amplitudes $B_n(0)$ for $n = 1, 2, \ldots$ will be nonzero, it is necessary to restrict R to the class $R = 1 + \epsilon\alpha^2$ in order to ensure that all the $B_n(\bar{t})$ for $n > 1$ decay exponentially as $t \to \infty$. Suppose, however, that we consider a *special* $h(x)$ consisting of the *single* harmonic $B_m(0) \sin mx$ for some $m > 1$. It is possible to construct a weakly nonlinear *bounded* solution for this initial condition with $R = m^2 + \epsilon\alpha$. However, this formal solution is not robust, because any infinitesimal perturbation having a lower harmonic will dominate and *grow exponentially*. Such a formal solution is possible due to the fact that the nonlinear term v_0^3 on the right-hand side of (6.1.115a) produces only *higher harmonics* of $\sin mx$ (but no subharmonics). We do not pursue these special solutions, as they are unphysical in practical applications.

Two-dimensional model equation

In the example (6.1.100)–(6.1.102), the spatial dependence of the solution was embodied in the discrete eigenfunctions $\sin nx$. We now consider a two-dimensional mathematical model proposed by Eckhaus (see Sec. 7.1 of [6.12]) to simulate the features encountered in hydrodynamic stability problems. In this model, the dependence of the solution on one of the spatial variables is given in terms of discrete eigenvalues, whereas the other variable ranges over $(-\infty, \infty)$ and requires a continuous spectrum.

We want to study

$$u_t - u_y u_{xx} = \frac{1}{R}(u_{xx} + u_{yy}) - u_{xxxx} \tag{6.1.125}$$

with the boundary conditions

$$u(x, 0, t; \epsilon) = 0, \quad u(x, 1, t; \epsilon) = 1; \quad t > 0 \tag{6.1.126}$$

and the requirement that u and its derivatives vanish as $|x| \to \infty$. The initial condition is

$$u(x, y, 0; \epsilon) = y + \epsilon h(x, y; \epsilon), \tag{6.1.127}$$

where $0 < \epsilon \ll 1$.

We set

$$u(x, y, t; \epsilon) = y + \epsilon v(x, y, t; \epsilon) \tag{6.1.128}$$

to derive the following problem for v:

$$v_t + \left(1 - \frac{1}{R}\right) v_{xx} - \frac{1}{R} v_{yy} + v_{xxxx} + \epsilon v_y v_{xx} = 0. \tag{6.1.129}$$

$$v(x, 0, t; \epsilon) = v(x, 1, t; \epsilon) = 0, \quad t > 0. \tag{6.1.130}$$

$$v, v_x, v_y, v_t, \ldots \to 0 \quad \text{as} \quad |x| \to \infty. \tag{6.1.131}$$

$$v(x, y, 0; \epsilon) = h(x, y; \epsilon). \tag{6.1.132}$$

Consider the linear problem ($\epsilon = 0$). The eigenfunctions for the y dependence of the solution are $\sin n\pi y$, $n = 1, 2, \ldots$, and we expand v in the series

$$v(x, y, t; 0) = \sum_{n=1}^{\infty} w_n(x, t) \sin n\pi y \tag{6.1.133}$$

and the w_n satisfy

$$\frac{\partial w_n}{\partial t} + \left(1 - \frac{1}{R}\right) \frac{\partial^2 w_n}{\partial x^2} + \frac{\partial^4 w_n}{\partial x^4} + \frac{n^2\pi^2}{R} w_n = 0, \tag{6.1.134}$$

subject to

$$w_n(x, 0) = 2 \int_0^1 h(x, y) \sin n\pi y \, dy. \tag{6.1.135}$$

To solve (6.1.134) one can use Fourier transforms in x to obtain

$$\frac{\partial W_n}{\partial t} = \sigma_n(k, R) W_n; \quad \sigma_n = k^2(1 - k^2) - \frac{n^2\pi^2 + k^2}{R}, \tag{6.1.136}$$

where $W_n(k, t), k =$ real, is the Fourier transform of $w_n(x, t)$. Thus, solutions grow or decay exponentially depending on whether $\sigma_n > 0$ or $\sigma_n < 0$, respectively. The maximum growth rate σ_n occurs for $n = 1$ for any k, i.e., the linearly most unstable mode is $n = 1$, and henceforth we restrict attention to this mode. The neutral stability boundary for R as a function of k is obtained by setting $\sigma_1 = 0$. This curve is shown in Figure 6.1.1 for the interval $0 < k < 1$. For $|k| > 1$, all solutions are stable for $R > 0$. Also, since σ_1 is even in k, we only need to consider $k > 0$.

The minimum value of R, denoted by R_c, occurs at $k = k_c$, where

$$k_c = \left[\left(\pi^4 + \pi^2\right)^{1/2} - \pi^2\right]^{1/2} = 0.6985\ldots, \tag{6.1.137a}$$

$$R_c = \left[\frac{\pi^2 + k_c^2}{k_c^2(1 - k_c^2)}\right] = 41.45\ldots. \tag{6.1.137b}$$

Let us first study the effect of small nonlinearity and small instability for the linearly most unstable mode $n = 1$ and the *fixed* wave number $k = k_c$. We will later consider the effect of allowing k to deviate slightly from k_c.

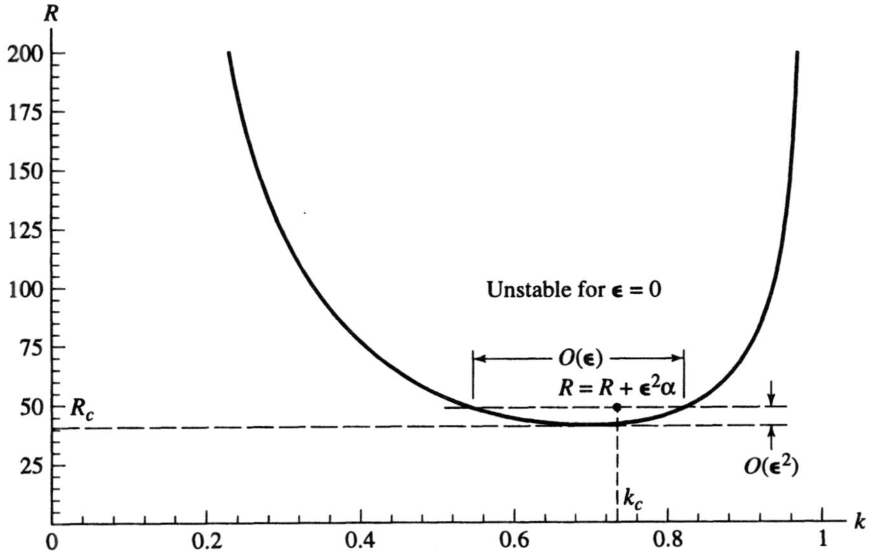

FIGURE 6.1.1. Linear Stability Boundary for $n = 1$

If we assume, as in the example of (6.1.103), that $R = R_c + \epsilon^a \alpha$, and expand v in the form

$$v(x, y, t; \epsilon) = v_0(x, y, t, \bar{t}) + \epsilon^\gamma v_1(x, y, t, \bar{t})$$

with $\bar{t} = \epsilon^\beta t$, (6.1.129) gives

$$L(v_0) + \epsilon^\gamma L(v_1) + \epsilon^\beta \frac{\partial v_0}{\partial \bar{t}} + \epsilon^a \frac{\alpha}{R_c^2} \Delta v_0 + \epsilon \frac{\partial v_0}{\partial y} \frac{\partial^2 v_0}{\partial x^2} + \ldots = 0, \quad (6.1.138a)$$

where

$$L = \frac{\partial}{\partial t} + \left(1 - \frac{1}{R_c}\right) \frac{\partial^2}{\partial x^2} - \frac{1}{R_c} \frac{\partial^2}{\partial y^2} + \frac{\partial^4}{\partial x^4} \quad (6.1.138b)$$

and

$$\Delta = \frac{\partial^2}{\partial x^2} + \frac{\partial^2}{\partial y^2}. \quad (6.1.138c)$$

The richest equations correspond to the choice $\gamma = \beta = a = 1$, which introduces the effects of slow variations with respect to $\bar{t}$, small instability, and small nonlinearity simultaneously to $O(\epsilon)$. We will show next that this choice does not result in a bounded solution. It suffices to consider the special initial condition $h(x, y) = \cos k_c x \sin \pi y$. The solution for v_0 is then given by

$$v_0(x, y, t, \bar{t}) = A(\bar{t}) \cos k_c x \sin \pi y, \quad \bar{t} = \epsilon t, \quad (6.1.139)$$

where we are ignoring higher harmonics in both n and k_c. We then find that v_1 satisfies (see (6.1.138))

$$\begin{aligned} L(v_1) &= -\frac{\partial v_0}{\partial \bar{t}} - \frac{\alpha}{R_c^2} \Delta v_0 - \frac{\partial v_0}{\partial y} \frac{\partial^2 v_0}{\partial x^2} \\ &= -\left[\frac{dA}{d\bar{t}} - \frac{\alpha}{R^2}(k_c^2 + \pi^2) A\right] \cos k_c x \sin \pi y + \ldots, \end{aligned} \quad (6.1.140)$$

where ... indicates terms that have no secular contribution to v_1. The term exhibited on the right-hand side of (6.1.140) is a homogeneous solution; therefore its contribution to v_1 is secular. To remove this term, we must set

$$\frac{dA}{d\bar{t}} - \frac{\alpha}{R_c^2}(k_c^2 + \pi^2) A = 0, \quad (6.1.141)$$

which predicts that A_1 *grows* exponentially on the $\bar{t} = \epsilon t$ scale.

In this example, the choice of the various scalings that corresponds to the richest equation for v_1 does not result in a stable v_1. We see that we cannot allow $R - R_c$ and β to be $O(\epsilon)$, as this choice leads to (6.1.141). On the other hand, the nonlinear term $-(\partial v_0/\partial y)(\partial^2 v_0/\partial x^2)$ on the right-hand side of (6.1.140) does not have a secular contribution in v_1. Therefore, we must take $R - R_c$ and β to be $O(\epsilon^2)$ and maintain $\gamma = 1$. We set $R = R_c + \epsilon^2 \alpha, \bar{t} = \epsilon^2 t$ and expand

$$v(x, y, t; \epsilon) = \sum_{i=0}^{2} \epsilon^i v_i(x, y, t, \bar{t}) + O(\epsilon^3). \quad (6.1.142)$$

Now, the equations governing v_0, v_1, and v_2 are

$$L(v_0) = 0, \tag{6.1.143a}$$

$$L(v_1) = -\frac{\partial v_0}{\partial y}\frac{\partial^2 v_0}{\partial x^2}, \tag{6.1.143b}$$

$$L(v_2) = -\frac{\partial v_0}{\partial \bar{t}} - \frac{\partial v_0}{\partial y}\frac{\partial^2 v_1}{\partial x^2} - \frac{\partial v_1}{\partial y}\frac{\partial^2 v_0}{\partial x^2} - \frac{\alpha}{R_c^2}\Delta v_0. \tag{6.1.143c}$$

We again restrict attention to the simple initial condition $h(x, y) = \cos k_c x \sin \pi y$, and we consider only the corresponding term in v_0:

$$v_0 = A_0(\bar{t}) \cos k_c x \sin \pi y. \tag{6.1.144}$$

The solution of (6.1.143b) then has the form

$$v_1 = \left[A_1(\bar{t}) \cos k_c x + B_1(\bar{t}) \sin k_c x\right] \sin \pi y + c_1 A_0^2 \sin 2\pi y + c_2 A_0^2 \cos 2k_c x \sin 2\pi y, \tag{6.1.145}$$

where

$$c_1 = \frac{k_c^2 R_c}{16\pi}, \qquad c_2 = \frac{k_c^2 \pi R_c}{16[(\pi^2 + k_c^2) - R_c k_c^2(1 - k_c^2)]} \tag{6.1.146}$$

and $A_1(\bar{t})$, $B_1(\bar{t})$ are functions to be defined by consistency conditions on v_3.

The terms multiplied by A_1 and B_1 do not have a secular contribution to the solution for v_2; hence we only keep track of the terms multiplied by c_1 and c_2 as well as v_0. We then find that (6.1.143c) becomes

$$L(v_2) = -\left[\frac{dA_1}{d\bar{t}} - \frac{\alpha}{R_c^2}(k_c^2 + \pi^2)A_1 + \delta A_1^3\right] \cos k_c x \sin \pi y + \dots, \tag{6.1.147}$$

where $\dots$ indicates terms that have no secular contribution to v_2, and δ is the *positive* constant

$$\delta = \frac{k_c^4 R_c}{32}\left[\frac{\pi^2 + 2k_c^2 - 2R_c k_c^2(1 - k_c^2)}{\pi^2 + k_c^2 - R_c k_c^2(1 - 4k_c^2)}\right] = 0.5138\dots. \tag{6.1.148}$$

Therefore, A_1 satisfies the amplitude equation

$$\frac{dA_1}{d\bar{t}} - \frac{\alpha}{R_c^2}(k_c^2 + \pi^2)A_1 + \delta A_1^3 = 0. \tag{6.1.149a}$$

It has the steady-state solution

$$A_1^2 \to \frac{\alpha(k_c^2 + \pi^2)}{\delta R_c^2} = \frac{32\alpha(k_c^2 + \pi^2)[\pi^2 + k_c^2 - R_c k_c^2(1 - 4k_c^2)]}{R_c^3 k_c^4[\pi^2 + 2k_c^2 - 2R_c k_c^2(1 - k_c^2)]}. \tag{6.1.149b}$$

Let us now consider a more general initial condition with wave number k close to, but not exactly equal to, k_c. We see from Figure 6.1.1 that for any given $R = R_c + O(\epsilon^2)$, there is an interval in k centered at $k = k_c$ and of extent $k - k_c = O(\epsilon)$

for which solutions are linearly unstable. Thus, let us choose $R = R_c + \epsilon^2\alpha$ as before, and consider the initial condition $k(x, y; \epsilon) = \cos kx \sin \pi y$, where $k = k_c + \epsilon\kappa$. Here κ is a positive or negative constant independent of ϵ. The choice $\kappa = 0$ corresponds to the case just studied. We may write $h(x, y; \epsilon)$ in the form

$$h(x, y; \epsilon) = [\cos k_c x \cos \kappa\tilde{x} - \sin k_c x \sin \kappa\tilde{x}] \sin \pi y, \tag{6.1.150}$$

where $\tilde{x} = \epsilon x$.

Now, we expect the solution to depend also on the slow scale $\tilde{x}$, and we expand v as

$$v(x, y, t; \epsilon) = \sum_{i=0}^{2} \epsilon^i v_i(x, y, t, \tilde{x}, \tilde{t}, \bar{t}) + O(\epsilon^3). \tag{6.1.151}$$

The need for the slow scale $\tilde{t} = \epsilon t$ will soon become evident. A straightforward calculation of the various derivatives (that we do not list) leads to the following equations for v_0, v_1, and v_2:

$$L(v_0) = 0, \tag{6.1.152a}$$

$$L(v_1) = -\frac{\partial v_0}{\partial \tilde{t}} - 2\left(1 - \frac{1}{R_c}\right)\frac{\partial^2 v_0}{\partial x \partial \tilde{x}} - \frac{\partial v_0}{\partial y}\frac{\partial^2 v_0}{\partial x^2} - 4\frac{\partial^4 v_0}{\partial x^3 \partial \tilde{x}}, \tag{6.1.152b}$$

$$\begin{aligned} L(v_2) = &-\frac{\partial v_1}{\partial \tilde{t}} - \frac{\partial v_0}{\partial \bar{t}} - 2\left(1 - \frac{1}{R_c}\right)\frac{\partial^2 v_1}{\partial x \partial \tilde{x}} - \left(1 - \frac{1}{R_c}\right)\frac{\partial^2 v_0}{\partial \tilde{x}^2} \\ &- \frac{\partial v_0}{\partial y}\frac{\partial^2 v_1}{\partial x^2} - 2\frac{\partial v_0}{\partial y}\frac{\partial^2 v_0}{\partial x \partial \tilde{x}} - \frac{\partial v_1}{\partial y}\frac{\partial^2 v_0}{\partial x^2} \\ &- \frac{\alpha}{R_c^2}\Delta v_0 - 4\frac{\partial^4 v_1}{\partial x^3 \partial \tilde{x}} - 6\frac{\partial^4 v_0}{\partial x^2 \partial \tilde{x}^2}. \end{aligned} \tag{6.1.152c}$$

We write the solution of (6.1.152a) using complex notation

$$v_0 = [\rho(\tilde{x}, \tilde{t}, \bar{t})e^{ik_c x} + \rho^* e^{-ik_c x}] \sin \pi y, \tag{6.1.153}$$

where ρ^* denotes the complex conjugate of ρ. Thus, if we express

$$\rho = \rho_1 + i\rho_2, \quad \rho^* = \rho_1 - i\rho_2, \tag{6.1.154a}$$

where ρ_1 and ρ_2 are real, we have the initial conditions

$$\rho_1(\tilde{x}, 0, 0) = \frac{1}{2}\cos \kappa\tilde{x}, \quad \rho_2(\tilde{x}, 0, 0) = \frac{1}{2}\sin \kappa\tilde{x}. \tag{6.1.154b}$$

We use the expression in (6.1.153) to compute the right-hand side of (6.1.152b) to obtain

$$\begin{aligned} L(v_1) = &\left(-\frac{\partial \rho}{\partial \tilde{t}} - i\nu\frac{\partial \rho}{\partial \tilde{x}}\right)e^{ik_c x}\sin \pi y + \left(-\frac{\partial \rho^*}{\partial \tilde{t}} + i\nu\frac{\partial \rho}{\partial \tilde{x}}\right)e^{-ik_c x}\sin \pi y \\ &+ \frac{\pi}{2}k_c^2(\rho^2 e^{2ik_c x} + 2\rho\rho^* + \rho^{*2} e^{-2ik_c x})\sin 2\pi y, \end{aligned} \tag{6.1.155a}$$

where

$$\nu = k_c \left[2\left(1 - \frac{1}{R_c}\right) - 4k_c^2 \right]. \qquad (6.1.155b)$$

Removing the coefficients of the homogeneous solutions $e^{\pm ik_c x} \sin \pi y$ requires

$$\rho_{\tilde{t}} + i\nu\rho_{\tilde{x}} = 0, \quad \rho^*_{\tilde{t}} - i\nu\rho^*_{\tilde{x}} = 0. \qquad (6.1.156)$$

These conditions define the $\tilde{x}, \tilde{t}$ dependence of ρ and ρ^*

$$\rho = \overline{\rho}(\psi, \bar{t}), \quad \psi = \tilde{x} - i\nu\tilde{t}, \qquad (6.1.157a)$$

$$\rho^* = \overline{\rho}^*(\psi^*, \bar{t}), \quad \psi^* = \tilde{x} + i\nu\tilde{t}. \qquad (6.1.157b)$$

This result exhibits the need for including a $\tilde{t}$ dependence in the solution.

We now solve what remains of (6.1.155a) to obtain

$$v_1 = v_{1_H} + [D_1\overline{\rho}\,\overline{\rho}^* + D_2(\overline{\rho}^2 e^{2ik_c x} + \overline{\rho}^{*2} e^{-2ik_c x})] \sin 2\pi y, \qquad (6.1.158)$$

where v_{1_H} is the homogeneous solution, and the two constants D_1 and D_2 are

$$D_1 = \frac{k_c^2 R_c}{4\pi} = 4c_1, \quad D_2 = \frac{\pi k_c^2 R_c}{8[\pi^2 + k_c^2 - R_c k_c^2(1 - k_c^2)]} = 2c_2. \quad (6.1.159)$$

We shall ignore v_{1_H} in our calculations of the right-hand side of (6.1.152c) because removing secular terms involving v_{1_H} will give two conditions analogous to (6.1.156) for the dependence of the complex amplitudes in v_{1_H} on the $\tilde{x}$ and $\tilde{t}$ scales. These amplitudes are only needed if we wish to compute the solution to $O(\epsilon)$. For the purposes of this illustrative example, we confine our attention to defining only v_0 completely.

Using the expression in (6.1.158) (with $v_{1_H} = 0$) in (6.1.152c) and requiring the coefficients of $e^{\pm ik_c x} \sin \pi y$ to vanish gives the following two conditions on the dependence of $\overline{\rho}$ on $\bar{t}$ and ψ and of $\overline{\rho}^*$ on $\bar{t}$ and ψ^*:

$$\frac{\partial \overline{\rho}}{\partial \bar{t}} + (1 - 6k_c^2)\frac{\partial^2 \overline{\rho}}{\partial \psi^2} - \frac{\alpha}{R_c^2}(k_c^2 + \pi^2)\overline{\rho} + \pi k_c^2(D_1 - D_2)\overline{\rho}^2\overline{\rho}^* = 0, \quad (6.1.160a)$$

$$\frac{\partial \overline{\rho}^*}{\partial \bar{t}} + (1 - 6k_c^2)\frac{\partial^2 \overline{\rho}^*}{\partial \psi^*} - \frac{\alpha}{R_c^2}(k_c^2 + \pi^2)\overline{\rho}^* + \pi k_c^2(D_1 - D_2)\overline{\rho}^{*2}\overline{\rho} = 0. \quad (6.1.160b)$$

These two equations are formally identical (they are complex conjugates). If we let Φ denote $\overline{\rho}$ ($\overline{\rho}^*$) and ξ denote ψ (ψ^*), we have

$$\frac{\partial \Phi}{\partial \bar{t}} + (1 - 6k_c^2)\frac{\partial^2 \Phi}{\partial \xi^2} - \frac{\alpha}{R_c^2}(k_c^2 + \pi^2)\Phi + \pi k_c^2(D_1 - D_2)|\Phi|^2\Phi = 0. \quad (6.1.161)$$

Equation (6.1.161) is often referred to as a *Ginzburg-Landau* equation. It is easily seen that for $\kappa = 0$ we recover the result in (6.1.149a). The reader will find a general discussion of various aspects of (6.1.161) as well as a historical perspective in [6.10].

Problems

1. Consider the solution of (6.1.13) to $O(\epsilon)$ for the initial conditions (6.1.19).
   a. Show that when we use (6.1.30) for u_1, the initial conditions (6.1.23) and (6.1.29) imply that

$$A_1(0) = -B_1(0) = -\frac{3\rho_0^3}{16(1+k^2)}, \tag{6.1.162a}$$

$$A_3(0) = \frac{\rho_0^3}{64}\left[1 + 3\left(\frac{1+k^2}{1+9k^2}\right)^{1/2}\right], \tag{6.1.162b}$$

$$B_3(0) = \frac{\rho_0^3}{64}\left[1 - 3\left(\frac{1+k^2}{1+9k^2}\right)^{1/2}\right], \tag{6.1.162c}$$

$$A_n(0) = B_n(0) = 0, \quad n \neq 1, \quad n \neq 3, \tag{6.1.162d}$$

$$\alpha_n(0) = \beta_n(0) = 0, \quad n = 1, 2, \ldots. \tag{6.1.162e}$$

   b. Calculate the right-hand side of (6.1.21c) and isolate the terms that satisfy the homogeneous equation. Remove these terms to conclude that all the A_n and B_n are constants equal to their initial values. Hence, the only nonvanishing terms in the infinite series in (6.1.30) correspond to $n = 1$ and $n = 3$. Show that

$$\alpha_1' = \phi', \quad \beta_1' = -2\phi', \tag{6.1.163a}$$

$$\alpha_3' = -\frac{3\rho_0^2}{4\sqrt{1+9k^2}} = \beta_3', \tag{6.1.163b}$$

$$\alpha_n' = \beta_n' = 0, \quad n \neq 1, \quad n \neq 3. \tag{6.1.163c}$$

   c. Solve (6.1.21c) with the terms that remain on the right-hand side and show that boundedness of u_2 on the $\tilde{t}$ scale requires that we choose ω_2 as in (6.1.32).

2. Consider the solution of (6.1.13) with initial conditions

$$u(x, 0; \epsilon) = A \sin kx + B \sin \ell x, \tag{6.1.164a}$$

$$u(x, 0; \epsilon) = -A(1+k^2)^{1/2} \cos kx - B(1+\ell^2)^{1/2} \cos \ell x, \tag{6.1.164b}$$

   where A, B, k, and ℓ are constants. Since the solution must involve at least two waves, each with a different frequency shift, we cannot use a single fast time scale as in (6.1.20). This is analogous to the situation discussed in Sec. 4.3.6 for the pair of coupled weakly nonlinear oscillators. Guided by the approach used there, assume an expansion for u in the form

$$u(x, t; \epsilon) = u_0(x, t_0, t_1, \ldots) + \epsilon u_1(x, t_0, t_1, \ldots) + O(\epsilon^2),$$

   where $t_0 = t, t_1 = \epsilon t, \ldots, t_n = \epsilon^n t$.
   a. Assume u_0 in the form

$$u_0(x, t_0, t_1, \ldots) = a_0 \sin(kx - \omega t_0 + \phi_0) + b_0 \sin(\ell x - \mu t_0 + \psi_0),$$

where $\omega = (1 + k^2)^{1/2}$ and $\mu = (1 + \ell^2)^{1/2}$, and a_0, b_0, ϕ_0, and ψ_0 depend on the slow scales $t_1, t_2, \ldots$. Determine the dependence of a_0, b_0, ϕ_0, and ψ_0 on t_1 by requiring u_1 to be bounded with respect to t_1. Show that small divisors occur if $|k/\ell| \approx 1$. If $k = \ell$ exactly, we actually have only one initial wave, and the solution reduces to that discussed in Sec. 6.1.1.

b. Consider the solution for u_2 and determine the dependence of a_0, b_0, ϕ_0, and ψ_0 on t_2 and the dependence of u_1 on t_1.

c. In (6.1.164) set $k = k_0$ and $\ell = k_0 + \epsilon\kappa$, where $k_0 (\neq 0)$ and κ are arbitrary finite constants independent of ϵ. Use trigonometric identities to express (6.1.164) for this case in the form of a *single* slowly varying wave

$$u(x, 0; \epsilon) = \rho_0(\tilde{x}) \sin[k_0 x + \phi_0(\tilde{x})], \tag{6.1.165a}$$
$$u_t(x, 0; \epsilon) = -\rho_0(\tilde{x})(1 + k_0^2)^{1/2} \cos[k_0 x + \phi_0(\tilde{x})], \tag{6.1.165b}$$

where $\tilde{x} = \epsilon x$, and ρ_0 and ϕ_0 are periodic functions of $\tilde{x}$, which reduce to constants as $\kappa \to 0$. Show that a multiple-scale expansion for u as a function of x, t, $\tilde{x}$, and $\tilde{t} = \epsilon t$ implies that, to $O(1)$, u is a wave that varies slowly in space and time, in the form (6.1.73). Show also that $k(\tilde{x}, \tilde{t})$, $\rho(\tilde{x}, \tilde{t})$, and $\phi(\tilde{x}, \tilde{t})$ in (6.1.73) obey (6.1.79), (6.1.82), and (6.1.85), respectively, with $K' = 0$, i.e., $K = k_0 =$ constant. Thus, the near resonant interaction between two waves for this problem merely corresponds to a periodic variation of the amplitude with

$$\xi = \tilde{x} - k_0\tilde{t}/\sqrt{1 + k_0^2}$$

as given by $\rho = \rho_0(\xi)$. The phase has a corresponding periodic variation, $\phi_0(\xi)$, plus a secular part as given by (6.1.86).

d. For the case of three initial waves with wave numbers k, ℓ, and m, show that resonance occurs only if any two of these waves have nearly equal wave number. Set $k = k_0$, $\ell = \ell_0$, and $m = \ell_0 + \epsilon\lambda$, where k_0, ℓ_0, and λ are three arbitrary constants independent of ϵ and $k_0 \neq \ell_0$. Use a multiple-scale expansion as in part b to derive the solution to $O(1)$. Show that for the special case $k_0 \approx \ell_0$ the problem reduces to that in part b.

3. Consider the weakly nonlinear Korteweg–de Vries equation

$$u_t + \left(1 + \frac{3\epsilon}{2} u\right) u_x + \frac{\delta^2}{6} u_{xxx} = 0 \tag{6.1.166}$$

with initial condition

$$u(x, 0; \epsilon) = \rho_0 \sin(kx + \phi_0), \tag{6.1.167}$$

where ρ_0, k, and ϕ_0 are prescribed constants. Compute the multiple-scale expansion of the solution in the form (6.1.20) to $O(1)$ and determine ω_2. In particular, show that

$$u_0 = \rho_0 \sin(\theta_1^+ + \phi_0), \tag{6.1.168a}$$
$$u_1 = -\frac{3\rho_0^2}{4\delta^2 k^2} \left[\cos 2(\theta_1^+ + \phi_0) - \cos(\theta_2^+ + 2\phi_0)\right], \tag{6.1.168b}$$

where

$$\theta_n^+ = nkx + \left(-nkt^+ + k^3 n^3 \frac{\delta^2}{6} t^+\right), \tag{6.1.169a}$$

$$t^+ = 1 + \epsilon^2 \frac{-27\rho_0^2}{8k^2\delta^2(\delta^2k^2 - 6)}. \tag{6.1.169b}$$

4. Having removed the terms proportional to $\cos\psi$ and $\sin\psi$ from the right-hand side of (6.1.74), solve what remains to compute

$$u_1 = \rho_1(\tilde{x}, \tilde{t}) \sin\psi_1 - \frac{\rho^3}{32} \sin 3\psi, \quad \psi_1 = \theta + \phi_1(\tilde{x}, \tilde{t}). \tag{6.1.170}$$

Consider (6.1.71c) for u_2 and derive the equations governing ρ_1 and ψ. Solve these and comment on the solution to $O(\epsilon)$.

5. Consider the weakly nonlinear wave equation

$$u_{tt} - u_{xx} + u$$

$$+ \frac{\epsilon}{2\pi} \int_{-\infty}^{\infty} \int_{-\infty}^{\infty} u(x - \xi, t; \epsilon) u(\xi - \eta, t; \epsilon) u(\eta, t; \epsilon) d\xi d\eta = 0, \tag{6.1.171}$$

where the nonlinear term is in convolution form. Fourier transformation of this equation gives

$$U_{tt} + (1 + k^2)U + \epsilon U^3 = 0, \tag{6.1.172}$$

where

$$U(k, t; \epsilon) = \frac{1}{\sqrt{2\pi}} \int_{-\infty}^{\infty} e^{-ikx} u(x, t; \epsilon) dx \tag{6.1.173}$$

is the Fourier transform of u.

Choose the initial conditions

$$u(x, 0; \epsilon) = Ae^{-|x|}, \quad A = \text{constant}, \tag{6.1.174a}$$

$$u_t(x, 0; \epsilon) = 0, \tag{6.1.174b}$$

which have Fourier transforms

$$U(k, 0; \epsilon) = \sqrt{\frac{2}{\pi}} \frac{A}{1 + k^2}, \tag{6.1.175a}$$

$$U_t(k, 0; \epsilon) = 0. \tag{6.1.175b}$$

Since all solutions of (6.1.172) subject to (6.1.175) are periodic, expand

$$U(k, t; \epsilon) = U_0(k_1, t^+) + \epsilon U_1(k; t^+) + O(\epsilon^2) \tag{6.1.176}$$

in terms of the strained coordinate

$$t^+ = (1 + \epsilon\omega_1 + \ldots)t \tag{6.1.177}$$

and derive U_0, U_1, and ω_1. Invert (6.1.176) to compute the asymptotic expansion of u in the form

$$\begin{aligned} u(x,t;\epsilon) &= \frac{1}{\sqrt{2\pi}}\int_{-\infty}^{\infty} e^{ikx}U(k,t;\epsilon)dk \\ &= \frac{1}{\sqrt{2\pi}}\int_{-\infty}^{\infty} e^{ikx}U_0(k,t^+)dk + \frac{\epsilon}{\sqrt{2\pi}}\int_{-\infty}^{\infty} e^{ikx}U_1(k,t^+)dk \\ &\quad + O(\epsilon^2), \end{aligned} \tag{6.1.178}$$

and discuss the behavior of the solution.

6. Study the weakly nonlinear stability of

$$u_t - u + uu_x = \frac{1}{R}u_{xx}; \quad R = \text{constant} > 0; \quad 0 \le x \le \pi; \quad 0 \le t \tag{6.1.179}$$

with the boundary condition (6.1.101) and initial condition (6.1.102).

6.2 Weakly Nonlinear Conservation Laws

In this section, we study the weakly nonlinear system

$$\frac{\partial U_i}{\partial t} + \lambda_i \frac{\partial U_i}{\partial x} + \sum_{j=1}^{n} C_{ij}U_j =$$

$$\epsilon\phi_i\left(U_1,\ldots,U_n,\frac{\partial U_1}{\partial x},\ldots,\frac{\partial U_n}{\partial x},\frac{\partial^2 U_1}{\partial x^2}\ldots\right) + O(\epsilon^2), \quad i = 1,\ldots,n \tag{6.2.1}$$

for constant λ_i and C_{ij}. In Sec. 6.2.1, we show that a special case of this system results from perturbing a hyperbolic system of conservation laws near a uniform solution and then introducing characteristic dependent variables. One can also obtain (6.2.1) for $n = 2$ by decomposing a weakly nonlinear wave equation into two first-order equations as discussed in Sec. 6.2.3. Other physical examples leading to (6.2.1) are discussed throughout this section.

In contrast to the approach in Sec. 6.1, we use the solution of the unperturbed problem in its general form rather than expressing it as a series of harmonics. In the remainder of this section, we implement this approach for various physical problems that correspond to special values of the constants λ_i, C_{ij}, and the perturbation functions ϕ_i.

6.2.1 Characteristic Dependent Variables

To simplify the derivations, let us restrict attention to the case $n = 2$. The results we derive here also hold for $n > 2$. Consider the vector conservation law in divergence form

$$\mathbf{p}_t + \mathbf{q}_x = \mathbf{s}, \tag{6.2.2}$$

where $\mathbf{p} = (p_1(u_1, u_2), p_2(u_1, u_2))$ is the conserved quantity that we assume to be a function of the two dependent variables $u_1(x, t)$ and $u_2(x, t)$. The flux is denoted by $\mathbf{q} = (q_1(u_1, u_2), q_2(u_1, u_2))$, and the source term is $\mathbf{s} = (s_1(u_1, u_2), s_2(u_1, u_2); \mu)$. We let the source vector depend on the constant μ, a dimensionless parameter that will be defined for each physical application. The more general problem where $\mathbf{s}$ also depends on x, as in models with axial or spherical symmetry, is more difficult and will not be considered.

If we evaluate $\mathbf{p}_t$ and $\mathbf{q}_x$ in terms of the partial derivatives $\mathbf{u}_t = \left(\frac{\partial u_1}{\partial t}, \frac{\partial u_2}{\partial t}\right)$ and $\mathbf{u}_x = \left(\frac{\partial u_1}{\partial x}, \frac{\partial u_2}{\partial x}\right)$, we have

$$P\mathbf{u}_t + Q\mathbf{u}_x = \mathbf{s}, \tag{6.2.3}$$

where P and Q are the Jacobian matrices

$$P = \frac{\partial(p_1, p_2)}{\partial(u_1, u_2)}, \quad Q = \frac{\partial(q_1, q_2)}{\partial(u_1, u_2)}. \tag{6.2.4}$$

Multiplying (6.2.3) by P^{-1} gives

$$\mathbf{u}_t + A(\mathbf{u})\mathbf{u}_x = \mathbf{r}, \tag{6.2.5}$$

where

$$A = P^{-1}Q, \quad \mathbf{r} = P^{-1}\mathbf{s}. \tag{6.2.6}$$

We will restrict attention to the *strictly hyperbolic* problem for which the eigenvalues of A are real and distinct over the solution domain. We wish to study solutions of (6.2.5) that are close to a uniform state defined by the *constant* vector $\mathbf{v} = (v_1(\mu), v_2(\mu))$. Thus, v_1 and v_2 are solutions of the pair of algebraic equations

$$r_1(v_1, v_2; \mu) = 0, \quad r_2(v_1, v_2; \mu) = 0. \tag{6.2.7}$$

If $\mathbf{s} = 0$, the uniform state corresponds to *arbitrary* constants v_1 and v_2.

One possible means of generating a solution close to the uniform state is to prescribe initial conditions that deviate slightly from this state. Thus, we consider the initial-value problem for (6.2.5)

$$u_i(x, 0; \epsilon, \mu) = v_i(\mu) + \epsilon w_i^*(x) + O(\epsilon^2), \quad i = 1, 2, \tag{6.2.8}$$

where ϵ is a small parameter and the w_i^* are prescribed *bounded* functions of x. In view of the fact that the v_i are *exact* uniform solutions of (6.2.5), we restrict the $w_i^*(x)$ to have zero average, i.e.,

$$\lim_{\ell\to\infty} \frac{1}{2\ell} \int_{-\ell}^{\ell} w_i^*(x)dx = 0. \tag{6.2.9a}$$

For 2ℓ-periodic initial data, w_i^*, this reduces to

$$\int_{-\ell}^{\ell} w_i^*(x)dx = 0. \tag{6.2.9b}$$

For future reference, we note that if solutions of (6.2.5) cease to exist in the strict sense (as when characteristics of the same family cross), we may introduce shocks that are consistent with (6.2.2). Such shocks propagate with speed $C_s = dx/dt$ given by (see 3.1.142)

$$C_s = \frac{[q_i]}{[p_i]}, \quad i = 1, 2, \tag{6.2.10}$$

where [ ] denotes the jump in a quantity across a shock. In particular, if we denote the values of u_i just ahead and just behind the shock by u_i^+ and u_i^-, respectively, the pair of conditions (6.2.10) reduce to *two* algebraic equations linking the five quantities C_s, u_1^+, u_1^-, u_2^+, and u_2^-. We will refer to the exact shock conditions (6.2.10) in Secs. 6.2.4–6.2.6, where we discuss examples that admit shocks.

Let us introduce the rescaled dependent variables w_i defined by

$$u_i(x, t; \epsilon, \mu) = v_i(\mu) + \epsilon w_i(x, t; \epsilon, \mu), \quad i = 1, 2. \tag{6.2.11}$$

If we substitute (6.2.11) for the u_i in (6.2.5) and expand for small ϵ, we obtain the following vector equation for $\mathbf{w} = (w_1, w_2)$ correct to $O(\epsilon)$:

$$\mathbf{w}_t + A^{(0)}(\mu)\mathbf{w}_x + B(\mu)\mathbf{w} = \epsilon\mathbf{f}(\mathbf{w}, \mathbf{w}_x; \mu) + O(\epsilon^2), \tag{6.2.12}$$

where

$$A_{ij}^{(0)}(\mu) = A_{ij}(v_1(\mu), v_2(\mu)), \quad i, j = 1, 2, \tag{6.2.13a}$$

$$B_{ij}(\mu) = -\frac{\partial r_i}{\partial u_j}(v_1(\mu), v_2(\mu); \mu); \quad i, j = 1, 2, \tag{6.2.13b}$$

$$\begin{aligned} f_i(\mathbf{w}, \mathbf{w}_x; \mu) = \sum_{j=1}^{2}\sum_{k=1}^{2}\Bigg[&\frac{1}{2}\frac{\partial^2 r_i}{\partial u_j \partial u_k}(v_1(\mu), v_2(\mu); \mu) w_j w_k \\ &- \frac{\partial A_{ik}}{\partial u_j}(v_1(\mu), v_2(\mu)) w_j \frac{\partial w_k}{\partial x}\Bigg], \quad i = 1, 2. \end{aligned} \tag{6.2.13c}$$

Thus, in (6.2.12) the matrix components of $A^{(0)}$ and B and the coefficients of the quadratic terms in $\mathbf{f}$ are all constants involving the parameter μ. Note also that B is identically equal to zero, and the parameter μ is absent if the source terms are absent from (6.2.1), i.e., $\mathbf{s} \equiv 0$. Most of our discussion in this section is restricted to this case. The effect of source terms is considered in Sec. 6.2.6.

Now we transform dependent variables to characteristic form using a basis of eigenvectors for $A^{(0)}$. For a more detailed discussion, see Sec. 4.5.3 of [6.17]. The eigenvalues of $A^{(0)}$ are given by

$$\lambda_{1,2}(\mu) = \frac{1}{2}\left[A_{11}^{(0)} + A_{22}^{(0)} \pm \sqrt{(A_{11}^{(0)} - A_{22}^{(0)})^2 + 4A_{12}^{(0)}A_{21}^{(0)}}\right] \tag{6.2.14}$$

and are assumed to be real and distinct, i.e.,

$$(A_{11}^{(0)} - A_{12}^{(0)})^2 + 4A_{12}^{(0)}A_{21}^{(0)} > 0. \tag{6.2.15}$$

Note $\lambda_1 > \lambda_2$. A linear transformation to a basis of eigenvectors may be defined by

$$\mathbf{w} = W\mathbf{U}, \tag{6.2.16}$$

where $\mathbf{U} = (U_1, U_2)$, and W is the matrix

$$W(\mu) = \begin{pmatrix} -A_{12}^{(0)} & -A_{12}^{(0)} \\ A_{11}^{(0)} - \lambda_1 & A_{11}^{(0)} - \lambda_2 \end{pmatrix}. \tag{6.2.17a}$$

The inverse of W, denoted by V, is then given by

$$V(\mu) = \frac{1}{A_{12}^{(0)}(\lambda_2 - \lambda_1)} \begin{pmatrix} A_{11}^{(0)} - \lambda_2 & A_{12}^{(0)} \\ \lambda_1 - A_{11}^{(0)} & -A_{12}^{(0)} \end{pmatrix}. \tag{6.2.17b}$$

The equation for $\mathbf{U}$ is obtained by substituting (6.2.16) for $\mathbf{w}$ in (6.2.12):

$$W\mathbf{U}_t + A^{(0)}W\mathbf{U}_x + BW\mathbf{U} = \epsilon\mathbf{f}(W\mathbf{U}, W\mathbf{U}_x; \mu) + O(\epsilon^2), \tag{6.2.18}$$

then multiplying this by V to find

$$\mathbf{U}_t + \Lambda\mathbf{U}_x + C\mathbf{U} = \epsilon\boldsymbol{\phi}(\mathbf{U}, \mathbf{U}_x; \mu). \tag{6.2.19a}$$

This has the component form

$$\frac{\partial U_i}{\partial t} + \lambda_i \frac{\partial U_i}{\partial x} + \sum_{j=1}^{2} C_{ij}U_j = \epsilon\phi_i\left(U_1, U_2, \frac{\partial U_1}{\partial x}, \frac{\partial U_2}{\partial x}; \mu\right), \quad i = 1, 2, \tag{6.2.19b}$$

where

$$\Lambda(\mu) = VAW = \begin{pmatrix} \lambda_1 & 0 \\ 0 & \lambda_2 \end{pmatrix}, \tag{6.2.20a}$$

$$C(\mu) = VBW, \tag{6.2.20b}$$

$$\boldsymbol{\phi} = V\mathbf{f}(W\mathbf{U}, W\mathbf{U}_x; \mu). \tag{6.2.20c}$$

Equation (6.2.19) is in the standard form (6.2.1) for $n = 2$. The fact that Λ is diagonal follows directly from the choice (6.2.16) of a basis of eigenvectors. Note again that the C_{ij} all vanish if the source vector $\mathbf{s}$ is absent in (6.2.3). The components of $\boldsymbol{\phi}$ in terms of the U_i and $(\partial U_i/\partial x)$ can be calculated using the expression (6.2.13b) for the f_i. We find that ϕ_1 and ϕ_2, the components of $\boldsymbol{\phi}$, are given by the quadratic form

$$\phi_i = \sum_{j=1}^{2}\sum_{k=1}^{2}\left[F_{ijk}(\mu)U_j\frac{\partial U_k}{\partial x} + G_{ijk}(\mu)U_jU_k\right], \tag{6.2.21}$$

where

$$F_{ijk}(\mu) = -\sum_{r=1}^{2}\sum_{s=1}^{2}\gamma_{irs}(\mu)W_{rj}W_{sk}, \tag{6.2.22a}$$

$$\gamma_{irs}(\mu) = \sum_{m=1}^{2} \frac{\partial A_{ms}}{\partial u_r}(v_1(\mu), v_2(\mu))V_{im}, \tag{6.2.22b}$$

and

$$G_{ijk}(\mu) = \frac{1}{2}\sum_{r=1}^{2}\sum_{s=1}^{2} \beta_{irs}(\mu)W_{rj}W_{sk}, \tag{6.2.23a}$$

$$\beta_{irs}(\mu) = \beta_{isr}(\mu) = \sum_{m=1}^{2} \frac{\partial^2 r_m}{\partial u_r \partial u_s}(v_1(\mu), v_2(\mu))V_{im}. \tag{6.2.23b}$$

Once the uniform state is established, the constants F_{ijk} and G_{ijk} are easily computed in terms of μ. Thus, in addition to ϵ, (6.2.19) involves the 20 parameters λ_i, C_{ij}, F_{ijk}, and G_{ijk}.

The initial conditions (6.2.8) imply that the U_i must satisfy

$$U_i(x, 0; \epsilon, \mu) = \sum_{j=1}^{2} V_{ij}(\mu)w_j^*(x) + O(\epsilon) \equiv U_i^*(x) + O(\epsilon). \tag{6.2.24}$$

In some applications, we wish to study (6.2.19) in a small neighborhood of a critical value of $\mu = \mu_0$. For example, the parameters in (6.2.19) evaluated at $\mu = \mu_0$ might correspond to neutral stability of the linear ($\epsilon = 0$) problem. In such cases, it is useful to expand (6.2.19) further.

We choose

$$\mu = \mu_0 + \epsilon\mu_1 \tag{6.2.25}$$

and expand the λ_i, C_{ij}, and ϕ_i as follows:

$$\lambda_i = \lambda_i^{(0)}(\mu_0) + \epsilon\lambda_i^{(1)}(\mu_0) + O(\epsilon^2), \tag{6.2.26a}$$

$$C_{ij} = C_{ij}^{(0)}(\mu_0) + \epsilon C_{ij}^{(1)}(\mu_0) + O(\epsilon^2), \tag{6.2.26b}$$

$$\phi_i = \phi_i^{(0)} + O(\epsilon), \tag{6.2.26c}$$

where

$$\lambda_i^{(1)}(\mu_0) = \lambda_i'(\mu_0)\mu_1, \quad C_{ij}^{(1)}(\mu_0) = C_{ij}'(\mu_0)\mu_1, \tag{6.2.27}$$

$$\phi_i^{(0)} = \sum_{j=1}^{2}\sum_{k=1}^{2} F_{ijk}(\mu_0)U_j\frac{\partial U_k}{\partial x} + G_{ijk}(\mu_0)U_jU_k. \tag{6.2.28}$$

The governing system to $O(\epsilon)$ now becomes

$$\frac{\partial U_i}{\partial t} + \lambda_i^{(0)}\frac{\partial U_i}{\partial x} + \sum_{j=1}^{2} C_{ij}^{(0)}U_j = \epsilon\left[-\lambda_i^{(1)}\frac{\partial U_i}{\partial x} - \sum_{j=1}^{2} C_{ij}^{(1)}U_j + \phi_i^{(0)}\right], \quad i = 1, 2. \tag{6.2.29}$$

An example, channel flow

We consider the flow of shallow water down a slightly inclined open channel, as discussed in [6.30], to illustrate many of the ideas in this section. If one accounts for the friction at the boundaries with an empirical body force term proportional to the square of the flow speed, the laws of mass and momentum conservation take on the following form corresponding to (6.2.2) in appropriate dimensionless variables (see section 5.5.1 of [6.17]):

$$h_t + (uh)_x = 0, \tag{6.2.30a}$$

$$(uh)_t + \left(u^2 h + \frac{h^2}{2}\right)_x = h - \frac{u^2}{F^2}. \tag{6.2.30b}$$

As shown in Figure 6.2.1, h is the free surface height normal to the channel bottom, and u is the flow speed parallel to the channel bottom averaged over h. The first and second terms on the right-hand side of (6.2.30b) represent the gravitational and friction forces acting on a column of water of width dx. The Froude number, F, is the constant dimensionless speed of undisturbed flow. This is the uniform state $h = 1, u = F$, where the gravitational and friction forces are in perfect balance. The choice of horizontal and vertical length scales and the time scale that led to the above dimensionless form implies that $F = (\tan s/D)^{1/2}$, where s is the channel slope and D is the friction coefficient.

For this example $u_1 = h$, $u_2 = u$, the parameter μ is the Froude number F, and the various components in (6.2.2) are $p_1 = u_1$, $q_1 = p_2 = u_1 u_2$, $q_2 = u_1 u_2^2 + u_1^2/2$, $s_1 = 0$, $s_2 = u_1 - u_2^2/\mu^2$. The components of A and $\mathbf{r}$ in (6.2.5) are then found to be $A_{11} = A_{22} = u_2$, $A_{12} = u_1$, $A_{21} = 1$, $r_1 = 0$, and $r_2 = 1 - u_2^2/u_1\mu^2$. The initial conditions (6.2.11) consist of small perturbations to the uniform state $u_1 = v_1 = 1$, $u_2 = v_2 = \mu$. The eigenvalues of A are $u_2 \pm \sqrt{u_1}$, and these are real and distinct for a physically realistic solution where the free surface height u_1 is positive.

It follows from (6.2.13) that the components in (6.2.12) are

$$A_{ij}^{(0)}(\mu) = \begin{pmatrix} \mu & 1 \\ 1 & \mu \end{pmatrix}, \quad B_{ij}(\mu) = \begin{pmatrix} 0 & 0 \\ -1 & \frac{2}{\mu} \end{pmatrix}. \tag{6.2.31}$$

$$f_1 = -w_1 \frac{\partial w_2}{\partial x} - w_2 \frac{\partial w_1}{\partial x}, \tag{6.2.32a}$$

$$f_2 = -w_2 \frac{\partial w_2}{\partial x} - w_1^2 + \frac{2}{\mu} w_1 w_2 - \frac{1}{\mu^2} w_2^2. \tag{6.2.32b}$$

The eigenvalues of $A_{ij}^{(0)}$ are

$$\lambda_1^{(0)} = \mu + 1; \quad \lambda_2^{(0)} = \mu - 1. \tag{6.2.33}$$

Using (6.2.17), we obtain

$$W_{ij}(\mu) = \begin{pmatrix} -1 & -1 \\ -1 & 1 \end{pmatrix}; \quad V_{ij}(\mu) = \begin{pmatrix} -\frac{1}{2} & -\frac{1}{2} \\ -\frac{1}{2} & \frac{1}{2} \end{pmatrix}, \tag{6.2.34}$$

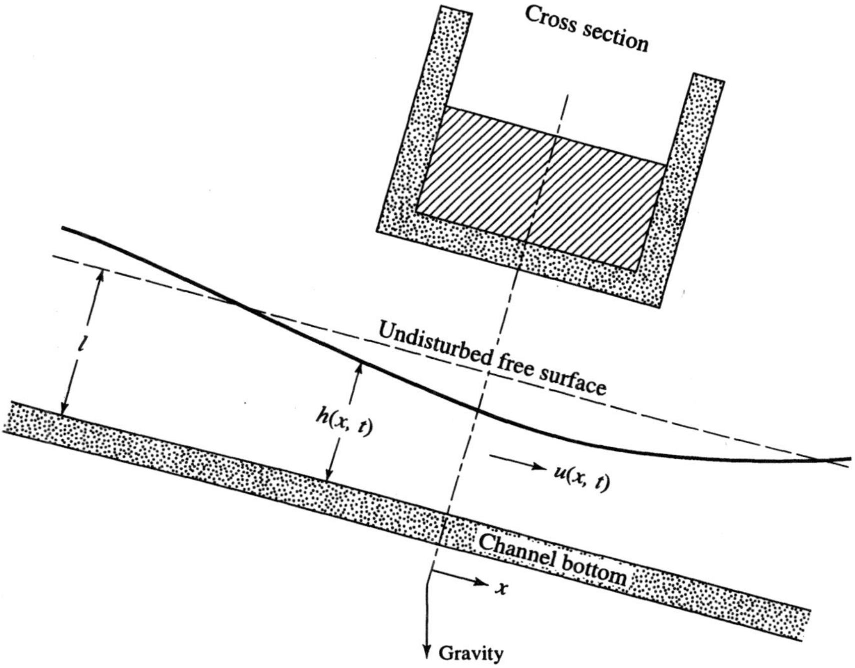

FIGURE 6.2.1. Shallow Water Flow Down an Inclined Open Channel; the Uniform State is $u = F, h = 1$

and it follows from (6.2.20b) that

$$C_{ij}(\mu) = \begin{pmatrix} -\frac{1}{2} + \frac{1}{\mu} & -\frac{1}{2} - \frac{1}{\mu} \\ \frac{1}{2} - \frac{1}{\mu} & \frac{1}{2} + \frac{1}{\mu} \end{pmatrix}. \tag{6.2.35}$$

Finally, using (6.2.21)–(6.2.23), we find

$$\phi_1 = \frac{3}{2} U_1 \frac{\partial U_1}{\partial x} - \frac{1}{2} U_1 \frac{\partial U_2}{\partial x} - \frac{1}{2} U_2 \frac{\partial U_1}{\partial x} - \frac{1}{2} U_2 \frac{\partial U_2}{\partial x} + \frac{(\mu - 1)^2}{2\mu^2} U_1^2 + \frac{\mu^2 - 1}{\mu^2} U_1 U_2 + \frac{(\mu + 1)^2}{2\mu^2} U_2^2, \tag{6.2.36a}$$

$$\phi_2 = \frac{1}{2} U_1 \frac{\partial U_1}{\partial x} + \frac{1}{2} U_1 \frac{\partial U_2}{\partial x} + \frac{1}{2} U_2 \frac{\partial U_1}{\partial x} - \frac{3}{2} U_2 \frac{\partial U_2}{\partial x} - \frac{(\mu - 1)^2}{2\mu^2} U_1^2 - \frac{\mu^2 - 1}{\mu^2} U_1 U_2 - \frac{(\mu + 1)^2}{2\mu^2} U_2^2. \tag{6.2.36b}$$

Equations (6.2.33), (6.2.35), and (6.2.36) define the standard form (6.2.19) for this example.

As we will show, solutions of the linear problem ($\epsilon = 0$) for this example are unstable if $\mu > 2$. Thus, for a weakly nonlinear stability analysis, we set $\mu = 2 + \epsilon\mu_1$, where μ_1 is a positive $O(1)$ constant. Expanding the λ_i, C_{ij}, and ϕ_i as in (6.2.26) gives the following system to $O(\epsilon)$:

$$\frac{\partial U_1}{\partial t} + 3\frac{\partial U_1}{\partial x} - U_2 = \epsilon\left[-\mu_1\frac{\partial U_1}{\partial x} + \frac{\mu_1}{4}(U_1 - U_2) + \frac{3}{4}U_1\frac{\partial U_1}{\partial x}\right.$$
$$- \frac{1}{2}U_1\frac{\partial U_2}{\partial x} - \frac{1}{2}U_2\frac{\partial U_1}{\partial x} - \frac{1}{2}U_2\frac{\partial U_2}{\partial x} + \frac{1}{8}U_1^2 + \frac{3}{4}U_1U_2$$
$$\left. + \frac{9}{8}U_2^2\right], \tag{6.2.37a}$$

$$\frac{\partial U_2}{\partial t} + \frac{\partial U_2}{\partial x} + U_2 = \epsilon\left[-\mu_1\frac{\partial U_2}{\partial x} + \frac{\mu_1}{4}(U_2 - U_1) + \frac{1}{2}U_1\frac{\partial U_1}{\partial x}\right.$$
$$+ \frac{1}{2}U_1\frac{\partial U_2}{\partial x} + \frac{1}{2}U_2\frac{\partial U_1}{\partial x} - \frac{3}{2}U_2\frac{\partial U_2}{\partial x} - \frac{1}{8}U_1^2 - \frac{3}{4}U_1U_2$$
$$\left. - \frac{9}{8}U_2^2\right]. \tag{6.2.37b}$$

We discuss the multiple-scale expansion for this system in Sec. 6.2.6.

6.2.2 The Linear Problem ($\epsilon = 0$): Stability, General Solution

Here we study the stability and explicit form of the unperturbed problem (6.2.19), i.e., the linear system

$$\frac{\partial U_i}{\partial t} + \lambda_i\frac{\partial U_i}{\partial x} + \sum_{j=1}^{2} C_{ij}U_j = 0, \quad i = 1, 2, \tag{6.2.38a}$$

subject to the initial conditions

$$U_i(x, 0) = U_i^*(x), \quad i = 1, 2. \tag{6.2.38b}$$

If source terms are absent to $O(1)$, i.e., all the C_{ij} vanish, the linear solution is just

$$U_1 = U_1^*(x - \lambda_1 t); \quad U_2 = U_2^*(x - \lambda_2 t), \tag{6.2.39}$$

which is neutrally stable as the U_i maintain their initial values along the characteristics $x - \lambda_i t = \xi_i$ = constant, respectively. Most of our work in this section and in the literature concerns this case.

Stability of the general problem

To analyze the general case, it is convenient to combine the two first-order equations (6.2.19b) into the single second-order equation for either U_1 or U_2

$$\left(\frac{\partial}{\partial t} + \lambda_1\frac{\partial}{\partial x}\right)\left(\frac{\partial}{\partial t} + \lambda_2\frac{\partial}{\partial x}\right)U_i + \left((C_{11} + C_{22})\frac{\partial}{\partial t} + \right.$$

$$(C_{11}\lambda_2 + C_{22}\lambda_1)\frac{\partial}{\partial x}\bigg) U_i + (C_{11}C_{22} - C_{12}C_{21})U_i = 0, \quad i = 1, 2. \quad (6.2.40)$$

The stability of this equation for the special case $C_{11}C_{22} - C_{12}C_{12} = 0$ is discussed in Sec. 3.1 of [6.27], where the following conditions are derived for solutions to decay as $t \to \infty$:

$$\lambda_2 < \frac{C_{11}\lambda_2 + C_{22}\lambda_1}{C_{11} + C_{22}} < \lambda_1, \quad C_{11} + C_{22} > 0. \quad (6.2.41a)$$

These conditions are equivalent to

$$C_{11} \geq 0 \quad \text{and} \quad C_{22} \geq 0, \quad (6.2.41b)$$

and they also follow by introducing the characteristic independent variables $\xi_i = x - \lambda_i t$ and considering the propagation of discontinuities along characteristics (see (3.1.67)–(3.1.71)).

It is instructive to consider the general case (6.2.40) using the standard approach. We look for solutions of the form $U_i \sim \exp(ikx + \sigma t)$, where k is a real constant and σ is a complex constant. We envision representing the general solution as a continuous superposition for $-\infty < k < \infty$ as would result from using Fourier transforms to solve (6.2.38). Substituting the assumed form for U_i into (6.2.40) gives the following quadratic equation for σ:

$$\begin{aligned}&\sigma^2 + [C_{11} + C_{22} + ik(\lambda_1 + \lambda_2)]\sigma \\ &\quad + [C_{11}C_{22} - C_{12}C_{21} - k^2\lambda_1\lambda_2 + ik(C_{11}\lambda_2 + C_{22}\lambda_1)] = 0. \quad (6.2.42)\end{aligned}$$

For stability, we must have the real parts of the two roots of (6.2.42) nonpositive, and this is easily seen to require

$$-(C_{11} + C_{22}) + [(a^2 + b^2)^{1/2} + a]^{1/2} \leq 0, \quad (6.2.43a)$$

$$-(C_{11} + C_{22}) - [(a^2 + b^2)^{1/2} + a]^{1/2} \leq 0, \quad (6.2.43b)$$

where

$$a = \frac{1}{2}(C_{11} - C_{22})^2 + 2C_{12}C_{21} - \frac{k^2}{2}(\lambda_1 - \lambda_2)^2, \quad (6.2.44a)$$

$$b = k(\lambda_1 - \lambda_2)(C_{11} - C_{22}). \quad (6.2.44b)$$

For $|k| \to \infty$, the two conditions (6.2.43) reduce to

$$-(C_{11} + C_{22}) + |C_{11} - C_{22}| \leq 0, \quad (6.2.45a)$$

$$-(C_{11} + C_{22}) - |C_{11} - C_{22}| \leq 0, \quad (6.2.45b)$$

and these combine to give (6.2.41b).

In the limit $k \to 0$, (6.2.43) give

$$-(C_{11} + C_{22}) + [(C_{11} - C_{22})^2 + 4C_{12}C_{21}]^{1/2} \leq 0, \quad (6.2.46a)$$

$$-(C_{11} + C_{22}) - [(C_{11} - C_{22})^2 + 4C_{12}C_{21}]^{1/2} \leq 0. \quad (6.2.46b)$$

If the conditions (6.2.41b) hold, (6.2.46b) is automatically satisfied, and (6.2.46a) gives the additional condition

$$C_{11}C_{22} \geq C_{12}C_{21}. \tag{6.2.47}$$

That the condition (6.2.47) is necessary for long waves ($k \to 0$) is physically obvious from (6.2.40), which reduces to the oscillator equation

$$\frac{\partial^2 U_i}{\partial t^2} + (C_{11} + C_{22})\frac{\partial U_i}{\partial t} + (C_{11}C_{22} - C_{12}C_{21})U_i = 0 \tag{6.2.48}$$

if x derivatives are ignored. With $C_{11} \geq 0$ and $C_{22} \geq 0$, we are assured that the damping coefficient is non-negative. The requirement (6.2.47) implies that the restoring force opposes the displacement.

A more careful analysis of the stability condition (6.2.43a) shows that for k sufficiently large it is possible to violate the *limiting* condition (6.2.47) but still satisfy the *exact* condition (6.2.43a) as long as $C_{11} \geq 0$ and $C_{22} \geq 0$. For example, with $C_{11} = C_{22} = 0$, $C_{12} = C_{21} = 1$, (6.2.47) is violated, but if we have $\lambda_1 = 3$, $\lambda_2 = 1$, condition (6.2.43a) holds as long as $k \geq 1/2$, for which $a \geq 0$. Thus, the problem with the above choices of C_{ij} and λ_i that violate (6.2.47) is stable as long as $k \geq 1/2$. In summary, the necessary stability conditions *for all k* are

$$C_{11} \geq 0, \quad C_{22} \geq 0; \quad C_{11}C_{22} - C_{12}C_{21} \geq 0. \tag{6.2.49}$$

Henceforth, we will refer to these as conditions (1), (2), and (3) respectively.

For the example of channel flow (see (6.2.30), (6.2.35)), condition (1) gives $\mu \leq 2$. Condition (2) gives $\mu \geq -2$, and condition (3) is satisfied with the equal sign since $C_{11}C_{22} - C_{12}C_{21} \equiv 0$. Since μ must be positive, the stability requirement is embodied in condition (1): $\mu \leq 2$.

We conclude this discussion of the stability of (6.2.38) by pointing out that we do not restrict attention to the value of k that is linearly most unstable as in Sec. 6.1.3. The main reason is that our weakly nonlinear solutions exhibit wave steepening and shock formation to leading order (as we will show in various examples). Thus, these solutions will involve all the harmonics of a given initial disturbance; the higher harmonics do not decay. Because of this feature, it is essential to have available a *general* solution of the unperturbed problem, regardless of the choice of initial disturbance. The ease with which our perturbation analysis for the weakly nonlinear problem (6.2.19) can be implemented will therefore depend on the form of this general solution, discussed next.

General solution

To solve (6.2.38), we introduce the characteristic independent variables

$$\xi_i = x - \lambda_i t; \quad i = 1, 2 \tag{6.2.50}$$

to obtain

$$\kappa\frac{\partial \overline{U}_1}{\partial \xi_2} + C_{11}\overline{U}_1 + C_{12}\overline{U}_2 = 0, \tag{6.2.51a}$$

$$-\kappa \frac{\partial \overline{U}_2}{\partial \xi_1} + C_{21}\overline{U}_1 + C_{22}\overline{U}_2 = 0, \tag{6.2.51b}$$

where we have introduced the notation

$$\kappa = \lambda_1 - \lambda_2 > 0, \quad \overline{U}_i(\xi_1, \xi_2) = U_i\left(\frac{\lambda_1\xi_2 - \lambda_2\xi_1}{\kappa}, \frac{\xi_2 - \xi_1}{\kappa}\right). \tag{6.2.52}$$

The initial conditions (6.2.38b) become

$$\overline{U}_i(\xi_2, \xi_2) = U_i^*(\xi_2), \quad i = 1, 2. \tag{6.2.53}$$

The solution for the case where all the C_{ij} vanish is the d'Alembert solution (6.2.39). Here the linear solution is always neutrally stable. Examples of weakly nonlinear problems in this class are discussed in Secs. 6.2.3–6.2.5. The case where one of the $C_{ij}, i \neq j$ vanishes is also elementary. With no loss of generality, we consider the case $C_{21} = 0$, $C_{12} \neq 0$. The example of channel flow with neutral stability ($\mu = 2$) is in this class (see (6.2.37) with $\epsilon = 0$). The weakly nonlinear weakly unstable case is discussed in Sec. 6.2.6. If $C_{21} = 0$, (6.2.51b) decouples from (6.2.51a), and we can solve for $\overline{U}_2$ in the form

$$\overline{U}_2(\xi_1, \xi_2) = U_2^*(\xi_2) \exp \frac{C_{22}}{\kappa}(\xi_1 - \xi_2). \tag{6.2.54a}$$

Using this result in (6.2.51a) and integrating gives

$$\overline{U}_1(\xi_1, \xi_2) = U_1^*(\xi_1) \exp \frac{C_{11}}{\kappa}(\xi_1 - \xi_2)$$

$$-\frac{C_{12}}{\kappa}\left[\exp\left(\frac{C_{22}\xi_1 - C_{11}\xi_2}{\kappa}\right)\right]\int_{\xi_1}^{\xi_2} U_2^*(s) \exp\left(\frac{C_{11} - C_{22}}{\kappa}s\right) ds. \tag{6.2.54b}$$

Note that with $C_{21} = 0$, the stability condition (3) is automatically satisfied if conditions (1) and (2) hold; solutions decay as $t \to \infty$ as long as $C_{11} > 0$ and $C_{22} > 0$. For example, if $U_2^* = A \sin kx$, the solution (6.2.54), expressed in terms of x and t, becomes

$$U_2 = A \sin k(x - \lambda_2 t) \exp(-C_{22}t), \tag{6.2.55a}$$

$$U_1 = U_1^*(x - \lambda_1 t)\exp(-C_{11}t) - A\frac{C_{12}}{\kappa}\left[\exp(-C_{22}t)\sin[k(x - \lambda_2 t) - \alpha]\right.$$

$$\left. + \exp(-C_{11}t)\sin[k(x - \lambda_1 t) - \alpha]\right], \tag{6.2.55b}$$

where α is the constant phase shift $\alpha = \tan^{-1}(k\kappa/(C_{11} - C_{22}))$. The exponential decay of the solution for arbitrary k (as long as $C_{11} > 0$, $C_{22} > 0$) is evident.

The general problem with $C_{12} \neq 0$, $C_{21} \neq 0$ is also solvable explicitly but is somewhat more complicated than the case just discussed. The interested reader can find this result in [6.20]. As no examples of weakly nonlinear problems in this class have been worked out, we omit the derivation of the linear solution.

6.2.3 Weakly Nonlinear Wave Equation: Examples with $n = 2$, $C_{ij} = 0$, $\phi_i(U_1, U_2)$

General weakly nonlinear wave equation

In this section we study (6.2.1) for the case $n = 2$, $C_{ij} = 0$ and ϕ_i independent of $(\partial U_i/\partial x)$. For this special case, the origin of (6.2.1) as a pair of perturbed conservation laws is not very interesting. However, we show next that (6.2.1) also arises by expressing the weakly nonlinear wave equation

$$u_{tt} - u_{xx} + \epsilon H(u_t, u_x) = 0, \tag{6.2.56}$$

as a pair of first-order equations. To do so, we set

$$u_x = U_1 + U_2, \quad u_t = -U_1 + U_2. \tag{6.2.57}$$

It then follows that

$$\frac{\partial U_1}{\partial t} + \frac{\partial U_2}{\partial t} + \frac{\partial U_1}{\partial x} - \frac{\partial U_2}{\partial x} = 0 \tag{6.2.58a}$$

for consistency ($u_{xt} - u_{tx} = 0$). Equation (6.2.56) then gives the second component

$$-\frac{\partial U_1}{\partial t} + \frac{\partial U_2}{\partial t} - \frac{\partial U_1}{\partial x} - \frac{\partial U_2}{\partial x} = -\epsilon H(-U_1 + U_2, U_1 + U_2). \tag{6.2.58b}$$

Adding and subtracting (6.2.58a), (6.2.58b) gives the desired result

$$\frac{\partial U_1}{\partial t} + \frac{\partial U_1}{\partial x} = \frac{\epsilon}{2} H(-U_1 + U_2, U_1 + U_2), \tag{6.2.59a}$$

$$\frac{\partial U_2}{\partial t} - \frac{\partial U_2}{\partial x} = -\frac{\epsilon}{2} H(-U_1 + U_2, U_1 + U_2). \tag{6.2.59b}$$

For general initial conditions,

$$u(x, 0; \epsilon) = u_0(x); \quad u_t(x, 0; \epsilon) = v_0(x), \tag{6.2.60}$$

we find

$$U_1(x, 0; \epsilon) = \frac{u_0'(x) - v_0(x)}{2} \equiv U_1^*(x), \tag{6.2.61a}$$

$$U_2(x, 0; \epsilon) = \frac{u_0'(x) + v_0(x)}{2} \equiv U_2^*(x). \tag{6.2.61b}$$

Since H does not involve the $(\partial U_i/\partial x)$, solutions are shock-free.

We assume a multiple-scale expansion in terms of the two characteristic independent variables $\xi_1 = x - t$, $\xi_2 = x + t$ and the slow time $\tilde{t} = \epsilon t$

$$U_i(x, t; \epsilon) = U_{i0}(\xi_1, \xi_2, \tilde{t}) + \epsilon U_{i1}(\xi_1, \xi_2, \tilde{t}) + O(\epsilon^2). \tag{6.2.62}$$

It should be noted that in this section we will only derive U_{i0} explicitly. If the higher-order terms U_{i1}, etc. are to be calculated, one must, in general, also include a dependence on $\epsilon^2 t$, etc., in the expansion. For the assumed multiple-scale expansion (6.2.62), derivatives are given by

$$\frac{\partial U_i}{\partial t} = -\frac{\partial U_{i0}}{\partial \xi_1} + \frac{\partial U_{i0}}{\partial \xi_2} + \epsilon\left(-\frac{\partial U_{i1}}{\partial \xi_1} + \frac{\partial U_{i1}}{\partial \xi_2} + \frac{\partial U_{io}}{\partial \tilde{t}}\right) + O(\epsilon^2), \quad (6.2.63)$$

$$\frac{\partial U_i}{\partial x} = \frac{\partial U_{i0}}{\partial \xi_1} + \frac{\partial U_{i0}}{\partial \xi_2} + \epsilon\left(\frac{\partial U_{i1}}{\partial \xi_1} + \frac{\partial U_{i1}}{\partial \xi_2}\right) + O(\epsilon^2). \quad (6.2.64)$$

The equations and initial conditions governing U_{10} and U_{20} are then obtained from (6.2.59) and (6.2.61) in the form

$$2\frac{\partial U_{10}}{\partial \xi_2} = 0, \quad U_{10}(\xi_1, \xi_1, 0) = U_1^*(\xi_1), \quad (6.2.65a)$$

$$-2\frac{\partial U_{10}}{\partial \xi_1} = 0, \quad U_{20}(\xi_2, \xi_2, 0) = U_2^*(\xi_2). \quad (6.2.65b)$$

Thus,

$$U_{10}(\xi_1, \xi_2, \tilde{t}) = f_0(\xi_1, \tilde{t}), \quad f_0(\xi_1, 0) = U_1^*(\xi_1), \quad (6.2.66a)$$

$$U_{20}(\xi_1, \xi_2, \tilde{t}) = g_0(\xi_2, \tilde{t}), \quad g_0(\xi_2, 0) = U_2^*(\xi_2). \quad (6.2.66b)$$

To derive the evolution equations for f_0 and g_0 we consider the terms of $O(\epsilon)$. Using (6.2.66) in the right-hand side of the equations governing U_{11} and U_{21}, we obtain

$$2\frac{\partial U_{11}}{\partial \xi_2} = -\frac{\partial f_0}{\partial \tilde{t}} + \frac{1}{2}H(-f_0 + g_0, f_0 + g_0), \quad (6.2.67)$$

$$-2\frac{\partial U_{21}}{\partial \xi_1} = -\frac{\partial g_0}{\partial \tilde{t}} - \frac{1}{2}H(-f_0 + g_0, f_0 + g_0). \quad (6.2.68)$$

The initial conditions are

$$U_{11}(\xi_1, \xi_1, 0) = U_{21}(\xi_2, \xi_2, 0) = 0. \quad (6.2.69)$$

We now consider the solution for U_{11} and U_{21} in order to isolate inconsistent terms and derive evolution equations for f_0 and g_0. To fix ideas, let us restrict attention to periodic initial data. In particular, assume that $U_1^*(x)$ and $U_2^*(x)$ are 2ℓ-periodic functions with zero average. It is then easy to prove that the exact solution is also 2ℓ-periodic in x. To see this, we note that we may solve (6.2.59) by "time-stepping" U_1 along the ξ_2 characteristic and U_2 along the ξ_1 characteristic. Thus, the outcome for U_1 and U_2 at any given point (x, t) is *exactly the same* at the points $(x + 2n\ell, t)$, $n = \pm 1, \pm 2, \ldots$. Periodicity of the exact solution in x then implies periodicity of f_0 and g_0 with respect to ξ_1 and ξ_2, respectively.

In the integration of (6.2.67) for U_1 with respect to ξ_2, an inconsistent term proportional to ξ_2 will be contributed by any term on the right-hand side that is *independent* of ξ_2, i.e., by a function of ξ_1 and $\tilde{t}$ only. The leading term $-(\partial f_0/\partial \tilde{t})$ on the right-hand side of (6.2.67) is such a term and must be removed. We must also identify and remove all the terms in $H/2$ that are independent of ξ_2. Since f_0 and g_0 are 2ℓ-periodic functions of their respective fast scales, $H(-f_0 + g_0, f_0 + g_0)$ is a 2ℓ-periodic function of ξ_1 and ξ_2. The Fourier series of $H(-f_0 + g_0, f_0 + g_0)$

with respect to ξ_2 will, in general, have an average term that must be removed also. Similar remarks apply to (6.2.68). Therefore, we remove inconsistent terms by setting

$$\frac{\partial f_0}{\partial \tilde{t}} - \frac{1}{2}\langle H\rangle(\xi_1, \tilde{t}) = 0, \tag{6.2.70a}$$

$$\frac{\partial g_0}{\partial \tilde{t}} + \frac{1}{2}\langle H\rangle(\xi_2, \tilde{t}) = 0, \tag{6.2.70b}$$

where the averages $\langle H\rangle(\xi_1, \tilde{t})$ and $\langle H\rangle(\xi_2, \tilde{t})$ are defined by

$$\langle H\rangle(\xi_1, \tilde{t}) = \frac{1}{2\ell}\int_{-\ell}^{\ell} H(-f_0(\xi_1, \tilde{t}) + g_0(\xi_2, \tilde{t}), f_0(\xi_1, \tilde{t}) + g_0(\xi_2, \tilde{t}))d\xi_2, \tag{6.2.71a}$$

$$\langle H\rangle(\xi_2, \tilde{t}) = \frac{1}{2\ell}\int_{-\ell}^{\ell} H(-f_0(\xi_1, \tilde{t}) + g_0(\xi_2, \tilde{t}), f_0(\xi_1, \tilde{t}) + g_0(\xi_2, \tilde{t}))d\xi_1. \tag{6.2.71b}$$

This result was first derived in [6.6] proceeding directly from (6.2.56). In [6.11] it is proven that for periodic initial data, f_0 and g_0 given by (6.2.70) are the asymptotic terms of $O(1)$ for U_1 and U_2, respectively, uniformly in the interval $0 \le t \le T(\epsilon) = O(\epsilon^{-1})$.

The extension of (6.2.70) to the case of more general *bounded* initial data is also possible. Here, we define the averages H_1 and H_2 as follows:

$$\langle H\rangle(\xi_1, \tilde{t}) = \lim_{\ell\to\infty} \frac{1}{2\ell}\int_{-\ell}^{\ell} H d\xi_2, \tag{6.2.72a}$$

$$\langle H\rangle(\xi_2, \tilde{t}) = \lim_{\ell\to\infty} \frac{1}{2\ell}\int_{-\ell}^{\ell} H d\xi_1. \tag{6.2.72b}$$

Of course, in this case we must assume that the integrals in (6.2.72) exist.

Cubic damping, waves in one direction

To illustrate these ideas, let us consider the special case $H = u_t^3$. We then find

$$H(-f_0+g_0, f_0+g_0) = (-f_0+g_0)^3 = -f_0^3+3f_0^2g_0-3f_0g_0^2+g_0^3. \tag{6.2.73}$$

The evolution equations (6.2.70) become

$$\frac{\partial f_0}{\partial \tilde{t}} + \frac{1}{2}f_0^3 - \frac{3}{2}\langle g_0\rangle(\tilde{t})f_0^2 + \frac{3}{2}\langle g_0^2\rangle(\tilde{t})f_0 - \frac{1}{2}\langle g_0^3\rangle(\tilde{t}) = 0, \tag{6.2.74a}$$

$$\frac{\partial g_0}{\partial \tilde{t}} + \frac{1}{2}g_0^3 - \frac{3}{2}\langle f_0\rangle(\tilde{t})g_0^2 + \frac{3}{2}\langle f_0^2\rangle(\tilde{t})g_0 - \frac{1}{2}\langle f_0^3\rangle(\tilde{t}) = 0, \tag{6.2.74b}$$

where for a function $h(\xi_i, \tilde{t})$, $i = 1$ or 2, $\langle h\rangle(\tilde{t})$ denotes the average of h over one period in ξ_1 or ξ_2, and is therefore a function of $\tilde{t}$ only

$$\langle h\rangle(\tilde{t}) = \frac{1}{2\ell}\int_{-\ell}^{\ell} h(\xi_i, \tilde{t})d\xi_i, \quad i = 1, 2. \tag{6.2.75}$$

The solution of the coupled system (6.2.74) for *general* periodic initial data $f_0(\xi_1, 0)$ and $g_0(\xi_2, 0)$ with zero average $(\langle f_0 \rangle(0) = \langle g_0 \rangle(0) = 0)$ is difficult. However, if the initial data are restricted further to satisfynecessary conditions for the vanishing of odd-powered averages for

$$\langle f_0^{2n+1} \rangle(0) = \langle g_0^{2n+1} \rangle(0) = 0, \quad n = 0, 1, 2, \ldots, \tag{6.2.76}$$

then the system (6.2.74) simplifies to (see Problem 5)

$$\frac{\partial f_0}{\partial \tilde{t}} + \frac{1}{2} f_0^3 + \frac{3}{2} \langle g_0^2 \rangle(\tilde{t}) f_0 = 0, \tag{6.2.77a}$$

$$\frac{\partial g_0}{\partial \tilde{t}} + \frac{1}{2} g_0^3 + \frac{3}{2} \langle f_0^2 \rangle(\tilde{t}) g_0 = 0, \tag{6.2.77b}$$

because $\langle f_0 \rangle(\tilde{t})$, $\langle g_0 \rangle(\tilde{t})$, $\langle f_0^3 \rangle(\tilde{t})$, and $\langle g_0^3 \rangle(\tilde{t})$ all vanish identically. Although this system is still coupled, we can solve it explicitly for certain initial conditions. Note that the restriction (6.2.76) excludes certain simple multiharmonic initial data having a zero average, such as $f_0(x, 0) = \cos x + \cos 2x$, for which $\langle f_0^3 \rangle(0) = 3/4$.

Consider first the simple initial-value problem

$$u(x, 0; \epsilon) = A \sin kx, \quad u_t(x, 0; \epsilon) = -Ak \cos kx, \tag{6.2.78}$$

where A and k are constants. We note that for $\epsilon = 0$ the solution of (6.2.56) is the traveling wave $u = A \sin k(x - t)$, i.e., $U_1 = (u_x - u_t)/2 = Ak \cos k(x - t)$, $U_2 = (u_x + u_t)/2 = 0$. The initial conditions (6.2.66) for f_0 and g_0 are

$$f_0(\xi_1, 0) = Ak \cos k\xi_1, \quad g_0(\xi_2, 0) = 0, \tag{6.2.79}$$

and we see that the restriction (6.2.76) is trivially satisfied. Since $g_0(\xi_2, 0) = 0$, (6.2.77b) implies $g_0(\xi_2, \tilde{t}) = 0$, and (6.2.77a) simplifies to

$$\frac{\partial f_0}{\partial \tilde{t}} + \frac{1}{2} f_0^3 = 0.$$

The solution satisfying the initial condition (6.2.79) is

$$f_0(\xi_1, \tilde{t}) = \frac{Ak \cos k\xi_1}{[1 + A^2 k^2 \tilde{t} \cos^2 k\xi_1]^{1/2}}. \tag{6.2.80}$$

To compute u to leading order, we note from (6.2.57) that $u_x = f_0 + O(\epsilon)$. Therefore, integrating (6.2.80) with respect to ξ_1 and imposing the initial condition (6.2.78) for $u(x, 0; \epsilon)$ gives

$$u = \frac{1}{k\tilde{t}^{1/2}} \sin^{-1} \left\{ \left[\frac{A^2 k^2 \tilde{t}}{1 + A^2 k^2 \tilde{t}} \right]^{1/2} \sin k\xi_1 \right\} + O(\epsilon), \tag{6.2.81}$$

where the arcsine is in the interval $\left(-\frac{\pi}{2}, \frac{\pi}{2}\right)$. We see that the amplitude decays like $\tilde{t}^{-1/2}$.

Having removed the inconsistent terms from the right-hand sides of (6.2.67)–(6.2.68), and noting that $g_0 = 0$, U_{11} and U_{21} satisfy the homogeneous equations.

Therefore, $U_{11} = f_1(\xi_1, \tilde{t})$; $U_{21} = g_1(\xi_2, \tilde{t})$ in this case. The evolution equations for f_1 and g_1 are obtained by consistency arguments on the solution to $O(\epsilon^2)$. The details can be found in [6.6] and are not given here. We only point out that both f_1 and g_1 are present in the $O(\epsilon)$ solution.

To appreciate the computational complexity that we have avoided by deriving the explicit formula (6.2.81) instead of its Fourier series, let us work out the series expansion for the $O(1)$ solution. First, we expand f_0 in a Fourier cosine series and then integrate the result term by term to obtain

$$u = \sum_{n=1}^{\infty} B_{2n-1}(\tilde{t}) \sin(2n-1)k\xi_1, \tag{6.2.82a}$$

where

$$B_{2n-1} = \frac{2A}{(2n-1)\pi} \int_0^{\pi} \frac{\cos\xi \cos(2n-1)\xi}{[1 + A^2k^2\tilde{t}\cos^2\xi]^{1/2}}\, d\xi. \tag{6.2.82b}$$

Each B_{2n-1} can be expressed as $(1 + A^2k^2\tilde{t})^{-1/2}$ times a power series in $z^2 = A^2k^2\tilde{t}/(1 + A^2k^2\tilde{t})$. For example,

$$B_1(\tilde{t}) = \frac{A}{(1 + A^2k^2\tilde{t})^{1/2}} \left[1 + \frac{z^2}{8} + \frac{3}{64}z^4 + \dots \right]. \tag{6.2.83}$$

Cubic damping, waves in both directions

Consider now the solution for $H = u_t^3$ with the initial conditions

$$u(x, 0; \epsilon) = 2\sin x, \quad u_t(x, 0; \epsilon) = 0. \tag{6.2.84a}$$

The initial conditions for f_0 and g_0 are

$$f_0(\xi_1, 0) = \sin\xi_1, \quad g_0(\xi_2, 0) = \sin\xi_2, \tag{6.2.84b}$$

and again they satisfy the restrictions (6.2.76). In this case, the pair of equations (6.2.77) are coupled, but, because of our choice for $u_t(x, 0; \epsilon)$, they are of the same form. Consider (6.2.77a) and set $\langle g_0^2 \rangle$ = constant = $c/3$ temporarily. Integrating the result gives

$$f_0 = \frac{c^{1/2}e^{-c\tilde{t}/2}}{[F(\xi_1) - e^{-c\tilde{t}}]^{1/2}}, \tag{6.2.85}$$

where F is an arbitrary function of ξ_1.

Having found the form of the solution for $\langle g_0^2 \rangle$ = constant, let us assume that the actual solution with $\langle g_0^2 \rangle(\tilde{t})$ has the same structure by introducing unknown functions of $\tilde{t}$ whenever c appears, i.e.,

$$f_0(\xi_1, \tilde{t}) = \frac{\lambda(\tilde{t})}{[F(\xi_1) + \phi(\tilde{t})]^{1/2}}, \tag{6.2.86}$$

where λ and ϕ are unknowns. Substituting (6.2.86) into (6.2.77a) shows that the assumed form is indeed a solution if λ and ϕ satisfy

$$2\lambda'(\tilde{t}) + 3\langle g_0^2 \rangle(\tilde{t})\lambda(\tilde{t}) = 0, \tag{6.2.87a}$$

$$\phi'(\tilde{t}) = \lambda^2(\tilde{t}). \tag{6.2.87b}$$

In view of the symmetry of (6.2.77) and our initial conditions on f_0 and g_0, we have g_0 in the form

$$g_0(\xi_2, \tilde{t}) = \frac{\lambda(\tilde{t})}{[F(\xi_2) + \phi(\tilde{t})]^{1/2}}. \tag{6.2.88}$$

Equations (6.2.86) and (6.2.88) satisfy the initial conditions (6.2.84b) if $F_0(\xi_i) = \sec^2 \xi_i$, $\lambda(0) = 1$ and $\phi(0) = 0$.

Thus,

$$f_0 = \frac{\lambda(\tilde{t}) \cos \xi_1}{[1 - \phi(\tilde{t}) \cos^2 \xi_1]^{1/2}}, \quad g_0 = \frac{\lambda(\tilde{t}) \cos \xi_2}{[1 + \phi(\tilde{t}) \cos^2 \xi_2]^{1/2}}. \tag{6.2.89}$$

We now calculate $\langle g_0^2 \rangle(\tilde{t})$ using the above expression for g_0 in the defining relation (6.2.75) with $h_i = g_0^2$:

$$\begin{aligned} \langle g_0^2 \rangle(\tilde{t}) &= \frac{\lambda^2(\tilde{t})}{2\pi} \int_{-\pi}^{\pi} \frac{\cos^2 \xi_2 d\xi_2}{1 + \phi^2(\tilde{t}) \cos^2 \xi_2} \\ &= \frac{\lambda^2(\tilde{t})}{\phi(\tilde{t})} - \frac{\lambda^2(\tilde{t})}{\phi(\tilde{t})\sqrt{1 + \phi(\tilde{t})}}. \end{aligned} \tag{6.2.90}$$

Using this expression for $\langle g_0^2 \rangle(\tilde{t})$ in (6.2.87a) gives

$$-\frac{2}{3}\frac{\lambda'}{\lambda} = \frac{\phi'}{\phi} - \frac{\phi'}{\phi\sqrt{1+\phi}}, \tag{6.2.91}$$

which upon integration and use of (6.2.87b) results in

$$\log \lambda^{-2/3} = \log(\phi')^{-1/3} = \log \frac{\phi[\sqrt{1+\phi} + 1]}{\sqrt{1+\phi} - 1} + \text{constant}. \tag{6.2.92}$$

Therefore,

$$\phi' = 2^6[1 + \sqrt{1+\phi}]^{-6}. \tag{6.2.93}$$

If we introduce the new variable

$$m(\tilde{t}) = 1 + \sqrt{1+\phi}, \quad m(0) = 2, \tag{6.2.94}$$

we can integrate (6.2.93) to find

$$m^8(\tilde{t}) - \frac{8}{7} m^7(\tilde{t}) = 2^8 \left(\tilde{t} + \frac{3}{7}\right). \tag{6.2.95}$$

The solution of this algebraic equation gives $m(\tilde{t})$, and (6.2.87b) gives λ in terms of m

$$\lambda(\tilde{t}) = \sqrt{\phi'} = \frac{2^3}{m(\tilde{t})}. \tag{6.2.96}$$

This defines f_0 and g_0 in terms of $m(\tilde{t})$. To compute u to leading order, we integrate $f_0 + g_0$ with respect to x and find

$$u = \frac{\lambda(\tilde{t})}{\sqrt{\phi(\tilde{t})}} \left\{ \sin^{-1}\left[\left(\frac{\phi(\tilde{t})}{1+\phi(\tilde{t})} \right)^{1/2} \sin\xi_1 \right] + \sin^{-1}\left[\left(\frac{\phi(\tilde{t})}{1+\phi(\tilde{t})} \right)^{1/2} \sin\xi_2 \right] \right\} + O(\epsilon). \tag{6.2.97}$$

The simplicity and symmetry of the perturbation function and the initial conditions were crucial in calculating this explicit result. It is not always possible to solve the coupled system (6.2.70) analytically for more general perturbation functions and periodic initial conditions.

For initial conditions with compact support (i.e., f_0 and g_0 both equal zero initially outside a bounded interval), the evolution equations simplify considerably. For example, with $H = u_t^3$, the expressions in (6.2.74) no longer contain any of the averages. More interesting examples for isolated initial disturbances are discussed in Secs. 6.2.3–6.2.5. If the initial conditions are bounded, nonperiodic functions on the entire interval, the evolution equations can be solved analytically in certain cases (see Problem 2).

The expansion procedure we have outlined also applies (with slight modifications) to boundary-value problems with homogeneous boundary conditions ([6.21] and Problem 2), to signaling problems on the semi-infinite interval ([6.5] and Problem 6), and to the solution of elliptic weakly nonlinear equations in terms of complex characteristics, [6.4].

6.2.4 Shallow Water Flow; Examples with $n = 2$, $C_{ij} = 0$, $\phi_i(U_1, U_2, \frac{\partial U_1}{\partial x}, \frac{\partial U_2}{\partial x}, \ldots)$

In this section, we use the approximate equations for shallow water flow in two dimensions to illustrate various features of solutions for the case where the C_{ij} matrix in (6.2.1) vanishes, and where the ϕ_i depend on first and higher derivatives of the U_i.

Governing equations for shallow water flow

We consider an ideal fluid where the effects of viscosity, surface tension, and density variation are ignored. In the shallow water limit, where the characteristic length scale of a horizontal disturbance is assumed to be very large relative to the undisturbed free surface height, we ignore vertical fluid motion. The state of the fluid is then defined in terms of the dimensionless free surface height, $h(x, t)$, and the vertically averaged dimensionless horizontal speed, $u(x, t)$. We allow for specified small bottom surface variations, $y_b = \epsilon b(x, t)$, in a coordinate system where the undisturbed flow ($\epsilon = 0$) is defined by $u = 0$ and $h = 1$. The geometry is sketched in Figure 6.2.2.

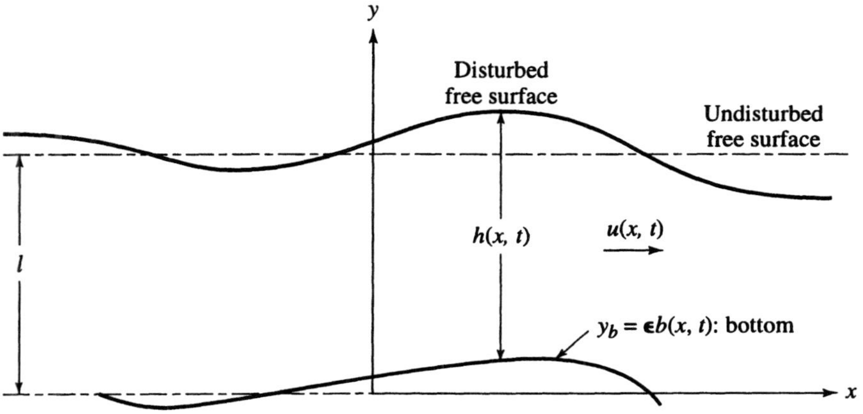

FIGURE 6.2.2. Geometry

The conservation laws of mass and momentum (see chapter 3 of [6.17]) are then given by (see (6.2.2))

$$\text{mass:} \qquad h_t + (uh)_x = 0, \tag{6.2.98a}$$

$$\text{momentum:} \qquad (uh)_t + (u^2h + h^2/2)_x + \epsilon h b_x = 0. \tag{6.2.98b}$$

The associated shock conditions are (see (6.2.10))

$$C_s = \frac{[uh]}{[h]} = \frac{[u^2h + h^2/2]}{[uh]}, \tag{6.2.99}$$

where [ ] denotes the jump in a variable across a shock.

Simplifying (6.2.98b) using (6.2.98a), we find (see (6.2.5))

$$h_t + uh_x + hu_x = 0, \tag{6.2.100a}$$

$$u_t + h_x + uu_x = -\epsilon b_x. \tag{6.2.100b}$$

The exact (deep water) formulation involves the additional parameter δ, the ratio of the undisturbed free surface height to the horizontal disturbance wavelength. One can show that in the limit $\delta \to 0$, the exact problem indeed gives (6.2.100). In order to account for the leading contribution for $\delta \neq 0$, one must add the following $O(\delta^2)$ terms to the right-hand side of (6.2.100b):

$$h_t + uh_x + hu_x = 0. \tag{6.2.101a}$$

$$u_t + h_x + uu_x = -\epsilon b_x - \frac{1}{3}\delta^2 h_{xtt} - \frac{1}{2}\delta^2 \epsilon b_{xtt}. \tag{6.2.101b}$$

This so-called *Boussinesq approximation* is derived in detail for the special case $b \equiv 0$ in section 5.2.4 of [6.18]; the added term due to a variable bottom is given in [6.19].

In some applications, e.g., a river flowing over an isolated bump on the bottom, the conditions at upstream infinity in a coordinate system fixed to the bump are $u = F =$ constant, $h = 1$. In this coordinate system, the bump is defined by $b = B(\overline{x})$ on $0 \leq \overline{x} \leq 1$ and $b = 0$ otherwise. The Galilean transformation

$$\overline{x} = x + Ft, \quad \overline{t} = t, \tag{6.2.102a}$$

$$u = \overline{u} - F, \quad h = \overline{h} \tag{6.2.102b}$$

then transforms (6.2.101) to

$$\overline{h}_{\overline{t}} + \overline{u}\overline{h}_{\overline{x}} + \overline{h}\overline{u}_{\overline{x}} = 0, \tag{6.2.103a}$$

$$\overline{u}_{\overline{t}} + \overline{h}_{\overline{x}} + \overline{u}\,\overline{u}_{\overline{x}} = -\epsilon B'(\overline{x}) - \frac{1}{3}\delta^2[F^2 h_{\overline{x}\overline{x}\overline{x}} + 2Fh_{\overline{x}\overline{x}\overline{t}} + h_{\overline{x}\overline{t}\overline{t}}] - \frac{1}{2}\delta^2\epsilon F^2 B'''(\overline{x}). \tag{6.2.103b}$$

Surface disturbances over a flat bottom, Korteweg–de Vries equation

The simplest problem consists of the special case $b \equiv 0$ in (6.2.101) and an initial surface and velocity disturbance of amplitude $O(\epsilon)$ to the quiescent state $u = 0$, $h = 1$. It is shown in [6.18] that the choice $\delta = O(\epsilon^{1/2})$ results in the richest limit, and we accordingly set $\delta = \kappa\epsilon^{1/2}$ in (6.2.101). Here κ is an arbitrary constant independent of ϵ (not related to the κ in (6.2.52)). The initial disturbance is specified in the form

$$h(x, 0; \epsilon) = 1 + \epsilon h^*(x); \quad u(x, 0; \epsilon) = \epsilon u^*(x). \tag{6.2.104}$$

With $u_1 = h$, $u_2 = u$, (6.2.11) give $u_1 = 1 + \epsilon w_1(x, t; \epsilon, \kappa)$, $u_2 = \epsilon w_2(x, t; \epsilon, \kappa)$, and (6.2.12) becomes

$$\frac{\partial w_1}{\partial t} + \frac{\partial w_2}{\partial x} = -\epsilon\frac{\partial}{\partial x}(w_1 w_2), \tag{6.2.105a}$$

$$\frac{\partial w_2}{\partial t} + \frac{\partial w_1}{\partial x} = -\epsilon\frac{\partial}{\partial x}\left(\frac{w_2^2}{2} + \frac{\kappa^2}{3}\frac{\partial^2 w_1}{\partial t^2}\right). \tag{6.2.105b}$$

The initial conditions are $w_1(x, 0; \epsilon) = h^*(x)$, $w_2(x, 0; \epsilon) = u^*(x)$.

The eigenvalues of $A^{(0)}$ are $\lambda_{1,2} = \pm 1$, and the linear transformation to the U_1, U_2 variables is the same as in the example of channel flow, i.e., $w_1 = -U_1 - U_2$, $w_2 = -U_1 + U_2$. The standard form system (6.2.19) is then easily calculated

$$\frac{\partial U_1}{\partial t} + \frac{\partial U_1}{\partial x} = \epsilon\frac{\partial}{\partial x}\left[\frac{3}{4}U_1^2 - \frac{1}{2}U_1U_2 - \frac{1}{4}U_2^2 - \frac{\kappa^2}{6}\frac{\partial^2}{\partial t^2}(U_1 + U_2)\right], \tag{6.2.106a}$$

$$\frac{\partial U_2}{\partial t} - \frac{\partial U_2}{\partial x} = \epsilon \frac{\partial}{\partial x}\left[\frac{1}{4}U_1^2 + \frac{1}{2}U_1U_2 - \frac{3}{4}U_2^2 + \frac{\kappa^2}{6}\frac{\partial^2}{\partial t^2}(U_1 + U_2)\right]. \tag{6.2.106b}$$

The initial conditions are

$$U_1(x, 0; \epsilon) = -\frac{1}{2}[h^*(x) + u^*(x)] \equiv U_1^*(x), \tag{6.2.107a}$$

$$U_2(x, 0; \epsilon) = \frac{1}{2}[-h^*(x) + u^*(x)] \equiv U_2^*(x). \tag{6.2.107b}$$

We expand the U_i as in (6.2.62) to find

$$U_{10} = f_0(\xi_1, \tilde{t}); \quad U_{20} = g_0(\xi_2, \tilde{t}) \tag{6.2.108}$$

as in (6.2.66), with initial conditions

$$f_0(\xi_1, 0) = U_1^*(\xi_1); \quad g_0(\xi_2, 0) = U_2^*(\xi_2). \tag{6.2.109}$$

The equations for U_{11} and U_{21} that result from (6.2.106) are then obtained as follows

$$2\frac{\partial U_{11}}{\partial \xi_2} = -\frac{\partial f_0}{\partial \tilde{t}} + \left(\frac{\partial}{\partial \xi_1} + \frac{\partial}{\partial \xi_2}\right)\left[\frac{3}{4}f_0^2 - \frac{1}{2}f_0g_0 - \frac{1}{4}g_0^2 - \frac{\kappa^2}{6}\left(-\frac{\partial}{\partial \xi_1} + \frac{\partial}{\partial \xi_2}\right)^2(f_0 + g_0)\right], \tag{6.2.110a}$$

$$-2\frac{\partial U_{21}}{\partial \xi_1} = -\frac{\partial g_0}{\partial \tilde{t}} + \left(\frac{\partial}{\partial \xi_1} + \frac{\partial}{\partial \xi_2}\right)\left[\frac{1}{4}f_0^2 + \frac{1}{2}f_0g_0 - \frac{3}{4}g_0^2 + \frac{\kappa^2}{6}\left(-\frac{\partial}{\partial \xi_1} + \frac{\partial}{\partial \xi_2}\right)^2(f_0 + g_0)\right]. \tag{6.2.110b}$$

Simplifying the right-hand sides, we obtain

$$2\frac{\partial U_{11}}{\partial \xi_2} = \left(-\frac{\partial f_0}{\partial \tilde{t}} + \frac{3}{2}f_0\frac{\partial f_0}{\partial \xi_1} - \frac{\kappa^2}{6}\frac{\partial^3 f_0}{\partial \xi_1^3}\right) - \frac{1}{2}(f_0 + g_0)\frac{\partial g_0}{\partial \xi_2} - \frac{1}{2}g_0\frac{\partial f_0}{\partial \xi_1} - \frac{\kappa^2}{6}\frac{\partial^3 g_0}{\partial \xi_2^3}, \tag{6.2.111a}$$

$$-2\frac{\partial U_{21}}{\partial \xi_1} = \left(-\frac{\partial g_0}{\partial \tilde{t}} - \frac{3}{2}g_0\frac{\partial g_0}{\partial \xi_2} + \frac{\kappa^2}{6}\frac{\partial^3 g_0}{\partial \xi_2^3}\right) + \frac{1}{2}(f_0 + g_0)\frac{\partial f_0}{\partial \xi_1} + \frac{1}{2}f_0\frac{\partial g_0}{\partial \xi_2} - \frac{\kappa^2}{6}\frac{\partial^3 f_0}{\partial \xi_1^3}. \tag{6.2.111b}$$

The first expression in parentheses on the right-hand side of (6.2.111a) is independent of ξ_2. Therefore, upon integration with respect to ξ_2 it contributes an inconsistent term proportional to ξ_2 in the solution for U_{11}. Similarly, the first expression in parentheses on the right-hand side of (6.2.111b) contributes an inconsistent term proportional to ξ_1. Removing these inconsistent terms gives the *decoupled* evolution equations

$$\frac{\partial f_0}{\partial \tilde{t}} - \frac{3}{2} f_0 \frac{\partial f_0}{\partial \xi_1} + \frac{\kappa^2}{6} \frac{\partial^3 f_0}{\partial \xi_1^3} = 0, \tag{6.2.112a}$$

$$\frac{\partial g_0}{\partial \tilde{t}} + \frac{3}{2} g_0 \frac{\partial g_0}{\partial \xi_2} - \frac{\kappa^2}{6} \frac{\partial^3 g_0}{\partial \xi_2^3} = 0. \tag{6.2.112b}$$

Because we only have quadratic nonlinearities in (6.2.105), we do not encounter a coupling between the f_0 and g_0 equations as in (6.2.74); the remaining terms on the right-hand sides of (6.2.111) are perfectly consistent as long as the U_i^* are bounded integrable functions. In fact, we can now integrate what remains of (6.2.111) to obtain

$$\begin{aligned} U_{11}(\xi_1, \xi_2, \tilde{t}) = {} & f_1(\xi_1, \tilde{t}) - \frac{1}{4} f_0 g_0 - \frac{1}{8} g_0^2 \\ & - \frac{1}{4} \frac{\partial f_0}{\partial \xi_1} G_0(\xi_2, \tilde{t}) - \frac{\kappa^2}{12} \frac{\partial^2 g_0}{\partial \xi_2^2}. \end{aligned} \tag{6.2.113}$$

$$\begin{aligned} U_{21}(\xi_1, \xi_2, \tilde{t}) = {} & g_1(\xi_1, \tilde{t}) - \frac{1}{4} f_0 g_0 - \frac{1}{8} f_0^2 \\ & - \frac{1}{4} \frac{\partial g_0}{\partial \xi_2} F_0(\xi_1, \tilde{t}) - \frac{\kappa^2}{12} \frac{\partial^2 f_0}{\partial \xi_1^2}, \end{aligned} \tag{6.2.114}$$

where $f_1(\xi_1, \tilde{t})$ and $g_1(\xi_2, \tilde{t})$ are new unknowns to be determined by consistency conditions on the solution to $O(\epsilon^2)$, and G_0, F_0 are the indefinite integrals of g_0 and f_0 with respect to ξ_2 and ξ_1, respectively,

$$\frac{\partial G_0}{\partial \xi_2} = g_0(\xi_2, \tilde{t}), \quad \frac{\partial F_0}{\partial \xi_1} = f_0(\xi_1, \tilde{t}). \tag{6.2.115}$$

This result is independent of the nature (periodic or not) of the initial conditions. We point out again that if the solution for U_1 and U_2 is to be computed to $O(\epsilon)$, i.e., if we wish to determine f_1 and g_1, we should include a dependence on the second slow time $t_2 = \epsilon^2 t$. The calculations are straightforward but tedious and are not discussed here.

Let us now examine the evolution equations (6.2.112). These are to be solved individually for the initial conditions (6.2.109). Each of (6.2.112) is a Korteweg–de Vries equation, and its exact solution is possible for initial data that decay sufficiently fast as $|\xi_i| \to \infty$. The solution procedure, known as *inverse scattering theory*, was developed in 1967 in [6.14]. It has been studied extensively for a number of different equations since then. For example, see the discussion in [6.1], [6.8], and [6.27]. A discussion of this theory would take us too far afield and is omitted. Without going into the details of the solution for f_0 and g_0, we have already a remarkable result in the decoupled system (6.2.112). This result shows that the f_0 and g_0 waves evolve *independently*, each obeying its own nonlinear equation. Equation (6.2.112a) for f_0 defines a wave that propagates to the right with unit speed in the xt frame (as exhibited by the dependence of f_0 on ξ_1) and changes slowly in time (as exhibited by the dependence of f_0 on $\tilde{t}$). The g_0 wave propagates to the left in exactly the same way. The solution for h and u to $O(\epsilon)$ involves both waves and has the form

$$h(x, t; \epsilon) = 1 - \epsilon[f_0(\xi_1, \tilde{t}) + g_0(\xi_2, \tilde{t})] + O(\epsilon^2), \tag{6.2.116a}$$

$$u(x, t; \epsilon) = \epsilon[-f_0(\xi_1, \tilde{t}) + g_0(\xi_2, \tilde{t})] + O(\epsilon^2). \tag{6.2.116b}$$

It is interesting to consider the special case where one of the waves, say g_0, is absent and to express the remaining evolution equation in terms of the original h, u variables as functions of x and t. We see that $g_0(\xi_2, \tilde{t}) \equiv 0$ if we have $g_0(\xi_2, 0) = 0$, i.e., by choosing $u^*(x) = h^*(x)$ (see (6.2.107b)). If we ignore terms of order ϵ^2, (6.2.116) give $f_0(\xi_1, \tilde{t}) = -(h-1)/\epsilon = -u/\epsilon$. Let us now transform independent variables from $(\xi_1, \tilde{t})$ to (x, t). We have

$$x = \xi_1 + \frac{\tilde{t}}{\epsilon}, \quad t = \frac{\tilde{t}}{\epsilon}. \tag{6.2.117}$$

Therefore,

$$\frac{\partial}{\partial \xi_1} = \frac{\partial}{\partial x}, \quad \frac{\partial^3}{\partial \xi_1^3} = \frac{\partial^3}{\partial x^3}, \quad \frac{\partial}{\partial \tilde{t}} = \frac{1}{\epsilon}\frac{\partial}{\partial x} + \frac{1}{\epsilon}\frac{\partial}{\partial t}, \tag{6.2.118}$$

and (6.2.112a) becomes (after multiplying by $-\epsilon^2/2$ and using $\kappa^2 = \delta/\epsilon$)

$$h_t + \frac{1}{2}(3h - 1)h_x + \frac{\delta^2}{6} h_{xxx} = 0 \tag{6.2.119a}$$

or (see (4.1.46))

$$u_t + \left(\frac{3}{2}u + 1\right) u_x + \frac{\delta^2}{6} u_{xxx} = 0. \tag{6.2.119b}$$

This is the form usually found in the literature for the Korteweg–de Vries equation. The transformation $\frac{3}{2}u + 1 = \frac{3}{2}w$ or $\frac{3}{2}h - \frac{1}{2} = \frac{3}{2}w$ takes each of (6.2.119) to the generic form

$$w_t + \frac{3}{2} w w_x + \frac{\delta^2}{6} w_{xxx} = 0. \tag{6.2.120}$$

Isolated surface disturbance over a flat bottom: $\delta \ll \epsilon^{1/2}$, solutions with shocks

If we restrict attention to the special case $\delta \ll \epsilon^{1/2} (\kappa \ll 1)$, then the third derivative terms in the evolution equations are absent, and the exact solution is straightforward. Let us consider the initial-value problem studied in Sec. 8.4.4 of [6.17]. We take the continuous, piecewise linear, surface disturbance

$$h^*(x) = \begin{cases} 2x - 1, & 0 \le x \le 1/2 \\ -1 - 2x, & -1/2 \le x \le 0 \\ 0, & |x| \ge 1/2 \end{cases} \tag{6.2.121}$$

and $u^*(x) = 0$. Thus, initially, the water is at rest, and the free surface has a triangular depression over $-1/2 \leq x \leq 1/2$. From (6.2.107), (6.2.109) we see that

$$f_0(\xi_1, 0) = -\frac{1}{2}h_1^*(\xi_1), \quad g_0(\xi_2, 0) = -\frac{1}{2}h_1^*(\xi_2). \tag{6.2.122}$$

The evolution equations for f_0 and g_0 in the limiting case $\kappa = 0$ are the decoupled quasilinear first-order system

$$\frac{\partial f_0}{\partial \tilde{t}} - \frac{3}{2}f_0\frac{\partial f_0}{\partial \xi_1} = 0, \quad \frac{\partial g_0}{\partial \tilde{t}} + \frac{3}{2}g_0\frac{\partial g_0}{\partial \xi_2} = 0. \tag{6.2.123}$$

We need only solve the equation for f_0, as the solution for g_0 follows by symmetry from $g_0(\xi_2, \tilde{t}) = f_0(-\xi_2, \tilde{t})$.

In general, as well as for the example (6.2.121), characteristics of (6.2.123) will cross, and we need to introduce shocks that are consistent with the exact shock conditions (6.2.99). If we denote $[h] = h^+ - h^-$, $[u] = u^+ - u^-$ and use these in (6.2.99) we have two relations linking the five quantities C_s, u^+, u^-, h^+, and h^-. Eliminating u^+ from these two relations gives the condition

$$C_s^2 - 2C_s u^- + (u^-)^2 - \frac{h^+(h^+ + h^-)}{2h^-} = 0. \tag{6.2.124}$$

Let us expand C_s in a series in powers of ϵ

$$C_s = C_0 + \epsilon C_1 + O(\epsilon^2). \tag{6.2.125}$$

Since $C_s = (dx/dt)$ along a shock, we have

$$C_s = \frac{dx}{dt} = \begin{cases} \dfrac{d}{dt}(\xi_1 + t) = \epsilon\dfrac{d\xi_1}{d\tilde{t}} + 1 & \text{for } f_0 \\ \dfrac{d}{dt}(\xi_2 - t) = \epsilon\dfrac{d\xi_2}{d\tilde{t}} - 1 & \text{for } g_0. \end{cases} \tag{6.2.126}$$

Equating (6.2.125) and (6.2.126) gives $C_0 = \pm 1$. As expected, the shock speed equals the characteristic speed to leading order. To $O(\epsilon)$, we have $C_1 = (d\xi_1/d\tilde{t})$ and $C_1 = (d\xi_2/d\tilde{t})$ for f_0 and g_0, respectively.

Since the f_0 and g_0 disturbances evolve independently, we may set $g_0 = 0$ in calculating the jump condition for f_0. In this case, (6.2.116) gives

$$h^\pm = 1 - \epsilon f_0^\pm + O(\epsilon^2), \quad u^- = -\epsilon f_0^- + O(\epsilon^2). \tag{6.2.127}$$

Substituting these expansions and $C_s = 1 + \epsilon(d\xi_1/d\tilde{t}) + O(\epsilon^2)$ into (6.2.124) gives

$$\frac{d\xi_1}{d\tilde{t}} = -\frac{3}{4}(f_0^+ + f_0^-). \tag{6.2.128a}$$

Similarly, the shock condition for g_0 is

$$\frac{d\xi_2}{d\tilde{t}} = \frac{3}{4}(g_0^+ + g_0^-). \tag{6.2.128b}$$

Having derived the shock conditions in terms of the $\xi_1, \tilde{t}$ variables, we can now calculate the solution. The characteristic equations are

$$\frac{d\tilde{t}}{ds} = 1, \quad \frac{d\xi_1}{ds} = -\frac{3}{2} f_0, \quad \frac{df_0}{ds} = 0. \tag{6.2.129}$$

Solving these subject to $f_0(\xi_1, 0) = -\frac{1}{2} h^*(\xi_1)$ gives the following one-parameter family

$$\xi_1 = \frac{3}{4} h^*(\xi)\tilde{t} + \xi, \quad f_0 = -\frac{1}{2} h^*(\xi), \tag{6.2.130}$$

where ξ is the parameter that fixes each characteristic. As seen in Figure 6.2.3(a), the straight characteristics emerging from the interval $-1/2 \leq \xi_1 \leq 0, \tilde{t} = 0$, all intersect at the point $\xi_1 = -1/2, \tilde{t} = 2/3$. Therefore, a strict (shock-free) solution exists only for $\tilde{t} < 2/3$. This is given by

$$f_0(\xi_1, \tilde{t}) = \begin{cases} \dfrac{1-2\xi_1}{2+3\tilde{t}}, & -\dfrac{3}{4}\tilde{t} \leq \xi_1 \leq \dfrac{1}{2} \\ \dfrac{1+2\xi_1}{2-3\tilde{t}}, & -\dfrac{1}{2} \leq \xi_1 \leq -\dfrac{3}{4}\tilde{t} \\ 0, & |\xi_1| \geq \dfrac{1}{2}. \end{cases} \tag{6.2.131}$$

For $\tilde{t} \geq 2/3$ we introduce the shock satisfying (6.2.128a), where $f_0^+ = (1 - 2\xi_1)/(2+3\tilde{t})$ and $f_0^- = 0$. The solution of (6.2.128a) subject to the initial condition $\xi_1 = -1/2$ at $\tilde{t} = 2/3$ defines the shock

$$\xi_1 = \frac{1}{2} - \frac{1}{2}(2 + 3\tilde{t})^{1/2} \equiv S(\tilde{t}). \tag{6.2.132}$$

Note that this shock becomes stationary, $(d\xi_1/d\tilde{t}) \to 0$, as $\tilde{t} \to \infty$. As indicated in Figure 6.2.3(b), the solution for f_0 for $\tilde{t} > 2/3$ is $f_0 = 0$ to the left of the shock $S(\tilde{t})$ and $f_0 = f_0^+$ to the right of $S(\tilde{t})$.

For a more general concave function $h^*(x)$ with $h^*(\pm 1/2) = 0, h^{*\prime\prime} > 0$, e.g., $h^* = 4x^2 - 1$, the characteristics emerging from $-\frac{1}{2} \leq \xi_1 \leq 0$ do not intersect at a point. This situation is illustrated later for the example of an isolated bottom disturbance.

Consider the qualitative behavior of the solution for h to $O(\epsilon)$. The initial surface disturbance h^* splits into two identical waves, one propagating to the right $(-\epsilon f_0)$ and one propagating to the left $(-\epsilon g_0)$. If $t = O(1)$, i.e., $\tilde{t} = O(\epsilon)$, we may ignore the $\tilde{t}$ dependence in f_0 and g_0 and have the result predicted by linear theory; the f_0 and g_0 waves, respectively, propagate unchanged with speed ± 1 in x. Over longer periods, i.e., $\tilde{t} = O(1)$, the front parts of the f_0 and g_0 waves steepen, and a shock is formed at $t = 2/3\epsilon$. This nonlinear evolution occurs over the $\tilde{t}$ scale, and the two waves do not interact.

The shock trajectories in the xt plane are defined to $O(1)$ only for $\tilde{t} = O(1)$. Expressing (6.2.132) in terms of the x, t variables gives

$$x = t + \frac{1}{2} - \frac{1}{2}(2 + 3\epsilon t)^{1/2} + O(\epsilon), \quad t > 2/3\epsilon. \tag{6.2.133}$$

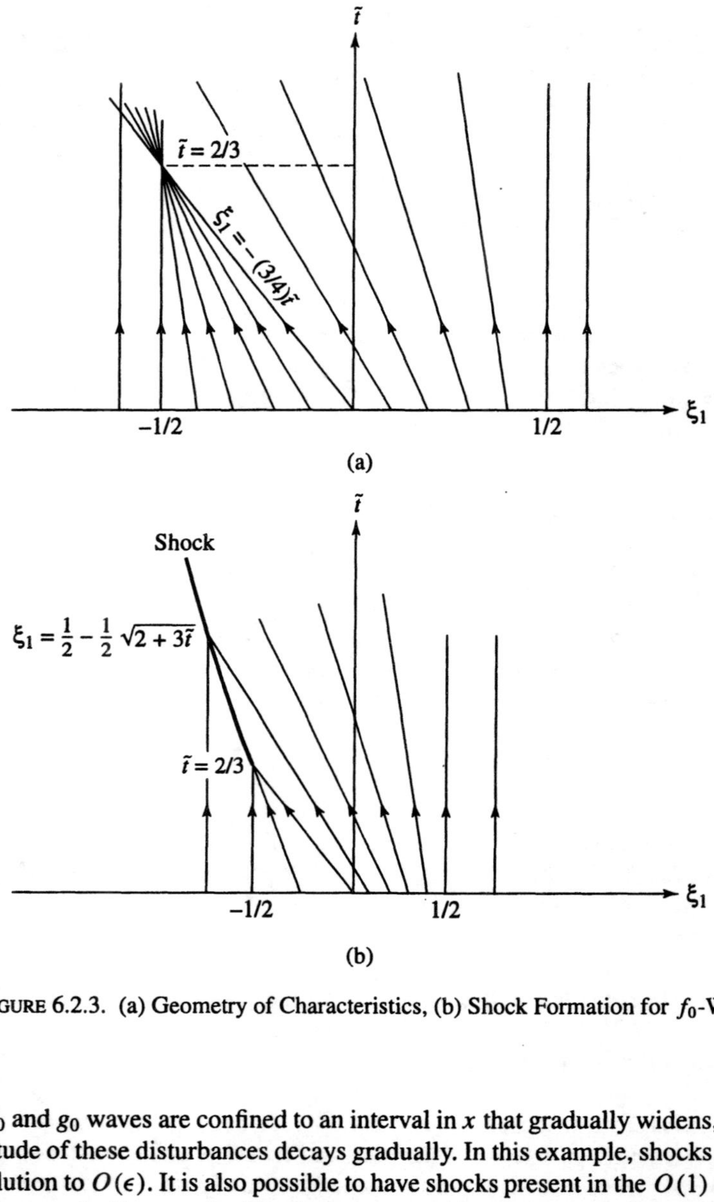

FIGURE 6.2.3. (a) Geometry of Characteristics, (b) Shock Formation for f_0-Wave

The f_0 and g_0 waves are confined to an interval in x that gradually widens, and the amplitude of these disturbances decays gradually. In this example, shocks arose in the solution to $O(\epsilon)$. It is also possible to have shocks present in the $O(1)$ solution as, for example, if we study disturbances to a uniformly propagating $O(1)$ shock. A number of cases are studied in detail in [6.29].

As pointed out in [6.17], we cannot regard the limiting solution we have found here for $\kappa = 0$ as an outer limit of the Korteweg–de Vries equation (6.2.112); the third derivative term multiplied by κ precludes an interior-layer solution of (6.2.112) in a small (relative to κ) neighborhood of a shock. This behavior is in contrast to that for Burgers' equation, where the small parameter multiplies a second derivative term. Conservation laws with small dissipation lead to evolution equations of Burgers' type as discussed, for example, in [6.13] and in Sec. 6.2.5.

Isolated bottom disturbance: $F \neq 1$

We now consider the effect of an isolated bottom disturbance, e.g., a stationary bump, in a coordinate system where the undisturbed flow is $h = 1$ and $u = F =$ constant > 0. This problem was studied in [6.19] for the case where the Froude number F is either a constant or a slowly varying function of time to illustrate the solution behavior near the critical value $F = 1$. Here, we restrict attention to the simpler case $F =$ constant for which the flow equations are given by (6.2.103). To simplify the notation, we drop the overbars and denote the independent variables by x and t in (6.2.103).

At time $t = 0^-$ the water is stationary ($u = 0$), and the surface height measured from the bottom is given by $h = 1 - \epsilon B$, where B is an isolated bump, e.g.,

$$B(x) = \begin{cases} 1 - 4x^2, & |x| \leq 1/2 \\ 0, & |x| \geq 1/2. \end{cases} \tag{6.2.134}$$

At time $t = 0^+$, we set the entire body of water in motion to the right, i.e.,

$$h(x, 0^+; \epsilon) = 1 - \epsilon B(x), \quad u(x, 0^+; \epsilon) = F. \tag{6.2.135}$$

Denoting $u_1 = h$, $u_2 = u$, $u_1 = 1 + \epsilon w_1$, and $u_2 = F + \epsilon w_2$, we obtain the following equations for w_1 and w_2 from (6.2.103):

$$\frac{\partial w_1}{\partial t} + F\frac{\partial w_1}{\partial x} + \frac{\partial w_2}{\partial x} = -\epsilon\frac{\partial}{\partial x}(w_1 w_2), \tag{6.2.136a}$$

$$\frac{\partial w_2}{\partial t} + \frac{\partial w_1}{\partial x} + F\frac{\partial w_2}{\partial x} = -B'(x)$$

$$-\epsilon\frac{\partial}{\partial x}\left[\frac{w_2^2}{2} + \frac{\kappa^2}{2}F^2 B''(x) + \frac{\kappa^3}{3}D(w_1)\right]. \tag{6.2.136b}$$

Here D is the second-order operator

$$D \equiv F^2\frac{\partial^2}{\partial x^2} + 2F\frac{\partial^2}{\partial x \partial t} + \frac{\partial^2}{\partial t^2}, \tag{6.2.137}$$

and we have set $\delta^2 = \kappa^2\epsilon$. The initial conditions are

$$w_1(x, 0; \epsilon, \kappa, F) = -B(x), \quad w_2(x, 0; \epsilon, \kappa, F) = 0. \tag{6.2.138}$$

The eigenvalues of $A^{(0)}$ (see (6.2.33)) are $\lambda_1 = F + 1$, $\lambda_1 = F - 1$, and we have the same transformation, (6.2.34), as for channel flow, i.e., $w_1 = -U_1 - U_2$, $w_2 = -U_1 + U_2$. The equations for U_1 and U_2 that result from (6.2.136) are

$$\frac{\partial U_1}{\partial t} + (F + 1)\frac{\partial U_1}{\partial x} - \frac{1}{2}B'(x) = \frac{\epsilon}{2}\frac{\partial}{\partial x}\left[\frac{3}{2}U_1^2 - U_1 U_2 - \frac{1}{2}U_2^2\right.$$

$$+ \frac{\kappa^2}{2} F^2 B''(x) - \frac{\kappa^2}{3} D(U_1 + U_2)\Bigg] \equiv \epsilon \frac{\partial}{\partial x} R_1(U_1, U_2, x), \quad (6.2.139a)$$

$$\frac{\partial U_2}{\partial t} + (F-1)\frac{\partial U_2}{\partial x} + \frac{1}{2} B'(x) = \frac{\epsilon}{2}\frac{\partial}{\partial x}\Bigg[\frac{1}{2} U_1^2 + U_1 U_2 - \frac{3}{2} U_2^2$$

$$- \frac{\kappa^2}{2} F^2 B''(x) + \frac{\kappa^2}{3} D(U_1 + U_2)\Bigg] \equiv \epsilon \frac{\partial}{\partial x} R_2(U_1, U_2, x). \quad (6.2.139b)$$

This is essentially in the standard form (6.2.19) with the added terms $\pm B'(x)/2$ to the left-hand sides and the occurrence of $B''(x)$ in the right-hand sides. The initial conditions are

$$U_1(x, 0; \epsilon, \kappa, F) = \frac{B(x)}{2}, \quad U_2(x, 0; \epsilon, \kappa, F) = \frac{B(x)}{2}. \quad (6.2.140)$$

We now expand U_1 and U_2

$$U_i(x, t; \epsilon, \kappa, F) = U_{i0}(\xi_1, \xi_2, \tilde{t}; \kappa, F) + \epsilon U_{i1}(\xi_1, \xi_2, \tilde{t}; \kappa, F) + O(\epsilon^2), \quad i = 1, 2, \quad (6.2.141)$$

where $\xi_1 = x - (F+1)t, \xi_2 = x - (F-1)t$. Derivatives transform as follows:

$$\frac{\partial}{\partial x} = \frac{\partial}{\partial \xi_1} + \frac{\partial}{\partial \xi_2}, \quad \frac{\partial}{\partial t} = -(F+1)\frac{\partial}{\partial \xi_1} - (F-1)\frac{\partial}{\partial \xi_2} + \epsilon \frac{\partial}{\partial \tilde{t}}. \quad (6.2.142)$$

Therefore, the U_{i0} and U_{i1} satisfy

$$2\frac{\partial U_{10}}{\partial \xi_2} - \frac{1}{2} B'(x) = 0, \quad -2\frac{\partial U_{20}}{\partial \xi_1} - \frac{1}{2} B'(x) = 0, \quad (6.2.143)$$

$$2\frac{\partial U_{11}}{\partial \xi_2} = -\frac{\partial U_{10}}{\partial \tilde{t}} + \left(\frac{\partial}{\partial \xi_1} + \frac{\partial}{\partial \xi_2}\right) R_1(U_{10}, U_{20}, x), \quad (6.2.144a)$$

$$-2\frac{\partial U_{21}}{\partial \xi_1} = -\frac{\partial U_{20}}{\partial \tilde{t}} + \left(\frac{\partial}{\partial \xi_1} + \frac{\partial}{\partial \xi_2}\right) R_2(U_{10}, U_{20}, x). \quad (6.2.144b)$$

The solution of (6.2.143) is easily found in the form

$$U_{10} = f_0(\xi_1, \tilde{t}; \kappa, F) + \frac{B(x)}{2(F+1)}, \quad (6.2.145a)$$

$$U_{20} = g_0(\xi_2, \tilde{t}; \kappa, F) - \frac{B(x)}{2(F-1)}. \quad (6.2.145b)$$

The initial conditions (6.2.140) imply that the unknown functions f_0 and g_0 must satisfy

$$f_0(x, 0; \kappa, F) = \frac{FB(x)}{2(F+1)}; \quad g_0(x, 0; \kappa, F) = \frac{FB(x)}{2(F-1)}. \quad (6.2.146)$$

We note the singularity in U_{20} for the critical Froude number $F = 1$. If $F \approx 1$, our results break down and we need a different expansion, as will be discussed. The expansion (6.2.141) is thus restricted to values of F such that $F - 1 = O_s(1)$.

To derive the evolution equations for f_0 and g_0, we examine the solutions for U_{11} and U_{21}. The details are entirely analogous to those for the previous example (surface disturbance) and are not repeated. In fact, we find *exactly* the same evolution equations (6.2.112) in this case. Once the evolution equations are solved for f_0 and g_0, we can express the solution for the physical variables h and u in the form

$$h(x, t; \epsilon, \kappa, F) = 1 + \epsilon \left[\frac{B(x)}{F^2 - 1} - f_0 - g_0 \right] + O(\epsilon^2), \qquad (6.2.147a)$$

$$u(x, t; \epsilon, \kappa, F) = F + \epsilon \left[-\frac{FB(x)}{F^2 - 1} - f_0 + g_0 \right] + O(\epsilon^2). \quad (6.2.147b)$$

Thus, the solution to $O(\epsilon)$ consists of three components: a stationary disturbance over the bump plus waves propagating to the right (f_0) and left (g_0).

If we set $\kappa \equiv 0$, the evolution equations (6.2.123) that result can be solved explicitly for the initial conditions (6.2.146). We find

$$f_0 = \frac{F}{2(F+1)} B(\overline{\xi}_1), \quad g_0 = \frac{F}{2(F-1)} B(\overline{\xi}_2), \qquad (6.2.148)$$

where $\overline{\xi}_1$ and $\overline{\xi}_2$ are constants along the ξ_1 and ξ_2 characteristics, respectively. The characteristics are defined implicitly by

$$\overline{\xi}_1 - \xi_1 - \frac{3F}{4(F+1)} B(\overline{\xi}_1)\tilde{t} = 0, \qquad (6.2.149a)$$

$$\overline{\xi}_2 - \xi_2 + \frac{3F}{4(F-1)} B(\overline{\xi}_1)\tilde{t} = 0. \qquad (6.2.149b)$$

For a given $B(x)$, one solves (6.2.149a) for $\overline{\xi}_1$ in terms of ξ_1 and $\tilde{t}$ and then substitutes this expression into the first equation (6.2.148) to obtain f_0 as a function of ξ_1 and $\tilde{t}$. Similarly, the solution of (6.2.149b) for $\overline{\xi}_2$, when used in the second equation (6.2.148), defines g_0 as a function of ξ_2 and $\tilde{t}$.

The solution for $\overline{\xi}_1$ in terms of ξ_1 and $\tilde{t}$ is unique as long as the one-parameter family of straight lines, (6.2.149a), in the $\xi_1\tilde{t}$ plane does not envelop. A necessary condition for envelopment to occur is that

$$\frac{dB}{d\overline{\xi}_1} = \frac{4(F+1)}{3F\tilde{t}} \qquad (6.2.150)$$

have a real solution for $\overline{\xi}_1$ as a function of $\tilde{t}$ for $\tilde{t} > 0$. Thus, the $\overline{\xi}_1$ characteristics will envelop only if $B' > 0$. Similar remarks apply for the $\overline{\xi}_2$ characteristics. In particular, these envelop if $B' < 0$. When characteristics envelop, we must introduce shocks as discussed later.

Since $B = 0$ if $|x| \geq 1/2$, the three components of the solution: the stationary disturbance over the bump and the f_0, g_0 waves, eventually separate if $F \neq 1$. (We will see that a more precise condition for the separation of the three components is $|F - 1|\epsilon^{-1/2} > \sqrt{3/2}$.) In particular, for sufficiently large $\tilde{t}$, all that remains of the solution over the unit interval $|x| \leq 1/2$ is the *stationary* disturbance

$$h_s(x; \epsilon, F) = 1 + \epsilon \frac{B(x)}{F^2 - 1} + O(\epsilon^2), \tag{6.2.151a}$$

$$u_s(x; \epsilon, F) = F - \epsilon \frac{FB(x)}{F^2 - 1} + O(\epsilon^2). \tag{6.2.151b}$$

It is interesting to verify that the above is the correct asymptotic expansion of the exact steady solution of (6.2.103) with $\kappa = 0$. If we set $\partial/\partial t = 0$, $\kappa = 0$ in (6.2.103) and assume the boundary conditions $u = F$, $h = 1$ at $x = -1/2$, we calculate the following exact integrals:

$$uh = F, \quad \frac{u^2}{2} + h + \epsilon B = 1 + \frac{F^2}{2}. \tag{6.2.152}$$

If these two algebraic relations have real solutions $h_s(x; \epsilon, F)$ and $u_s(x; \epsilon, F)$, these represent a steady state over $|x| \leq 1/2$. Eliminating h from (6.2.152) gives the cubic

$$G(u_s; \epsilon, F) \equiv \frac{u_s^3}{2} - \left[\frac{F^2}{2} + 1 - \epsilon B(x)\right] u_s + F = 0. \tag{6.2.153}$$

The stationary points corresponding to $(\partial G/\partial u_s) = 0$ are

$$u_s^{(0)} = \pm\left[\frac{2}{3}\left(\frac{F^2}{2} + 1 - \epsilon B\right)\right]^{1/2}. \tag{6.2.154}$$

It then follows that (with $B_{\max} = 1$) the critical values of F between which (6.2.153) has no positive real root satisfy

$$\frac{F_c^2}{2} - F_c^{2/3} + \frac{2}{3}(1 - \epsilon) = 0 \tag{6.2.155a}$$

or

$$F_c = 1 \pm \epsilon^{1/2}\sqrt{\frac{3}{2}} + O(\epsilon). \tag{6.2.155b}$$

For $F > 1 + \epsilon^{1/2}\sqrt{3/2} + O(\epsilon)$, (6.2.153) has two real positive roots for u_s; these are conveniently characterized by their behavior as $F \to \infty$. It is easily seen that the smaller positive root has the behavior $u \sim 2/F$, which implies $h \sim F^2/2$ and therefore does not satisfy the boundary conditions at $x = \pm 1/2$. Conversely, the larger root given by (6.2.151b) satisfies (6.2.153) to $O(\epsilon)$, and we conclude that (6.2.151) is the asymptotic expansion of the exact steady solution (6.2.152).

When the characteristics (6.2.149) envelop, we need to introduce appropriate shocks for the f_0 and g_0 evolution equations. The analysis here is essentially the same as for the case discussed following (6.2.123). We observe that the two shock conditions (6.2.99) are invariant under the Galilean transformation (6.2.102). Thus,

(6.2.99) also hold for the system (6.2.103). Now

$$C_s = \frac{dx}{dt} = \begin{cases} (F+1) + \epsilon \dfrac{d\xi_1}{d\tilde{t}} & \text{for } f_0 \\ (F-1) + \epsilon \dfrac{d\xi_2}{d\tilde{t}} & \text{for } g_0, \end{cases} \tag{6.2.156}$$

and using (6.2.124) we obtain the same shock conditions as (6.2.128).

To illustrate ideas, consider the disturbance due to the parabolic bump $B(x)$ defined by (6.2.134). We compute the solution for f_0 in parametric form (see (6.2.148)–(6.2.149)):

$$f_0 = \frac{F(1-4\bar{\xi}_1^2)}{2(F+1)}, \quad \bar{\xi}_1 - \xi_1 - \frac{3F\tilde{t}}{4(F+1)}(1-4\bar{\xi}_1^2) = 0. \tag{6.2.157}$$

To express f_0 as a function of ξ_1 and $\tilde{t}$, we solve the quadratic expression in the second equation (6.2.157) for $\bar{\xi}_1$ and use the root (the other root is spurious)

$$\bar{\xi}_1 = \left[-1 + \sqrt{1 + 16c\tilde{t}(\xi_1 + c\tilde{t})}\right]/8c\tilde{t}, \tag{6.2.158}$$

where $c = 3F/4(F+1)$. Using (6.2.158) in the first equation (6.2.157) gives

$$f_0 = \left[\sqrt{1 + 16c\tilde{t}(\xi_1 + c\tilde{t})} - 1 - 8\xi_1 c\tilde{t}\right]/12c\tilde{t}^2, \tag{6.2.159}$$

and we verify that as $\tilde{t} \to 0$ this result tends to the correct initial condition (6.2.146). The solution is unique in the interval $0 \leq \tilde{t} \leq \tilde{t}_0 \equiv (F+1)/3F$. As $\tilde{t}$ increases, the portion of the wave lying between the characteristics $\bar{\xi}_1 = -0.5$ and $\bar{\xi}_1 = 0$ steepens, as shown in Figure 6.2.4. At the point $\xi_1 = -1/2, \tilde{t} = \tilde{t}_0$ (denoted by A), the characteristics begin to envelop. At this point, we introduce the shock defined by (6.2.128a) with f_0^+ given by (6.2.159) and $f_0^- = 0$. This differential equation defines a unique shock ABC starting from A and having the shape shown in Figure 6.2.3 for $F = 2$. Initially, the jump in f_0 at A is zero; it increases monotonically and reaches a maximum at the inflection point B. This is the point where the characteristic emanating from the origin intersects the shock. As $\tilde{t}$ increases beyond B, the jump in f_0 decreases monotonically and tends to zero as $\tilde{t} \to \infty$. It also follows from (6.2.128a) that the shock curve for $-\xi_1$ grows at a rate proportional to $\tilde{t}^{1/2}$ as $\tilde{t} \to \infty$.

The results for g_0 can be derived from the above by noting that the two problems are equivalent if we transform $\bar{\xi}_1 \to -\bar{\xi}_2, \xi_1 \to -\xi_2, \tilde{t} \to \tilde{t}, c \to 3F/4(F-1)$. Thus, aside from the change in the value of the constant c, the behavior of the g_0 wave is essentially found by reflecting the curves in Figure 6.2.3 relative to the $\tilde{t}$ axis. A numerical verification of these results is given in Sec. 3.1.2 of [6.19].

Isolated bottom disturbance $F \approx 1$

The basic reason for the nonuniformity in the solution (6.2.147) as $F \to 1$ can be traced to the contradictory implicit assumptions in this case that: (i) surface height

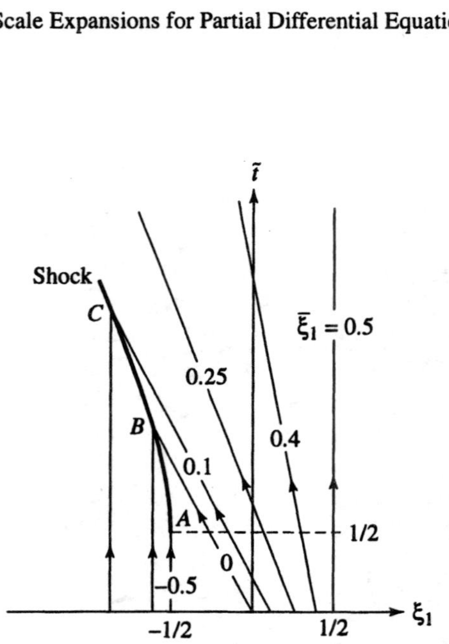

FIGURE 6.2.4. Shock Formation for the f_0-Wave, $F = 2$

and speed disturbances are of the same order, $O(\epsilon)$, as the bottom disturbance, and (ii) these disturbances propagate with speeds close to the characteristic speeds $F+1$ and $F-1$. For $F \approx 1$, the $F-1$ characteristic speed is small, i.e., the associated disturbance remains nearly stationary over the bump. This, in turn, implies that perturbations over the bump grow with time in contradiction to the assumed order of the free surface and speed perturbations. The above situation is very similar to the breakdown of supersonic small disturbance theory when the Mach number M is close to unity. See Problem 7 and chapter 2 of [6.7].

In order to establish the appropriate scales when $F \approx 1$, we set

$$F = 1+\epsilon^{\lambda}F^*, \quad \delta = \epsilon^{\alpha}\kappa^*, \quad h = 1+\epsilon^{\beta}w_1^*, \quad u = 1+\epsilon^{\lambda}F^*+\epsilon^{\beta}w_2^*, \tag{6.2.160}$$

where λ, α, and β are unknown positive constants to be determined, and F^*, κ^* are arbitrary $O(1)$ constants. The equations that follow from (6.2.103) (dropping overbars) for w_1^* and w_2^* are

$$\frac{\partial w_1^*}{\partial t} + \frac{\partial w_1^*}{\partial x} + \frac{\partial w_2^*}{\partial x} = -\epsilon^{\lambda}F^*\frac{\partial w_1^*}{\partial x} - \epsilon^{\beta}\frac{\partial}{\partial x}(w_1^* w_2^*), \tag{6.2.161a}$$

$$\frac{\partial w_2^*}{\partial t} + \frac{\partial w_1^*}{\partial x} + \frac{\partial w_2^*}{\partial x} = -\epsilon^{\lambda}F^*\frac{\partial w_2^*}{\partial x} - \epsilon^{1-\beta}B'(x) - \frac{\partial}{\partial x}\left[\frac{\epsilon^{\beta}}{2}w_2^{*^2} + \epsilon^{2\alpha}\frac{\kappa^{*^2}}{3}D^*(w_1^*)\right] + O\left(\epsilon^{2\alpha+\lambda}\right) + O\left(\epsilon^{2\alpha+1-\beta}\right), \tag{6.2.161b}$$

where (see (6.2.137))

$$D^* = \frac{\partial^2}{\partial x^2} + 2\frac{\partial^2}{\partial x \partial t} + \frac{\partial^2}{\partial t^2}.$$

We see that the richest equations result for $\lambda = \beta = 1 - \beta = 2\alpha$, i.e., $\lambda = \beta = 1/2$ and $\alpha = 1/4$. With this choice, we obtain the following governing system in the standard form (6.2.19) for $w_1^* = -U_1^* - U_2^*$, $w_2^* = -U_1^* + U_2^*$:

$$\frac{\partial U_1^*}{\partial t} + 2\frac{\partial U_1^*}{\partial x} = \frac{\epsilon^{1/2}}{2}\frac{\partial}{\partial x}\left[-2F^*U_1 + B(x) + \frac{3}{2}U_1^{*^2} - U_1^*U_2^* - \frac{1}{2}U_2^{*^2} - \frac{\kappa^{*^2}}{3}D^*(U_1^* + U_2^*)\right] + O(\epsilon) \equiv \epsilon^{1/2}\frac{\partial}{\partial x}R_1^*(U_1^*, U_2^*, x) + O(\epsilon), \tag{6.2.162a}$$

$$\frac{\partial U_2^*}{\partial t} = \frac{\epsilon^{1/2}}{2}\frac{\partial}{\partial x}\left[-2F^*U_2^* - B(x) + \frac{1}{2}U_1^* + U_1^*U_2^* - \frac{3}{2}U_2^{*^2} + \frac{\kappa^{*^2}}{3}D(U_1^* + U_2^*)\right] + O(\epsilon) = \epsilon^{1/2}\frac{\partial}{\partial x}R_2^*(U_1^*, U_2^*, x) + O(\epsilon). \tag{6.2.162b}$$

The initial conditions (6.2.135) imply that U_1^* and U_2^* are $O(\epsilon^{1/2})$ initially.

We expand U_1^* and U_2^* in multiple-scale form

$$U_i^*(x, t; \epsilon, \kappa^*, F^*) = U_{i0}^*(\xi_1^*, \xi_2^*, t^*; \kappa^*, F^*) + \epsilon^{1/2}U_{i1}(\xi_1^*, \xi_2^*, t^*; \kappa^*, F^*) + O(\epsilon), \tag{6.2.163}$$

where $\xi_1^* = x - 2t$, $\xi_2^* = x$, and $t^* = \epsilon^{1/2}t$. Equations (6.2.162) give

$$2\frac{\partial U_{10}^*}{\partial \xi_2^*} = 0, \quad -2\frac{\partial U_{20}^*}{\partial \xi_1^*} = 0. \tag{6.2.164}$$

$$2\frac{\partial U_{11}^*}{\partial \xi_2^*} = -\frac{\partial U_{10}^*}{\partial t^*} + \left(\frac{\partial}{\partial \xi_1^*} + \frac{\partial}{\partial \xi_2^*}\right)R_1^*(U_{10}^*, U_{20}^*, x), \tag{6.2.165a}$$

$$-2\frac{\partial U_{21}^*}{\partial \xi_1^*} = -\frac{\partial U_{20}^*}{\partial t^*} + \left(\frac{\partial}{\partial \xi_1^*} + \frac{\partial}{\partial \xi_2^*}\right)R_2^*(U_{10}^*, U_{20}^*, x). \tag{6.2.165b}$$

When the solution $U_{10}^* = f_0^*(\xi_1^*, t^*; \kappa^*, F^*)$, $U_{20}^* = g_0^*(\xi_2^*, t^*; \kappa^*, F^*)$ of (6.2.164) is used in (6.2.165) and inconsistent terms are removed, we obtain the evolution equations

$$\frac{\partial f_0^*}{\partial t^*} + \left(F^* - \frac{3}{2}f_0^*\right)\frac{\partial f_0^*}{\partial \xi_1^*} + \frac{\kappa^{*^2}}{6}\frac{\partial^3 f_0^*}{\partial \xi_1^{*^3}} = 0, \tag{6.2.166a}$$

$$\frac{\partial g_0^*}{\partial t^*} + \left(F^* + \frac{3}{2}g_0^*\right)\frac{\partial g_0^*}{\partial \xi_2^*} - \frac{\kappa^{*^2}}{6}\frac{\partial^3 g_0^*}{\partial \xi_2^{*^3}} = B'(x). \tag{6.2.166b}$$

The solution of (6.2.166a) with $f_0^*(\xi_1^*, 0; \kappa^*, F^*) = 0$ is $f_0^* \equiv 0$. Equation (6.2.166b) is a *forced* Korteweg–de Vries equation which must be solved numerically for the initial condition $g_0^*(\xi_2^*, 0; \kappa^*, F^*) = 0$. Since $f_0^* = 0$, h and u depend only on g_0^* to $O(\epsilon^{1/2})$ and are given by

$$h = 1 - \epsilon^{1/2} g_0^*(x, t^*; \kappa^*, F^*) + O(\epsilon),$$
$$u = 1 + \epsilon^{1/2}[F^* + g_0^*(x, t^*; \kappa^*, F^*) + O(\epsilon).$$

Expressing (6.2.166b) in terms of u and h as functions of x, t gives

$$h_t + \left(F + \frac{1}{2} - \frac{3}{2}h\right) h_x - \frac{\delta^2}{6} h_{xxx} = -y_b'(x), \qquad (6.2.167a)$$

$$u_t - \left(1 + \frac{F}{2} - \frac{3}{2}u\right) u_x - \frac{\delta^2}{6} u_{xxx} = y_b'(x) \qquad (6.2.167b)$$

to leading order. For further details, including solutions with $\kappa^* = 0$, comparisons with numerical calculations, and other references to this problem, the reader is referred to [6.19].

Signaling problem

We conclude our discussion of shallow water waves with the problem of an idealized wavemaker introducing a small disturbance at one end of a semi-infinite body of water of constant depth at rest. Thus, the governing equations are (6.2.98) with $b = 0$. The initial conditions are $h = 1, u = 0$ at $t = 0$ $x > 0$, and the boundary condition representing the idealized wavemaker is

$$u(\epsilon x_w(t), t; \epsilon) = \epsilon \dot{x}_w(t), \quad t > 0. \qquad (6.2.168)$$

Here $\epsilon x_w(t)$ is the horizontal displacement of the wavemaker, assumed to be small, and (6.2.168) states that the horizontal component of the water velocity at the wavemaker must equal the wavemaker velocity.

The transformation to standard form proceeds as for (6.2.106), and if we set $\kappa = 0$ for simplicity, we have

$$\frac{\partial U_1}{\partial t} + \frac{\partial U_1}{\partial x} = \epsilon \frac{\partial}{\partial x}\left[\frac{3}{4}U_1^2 - \frac{1}{2}U_1 U_2 - \frac{1}{4}U_2^2\right], \qquad (6.2.169a)$$

$$\frac{\partial U_2}{\partial t} - \frac{\partial U_2}{\partial x} = \epsilon \frac{\partial}{\partial x}\left[\frac{1}{4}U_1^2 + \frac{1}{2}U_1 U_2 - \frac{3}{4}U_2^2\right]. \qquad (6.2.169b)$$

The initial conditions are

$$U_1(x, 0; \epsilon) = 0; \quad U_2(x, 0; \epsilon) = 0, \quad x > 0, \qquad (6.2.170)$$

and the wavemaker boundary condition becomes

$$-U_1(\epsilon x_w(t), t; \epsilon) + U_2(\epsilon x_w(t), t; \epsilon) = \dot{x}_w(t). \qquad (6.2.171)$$

For the signaling problem, it is more convenient to use $\zeta_1 = t - x = -\xi_1$ instead of ξ_1. More importantly, the slow variable is now $\tilde{x} = \epsilon x$ instead of ϵt.

Thus, we expand U_i in the following multiple-scale form:

$$U_i(x,t;\epsilon) = U_{i0}(\zeta_1,\zeta_2,\tilde{x}) + \epsilon U_{i1}(\zeta_1,\zeta_2,\tilde{x}) + O(\epsilon^2), \tag{6.2.172}$$

where

$$\zeta_1 = t - x, \quad \zeta_2 = t + x. \tag{6.2.173}$$

Derivatives transform according to

$$\frac{\partial}{\partial x} = -\frac{\partial}{\partial \zeta_1} + \frac{\partial}{\partial \zeta_2} + \epsilon \frac{\partial}{\partial \tilde{x}}, \quad \frac{\partial}{\partial t} = \frac{\partial}{\partial \zeta_1} + \frac{\partial}{\partial \zeta_2}, \tag{6.2.174}$$

and we find the following equations governing the U_{i0} and U_i:

$$2\frac{\partial U_{10}}{\partial \zeta_2} = 0, \quad 2\frac{\partial U_{20}}{\partial \zeta_1} = 0. \tag{6.2.175}$$

$$2\frac{\partial U_{11}}{\partial \zeta_2} = -\frac{\partial U_{10}}{\partial \tilde{x}} + \left(\frac{\partial}{\partial \zeta_2} - \frac{\partial}{\partial \zeta_1}\right)\left(\frac{3}{4}U_{10}^2 - \frac{1}{2}U_{10}U_{20} - \frac{1}{4}U_{20}\right), \tag{6.2.176a}$$

$$2\frac{\partial U_{21}}{\partial \zeta_1} = \frac{\partial U_{20}}{\partial \tilde{x}} + \left(\frac{\partial}{\partial \zeta_2} - \frac{\partial}{\partial \zeta_1}\right)\left(\frac{1}{4}U_{10}^2 + \frac{1}{2}U_{10}U_{20} - \frac{3}{4}U_{20}^2\right). \tag{6.2.176b}$$

We express the solution of (6.2.175) in the form

$$U_{10} = f_0(\zeta_1,\tilde{x}); \quad U_{20} = g_0(\zeta_2,\tilde{x}), \tag{6.2.177}$$

where the initial conditions are

$$f_0(-x,\tilde{x}) = 0, \quad g_0(x,\tilde{x}) = 0, \quad x > 0. \tag{6.2.178}$$

The boundary condition (6.2.171) gives

$$-U_{10}(t,t,0) + U_{20}(t,t,0) = \dot{x}_w(t), \quad t > 0, \tag{6.2.179}$$

i.e.,

$$-f_0(t,0) + g_0(t,0) = \dot{x}_w(t), \quad t > 0. \tag{6.2.180}$$

When we substitute the expressions given by (6.2.177) into the right-hand sides of (6.2.176) and remove inconsistent terms in U_{11} and U_{21}, we find the following evolution equations for f_0 and g_0:

$$\frac{\partial f_0}{\partial \tilde{x}} + \frac{3}{2}f_0\frac{\partial f_0}{\partial \zeta_1} = 0, \tag{6.2.181a}$$

$$\frac{\partial g_0}{\partial \tilde{x}} - \frac{3}{2}g_0\frac{\partial g_0}{\partial \zeta_2} = 0. \tag{6.2.181b}$$

Consider (6.2.181b) first. The transformation from the (x,t) to $(\zeta_2,\tilde{x})$ variables is given by $\zeta_2 = t + x$, $\tilde{x} = \epsilon x$. Thus, the $x > 0$ axis ($t = 0$) corresponds to the line $\zeta_2 = \tilde{x}/\epsilon$, $\tilde{x} > 0$, whereas the $t > 0$ axis ($x = 0$) corresponds to the line $\zeta_2 > 0$, $\tilde{x} = 0$. The domain $x > 0$, $t > 0$ is therefore the triangular region:

$\zeta_2 > 0, 0 < \tilde{x} < \epsilon\zeta_2$, as shown in Figure 6.2.5(a). The solution of (6.2.181b) subject to the initial condition $g_0(x, \tilde{x}) = 0, x > 0$ gives $g_0(\zeta_2, \tilde{x}) \equiv 0$. In particular, $g_2(t, 0) = 0$ for $t > 0$ and the boundary condition (6.2.180) reduces to

$$f_0(t, 0) = -\dot{x}_w(t), \quad t > 0. \tag{6.2.182}$$

Now we can solve (6.2.181a). The positive x axis maps to the line $\zeta_1 = -\tilde{x}/\epsilon$, $\tilde{x} > 0$ in the $\tilde{x}\zeta_1$ plane, whereas the positive t axis maps to the positive ζ_1 axis. The domain of interest is thus $\tilde{x} > 0$ for $\zeta_1 > 0$, and $\tilde{x} > -\epsilon\zeta_1$ for $\zeta_1 < 0$, as shown in Figure 6.2.5(b). The solution of (6.2.181a) is $f_0(\zeta_1, \tilde{x}) \equiv 0$ for $\zeta_1 < 0$, $\tilde{x} > 0$.

In order to derive an explicit result for $\zeta_1 > 0$, let us choose the monotone wavemaker displacement

$$x_w(t) = \begin{cases} t - \dfrac{t^2}{2}, & 0 < t \le 1 \\ \dfrac{1}{2}, & t \ge 1. \end{cases} \tag{6.2.183}$$

We then find

$$f_0(\zeta_1, \tilde{x}) = \frac{\zeta_1 - 1}{1 + 3\tilde{x}/2}. \tag{6.2.184}$$

The characteristics emanating from the $\zeta_1 > 0$ axis cross with the horizontal characteristics for $\zeta_1 < 0$, and we need to introduce a shock at the origin. The correct shock condition for (6.2.181a) is derived from the exact shock conditions (6.2.99) just as in (6.2.128). Along a shock in the $\tilde{x}\zeta_1$ plane, we have

$$\frac{d\zeta_1}{d\tilde{x}} = \frac{d\zeta_1/dt}{\epsilon dx/dt} = \frac{1 - (dx/dt)}{\epsilon(dx/dt)}. \tag{6.2.185a}$$

Therefore, $(d\zeta_1/d\tilde{x})$ along a shock is given by

$$\left(\frac{d\zeta_1}{d\tilde{x}}\right) = \frac{1 - C_s}{\epsilon C_s}, \tag{6.2.185b}$$

where $C_s = (dx/dt)$ is the shock speed in the xt plane. Solving (6.2.185b) gives

$$C_s = 1 - \epsilon\left(\frac{d\zeta_1}{d\tilde{x}}\right) + O(\epsilon^2). \tag{6.2.186}$$

If we use this expression for C_s in (6.2.124) and recall that $h^\pm$ and $u^\pm$ are given by (6.2.127), we find

$$\left(\frac{d\zeta_1}{d\tilde{x}}\right) = \frac{3}{4}(f_0^+ + f_0^-). \tag{6.2.187}$$

For the choice of x_w given by (6.2.183), we have $f_0^+ = 0$, and f_0^- is given by (6.2.184). Solving (6.2.187) for this case (and the initial condition $\zeta_1 = 0$ at $\tilde{x} = 0$) gives the parabolic shock

$$\zeta_1 = 1 - \sqrt{1 + 3\tilde{x}/2}. \tag{6.2.188a}$$

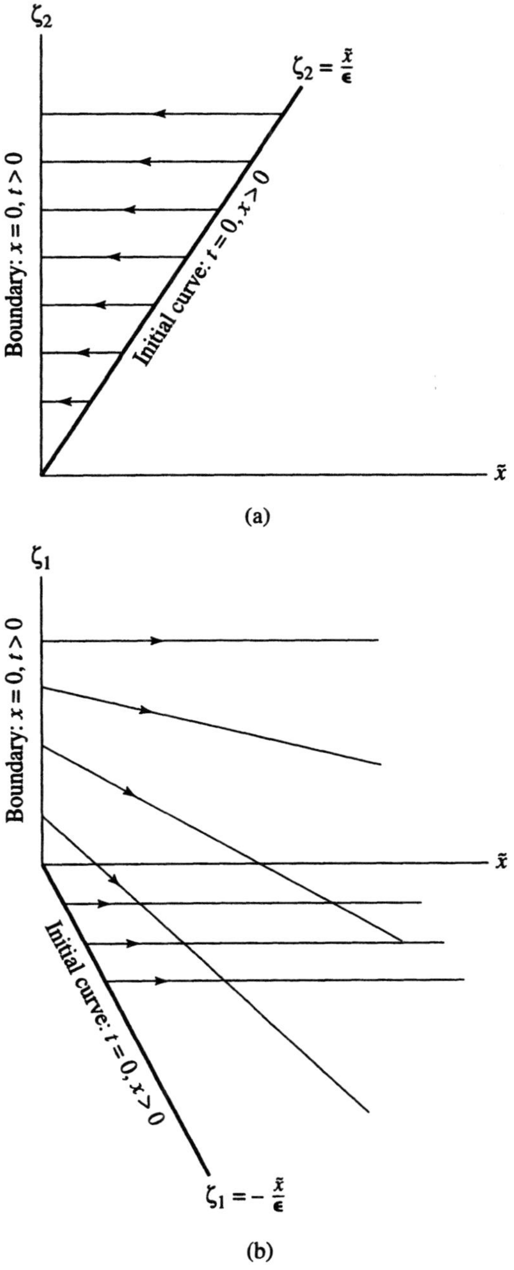

FIGURE 6.2.5. (a) $\tilde{x}\zeta_2$ plane, (b) $\tilde{x}\zeta_1$ plane

In the xt plane, the shock is given by

$$t = x + 1 - \sqrt{1 + 3\epsilon x/2}. \tag{6.2.188b}$$

Figure 6.2.6 shows the shock in both planes; Figure 6.2.6(b) is derived for $\epsilon = 0.5$. The solution for h and u behind the shock to $O(\epsilon)$ follows using (6.2.116):

$$h(x, t; \epsilon) = 1 + \epsilon \frac{1 + x - t}{1 + 3\epsilon x/2} + O(\epsilon^2), \tag{6.2.189a}$$

$$u(x, t; \epsilon) = \epsilon \frac{1 + x - t}{1 + 3\epsilon x/2} + O(\epsilon^2). \tag{6.2.189b}$$

For our choice of x_w, no other shocks are needed. However, if the wavemaker were to reaccelerate for $t > 1$, characteristics would again cross, and a new shock would be needed.

6.2.5 *Gas Dynamics, An Example with $n = 3$, $C_{ij} = 0$*

Equations of motion

We consider the one-dimensional flow of an ideal gas, i.e., a gas where the pressure p is related to the temperature θ and density ρ by the equation of state

$$p = \rho\theta. \tag{6.2.190}$$

Here p, ρ, and θ are made dimensionless by dividing by their ambient values, p_0, ρ_0, and θ_0 respectively. The equations of mass, momentum, and energy conservation are then given by (e.g., see (3.63) of [6.17])

$$\rho_t + (\rho u)_x = 0, \tag{6.2.191a}$$

$$\rho u_t + \rho u u_x + \frac{p_x}{\gamma} = \frac{4}{3\text{Re}} u_{xx}, \tag{6.2.191b}$$

$$\rho\theta_t + \rho u\theta_x + (\gamma - 1)p u_x = \frac{4\gamma(\gamma - 1)}{3\text{Re}} u_x^2 + \frac{\gamma}{\text{Re Pr}} \theta_{xx}. \tag{6.2.191c}$$

We have simplified the divergence form of these equations (see (3.56) of [6.17]) to obtain the above formulation. The dimensionless speed (normalized by the ambient sound speed c_0) is u, and the dimensionless parameters γ, Re, and Pr are given by

$$\gamma = \frac{C_p}{C_v} = \frac{\text{Specific heat at constant pressure}}{\text{Specific heat at constant volume}}, \tag{6.2.192a}$$

$$\text{Re} = \frac{c_0 L_0}{\nu_0} = \frac{c_0^2 T_0}{\nu_0} = \text{Reynolds number}, \tag{6.2.192b}$$

$$\text{Pr} = \frac{\mu C_p}{\lambda} = \text{Prandtl number}. \tag{6.2.192c}$$

The Reynolds number is defined in terms of the ambient kinematic viscosity ν_0, the ambient sound speed $c_0 = (\gamma p_0/\rho_0)^{1/2}$, and either a characteristic length L_0 (initial value problem) or $c_0 T_0$ (signaling problem), where T_0 is a characteristic time

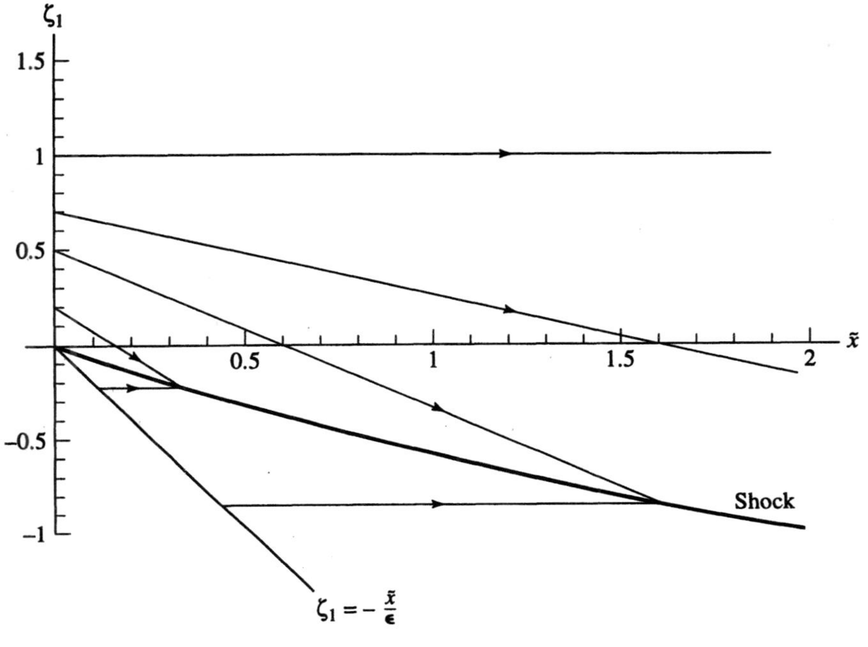

(a)

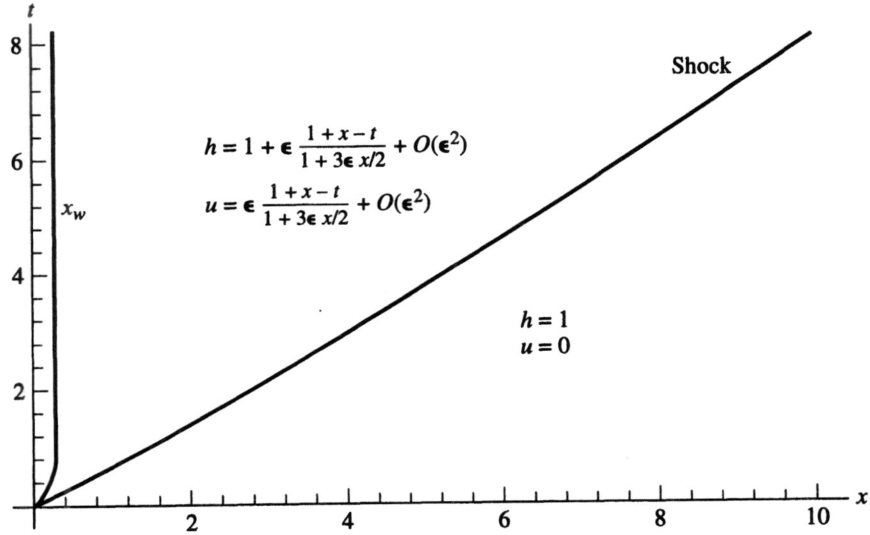

(b)

FIGURE 6.2.6. Shock trajectory (a) $\tilde{x}\zeta_1$ Plane; (b) xt Plane, $\epsilon = 0.5$

scale for the signal. In (6.2.192c), μ is the coefficient of viscosity, $\mu = \nu_0 \rho_0$, and λ is the thermal conductivity. Note the $(1/\gamma)$ factor multiplying p_x in (6.2.191b) that arises from our choice of dimensionless variables. Also, the dimensionless sound speed c for the flow is given by

$$c = \sqrt{\frac{p}{\rho}}. \tag{6.2.193}$$

There are various choices of three independent state variables, e.g., (u, ρ, p), (u, ρ, θ), (u, p, c), etc. The example we study is most conveniently formulated using (u, p, s), where s is the dimensionless entropy

$$s = \log\left(\frac{p}{\rho^\gamma}\right). \tag{6.2.194}$$

In order to facilitate the transformation to these variables, we introduce the notation

$$D \equiv \frac{\partial}{\partial t} + u\frac{\partial}{\partial x}, \tag{6.2.195}$$

and write (6.2.191a) and (6.2.191c) in the form

$$D(\rho) + u\rho_x = 0, \tag{6.2.196a}$$

$$\rho D(\theta) + (\gamma - 1)pu_x = \frac{4\gamma(\gamma - 1)}{3\mathrm{Re}}u_x^2 + \frac{\gamma}{\mathrm{Re\,Pr}}\theta_{xx} \equiv M. \tag{6.2.196b}$$

If we express p in terms of ρ and θ using (6.2.190), we have

$$D\left(\frac{p}{\rho^\gamma}\right) = D(\rho^{1-\gamma}\theta) = \rho^{1-\gamma}D(\theta) + (1-\gamma)\rho^{-\gamma}\theta D(\rho).$$

Using (6.2.196a) for $D(\rho)$ and $\theta = p/\rho$ gives

$$D\left(\frac{p}{\rho^\gamma}\right) = \rho^{1-\gamma}D(\theta) - (1-\gamma)\rho^{-\gamma}pu_x.$$

Now, using (6.2.196b) for $D(\theta)$ gives the energy equation

$$D\left(\frac{p}{\rho^\gamma}\right) = \rho^{-\gamma}M. \tag{6.2.197}$$

Next, we set $(p/\rho^\gamma) = e^s$ and find that (6.2.196a) transforms to

$$D(p) + \gamma pu_x = pD(s). \tag{6.2.198}$$

The energy equation (6.2.197) becomes

$$D(s) = \frac{M}{p}, \tag{6.2.199}$$

and using this expression for $D(s)$ in (6.2.198) gives

$$D(p) + \gamma pu_x = M. \tag{6.2.200}$$

Finally, the momentum equation (6.2.191b) becomes

$$D(u) + \frac{p_x}{\gamma \rho(p, s)} = \frac{4}{3\text{Re}} u_{xx}, \tag{6.2.201}$$

where $\rho(p, s)$ is obtained from (6.2.194) in the form

$$\rho(p, s) = (pe^{-s})^{1/\gamma}. \tag{6.2.202}$$

Also, since M involves θ, we must express θ as a function of p, s using (6.2.190) and (6.2.202)

$$\theta(p, s) = \frac{p}{\rho(p, s)}. \tag{6.2.203}$$

We rearrange equations (6.2.199)–(6.2.201) for the vector (u, p, s) to obtain the standard form (6.2.5) with

$$A(\mathbf{u}) = \begin{pmatrix} u & \dfrac{1}{\gamma\rho} & 0 \\ \gamma p & u & 0 \\ 0 & 0 & u \end{pmatrix}, \quad \mathbf{r} = \begin{pmatrix} \dfrac{4}{3\text{Re}} u_{xx} \\ M \\ M/p \end{pmatrix}. \tag{6.2.204}$$

Inviscid, non-heat-conducting flow

If the flow is inviscid ($\nu_0 = 0$ or Re $\to \infty$) and non-heat-conducting ($\lambda = 0$ or $Pr \to \infty$), then $\mathbf{r} = 0$, and the system reduces to

$$\mathbf{u}_t + A(\mathbf{u})\mathbf{u}_x = 0, \tag{6.2.205}$$

which is hyperbolic. The three characteristics are $\lambda_1 = u+c$, $\lambda_2 = u-c$, and $\lambda_3 = 0$. If, in addition, the entropy is constant ($s = 0$), we have two dependent variables (u, p). Such a flow may be generated, for example, by initial conditions with $s = 0$ throughout the domain. It then follows that s remains constant throughout the flow, except across shocks where the change in entropy is proportional to the cube of the change in pressure (or speed). For details, see the discussion in Sec. 5.3 of [6.17]. In particular, for a small disturbance theory, we may ignore entropy variations to $O(\epsilon^2)$.

The equations governing *isentropic* flow are formally analogous to those for shallow water flow discussed in Sec. 6.2.4. Using (p, u) as the two dependent variables, we have ($p = \rho^\gamma$)

$$p_t + up_x + \gamma p u_x = 0, \tag{6.2.206a}$$

$$u_t + \frac{1}{\gamma p^{1/\gamma}} p_x + uu_x = 0. \tag{6.2.206b}$$

Thus (see (6.2.100)), we may identify $u \to u$, $p \to \sqrt{h}$, $\gamma \to 2$. An equivalent formulation in terms of (c, u) variables gives

$$c_t + uc_x + \frac{\gamma - 1}{2} cu_x = 0, \tag{6.2.207a}$$

$$u_t + \frac{2}{\gamma - 1} cc_x + uu_x = 0. \tag{6.2.207b}$$

Now, we identify $c \to \sqrt{h}$.

The shock condition that replaces (6.2.124) is now given by (see Sec. 5.3 of [6.17])

$$C_s^2 - \frac{\gamma+1}{2} u^- C_s - 1 = 0 \tag{6.2.208}$$

for the case where $u^+ = 0$, $c^+ = 1$. Thus, the analogy does not carry over to shocks and is only qualitative in this sense. Nevertheless, the mathematical details for solutions, including multiple-scale expansions, are identical to those discussed earlier for shallow water flow and are not repeated.

Transformation to standard form

Small disturbance theory corresponds to weak perturbations to the ambient state $u = 0$, $p = 1$, $s = 0$. Thus, we set

$$u = \epsilon w_1; \quad p = 1 + \epsilon w_2, \quad s = \epsilon w_3. \tag{6.2.209}$$

We also wish to consider flow with small viscosity, and we set

$$\frac{1}{\mathrm{Re}} = \sigma\epsilon, \quad \sigma = O(1). \tag{6.2.210}$$

Substituting (6.2.209)–(6.2.210) into (6.2.199)–(6.2.201) gives the following system governing the w_i to $O(\epsilon)$:

$$\frac{\partial w_1}{\partial t} + \frac{1}{\gamma}\frac{\partial w_2}{\partial x} = \epsilon\left[-w_1\frac{\partial w_1}{\partial x} - \frac{(w_3 - w_2)}{\gamma^2}\frac{\partial w_2}{\partial x} + \frac{4\sigma}{3}\frac{\partial^2 w_1}{\partial x^2}\right] + O(\epsilon^2), \tag{6.2.211a}$$

$$\frac{\partial w_2}{\partial t} + \gamma\frac{\partial w_1}{\partial x} = \epsilon\left\{-w_1\frac{\partial w_2}{\partial x} - \gamma w_2\frac{\partial w_1}{\partial x} + \sigma\left[(\gamma-1)\frac{\partial^2 w_2}{\partial x^2} + \frac{\partial^2 w_3}{\partial x^2}\right]\right\} + O(\epsilon^2), \tag{6.2.211b}$$

$$\frac{\partial w_3}{\partial t} = \epsilon\left\{-w_1\frac{\partial w_3}{\partial x} + \sigma\left[(\gamma-1)\frac{\partial^2 w_2}{\partial x^2} + \frac{\partial^2 w_3}{\partial x^2}\right]\right\} + O(\epsilon^2). \tag{6.2.211c}$$

The characteristics of the $\epsilon = 0$ problem are $\lambda_1 = 1$, $\lambda_2 = -1$, $\lambda_3 = 0$. We introduce the characteristic dependent variables

$$U_1 = \gamma w_1 + w_2, \quad U_2 = \gamma w_1 - w_2, \quad U_3 = w_3 \tag{6.2.212a}$$

with inverse

$$w_1 = \frac{U_1 + U_2}{2\gamma}, \quad w_2 = \frac{U_1 - U_2}{2}, \quad w_3 = U_3, \tag{6.2.212b}$$

and calculate the following system in the standard form (6.2.1)

$$\frac{\partial U_1}{\partial t} + \frac{\partial U_1}{\partial x} = \epsilon\left\{-\left(\frac{1+\gamma}{4\gamma}\right)\left[U_1\left(\frac{\partial U_1}{\partial x} + \frac{\partial U_2}{\partial x}\right) - U_2\frac{\partial U_2}{\partial x}\right]\right.$$

$$+ \frac{\gamma - 3}{4\gamma} U_2 \frac{\partial U_1}{\partial x} + \frac{U_3}{2\gamma} \left(\frac{\partial U_2}{\partial x} - \frac{\partial U_1}{\partial x} \right) + \frac{\sigma(1 + 3\gamma)}{6} \frac{\partial^2 U_1}{\partial x^2}$$

$$+ \left. \frac{\sigma(7 - 3\gamma)}{6} \frac{\partial^2 U_2}{\partial x^2} + \sigma \frac{\partial^2 U_3}{\partial x^2} \right\} + O(\epsilon^2), \tag{6.2.213a}$$

$$\frac{\partial U_2}{\partial t} - \frac{\partial U_1}{\partial x} = \epsilon \left\{ - \left(\frac{1 + \gamma}{4\gamma} \right) \left[U_2 \left(\frac{\partial U_1}{\partial x} + \frac{\partial U_2}{\partial x} \right) - U_1 \frac{\partial U_1}{\partial x} \right] \right.$$

$$+ \frac{\gamma - 3}{4\gamma} U_1 \frac{\partial U_2}{\partial x} + \frac{U_3}{2\gamma} \left(\frac{\partial U_2}{\partial x} - \frac{\partial U_1}{\partial x} \right) + \frac{\sigma(7 - 3\gamma)}{6} \frac{\partial^2 U_1}{\partial x^2}$$

$$+ \left. \frac{\sigma(1 + 3\gamma)}{6} \frac{\partial^2 U_2}{\partial x^2} - \sigma \frac{\partial^2 U_3}{\partial x^2} \right\} + O(\epsilon^2), \tag{6.2.213b}$$

$$\frac{\partial U_3}{\partial t} = \epsilon \left[-\frac{1}{2\gamma} (U_1 + U_2) \frac{\partial U_3}{\partial x} + \frac{\sigma}{2} (\gamma - 1) \left(\frac{\partial^2 U_1}{\partial x^2} - \frac{\partial^2 U_2}{\partial x^2} \right) + \sigma \frac{\partial^2 U_3}{\partial x^2} \right]$$

$$+ O(\epsilon^2). \tag{6.2.213c}$$

Expansion procedure

The system (6.2.213) is a special case of the general form (6.2.1) for $n = 3$, $C_{ij} = 0$. The ϕ_i are as in (6.2.21) with $n = 3$, $G_{ijk} = 0$. The ϕ_i also contain the second derivate terms proportional to $\partial^2 U_j / \partial x^2$. Thus, (6.2.213) is a special case of the general system

$$\frac{\partial U_i}{\partial t} + \lambda_i \frac{\partial U_i}{\partial x} = \epsilon \left[\sum_{j=1}^{3} \sum_{k=1}^{3} F_{ijk} U_j \frac{\partial U_k}{\partial x} + \sum_{j=1}^{3} D_{ij} \frac{\partial^2 U_j}{\partial x^2} \right], \quad i = 1, 2, 3. \tag{6.2.214}$$

The essential new feature for $n > 2$ is that there are more characteristics, $x - \lambda_i t = \xi_i =$ constant, than independent variables. Hence, it is not possible to use all the ξ_i as independent fast scales; we must select only two of these and express the rest in terms of the two selected. For $n = 3$, let us choose $\xi_1 = x - \lambda_1 t$ and $\xi_2 = x - \lambda_2 t$ as the two fast scales. It then follows that

$$\xi_3 = x - \lambda_3 t = \frac{1}{(\lambda_1 - \lambda_2)} \left[(\lambda_3 - \lambda_2)\xi_1 + (\lambda_1 - \lambda_3)\xi_2 \right] \equiv \alpha_1 \xi_1 + \alpha_2 \xi_2. \tag{6.2.215}$$

Note that since the λ_i are *distinct*, $\lambda_1 - \lambda_2 \neq 0$. Also, $\alpha_1 \neq 0, \alpha_2 \neq 0$. We then expand the solution of (6.2.214) in the multiple-scale form

$$U_i(x, t; \epsilon) = U_{i0}(\xi_1, \xi_2, \tilde{t}) + \epsilon U_{i1}(\xi_1, \xi_2, \tilde{t}) + O(\epsilon^2). \tag{6.2.216}$$

Derivatives transform as follows:

$$\frac{\partial}{\partial x} = \frac{\partial}{\partial \xi_1} + \frac{\partial}{\partial \xi_2}, \quad \frac{\partial}{\partial t} = -\lambda_1 \frac{\partial}{\partial \xi_1} - \lambda_2 \frac{\partial}{\partial \xi_2} + \epsilon \frac{\partial}{\partial \tilde{t}}. \tag{6.2.217}$$

Therefore,

$$\frac{\partial}{\partial t} + \lambda_i \frac{\partial}{\partial x} = (\lambda_i - \lambda_1)\frac{\partial}{\partial \xi_1} + (\lambda_i - \lambda_2)\frac{\partial}{\partial \xi_2}$$

$$= \begin{cases} (\lambda_1 - \lambda_2)\dfrac{\partial}{\partial \xi_2}, & \text{if } i = 1 \\ (\lambda_2 - \lambda_1)\dfrac{\partial}{\partial \xi_1}, & \text{if } i = 2 \\ (\lambda_3 - \lambda_1)\dfrac{\partial}{\partial \xi_1} + (\lambda_3 - \lambda_2)\dfrac{\partial}{\partial \xi_2}, & \text{if } i = 3. \end{cases} \tag{6.2.218}$$

Substituting these expansions into (6.2.214) gives the following equations governing the U_{i0} and U_{i1}:

$$(\lambda_i - \lambda_1)\frac{\partial U_{i0}}{\partial \xi_1} + (\lambda_i - \lambda_2)\frac{\partial U_{i0}}{\partial \xi_2} = 0, \quad i = 1, 2, 3, \tag{6.2.219}$$

$$(\lambda_i - \lambda_1)\frac{\partial U_{i1}}{\partial \xi_1} + (\lambda_i - \lambda_2)\frac{\partial U_{i1}}{\partial \xi_2} = \sum_{j=1}^{3}\sum_{k=1}^{3} F_{ijk} U_{j0}\left(\frac{\partial}{\partial \xi_1} + \frac{\partial}{\partial \xi_2}\right) U_{k0}$$

$$+ \sum_{j=1}^{3} D_{ij}\left(\frac{\partial}{\partial \xi_1} + \frac{\partial}{\partial \xi_2}\right)^2 U_{j0} - \frac{\partial U_{i0}}{\partial \tilde{t}}, \quad i = 1, 2, 3. \tag{6.2.220}$$

The solution of (6.2.219) is

$$U_{i0} = f_i(\xi_i, \tilde{t}), \quad i = 1, 2, 3, \tag{6.2.221}$$

where ξ_3 is the function of ξ_1 and ξ_2 defined by (6.2.215).

Using (6.2.219) to simplify the right-hand side of (6.2.220) gives

$$(\lambda_i - \lambda_1)\frac{\partial U_{i1}}{\partial \xi_1} + (\lambda_i - \lambda_2)\frac{\partial U_{i1}}{\partial \xi_2} = -\frac{\partial f_i}{\partial \tilde{t}} + \sum_{j=1}^{3}\sum_{k=1}^{3} F_{ijk} f_j \frac{\partial f_k}{\partial \xi_k}$$

$$+ \sum_{j=1}^{3} D_{ij}\frac{\partial^2 f_j}{\partial \xi_j^2}, \quad i = 1, 2, 3. \tag{6.2.222}$$

In preparation for deriving the evolution equations for the f_i, we write the three components of (6.2.222) and rearrange the terms on the right-hand sides as follows:

$$(\lambda_1 - \lambda_2)\frac{\partial U_{11}}{\partial \xi_2} = \left[-\frac{\partial f_1}{\partial \tilde{t}} + F_{111} f_1 \frac{\partial f_1}{\partial \xi_1} + D_{11}\frac{\partial^2 f_1}{\partial \xi_1^2}\right] + R_{12}(\xi_2, \tilde{t})$$

$$+ R_{13}(\xi_3, \tilde{t}) + \sum_{r=1}^{2}\Big[P_{11}^{(r)}(\xi_1, \tilde{t}) P_{12}^{(r)}(\xi_2, \tilde{t}) + Q_{11}^{(r)}(\xi_1, \tilde{t}) Q_{13}^{(r)}(\xi_3, \tilde{t})$$

$$+ S_{12}^{(r)}(\xi_2, \tilde{t}) S_{13}^{(r)}(\xi_3, \tilde{t})\Big], \tag{6.2.223a}$$

$$(\lambda_2 - \lambda_1)\frac{\partial U_{21}}{\partial \xi_1} = \left[-\frac{\partial f_2}{\partial \tilde{t}} + F_{222} f_2 \frac{\partial f_2}{\partial \xi_2} + D_{22}\frac{\partial^2 f_2}{\partial \xi_2^2}\right] + R_{21}(\xi_1, \tilde{t})$$

$$+ R_{23}(\xi_3, \tilde{t}) + \sum_{r=1}^{2} \Big[P_{21}^{(r)}(\xi_1, \tilde{t}) P_{22}^{(r)}(\xi_2, \tilde{t}) + Q_{21}^{(r)}(\xi_1, \tilde{t}) Q_{23}^{(r)}(\xi_3, \tilde{t})$$

$$+ S_{22}^{(r)}(\xi_2, \tilde{t}) S_{23}^{(r)}(\xi_2, \tilde{t}) \Big], \tag{6.2.223b}$$

$$(\lambda_3 - \lambda_1) \frac{\partial U_{31}}{\partial \xi_1} + (\lambda_3 - \lambda_2) \frac{\partial U_{31}}{\partial \xi_2} = \left[-\frac{\partial f_3}{\partial \tilde{t}} + F_{333} f_3 \frac{\partial f_3}{\partial \xi_3} + D_{33} \frac{\partial^2 f_3}{\partial \xi_3^2} \right]$$

$$+ R_{31}(\xi_1, \tilde{t}) + R_{32}(\xi_2, \tilde{t}) + \sum_{r=1}^{2} \Big[P_{31}^{(r)}(\xi_1, \tilde{t}) P_{32}^{(r)}(\xi_2, \tilde{t})$$

$$+ Q_{31}^{(r)}(\xi_1, \tilde{t}) Q_{33}^{(r)}(\xi_3, \tilde{t}) + S_{32}^{(r)}(\xi_2, \tilde{t}) S_{33}^{(r)}(\xi_3, \tilde{t}) \Big]. \tag{6.2.223c}$$

In particular, in (6.2.223a) we have denoted

$$R_{12} = F_{122} f_2 \frac{\partial f_2}{\partial \xi_2} + D_{12} \frac{\partial^2 f_2}{\partial \xi_2^2}, \quad R_{13} = F_{133} f_3 \frac{\partial f_3}{\partial \xi_3} + D_{13} \frac{\partial^2 f_3}{\partial \xi_3^2}$$

$$\begin{aligned} P_{11}^{(1)} P_{12}^{(1)} + P_{11}^{(2)} P_{12}^{(2)} &= F_{112} f_1 \frac{\partial f_2}{\partial \xi_2} + F_{121} f_2 \frac{\partial f_1}{\partial \xi_1} \\ Q_{11}^{(1)} Q_{13}^{(1)} + Q_{11}^{(2)} Q_{13}^{(2)} &= F_{113} f_1 \frac{\partial f_3}{\partial \xi_3} + F_{131} f_3 \frac{\partial f_1}{\partial \xi_1} \\ S_{12}^{(1)} S_{13}^{(1)} + S_{12}^{(2)} S_{13}^{(2)} &= F_{123} f_2 \frac{\partial f_3}{\partial \xi_3} + F_{132} f_3 \frac{\partial f_2}{\partial \xi_2}. \end{aligned} \tag{6.2.224}$$

Thus, the R_{ij} refer to functions of $(\xi_j, \tilde{t})$ occurring in the ith equation; the $P_{ij}\ P_{ik}$ are products of f_j with $\partial f_k/\partial \xi_k$ or f_k with $\partial f_j/\partial \xi_j$ occurring in the ith equation, etc.

Evolution equations, isolated initial data

Consider (6.2.223a) for U_{11}. The leading bracketed term on the right-hand side, $-(\partial f_1/\partial \tilde{t}) + F_{111} f_1 (\partial f_1/\partial \xi_1) + D_{11}(\partial^2 f_1/\partial \xi_1^2)$, consists of functions of ξ_1 and $\tilde{t}$ only. Upon integration with respect to ξ_2, this term will contribute an inconsistent component to U_{11} that is proportional to ξ_2. Now, as long as we ensure that the $f_i(\xi_i, \tilde{t})$ are bounded functions of their respective ξ_i with zero average, the terms R_{12}, R_{13}, and $P_{11}^{(r)}\ P_{12}^{(r)}$ are not troublesome. The remaining product terms $Q_{11}^{(r)}\ Q_{13}^{(r)}$ and $S_{12}^{(r)}\ S_{13}^{(r)}$ are also harmless if their average values over ξ_2 vanish. We denote these averages by

$$\langle Q_{11}^{(r)} Q_{13}^{(r)} \rangle(\xi_1, \tilde{t}) = \lim_{\ell \to \infty} \frac{1}{2\ell} \int_{-\ell}^{\ell} Q_{11}^{(r)}(\xi_1, \tilde{t}) Q_{13}^{(r)}(\alpha_1 \xi_1 + \alpha_2 \xi_2) d\xi_2, \tag{6.2.225a}$$

$$\langle S_{12}^{(r)} S_{13}^{(r)} \rangle(\xi_1, \tilde{t}) = \lim_{\ell \to \infty} \frac{1}{2\ell} \int_{-\ell}^{\ell} S_{12}^{(r)}(\xi_2, \tilde{t}) S_{13}^{(r)}(\alpha_1 \xi_1 + \alpha_2 \xi_2) d\xi_2. \tag{6.2.225b}$$

If the functions in the integrands are 2ℓ-periodic in the ξ_i variables, we omit the limit and choose ℓ as the half-period. We will see that for such periodic functions these averages may be nonzero functions of ξ_1 and $\tilde{t}$, in which case they must also be removed from the right-hand side of (6.2.223a). Notice incidentally that for problems with two waves ($n = 2$) the average terms (6.2.225) do not occur.

It is difficult to give necessary conditions on the initial values of f_i for the vanishing of these averages. A sufficient condition of some physical interest for the vanishing of the averages (6.2.225) is to have isolated initial data, i.e., the f_i initially vanish outside some bounded interval in x (compact support). A proof of this statement for the hyperbolic problem $D_{ij} \equiv 0$ follows easily from the observation that disturbances propagate along characteristics. Thus, if the initial data have compact support, the exact solution also has compact support for finite t. Therefore, the integrals in (6.2.225) remain finite as $\ell \to \infty$, and dividing by 2ℓ gives a zero average. The addition of diffusion (for physically consistent constants D_{ij}, as in gas dynamics) does not affect this outcome in the sense that along an exact solution the integrals in (6.2.225) still remain finite as $\ell \to \infty$. Similar remarks apply for (6.2.223b) and (6.2.223c).

We conclude that, for isolated initial data, the f_i satisfy the *decoupled* Burgers' type equations

$$\frac{\partial f_i}{\partial \tilde{t}} - F_{iii} f_i \frac{\partial f_i}{\partial \xi_i} = D_{ii} \frac{\partial^2 f_i}{\partial \xi_i^2}, \quad i = 1, 2, 3. \tag{6.2.226}$$

For the example of gas dynamics, we have $F_{111} = F_{222} = -(\gamma + 1)/4\gamma$, $F_{333} = 0$, $D_{11} = D_{22} = \sigma(1 + 3\gamma)/6$, $D_{33} = \sigma$. Thus f_1 and f_2 satisfy Burgers' equations

$$\frac{\partial f_i}{\partial \tilde{t}} + \frac{(\gamma + 1)}{4\gamma} f_i \frac{\partial f_i}{\partial \xi_i} = \frac{\sigma}{6}(1 + 3\gamma) \frac{\partial^2 f_i}{\partial \xi_i^2}, \quad i = 1, 2, \tag{6.2.227a}$$

and f_3 satisfies the linear diffusion equation

$$\frac{\partial f_3}{\partial \tilde{t}} = \sigma \frac{\partial^2 f_3}{\partial \xi_3^2}. \tag{6.2.227b}$$

Once the $f_i (i = 1, 2, 3)$ have been found, the velocity, pressure, and entropy are known to $O(\epsilon)$:

$$u = \frac{\epsilon}{2\gamma} \left[f_1(\xi_1, \tilde{t}) + f_2(\xi_2, \tilde{t}) \right] + O(\epsilon^2), \tag{6.2.228a}$$

$$p = 1 + \frac{\epsilon}{2} \left[f_1(\xi_1, \tilde{t}) - f_2(\xi_2, \tilde{t}) \right] + O(\epsilon^2), \tag{6.2.228b}$$

$$s = \epsilon f_3(\xi_3, \tilde{t}) + O(\epsilon^2). \tag{6.2.228c}$$

For example, if the initial conditions are chosen such that $f_2(x, 0) = 0$ for all x, then $f_2(\xi_2, \tilde{t}) \equiv 0$, and the velocity u is given by $\epsilon f_1/2\gamma$. Expressing the evolution equation for f_1 in terms of u, and using $x = \xi_1 + \tilde{t}/\epsilon$ and $t = \tilde{t}/\epsilon$ as independent variables (see (6.2.117)), gives the well-known result ($\text{Re}^{-1} = \sigma\epsilon$)

$$\frac{\partial u}{\partial t} + \left(1 + \frac{\gamma + 1}{2} u\right) \frac{\partial u}{\partial x} = \frac{1 + 3\gamma}{6\text{Re}} \frac{\partial^2 u}{\partial x^2}. \tag{6.2.229}$$

In particular, for $\sigma \ll 1$, i.e., $\frac{1}{\epsilon \mathrm{Re}} \ll 1$, the effect of the second derivative term in (6.2.229) is to smooth out the discontinuities in $O(\sigma)$ neighborhoods of shocks as discussed in Sec. 3.1.3.

Evolution equations, periodic disturbances, resonance

If the initial data are periodic with zero average (let us assume the period equals 2π for simplicity), products of the form $H_{jk}(\xi_k, \tilde{t})H_{j\ell}(\xi_\ell, \tilde{t})$, with $k \neq \ell$, must be considered as possible contributors of inconsistent terms in the equation for U_{1j}. As argued earlier (see the discussion following (6.2.69)), the periodicity of the initial data implies that the f_j are also periodic functions of their respective characteristic variables. For appropriate restrictions on the coefficients F_{ijk} and the initial data, the $f_j(\xi_j, \tilde{t})$ will have zero average over one period of their respective ξ_j. In this case, the f_j have the following complex Fourier series:

$$f_j(\xi_j, \tilde{t}) = \sum_{\substack{n=-\infty \\ n\neq 0}}^{\infty} c_{jn}(\tilde{t})e^{in\xi_j}, \quad j = 1, 2, 3. \tag{6.2.230}$$

Since the H_{jk} are either equal to f_k or to $\partial f_k/\partial \xi_k$ (see (6.2.224)), the products $H_{jk}H_{j\ell}$ have the double Fourier series

$$H_{jk}(\xi_k, \tilde{t})H_{j\ell}(\xi_\ell, \tilde{t}) = \sum_{\substack{n=-\infty \\ n\neq 0,}}^{\infty} \sum_{\substack{m=-\infty \\ m\neq 0}}^{\infty} d_{jnm}(t)e^{i(n\xi_k + m\xi_\ell)}. \tag{6.2.231}$$

The question now is under what circumstances $n\xi_k + m\xi_\ell$ is independent of ξ_j in the j-th equation (6.2.223) for U_{j1}.

Consider first (6.2.223a) for U_{11}. Each of the products $Q_{11}^{(r)}(\xi_1, \tilde{t})Q_{13}^{(r)}(\xi_3, \tilde{t})$ is a series of the form (6.2.231) with $j = 1, k = 1, \ell = 3$. In particular, the dependence on ξ_1 and ξ_3 is through a series of terms proportional to $\exp i(n\xi_1 + m\xi_3)$, where m and n range over all positive and negative integers (but not zero). We have $\exp i(n\xi_1 + m\xi_3) = \exp i[(n + m\alpha_1)\xi_1 + m\alpha_2\xi_2]$ when we use (6.2.215) for ξ_3. Since neither m nor α_2 vanish, we conclude that no term in the double Fourier series (6.2.231) for $Q_{11}\ Q_{13}$ is independent of ξ_2, i.e., $\langle Q_{11}Q_{13}\rangle(\xi_1, \tilde{t}) \equiv 0$. However, the products $S_{12}^{(r)}\ S_{13}^{(r)}$ consist of double Fourier series (6.2.231) with the terms $\exp i(n\xi_2 + m\xi_3) = \exp i[m\alpha_1\xi_1 + (n + m\alpha_2)\xi_2]$. For a given $\alpha_2 = (\lambda_1 - \lambda_3)/(\lambda_1 - \lambda_2)$, this term is independent of ξ_2 if there are integers m and n such that $n + m\alpha_2 = 0$, i.e., if the *resonance condition*

$$\frac{\lambda_1 - \lambda_3}{\lambda_1 - \lambda_2} = -\frac{n}{m} \tag{6.2.232}$$

is satisfied for positive or negative integers n and m. For example, in gas dynamics we saw that $\lambda_1 = 1, \lambda_2 = -1$ and $\lambda_3 = 0$. Therefore, $(\lambda_1 - \lambda_3)/(\lambda_1 - \lambda_2) = 1/2$, and all the terms d_{1nm} with $n/m = -1/2$ are independent of ξ_2. The nonzero average terms, which are functions of ξ_1 and $\tilde{t}$, are given by

$$\langle S_{12}^{(r)} S_{13}^{(r)} \rangle (\xi_1, \tilde{t}) = \frac{1}{2\pi} \int_{-\pi}^{\pi} S_{12}^{(r)}(\xi_2, \tilde{t}) S_{13}^{(3)}(\alpha_1 \xi_1 + \alpha_2 \xi_2, \tilde{t}) d\xi_2, \quad r = 1, 2. \tag{6.2.233}$$

Similar considerations apply to (6.2.223b) for U_{21}. The products $P_{21}^{(r)}\ P_{22}^{(r)}$ and $S_{22}^{(r)}\ S_{23}^{(r)}$ are harmless, but the products $Q_{21}^{(r)}\ Q_{23}^{(r)}$ have nonzero averages if the resonance condition

$$\frac{\lambda_3 - \lambda_2}{\lambda_1 - \lambda_2} = -\frac{n}{m} \tag{6.2.234}$$

is satisfied for integer values of m and n. Again, we note that in gas dynamics $(\lambda_3 - \lambda_2)/(\lambda_1 - \lambda_2) = 1/2$. Therefore, (6.2.234) holds for an infinite subset of the Fourier series for $Q_{21}^{(r)}\ Q_{23}^{(r)}$. The nonzero average terms are given by

$$\langle Q_{21}^{(r)} Q_{23}^{(r)} \rangle (\xi_2, \tilde{t}) = \frac{1}{2\pi} \int_{-\pi}^{\pi} Q_{21}^{(r)}(\xi_1, \tilde{t}) Q_{23}^{(r)}(\alpha_1 \xi_1 + \alpha_2 \xi_2, \tilde{t}) d\xi_1, \quad r = 1, 2. \tag{6.2.235}$$

Consider finally (6.2.223c), which is a linear first-order partial differential equation for U_{31}. The characteristic equations are

$$\frac{d\xi_1}{ds} = (\lambda_3 - \lambda_1), \qquad \frac{d\xi_2}{ds} = (\lambda_3 - \lambda_2), \tag{6.2.236a}$$

$$\begin{aligned} \frac{dU_{31}}{ds} &= \left[R_{33}(\alpha_1 \xi_1 + \alpha_2 \xi_2, \tilde{t}) + \sum_{r=1}^{2} P_{31}^{(r)}(\xi_1, \tilde{t}) P_{32}^{(r)}(\xi_2, \tilde{t}) \right] \\ &\quad + R_{32}(\xi_2, \tilde{t}) + R_{31}(\xi_1, \tilde{t}) + \sum_{r=1}^{2} \Big[Q_{31}^{(r)}(\xi_1, \tilde{t}) Q_{33}^{(r)}(\alpha_1 \xi_1 + \alpha_2 \xi_2, \tilde{t}) \\ &\quad + S_{32}^{(r)}(\xi_2, \tilde{t}) S_{33}^{(r)}(\alpha_1 \xi_1 + \alpha_2 \xi_2, \tilde{t}) \Big], \end{aligned} \tag{6.2.236b}$$

where R_{33} denotes the first set of bracketed terms on the right-hand side of (6.2.223c).

Integrating the two equations (6.2.236a) gives $\xi_1 = (\lambda_3 - \lambda_1)s +$ constant, $\xi_2 = (\lambda_3 - \lambda_2)s +$ constant. Therefore, $\xi_3 = [\alpha_1(\lambda_3 - \lambda_1) + \alpha_2(\lambda_3 - \lambda_2)]s +$ constant $=$ constant because the coefficient of s vanishes identically. Thus, all terms, such as R_{33}, that depend only on $\xi_3, \tilde{t}$ give rise to a contribution to U_{31} that is proportional to s and is inconsistent. Also, the sum of products $P_{31}^{(r)}\ P_{32}^{(r)}$ gives an inconsistent contribution to U_{31} (they do not depend on ξ_1 or ξ_2) if the resonance condition

$$\frac{\lambda_3 - \lambda_1}{\lambda_3 - \lambda_2} = -\frac{n}{m} \tag{6.2.237}$$

is satisfied for integer values of n and m, as in gas dynamics. Note that if the two resonance conditions (6.2.232) and (6.2.234) hold, the third condition (6.2.237) is automatically implied. In this case, the averages of $P_{31}^{(r)}\ P_{32}^{(r)}$ may be expressed in terms of integrals either with respect to ξ_1 or ξ_2. For example, using ξ_1 and expressing ξ_2 in terms of ξ_1 and ξ_3, we have

$$\langle P_{31}^{(r)} P_{32}^{(r)} \rangle (\xi_3, \tilde{t}) = \frac{1}{2\pi} \int_{-\pi}^{\pi} P_{31}^{(r)}(\xi_1, \tilde{t}) P_{32}^{(r)} \left(\frac{1}{\alpha_2} \xi_3 - \frac{\alpha_1}{\alpha_2} \xi_1, \tilde{t} \right) d\xi_1. \tag{6.2.238}$$

In summary, the evolution equations for the f_i satisfy the coupled system (if the characteristics satisfy the resonance conditions (6.2.232), (6.2.234), and (6.2.237))

$$\frac{\partial f_1}{\partial \tilde{t}} - F_{111} f_1 \frac{\partial f_1}{\partial \xi_1} - F_{123} \langle f_2 \frac{\partial f_3}{\partial \xi_3} \rangle (\xi_1, \tilde{t}) - F_{132} \langle f_3 \frac{\partial f_2}{\partial \xi_2} \rangle (\xi_1, \tilde{t}) = D_{11} \frac{\partial^2 f_1}{\partial \xi_1^2}, \tag{6.2.239a}$$

$$\frac{\partial f_2}{\partial \tilde{t}} - F_{222} f_2 \frac{\partial f_2}{\partial \xi_2} - F_{213} \langle f_1 \frac{\partial f_3}{\partial \xi_3} \rangle (\xi_2, \tilde{t}) - F_{231} \langle f_3 \frac{\partial f_1}{\partial \xi_1} \rangle (\xi_2, \tilde{t}) = D_{22} \frac{\partial^2 f_2}{\partial \xi_2^2}, \tag{6.2.239b}$$

$$\frac{\partial f_3}{\partial \tilde{t}} - F_{333} f_3 \frac{\partial f_3}{\partial \xi_3} - F_{312} \langle f_1 \frac{\partial f_2}{\partial \xi_2} \rangle (\xi_3, \tilde{t}) - F_{321} \langle f_2 \frac{\partial f_1}{\partial \xi_1} \rangle (\xi_3, \tilde{t}) = D_{33} \frac{\partial^2 f_3}{\partial \xi_2^2}. \tag{6.2.239c}$$

It should be kept in mind that $\xi_3 = \alpha_1 \xi_1 + \alpha_2 \xi_2$ throughout.

For the example of gas dynamics $F_{333} = F_{312} = F_{321} = 0$, $D_{33} = \sigma$. Thus, (6.2.239c) again reduces to the linear diffusion equation (6.2.222b) for f_3. Once this is solved, (6.2.239a) and (6.2.239b) give a pair of coupled equations for f_1 and f_2. These evolution equations were first derived for the inviscid problem $\sigma = 0$ in [6.23]. These equations have been studied extensively in the subsequent literature, but a discussion of these results is beyond the scope of this book.

6.2.6 *Channel Flow, An Example with $n = 2$, $C_{21} = O(\epsilon)$*

In our discussion of solutions of (6.2.1) so far, we have only considered examples where all the C_{ij} are absent from the leading approximation for the U_i. In this case, the unperturbed solution is neutrally stable on the t scale, and the weak nonlinearities determine the slow evolution of the two waves $U_1 = f_0(x - \lambda_1 t, \tilde{t}) + O(\epsilon)$ and $U_2 = g_0(x - \lambda_2 t, \tilde{t}) + O(\epsilon)$. If the C_{ij} are present to $O(1)$, the behavior of these waves on the t scale is determined by the stability conditions (6.2.49). Problems where one or both of the U_i grow exponentially on the t scale are inconsistent with the small disturbance assumption that led to (6.2.1). On the other hand, problems where *both* waves decay exponentially on the t scale are not interesting. We restrict attention to the case where one of the U_i is neutrally stable for ($\epsilon = 0$) and the other decays exponentially on the t scale. We then look for the effects of weak nonlinearity and weak instability on the solution over long times. The case where both U_i are neutrally stable for $\epsilon = 0$ is also interesting. It is discussed in [6.20].

In order to have one of the U_i decay exponentially on the t scale and to have the other be neutrally stable for $\epsilon = 0$, we set $C_{11} = O_s(1) > 0$ and $C_{22} = O_s(\epsilon) < 0$. If, in addition, we assume that C_{21} (or C_{12}) is small, the $\epsilon = 0$ problem has the elementary solution (6.2.54). More specifically, we assume the following behavior for the λ_i and C_{ij}:

$$\lambda_i = \lambda_i^{(0)} + \epsilon\lambda_i^{(1)} + O(\epsilon^2), \quad i = 1, 2, \tag{6.2.240a}$$

$$C_{11} = \epsilon C_{11}^{(1)} + O(\epsilon^2), \quad C_{11}^{(1)} = \frac{dC_{11}(\mu_0)}{d\mu}\mu_1, \tag{6.2.240b}$$

$$C_{12} = C_{12}^{(0)} + \epsilon C_{12}^{(1)} + O(\epsilon^2), \tag{6.2.240c}$$

$$C_{21} = \epsilon C_{21}^{(1)} + O(\epsilon^2), \tag{6.2.240d}$$

$$C_{22} = C_{22}^{(0)} + \epsilon C_{22}^{(1)} + O(\epsilon^2), \quad C_{22}^{(0)} > 0. \tag{6.2.240e}$$

The choice $C_{22}^{(0)} > 0$ ensures that U_2 decays exponentially on the fast scale. The sign of $C_{11}^{(1)}$ determines the stability condition (1) (see (6.2.49)). In particular, with $C_{11}^{(1)} < 0$, U_1 grows exponentially on the $\tilde{t}$ scale if nonlinearities are ignored. We also have $C_{11}C_{22} - C_{12}C_{21} = \epsilon(C_{11}^{(1)}C_{22}^{(0)} - C_{12}^{(0)}C_{21}^{(1)})$. Therefore, the sign of $C_{11}^{(1)}C_{22}^{(0)} - C_{12}^{(0)}C_{21}^{(1)}$ determines the stability condition (3).

For channel flow that is marginally unstable in the linear sense, we have $\mu = F$, $\mu_0 = 2$

$$F = 2 + \epsilon\alpha, \quad \alpha > 0, \tag{6.2.241a}$$

$$\lambda_1 = F + 1 = 3 + \epsilon\alpha, \quad \lambda_2 = F - 1 = 1 + \epsilon\alpha, \tag{6.2.241b}$$

$$C_{11} = -C_{21} = -\frac{1}{2} + \frac{1}{F} = -\epsilon\frac{\alpha}{4} + O(\epsilon^2), \tag{6.2.241c}$$

$$C_{12} = -C_{22} = -\frac{1}{2} - \frac{1}{F} = -1 + \frac{\epsilon\alpha}{4} + O(\epsilon^2). \tag{6.2.241d}$$

Thus,

$$\lambda_1^{(0)} = 3, \quad \lambda_2^{(0)} = 1, \quad \lambda_1^{(1)} = \lambda_2^{(1)} = \alpha, \tag{6.2.242a}$$

$$C_{11}^{(1)} = -C_{21}^{(1)} = -\frac{\alpha}{4}, \tag{6.2.242b}$$

$$C_{22}^{(0)} = -C_{12}^{(0)} = 1, \tag{6.2.242c}$$

$$C_{22}^{(1)} = -C_{12}^{(1)} = -\frac{\alpha}{4}. \tag{6.2.242d}$$

For this example, $C_{11}C_{22} - C_{12}C_{21} \equiv 0$. Therefore, the choice $\alpha > 0$ implies that only the second stability condition is violated.

With $C_{11}^{(0)} = C_{21}^{(0)} = 0$, the system (6.2.240) simplifies to

$$\frac{\partial U_i}{\partial t} + \lambda_i^{(0)}\frac{\partial U_i}{\partial x} + C_{i2}^{(0)}U_2 = \epsilon\left(-\lambda_i^{(1)}\frac{\partial U_i}{\partial x}\right.$$

$$- \sum_{j=1}^{2} C_{ij}^{(1)} U_j + \phi_i^{(0)} \Bigg) + O(\epsilon^2), \quad i = 1, 2. \tag{6.2.243}$$

Expansion procedure

As we only intend to compute the solution of (6.2.243) for the U_i to $O(1)$ explicitly, we assume an expansion in terms of the two characteristic fast scales $\xi_i = x - \lambda_i t$ and the single slow scale $\tilde{t} = \epsilon t$

$$U_i(x, t; \epsilon) = U_{i0}(\xi_1, \xi_2, \tilde{t}) + \epsilon U_{i1}(\xi_1, \xi_2, \tilde{t}) + O(\epsilon^2). \tag{6.2.244}$$

The equations governing U_{i0} and U_{i1} that result from (6.2.243) are

$$(\lambda_1^{(0)} - \lambda_2^{(0)}) \frac{\partial U_{10}}{\partial \xi_2} + C_{12}^{(0)} U_{20} = 0, \tag{6.2.245a}$$

$$-(\lambda_1^{(0)} - \lambda_2^{(0)}) \frac{\partial U_{20}}{\partial \xi_1} + C_{22}^{(0)} U_{20} = 0. \tag{6.2.245b}$$

$$(\lambda_1^{(0)} - \lambda_2^{(0)}) \frac{\partial U_{11}}{\partial \xi_2} + C_{12}^{(0)} U_{21} = - \frac{\partial U_{10}}{\partial \tilde{t}} - C_{11}^{(1)} U_{10} - C_{12}^{(1)} U_{20}$$
$$- \lambda_1^{(1)} \left(\frac{\partial U_{10}}{\partial \xi_1} + \frac{\partial U_{10}}{\partial \xi_2} \right) + \phi_1^{(0)}, \tag{6.2.246a}$$

$$-(\lambda_1^{(0)} - \lambda_2^{(0)}) \frac{\partial U_{21}}{\partial \xi_1} + C_{22}^{(0)} U_{21} = - \frac{\partial U_{20}}{\partial \tilde{t}} - C_{21}^{(1)} U_{10} - C_{22}^{(1)} U_{20}$$
$$- \lambda_2^{(1)} \left(\frac{\partial U_{20}}{\partial \xi_1} + \frac{\partial U_{20}}{\partial \xi_2} \right) + \phi_2^{(0)}. \tag{6.2.246b}$$

We first solve (6.2.245b) and then use this result to solve (6.2.245a) and obtain (see (6.2.54))

$$U_{10} = g_0(\xi_1, \tilde{t}) - \tilde{C}_{12} F_0(\xi_2, \tilde{t}) \exp(\tilde{C}_{22} \xi_1), \tag{6.2.247a}$$

$$U_{20} = f_0(\xi_2, \tilde{t}) \exp(\tilde{C}_{22} \xi_1), \tag{6.2.247b}$$

where we have introduced the following notation:

$$\tilde{C}_{12} = \frac{C_{12}^{(0)}}{\lambda_1^{(0)} - \lambda_2^{(0)}}, \quad \tilde{C}_{22} = \frac{C_{22}^{(0)}}{\lambda_1^{(0)} - \lambda_2^{(0)}}, \quad \frac{\partial F_0}{\partial \xi_2} = f_0(\xi_2, \tilde{t}). \tag{6.2.248}$$

The initial conditions (6.2.24) require that

$$U_1^*(x) = g_0(x, 0) - \tilde{C}_{12} F_0(x, 0) \exp(\tilde{C}_{22} x), \tag{6.2.249a}$$

$$U_2^*(x) = f_0(x, 0) \exp(\tilde{C}_{22} x). \tag{6.2.249b}$$

Henceforth, we restrict attention to the case where the $U_i^*(x)$ are 2ℓ-periodic functions with zero average. Our discussion also carries over to bounded initial data, but we do not give the details. This case is discussed for the example of channel flow in [6.30].

Once $g_0(\xi_1, \tilde{t})$ and $f_0(\xi_2, t)$ are defined, we can express the solution for the original physical variables $u_i(x, t; \epsilon)$ as follows, using (6.2.11), (6.2.16), and (6.2.247):

$$u_i(x, t; \epsilon) = v_i(\mu_0) + \epsilon\Big[v_i'(\mu_0)\mu_1 + \sum_{j=1}^{2} W_{ij}(\mu_0)U_{j0}(x - \lambda_1^{(0)}t, x - \lambda_2^{(0)}t, \tilde{t})\Big] + O(\epsilon^2), \quad i = 1, 2. \tag{6.2.250}$$

It will be more transparent to regroup the terms of order ϵ in (6.2.250) so as to separate the dependence of the solution on g_0 from that on f_0 and F_0 because, as we will verify, the terms depending on f_0 and F_0 decay rapidly. We introduce the notation

$$\overline{f}_0(\xi_2, \tilde{t}) = f_0(\xi_2, \tilde{t}) \exp(\tilde{C}_{22}\xi_2), \tag{6.2.251a}$$

$$\overline{F}_0(\xi_2, \tilde{t}) = F_0(\xi_2, \tilde{t}) \exp(\tilde{C}_{22}\xi_2). \tag{6.2.251b}$$

Equation (6.2.250) can now be written as

$$u_i(x, t; \epsilon) = v_i(\mu_0) + \epsilon\Big\{v_i'(\mu_0)\mu_1 + W_{i1}(\mu_0)g_0(x - \lambda_1^{(0)}t, \tilde{t}) + \Big[W_{i2}\overline{f}_0 - W_{i1}\tilde{C}_{12}\overline{F}_0\Big]\exp(-C_{22}^{(0)}t)\Big\} + O(\epsilon^2), \quad i = 1, 2. \tag{6.2.252}$$

We will show presently that $\overline{f}_0$ and $\overline{F}_0$ are periodic functions of ξ_2 and may, at most, grow exponentially with respect to $\tilde{t}$. Therefore, with $C_{22}^{(0)} > 0$, the terms in (6.2.252) multiplied by $\exp(-C_{22}^{(0)}t)$ decay rapidly with t leaving the contribution involving g_0 only.

Evolution equations

We first consider (6.2.246b) and use the solution (6.2.247) to evaluate the right-hand side. After some algebra, we can reduce the result to the form

$$\frac{\partial}{\partial \xi_1}\left(\frac{U_{21}}{E}\right) = -\frac{1}{(\lambda_1^{(0)} - \lambda_2^{(0)})}\Bigg\{\Bigg[-\frac{\partial f_0}{\partial \tilde{t}} + C_{21}^{(1)}\tilde{C}_{12}F_0 - (C_{22}^{(1)} + \lambda_2^{(1)}\tilde{C}_{22})f_0 - \lambda_2^{(1)}\frac{\partial f_0}{\partial \xi_2}\Bigg] + E^{-1}\Bigg[-C_{21}^{(1)}g_0 + F_{211}g_0\frac{\partial g_0}{\partial \xi_1} + G_{211}g_0^2\Bigg] + \ldots\Bigg\}. \tag{6.2.253}$$

Here ... indicates products of functions of $(\xi_1, \tilde{t})$ with functions of $(\xi_2, \tilde{t})$ that we do not list (see (3.14) of [6.20]), and we have introduced the notation

$$E(\xi_1) = \exp(\tilde{C}_{22}\xi_1). \tag{6.2.254}$$

The first group of terms in square brackets on the right-hand side of (6.2.253) is independent of ξ_1 and will, upon integration, lead to an inconsistent contribution proportional to $\xi_1 E(\xi_1)$ to U_{21}. These are the only terms independent of ξ_1 as long as $\langle g_0\rangle(\tilde{t}) \equiv 0$ (see (6.2.75) for the definition of $\langle g_0\rangle(\tilde{t})$). Note, in particular, that in the first group of terms in square brackets we have not included contributions from products such as $f_0 g_0$, $F_0 g_0$, and $g_0(\partial f_0/\partial \xi_2)$, precisely because we are assuming $\langle g_0\rangle(\tilde{t}) \equiv 0$. Later on, we shall derive the necessary condition for this to be true if $\langle g_0\rangle(0) = 0$. Thus, we find the following *linear* evolution equation for $f_0(\xi_2, \tilde{t})$:

$$\frac{\partial f_0}{\partial \tilde{t}} - C_{21}^{(1)}\tilde{C}_{12}F_0 + (C_{22}^{(1)} + \lambda_2^{(1)}\tilde{C}_{22})f_0 + \lambda_2^{(1)}\frac{\partial f_0}{\partial \xi_2} = 0. \tag{6.2.255}$$

We will study the general solution of (6.2.255) later.

The second group of terms in square brackets on the right-hand side of (6.2.253) is independent of ξ_2. Upon integration, this group of terms leads to a perfectly acceptable contribution depending on $(\xi_1, \tilde{t})$ for U_{21}. However, when this contribution is substituted into the left-hand side of (6.2.246a) and integrated with respect to ξ_2, it will lead to inconsistent terms proportional to ξ_2 in U_{11}. We anticipate this occurrence and keep track of the contribution to U_{21} of this second group of terms. More specifically, we observe that if $\langle g_0\rangle(\tilde{t}) \equiv 0$ and (6.2.255) is satisfied, U_{21} has the form

$$U_{21} = \frac{C_{21}^{(1)}}{\lambda_1^{(0)} - \lambda_2^{(0)}} E(\xi_1) \int^{\xi_1} \frac{g_0(s, \tilde{t})}{E(s)} ds - \frac{(F_{211}\tilde{C}_{22} + 2G_{211})E(\xi_1)}{2(\lambda_1^{(0)} - \lambda_2^{(0)})}$$

$$\cdot \int^{\xi_1} \frac{g_0^2(s, \tilde{t})}{E(s)} ds - \frac{F_{211}}{2(\lambda_1^{(0)} - \lambda_2^{(0)})} g_0^2(\xi_1, \tilde{t}) + K(\tilde{t})E(\xi_1) + \dots, \tag{6.2.256}$$

where $K(\tilde{t})$ is an arbitrary function of $\tilde{t}$. Here ... denotes terms that either depend on $(\xi_2, \tilde{t})$ or products of functions of $(\xi_1, \tilde{t})$ with functions of $(\xi_2, \tilde{t})$, all of which are consistent as long as U_{10} and U_{20} are bounded functions of ξ_1 and ξ_2.

Now, we transfer $C_{12}^{(0)} U_{21}$ (using the expression given by (6.2.256) for U_{21}) to the right-hand side of (6.2.246a) and isolate all the terms that are independent of ξ_2. Such terms give rise to inconsistent contributions proportional to ξ_2 in U_{11}. Setting the sum of these terms equal to zero gives the following evolution equation for g_0:

$$\frac{\partial g_0}{\partial \tilde{t}} + (\lambda_1^{(1)} - F_{111}g_0)\frac{\partial g_0}{\partial \xi_1} + C_{11}^{(1)}g_0 + \gamma g_0^2 + \tilde{C}_{12}C_{21}^{(1)} \int^{\xi_1} \frac{g_0(s, \tilde{t})}{E(s)} ds$$

$$+ \nu E(\xi_1) \int^{\xi_1} \frac{g_0^2(s, \tilde{t})}{E(s)} ds + K_1(\tilde{t})E(\xi_1) = 0, \tag{6.2.257}$$

where $K_1 = C_{12}^{(0)} K$, and we have introduced the notation

$$\gamma = -(G_{111} + \tilde{C}_{12}F_{211}/2), \quad \nu = -\tilde{C}_{12}(G_{211} + \tilde{C}_{22}F_{211}/2). \tag{6.2.258}$$

We note that for channel flow with $F = 2+\epsilon\alpha$, the evolution equations (6.2.255) for f_0 and (6.2.257) for g_0 reduce to

$$\frac{\partial f_0}{\partial \tilde{t}} + \alpha \frac{\partial f_0}{\partial \xi_2} + \frac{\alpha}{8} f_0 + \frac{\alpha}{4} F_0 = 0, \tag{6.2.259a}$$

$$\frac{\partial g_0}{\partial \tilde{t}} + \left(\alpha - \frac{3}{2} g_0\right) \frac{\partial g_0}{\partial \xi_1} - \frac{\alpha}{4} g_0 - \frac{\alpha}{8} e^{\xi_1/2} \int^{\xi_1} e^{-s/2} g_0(s, \tilde{t}) ds$$
$$+ K_1(\tilde{t}) e^{\xi_1/2} = 0. \tag{6.2.259b}$$

This result agrees with (3.11) and (3.12), respectively, of [6.30] once the notation is reconciled ($\xi_1 \to \xi, \xi_2 \to \eta, f_0 \to -f_1/2, g_0 \to -g_1/2, F_0 \to G, K_1 \to C_1$). Thus, for the problem of channel flow the term involving g_0^2 and the term involving the integral of g_0^2/E are absent from (6.2.257) because $\gamma = 0$ and $\nu = 0$. We will see later that $\gamma = \nu = 0$ satisfy the necessary condition for having $\langle g_0 \rangle(\tilde{t}) \equiv 0$.

Solution for f_0

For a given 2ℓ-periodic function $U_2^*(x)$ with zero average, we can express $f_0(x, 0)$ in the form (see (6.2.249b))

$$f_0(x, 0) = \exp(-\tilde{C}_{22} x) \sum_{n=1}^{\infty} \left(\alpha_n \cos \frac{n\pi x}{\ell} + \beta_n \sin \frac{n\pi x}{\ell}\right), \tag{6.2.260}$$

for known constants α_n and β_n. In view of this structure for f_0, we express the solution of (6.2.255) in the following series form:

$$f_0(\xi_2, \tilde{t}) = \exp(-\tilde{C}_{22}\xi_2) \sum_{n=1}^{\infty} \left[\tilde{\alpha}_n(\tilde{t}) \cos \frac{n\pi \xi_2}{\ell} + \tilde{\beta}_n(\tilde{t}) \sin \frac{n\pi \xi_2}{\ell}\right], \tag{6.2.261}$$

where $\tilde{\alpha}_n(0) = \alpha_n$ and $\tilde{\beta}_n(0) = \beta_n$. Thus, $\overline{f}_0(\xi_2, \tilde{t})$ is periodic (see (6.2.251a)). Integrating (6.2.261) defines F_0

$$F_0(\xi_2, \tilde{t}) = \exp(-\tilde{C}_{22}\xi_2) \sum_{n=1}^{\infty} \left[-\frac{\tilde{\alpha}_n \tilde{C}_{22} + \tilde{\beta}_n n\pi/\ell}{\tilde{C}_{22}^2 + n^2\pi^2/\ell^2} \cos \frac{n\pi\xi_2}{\ell} \right.$$
$$\left. + \frac{-\tilde{\beta}_n \tilde{C}_{22} + \tilde{\alpha}_n n\pi/\ell}{\tilde{C}_{22}^2 + n^2\pi^2/\ell^2} \sin \frac{n\pi\xi_2}{\ell} \right] + k(\tilde{t}), \tag{6.2.262}$$

where $k(\tilde{t})$ is an integration constant. If we now substitute (6.2.261), (6.2.262), and the expressions that result for $(\partial f_0/\partial \tilde{t})$ and $(\partial f_0/\partial \xi_2)$ into (6.2.255), we see that $k(\tilde{t}) \equiv 0$. Thus, $\overline{F}_0$ is also a periodic function of ξ_2, and this verifies the claim following (6.2.252) for the case of periodic initial data. We also find that the Fourier coefficients $\tilde{\alpha}_n$ and $\tilde{\beta}_n$ satisfy the linear system

$$\frac{d\tilde{\alpha}_n}{d\tilde{t}} + \mu_n \tilde{\alpha}_n + \gamma_n \tilde{\beta}_n = 0, \tag{6.2.263a}$$

$$\frac{d\tilde{\beta}_n}{d\tilde{t}} + \mu_n \tilde{\beta}_n - \gamma_n \tilde{\alpha}_n = 0, \tag{6.2.263b}$$

where μ_n and γ_n are the constants

$$\mu_n = C_{12}^{(1)} + \frac{C_{21}^{(1)}\tilde{C}_{12}\tilde{C}_{22}}{\tilde{C}_{22}^2 + n^2\pi^2/\ell^2}, \tag{6.2.264a}$$

$$\gamma_n = \frac{n\pi/\ell}{\tilde{C}_{22}^2 + n^2\pi^2/\ell^2}\left[C_{21}^{(1)}\tilde{C}_{12} - \lambda_2^{(1)}\left(\tilde{C}_{22}^2 + \frac{n^2\pi^2}{\ell^2}\right)\right]. \tag{6.2.264b}$$

The solution for $\tilde{\alpha}_n$ and $\tilde{\beta}_n$ that satisfies the initial conditions is

$$\tilde{\alpha}_n(\tilde{t}) = [\alpha_n \cos\gamma_n\tilde{t} - \beta_n \sin\gamma_n\tilde{t}]\exp(-\mu_n\tilde{t}), \tag{6.2.265a}$$

$$\tilde{\beta}_n(\tilde{t}) = [\beta_n \cos\gamma_n\tilde{t} + \alpha_n \sin\gamma_n\tilde{t}]\exp(-\mu_n\tilde{t}), \tag{6.2.265b}$$

and this defines $f_0(\xi_2, \tilde{t})$ and $F_0(\xi_2, \tilde{t})$ explicitly.

We note, as pointed out earlier, that the terms in the Fourier series for $\overline{f}_0$ and $\overline{F}_0$ are multiplied by the factor $\exp(-\mu_n\tilde{t})$, which grows exponentially on the $\tilde{t}$ scale if $\mu_n < 0$. However, since $C_{22}^{(0)} > 0$, this growth is suppressed by the factor $\exp(-C_{22}^{(0)}t)$ multiplying the terms in square brackets in (6.2.252). We also note that $\mu_n = O(1)$ for all n; actually, $\mu_n \to C_{22}^{(1)}$ as $n \to \infty$. Therefore, all the harmonics of $\overline{f}_0$ and $\overline{F}_0$ are multiplied by $\exp(-C_{22}^{(0)}t)$ and die out for $C_{22}^{(0)} > 0$. In fact, the terms in square brackets in (6.2.252) become $O(\epsilon)$ and may be ignored for $t \geq t_0 = O(|\log\epsilon|)$.

The equation for g_0

For 2ℓ-periodic initial data with zero average, we have shown that $\overline{F}_0(x, 0) = F_0(x, 0)\exp(\tilde{C}_{22}x)$ is 2ℓ-periodic with zero average. Therefore, (6.2.249a) implies that $g_0(x, 0)$ is also 2ℓ-periodic with zero average. This, in turn, implies that $g_0(\xi_1, \tilde{t})$ is 2ℓ-periodic in ξ_1 but does *not* guarantee that the average value of g_0 remains zero for all $\tilde{t} > 0$. Next we derive the necessary condition for this to be true, i.e., for $\langle g_0\rangle(0) = 0$ to imply that $\langle g_0\rangle(\tilde{t}) \equiv 0$ for $\tilde{t} > 0$.

We transform (6.2.257) to a second-order hyperbolic equation for g_0 that is free of integral terms by dividing it by $E(\xi_1)$, taking the partial derivative of the result with respect to ξ_1 and then canceling out $E(\xi_1)$. We find

$$\begin{aligned}\frac{\partial^2 g_0}{\partial\xi_1\partial\tilde{t}} &+ (\lambda_1^{(1)} - F_{111}g_0)\frac{\partial^2 g_0}{\partial\xi_1^2} + (C_{11}^{(1)} - \lambda_1^{(1)}\tilde{C}_{22})\frac{\partial g_0}{\partial\xi_1} \\ &+(2\gamma + F_{111}\tilde{C}_{22})g_0\frac{\partial g_0}{\partial\xi_1} - F_{111}\left(\frac{\partial g_0}{\partial\xi_1}\right)^2 - \tilde{C}_{22}\frac{\partial g_0}{\partial\tilde{t}} \\ &+\mathcal{D}g_0 + \mathcal{E}g_0^2 = 0,\end{aligned} \tag{6.2.266}$$

where we have introduced the notation

$$\mathcal{D} = \tilde{C}_{12}C_{21}^{(1)} - C_{11}^{(1)}\tilde{C}_{22}, \tag{6.2.267a}$$

$$\begin{aligned}\mathcal{E} &= \nu - \gamma\tilde{C}_{22} \\ &= \tilde{C}_{22}(G_{111} + \tilde{C}_{12}F_{211}/2) - \tilde{C}_{12}(G_{211} + \tilde{C}_{22}F_{211}/2).\end{aligned} \tag{6.2.267b}$$

Note that both $\mathcal{D}$ and $\mathcal{E}$ are zero for channel flow.

We decompose $g_0(\xi_1, \tilde{t})$ into its average and oscillatory parts and substitute this into (6.2.266). Most of the terms in (6.2.266) have no average part. In particular, the averages of $\partial^2 g_0/\partial\xi_1\partial\tilde{t}$, $\partial^2 g_0/\partial\xi_1^2$, $g_0\partial^2 g_0/\partial\xi_1^2 + (\partial g_0/\partial\xi_1)^2 = \partial(g_0\partial g_0/\partial\xi_1)/\partial\xi_1$, $\partial g_0/\partial\xi_1$, and $g_0\partial g_0/\partial\xi_1 = (1/2)\partial(g_0^2)/\partial\xi_1$ are all zero. The average of (6.2.266) thus reduces to

$$-\tilde{C}_{22}\frac{d}{d\tilde{t}}\langle g_0\rangle(\tilde{t}) + \mathcal{D}\langle g_0\rangle(\tilde{t}) + \mathcal{E}\langle g_0^2\rangle(\tilde{t}) = 0. \tag{6.2.268}$$

The third term, $\mathcal{E}\langle g_0^2\rangle(\tilde{t})$, in (6.2.268) is key in determining $\langle g_0\rangle(\tilde{t})$. If $\mathcal{E} = 0$, then (6.2.268) is a linear homogeneous equation for $\langle g_0\rangle(\tilde{t})$, and its solution is identically equal to zero for $\langle g_0\rangle(0) = 0$. We note that $\langle g_0^2\rangle(\tilde{t})$ only vanishes if $g_0(\xi_1, \tilde{t}) \equiv 0$. Thus, if $\mathcal{E} \neq 0$, we find upon evaluating (6.2.268) at $\tilde{t} = 0$ that $\frac{d}{d\tilde{t}}\langle g_0\rangle(0) = (\mathcal{E}/\tilde{C}_{22})\langle g_0^2\rangle(0) \neq 0$, which implies that $\langle g_0\rangle(\tilde{t}) \neq 0$ even if $\langle g_0\rangle(0) = 0$.

In deriving the evolution equations for f_0 and g_0, we have assumed that $\langle g_0\rangle(\tilde{t}) \equiv 0$. The above argument shows that a necessary condition for this to be true (if $\langle g_0\rangle(0) = 0$) is to have $\mathcal{E} = 0$. This condition, when written out explicitly in terms of the primitive variables of (6.2.5), becomes

$$\frac{2}{A_{12}}\left\{(A_{11} - A_{22})(C_{22} + C_{12})^2 + [(A_{11} - A_{22})^2 + 4A_{12}A_{21}]^{1/2}\right\} L(r_1)$$

$$= (C_{12} + C_{22})L(r_2), \tag{6.2.269}$$

where

$$L(r_i) = W_{11}^2\frac{\partial^2 r_i}{\partial u_1^2} + 2W_{11}W_{21}\frac{\partial^2 r_i}{\partial u_1\partial u_2} + W_{21}^2\frac{\partial^2 r_i}{\partial u_2^2}. \tag{6.2.270}$$

Here, the A_{ij}, C_{ij}, and W_{ij} are all evaluated for $\mu = \mu_0$. For the case of channel flow, (6.2.269) is satisfied because $r_1 = 0$ and $C_{12} + C_{22} = 0$.

Returning to (6.2.257), we use the periodicity condition on g_0, i.e., $g_0(\xi_1 + 2\ell, \tilde{t}) = g_0(\xi_1, \tilde{t})$ to determine $K_1(\tilde{t})$

$$K_1(\tilde{t}) = \frac{E(2\ell)}{1 - E(2\ell)}\int_{-\ell}^{\ell}\frac{\tilde{C}_{12}C_{21}^{(1)}g_0(s, \tilde{t}) + \nu g_0^2(s, \tilde{t})}{E(s)}\,ds. \tag{6.2.271}$$

For a given divergence form (6.2.2) of the original physical problem, there corresponds a shock condition for g_0 in the $\xi_1\tilde{t}$ plane. This shock condition then determines the appropriate divergence form for (6.2.257). The details of this calculation are entirely analogous to those leading to (6.2.128). It is shown in [6.20] and [6.30] that for channel flow the shock condition is

$$\frac{d\xi_1}{d\tilde{t}} = \alpha - \frac{3}{4}(g_0^+ + g_0^-). \tag{6.2.272}$$

Therefore, the correct divergence form for (6.2.259b) is

$$\frac{d}{d\tilde{t}}\left(\alpha - \frac{3}{2}g_0\right) + \frac{\partial}{\partial \xi_1}\left[\frac{1}{2}\left(\alpha - \frac{3}{2}g_0\right)^2\right] + \frac{3}{8}\alpha g_0$$

$$+ \frac{3}{16}\alpha e^{\xi_1/2}\left[\int_{-\ell}^{\xi_1} g_0(s,\tilde{t})e^{-s/2}ds + \frac{e^{\ell}}{1-e^{\ell}}\int_{-\ell}^{\ell} g_0 e^{-s/2}ds\right]. \tag{6.2.273}$$

This is the appropriate form to use for calculating g_0 numerically as shocks are captured automatically (see [6.30]).

For the general case (6.2.257), the divergence form analogous to (6.2.273) is

$$\frac{\partial}{\partial \tilde{t}}(\lambda_1^{(1)} - F_{111}g_0) + \frac{\partial}{\partial \xi_1}\left[\frac{1}{2}(\lambda_1^{(1)} - F_{111}g_0)^2\right] - F_{111}C_{11}^{(1)}g_0$$

$$- \gamma F_{111}g_0^2 - F_{111}E(\xi_1)\int_{-\ell}^{\xi_1}\frac{\tilde{C}_{12}C_{21}^{(1)}g_0(s,\tilde{t}) + \gamma\tilde{C}_{22}g_0^2(s,\tilde{t})}{E(s)}ds$$

$$- \frac{F_{111}E(2\ell)}{1-E(2\ell)}\int_{-\ell}^{\ell}\frac{\tilde{C}_{12}C_{21}^{(1)}g_0(s,\tilde{t}) + \gamma\tilde{C}_{22}g_0^2(s,\tilde{t})}{E(s)}ds. \tag{6.2.274}$$

Note that in view of the condition $\mathcal{E} = 0$, we have set $\nu = \gamma\tilde{C}_{22}$ in (6.2.274).

Numerical solutions of (6.2.273) and the exact system (6.2.30) given in [6.30] show excellent agreement. An arbitrary periodic initial disturbance evolves slowly and tends to a traveling wave called a *roll wave* as $\tilde{t} \to \infty$. Roll waves consist of a periodic pattern of shocks separating special continuous solutions of (6.2.30) in a uniformly translating frame. Numerical solutions of (6.2.274) for various parameter values are given in [6.20]. Roll waves also exist in this general case, and their asymptotic behavior can be computed using (6.2.274) for the case $\mathcal{D} = 0$. If we let $\mathcal{D}$ become progressively more negative, we observe roll waves that have progressively more shocks per period. For more details, we refer the reader to [6.20] and [6.30] and the references cited therein.

The effect of weak dissipation

Consider now the effect of having $\mathbf{f}$ in (6.2.12) depend also on $\mathbf{w}_{xx}$ linearly to model weak dissipation. Specifically, we assume that the f_i given in (6.2.13c) have the added terms

$$\Delta f_i = \sigma_i\frac{\partial^2 w_i}{\partial x^2}, \quad i = 1, 2, \tag{6.2.275}$$

where the σ_i are positive $O(1)$ constants (see (6.2.211)). Since the $O(1)$ problem for (6.2.12) is unaffected, we are still able to achieve the standard form (6.2.19), except the ϕ_i now have the added terms

$$\Delta\phi_i = \sum_{k=1}^{3}\sigma_{ik}\frac{\partial^2 U_k}{\partial x^2}, \quad i = 1, 2. \tag{6.2.276}$$

The constants σ_{ik} are defined in terms of the σ_i and the matrix components of W and V (see (6.2.17)) as follows:

$$\sigma_{ik} = \sum_{j=1}^{2} \sigma_j V_{ij} W_{jk}, \quad i, k = 1, 2. \tag{6.2.277}$$

We expand the U_i as in (6.2.244) and find that the U_{i0} are still given by (6.2.247). The $\phi_i^{(0)}$ on the right-hand sides of (6.2.246) now have the following added terms:

$$\Delta\phi_i^{(0)} = \sum_{k=1}^{2} \sigma_{ij} \mathcal{L}(U_{k0}), \tag{6.2.278}$$

where

$$\mathcal{L} \equiv \frac{\partial^2}{\partial \xi_1^2} + 2\frac{\partial^2}{\partial \xi_1 \partial \xi_2} + \frac{\partial^2}{\partial \xi_2^2}. \tag{6.2.279}$$

Proceeding as before, we remove the collection of terms independent of ξ_1 from the modified right-hand side of (6.2.253) to find the following evolution equation for f_0:

$$\frac{\partial f_0}{\partial \tilde{t}} + \tilde{C}_{12}(\sigma_{21}\tilde{C}_{22}^2 - C_{21}^{(1)})F_0 + [C_{22}^{(1)} + \tilde{C}_{22}(\lambda_2^{(1)} + 2\sigma_{21}\tilde{C}_{12}$$
$$-\sigma_{22}\tilde{C}_{22})]f_0 + (\lambda_2^{(1)} + \sigma_{21}\tilde{C}_{12} - 2\sigma_{22}\tilde{C}_{22})\frac{\partial f_0}{\partial \xi_2} = \sigma_{22}\frac{\partial^2 f_0}{\partial \xi_2^2}. \tag{6.2.280}$$

Thus, the effect of weak diffusion is to modify the coefficients of F_0, f_0 and $\partial f_0/\partial \xi_2$ in (6.2.255) and, more importantly, to introduce the second derivative term $\sigma_{22}\partial^2 f_0/\partial \xi_2^2$ on the right-hand side. The series form (6.2.261) for the solution of (6.2.280) remains appropriate, and we obtain equations formally analogous to (6.2.265) governing the coefficients. Again, the contribution of f_0 to the solution (6.2.252) is suppressed by the factor $\exp(-C_{12}^{(0)}t)$ for $C_{22}^{(0)} > 0$.

Returning to the amended (6.2.253), we isolate terms independent of ξ_2 and compute the following addition to U_{21} as given by (6.2.256):

$$\Delta U_{21} = -\frac{\sigma_{21}}{\lambda_1^{(0)} - \lambda_2^{(0)}}\left[\frac{\partial g_0}{\partial \xi_1} + \tilde{C}_{22}g_0 + \tilde{C}_{22}^2 E(\xi_1)\int^{\xi_1}\frac{g_0(s,\tilde{t})}{E(s)}ds\right]. \tag{6.2.281}$$

Including this addition to U_{21} in the modified (6.2.246a) and removing terms proportional to ξ_2 from the solution for U_{11} gives the following evolution equation for g_0 that generalizes (6.2.257):

$$\frac{\partial g_0}{\partial \tilde{t}} + (\lambda_1^{(1)} - \sigma_{21}\tilde{C}_{12} - F_{111}g_0)\frac{\partial g_0}{\partial \xi_1} + (C_{11}^{(1)} - \sigma_{21}\tilde{C}_{12}\tilde{C}_{22})g_0$$
$$+ \gamma g_0^2 + \tilde{C}_{12}(C_{21}^{(1)} - \sigma_{21}\tilde{C}_{22}^2)E(\xi_1)\int^{\xi_1}\frac{g_0(s,\tilde{t})}{E(s)}ds$$
$$+ \nu E(\xi_1)\int^{\xi_1}\frac{g_0^2(s,\tilde{t})}{E(s)}ds + K_1(\tilde{t})E(\xi_1) = \sigma_{11}\frac{\partial^2 g_0}{\partial \xi_1^2}. \tag{6.2.282}$$

Again, we see that the dissipative terms in (6.2.12) modify the coefficients of the linear terms in (6.2.257) and introduce the second derivative term $\sigma_{11}\partial^2 g_0/\partial\xi_1^2$ to the right-hand side. We note that the coefficient σ_{12} does not occur in either evolution equation. The condition $\mathcal{E} = \nu - \gamma\tilde{C}_{22} = 0$, necessary for $\langle g_0\rangle(\tilde{t}) \equiv 0$, still holds, and we compute $K_1(\tilde{t})$ as in (6.2.271) in the form

$$K_1(\tilde{t}) = \frac{E(2\ell)}{1 - E(2\ell)} \int_{-\ell}^{\ell} \frac{\tilde{C}_{12}(C_{21}^{(1)} - \sigma_{21}\tilde{C}_{22}^2)g_0(s,\tilde{t}) + \nu g_0^2(s,\tilde{t})}{E(s)}\, ds. \tag{6.2.283}$$

The magnitude of σ_{11} relative to γ determines the relative importance of weak dissipation and weak nonlinearity. Recall that we have assumed at the outset that dissipation is small by having the σ_i multiplied by ϵ in the right-hand side of (6.2.12). As for Burgers' equation with very small dissipation, solutions of the limiting equation (6.2.257) are outer limits of solutions of (6.2.282) as $\sigma_{11} \to 0^+$ and $\sigma_{21} \to 0^+$ everywhere away from shocks. The effect of the second derivative term on the right-hand side of (6.2.282) is to smooth out the solution of (6.2.257) in an $O(\sigma_{11})$ neighborhood of a shock. We therefore conclude that small dissipation does not essentially alter the qualitative behavior of solutions; shocks are smoothed and various parameters occurring in (6.2.257) are modified by the terms involving σ_{11} and σ_{21}.

The situation is fundamentally different if the dissipative terms added to (6.2.12) are $O(1)$. Such terms alter the $O(1)$ problem; in particular, it is no longer hyperbolic, and its general solution is significantly more complicated than (6.2.247). A multiple-scale expansion of this solution for the case of weak instability, $\mu - \mu_0 = O(\epsilon)$, is difficult and, to our knowledge, has not been worked out. What is well known is the technique discussed in Sec. 6.1.3 for solving the class of *very weakly unstable* problems, $\mu - \mu_0 = O(\epsilon^2)$, if the linear solution has a nonzero initial wave number, k_c, that is most unstable.

Problems

1. Show that $g_0(\xi_2, \tilde{t}) \equiv 0$ in the solution of (6.2.56) as long as the initial conditions are in the form

$$u(x, 0; \epsilon) = \rho(x), \quad u_t(x, 0; \epsilon) = -\rho'(x), \tag{6.2.284}$$

and the following condition on H holds:

$$\lim_{\ell\to\infty} \frac{1}{2\ell} \int_{-\ell}^{\ell} H(-f_0(\xi_1,\tilde{t}), f_0(\xi_1,\tilde{t}))d\xi_1 = 0.$$

2. Calculate the solution of (6.2.56) for $H = u_t^3$ to $O(1)$ and the following pairs of initial conditions:

a.

$$u(x, 0; \epsilon) = Ae^{-k|x|}, \; A = \text{constant}, k = \text{constant} > 0, \tag{6.2.285a}$$

$$u_t(x, 0; \epsilon) = -Ak\text{sgn}(x)e^{-k|x|}. \tag{6.2.285b}$$

b.

$$u(x, 0; \epsilon) = 2e^{-|x|}, \quad u_t(x, 0; \epsilon) = 0. \tag{6.2.286}$$

3. Consider the wave equation (6.2.56) with the nonlinear term

$$H = -\alpha u_t + \beta u_t^3, \tag{6.2.287}$$

where α and β are positive constants.

a. For the initial values in (6.2.78), show that f_0 is given by

$$f_0(\xi_1, \tilde{t}) = \frac{Ak \cos k\xi_1}{[(1 - e^{-\beta\tilde{t}})(\alpha/\beta)A^2k^2 \cos^2 k\xi_1 + e^{-\beta\tilde{t}}]^{1/2}}. \tag{6.2.288}$$

Therefore,

$$u = \frac{1}{k}\left(\frac{\beta/\alpha}{1 - e^{-\beta\tilde{t}}}\right)^{1/2} \sin^{-1}\left\{\left[\frac{(1 - e^{-\beta\tilde{t}})\alpha k^2 A^2}{\alpha k^2 A^2 + (\beta - \alpha k^2 A^2)e^{-\beta\tilde{t}}}\right]^{1/2} \cdot \sin k\xi_1\right\} + O(\epsilon), \tag{6.2.289}$$

where the arcsine is in the interval $\left(-\frac{\pi}{2}, \frac{\pi}{2}\right)$.

b. Show that as $\tilde{t} \to \infty$, u, to leading order, tends to the "saw-tooth" wave with slope $\pm\sqrt{\beta/\alpha}$. In this limit, the amplitude $\frac{1}{k}\sqrt{\beta/\alpha}$ does not depend on the initial amplitude A.

c. Now consider the initial values (6.2.84a). Show that the solution for u to leading order is again given by (6.2.97), where λ and m are now given by

$$\lambda(\tilde{t}) = 8e^{\beta\tilde{t}/2}m^{-3}(\tilde{t}), \tag{6.2.290}$$

$$m^8 - \frac{8}{7}m^7 = 2^8\left[\frac{\alpha}{\beta}(e^{\beta\tilde{t}} - 1) + \frac{3}{7}\right]. \tag{6.2.291}$$

4. We wish to study the initial and boundary-value problem for (6.2.56) with the homogeneous boundary conditions

$$u(0, t; \epsilon) = u(\ell, t; \epsilon) = 0, \quad t > 0. \tag{6.2.292}$$

As pointed out in [6.21], we can extend the definition of u over the entire x axis by requiring u to be an odd 2ℓ-periodic function

$$u(-x, t; \epsilon) = -u(x, t; \epsilon). \tag{6.2.293}$$

$$u(x + 2n\ell, t; \epsilon) = u(x, t; \epsilon), \quad n = \pm 1, \pm 2, \ldots. \tag{6.2.294}$$

Condition (6.2.293) implies that

$$u_t(-x, t; \epsilon) = -u_t(x, t; \epsilon), \quad u_x(-x, t; \epsilon) = u_x(x, t; \epsilon). \tag{6.2.295}$$

Therefore, the definition of H must be extended as

$$H(-u_t, u_x) = -H(u_t, u_x) \tag{6.2.296}$$

in order that the extended solutions satisfy (6.2.56) for all x. Note that this requirement is automatically satisfied if H is an odd function of u_t. If it is even in u_t, we must use (6.2.296) and change the sign of H in $(-\ell, 0)$.

Now, since u is odd and periodic, the definitions of U_1 and U_2 in (6.2.57) imply that

$$g_0(\xi_2, \tilde{t}) = f_0(-\xi_2, \tilde{t}). \tag{6.2.297}$$

The condition (6.2.297) then implies that the two evolution equations (6.2.70) reduce to

$$\frac{\partial f_0}{\partial \tilde{t}} - \frac{1}{4\ell}\int_{-\ell}^{\ell} H(-f_0(\xi_1, \tilde{t}) + f_0(-\xi_2, t), f_0(\xi_1, \tilde{t}) + f_0(-\xi_2, \tilde{t}))d\xi_2 = 0. \tag{6.2.298}$$

Use the results above to calculate the solution to $O(1)$ for

$$u_{tt} - u_{xx} + \epsilon(-u_t + \frac{1}{3}u_t^3) = 0, \quad 0 \le x \le \pi. \tag{6.2.299}$$

$$u(x, 0; \epsilon) = 0, \quad u_t(x, 0; \epsilon) = \sin x. \tag{6.2.300}$$

$$u(0, t; \epsilon) = u(\pi, t; \epsilon) = 0, \quad t > 0. \tag{6.2.301}$$

In particular, derive the asymptotic form of the solution for $t \to \infty$ [6.21].

5. We wish to show that for 2ℓ-periodic initial data $f_0(\xi_1, 0)$ and $g_0(\xi_2, 0)$ that also satisfy the condition (6.2.76) we have $\langle f_0\rangle(\tilde{t}) = \langle g_0\rangle(\tilde{t}) = \langle f_0^3\rangle(\tilde{t}) = \langle g_0^3\rangle(\tilde{t}) = 0$. Take the average of (6.2.74a) and (6.2.74b) over one period and conclude that

$$\frac{d}{d\tilde{t}}\left[\langle f_0\rangle(\tilde{t})\right]_{\tilde{t}=0} = \frac{d}{d\tilde{t}}\left[\langle g_0\rangle(\tilde{t})\right]_{\tilde{t}=0} = 0. \tag{6.2.302}$$

Multiply (6.2.74a) by $3f_0^2$, (6.2.74b) by $3g_0^2$, and then average the resulting equations to show that

$$\frac{d}{d\tilde{t}}\left[\langle f_0^3\rangle(\tilde{t})\right]_{\tilde{t}=0} = \frac{d}{d\tilde{t}}\left[\langle g_0^3\rangle(\tilde{t})\right]_{\tilde{t}=0} = 0. \tag{6.2.303}$$

Use induction to conclude that all the derivatives of $\langle f_0\rangle(\tilde{t})$, $\langle g_0\rangle(\tilde{t})$, $\langle f_0^3\rangle(\tilde{t})$, and $\langle g_0^3\rangle(\tilde{t})$ vanish at $\tilde{t} = 0$. Hence, $\langle f_0\rangle(\tilde{t}) = \langle g_0\rangle(\tilde{t}) = \langle f_0^3\rangle(\tilde{t}) = \langle g_0^3\rangle(\tilde{t}) = 0$, and (6.2.77) follows from (6.2.74) under the assumption (6.2.76) and 2ℓ-periodic initial values for f_0 and g_0.

6. Consider the signaling problem

$$u(0, t; \epsilon) = \sin \omega t, \quad \omega = \text{ constant}, \ t > 0, \tag{6.2.304a}$$

$$u(x, 0; \epsilon) = 0, \tag{6.2.304b}$$

for the wave equation

$$u_{tt} - u_{xx} + \epsilon u_t^3 = 0. \tag{6.2.305}$$

Solve the system (6.2.59) for this case using the multiple-scale expansionsignaling problem for

$$U_i(x, t; \epsilon) = U_{i0}(\zeta_1, \zeta_2, \tilde{x}) + \epsilon U_{i1}(\zeta_1, \zeta_2, \tilde{x}) + O(\epsilon^2), \qquad (6.2.306)$$

where

$$\zeta_1 = t - x, \quad \zeta_2 = t + x, \quad \tilde{x} = \epsilon x. \qquad (6.2.307)$$

7. The velocity potential $\Phi(x, y; \epsilon, M)$ for steady two-dimensional supersonic flow of an inviscid perfect gas is

$$(a^2 - \Phi_x^2)\Phi_{xx} - 2\Phi_x\Phi_y\Phi_{xy} + (a^2 - \Phi_y^2)\Phi_{yy} = 0. \qquad (6.2.308)$$

Here a is the dimensionless local sound speed related to Φ according to

$$a^2 = 1 + \frac{M^2(\gamma - 1)}{2} - \frac{(\gamma - 1)}{2}(\Phi_x^2 + \Phi_y^2), \qquad (6.2.309)$$

where $M > 1$ is the Mach number (dimensionless flow speed) at $x = -\infty$, and γ is the constant ratio of specific heats.

We consider flow over an airfoil that is symmetric; thus we need only consider $y \geq 0$. The upper surface of the airfoil is defined in dimensionless form by

$$y = \begin{cases} \epsilon F(x), & 0 \leq x \leq 1,\ F(0) = F(1) = 0 \\ 0, & x \leq 0, x \geq 1. \end{cases} \qquad (6.2.310)$$

Here ϵ measures the half-thickness of the airfoil, and we shall assume $0 < \epsilon \ll 1$.

The boundary condition representing flow tangency to the airfoil is

$$\Phi_y(x, \epsilon F(x); \epsilon, M) = \epsilon F'(x)\Phi_x(x, \epsilon F(x); \epsilon, M), \quad 0 \leq x \leq 1, \quad (6.2.311)$$

and symmetry requires

$$\Phi_y(x, 0; \epsilon, M) = 0, \quad x < 0, \quad x > 1. \qquad (6.2.312)$$

The boundary condition at upstream infinity is

$$\Phi \to Mx, \text{ as } x \to -\infty. \qquad (6.2.313)$$

For more details of the above formulation, see section 2.4 of [6.7].

a. Set $\Phi = M(x + \epsilon\phi(x, y; \epsilon, M))$ and show that a is given by

$$a^2 = 1 - \epsilon(\gamma - 1)M^2\phi_x - \epsilon^2\frac{(\gamma - 1)}{2}M^2(\phi_x^2 + \phi_y^2) \qquad (6.2.314)$$

and that (6.2.308) becomes

$$\phi_{yy} - (M^2 - 1)\phi_{xx} = \epsilon M^2[(\gamma + 1)\phi_x\phi_{xx} + 2\phi_y\phi_{xy} + (\gamma - 1)\phi_x\phi_{yy}]$$

$$+\epsilon^2 M^2\left[\frac{(\gamma - 1)}{2}(\phi_x^2 + \phi_y^2)(\phi_{xx} + \phi_{yy}) + \phi_x^2\phi_{xx} + 2\phi_x\phi_y\phi_{xy} + \phi_y^2\phi_{yy}\right]. \qquad (6.2.315)$$

Thus, for ϵ small and $M > 1$, this is a weakly nonlinear wave equation.

The appropriate shock condition for small ϵ is

$$\left(\frac{dy}{dx}\right)_{\text{shock}} = \pm\frac{1}{\sqrt{M^2-1}} \pm \epsilon\frac{(\gamma+1)}{4}\frac{M^2}{(M^2-1)^2}\phi_y^+ + O(\epsilon), \tag{6.2.316}$$

where ϕ_y^+ is the value of ϕ_y just downstream of the shock, and it is assumed that $\phi_x^- = \phi_y^- = 0$.

b. Set $\overline{y} = \sqrt{M^2-1}y$, $\phi_x = U_1 + U_2$, $\phi_{\overline{y}} = -U_1 + U_2$, and transform (6.2.315) to the standard form (6.2.1) with $\lambda_1 = 1, \lambda_2 = -1, C_{ij} = 0$.

c. Expand U_1 and U_2 in the multiple-scale form

$$U_i = U_{i0}(\xi_1, \xi_2, \tilde{y}) + \epsilon U_{i1}(\xi_1, \xi_2, \tilde{y}) + O(\epsilon^2), \tag{6.2.317}$$

where $\xi_1 = x - \overline{y}, \xi_2 = x + \overline{y}, \tilde{y} = \epsilon\overline{y}, U_{10} = f_0(\xi_1, \tilde{y}), U_{20} = g_0(\xi_2, \tilde{y})$. Show that f_0 and g_0 satisfy the evolution equations

$$2\frac{\partial f_0}{\partial \tilde{y}} + \frac{M^4(\gamma+1)}{M^2-1}f_0\frac{\partial f_0}{\partial \xi_1} = 0, \tag{6.2.318a}$$

$$2\frac{\partial g_0}{\partial \tilde{y}} - \frac{M^4(\gamma+1)}{M^2-1}g_0\frac{\partial g_0}{\partial \xi_2} = 0. \tag{6.2.318b}$$

The failure of this result for $M \approx 1$ is evident. The appropriate expansion procedure for $M \approx 1$ is discussed in [6.7].

d. Use the boundary conditions to show that for $y \geq 0$, $g_0 \equiv 0$ and f_0 satisfies

$$f_0(x,0) = \begin{cases} -F'(x)/\sqrt{M^2-1}, & 0 \leq x \leq 1 \\ 0, & x < 0, x > 1. \end{cases} \tag{6.2.319}$$

Solve (6.2.318a) for the case $F(x) = 4x(1-x)$, and calculate the shocks that emanate from the airfoil leading and trailing edges.

e. Observe that the characteristics of (6.2.318a) are the straight lines in the xy plane

$$x - \sqrt{M^2-1}y - \epsilon\frac{M^4(\gamma+1)}{2(M^2-1)^{3/2}}yf_0(\xi) = \xi = \text{constant}, \tag{6.2.320}$$

which may be intepreted as "strained" characteristic coordinates

$$x - \sqrt{M^2-1}\left(1 + \epsilon\frac{M^4(\gamma+1)f_0}{2(M^2-1)^2} + O(\epsilon^2)\right)y = \text{constant}. \tag{6.2.321}$$

Show that the solution for U_1 to $O(\epsilon)$ *cannot* be expressed in terms of the one strained coordinate (6.2.321).

A more general version of this problem, where the oncoming Mach number varies slowly with altitude, is also interesting. In this case, a supersonic disturbance originating at the airfoil may become sonic. The appropriate expansion and matching procedures are discussed in [6.25].

8. Consider shallow water flow of depth $h = 1$ and speed $u = F > 0$ over a flat bottom ($b \equiv 0$) for $t < 0$. At $t = 0^+$, the bottom over $0 < x < 1$

drops a distance ϵ due to an earthquake. Calculate the solution for $h(x, t; \epsilon)$ and $u(x, t; \epsilon)$ to $O(\epsilon)$.

6.3 Multiple-Scale Homogenization

Partial differential equations that involve rapidly varying terms are found in many applications. In this section, we consider two simple examples to illustrate ideas. Because both examples contain only one fast independent variable, the analysis is considerably simplified as the solution to each order obeys an ordinary differential operator that is easily solved. Extension of the method to problems with more than one fast scale is possible but is not discussed.

In Section 3.3.9, an effective thermal conductivity was found by using limit process expansions for compact inclusions. When the nonuniformities can be approximated by a continuous but rapidly varying inhomogeneity, it is again possible to obtain an effective conductivity and a homogenized equation by using multiple-scale expansions.

6.3.1 One-Dimensional Steady Heat Conduction

The simple problem to be studied first is one-dimensional steady heat conduction. The heat flux q_x in the x direction is given by

$$q_x(x) = -k\left(x, \frac{x}{\epsilon}\right)\frac{\partial T}{\partial x}, \tag{6.3.1}$$

where T is the temperature and k is the thermal conductivity. Here the characteristic length of the global problem is taken to be unity. The thermal conductivity shows a slow variation on the x scale and a rapid variation on the scale (x/ϵ), where ϵ is the small parameter. An asymptotic description is sought for small ϵ. In the absence of heat sources, the basic equation for heat conservation is

$$\frac{\partial q_x}{\partial x} = 0 = \frac{\partial}{\partial x}\left(k\left(x, \frac{x}{\epsilon}\right)\frac{\partial T}{\partial x}\right), \qquad 0 < x < 1, \tag{6.3.2}$$

where k is bounded. In most cases, the rapid variations are periodic or random. Typical boundary conditions could be

$$T(0) = T_L, \qquad T(1) = 0. \tag{6.3.3}$$

An asymptotic expansion for T can be expressed in terms of a fast variable

$$x^* = \frac{x}{\epsilon}$$

and a slow variable $x = \epsilon x^*$ based on the global scale.

We regard $T(x; \epsilon)$ as a function of x, x^*, and ϵ and expand it in the usual form

$$T(x; \epsilon) = T_0(x, x^*) + \epsilon T_1(x, x^*) + \epsilon^2 T_2(x, x^*) + \cdots. \tag{6.3.4}$$

Thus, (6.3.2) gives

$$\left(\frac{1}{\epsilon}\frac{\partial}{\partial x^*}+\frac{\partial}{\partial x}\right)\left\{k(x,x^*)\left(\frac{1}{\epsilon}\frac{\partial T_0}{\partial x^*}+\frac{\partial T_1}{\partial x^*}+\epsilon\frac{\partial T_2}{\partial x^*}+\frac{\partial T_0}{\partial x}\right.\right.$$
$$\left.\left.+\epsilon\frac{\partial T_1}{\partial x}+\cdots\right)\right\}=0.$$

Collecting the terms of $O\left(\epsilon^{-2}\right)$, $O\left(\epsilon^{-1}\right)$, and $O(1)$, respectively, we have to study the sequence of approximating equations

$$\frac{\partial}{\partial x^*}\left(k(x,x^*)\frac{\partial T_0}{\partial x^*}\right)=0. \tag{6.3.5}$$

$$\frac{\partial}{\partial x^*}\left(k(x,x^*)\frac{\partial T_1}{\partial x^*}\right)=-\frac{\partial}{\partial x^*}\left(k(x,x^*)\frac{\partial T_0}{\partial x}\right)-\frac{\partial}{\partial x}\left(k(x,x^*)\frac{\partial T_0}{\partial x^*}\right). \tag{6.3.6}$$

$$\frac{\partial}{\partial x^*}\left(k(x,x^*)\frac{\partial T_2}{\partial x^*}\right)=-\frac{\partial}{\partial x^*}\left(k(x,x^*)\frac{\partial T_1}{\partial x}\right)-\frac{\partial}{\partial x}\left(k(x,x^*)\frac{\partial T_1}{\partial x^*}\right)$$
$$-\frac{\partial}{\partial x}\left(k(x,x^*)\frac{\partial T_0}{\partial x}\right). \tag{6.3.7}$$

Equation (6.3.5) gives

$$\frac{\partial T_0}{\partial x^*}=\frac{B_0(x)}{k(x,x^*)}$$

$$T_0(x,x^*)=B_0(x)\int_0^{x^*}\frac{d\xi}{k(x,\xi)}+\theta_0(x), \tag{6.3.8}$$

where B_0 and θ_0 are arbitrary at this stage.

Now, we assume that the average of k^{-1} on the fast scale exists, i.e.,

$$\langle k^{-1}\rangle(x)\equiv\lim_{x^*\to\infty}\frac{1}{x^*}\int_0^{x^*}\frac{d\xi}{k(x,\xi)} \tag{6.3.9}$$

is a bounded function of x. The notation $\langle k^{-1}\rangle(x)$ means the average with respect to the fast scale resulting in a function of the slow scale x. If k is periodic in x^*, the average can be calculated over a period, which, of course, is much smaller than the global scale (see (6.2.75)). We set $B_0(x)=0$ to prevent T_0 from growing linearly on the fast scale, as this would make the expansion (6.3.4) inconsistent. Thus,

$$T_0(x,x^*)=\theta_0(x). \tag{6.3.10}$$

Next, (6.3.6) reads

$$\frac{\partial}{\partial x^*}\left(k(x,x^*)\frac{\partial T_1}{\partial x^*}\right)=-\frac{d\theta_0}{dx}\frac{\partial k}{\partial x^*} \tag{6.3.11}$$

or

$$k(x, x^*)\frac{\partial T_1}{\partial x^*} = -k(x, x^*)\frac{d\theta_0}{dx} + B_1(x),$$

$$T_1(x, x^*) = -x^*\frac{d\theta_0}{dx} + B_1(x)\int_0^{x^*}\frac{d\xi}{k(x, \xi)} + \theta_1(x). \tag{6.3.12}$$

Again, to prevent growth of $T_1(x, x^*)$ on the x^* scale, we have the requirement (see (6.3.9))

$$B_1(x) = \frac{1}{< k^{-1} > (x)}\frac{d\theta_0}{dx}. \tag{6.3.13}$$

Thus,

$$T_1(x, x^*) = \frac{d\theta_0}{dx}\left\{\frac{1}{< k^{-1} > (x)}\int_0^{x^*}\frac{d\xi}{k(x, \xi)} - x^*\right\} + \theta_1(x). \tag{6.3.14}$$

Note that

$$\frac{\partial T_1}{\partial x} = \frac{\partial}{\partial x}\left\{\frac{d\theta_0}{dx}\left(\frac{1}{< k^{-1} > (x)}\int_0^{x^*}\frac{d\xi}{k(x, \xi)} - x^*\right)\right\} + \frac{d\theta_1}{dx} \tag{6.3.15}$$

and

$$\frac{\partial T_1}{\partial x^*} = \frac{d\theta_0}{dx}\left\{\frac{1}{k(x, x^*) < k^{-1} > (x)} - 1\right\}. \tag{6.3.16}$$

Now, to get the homogenized equation for $\theta_0(x)$, it is necessary to consider the behavior of $T_2(x, x^*)$. From (6.3.7),

$$\frac{\partial}{\partial x^*}\left(k(x, x^*)\frac{\partial T_2}{\partial x^*}\right) = -\frac{\partial}{\partial x^*}\left(k(x, x^*)\frac{\partial T_1}{\partial x}\right) - \frac{\partial}{\partial x}\left(\frac{d\theta_0}{dx}\left(\frac{1}{< k^{-1} > (x)} - k(x, x^*)\right)\right) - \frac{\partial}{\partial x}\left(k(x, x^*)\frac{d\theta_0}{dx}\right)$$

$$= -\frac{\partial}{\partial x^*}\left(k(x, x^*)\frac{\partial T_1}{\partial x}\right) - \frac{d}{dx}\left(\frac{1}{< k^{-1} > (x)}\frac{d\theta_0}{dx}\right). \tag{6.3.17}$$

Integration gives

$$k(x, x^*)\frac{\partial T_2}{\partial x^*} = -k(x, x^*)\frac{\partial T_1}{\partial x} - x^*\frac{d}{dx}\left(\frac{1}{< k^{-1} > (x)}\frac{d\theta_0}{dx}\right) + B_3(x). \tag{6.3.18}$$

It can be seen from (6.3.15) that $(\partial T_1/\partial x)$ does not grow on the fast scale like x^* so that to prevent growth it is again necessary that

$$\frac{d}{dx}\left(\frac{1}{< k^{-1} > (x)}\frac{d\theta_0}{dx}\right) = 0. \tag{6.3.19}$$

This is the homogenized equation

$$\frac{d}{dx}\left(k_{eff}(x)\frac{d\theta_0}{dx}\right) = 0 \tag{6.3.20}$$

with the effective thermal conductivity, $k_{eff}(x)$,

$$\frac{1}{k_{eff}(x)} = \left\langle k^{-1}\right\rangle(x) = \lim_{x^*\to\infty}\int_0^{x^*}\frac{d\xi}{k(x,\xi)}. \tag{6.3.21}$$

The asymptotic approximation to (6.3.2) is (6.3.20), and the boundary conditions (6.3.3) can be satisfied by $\theta_0(x)$ to obtain the first approximation to the heat flux

$$q_x = \text{constant} = -k_{eff}(x)\frac{d\theta_0}{dx}. \tag{6.3.22}$$

The rule for calculating $k_{eff}(x)$ is the usual rule for resistances in series since resistance $\approx$ (1/conductivity). The original problem can be solved exactly, and it can be shown that the asymptotic solution just constructed is correct.

6.3.2 *Two-Dimensional Steady Heat Conduction*

A simple generalization is to the case where there is a two-dimensional temperature field, $T(x, y)$, but the conductivity still depends on only one fast variable. The local heat flux is isotropic, and the heat flux vector has the components

$$q_x = -k(x, x^*, y)\frac{\partial T}{\partial x}, \qquad q_y = -k(x, x^*, y)\frac{\partial T}{\partial y}. \tag{6.3.23}$$

The basic conservation equation is

$$\frac{\partial}{\partial x}\left(k(x, x^*, y)\frac{\partial T}{\partial x}\right) + \frac{\partial}{\partial y}\left(k(x, x^*, y)\frac{\partial T}{\partial y}\right) = 0. \tag{6.3.24}$$

Suitable conditions, such as prescribed temperature, are provided on the boundary of a domain in (x, y). The asymptotic expansion in (x, x^*, y) now reads

$$T(x, y; \epsilon) = T_0(x, x^*, y) + \epsilon T_1(x, x^*, y) + \epsilon^2 T_2(x, x^*, y) + \cdots. \tag{6.3.25}$$

The first two approximate equations are basically the same as (6.3.5) and (6.3.6) since the y dependence only enters in the $O(1)$ equation. The resulting solutions are

$$T_0(x, x^*, y) = \theta_0(x, y). \tag{6.3.26}$$

$$T_1(x, x^*, y) = \frac{\partial \theta_0}{\partial x}\left\{\frac{1}{< k^{-1} > (x, y)}\int_0^{x^*}\frac{d\xi}{k(x, \xi, y)} - x^*\right\} + \theta_1(x, y), \tag{6.3.27}$$

where

$$< k^{-1} > (x, y) = \lim_{x^*\to\infty}\int_0^{x^*}\frac{d\xi}{k(x, \xi, y)}.$$

The $O(1)$ equation now reads

$$\frac{\partial}{\partial x^*}\left(k(x, x^*, y)\frac{\partial T_2}{\partial x^*}\right) = -\frac{\partial}{\partial x^*}\left(k(x, x^*, y)\frac{\partial T_1}{\partial x}\right) - \frac{\partial}{\partial x}\left(k(x, x^*, y)\frac{\partial T_1}{\partial x^*}\right)$$

$$-\frac{\partial}{\partial x}\left(k(x, x^*, y)\frac{\partial T_0}{\partial x}\right) - \frac{\partial}{\partial y}\left(k(x, x^*, y)\frac{\partial T_0}{\partial y}\right). \tag{6.3.28}$$

As before,

$$\begin{aligned}\frac{\partial}{\partial x^*}\left(k(x, x^*, y)\frac{\partial T_2}{\partial x}\right) = &-\frac{\partial}{\partial x^*}\left(k(x, x^*, y)\frac{\partial T_1}{\partial x}\right)\\ &-\frac{\partial}{\partial x}\left\{\frac{1}{< k^{-1} > (x, y)}\frac{\partial \theta_0}{\partial x}\right\}\\ &-\frac{\partial}{\partial y}\left(k(x, x^*, y)\frac{\partial \theta_0}{\partial y}\right).\end{aligned}$$

Integration gives

$$k(x, x^*, y)\frac{\partial T_2}{\partial x} = -k(x, x^*, y)\frac{\partial T_1}{\partial x} - x^*\frac{\partial}{\partial x}\left\{\frac{1}{< k^{-1} > (x, y)}\frac{\partial \theta_0}{\partial x}\right\}$$

$$-\frac{\partial}{\partial y}\left(\int_0^{x^*}\frac{\partial \theta_0}{\partial y}k(x, \xi, y)d\xi\right). \tag{6.3.29}$$

Assume the existence of

$$\lim_{x^*\to\infty}\frac{1}{x^*}\int_0^{x^*}k(x, \xi, y)d\xi \equiv < k > (x, y). \tag{6.3.30}$$

As before, $(\partial T_2/\partial x)$ does not grow like x^* so that to prevent rapid growth on the x^* scale, it is necessary that

$$\frac{\partial}{\partial x}\left\{\frac{1}{< k^{-1} > (x, y)}\frac{\partial \theta_0}{\partial x}\right\} + \frac{\partial}{\partial y}\left(< k > (x, y)\frac{\partial \theta_0}{\partial y}\right) = 0. \tag{6.3.31}$$

This is the homogenized version of (6.3.24), which can be used to satisfy the global boundary conditions. The heat flux components are

$$q_x = -\frac{1}{< k^{-1} > (x, y)} \frac{\partial \theta_0}{\partial x}, \qquad q_y = - < k > (x, y) \frac{\partial \theta_0}{\partial y}. \quad (6.3.32)$$

The effective conductivity in the x and y directions is different so that the medium is "globally" anisotropic, in contrast to its local behavior. The rule for conductivity in the y direction, normal to the direction of rapid change, is essentially that for resistances in parallel.

An approach similar to the one presented here was given by J. B. Keller [6.15]. Another approach to these problems that goes more deeply into various mathematical points appears in [6.2], and there are many earlier references.

References

6.1. M.J. Ablowitz and H. Segur, *Solitons and the Inverse Scattering Method*, SIAM, Philadelphia, 1981.

6.2. A. Bensoussan, J.-L. Lions, and G. Papanicolaou, *Asymptotic analysis for periodic structures*, Studies in Mathematics and Its Applications 5, North-Holland, Amsterdam, 1978.

6.3. S.C. Chikwendu, "Non-linear wave propagation solutions by Fourier transform perturbation," *Int. J. Non-linear Mechanics*, **16**, 1981, pp. 117–128.

6.4. S.C. Chikwendu, "Asymptotic solutions of some weakly nonlinear elliptic equations," *SIAM J. Appl. Math.*, **31**, 1976, pp. 286–303.

6.5. S.C. Chikwendu and C.V. Easwaran, "Multiple-scale solution of initial-boundary value problems for weakly nonlinear wave equations on the semi-infinite line," *SIAM J. Appl. Math.*, **52**, 1992, pp. 946–958.

6.6. S.C. Chikwendu and J. Kevorkian, "A perturbation method for hyperbolic equations with small nonlinearities," *SIAM J. Appl. Math.*, **22**, 1972, pp. 235–258.

6.7. J.D. Cole and L.P. Cook, *Transonic Aerodynamics*, North-Holland, Amsterdam, 1986.

6.8. P.G. Drazin, *Solitons*, Cambridge University Press, Cambridge, 1983.

6.9. P.G. Drazin and W.H. Reid, *Hydrodynamic Stability*, Cambridge University Press, Cambridge, 1991.

6.10. W. Eckhaus, "On modulation equations of the Ginzburg-Landau type," International Conference on Industrial and Applied Mathematics, R.E. O'Malley, Jr., Ed., Society for Industrial and Applied Mathematics, Philadelphia, 1991, pp. 83–98.

6.11. W. Eckhaus, "New approach to the asymptotic theory of nonlinear oscillations and wave-propagation," *J. Math. Anal. Appl.*, **49**, 1975, pp. 575–611.

6.12. W. Eckhaus, *Studies in Nonlinear Stability Theory*, Springer Tracts in Natural Philosophy, vol. 6, Springer-Verlag, Berlin, 1965.

6.13. C.L. Frenzen and J. Kevorkian, "A review of multiple scale and reductive perturbation methods for deriving uncoupled evolution equations," *Wave Motion*, **7**, 1985, pp. 25–42.

6.14. C.S. Gardner, J.M. Greene, M.D. Kruskal, and R.M. Miura, "Method for solving the Korteweg–de Vries equation," *Phys. Rev. Lett.*, **19**, 1967, pp. 1095–1097.

6.15. J.B. Keller, *D'Arcy's law for flow in porous media and the two-space method*, Nonlinear Partial Differential Equations in Engineering and Applied Sciences, R.L.

Sternberg, A.J. Kalmowski, and J.S. Papadokis, Eds., Marcel Dekker, New York, 1980.

6.16. J.B. Keller and S. Kogelman, "Asymptotic solutions of initial value problems for nonlinear partial differential equations," *SIAM J. Appl. Math.*, **18**, 1970, pp. 748–758.

6.17. J. Kevorkian, *Partial Differential Equations: Analytical Solution Techniques*, Chapman and Hall, New York, London, 1990, 1993.

6.18. J. Kevorkian and J.D. Cole, *Perturbation Methods in Applied Mathematics*, Springer-Verlag, New York, 1981.

6.19. J. Kevorkian and J. Yu, "Passage through the critical Froude number for shallow water waves over a variable bottom," *J. Fluid Mech.*, **204**, 1989, pp. 31–56.

6.20. J. Kevorkian, J. Yu, and L. Wang, "Weakly nonlinear waves for a class of linearly unstable hyperbolic conservation laws with source terms," *SIAM J. Appl. Math.*, **55**, 1995, pp. 446–484.

6.21. R.W. Lardner, "Asymptotic solutions of nonlinear wave equations using the methods of averaging and two-timing," *Q. Appl. Math.*, **35**, 1977, pp. 225–238.

6.22. J.C. Luke, "A perturbation method for nonlinear dispersive wave problems," *Proc. Royal Soc. London A*, **292**, 1966, pp. 403–412.

6.23. A. Majda and R. Rosales, "Resonantly interacting weakly nonlinear hyperbolic waves, I. A single space variable," *Stud. Appl. Math.*, **71**, 1984, pp. 149–179.

6.24. B.J. Matkowsky, "A simple nonlinear dynamic stability problem," *Bull. Am. Math. Soc.*, **76**, 1970, pp. 620–625.

6.25. G. Pechuzal and J. Kevorkian, "Supersonic-transonic flow generated by a thin airfoil in a stratified atmosphere," *SIAM J. Appl. Math.*, **33**, 1977, pp. 8–33.

6.26. R. Srinivasan, "Asymptotic solution of the weakly nonlinear Schrödinger equation with variable coefficients," *Stud. Appl. Math.*, **84**, 1991, pp. 145–165.

6.27. G.B. Whitham, *Linear and Nonlinear Waves*, Wiley, New York, 1974.

6.28. G.B. Whitham, "A general approach to linear and nonlinear dispersive waves," *J. Fluid Mech.*, **22**, 1965, pp. 273–283.

6.29. J. Yu and J. Kevorkian, "The interaction of a strong bore with small disturbances in shallow water," *Stud. Appl. Math.*, **91**, 1994, pp. 247–273.

6.30. J. Yu and J. Kevorkian, "Nonlinear evolution of small disturbances into roll waves in an inclined open channel," *J. Fluid Mech.*, **243**, 1992, pp. 575–594.

Index

Applied Mathematical Sciences

(continued from page ii)

61. *Sattinger/Weaver:* Lie Groups and Algebras with Applications to Physics, Geometry, and Mechanics.
62. *LaSalle:* The Stability and Control of Discrete Processes.
63. *Grasman:* Asymptotic Methods of Relaxation Oscillations and Applications.
64. *Hsu:* Cell-to-Cell Mapping: A Method of Global Analysis for Nonlinear Systems.
65. *Rand/Armbruster:* Perturbation Methods, Bifurcation Theory and Computer Algebra.
66. *Hlavácek/Haslinger/Necasl/Lovísek:* Solution of Variational Inequalities in Mechanics.
67. *Cercignani:* The Boltzmann Equation and Its Applications.
68. *Temam:* Infinite Dimensional Dynamical Systems in Mechanics and Physics.
69. *Golubitsky/Stewart/Schaeffer:* Singularities and Groups in Bifurcation Theory, Vol. II.
70. *Constantin/Foias/Nicolaenko/Temam:* Integral Manifolds and Inertial Manifolds for Dissipative Partial Differential Equations.
71. *Catlin:* Estimation, Control, and the Discrete Kalman Filter.
72. *Lochak/Meunier:* Multiphase Averaging for Classical Systems.
73. *Wiggins:* Global Bifurcations and Chaos.
74. *Mawhin/Willem:* Critical Point Theory and Hamiltonian Systems.
75. *Abraham/Marsden/Ratiu:* Manifolds, Tensor Analysis, and Applications, 2nd ed.
76. *Lagerstrom:* Matched Asymptotic Expansions: Ideas and Techniques.
77. *Aldous:* Probability Approximations via the Poisson Clumping Heuristic.
78. *Dacorogna:* Direct Methods in the Calculus of Variations.
79. *Hernández-Lerma:* Adaptive Markov Processes.
80. *Lawden:* Elliptic Functions and Applications.
81. *Bluman/Kumei:* Symmetries and Differential Equations.
82. *Kress:* Linear Integral Equations.
83. *Bebernes/Eberly:* Mathematical Problems from Combustion Theory.
84. *Joseph:* Fluid Dynamics of Viscoelastic Fluids.
85. *Yang:* Wave Packets and Their Bifurcations in Geophysical Fluid Dynamics.
86. *Dendrinos/Sonis:* Chaos and Socio-Spatial Dynamics.
87. *Weder:* Spectral and Scattering Theory for Wave Propagation in Perturbed Stratified Media.
88. *Bogaevski/Povzner:* Algebraic Methods in Nonlinear Perturbation Theory.
89. *O'Malley:* Singular Perturbation Methods for Ordinary Differential Equations.
90. *Meyer/Hall:* Introduction to Hamiltonian Dynamical Systems and the N-body Problem.
91. *Straughan:* The Energy Method, Stability, and Nonlinear Convection.
92. *Naber:* The Geometry of Minkowski Spacetime.
93. *Colton/Kress:* Inverse Acoustic and Electromagnetic Scattering Theory.
94. *Hoppensteadt:* Analysis and Simulation of Chaotic Systems.
95. *Hackbusch:* Iterative Solution of Large Sparse Systems of Equations.
96. *Marchioro/Pulvirenti:* Mathematical Theory of Incompressible Nonviscous Fluids.
97. *Lasota/Mackey:* Chaos, Fractals, and Noise: Stochastic Aspects of Dynamics, 2nd ed.
98. *de Boor/Höllig/Riemenschneider:* Box Splines.
99. *Hale/Lunel:* Introduction to Functional Differential Equations.
100. *Sirovich (ed):* Trends and Perspectives in Applied Mathematics.
101. *Nusse/Yorke:* Dynamics: Numerical Explorations.
102. *Chossat/Iooss:* The Couette-Taylor Problem.
103. *Chorin:* Vorticity and Turbulence.
104. *Farkas:* Periodic Motions.
105. *Wiggins:* Normally Hyperbolic Invariant Manifolds in Dynamical Systems.
106. *Cercignani/Illner/Pulvirenti:* The Mathematical Theory of Dilute Gases.
107. *Antman:* Nonlinear Problems of Elasticity.
108. *Zeidler:* Applied Functional Analysis: Applications to Mathematical Physics.
109. *Zeidler:* Applied Functional Analysis: Main Principles and Their Applications.
110. *Diekmann/van Gils/Verduyn Lunel/Walther:* Delay Equations: Functional-, Complex-, and Nonlinear Analysis.
111. *Visintin:* Differential Models of Hysteresis.
112. *Kuznetsov:* Elements of Applied Bifurcation Theory.
113. *Hislop/Sigal:* Introduction to Spectral Theory: With Applications to Schrödinger Operators.
114. *Kevorkian/Cole:* Multiple Scale and Singular Perturbation Methods.
115. *Taylor:* Partial Differential Equations I, Basic Theory.
116. *Taylor:* Partial Differential Equations II, Qualitative Studies of Linear Equations.
117. *Taylor:* Partial Differential Equations III, Nonlinear Equations.